Integrating Human Health into Urban and Transport Planning

Mark Nieuwenhuijsen • Haneen Khreis
Editors

Integrating Human Health into Urban and Transport Planning

A Framework

Editors
Mark Nieuwenhuijsen
Barcelona Institute for Global Health
ISGlobal
Barcelona, Spain

Haneen Khreis
Texas A&M Transportation Institute, Center for Advancing Research in Transportation Emissions, Energy, and Health
College Station, TX, USA

ISBN 978-3-319-74982-2 ISBN 978-3-319-74983-9 (eBook)
https://doi.org/10.1007/978-3-319-74983-9

Library of Congress Control Number: 2018942501

Printed on acid-free paper

This Springer imprint is published by the registered company Springer International Publishing AG part of Springer Nature.
The registered company address is: Gewerbestrasse 11, 6330 Cham, Switzerland

Preface

The world is urbanising rapidly and cities are not only centres of innovation and wealth creation but also hot spots of air pollution and noise, heat islands and lack green space and physical activity, partly due to poor urban and transport planning. Recent studies show large impacts from urban and transport planning on health, and urgent action is needed to reduce these negative health impacts.

The urban environment is a complex interlinked system. Decision-makers need not only better data on the complexity of factors in environmental and developmental processes affecting human health, but also an enhanced understanding of the linkages between these factors and effects to be able to know at which level to target their actions most effectively. In recent years, there also has been a shift from trying to change at the national level to more comprehensive and ambitious actions being developed and implemented at the regional and local levels. Cities have come to the forefront of providing solutions for environmental issues such as climate change, which has co-benefits on health, but yet need better knowledge for wider health-centric action.

This volume brings together the world's leading experts on urban and transport planning, environmental exposures, physical activity, health and health impact assessment to discuss challenges and solutions in cities and provide a conceptual framework and work programme for actions and future research needs. It presents the current evidence base, benefits and case studies of better integrating health and the environment into urban development and transport planning, a conceptual framework describing this, and strategies for future action and research.

Barcelona, Spain — Mark Nieuwenhuijsen
College Station, TX, USA — Haneen Khreis

Contents

Part I
Introduction and Key Concepts

Chapter 1
Urban and Transport Planning, Environment and Health

Mark Nieuwenhuijsen and Haneen Khreis

1.1 Introduction

1.1.1 The Era of the City

Over 50% of people worldwide live in cities, and this figure will increase to up to 70% over the next 20 years (United Nations 2014). The United Nations projects that nearly all global population growth from thc ycars 2016 to 2030 will be absorbed by cities, about 1.1 billion new urbanites over the next 14 years. While 50 years ago there were only 3 megacities (>10 m inhabitants), Tokyo, Osaka and New York-Newark, today there are 28 megacities, and by 2030 there will be 41 (United Nations 2015a). In recent years, attention has turned to cities, not only because of the multiple challenges cities present but also because of the potential solutions cities could provide to pressing societal problems such as poverty, climate change and disease. The United Nations Habitat New Urban Agenda (UN Habitat 2016) and the Sustainable Development Goals (SDG 2015) have provided new impetus into the

M. Nieuwenhuijsen (✉)
ISGlobal-CREAL, Parc de Recerca Biomèdica de Barcelona – PRBB, Barcelona, Spain

ISGlobal, Centre for Research in Environmental Epidemiology (CREAL), Barcelona, Spain

Universitat Pompeu Fabra (UPF), Barcelona, Spain

CIBER Epidemiologia y Salud Publica (CIBERESP), Madrid, Spain
e-mail: mark.nieuwenhuijsen@isglobal.org

H. Khreis
ISGlobal, Centre for Research in Environmental Epidemiology (CREAL), Barcelona, Spain

Universitat Pompeu Fabra (UPF), Barcelona, Spain

CIBER Epidemiologia y Salud Publica (CIBERESP), Madrid, Spain

Texas A&M Transportation Institute (TTI), College Station, TX, USA

Institute for Transport Studies, University of Leeds, Leeds, UK

M. Nieuwenhuijsen, H. Khreis (eds.), *Integrating Human Health into Urban and Transport Planning*, https://doi.org/10.1007/978-3-319-74983-9_1

urban development agenda and its linkages to human health. New organizations like C40 (www.C40.org), led by city mayors, have taken the lead in the fight against climate change.

Cities have long been known to be society's predominant engine of innovation and wealth creation, but they are also a main source of pollution, crime and disease (Bettencourt et al. 2007). Today, only 600 urban centres generate about 60% of the global GDP (Dobbs et al. 2011). The top 20 metropolitan areas in the United States *contribute* 52% of the total country's *GDP*. Also, there is a strong positive correlation between the degree of urbanization of a country and its per capita income that has long been recognized. Double a city's population, and its economic productivity goes up by 130% (Bettencourt et al. 2007; Bettencourt and West 2010). Resource sharing, quicker and better matching and more learning have been suggested as the drivers of the higher productivity of cities. Increasing urban population density is favourable as it gives residents greater opportunity for face-to-face interaction. Productivity and innovation tend to be higher in cities leading to higher wages and improved standards of living. As such, cities attract more workers resulting in positive loops of growth, productivity and innovation.

1.1.2 Beyond the Benefits of City Living

Unfortunately, cities are also hotspots for a number of adverse environmental exposures and lifestyles which are detrimental to human health (Khreis et al. 2017). Current urban and transport processes in cities have been less than optimal, creating air pollution, noise, heat islands, lack of green space and sedentary behaviour, to name a few (Nieuwenhuijsen 2016). As a result, within and between cities, there is considerable variation in the levels of environmental exposures, such as air pollution, noise, temperature and green space, and in physical activity levels and motor vehicle crashes, partly due to urban and transport planning practices (Nieuwenhuijsen 2016).

These exposures are associated with multiple health endpoints (Khreis et al. 2016). Air pollution (Beelen et al. 2014; Héroux et al. 2015; Khreis et al. 2017), noise (Basner et al. 2014; Halonen et al. 2015) and temperature (Gasparrini et al. 2015) cause adverse health outcomes including increased morbidity and premature mortality. Green space has predominantly been associated with positive health outcomes (Gascon et al. 2016; Hartig et al. 2014; Nieuwenhuijsen et al. 2017) but also some negative impacts such as urban sprawl, gentrification and spread of infectious diseases (Cucca 2012; Hartig et al. 2014; Lõhmus and Balbus 2015). Physical activity has many health benefits (Woodcock et al. 2011).

The global burden of disease attributable to these exposures is large. Worldwide, approximately 3–4 million deaths each year are attributable to ambient air pollution and 2.1 million deaths to insufficient physical activity (Forouzanfar et al. 2015). Motor vehicle crashes cause over 1.5 million global deaths annually and 79.6 million healthy years of life lost (Bhalla et al. 2014). A recent study showed that 20%

Table 1.1 World trends affecting cities

Population growth
Economic growth
Urbanization
Technological development
Ageing population
Rising inequalities
Climate change

of premature mortality in a city like Barcelona is related to urban and transport planning-related exposures, including air pollution, noise, temperature, green space and physical activity, not meeting international exposure-level guidelines (Mueller et al. 2017).

Further, global challenges that are linked to urban and transport planning in cities are climate change, rising inequalities and ageing populations (Table 1.1). Climate change and emerging environmental risks have been identified as "the biggest global health threat of the twenty-first century" (Costello et al. 2009) and have put further pressures on cities. Mean surface temperature is expected to increase heat stress and flooding, as well as indirect health impacts mediated through infectious diseases, air quality and food security (Costello et al. 2009; Patz et al. 2016). Many coastal cities are at risk of flooding, and the increasing temperature exacerbates heat island effects.

Furthermore, there is concern of the impact of increasing social and economic inequality in cities on health. The World Health Organization's Commission on the Social Determinants of Health (2008) noted that "social injustice is killing people on a grand scale" (p. 26), which is acutely apparent in cities (WHO and UN Habitat 2010). In many cities, poorer people live in areas where there is more pollution and also where there is poorer access to public transport. In some low-income countries, the majority of people in cities live in informal settlements where they are exposed to greater health risks from environmental exposures and insecurity and where there is a lack of access to basic services such as safe water and sanitation and.

Also today, in cities there are people from different cultural and ethnic backgrounds, religions and social statuses. They are increasingly diverse and this is reinforced by the impact of immigration. Cities around the world are discussing these changes, with particular intensity in the European Union due to the recent influx of refugees. Cities have to deal with the great challenges of absorbing diversity and counteracting inequality and spatial segregation.

Finally, the trend of an ageing population provides further challenges. By 2050, the global population aged 60 years or over is projected to become more than double that of 2015, reaching nearly 2.1 billion, and the number of people aged 80 and over is growing even faster (UN 2015b). This ageing population presents specific urban planning and mobility challenges and echoes calls for city design to adapt to older populations making sure they do not become isolated or are at increased risks of environmental hazard.

Further economic growth and new technological developments in cities may help to address some of the challenges faced by cities. Many of these challenges need appropriate financial resources and adequate governance structures to address them. There needs to be policies in place to improve urban and transport planning with the aim to improve health and reduce inequalities. Cities are often eager to brand themselves as technologically innovative, smart and connected, and technological advances could address a number of challenges and make cities more efficient and provide cost savings that could be invested elsewhere, but they may also provide new challenges. Examples of these are the introduction of electric vehicles and autonomous vehicles (or the combination of the two) that have great potential to reduce the pollution levels and the number of vehicles in the city, respectively, but may provide concerns regarding the electric supply and privacy, respectively. They also may solve only part of the problem if there is no good legislation is in place.

1.2 Linking Urban and Transport Planning, Environment and Health

From a health point of view, urban and transport planning and policies in cities have often been deleterious to health. There is, however, another plausible reality, one where urban and transport planning and policy become promoters and drivers of good public health, creating healthier and more sustainable communities. But how can we make cities that promote health? That may start from the understanding and acknowledgement of the clear link between land use, behaviour, exposures, morbidity and mortality (Fig. 1.1).

Urban planners often describe land use in terms of the five Ds: density, diversity, design, destination accessibility and distance to transit (Ewing and Cervero 2010). Higher population and development density, for example, could lead to shorter distances because destinations become closer to origins. Trips of shorter distances are easier and more convenient to walk or cycle. Diversity is a measure of the land use mix including homes, shops, schools and workplaces in an area. When there are more shops around the home, people may travel shorter and use transport modes other than the car to get to a shop. Design describes the overall infrastructure, and a good design encourages public and active transportation and discourages the use of cars. Destination accessibility is a measure for how accessible places are, while distance to transit expresses the shortest distance to a bus stop or railway station. Both will encourage the use of public and active transportation when the destination accessibility is higher and the distance to public transport shorter. Greater density, diversity and destination accessibility, better design and shorter distance to transit are characteristics of so-called compact cities, an example of which is the Barcelona City, where the density, diversity and destination accessibility are high and distance

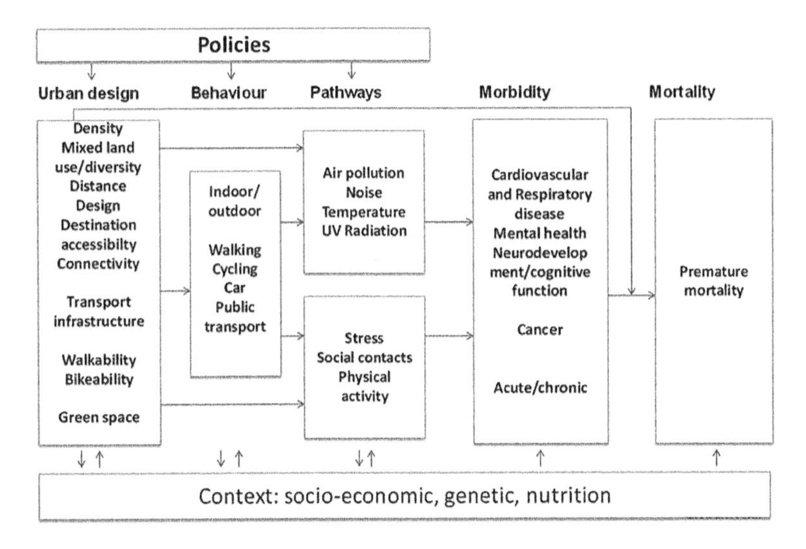

Fig. 1.1 Conceptual framework of links between urban and transport planning, environmental exposures, physical activity and health (After Nieuwenhuijsen 2016)

to transit is low. Conversely, a city like Atlanta is a sprawling city, where the opposite is true.

The above land use characteristics influence transport mode choice, travel demand and traffic levels and, in vicious cycles, also influence the investment required to upscale and maintain transport systems that can accommodate the current situation. Large investment in infrastructure for cars has led to car-dominated cities, where, in many cities or neighbourhoods, it has become hard to imagine not having any cars around. This has often happened at the expense of investments in public and active transportation, including our most basic and easiest form of mobility: walking. In many cities, public space is dominated by cars. More car-accommodating infrastructure such as roads and parking lots leads to greater use of cars and to higher levels of environmental stressors such as air pollution, noise and heat islands and to less active mobility and physical activity. Furthermore, infrastructure for cars takes up a large amount of the often already-limited public space in cities that can be used for other purposes such as green space or public space for people to meet, interact, relax or simply walk. These exposures and lifestyles are detrimental to health and increase morbidity and premature mortality.

These trends are happening at a time of mechanization and automation, when the majority of employment opportunities are sedentary in nature as compared to positions involving manual labour prior to the industrial revolution. Too many people step in their car at home, drive to work, sit at a desk the entire day and drive back to sit at home. And often, pressure on time leaves little space for individuals to go to the gym or for a walk to get some physical activity, not to mention the need to pay for physical activity.

On the other side, a move away from car infrastructure to infrastructure for public and/or active transportation can lead to an increase in use of public and/or active transportation and reduce air pollution, noise, heat island effects and stress while increasing physical activity. This will reduce the morbidity and premature mortality burden. Furthermore a better provision of green infrastructure may provide opportunities for walking, relaxation and social contacts and thereby increase physical activity, reduce stress, improve restoration and reduce morbidity and premature mortality (Nieuwenhuijsen, Khreis, Triguero-Mas, et al. 2017).

While many policymakers continue to advance the justification and merit of motorized transport and urban planning policies, growing evidence, which we synthesize in this book, strongly suggests that such policies not only help facilitate a self-reinforcing cycle of car-reliant travel and urban sprawl but, importantly, continue to discount numerous health impacts related with transport and urban planning and policy.

1.3 Assessing the Health Impacts of Urban and Transport Planning

In recent years, numerous health impact assessment studies have started to quantify the potential health impacts of urban and transport planning and policy in cities. For most, if not all, of the studies, we see clear health impacts from different urban and transport scenarios suggesting that there is a need to make changes in cities to improve health, primarily by land use changes, and move towards public and active transportation.

For example, Reisi et al. (2016) evaluated three urban planning scenarios in Melbourne for 2030: a base case scenario based on governmental plans, fringe focus scenario based on expansive urban development patterns and activity centres scenario based on compact urban development patterns. The authors showed that the compact urban development scenario resulted in the least greenhouse gases and other emissions, as well as a reduction of mortality when compared to the other scenarios. Stevenson et al. (2016) estimated the population health impacts arising from alternative land use and transport policy initiatives in six major cities. Land use changes were modelled to reflect a compact city in which land use density and diversity were increased and distances to public transport were reduced with the objective to reduce private motorized transport and promote a modal shift to walking, cycling and public transport. The modelled compact city scenario resulted in health gains across all cities.

Some studies have evaluated the impacts of specific transport policy measures in cities. Woodcock et al. (2009) estimated the health impacts of alternative transport scenarios for two settings: London, UK, and Delhi, India. The authors found that a combination of active travel and lower-emission motor vehicles would give the largest benefits. Creutzig et al. (2012) evaluated scenarios of increasingly ambitious policy packages, reducing greenhouse gas emissions from urban transport by up to 80% from 2010 to 2040. Based on stakeholder interviews and data analysis, the main target was a modal shift from motorized individual transport to public transit and nonmotorized individual transport (walking and cycling) in four European cities (Barcelona, Malmö, Sofia and Freiburg). The authors reported significant concurrent co-benefits of better air quality, reduced noise, less traffic-related injuries and deaths and increased physical activity, alongside less congestion and monetary fuel savings. Xia et al. (2015) estimated that the shifting of 40% of vehicle kilometres travelled to alternative transport in Adelaide, South Australia, would reduce annual average particulate matter preventing 13 deaths a year and 118 disability-adjusted life years. Further health benefits would be obtained from improved physical fitness through, active transport, and these are much larger preventing 508 deaths a year and 6569 disability-adjusted life years. Similarly, a range of health impact assessment studies evaluating mortality and other health impacts of increases in active transport were recently reviewed and estimated considerable reductions in premature deaths and a reduced burden of morbidity with the most benefits attributable to increases in physical activity and only low increased risks of motor vehicles crashes and air pollution for those who switched to public and active transportation (Mueller et al. 2015; Tainio et al. 2016).

1.4 Paradigms of Cities

Over the past decades, numerous paradigms have evolved which have labelled cities in a number of different ways such as active cities, healthy cities, livable cities, resilient cities, smart cities and sustainable cities. These models are increasingly being envisioned and advocated to improve different aspects of cities including environmental conditions, public health, livability and economic performance. Some of these paradigms will be covered later in this book, but what is prominent is that there often are no rigid definitions or frameworks for these models. Many of these models are still developed in silos without much integration across the different relevant disciplines and perhaps with the lack of multidisciplinary teams.

A sustainable city is often defined as a city designed with consideration of local and global environmental impact, guided by urban sustainability visions and targets. People in a sustainable city try to minimize their required inputs of energy, water and food and waste output of heat, air and water pollution and greenhouse gas emissions. This may also have a beneficial impact on health. Although a seemingly simple and an alluring proposition, there yet is no completely agreed upon definition for what a sustainable city should be. Both conceptually and analytically, the sustainable city is challenging to define and delineate. Generally, developmental experts agree that a sustainable city should meet the needs of the present without compromising the ability of future generations to meet their own needs. This idea is general and ambiguous, and it leads to a great deal of variation in how cities carry out attempts to become more sustainable (Joss 2015; Magilavy 2011).

A livable city may be defined as a city that has robust and complete neighbourhoods, accessibility and sustainable mobility, a diverse and resilient local economy, vibrant public spaces and affordability. Livable cities enhance the wellbeing of all inhabitants, strengthen community, improve social and physical health and increase civic engagement by sustainably reshaping the built environment of our cities, suburbs, towns and villages (Crowhurst Lennard, Chap. 4).

A healthy city is a city with a commitment to health and a process and structure to achieve it. A healthy city is one that continually creates and improves its physical and social environments and expands the community resources that enable people to mutually support each other in performing all the functions of life and developing to their maximum potential. A Healthy Cities approach seeks to put health high on the political and social agenda of cities and to build a strong movement for public health at the local level. It strongly emphasizes equity, participatory governance and solidarity, intersectoral collaboration and action to address the determinants of health. Successful implementation of this approach requires innovative action addressing all aspects of health and living conditions and extensive networking between cities. This entails explicit political commitment, leadership, institutional change and intersectoral partnerships (WHO 2017, Tsouros, Chap. 5).

A resilient city is a city with the ability to persevere in the face of emergency, to continue its core mission despite daunting challenges. Resilience refers to equipping cities to face future shocks and stresses, for example, from climate change and depleted oil and fuel sources, and make it through crises. Thus a resilient city takes into consideration appropriate built form and physical infrastructure to be more resilient to the physical, social and economic challenges that come with, for example, depleting carbon-based fuels and climate change (Evans 2017).

1.5 Policies

Cities provide good opportunities for policy change as cities have direct local accountability and are generally more agile to act than national governments. However, cities are multicausal and complex systems, and strong political will and governance are needed to make changes happen. The infrastructure investments and the policies cities will adopt in the next years will lock the world into one path or another.

Environmental factors are highly modifiable, and environmental interventions at the community level, such as changes in urban and transport planning, have been shown to be promising and more effective than interventions at the individual level (Chokshi and Farley 2012). Further, changes in the urban environment are long lasting and are arguably permanent, while behavioural change campaigns and programmes are rarely maintained (Saelens et al. 2003). In order to implement interventions for the urban environments, however, decision-makers need to have a good understanding of the linkages between urban and transport planning, environmental exposures, behaviour and human health. To be able to know at which level and to what extent actions can be targeted effectively, decision-makers need a quantification of these impacts and linkages. They also need recommendations and quantification of the impacts associated with a range of available, plausible and feasible policy measures (or packages of measures) that can effectively address current issues. Without the latter, the more effective measures may be overlooked (May et al. 2016).

Various ideas and measures have been proposed to promote healthy urban living including the greening of cities (Khreis et al. 2016; Nieuwenhuijsen et al. 2017) and moving away from car-dominated cities towards car-free cities (Nieuwenhuijsen and Khreis 2016). Giles-Corti et al. (2016) recently identified eight integrated regional and local interventions that, when combined, encourage walking, cycling and public transport use while reducing private motor vehicle use. These interventions were destination accessibility, equitable distribution of employment across cities, managing demand by reducing the availability and increasing the cost of parking, designing pedestrian-friendly and cycling-friendly transport networks, achieving optimum levels of residential density, reducing distance to public transport and enhancing the desirability of active travel modes (e.g. by creating safe attractive neighbourhoods and safe, affordable and convenient public transport). They stated that together, these interventions will create healthier and more sustainable compact cities that reduce the environmental, social and behavioural risk factors that affect lifestyle choices, levels of traffic, environmental pollution, noise and crime.

Others have suggested that technological advances may improve the current conditions in cities such as the introduction of the electric cars or (shared) autonomous vehicles or the combination of the two. The scale of challenges faced in achieving long-term sustainability of the urban and transport infrastructure suggests that technological improvements, albeit important, will not be sufficient to solve transport-related health challenges. In fact, technological improvements have been shown counterproductive in instances such as the failure of the massive technology change from petrol towards diesel vehicles, initiated by the European car and oil industries, to mitigate climate change impacts, the subsequent Volkswagen diesel emission scandal (Schiermeier 2015) and its implications to public health (Nieuwenhuijsen and Khreis 2016). Emerging technologies including autonomous vehicles and electric or low emission vehicles, which may also have confounding effects that impact other sectors. Electric cars, for example, are only likely to address some of the challenges we have in cities such as air pollution and noise, to some extent, but not others like physical activity and space allocation, while the use and impact of autonomous vehicles are largely unpredictable. Autonomous vehicles have the potential to dramatically change how we commute and perceive the journey in the car and potentially change perceptions on where we can live. The commute in an autonomous vehicle may become part of the working day, as people sit back in their vehicles and perform tasks other than the driving. This may drive an important proportion of people to live further away from work, leading to more urban sprawl and reinforcing car dependence. Shared electric autonomous vehicles have the potential to reduce the number of vehicles on the road, if the number of journeys does not change dramatically and therefore allow public space to be used in a health-promoting manner, for example, by increasing green and public space. However, this will depend again on pricing, consumer behaviour and available alternatives.

Instead of relying on technological improvements to drive change, changing the urban environment, providing for public transport and active travel alongside behavioural and societal transformations will need to be put in place (Khreis et al. 2016). Making changes and proposing new policies is challenging as there is much vested interest and many stakeholders involved. There is also a further lack of tools that can support decision-makers, practitioners and researchers in policy options generation. Other barriers to good practice include poor and conflicting institutional coordination, unsupportive regulatory frameworks, biases in financing, poor data quality, lack of evidence of the performance of specific solutions, limited public support, lack of experience in stakeholder involvement and lack of political resolve (European Conference of Ministers of Transport 2006; May et al. 2017).

Therefore, while proposing changes, it is important to build an effective dialogue with the population that produces environmental stressors and is impacted at the same time, ensuring in this way public awareness and acceptance, as some measures can be restrictive in nature and therefore be politically unpopular (e.g. vehicle restricted areas and congestion charging zones). Participation should also include a variety of professionals and stakeholders that are already acting or can act towards improving health in cities and involve citizens in the development of future urban and transport scenarios. Researchers and practitioners can work together and provide better data and evidence of the health performance of specific solutions (Nieuwenhuijsen et al. 2017). A cultural shift and a reallocation of funding streams at the policy level are needed to include health assessments in proposed transport and urban development projects. The mitigation of the adverse health impacts associated with a proposed transport or urban project should be considered and dictated by clear policy and guidance as one of the objectives projects need to achieve. The lack of substantive influence for health in the transport and urban planning agenda may be traced back to the lack of clarity in policy and guidance (Khreis et al. 2016).

Finally, in cities and related research communities, there are often silos of urban planning and development, mobility and transport, parks and green space, environmental departments and (public) health departments that do not work together well enough, while multi-sectorial and systemic approaches are needed to tackle the multi-faceted environmental and health problems (Nieuwenhuijsen 2016). A systemic approach should now be the standard to avoid the flawed sector-centric planning and decision-making atmosphere. Unfortunately, these are complex challenges, the solutions to which cannot be prescribed or dictated. However, through this book, we hope to start a dialogue on these issues and their potential solutions.

1.6 This Book

In this book, we bring together experts from different sectors and disciplines to start a more holistic dialogue on how to improve and integrate urban and transport planning and health in cities. After the introduction chapter, the book starts off with a chapter on the New Urban Agenda and the Sustainable Development goals (Chap. 2), followed by concepts/paradigms of modern cities such as sustainable cities (Chap. 3), livable cities (Chap. 4), healthy cities (Chap. 4) and the human ecology (Chap. 5). This section is followed by a section on planning and development of cities providing some specific examples on how to create cities for people (Chap. 7), how to create car-free cities (Chap. 11) and the Superblocks initiative as a potential health-conducive urban model (Chap. 8). This section also includes a chapter with a description of the possible models that can be used in cities for city planning (Chap. 10) and a chapter on challenges of informal settlements (Chap. 9). The section ends with a chapter on the planning for healthy cities (Chap. 12). The following section deals with accessibility and mobility issues and their interactions with human health. Links between land use and mobility are described (Chaps. 13 and 14), followed by chapters on indicators and decision-making to advance health considerations within the transport agenda (Chap. 15) and policy measure generation and selection for healthy urban mobility options (Chap. 16). This section ends with a thought-provoking discussion about transport planners in their profession and how they may affect transport planning in the first instance (Chap. 17). The following section describes the impacts of urban and transport planning on our environment, motor vehicle crashes, physical activity and health. The chapters address specific health-relevant exposures including physical activity (Chap. 18), motor vehicle crashes (Chap. 19), green space (Chap. 20), air pollution (Chap. 21), noise (Chap. 22) and urban temperatures (Chap 23). This section is followed by a section on society, participation and health implications with chapters on social cohesion (Chap. 24), inequalities and health (Chap. 25), citizens' participation (Chap. 25) and citizens' science (Chaps. 26 and 27). The final section deals with policy and health impact assessment and includes one chapter on the co-benefits of climate change measures on health (Chap. 28), followed by two chapters on health impact assessment and its role in shaping policies to make cities healthier (Chaps. 29 and 30). Furthermore, this section includes chapters on barriers and enablers of policies (Chap. 31), the translation of evidence into practice (Chap. 32) and the use of conceptual models for policy implementation (Chap. 33).

This book is not a final word on these issues but is the start of a holistic dialogue that shows and documents the current state of the art of thinking in the different communities, the language used and the potential solutions that may lock the world's cities in a health-promoting path.

References

Basner, M., Babisch, W., Davis, A., Brink, M., Clark, C., Janssen, S., & Stansfeld, S. (2014). Auditory and non-auditory effects of noise on health. *Lancet, 383*(9925), 1325–1332.

Beelen, R., Raaschou-Nielsen, O., Stafoggia, M., Andersen, Z. J., Weinmayr, G., Hoffmann, B., et al. (2014). Effects of long-term exposure to air pollution on natural-cause mortality: An analysis of 22 European cohorts within the multicentre ESCAPE Project. *Lancet, 383*, 785–795.

Bettencourt, L. M., Lobo, J., Helbing, D., Kühnert, C., & West, G. B. (2007). Growth, innovation, scaling, and the pace of life in cities. *PNAS, 104*, 7301–7306.

Bettencourt, L. M., & West, G. (2010). A unified theory of urban living. *Nature, 467*, 912–913. Retrieved from http://www.nature.com/nature/journal/v467/n7318/abs/467912a.html.

Bhalla, K., Shotten, M., Cohen, A., Brauer, M., Shahraz, S., Burnett, R., et al. (2014). *Transport for health: The global burden of disease from motorized road transport*. Washington: World Bank Group. Retrieved from http://documents.worldbank.org/curated/en/2014/01/19308007/transport-health-global-burden-disease-motorized-road-transport.

Chokshi, D. A., & Farley, T. A. (2012). The cost-effectiveness of environmental approaches to disease prevention. *The New England Journal of Medicine, 367*, 295–297.

Costello, A., Abbas, M., Allen, A., Ball, S., Bell, S., Bellamy, R., et al. (2009). Managing the health effects of climate change: Lancet and University College London Institute for Global Health Commission. *Lancet, 373*, 1693–1733.

Creutzig, F., Mühlhoff, R., & Römer, J. (2012). Decarbonizing urban transport in European cities: Four cases show possibly high co-benefits. *Environmental Research Letters, 7*(4), 044042. Retrieved from http://iopscience.iop.org/article/10.1088/1748-9326/7/4/044042/meta.

Cucca, R. (2012). The unexpected consequences of sustainability: Green cities between innovation and ecogentrification. *Sociologica, 6*(2), 1–21.

Dobbs, R., Smit, S., Remes, J., Manyika, J., Roxburgh, C., & Restrepo, A. (2011). *Urban world: Mapping the economic power of cities*. New York: McKinsey Global Institute.

European Conference of Ministers of Transport. (2006). *Sustainable urban travel: Implementing sustainable urban travel policies: Applying the 2001 key messages*. Paris: ECMT.

Evans. (2017). Retrieved from https://www.thebalance.com/resilient-city-definition-and-urban-design-principles-3157826.

Ewing, R., & Cervero, R. (2010). Travel and the built environment: A meta-analysis. *Journal of the American Planning Association, 76*(3), 265–294. Retrieved from http://www.tandfonline.com/doi/abs/10.1080/01944361003766766.

Forouzanfar, M. H., Alexander, L., Anderson, H. R., Bachman, V. F., Biryukov, S., Brauer, M., et al. (2015). Global, regional, and national comparative risk assessment of 79 behavioural, environmental and occupational, and metabolic risks or clusters of risks in 188 countries, 1990–2013: A systematic analysis for the Global Burden of Disease Study 2013. *Lancet, 386*(10010), 2287–2323.

Gascon, M., Triguero-Mas, M., Martínez, D., Dadvand, P., Forns, J., Plasència, A., & Nieuwenhuijsen, M. J. (2016). Green space and mortality: A systematic review and meta-analysis. *Environment International, 2*(86), 60–67.

Gasparrini, A., Guo, Y., Hashizume, M., Lavigne, E., Zanobetti, A., Schwartz, J., et al. (2015). Mortality risk attributable to high and low ambient temperature: A multicountry observational study. *Lancet, 386*(9991), 369–375.

Giles-Corti, B., Vernez-Moudon, A., Reis, R., Turrell, G., Dannenberg, A. L., Badland, H., et al. (2016). City planning and population health: A global challenge. *The Lancet, 388*(10062), 2912–2924.

Halonen, J., Hansell, A., Gulliver, J., Morley, D., Blangiardo, M., Fecht, D., et al. (2015). Road traffic noise is associated with increased cardiovascular morbidity and mortality and all-cause mortality in London. *European Heart Journal, 36*, 2653–2661. https://doi.org/10.1093/eurheartj/ehv216.

Hartig, T., Mitchell, R., de Vries, S., & Frumkin, H. (2014). Nature and health. *Annual Review of Public Health, 35*, 207–228.

Héroux, M. E., Anderson, H. R., Atkinson, R., Brunekreef, B., Cohen, A., Forastiere, F., et al. (2015). Quantifying the health impacts of ambient air pollutants: Recommendations of a WHO/Europe Project. *International Journal of Public Health, 60*(5), 619–627.

Joss, S. (2015). *Sustainable cities: Governing for urban innovation*. London: Palgrave Macmillan.

Khreis, H., Warsow, K. M., Verlinghieri, E., Guzman, A., Pellecuer, L., Ferreira, A., et al. (2016). The health impacts of traffic-related exposures in urban areas: Understanding real effects, underlying driving forces and co-producing future directions. *Journal of Transport & Health, 3*(3), 249–267.

Khreis, H., Kelly, C., Tate, J., Parslow, R., Lucas, K., & Nieuwenhuijsen, M. (2017). Exposure to traffic-related air pollution and risk of development of childhood asthma: A systematic review and meta-analysis. *Environment International, 100*, 1–31.

Khreis, H., van Nunen, E., Mueller, N., Zandieh, R., & Nieuwenhuijsen, M. J. (2017). Commentary: How to create healthy environments in cities. *Epidemiology, 28*(1), 60–62.

Lõhmus, M., & Balbus, J. (2015). Making green infrastructure healthier infrastructure. *Infection ecology & epidemiology, 5*, PMCID: PMC4663195.

Magilavy, B. (2011). Sustainability plan. Sustainable city. Retrieved December 6, 2011.

May, A. D., Khreis, H., & Mullen, C. (2016). *Option generation for policy measures and packages: The role of the KonSULT knowledgebase*. Proceedings 14th World Conference on Transport Research, Shanghai.

May, A., Boehler-Baedeker, S., Delgado, L., Durlin, T., Enache, M., & van der Pas, J. W. (2017). Appropriate national policy frameworks for sustainable urban mobility plans. *European Transport Research Review, 9*(1), 7.

Mueller, N., Rojas-Rueda, D., Cole-Hunter, T., de Nazelle, A., Dons, E., Gerike, R., et al. (2015). Health impact assessment of active transportation: A systematic review. *Preventive Medicine, 76*, 103–114. Retrieved from http://www.sciencedirect.com/science/article/pii/S0091743515001164.

Mueller, N., Rojas-Rueda, D., Basagaña, X., Cirach, M., Cole-Hunter, T., Dadvand, P., et al. (2017). Urban and transport planning related exposures and mortality: A health impact assessment for cities. *Environmental Health Perspectives, 125*(1), 89–96.

Nieuwenhuijsen, M. J. (2016). Urban and transport planning, environmental exposures and health-new concepts, methods and tools to improve health in cities. *Environmental Health, 15*(1), S38.

Nieuwenhuijsen, M. J., & Khreis, H. (2016). Car free cities: Pathway to healthy urban living. *Environment International, 94*, 251–262.

Nieuwenhuijsen, M. J., Khreis, H., Triguero-Mas, M., Gascon, M., & Dadvand, P. (2017). Fifty shades of green: Pathway to healthy urban living. *Epidemiology, 28*, 63–71.

Nieuwenhuijsen, M. J., Khreis, H., Verlinghieri, E., Mueller, N., & Rojas-Rueda, D. (2017). Participatory quantitative health impact assessment of urban and transport planning in cities: A review and research needs. *Environment International, 103*, 61–72.

Patz, J., Frumkin, H., Holloway, T., Vimont, D., & Haines, A. (2016). Climate change challenges and opportunities for global health. *JAMA, 53726*, 1565–1580.

Reisi, M., Aye, L., Rajabifard, A., & Ngo, T. (2016). Land-use planning: Implications for transport sustainability. *Land Use Policy, 50*, 252–261. Retrieved from http://www.sciencedirect.com/science/article/pii/S0264837715002896.

Saelens, B. E., Sallis, J. F., & Frank, L. D. (2003). Environmental correlates of walking and cycling: Findings from the transportation, urban design, and planning literatures. *Annals of Behavioral Medicine, 25*(2), 80–91.

SDG. (2015). Retrieved November 8, 2016, from https://sustainabledevelopment.un.org/?menu=1300

Stevenson, M., Thompson, J., de Sá, T. H., Ewing, R., Mohan, D., McClure, R., et al. (2016). Land use, transport, and population health: Estimating the health benefits of compact cities.

The Lancet, 388(10062), 2925–2935. Retrieved from http://www.sciencedirect.com/science/article/pii/S0140673616300678.
Tainio, M., de Nazelle, A. J., Götschi, T., Kahlmeier, S., Rojas-Rueda, D., Nieuwenhuijsen, M. J., et al. (2016). Can air pollution negate the health benefits of cycling and walking? *Preventive Medicine, 87*, 233–236. Retrieved from http://www.sciencedirect.com/science/article/pii/S0091743516000402.
UN Habitat. (2016). Retrieved November 8, 2016, from https://habitat3.org/the-new-urban-agenda/.
United Nations. (2014). *World urbanization prospects. The 2014 revision, highlights*. Retrieved from http://esa.un.org/unpd/wup/Highlights/WUP2014-Highlights.pdf.
UN. (2015a). *World urbanization prospects: The 2014 revision (ST/ESA/SER.A/366)*. United Nations, Department of Economic and Social Affairs, Population Division.
UN. (2015b). *World population ageing 2015* (ST/ESA/SER.A/390). United Nations, Department of Economic and Social Affairs, Population Division.
WHO. (2017). Retrieved November 6, 2017, from http://www.euro.who.int/en/health-topics/environment-and-health/urban-health/activities/healthy-cities/who-european-healthy-cities-network/what-is-a-healthy-city.
Woodcock, J., Edwards, P., Tonne, C., Armstrong, B. G., Ashiru, O., Banister, D., et al. (2009). Public health benefits of strategies to reduce greenhouse-gas emissions: Urban land transport. *The Lancet, 374*(9705), 1930–1943. Retrieved from http://www.sciencedirect.com/science/article/pii/S0140673609617141.
Woodcock, J., Franco, O. H., Orsini, N., & Roberts, I. (2011). Non-vigorous physical activity and all-cause mortality: Systematic review and meta-analysis of cohort studies. *International Journal of Epidemiology, 40*(1), 121–138.
WHO. (2008). Commission on Social Determinants of Health *Closing the gap in a generation: Health equity through action on the social determinants of health (Final Report, Executive Summary)*. Retrieved from http://whqlibdoc.who.int/hq/2008/WHO_IER_CSDH_08.1_eng.pdf.
WHO and UN Habitat. (2010). *Hidden cities: Unmasking and overcoming health inequities in urban settings*. Geneva: WHO.
Xia, T., Nitschke, M., Zhang, Y., Shah, P., Crabb, S., & Hansen, A. (2015). Traffic-related air pollution and health co-benefits of alternative transport in Adelaide, South Australia. *Environment International, 74*, 281–290. Retrieved from http://www.sciencedirect.com/science/article/pii/S0160412014002980.

Chapter 2
Health, Sustainable Development Goals and the New Urban Agenda

Joan Clos and Rosa Surinach

2.1 Introduction

Urbanization is one of the few human and social processes that spontaneously takes place as a result of government policies, including those in neighbouring regions, in the case of migration-related urbanization. Urbanization has both positive and negative implications. If managed effectively, it can result in thriving, productive and healthy cities, or, conversely, it can result in suffering and inequity on a large scale. We see this in our growing cities and in the projected growth of our combined urban population from 50% currently to around 70% in the next 30 years. This is why urbanization has become one of the most important global trends of the twenty-first century, and according to OECD recent research, its impacts will be more keenly felt than climate change in the immediate future. Investing in good urbanization is a guarantee of prosperity and development particularly in developing countries, where major urban transformations will inevitably take place. But how do we achieve good urbanization and, more importantly, how do we ensure urbanization results in good health and wellbeing?

The *UN Global Sample of Cities*, a UN-led initiative, was the first scientific analysis of world urbanization based on satellite images of a representative sample of 200 of the world's 4231 cities in 2010. The analysis revealed that current urban practices are unsustainable, even as they remain the main driver of economic development. Cities are increasingly less planned, leading to spontaneous urbanization (and urban poverty growth), which, in turn, decreases the quality of life for millions. The density of cities has also declined by 52.5% and 37.5% in developed and developing countries, respectively. Such urban sprawl and reduced density is a result of a

J. Clos (✉) · R. Surinach (✉)
United Nations Human Settlements Programme (UN-Habitat), Under-Secretary General, United Nations, Nairobi, Kenya
e-mail: jclos.mat@gmail.com; surinach@un.org

M. Nieuwenhuijsen, H. Khreis (eds.), *Integrating Human Health into Urban and Transport Planning*, https://doi.org/10.1007/978-3-319-74983-9_2

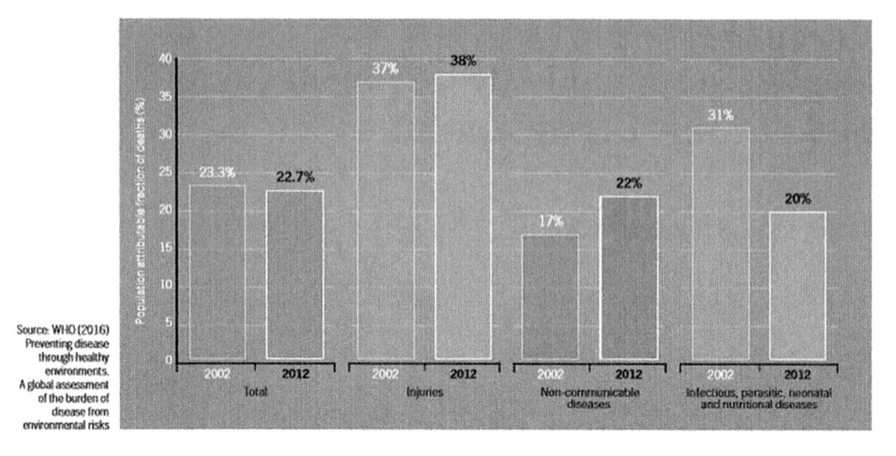

Fig. 2.1 Trend in the proportion of deaths attributable to the environment by disease group, 2002–2012

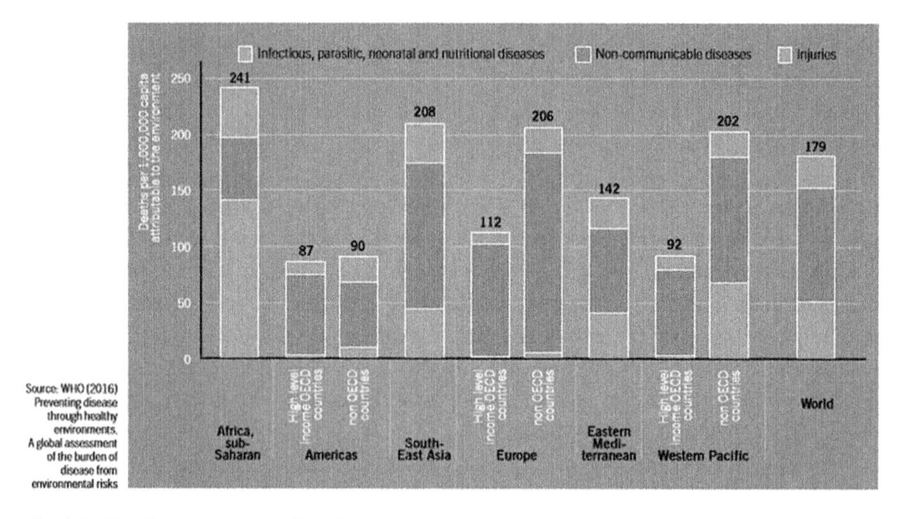

Fig. 2.2 Deaths per capita attributable to the environment, by region and disease group, 2012

change in lifestyle and has significant consequences for urban health both in terms of disease transmission and poor lifestyle.

A recent review of environmental health (WHO 2016a, b, c, d) has clearly shown the changing face of environmentally influenced diseases (see Figs. 2.1 and 2.2). In 2012, it was estimated that 12.6 million deaths were caused by poor environment, which influenced air, food and water quality. This equates to 23% of the global population. Further disaggregation shows that this figure raises to 26% of children under 5 years old and 25% of adults between the ages of 50 and 75. The difference between the total impacts for men and women is 2% extra for men, likely due to increased occupational risk, as the percentage of men employed is 50% higher than women. It is clear that most of these deaths are entirely preventable and that they could be greatly reduced by limiting exposure to multiple risk factors.

In terms of disease profiles, we see some patterns emerging. In many regions, we see a shift away from infectious, parasitic and nutritional diseases towards NCDs. This is mainly attributed to exposure to chemicals, poor air quality and unhealthy lifestyles. It does, however, give the clear message that in sub-Saharan Africa, for example, communicable diseases (mainly infectious, parasitic and nutritional diseases) will continue to exert a heavy health burden.

The current trends in urbanisation, when combined with epidemiological evidence, show an intensification of both infectious and non-communicable diseases in urban areas.

Historically, the evidence that poor environment is responsible for ill health is not new. The Report of the Sanitary Commission of Massachusetts (Shattuck 1850) not only clearly outlines the health impacts of poor environment, but it also compares Massachusetts with other cities worldwide. It also quantifies very elegantly the economic impact of early death of the breadwinner in a family and the costs of supporting those left behind. It is perhaps not surprising that this report was republished and considered essential reading for public health officials in 1948. The report has stood the test of time and is equally valid today, especially when its principles are applied to the developing country cities of today.

Moving to the present day and considering sustainable urbanisation, health is well articulated in the New Urban Agenda (NUA) which was negotiated in Quito, Ecuador, in 2016. The NUA has a *collective vision and political commitment to promote and realise sustainable urban development. It is a historic opportunity to leverage the key role of cities and human settlements as drivers of sustainable development.*[1] Its significance is that it is an action-oriented guide for national, subnational and local governments and that it is composed of five pillars to effectively address the complex challenges of urbanisation. These are national urban policies, rules and regulations, urban planning and design, financing urbanisation and local level implementation (UN-Habitat 2006).

The New Urban Agenda adopted at Habitat III differs from previous global conferences in its recognition of the fundamental linkages between health and sustainable urban development. It further clarifies the importance of these linkages and that health is not only about the provision of health-care services but that it also reflects decades of experience and advances in our understanding of how the shape and form of urban development influence the health of city residents. The NUA recognises that effective urban planning, infrastructure development and governance can mitigate risks and promote the health and wellbeing of urban populations. There is, however, a need to further address how specific urban policies can contribute to reducing disease burdens.

[1] Article 22 of the New Urban Agenda

2.2 Urban Policy and Its Contribution to Reducing Infectious and Non-communicable Diseases

Cities have the huge responsibility and opportunity of promoting healthier urban environments. They can do this through taxation and advocacy and by encouraging citizens to adopt a healthier lifestyle. Good examples include the control of tobacco, alcohol use and poor or "junk food". For infectious diseases, cities can provide incentives to improve access to cleaner water, sanitation, solid waste management and energy and through improved housing.

The global pandemic of non-communicable diseases, which include heart diseases, strokes, cancers and respiratory diseases, is the largest cause of death globally. In India, for example, cardiovascular diseases and cancer are the two leading causes of death. The design of urban space and planning policy greatly influence the main risk factors such as the lack of access to public transport and safe spaces for walking and cycling and increased use of motor vehicles which promote sedentary attitudes in addition to exposure to traffic-related air pollution. Citizens often have little control over these issues and certainly a limited ability to influence policy. When coupled with personal risk factors such as poor diet and alcohol and tobacco use, the result is a huge economic burden both in terms of lost productivity and health-care costs.

Considering infectious diseases, the lack of access to a good water supply and sanitation results in high levels of water-borne diseases including cholera and typhoid. Poor garbage or solid waste management causes blocked drainage channels which promote the spread of vector-borne diseases such as dengue, Zika and chikungunya, which are emerging disease threats in rapidly unplanned urban areas.

2.3 Urban Health Risks from Poor Air Quality and Extreme Climate Events

Air pollution is a major cause of NCDs. Vulnerable populations are often at risk from a variety of sources which may expose them to pollutants, including vehicle emissions and industrial discharges. In a recent analysis of data from 3000 cities around the world, 80% of them did not meet the recommended WHO air quality guidelines, a critical parameter being the presence of fine particulate matter, produced mainly by vehicle diesel-engines. 60% of European cities, 20% of North American cities and 40% of high-income Asian cities fail to meet adequate air quality standards. Effective planning policies can significantly reduce this problem (WHO 2016a, b, c, d).

Air pollution affects everyone in a city, both the rich and the poor. However, marginalised populations can often get the best benefits from air pollution mitigation policies. Many dwellers in low-income communities are exposed to multiple risk factors, not only air pollution from traffic, industrial plants and waste dumpsites but also indoor air pollution from poor cooking and heating and additional exposures in the workplace. Policies that support use of cleaner fuels for cooking, etc. can greatly assist in this respect. Women and children, as the most intensive users of housing, are generally in a more vulnerable situation.

The increasing effects of climate change are now very evident. Many people, especially in poorer communities, are forced to live in areas which are prone to flood risk and landslides. They lack access to green and blue spaces, they do not benefit from the reduced heat island effect and they lack access to areas free from pollution or places to exercise. In addition to physical health, wellbeing and, in particular, mental health are compromised when individuals do not have access to such facilities (Campbell-Lendrum and Corvalan 2007).

Urban design can help to reduce inequities both in service provision and health status and lead to more inclusive and productive urban societies.

2.4 Translating the Evidence into Good Policy

The body of evidence is considerable, but there is still room for improvement in terms of policy development. Despite the fact that it is likely that discrete, stepwise improvements in health care are achieved, there is the need to provide resource-constrained urban administrations with tools to help determine the disaggregated risks from multiple sources. To put it more simply, if you need to know where the most impact can be made for a specific investment, a better understanding of the discrete health benefits and their spatial disaggregation is critically important. The intra-urban differential then becomes very important.

In some cases, scientific research has greatly assisted policy. The case of cycling in polluted urban environments is a good example. For outdoor air $PM_{2.5}$ concentrations of 90 μm/m^3, the health benefit gained from exercise exceeds the health risks of inhaling particulates. However, if the $PM_{2.5}$ concentration exceeds 160 μm/m^3, the risks from air pollution are greater than the benefit of exercise (Tainio et al. 2016).

Concerning infectious diseases related to poor water and sanitation provision, it has to be stated that although the bulk of opinion shows that significant advances have been made where infrastructure provision has improved, there is a need for a more exhaustive research to quantify the incremental benefits from improved water, sanitation and waste management.

2.5 Focusing on the Most Vulnerable Populations

There is clear evidence that socio-economic deprivation and health inequity are inextricably linked. Early death and disability are significantly higher amongst slum and informal settlement communities. The current migration patterns have greatly exacerbated slum formation due to civil strife. In fact, in the Middle East region, the proportion of urban dwellers living in slums has increased, as compared to a decrease in other areas. Although the overall trends are declining, slum dwellers are increasing in terms of absolute numbers. The national statistics also mask the trends in some rapidly expanding areas such as the smaller urban centres in many sub-Saharan African countries (WHO and UN-Habitat 2016).

2.6 Critical Constraints to Delivering Improved Health

2.6.1 *Understanding Urban Form*

Perhaps one of the biggest hurdles to designing improved health systems is the failure to understand that the definition "urban" covers a range of possible situations depending on where you are in the world. It is often only an arbitrary or an administrative definition, and it has little bearing on the population demographics. Many settlements fall somewhere in between megacities and rural homesteads. These can be small towns, large villages, secondary urban centres or, indeed, conurbations. These settings have the population densities and the lack of access to basic services commensurate with the urban dwellings in megacities, but they suffer from the inefficiency of local authorities, who are severely resourced constrained and typically not fully supported by the central government and line ministries. A better understanding of the type of urban settlement will also be a critical factor in determining:

- Access to primary health-care facilities
- The potential to improve one's lifestyle
- The opportunities for community cohesion/engagement to support more formal health interventions

Assuming there is a better comprehension of the urban form, there is a need to fully understand that intra-urban health differentials exist. National estimates, or indeed city-wide data, may hide huge differentials between the wealthier and their poorer neighbours. Formal government structures have neither the resources nor the outreach to collect this information, and this is perhaps where local authority partnerships with civil society groups show a great deal of promise. Lower socio-economic residents are often exposed to multiple risks, as they lack services, they face risk of injury at their place of employment, and they live in areas prone to flooding or other risks and higher environmental stress.

2.6.2 *Multisectoral Approaches to Support Health Interventions*

It is generally being recognised that health systems delivery needs support from those outside the health sector if advances in global health are to be made. This is particularly important with respect to preventative approaches to health care, driven by poor environment, for example. The approach has been documented in areas such as malaria management (RBM 2013), but it has yet to reach its full potential. There is some evidence that it may be easier to promote the uptaking of multisectoral approaches in more resource-constrained urban areas. This may be driven out of necessity, where strong local leaders are convinced of the benefits, and, perhaps, at times of disease epidemics. The reason that the uptaking of multisectoral approaches is slow is also due to ignorance and fear and due to the concern that budget cuts or staff redundancy may result. In poorer cities, where resources are constrained, conditions will most likely encourage resourceful officials to adopt innovative approaches.

2.7 Urban Policies that Support Improved Public Health

2.7.1 Transport and Mobility

The provision of safe, affordable and accessible transport is a key element of the Sustainable Development Goals (SDGs) Target 11.2. It seeks to expand public transportation making it easier for vulnerable groups to use. Good mass transportation, high density and compact urban design go hand in hand. Efficient planning will result in lower travel times and costs, especially if the right mix of walking and cycling facilities are combined with public transport and rapid transit systems. This must be done in a way that reduces hazards for pedestrians and cyclists from motorised transport, which has been shown to greatly reduce the prevalence of NCDs. Encouraging less private motor vehicle use impacts positively on air and noise pollution. Public transport can also help to promote equity in access to employment, educational and social amenities.

Recent evidence indicates that significant health gains can be made from restricting the use of diesel engines for both commercial and private vehicles in dense urban systems. The exhaust gases are carcinogenic and emit a greater proportion of small particulate matter which is associated with increased stroke, heart and respiratory disease and early death (WHO 2003).

For example, traffic congestion in Mexico City needed urgent action to reduce congestion and improve air quality. Over the past 10 years, improved bus rapid transport systems serving around one million passengers a day encouraged 10% of the population to leave their cars at home. The reductions in air pollution resulted in the saving of 6100 days of lost work, together with reduced cases of bronchitis and deaths (WHO and UN-Habitat 2016). Additionally, an innovative bike-sharing programme was developed, and it was integrated with the metro rail and bus systems, offering one payment for all. Greater access to parks and green spaces was also made possible.

2.7.2 Energy

Inefficient energy use in homes in urban areas contributes significantly to both indoor and outdoor air pollution and greenhouse gas generation (WHO 2016c). A change in energy systems from fossil fuel to renewable sources will reduce these emissions and indirectly impact on health both in terms of extreme climate events and global warming.

It is reported that in Indian cities, close to a third of outdoor air pollution comes from indoor sources. The use of biomass fuels in Southeast Asia and African domestic environment is popular and results in more than one million deaths from chronic obstructive pulmonary disease (COPD). It is also responsible for half of all deaths from childhood pneumonia. Policies promoting the use of clean efficient cook stoves and similar devices for heating and lighting can avert a large number of these deaths. The use of photovoltaic lighting, for example, in addition to improved air quality, is much less risky in respect of injuries. Recent public-private partnerships (PPPs) have seen the cost of such solutions greatly reduced to a level at which they are affordable for the poor (WHO 2016c).

For example, a 2008 scheme in Cape Town, South Africa, where PPPs were used to initiate a housing improvement programme in a low-income area Khayelitsha Township, has had significant health outcomes. Houses have been upgraded with insulation, compact fluorescent lamps and solar heating, which has reduced energy costs by approximately USD 110 per household and 2.8 t of CO_2 per household per year. Huge reductions in respiratory diseases were also shown.

2.7.3 Water Sanitation and Waste Management

Not having access to clean and safe water and to sanitation facilities ranks as one of the highest causes of early death, particularly in children. There are not only risks from gastrointestinal diseases but also of many diseases related to water availability, such as the so-called water washed diseases, which include skin infections and trachoma. The results, when combined with malnutrition, can kill very young children rapidly. In addition to early death, many of the more chronic water-borne diseases cause anaemia. Chronic helminth infections result in sick children whose learning capacities are affected.

Solid waste management is also becoming increasingly important, as it is a source of local contamination, it pollutes water resources and it contributes to air pollution through the burning of waste and spontaneous combustion at landfilling sites. The collection and recycling of wastes is a popular occupation amongst young children, who reside at refuse tips and make a living from separating and selling valuable materials. This is often carried out in an unsafe manner, with little protection. Increased cases of chronic respiratory infections, chloro-acne and other diseases are common. Discarded waste also serves as a reservoir for disease vectors, such as rats, sandflies and mosquitos. This results in increased incidence of vector-borne diseases such as plague and leishmaniasis. More recently, there is evidence that Zika vectors bred in discarded refuse (PAHO 2016).

Although in the developing world much reusable waste is recycled, the remaining organic fraction is frequently discarded. This contributes to greenhouse gas emissions in the form of methane from landfills and wastewater sludge and biomass. Energy from waste needs to be better utilised as much of the organic material could be used to generate biogas for power generation.

In Nairobi, Kenya, between 2000 and 2012, a coordinated effort between government ministries, development agencies and civil society greatly improved access to water and sanitation in the city slums. Piped water access increased from 3% to 60% and the use of water-borne sanitation increased by six times. As a result, there have been marked declines in childhood death from diarrhoea. The results clearly demonstrated that removing the excreta from the immediate living environment is perhaps the most critical step in breaking the disease transmission route (WHO and UN-Habitat 2016).

2.7.4 Land Use Planning and Design and Provision of Green Space

Land use planning plays a critical role in health, as green spaces promote exercise and, hence, healthy lifestyles. But perhaps most importantly, green spaces have a great impact on mental health and wellbeing. This is an often neglected area of health which is responsible for a significant proportion of health care and social costs.

Everyone benefits from a well-designed city. Neighbourhood development strategies foster local traders and other services and amenities, which reduce transport burdens themselves. Careful design of urban spaces can reduce segregation and isolation and encourage interaction across different income groups and generations. Recent experiments, linking elderly residents in care homes with primary school children in Australia and the UK, noted significant improvements in the quality of life for the elderly.

Greening strategies designed to improve urban spaces are being recognised as a cost-effective intervention, compared to changes in infrastructure. The use of tree belts to filter air and dust and to reduce urban heat islands is well documented (FAO 2016). Tree planting can reduce energy costs by providing shade and so a temperature reduction in the need for climate control. In arid zones, they can be used to improve water retention and to act as wind breaks. Greening can also improve the social spaces where people meet, as they induce. Recycled wastewater can be used in urban food production for local consumption or to enhance lawns and verges. If the wastewater is used for non-food crops, standards of treatment can be relaxed. Tree-lined streets encourage pedestrian traffic and attract birds and other wildlife.

2.7.5 *Food and Nutrition*

Urban design and access to food systems can sometimes mean that obesity and stunting can co-exist in the same low-income area. If you live in a poor neighbourhood where there is little fresh food available, you are more likely to have a poor diet, rich in sugars and fats and highly processed food. This, in turn, means a greater likelihood of obesity. The proximity of fast-food outlets to schools and colleges has shown that institutions closer than 400 m show greater incidences of childhood obesity.

Effective planning can promote local food production, stimulate urban agriculture and enhance the possibility of recycling of waste to further promote food production. Much food is wasted during transportation, and supermarkets in wealthy areas of the city regularly throw food away whilst poor communities in the same city have insufficient food.

2.7.6 *Housing*

The method of construction and the contents and furnishings in many houses constitute a significant health risk. Aside from accidents from fires in cooking and heating, exposure to toxic substances such as lead in paint, asbestos, etc. mean that such risks kill more people than road traffic accidents (WHO 2011). Pest infestation can cause food poisoning, and overcrowding carries a major risk of mental illness. Poor housing is also expensive to run. Homes without insulation use too much energy, promoting poor ventilation and causing mould and associated respiratory illness. Conversely, ventilation may be very important to reduce risk from toxic building materials of naturally occurring substances such as radon gas (WHO 2011).

2.7.7 *Slum Upgrading*

Slum upgrading offers a great opportunity to reduce urban inequities, especially health inequities. Slums are frequently the "dormitories" where a large proportion of the urban workforce reside. Many support families in the rural hinterland, and as such, they may or may not have an interest in improving their living conditions. After all, if rents can be kept as low as possible, repatriating money to rural-based families can be maximised. Slums are increasingly being seen as an important part of the city, often cited close to the cities business districts. Intra-urban slums are often established on vacant sites. It is often not realised that many slums are complete communities that offer a parallel range of services found in the formal city. Integrating not just the physical infrastructure, but also breaking the social divide, can greatly enhance the overall economic gains. Basic standards with respect to housing need to be established, and supportive infrastructure such as water, sanitation, road networks, power, etc. must be developed. Many poor communities pay for services that the rich get for free. Improving slums and, in particular, reducing poverty and increasing security benefit the whole city and lift an otherwise frightening reputation that slum populations in cities may have for outsiders, who see them as no-go areas where safety and security are compromised.

2.8 The SDGs, the New Urban Agenda and Health

The SDGs present a real opportunity to bring change and, for the first time, to look at goals shared amongst all citizens of the world. It is true that the unfinished business of the Millennium Development Goals needs to be completed, but it is also clear the world is very much at risk as many rapidly industrialising economies exert a toll on the environment, with dramatic consequences if they continue without some course correction. The SDGs and the NUA are closely linked and mutually supportive. Health is a cornerstone of both.

2.8.1 Setting the Health-Based Targets for Clean Air, Water and Energy

The SDGs aim to protect the planet and to reduce the impact of human kind, but the consequences on health pose a more immediate threat. International standards for the SDGs targets need to be developed in a way to ensure protection without unnecessarily imposing a burden on countries, both from the costs of monitoring systems and in setting targets that are both impractical and unachievable.

Recent work on SDG 6 on access to water and sanitation by UN-Habitat and by the WHO (2015) has indicated that a progressive approach to monitoring is the preferred way ahead. The idea is that modest monitoring frameworks can be adopted at the outset, with the more in-depth systems being used as countries better see how monitoring can contribute to national processes.

2.8.2 Urban Polices and Their Health Implications

The health sector needs to provide better evidence-based guidance on the impact of urban policies, building on the work on its guidelines on drinking water, household energy, air quality, etc. The guidelines for housing are currently under development (WHO 2016d). These documents will provide practical guidance to local authorities to help in setting local bye-laws and other policies. Some areas need further development including waste management. The WHO "health in all policies" framework provides a suitable way to address health inequities in the urban setting.

2.8.3 *Reducing the Health Costs Through Good Urban Design*

More evidence is needed on the opportunities to reduce health costs by improved urban design and management. It is very likely that if current lifestyle patterns continue, the health-care cost burden, even for the best systems in the world, will collapse. The implications of the obesity epidemic in children can result in chronic ill health when they reach adulthood. If the current threats from drug resistance come to fruition, diseases that are now easily treated could become killers. Scarce resources could be stretched to their limits. Urban design for good health outcomes will be a popular tool to help combat disease.

Further analysis on the costs of inaction also needs to be done. Cost-benefit analyses drawing on the large databases of health services can be undertaken. New schemes can use such tools to do health economic assessments.

2.8.4 *Monitoring and Tracking Health Impacts in Cities*

Urban inequities, especially in low- and middle-income countries, will need to be further highlighted and better understood. Differentiated approaches to health care, advocacy campaigns and resources to tackle particular health problems can be targeted to those who are most needy, further reducing health costs. As more complex models will allow socio-economic strata to be studied more closely, spatial analysis of health-care patterns will become more important.

2.8.5 *Community Engagement in Supporting Improved Health Outcomes*

Currently, the huge potential locked up in communities to contribute to urban health improvements remains under exploited. Not only can communities play a key role in monitoring health patterns, but they can also assist in the improved delivery of services. Civil society has much expertise in organising advocacy campaigns and providing oversight on access to official facilities such as health-care centres, etc. Tapping this potential will need good local leaders with a vision of improved health and productivity for their cities. Matching capacity will also be needed within local authorities. Multi-stakeholder partnerships, used successfully in other sectors, can be used by the health sector. Sometimes, the efforts will translate into improved local legislation.

2.9 Conclusions

The unprecedented rates of urbanisation will bring significant changes to the health status of urban residents, and when coupled with climate change, many of the effects will be amplified, specifically heat and water stress. In developed economies, the elderly population is increasing, and, in the developing world, younger populations are at risk of poor lifestyle choices.

Not only inadequate urbanisation will increase the possibilities of communicable disease transmission, but poorly designed urban spaces will discourage healthy lifestyles and put urban populations at higher risk of many NCDs. The provision and cost of basic services, including access to primary health care, are linked to urbanisation density. If urbanisation is "inefficient", then the costs of provision of services are going to be much higher.

The disease threats fall into the two basic categories of the NCDs and communicable diseases. Countries at all levels of development will not escape the effects of these epidemics, and there are a variety of new threats which have the potential to cripple even the best health systems.

There is a huge need for disaggregation of urban health data to understand the intra-urban differences. This, when overlaid with socio-economic data and physical infrastructure and topology, can provide the information to support decision-making at the local level.

In terms of the priorities, the developing world is at risk from both communicable diseases *and* NCDs. Communicable diseases still exert a heavy toll, such as those resulting from contaminated water and lack of sanitation and vector-borne diseases (mainly malaria). However, even in the poorest economies, poor lifestyles are killing in similar proportions. For more developed economies, NCDs are the biggest killers, with mental health conditions deteriorating, including for displaced persons and refugees.

There is still a strong focus on curing the sick rather than prevention. Globally, approximately one third of diseases are the result of environmental factors, with children being some of the most vulnerable. If we consider the multiple risk factors that some communities live in (e.g. slums), the figures are most likely much higher.

Future health burden will never be reduced without an increased effort on prevention of disease through resilient, liveable and healthy cities, where access to basic services and reduction of environmental risks reduce disease and early death. Ageing populations will become a drain on health services if we don't ensure citizens have a healthy and active life. Threats from overuse of antibiotics are also an issue. If resistant strains develop, everyday diseases could kill.

As a priority, urgent analyses using the SDGs framework (and the UN-Habitat sample cities) of the health status in cities, disaggregated and overlaid with socio-economic profiles are needed. A city-city analysis of the problems and solutions driven by the urban authorities is needed alongside the realisation that many of the interventions that frame and support health improvements are needed to stem from outside the health sector. Localised plans of action for "sanitary revolutions" to reduce communicable diseases and campaigns to combat NCDs, led by urban design and continuous monitoring, need urgent attention.

References

Campbell-Lendrum, D., & Corvalan, C. (2007). Climate change and developing-country cities: Implications for environmental health and equity. *Journal of Urban Health, 84*(3 Suppl), i109–i117.

FAO. (2016). *Benefits of urban trees*. Rome: Food and Agriculture Organization. Retrieved October 20, 2017, from http://www.fao.org/resources/infographics/infographics-details/en/c/411348/.

PAHO. (2016). Retrieved October 20, 2017, from http://www.paho.org/hq/index.php?option=com_docman&task=doc_view&andItemid=270andgid=37114andlang=en.

RBM. (2013). *Multisectoral action framework for malaria, RBM partnership*. Retrieved October 20, 2017, from http://www.rollbackmalaria.org/files/files/about/MultisectoralApproach/Multisectoral-Action-Framework-for-Malaria.pdf.

SHATTUCK. (1850). *The 1850 report of the Sanitary Commission of the State of Massachusetts*. Retrieved October 20, 2017, from www.deltaomega.org/documents/shattuck.pdf.

Tainio, M., de Nazelle, A. J., Götschi, T., Kahlmeier, S., Rojas-Rueda, D., Nieuwenhuijsen, M. J., et al. (2016). Can air pollution negate the health benefits of cycling and walking? *Preventive Medicine, 87*, 233–236.

UN-Habitat. (2016). *The New Urban Agenda*. Retrieved October 20, 2017, from http://habitat3.org/the-new-urban-agenda/.

UN-Habitat and WHO. (2015). *Report of a meeting on GEMI*. Geneva: World Health Organization.

WHO. (2003). *Investing in mental health*. Geneva: World Health Organization.

WHO. (2011). *Health in the green economy: Health co-benefits of climate change mitigation – Housing sector*. Geneva: World Health Organization.

WHO. (2016a). Preventing disease through healthy environments. In A. Prüss-Ustün, J. Wolf, C. Corvalán, R. Bos, & M. Neira (Eds.), *A global assessment of the burden of disease from environmental risks (WHO, 2016)*. Geneva: World Health Organization.

WHO. (2016b). *Air pollution levels rising in many of the worlds' poorest cities*. Geneva: WHO.

WHO. (2016c). *Burning opportunity: Clean household energy for health, sustainable development, and wellbeing of women and children* (p. 2016). Geneva: World Health Organization.

WHO. (2016d). *Health and sustainable development – Housing indicators, guidance and tools*. World Health Organization. Retrieved October 20, 2017, from http://www.who.int/sustainable-development/housing/indicators/en/.

WHO and UN-Habitat. (2016). *Global report on urban health: Equitable, healthier cities for sustainable development*. Geneva: World Health Organization and UN-Habitat.

Chapter 3
Sustainable Cities, Policies and Healthy Cities

Garett Sansom and Kent E. Portney

3.1 Introduction: Linking Sustainability Policies to Public Health

While the concept of sustainable development has been used by certain private, public, and academic institutions for generations, it was not fully defined nor did it enter into common usage until the 1987 World Commission on Environment and Development and the subsequent report entitled *Our Common Future*. Here we rely on the definition most commonly associated today with sustainable practices that "Sustainable development is development that meets the needs of the present without compromising the ability of future generations to meet their own needs" (WCED 1987, p. 8). Despite this abstract definition, it can lead to fairly specific policy requirements.

In pursuit of attaining a sustainable society, further requirements have been placed on this concept; particularly, that environmental stewardship must be maintained, economic growth and development should be preserved, and equity should be explicitly sought. These three core concepts of sustainability—environment, economy, and equity—lend themselves to serving the well-being of individuals. Indeed, we proceed under the expectation that the pursuit of sustainability is largely undertaken for the purpose of improving and protecting the health of people and the

G. Sansom (✉)
Institute for Sustainable Communities, Texas A&M University, College Station, TX, USA

Texas A&M University, College Station, TX, USA
e-mail: gsansom@arch.tamu.edu

K. E. Portney (✉)
Texas A&M University, College Station, TX, USA

Institute for Science, Technology and Public Policy, The Bush School of Government and Public Service, College Station, TX, USA
e-mail: kportney@tamu.edu

M. Nieuwenhuijsen, H. Khreis (eds.), *Integrating Human Health into Urban and Transport Planning*, https://doi.org/10.1007/978-3-319-74983-9_3

ecosystems on which they rely. Here we look at the policies and programs US cities adopt and implement in pursuit of sustainability to see, at a most basic level, whether there is evidence that these policies and programs are at all related to public health. As discussed below, these sustainability policies and programs largely focus on protecting and improving the biophysical environment of the city, although they often use these policies in pursuit of economic growth and development.

This chapter addresses a specific question concerning the relationship between city sustainability policies and public health outcomes. There is little doubt that a central rationale underlying the sustainability programs and initiatives in cities is rooted in some conception of public health. What we mean by this is that advocates of urban sustainability often offer the argument that by pursuing sustainability policies, cities will improve the health of their populations. The dominant view of sustainability, of course, is related to the quality of the biophysical environment, and cities' policies in pursuit of sustainability promise to improve that environment. But many advocates of urban sustainability go beyond efforts to protect and improve that environment, suggesting that the ultimate purpose is to improve the health and well-being of cities' respective populations. Embedded in this notion is the idea that sustainability seeks to achieve greater equity in outcomes, especially public health outcomes. Curiously, while there are many studies of cities' sustainability policies and programs, there are very few efforts to empirically tie these policies to health outcomes. In short, we know much more about why some cities are more likely to pursue sustainability policies than we know about what these policies achieve. Comprehensive and definitive answers to this type of question are well beyond the scope of this chapter. What we wish to accomplish here is to explain the conceptual connections between sustainability policies and programs on one hand and public health outcomes on the other. We then examine the empirical patterns of relationship between sustainability and public health outcomes across the largest cities in the United States. Far from being definitive, this chapter seeks to begin a conversation about what kinds of public health outcomes and results seem to be produced when cities adopt and implement sustainability policies. Embedded in this is a call for better city-specific measures of public health outcomes to facilitate future research.

This chapter proceeds with a brief historical review of the development of public health to argue that the time is ripe for public policies seeking to achieve greater sustainability to be understood, at least in part, as efforts to improve health and well-being of city populations. It then looks explicitly at how the pursuit of sustainability is conceptually connected to community health and healthy communities. It then turns its attention to a preliminary investigation into the empirical links between city sustainability efforts and the improvement of public health outcomes, with special reference to obesity. As will be discussed, obesity has proven to be especially difficult for public health professionals to adequately address on a population level. If sustainable policies can be correlated with a reduction, or slowing, of the hard case of obesity, it will represent a significant potential for future interventions.

3.2 The Epidemiologic Transition and Modern Health Concerns

Currently, within the developed world, health concerns differ significantly from those of our ancestors. An examination of the history for the causes of mortality and morbidity within the United States shows a dramatic change from largely infectious and preventable diseases to chronic noninfectious conditions. All industrialized nations have undergone this dynamic population shift and witnessed the transition from a state of high mortality and birth rates to relatively low mortality and birth rates. This characteristic difference in disease distribution of a population from acute infectious conditions to chronic degenerative diseases, as the age pattern of mortality goes from younger to older individuals, is known as the epidemiologic transition.

The epidemiologic transition, first posited by Abdel Omran (1971) and expanded upon by modern scholars, outlines the stages that nations travel through in regard to health outcomes. Analysis of historical data on population health, as well as an examination of current global conditions, reveals four clear transitional stages. The importance of understanding these provisional steps is underscored by the realization that strategies for creating healthy communities differ dramatically depending upon where along the transition a group falls.

Since the advent of agriculture, the majority of the human experience has been within the first stage, referred to as the *Age of Pestilence and Famine.* This stage is characterized by high and fluctuating mortality rates and extended periods of hardships. Infant mortality is high, as are birth rates, and the average life expectancies typically range between 30 and 40 years old. The systematic theory of population growth, first proposed by the economist Thomas Robert Malthus, claims that instances of epidemics create a balance by providing a "check" on unsustainable growth patterns (Gilbert 2008). An illustrative example of this population check was witnessed during the devastating Eurasian pandemic known as the Black Death.

The Black Death, a phrase coined centuries after the event, took place between 1346 and 1353. This disease swept across western Asia, the Middle East, North Africa, and Europe, causing a catastrophic loss of life for urban and rural citizens. Estimates range as high as 200 million deaths were experienced during these years (Benedictow 2004). As trade increased along the Silk Road during this time, the pathogen *Yersinia pestis bacterium* was carried by fleas, which lived atop black rats (Bos et al. 2011). These vector species easily disseminated the pathogen across nations.

The second transitional stage is referred to as the *Age of Receding Pandemics.* While populations still experience local, or even far-reaching, disease clusters, the distinguishing difference is that these events will not stifle the decline in mortality or the increase in life expectancy. Population growth is not only sustained but begins to exhibit exponential growth. Urban growth during this time often results in heavily crowded city centers, which become the primary location for disease outbreaks.

Perhaps the most famous epidemic during this stage is the 1854 Cholera outbreak in London.

In the mid-seventeenth century, London had become overcrowded, poorly managed, and heavily polluted by human waste and industrial effluent. This was not something lost on those living or visiting the city, it was often referred to as "The Great Stink." In the words of Charles Dickens "I can certify that the offensive smells, even in that short whiff, have been of a most head-and-stomach-distending nature" (Dickens 2011). It is perhaps not surprising that within these conditions there was a severe outbreak of cholera in the city, claiming the lives of 616 individuals. It was traced back to water being drawn from a pump on Broad Street and was famously investigated by John Snow, known as the father of epidemiology, wherein he hypothesized that the water, not the foul-smelling air, caused the illness (Paneth, Vinten-Johansen, Brody, & Rip 2011).

The *Age of Degenerative and Human Created Diseases* is the third stage of this transition. The fourth stage is an extension of this with delayed onset of these degenerative illnesses. A marked increase in the average life expectancy can be expected during these ages. With longevity comes a host of chronic health conditions associated with old age. Death from infectious disease becomes increasingly rare as cardiovascular disease, cognitive disorders, and malignancies become increasingly common. Developed countries have transitioned into the fourth stage over the last few generations.

The United States has witnessed a reversal on health concerns as pandemics gave way to degenerative diseases and anthrophonic causes of mortality. According to the Centers for Disease Control and Prevention (CDC), chronic conditions, such as heart disease and cancer, comprised the majority of causes of mortality in 2010. In 1900, influenza and tuberculosis were the top two causes of death. Mortality rates for heart disease in 2010 were 192.9 per 100,000 and cancer was 185.9 per 100,000; combined they accounted for over 63 percent of all deaths, but less than 20 percent died of these causes in 1900. Forty-six percent of individuals died in 1900 from infectious diseases compared to less than three percent in 2010; Figure. 3.1 (National Center for Health Statistics 2016) further illustrates this change.

While success in vaccination campaigns, hygienic practices, access to health care and screening programs, as well as a variety of clinical advancements over the years, has protected populations from the historical causes of mortality, we are now faced with new challenges for which our standard model of care is ineffectual in meeting. Within the United States, the overall prevalence of obesity among adults has risen to 37.9 percent in 2014, up from 30.5 percent in 2000. Obesity disproportionality affects Hispanic, 42.5 percent, and African American, 48.1 percent, communities. This trend is correlated with rises in cardiovascular disease, diabetes, and cancer. The Trust for America's Health has said that "obesity remains one of the biggest threats to the health of our children and our country…" (TFAH 2016, p. 7).

Perhaps even more alarming is the increase in youth and adolescent chronic conditions. The World Health Organization (WHO) has declared that childhood obesity is one of the most serious public health issues in the twenty-first century. WHO researchers in the Department of Nutrition for Health and Development have

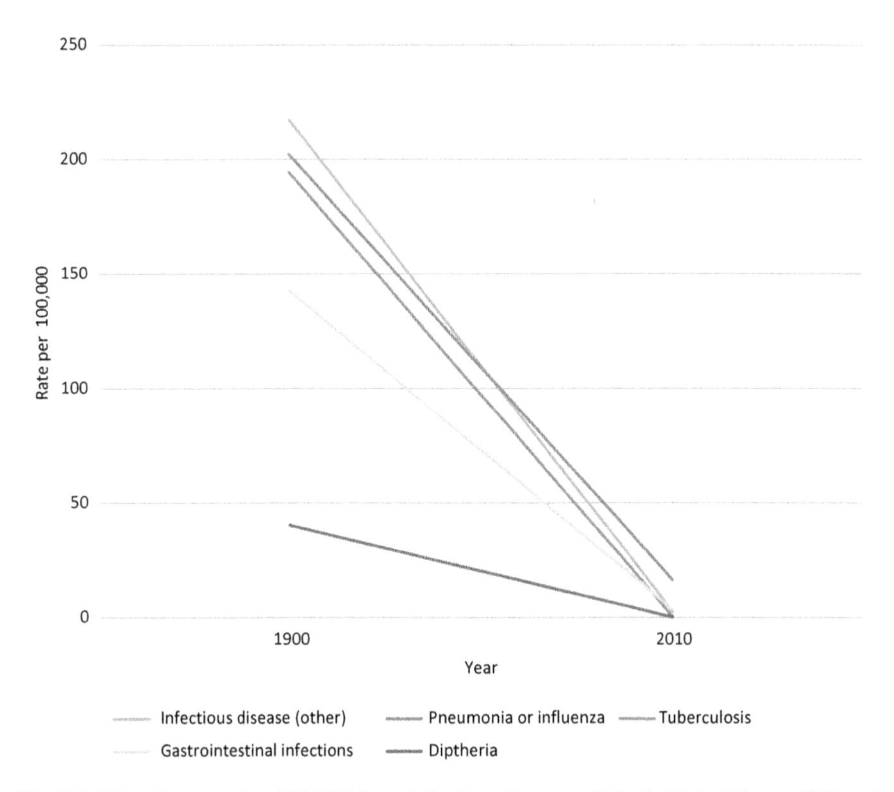

Fig. 3.1 Mortality rates (per 100,000) from infectious diseases within the United States, 1900 and 2010. Source: National Center for Health Statistics 2016

estimated that there are over 43 million obese preschool-aged children globally, a 60 percent increase since 1990 (de Onis, Blossner, & Borghi 2010). In the United States, an estimated 17 percent of US children and adolescents aged 2–19 years are obese, and another 16 percent are overweight in 2014, compared to only 5 percent obese and 10 percent overweight in 1974 (Fryar, Carroll, & Ogden 2016). There are already an estimated 2.8 million deaths annually associated with obesity; as the current generation ages, this number is expected to rise in the coming decades (WHO 2017).

Risk factors for illness have also changed. Roughly one in every four diseases diagnosed globally can be attributed to environmental exposures that are preventable. Furthermore, one in every three diseases in children under the age of five is caused by environmental exposures. Changes in our built environment have also changed risks, and nearly one million deaths annually can be attributed to road traffic accidents and unintentional injuries (Prüss-Ustün, Wolf, Corvalán, Bos, & Neira 2016).

This section is intended to illustrate the changing causes of morbidity and mortality, as well as the shift in exposures of concern. In 2008, for the first time since the dawn of civilization, the majority of earth's inhabitants lived in an urban center in place of rural locations. As industrial growth continues across the globe, and large migrations of individuals move into areas of increasing density, a new public health paradigm needs to be established. Current approaches to health interventions are often rooted in a framework that does not adequately represent current conditions. Creating improved health outcomes and emotionally fulfilling lives will inexorably be linked to the natural and built environment, access to healthcare and healthy living options, and equitable distributions of risks and benefits, namely, by establishing sustainable communities well poised to create a culture of health.

3.3 Pursuing Sustainability as a Strategy for Addressing Current Chronic and Degenerative Diseases

The idea that sustainability can be an appropriate vehicle for advancing public health is not new. Although earlier conceptions of this linkage focused mainly on the pursuit of sustainability as a way of reducing exposures to environmental hazards and toxins, more recent conceptions have been somewhat more expansive. The previous section explored many of the impressive efforts in curbing infectious diseases but also the sobering statistics for the rise of chronic and degenerative diseases. The silver lining for many chronic ailments—such as cardiovascular disease, cancer, stroke, and diabetes—is that they are preventable through nonclinical lifestyle changes. However, many conditions are associated with, or exacerbated by, environmental conditions such as air pollution. These environmental conditions are often outside of the individuals, control to change. These difficult to avoid exposures are associated with asthma, emphysema, and chronic obstructive pulmonary disease (COPD). Introduction of truly sustainable communities will address each of these to create healthy populations.

The rise in obesity rates within the United States paints a concerning picture for the future health of citizens. Obesity is highly correlated to a number of expensive and potentially debilitating diseases, such as type II diabetes and colon cancer, in addition to emotional and social harm. Despite having a firm understanding for the causes of obesity, namely, living a sedentary lifestyle and overconsuming calorie-dense foods, no efforts have been successful in stopping this increasing trend. The failure of standard approaches to health interventions is illustrative for the need to adopt a new approach with sustainability in mind.

The connection between the pursuit of sustainability and achievement of public health goals in American cities has been well documented. Arguing that the connection is finding its way into practice, for example, Jason Corburn (2009) describes how the city of San Francisco has made significant strides in planning for sustainability in a way that readily accommodates public health and healthy living goals.

His vision is that decentralized and resident-engaged planning facilitates the goals of achieving a more sustainable and equitable biophysical environment and public health outcomes.

3.4 Governance and Environmental Justice

Equitable distribution of diseases, and the burden of environmental pollution, is not realized within most urban centers in the United States. Research in minority communities has conclusively shown that poor minority populations shoulder an undue burden of exposure to industrial buildings, waste facilities, and urban pollution compared to majority populations. Further, low-income and marginalized communities, regardless of race, are more likely to live in areas characterized by poorer environmental conditions. Attempts to change or mitigate these unequal exposures is referred to as environmental justice.

Professor Robert Bullard, known as the father of environmental justice, in 1990 published his landmark book *Dumping in Dixie: Race, Class, and Environmental Quality*, which brought to light the poor environmental conditions that many vulnerable communities experienced in the United States. Born as an extension of the civil rights movements, proponents of environmental justice point to the established inequalities of exposure to pollutants and their related health outcomes, as well as the lack of representation within governing boards or staff of major national non-governmental organizations, such as the Sierra Club.

Identifying the challenges associated with marginalized groups has proven to be easier than showcasing a series of solutions. This twofold issue—that of representation and local conditions—may be accomplished through revitalizing local participation in governance. Empirical evidence suggests that support for, and inclusion of, environmental advocacy groups is associated with sustainability programs within urban centers throughout the United States (Berry & Portney 2013). Increasing local capacity is greatly impacted through empowered groups that are active in pursuing local interests.

Citizen involvement in many facets of government action has been greatly influenced and enriched through participation with stakeholders, interest groups, and individuals. Research on public health and hazard programs show particular promise in recent decades. Implementing high-quality, evidence-based response, health interventions, and preparedness programs has been shown to reduce human and ecological harm in communities within the United States. Program evaluations have indicated that participation from the community in pre-disaster planning and education outreach efforts is a critical facet for improving results.

Having a truly representative body of officeholders in federal, state, county, and local offices is also critical in addressing the needs from groups who traditionally had little voice in change. There is a startling gap between the amount of traditionally relegated groups, such as women and people of color, within the United States compared to how many hold positions in the government. For instance, women, who comprise 51 percent of the population, make up 29 percent of officeholders. Over 37 percent of individuals are people of color, but only 10 percent are officeholders.

3.5 The Natural and Built Environment

One would be hard-pressed to locate a more impressive accomplishment of human ingenuity than our built environment. Past generations would be utterly unprepared for the vast geographic footprint of modern cities, the interconnected highway and road systems, or the towering concrete buildings. Practically every moment, from waking to sleep, individuals are living in an environment that has been crafted from the natural world in a manner that is more comfortable and convenient for humans. Traveling across great distances requires almost no physical exertion; furthermore the trip will likely be made more pleasant through artificial heating and cooling. Within developed countries, careers often require sitting for great lengths of time in front of computers or other electric devices. While these changes have brought about great benefits to productivity, comfort, and development of knowledge, humans did not evolve from our modern lifestyles. More recent trends in public health and urban planning try to incorporate the benefits from our natural world, as well as modern development for sustainable urban growth.

Sustainable development can be envisioned as a unique approach that meets the needs of the present without compromising the ability of future generations to meet their own needs. It further encourages access to greenspace, walkability, and spaces to promote mental health. Sustainable development is approached on many facets—economic growth, public health, cultural and social development, and environmental preservation, just to name a few.

It would be difficult to overstate the impact that the built environment has on the well-being of individuals and communities. Traditional city planning and development within the United States has been primarily concerned with business development, trade, vehicle transportation networks, commerce and industrial placement, and public works systems. Urban planning has also provided highly successfully water purification systems and effluent disposal. However, there have been negative outcomes, which can be seen with air pollution, fossil fuel depletion, and land use contamination. Sustainability on the city level should include a long-term assessment of the environmental, economic, and equity impact of policy and development decisions. This approach is how sustainability can begin to address the modern predicament of health concerns.

The pursuit of sustainability offers a plausible solution for anthropogenic causes of diseases, as well as improved opportunities for preventing and mitigating the impacts from chronic conditions. The link between energy production, pollution, and health outcomes has been well documented. In 2015, President Barack Obama and the EPA announced the Clean Power Plan—a program put in place to take strong action against carbon pollution from power plants in the United States. EPA estimates that the implementation of this program will result in public health and climate benefits worth between $55 and $93 billion per year by 2030 and cost between $7.3 and $8.8 billion. Furthermore, the reduction on particle pollution is also projected to lead to a reduction in up to 3500 premature deaths, 90,000 asthma attacks in children, and 300,000 missed school and work days (United States Environmental Protection Agency 2017). This can serve as an initial step to promote sustainability through environmental stewardship and equitable reductions in risky exposures.

Improving the infrastructure, landscape design, and building management to encourage physical recreation and provide access to green walkable spaces provides a sustainable solution for improving health and reducing morbidity. Walkable neighborhoods reduce obesity and diabetes rates regardless of age, gender, or socioeconomic status (Rundle & Heymsfield 2016). Researchers have even found that those living in areas characterized by dense greenspace and natural environments have reduced mortality compared to those living in areas devoid of greenspace. Cognitive function has been shown to improve as well (Nieuwenhuijsen, Khreis, Triguero-Mas, Gascon, & Dadvand 2017). While the environmental, physical, and mental health benefits are clearly documented, the value that green space has for the health of the overall community can be tremendous. Civic engagement facilities, recreational and natural parks, as well as outdoor recreation facilities can have clear benefits for fostering positive social ties and a sense of place and can build social capital among individuals and groups.

Creating a monumental change in how health is perceived and maintained in developed countries, by adopting a new culture of health, is best served through adoption of sustainability. Mental, physical, and social well-being have been shown to be linked to the concepts of sustainability, what is required not is nationwide buy-in to these well-defined and accomplishable goals to meet the needs of the twenty-first century.

3.6 Sustainable Cities' Policies and Public Health in the United States

The idea of sustainable cities has been well represented in the United States. Over the last 20 years, many cities have created significant sustainability plans, often as a result of their long-term strategic planning processes. According to one estimate, by 2015 at least 50 of the largest 55 cities operate under a sustainability plan. Most such cities engage in making public policies and managing city programs in ways that are consistent with trying to achieve greater sustainability, environmental quality and equity, and energy efficiency. Cities that seem to take the pursuit of sustainability more seriously have been shown to engage in efforts to plan and implement policies on renewable energy and climate protection, public transit, waste reduction, water conservation, protection of environmentally sensitive land, green building, and dozens of other programs. Many of these cities explicitly include efforts to affect a variety of public health outcomes, from reducing exposures to toxics through asbestos and lead paint remediation to encouraging exercise through bicycle ridership programs and to promoting locally grown produce through community gardens and farmers markets and many other programs. As noted earlier, Corburn (2009) has documented these efforts in San Francisco, and similar efforts have been made in a wide array of cities including Seattle, Portland, New York, Philadelphia, Chicago, Los Angeles, and many others. Indeed, many of the sustainability plans adopted by cities contain explicit chapters dedicated to achieving targeted public health outcomes. For example, New York City's OneNYC Plan, the city's comprehensive strategic sustainability and resiliency plan that has taken the place of its predecessor PlaNYC, focuses on specific goals for reducing infant mortality, obesity through increased physical activity, and childhood asthma and improved nutrition in public schools and access to healthcare and supporting social services, among others (NYC 2016, pp. 73–86). Seattle's comprehensive plan, once titled "Toward a Sustainable Seattle" (Seattle 2015), incorporates indicators and goals related to "community well-being," including access to healthcare, reductions in infectious diseases, infant mortality, obesity, and exposure to secondhand tobacco smoke, among many others (Seattle 2017, pp. 153–155).

3.6.1 The Empirical Issues

The idea that when cities pursue sustainability as a matter of public policy that is effectively improving the health of their populations is tantalizing. While there is evidence that sustainability policies do in fact protect and improve the quality of the biophysical environment, neither the policies nor the environmental outcomes have been shown to be related to public health. Despite the logic of the expectation that these should be related, the evidence is lacking. So, this analysis examines the simple hypothesis, implied by the logic, that US cities electing to aggressively pursue sustainability policies and programs have healthier populations than cities electing not to pursue sustainability. Specifically, we expect sustainable cities to have smaller numbers of people with chronic health problems.

3.6.2 The Dependent Variables: Chronic Health Outcomes

Measuring public health outcomes for cities presents a significant challenge. Very little city-specific data are available. The local data that are available tend to be for counties or for metropolitan areas rather than cities per se. For the purposes of this analysis, we focus on two measures of chronic public health issues. The first of these is the percentage of the adult population with body mass indexes higher than 30 in 2013 as reported by the Center for Disease Control and Prevention, representing a measure of chronic obesity, in the county where the city resides. The second, also a measure of obesity and related issue, is an independent "Fattest Cities in America

Index" created and reported by WalletHub for the largest 100 metropolitan areas. We used the index scores for the metropolitan areas containing 54 of the 55 largest US cities (no index value is reported for Fresno, California). This index is a composite of some 17 specific indicators derived from a variety of official sources, including the percent of overweight and obese adults, teenagers, and children, projected obesity rates, and percent of adults who are physically active, who eat fewer than one serving of fruits and vegetables a day, and who have high serum cholesterol, high blood pressure, and heart disease. It also includes three indicators of healthy lifestyles. These indicators are weighted and combined into a single index score for each metropolitan area where the "fattest city" (Jackson, Mississippi) has a score of 84.93 and the least fat city (Seattle-Tacoma-Bellevue, Washington) has a score of 51.93. Among the 54 cities analyzed here, the "fattest" is Memphis, Tennessee (the second fattest city overall), with a score of 82.78, and the leanest is Seattle.

These county and metropolitan area data are merely an approximation of the health of the residents of the city but do provide at least some insight into the health of the people in the respective areas. For the BMI measure, we simply obtained information for the county (or largest county) in which each city exists. Some cities, such as Philadelphia, are coterminous with the county. Many others, such as Jacksonville, Florida, or Boston, Massachusetts, have counties that are only slightly larger than the cities themselves. A small number of cities are split between two and three counties, and for the purposes of this analysis, we used health data for the largest county. A few cities share a county, such as Los Angeles and Long Beach, California, both of which are in Los Angeles County, and Arlington and Fort Worth, Texas, both of which are in Tarrant County. In these cases, both cities are characterized by the same county data. And New York City consists of multiple counties or boroughs, so the BMI data for this city represents an average across all the boroughs. The correlation between these two measures for 54 cities is .673 (significant at $p < 0.01$), indicating that they are likely measuring a common underlying health condition.

County-level data for asthma, heart disease, and age-adjusted diabetes were collected from the Centers for Disease Control and Prevention's Community Health Needs Assessment tool (CDC, 2017). These three outcomes were chosen due to their clear impact on human well-being, as well as proving to be very challenging for public health professionals attempting to curb the steady growth of these ailments. If sustainable policies can be shown to be associated with a reduction, or even slowing, in diagnosis of these conditions, it could have a strong effect on health interventions.

Despite policies aimed at reducing air pollution, as well as clinical advances, nothing has reduced asthma within the United States. The last decade has shown a slow but steady increase in the prevalence in asthma among adults and children alike. While many advances in pharmaceutical approaches to managing asthma have been successful, little is known about the causal mechanism in an asthma diagnosis.

Heart disease is the number one cause of death in the United States, and complications associated with diabetes are in the top ten. These two highly correlated outcomes were chosen due to their enormous impact on the lives of Americans, both in finances and health. Healthy eating campaigns, healthy lifestyle interventions, and schoolwide programs have all failed to slow the tremendous increase in the prevalence of these two outcomes.

3.6.3 The Independent Variables: City Sustainability Policies and Sustainability Policies

The estimation of the effects of city sustainability on public health outcomes requires measures of urban sustainability policies and program. Here, we rely on two independent measures. The content of these two measures is summarized in Table 3.1. First, we rely on the Sustainable Cities Policy Index created by Portney (2013, 2017), computed for the 55 largest US cities based on the policies and programs of cities in 2011. Each city was evaluated to determine whether or not it had adopted and implemented each of the 38 policies and programs. The index represents a composite additive index of the number of some 38 different specific sustainability-related policies or programs that each city has adopted and implemented. These index values range from seven in Wichita, Kansas, to 35 in Seattle, Washington, Portland, Oregon, and San Francisco, California. In other words, Seattle, Portland, and San Francisco have adopted and implemented 35 of the 38 specific policies and programs designed to achieve greater city sustainability.

The second measure of sustainability is the Siemens Green Cities Index (Siemens 2011), measuring both environmental quality and city commitment to sustainability programs, for 21 large US cities. This index is reported for the year 2012. The overall Green City Index values range from 83.8 in San Francisco to 28.4 in Detroit and provide relative weighted assessments of how well each city performs on some 31 indicators in 16 different categories including carbon dioxide emissions, energy consumption, land use, building efficiency, transportation efficiency, water quality, waste, air quality, and environmental governance. Higher values represent better environmental performance.

The Sustainable Cities Policy Index is exclusively focused on policies, while the Siemens Green Cities Index also focuses on aspects of sustainability results, such as reduced carbon dioxide emissions, amount of electricity consumed, the extent of urban sprawl, water consumption per capita, and others. These two measures—the Sustainable Cities Index and the Siemens Green Cities Index—are quite closely correlated, at .772, significant at .001 level. Since the Siemens Green Cities Index is reported only for 21 of the largest US cities, this correlation is based on this smaller number of cities rather than the full group of the 55 largest cities. The high correlation for the 21 cities suggests that both indexes are measuring the same underlying sustainability variable. Results of this analysis are presented below for both groups of cities.

Table 3.1 Comparison of two measures of city sustainability

Sustainable Cities' Policy Index Has the city adopted and implemented a policy to:	Siemens Green Cities Index At what level is the city's:
1. Pursue targeted or cluster green economic development	1. Total CO_2 emissions per dollar of GDP
2. Develop an eco-industrial park	2. Total CO_2 emissions per capita
3. Redevelop at least one brownfield	3. CO_2 emission strategy
4. Develop eco-villages, urban infill housing, or transit oriented housing	4. Total electricity consumption per dollar of GDP
5. Use zoning to delineate environmentally sensitive growth or protected areas	5. Total electricity consumption per capita
6. Plan land use comprehensively to include environmental issues	6. Commitment to promoting green energy
7. Provide tax or fee incentives for environmentally friendly development	7. Standardized percent of city area devoted to greenspace
8. Operate or sponsor intra city mass transit	8. Population density (number of inhabitants per square mile)
9. Place limits on downtown parking	9. Commitment to improving amount of greenspace
10. Create intra city HOV car pool lanes	10. Commitment to containing urban sprawl and brownfield redevelopment
11. Establish alternatively fueled city vehicle program (green fleet)	11. Number of LEED-certified green buildings
12. Create a bicycle ridership or bike-sharing program	12. Requirement for energy audits and monitoring
13. Establish a household solid waste recycling program	13. Commitment to retro-fitting building for energy efficiency
14. Provide industrial recycling	14. Commuting to work with public transit, walking, or biking
15. Create hazardous waste recycling program	15. Commitment to providing public transit options
16. Operate an air pollution reduction program (e.g., VOC reduction)	16. Average commute-to-work time
17. Mandate recycled product purchasing by city government	17. Commitment to public transit incentives
18. Create a superfund (non-brownfield) site remediation program	18. Total water consumption in gallons per person per day
19. Engage in asbestos abatement	19. Amount of water leakage
20. Conduct lead paint abatement	20. Commitment to water quality from main water sources
21. Reduce pesticide use (integrated pest management)	21. Stormwater management plan
22. Create urban garden/sustainable food system or agriculture program	22. Amount of solid waste recycled
23. Mitigate the heat island effect	23. Commitment to waste reduction
24. Green building program	24. NOx emissions per person per year
25. Green affordable/low-income housing program	25. SO_2 emissions per person per year

(continued)

Table 3.1 (continued)

Sustainable Cities' Policy Index Has the city adopted and implemented a policy to:	Siemens Green Cities Index At what level is the city's:
26. Commit to renewable energy by city government (renewable energy portfolio)	26. PM10 emission per person per year
27. Create an energy conservation program	27. Commitment to air emission reduction
28. Offer alternative (renewable) energy to consumers	28. Commitment to green action plan
29. Conserve water	29. Extensiveness of environmental management
30. Operate a sustainability indicator project in the previous 5 years	30. Involvement of general public in monitoring environmental performance
31. Assess progress toward achieving indicators within the previous 5 years	
32. Create an action plan to achieve sustainability indicators	
33. Establish a single city office, agency, or person responsible for implementing sustainability initiatives	
34. Integrate sustainability goals into a citywide comprehensive or general plan	
35. Involve city, county, and metropolitan planning council in sustainability decisions	
36. Explicitly involved mayor/chief executive officer in sustainability decisions	
37. Involve the business community in sustainability decisions	
38. Involve the general public in sustainability planning	

3.7 Analysis and Results

The expectation that these measures of city sustainability should be related to public health outcomes is examined in Table 3.2, which reports the bivariate correlations. We report these correlations based on the 55 largest cities when the Sustainable Cities Policy Index is used as the independent variable and based on 21 large cities when the Siemens Green Cities Index is used as the independent variable. We also provide the correlations using the Sustainable Cities Policy Index, but just for the 21 Siemens cities. It is immediately evident that these correlations are extremely high, and in the expected direction, for four of the five measures of public health. Both measures of city sustainability are closely correlated with obesity, heart disease, and diabetes, but not with asthma. Among the cities studied here, there is a very strong tendency for those with aggressive sustainability programs and efforts to have considerably healthier county populations with respect to two measures of obesity, a measure of the rate of heart disease and a measure of the rate of diabetes.

To take a closer look at the anatomy of two of these correlations, we present the bivariate scattergrams that underlie the correlations with the age-adjusted diabetes rate. The scattergram in Fig. 3.2 makes clear the bivariate relationship between the

Table 3.2 Bivariate correlations between sustainable cities' measures and health outcome measures

	Health outcome measure				
Sustainable cities measures	% of adults with BMI over 30 2013	"Fattest Cities in America" Index 2016	% of adults with heart disease 2013	% of adults with asthma 2013	Diabetes age-adjusted rate 2013
Sustainable Cities Policy Index (2011)	−.452** ($n = 55$ cities)	−.356** ($n = 54$ cities)	−.292* ($n = 55$ cities)	.180 ($n = 55$ cities)	−.324** ($n = 55$ cities)
Siemens Green Cities Index (2012)	−.356** ($n = 21$ cities)	−.405* ($n = 21$ cities)	−.566** ($n = 21$ cities)	−.248 ($n = 21$ cities)	−.601** ($n = 21$ cities)
Sustainable Cities Policy Index (2011) for 21 Siemens Cities	−.637** ($n = 21$ cities)	−.414* ($n = 21$ cities)	−.477* ($n = 21$ cities)	−.023 ($n = 21$ cities)	−.529** ($n = 21$ cities)

*$P < .05$; **$P < .01$

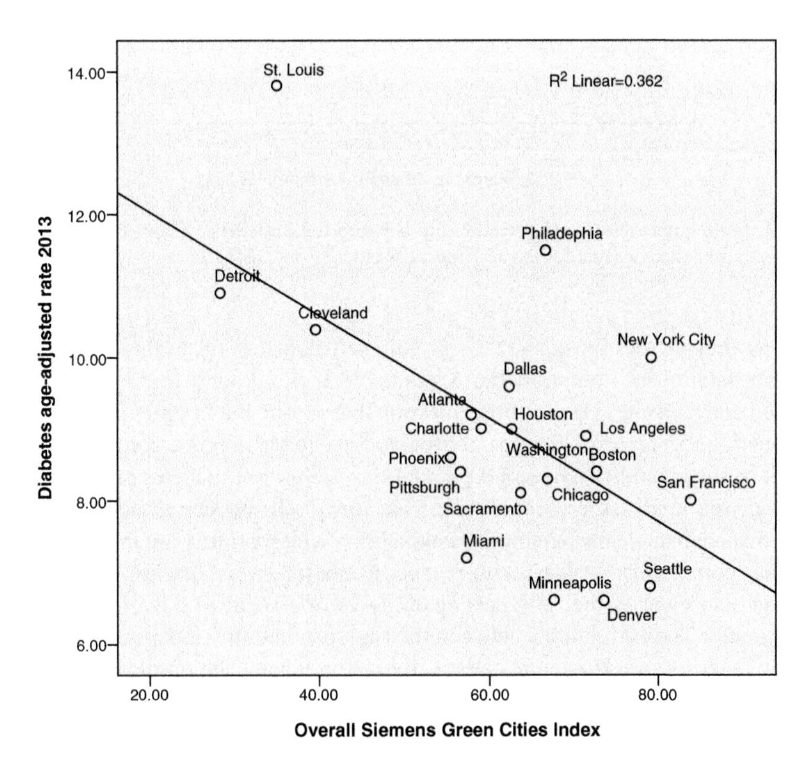

Fig. 3.2 Scattergram showing the relationship between the Siemens Green Cities Index and the age-adjusted diabetes rate for 21 large US cities ($r = -.601$)

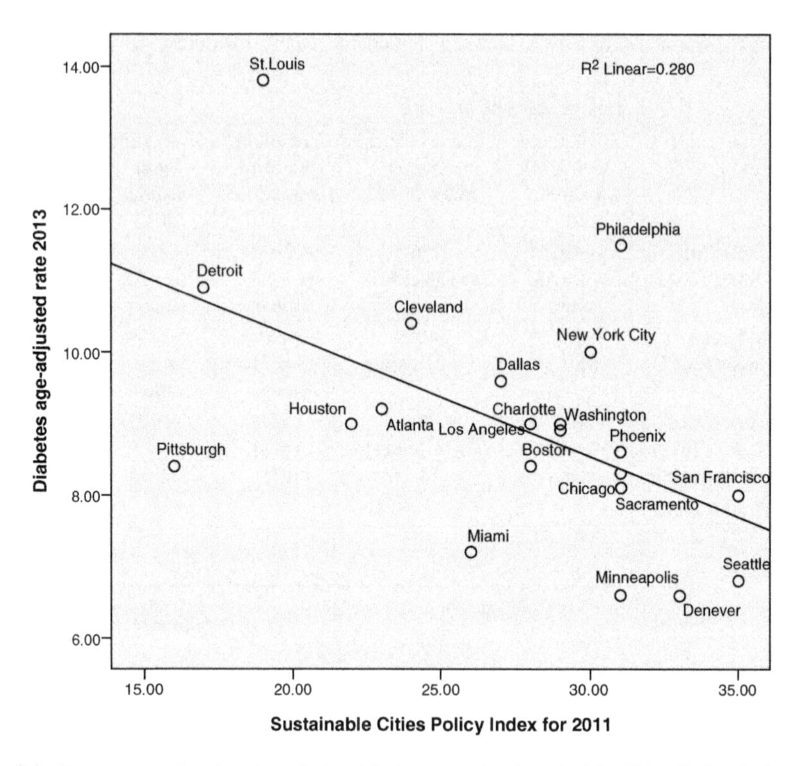

Fig. 3.3 Scattergram showing the relationship between the Sustainable Cities Policy Index and the age-adjusted diabetes rate for the 55 largest US cities ($r = -.324$)

Siemens Green Cities Index and the age-adjusted diabetes rate. Figure 3.3 shows the bivariate relationship between the Sustainable Cities Policy Index and the age-adjusted diabetes rate. There is little question that among the 21 cities in the Siemens Index and the 55 largest US cities, those that have made policy commitments to the pursuit of sustainability have adult populations with lower diabetes rates.

All of the analyses presented here essentially tell the same underlying story. Cities that have made the pursuit of sustainability a high priority are in counties with healthier populations, at least with respect to obesity, heart disease, and diabetes. What is also clear is that city sustainability efforts seem to have little effect on asthma rates. The correlations between the two sustainability independent variables and the measure of asthma are statistically insignificant. The implication of this is that city sustainability efforts probably do not adequately address asthma. Showing causality between an asthma diagnosis and environmental contaminants is difficult due to underlying confounding variables; however, recent research has shown a significant associations between traffic related pollution and asthma. This suggests that cities may be ineffectual at intervening to improve their air quality.

3.8 Discussion

This chapter has emphasized that adopting sustainable policies is uniquely qualified in tackling many modern public health concerns within developed countries. The analyses provided here take some initial steps that reveal an empirical link between sustainable policy adoption and health outcomes within the 55 largest US cities. This initial approach, while showcasing the broader implications of sustainable practices, is meant to continue a larger conversation with policy-makers, research groups, local stakeholders, and public health professionals on how to improve the health of individuals, as well environmental conditions. The core finding here is that among the largest US cities, those that are more aggressive in their pursuit of sustainability as a matter of public policy tend to be the same cities with positive public health outcomes. There are many opportunities to build upon, and improve, the approach presented here.

We do not wish to argue from these results that there is a causal connection. The potential presence of confounding factors may well influence the results of this bivariate analysis. While age was adjusted for when seeking correlates between our independent and dependent outcomes, many others proved beyond the scope of this chapter. Researchers have found that communities with high proportions of low socioeconomic status individuals and racial minority residents have multiple obstacles to overcome such as high obesity rates, the existence of food deserts, a lack of safe walkable streets, and cultural conditions (Wendell, Carlos, Jones, & Kraft 2006); this underscores the importance of including these variables in future analysis. This may also explain the lack of any significant findings within this report of sustainability efforts and an asthma diagnosis, as proximity to highways as well as the other mentioned confounding variables was not accounted for. There also may be a geographic disconnect between our variables, as health data is typically available at the county level and policy adoption is provided for cities. Those living within rural communities face different environmental conditions, as well as health and policy concerns, compared to their urban counterparts, and this difference should be accounted for.

Future research needs to be aimed to answer specific questions. For instance, which policies and programs seem better able to reduce heart disease, obesity, asthma rates, diabetes, and other public health outcomes? For example, when cities adopt and implement a bicycle ridership program, does this have an influence on health outcomes such as heart disease and obesity? When cities enact air emissions reductions, or climate protection programs, or a high-occupancy vehicle initiative, or a policy to support alternatively fueled vehicles, does this improve respiratory disease? While the goal for this line of research should be to provide clear policy prescriptions as a way of improving public health, intermediary projects need to be undertaken, specifically better ways to monitor, evaluate, and sustain current public health interventions and policy implications. Unfortunately there is much that we know we don't know about effective outcomes from local sustainability policy implementation. The results presented here, however, provide prima facie evidence that the city pursuit of sustainability may well represent an approach to achieving desired health outcomes.

References

Benedictow, O. (2004). *The black death 1346-1353: The complete history*. Suffolk, UK: Boydell Press.

Berry, J. M., & Portney, K. E. (2013). Sustainability and interest group participation in city politics. *Sustainability, 5*, 2077–2097.

Bos, K., Schuenemann, V., Golding, G., Burbano, H., Waglechner, N., Coombes, B., et al. (2011). A draft genome of *Yersinia Pestis* from victims of the black death. *Nature, 478*(7370), 506–510. https://doi.org/10.1038/nature10549.

Center for Disease Control and Prevention. 2017. *Assessment Tool: Diabetes and Obesity County Data*. Retrieved 2017, from https://www.cdc.gov/diabetes/data/county.html.

Corburn, J. (2009). *Toward the healthy city: People, places, and the politics of urban planning*. Cambridge, MA: MIT Press.

de Onis, M., Blossner, M., & Borghi, E. (2010). Global prevalence and trends of overweight and obesity among preschool children. *Am J Clin Nutr, 92*, 1257–1264.

Dickens, C. (2011). *Letters of Charles Dickens: 1833–1870*. Cambridge, UK: Cambridge University Press.

Fryar, C., Carroll, M., & Ogden, C. (2016). *Prevalence of overweight and obesity among children and adolescents aged 2–19 Years: United States, 1963–1965 through 2013–2014*. Hyattsville, MD: National Center for Health Statistics, Health E-Stats.

Gilbert, G. 2008. Introduction to Malthus T.R. *1798 An essay on the principle of population*. Oxford World's Classics reprint.

National Center for Health Statistics. (2016). *Health, United States, 2015: With special feature on racial and ethnic health disparities*. Hyattsville, MD: National Center for Health Statistics.

Nieuwenhuijsen, M. J., Khreis, H., Triguero-Mas, M., Gascon, M., & Dadvand, P. (2017). Fifty shades of green: Pathway to healthy urban living. *Epidemiology, 28*(1), 63–71.

NYC 2016. "Healthy Neighborhoods, Active Learning," *OneNYC 2016 Progress Report*. pp. 73–86. Retrieved from http://www1.nyc.gov/html/onenyc/downloads/pdf/publications/OneNYC-2016-Progress-Report.pdf.

Omran, A. (1971). The epidemiologic transition. *Milbank Memorial Fund Quarterly, 49*(4), 509–538.

Paneth, N., Vinten-Johansen, P., Brody, H., & Rip, M. (2011). A rivalry of foulness: official and unofficial investigations of the London cholera epidemic of 1854. *American Journal of Public Health., 88*(10), 1545–1553. https://doi.org/10.2105/ajph.88.10.1545.

Portney, Kent E. 2017. *Our Green Cities: 2011 sustainability rankings*. Retrieved from http://ourgreencities.com

Portney, K. E. (2013). *Taking sustainable cities seriously* (2nd ed.). Cambridge, MA: MIT Press.

Prüss-Ustün, A., Wolf, J., Corvalán, C., Bos, R., & Neira, M. (2016). *Preventing disease through healthy environments: A global assessment of the burden of disease from environmental risks*. Geneva: World Health Organization.

Rundle, A., & Heymsfield, S. (2016). Can walkable urban design play a role in reducing the incidence of obesity-related conditions? *JAMA., 315*(20), 2175–2177.

Seattle. 2017. *Seattle's comprehensive plan: citywide planning*. Retrieved from http://www.seattle.gov/Documents/Departments/OPCD/OngoingInitiatives/SeattlesComprehensivePlan/CouncilAdopted2016_CitywidePlanning.pdf.

Seattle. 2015. *City of Seattle comprehensive plan: Toward a sustainable Seattle*. Retrieved from https://www.seattle.gov/dpd/cs/groups/pan/@pan/documents/web_informational/dpdd016610.pdf.

Siemens. 2011. *U.S. and Canada Green City Index: Assessing the environmental performance of 27 Major US and Canadian Cities*. Retrieved from http://www.siemens.com/press/pool/de/events/2011/corporate/2011-06-northamerican/northamerican-gci-report-e.pdf.

TFAH. (2016). *The state of obesity: Better policies for a healthier America 2016*. Washington, D.C.: Trust for America's Health and Robert Wood Johnson Foundation. Retrieved from http://stateofobesity.org/files/stateofobesity2016.pdf.
United States Environmental Protection Agency. 2017. FACT SHEET: *Clean power plan by the numbers*. Retrieved 2017, from https://www.epa.gov/cleanpowerplan/fact-sheet-clean-power-plan-numbers.
WCED. (1987). World Commission on Environment and Development—The Brundtland Commission Report. In *Our common future*. New York: Oxford University Press.
Wendell, T., Carlos, P., Jones, L., & Kraft, K. (2006). Environmental justice: Obesity, physical activity, and healthy eating. *Journal of Physical Activity and Health, 3*(Suppl 1), S30–S54.
WHO (World Health Organization). 2017. *Obesity Situation and Trends: Global Health Observatory data*. Retrieved 2017, from http://www.who.int/gho/ncd/risk_factors/obesity_text/en/.

Chapter 4
Livable Cities: Concepts and Role in Improving Health

Suzanne H. Crowhurst Lennard

4.1 IMCL Principles of True Urbanism

The mission of the International Making Cities Livable (IMCL) movement founded in 1985 by Henry L. Lennard and myself is to enhance the well-being of all inhabitants, strengthen community, improve social and physical health, and increase civic engagement by sustainably reshaping the built environment of our cities, suburbs, towns, and villages (*International Making Cities Livable* n.d.-a).

Since 1985, IMCL has proposed a holistic vision for improving health and quality of life by reshaping the built environment. As Richard Jackson says: "the manner in which we design and build our communities – where we spend virtually our entire lives – has profound impacts on our physical, mental, social, environmental, and economic well-being" (Dannenberg et al. 2011, pg. xv). During the Obama presidency, the USA was fortunate to have a national urban policy agenda that promoted healthy communities and that recognized that "How a community is designed – including the layout of its roads, buildings and parks – has a huge impact on the health of its residents" (International Making Cities Livable n.d.-b). No doubt, under the subsequent regime, these priorities will be abandoned in favor of the GDP model.

IMCL's interdependent four-part strategy is called the PRINCIPLES OF TRUE URBANISM:

- Facilitate COMMUNITY SOCIAL LIFE
- Facilitate CONTACT WITH NATURE
- Facilitate INDEPENDENT MOBILITY
- Create a HOSPITABLE BUILT ENVIRONMENT

The following sections elaborate on each of these principles (Image 4.1).

S. H. C. Lennard (✉)
International Making Cities Livable, Portland, OR, USA
e-mail: suzanne.lennard@livablecities.org

M. Nieuwenhuijsen, H. Khreis (eds.), *Integrating Human Health into Urban and Transport Planning*, https://doi.org/10.1007/978-3-319-74983-9_4

Image 4.1 Healthy cities facilitate community social life. Photo credit: Suzanne Crowhurst Lennard

4.2 Facilitate Community Social Life

Key to achieving a high quality of life for all is the way we treat the public realm. The most essential task is to make it possible for people to come together, to form friendships and face-to-face social networks, and to develop social capital and community. It is the public places, streets, and squares that facilitate equitable social interactions.

Social Immune System: Social health is the foundation for physical health. The quality and quantity of social interaction and sense of belonging strongly influence both physical and mental health (House et al. 1988; Lomas 1998; Cohen 2004). We all need companionship and frequent face-to-face interaction with a wide circle of people who acknowledge us as human beings, share interests, and include us as a "member." These circles of friends and familiars form a "social immune system" to buffer stress, improve coping, and protect health. Integration in a social network produces positive psychological states (Cohen et al. 2000). Social circles "maintain, protect, promote and restore health" (Nestmann and Hurrelmann 1994).

Researchers in the Netherlands concerned with negative health effects of decrease in social contacts have called for "a more developed and detailed governmental policy to promote community" (De Vos 2003). Swedish authorities also noted: "Reducing social differences in health is a key public-health objective..." (Swedish National Institute of Public Health 2011). They emphasized: "...public health strategies that strengthen people's social networks...may have considerable potentials for health improvement, particularly for the most disadvantaged..." (Gele and Harsløf 2010).

Public Realm: We all need to be able to participate in a vibrant social life in public, but this cannot exist without a strong community and hospitable public spaces. A vibrant public realm encourages all to linger, share observations and perspectives, and get to know each other. It is essential in developing community and civic engagement and thus contributes to a more democratic way of life. As Martin Buber emphasized: "...architects must be set the task of also building for human contact, building surroundings that invite meeting and centers that shape meeting" (1967).

To create a hospitable public realm requires a mix of building uses—stores, workshops, or restaurants at street level, with dwellings and offices above. Residential buildings with windows and balconies overlooking the street or plaza create "eyes on the street" (Jacobs 1961), making the public realm safe and hospitable, a place where parents are comfortable letting their children roam.

To maintain human scale, buildings should be no higher than five stories, the height at which it is still possible to identify a face at the top window or call down to a child in the street (Gehl 2010). Even more valuable than the street in fostering social life is the plaza or square (Crowhurst Lennard and Lennard 2008). While the primary function of the street is movement, the primary function of the square is sojourn, to spend time and to socialize.

The Neighborhood Square, the Community's Living Room: A small neighborhood square is the most important element of a healthy neighborhood—and yet it is exactly this that is so lacking in modern cities. On a square, people's lives overlap; social interaction takes place while people shop or go about their daily lives. Centrally located traffic-free squares easily accessible by foot can provide multiple opportunities for all in the community to get to know others.

It is essential that the public realm include everyone—babies, toddlers, teens, youths, adults, and older people. It must involve people from all walks of life and sociocultural backgrounds, the poor and the well-to-do, integrating children into the complete community and building social support for elders and those less fortunate. Interaction in public builds community and social capital. It cements relations through repeated contact among inhabitants in multiple overlapping role relationships, thus strengthening the social fabric (International Making Cities Livable n.d.-c). As Lewis Mumford said, the public realm is the "ultimate expression of life in the city" (1961) and, according to Aristotle, is a fundamental requirement for citizens' well-being.

Social contact in the public realm is especially important for elders (Day 2008). Going shopping with friends, playing chess in the park, or meeting a friend on the street provides essential social interaction to maintain a sense of being valued by others.

Children need to grow up "within a web of sustained adult relationships," says Peter Benson (2006). At its best, the public realm is an incomparable teacher of social skills and attitudes (Lennard and Crowhurst Lennard 2000). The baby is introduced to members of the community, and the toddler learns how to address strangers as an equal, even before he learns how to speak. Social skills learned early shape a person's ability to maintain health and well-being throughout life and strengthen resilience.

Traffic-Free: To make the square safe for children and elders, and to allow the human voice to be heard, the square must be traffic-free—at least during those times when social life dominates and children are playing. The danger that a child might run into the traffic prevents a parent from allowing children to play freely on the square. Traffic noise inhibits conversation, especially for those with less-than-perfect hearing. Without the noise and potential dangers of traffic, all our senses open up. We can hear more clearly the voice of our companion. We are able to see and appreciate our surroundings more clearly. Freedom to be spontaneous and sociable on one's feet is an inestimable benefit of a traffic-free square. Air pollution and noise exposure from traffic are serious dangers to health, especially in places where people congregate (Nieuwenhuijsen and Khreis 2016).

Ideally, the square is not bounded by any street that carries traffic. Moving traffic and parked cars do not enliven a square as does an active building façade of small shops, cafes, and restaurants surmounted by homes with windows and balconies. The ease with which pedestrians can access shops within a traffic-free area has been shown by numerous studies to increase economic turnover (Monheim 1997; Nieuwenhuijsen and Khreis 2016; Lawlor n.d.).

Characteristics of Successful Neighborhood Squares: The neighborhood square needs to be a beautiful, lively place, surrounded by buildings housing some of the main commercial and civic activities. It should be a multifunctional space where markets and festivals take place. Cafes, shops, restaurants, and cultural and religious activities need to spill out onto the square from adjacent buildings (International Making Cities Livable n.d.-d).

A neighborhood or small-town square can only function if it is located at the heart of the neighborhood, at the crossing point of pedestrian routes, allowing everyone to mix and mingle. It must be surrounded by a compact, human scale and mixed-use urban fabric of shops and businesses, with people living over the shops, providing eyes on the street. A successful square cannot exist in a single-function zoned area.

A strong sense of place is conveyed by three special features: an outstanding building on the square, the experience of crossing a narrow threshold into an enclosed community "room," and the special treatment of the paving.

The size of a square must be appropriate to the size of the population served, and the events, and social life for which it is planned. For a square to function well as a gathering place for the community, it should be neither too large nor too small. It should be possible to easily recognize someone across the narrower dimension of the square. At a distance of 150 ft, clothing, gait, and general behavior can be recognized, allowing people to identify someone they know, and wave. Facial expression and emotion can be perceived at 115 ft (Gehl 2010). This distance allows a more reliable assumption as to the other person's readiness to engage in conversation. This also suggests that, to ensure the square is safe, by providing "eyes on the square," all parts of the square should be within 115 ft of some surrounding dwellings.

Buildings around a square must be human scale and correctly proportioned to the size of the square. Humans feel most comfortable when they can see a little sky within their normal sight lines. Since the normal angle of sight is approximately 50° above the horizontal, and one would want to have an area in the center of the square where a group of people may comfortably gather, face, and talk to one another, this indicates the maximum appropriate height of surrounding buildings.

Building façades should be designed to facilitate interaction between people in the public realm and those inside the building. Balconies, bay windows, articulated facades, and stoops greatly increase the number of potential interactions.

Small shops selling daily needs, especially food, attract many people to the square. They provide a more sociable shopping experience than a supermarket, and the shopkeeper gets to know regular customers. With shops on a square within a dense 10-min neighborhood, people shop more frequently, and this enhances the possibility for developing community.

More than anything else, food attracts people. Outdoor cafes and restaurants make the public space safe and hospitable, even into the night. While parents and grandparents talk, children can safely play nearby. Cafes allow people to spend more time in public, facilitating meetings and extended conversations late into the evening. This is especially valuable for elders.

For a square to be inviting, there must, of course, be places to sit without having to spend money, with a choice between sun and shade. Formal seating with backrests and armrests is especially helpful for elders. Seating needs to be arranged to facilitate eye contact between strangers, support discussion, or facilitate a game of chess among friends. Informal seating on steps, ledges, and walls support temporary sojourn and young people's need to perch and linger.

Public art offers an anchor in the public realm, drawing people together, offering pleasure, information, and a topic for conversation. Artworks that represent a neighborhood's traditional festival, legendary figures, historic locations, or events for which the neighborhood is known strengthen the community's shared sense of identity and pass the neighborhood's traditions down the generations. Representational works are beloved by children, especially when there is a story attached to them, when they offer opportunities for play and integrate water. Fountains are rejuvenating, refreshing, and calming.

The square may be the site of community festivals that create joy and well-being. They promote identity, pride, and community. It is especially important for children and young people to play important roles and be acknowledged for their contribution. A community that eats together on the square (as in Siena's neighborhood contrada) becomes one large family. It binds the community together and builds trust and a sense of well-being. Celebrating ethnic festivals together helps reduce social barriers.

There are six major neighborhood squares (campi) and numerous small neighborhood squares (campielli) in Venice, a few of which, such as Campo Santa Margherita and Campo San Giacomo dal'Orio, are still community gathering places. The Gracia district of Barcelona has ten neighborhood squares (plaças), most of which are very successful in supporting community life.

Main City Squares: Main squares at the city center may not be ideal for developing place-based social networks, but they can provide reasons for residents from all over the city to feel a shared identity, or attend a concert or festival together, and for visitors to enjoy the city's unique ambience.

A farmers' market on the square is immensely valuable as a source of healthy, fresh local food. The market is also one of the most powerful generators of social life. Farmers are skilled in interaction and act as "hosts." They know their regular customers, offer free samples, and make people feel valued. The well-to-do buy early in the day. The poor can often pick up free food at the end of the market. The farmers' market enhances children's sensory, cognitive, and social development, familiarizing them with the abundance and beauty of local produce and stimulating their appetite for fruit and vegetables. The daily market on Münsterplatz in Freiburg, Germany, is a powerful catalyst for social life in the city.

Image 4.2 Healthy cities facilitate contact with nature. Photo credit: Suzanne Crowhurst Lennard

Street entertainers are valuable on a main square. They cause "triangulation," breaking down the barriers between strangers (Whyte 1980). Suspicion and unease in the presence of strangers who seem very different from oneself are banished when everyone laughs together at the spectacle and exchanges smiles or comments—and laughter is good for one's health.

The main city square can be a catalyst for civic engagement. It is here that we become aware of shared experiences, common causes, and actions that we can take as "the people" to voice our opinions and together effect change, right wrongs, and improve our world. Daily political dialogue, civic engagement, and peaceful activism most frequently take place on main squares that combine city hall and a daily market. Elected officials have a visible presence in the community. They are known and recognized because they frequent the market. This immediacy and accessibility make political involvement personal for everyone who frequents the market. Padova, Italy, presents daily a supreme example of peaceful civic activism on the market places Piazza delle Erbe and Piazza della Frutta beside city hall (Image 4.2).

4.3 Facilitate Contact with Nature

Obviously, we must prevent pollution of our air by reducing vehicle emissions (Nieuwenhuijsen and Khreis 2016) and contamination of rivers and lakes as well as drinking water. Contamination can have immediate and long-term effects on our health and the health of all living things (Nieuwenhuijsen et al. 2017a, b). And besides, fresh air and clean water are immense sources of pleasure and well-being. But beyond that, research throughout Europe (Nieuwenhuijsen et al. 2017b) shows that protection of green and blue places in our cities is immensely valuable for the health and well-being of citizens. Much research supports what we would assume to be true—that parks and green areas support physical activity and are therefore good for health (Epstein et al. 2006; Fjortoft 2004). Green areas should support children's active life, with trees to climb and streams to play in. They should be designed to foster creativity, imagination, and social play, exploration, discovery, and vocabulary building.

What is important is that green areas are close at hand and easily accessible, especially by children (Richardson et al. 2013; Baranowski et al. 2000; Sallis et al. 1993). A very large park may not contribute so much to public health if it is not easily accessible to a large population; a smaller green space closer to home may benefit a far larger population. In Bristol, UK, for example, 90% of the population live within 300 m of parklands and waterways (Bristol 2015 European Green Capital 2015).

Research shows that play in nature improves concentration (Grahn et al. 1997; Wells 2000) and reduces and relieves attention deficit hyperactivity disorder (ADHD) (Kuo and Taylor 2004; Faber 2001a, b). Natural settings have been shown to encourage social play and cooperative relations (Kirkby 1989), protect emotional well-being (Huynh et al. 2013; Richardson et al. 2013), and promote emotional resilience (Wells and Evans 2003). Nature is particularly restorative for elders. Even a view of trees from your apartment improves emotional health, mental acuity, and productivity and reduces stress and violence (Kuo and Sullivan 2001).

Trees protect cardiovascular and respiratory health (Donovan et al. 2013). In neighborhoods with more, and diverse, trees, kids experience reduced asthma (Lovasi et al. 2013), and mothers are less likely to deliver undersized babies (Donovan et al. 2011; Dadvand et al. 2012). Trees and plants remove nitrogen dioxide and particulate matter from the air (Pugh et al. 2012).

Of course, we have known these things intuitively for millennia—the old European spa towns such as Montecatini or Baden always involve beautiful architecture in a beautiful natural setting—but now science is on board to back up this understanding.

A substantial level of health inequity exists between those living in well-to-do neighborhoods and those living in poor neighborhoods. One significant action to improve health in poor neighborhoods would be to provide more parks and trees there (Bienkowski 2015).

Community gardens are valuable in food deserts and poor underserved neighborhoods, not only because they help to provide essential healthy nutrition, which is especially valuable for children, but also because they help to build community networks, collaborative efforts, and a neighborhood "voice" (Girard et al. 2012; Wang et al. 2014). Direct engagement with a world that is alive, for example, by growing a garden, develops in children a sense of responsibility, and love for the earth, and sows the seeds of ecological awareness. Suburban neighborhoods have a far greater opportunity than do inner city neighborhoods to create community gardens.

Healthy, livable cities need green traffic-free commuter routes through natural ribbons throughout the city to remove the stress of commuting and reduce the day's tensions.

Cities need incidental nature in courtyards and tiny backyards, on balconies and roof gardens, growing up the walls and across the street to soften the hard

Image 4.3 Healthy cities facilitate independent mobility. Photo credit: Suzanne Crowhurst Lennard

environment, purify the air, and provide green views for urban dwellers. Nature provides habitat for wildlife, which also enhances our experience of the city. Beginning in 1972, Mayor Hahlweg was one of the first mayors to systematically promote green roofs, walls, and facades as well as green streets in Erlangen, Germany (Young 2012), for which he received the IMCL Award in 1991. Since 2016, Mayor Hidalgo has promoted a similar program in Paris (Cooke 2016).

Nature is attractive because it is complex and alive. The diversity of nature awakens the senses, improves concentration, and develops cognitive skills (Kahn and Kellert 2002). Children must be able to get up close with nature and develop "biophilia," so they will grow up to care for nature.

Architectural design that echoes the structural principles of nature—diversity within an overall unity, fractals, and increasing complexity—is found to be beautiful, and deeply attractive to humans, creating endorphins that increase mental and physical well-being (Kellert et al. 2008) (Image 4.3).

4.4 Facilitate Independent Mobility

To create a healthy, livable city requires a balanced transportation policy (Topp 1985) that encourages Active Living by Design, placing first priority on walking, second on biking, third on public transit, and lastly on the car (Active Living by Design n.d.). Taking the trips made by children, elders, the poor, and disabled as seriously as trips made by working adults helps to prioritize healthy, ecological transit modes.

Balanced transportation planning is not about movement of vehicles but about people and how we get around. The transportation planner must accommodate the varied trips that we all need to make—to school, work, shopping, the library, or theater—and make all trips as pleasant, economical, healthy, safe, comfortable, simple, and autonomous as possible. The day is past when transportation planning is simply planning for the car.

This policy requires reshaping suburban areas into human scale mixed-use neighborhoods so that destinations (school, shops, movie houses, library, restaurants, cafes, workplaces, community services, and public transit) are within a 10- or 20-min walking and biking radius from home for the majority of the population (Crowhurst Lennard and Lennard 1995; Ewing and Cervero 2010).

Children need to be able to walk to school and around their neighborhood, play out of doors, and meet friends and community members. They need free range within a safe territory to develop independence and spatial skills (Russell 2010; Malone 2007). They need to learn to orient themselves, identify landmarks, gauge distances, and create mental maps (Lynch 1997).

The City of Freiburg has excelled in accessibility for pedestrians, bicyclists, and public transit. They received the IMCL City of Vision Award in 1993 for their commitment to principles of livability and sustainability in all planning and urban design issues and have since received numerous awards for livability and as a child-friendly city.

More than simply wide sidewalks, what makes streets really safe is the presence of people on the streets, particularly familiars, and community members shopping and frequenting sidewalk cafes. Elders and the disabled in particular need traffic-free access to shops and services, cafes, and plazas where they can meet friends. Paving designs and textures help children, who are close to the ground, as well as the very elderly and the blind to find their way around.

Play streets are a—usually temporary—mechanism for transforming a street into a pedestrian-oriented street. They are valuable as a first step to increasing awareness of children's need to play out of doors and to more permanent solutions such as Wohnstrasse (traffic-calmed residential street) and pedestrian streets.

Pedestrian Routes: To maintain health and prevent obesity (Pappas et al. 2007), young people need at least 60 min of exercise everyday and adults at least 150 min every week (US DHHS 2008), and the best exercise is walking. Children, elders, and the disabled are safest on traffic-free pedestrian routes. These range from small family courts between rows of houses, and paths through parks and along rivers, to a network of pedestrian streets through the suburb and to the city center. The pedestrian network must be frequented by familiar adults and be adjacent to buildings that provide "eyes on the street."

It is not enough to have just one or two traffic-free streets—they need to be interlinked in a complete network, especially in the city center. The question is not so much how far are people willing to walk, as how interesting are the pedestrian ways, and how likely is one to run into friends or familiars.

Pedestrian streets may be accessible for vehicles, with restrictions: emergency vehicles have access when necessary; delivery vehicles are permitted at specific times when pedestrian activity is low; speed is limited to walking speed; and some pedestrian streets permit public transit—buses, streetcars, or light rail.

Traffic-Calmed Streets: Wohnstrasse, designated by the international blue sign, are easy to install in suburban areas. They are safe for children and elders. Pedestrians and vehicles have equal rights to the full width of the street. Children may play in the street. Vehicles must proceed at walking speed. They cannot use the street for through traffic, and parking is permitted only for delivery and residents. Kids, especially 10–14-year-olds, need adventure and independent exploration of their neighborhood, whether by skate boarding, by roller skating, or on foot. Any traffic on these streets must be at 20 mph or walking speed to make these activities safe.

Traffic limits the lives of children far more than the lives of adults. If a car traveling at 50 kph hits a child, the child's chance of survival is only 50%. If the car is traveling at 30 kph, the chance of survival increases to 90% (WHO n.d.). Pedestrian- and bike-related accidents are the leading cause of accidental deaths for children. Suburban traffic-calmed streets need wide sidewalks (at least 6 ft) buffered with a tree-planted verge, plantings, or rainwater gardens, and safe crossings achieved through raised crosswalks and necking. Traffic speed is reduced by roundabout, narrowed, and jogged traffic lanes.

Throughout the city and suburbs, we need to redesign arterials to give pedestrians priority on a continuous pedestrian network and slow the automobile to accommodate pedestrians.

Bicycle Networks: Children depend on the bike to explore further afield. Their increasing independence is essential to develop self-confidence. The bike provides a healthy transportation mode. We need a citywide network of bike paths that extend through suburban neighborhoods, as you find in Amsterdam, Copenhagen, Groningen, and Freiburg, and as is developing in Portland, Minneapolis, and other North American cities, so that it is possible to bike from a home in the suburb to work or school, shop in the city center, deliver goods inexpensively, or ride into the countryside at the weekend.

We cannot increase bike ridership unless we make bicycle networks really safe for people of all ages, for those who cannot afford a car, and even for very small children. The safest bike routes are those completely separated from traffic and separated from pedestrian routes. In the city, bicycles need, at the very least, clearly painted bike routes. The greatest care must be shown at the intersections. Dangers from right-turning vehicles can be minimized by bike lanes that are brightly painted and "bike boxes" in front of motorized vehicles at intersections. But the safest routes for bicyclists are buffered from moving or parked motorized vehicles by 3 ft painted buffer strip and a curb or planters (Andersen 2013). Seattle, Long Beach, and Indianapolis are now leading the way in the USA in protected cycle tracks.

Public Transportation: Rather than bussing kids to school, a better solution is to improve public transit. Older kids should be able to take transit on their own from a suburb to the town center. For parents to be comfortable with their children taking transit, parents themselves must use it to get to work or go shopping. Transit used

Image 4.4 Healthy cities create a hospitable urban fabric. Photo credit: François Schreuer

by everyone—rich and poor, old and young—is democratic and socially healthy. It also builds community. Those who regularly travel at the same time begin to recognize each other and develop a dialogue (Image 4.4).

4.5 A Hospitable Built Environment that Frames Social Life

Regional Planning: Our health is dependent on the prevention of sprawl. Cities and towns in a region must work together to focus new development within existing urban boundaries; establish guidelines for creating 10-min neighborhoods, locating housing close to jobs, schools, stores, and services; and prevent shopping malls and big box retail outside city limits where they destroy the city's economy and generate superfluous traffic. Leadership in this area has been taken by the Upper Rhine region where Germany, France, and Switzerland work together to achieve these goals.

Compact Urban Fabric: Strong and diverse social life, social and physical health, and commercial diversity flourish best within a compact, mixed-use urban fabric. Contiguous buildings form continuous walls enclosing streets and creating squares. A compact urban fabric makes it possible for a child to walk to school and for an elder to easily go shopping or run errands. It also makes viable a good transit system. When many errands and trips can be accomplished by foot within a small radius, community networks develop.

Urban Villages: For the most vulnerable among us, living in a complete community is essential for health and well-being. With familiar people on the street, children have the freedom they need to develop, and elders and those with disabilities have a supportive social network when needed. A complete community requires houses, apartments, schools, workplaces and shops, and a range of resources—natural play areas, library, shops, movie theater, etc.—within a compact 10-min walking radius. This is achieved in a city with a cellular structure of walkable suburbs and urban villages. The center of each suburb or urban village is a public place that brings the community together; the boundary of the suburb or urban village is visually defined by parks and natural features that are accessible and within walking distance for all.

Lewis Mumford called for us to plan cellular cities, where the neighborhood is a microcosm of the multifunctional core, containing diverse work opportunities, shopping, housing, and all necessary social and cultural infrastructure within a short radius. This goal is called in Germany "Die Stadt der kurzen Wege" (city of short distances), in England the "Urban Villages" approach, and in the USA "Complete Communities" or "Ten-Minute Neighborhoods." The French Quarter in Tübingen; Rieselfeld and Vauban in Freiburg, Germany; Poundbury in the UK; and Le Plessis Robinson, on the outskirts of Paris, are good examples.

Mixed Use Shop/Houses: A healthy, walkable lifestyle is dependent on the close proximity of living and working, shops, services, and social and cultural resources. At the center of the suburban core, the primary building block should be a shop, workshop, or restaurant at street level, with offices and dwellings above. Commercial activity at street level draws life onto the street and generates a business population with daytime jurisdiction. A high percentage of upper floors need to be residential, with windows that open, and balconies to provide, in Jane Jacobs' words, "eyes on the street" and interaction between residents and those on the street. The residential population provides nighttime jurisdiction, making even visitors aware that the street "belongs to" a community. The proximity of the private and public realms makes the public realm dynamic and hospitable and the private dwelling convenient. This is the way cities were built throughout the world until the twentieth century.

Human Scale Architecture: To create a safe and human scale public realm, buildings must be no higher than five stories, the height at which it is still possible to identify a face at the top window or call down to a child in the street. Windows and doors, balconies, and terraces give a building human scale. The street-level façade must emphasize interaction between interior and exterior with shop windows, steps, and porches. Interesting details and textures, especially close to the ground, enliven the walking experience and stimulate curiosity.

Integrating Varied Income Levels: It is important not to segregate the wealthy and the poor, if we are to retain our humanity. Gated compounds and luxury high-rise condos exacerbate distrust and resentment. We need neighborhoods with mixed incomes and housing that can accommodate a variety of income levels. Buildings can happily combine larger condos with views, rentable apartments with several bedrooms for families, and studios for young singles. If not always in the same building, it is nevertheless important that wealthy and poor live in the same neighborhood.

Inner Courtyards: In the continuous urban fabric of traditional cities, blocks are perforated by usable inner courtyards that allow air and sunlight to enter and narrow streets suitable for pedestrian ways. These provide sheltered semipublic outdoor areas, suitable for outdoor restaurants, gardens, and trees. In housing blocks, inner courtyards provide access to nature, a place for children to play, and sometimes, pedestrian shortcuts. Balconies and roof gardens open onto these quiet oases.

Neighborhood Squares and Main Squares: One major advantage of the continuous urban fabric, compared to isolated monumental buildings, is that urban fabric shapes the public realm, enabling the creation of neighborhood squares and main squares. Each neighborhood in the city needs to be structured around a community plaza that everyone in the neighborhood uses, and every city needs at least one major city square that fulfills the social and civic needs of the population. The new university town of Louvain-la-Neuve in Belgium, planned by Leon Krier, admirably achieves this goal.

An Urban Fabric that Integrates Nature: A healthy continuous urban fabric is not all raw concrete and glass. As much as possible, materials that are closer to nature and warmer to the touch (wood, stone, plaster, tile, brick) should be used, and nature should be encouraged to flourish across streets, up the walls, and to pour out of balconies, and across roofs, making the city green, comfortable, and biophilic.

The DNA of the City and Neighborhood: A healthy city embraces its own unique identity that grew from its historic roots and builds on its best-loved features. This identity of place allows the most diverse population to feel they have something in common they can all take pride in—a shared identity.

In order to fit into the context, new buildings must respect this "genetic code" (Crowhurst Lennard and Lennard 1995), reflecting at least some existing patterns or interpreting them in a contemporary idiom. Appropriate new infill buildings respect their neighbors and create a civil architectural dialogue, echoing some of the best characteristics of adjacent buildings. This mutual support from building to building illustrates a principle of civil interaction that subtly influences human interaction.

Beauty. The City as a Work of Art: Beauty is important to everyone but perhaps, as Mayor Joseph Riley emphasizes, especially to the disadvantaged. The enjoyment of beauty may be transformed in the body into "endorphins" that increase mental and physical well-being. This healing effect of a beautiful environment has been understood for millennia. Decoration, color, and interesting roofline engage the eye and the emotion, creating remembered landmarks that help people to find their way around.

Beauty in the built environment affects mental health, social health, and civic participation. In his research on urban aesthetics, Wolfgang Schulz (1994) observed: "bonds with the environment, the territory where one lives, reinforce the strength of the resident, help him endure burdens… A pleasant environment is perhaps much more important than we think." The principles of beauty—harmony in diversity, proportion, fit, and evidence of balanced growth—are prescriptions for how to live harmoniously in a community. A beautiful neighborhood is aesthetic as a whole. We cannot visualize "our" neighborhood if it sprawls into the rest of the city. We need distinct suburban neighborhoods, each with clear green boundaries, urban fabric, and main squares.

Above all, we need beautiful places, squares designed specifically for social life, as hospitable settings that invite everyone, that support children's play while elders talk, young people flirt, and people go shopping; places that residents feel is the most beautiful place on earth because that is where they meet friends, where they remember community festivals, and where they are recognized and valued; and places that epitomize all the most beautiful characteristics of their city and their community.

4.6 Why Are Cities Not Achieving These Health-Related Livability Goals?

In city-making, there are two competing value systems at work. The first is based on GDP. In this model, the city is seen as an economic engine. Its function is to fuel growth and raise standard of living. The second is based on the value of the quality of life. In this model, as outlined above, the function of the city is the "care and culture" (Mumford 1961) of human beings and of the earth.

The GDP Goal: The GDP model has governed the way cities have developed around the world in the twentieth century. Success is measured financially. The fastest way to grow the economy is through construction. Maximizing construction is therefore thought to be the solution for success.

There are major flaws in the GDP system. Fifty years ago, Lyndon Johnson criticized unbridled growth, which he declared led to soulless wealth. "He elucidated a new dream valuing quality of life above quantity of stuff" (De Graaf 2014). And Bobby Kennedy said on March 18, 1968, of the GDP "it measures everything in short, except that which makes life worthwhile" (Kennedy 1968).

Based on the drive to increase GDP, North America, Australia, and other parts of the world saw vast expanses of horizontal sprawl in the twentieth century. The negative side effects were noticeable from the beginning: social isolation and depression of those left "home alone" (first housewives, then children); destruction of social networks and civic engagement (Putnam 2000); unwalkability leading to obesity and related chronic diseases such as high cholesterol, high blood pressure, type -2 diabetes, sleep apnea, asthma, and liver disease (Frumkin et al. 2004); overdependence on the automobile and consumption of fossil fuels; etc.

Today, suburban sprawl is criticized and finally being reined in. However, the construction industry is innovative. To satisfy the increasing demand for growth and profits, the industry now focuses on an even more insidious, and vastly more profitable, form of overdevelopment—"vertical sprawl"—as Patrick Condon (International Making Cities Livable n.d.-e) calls it. This has taken hold of cities around the world—including Asia, Africa, and the Middle East. However, this may prove to be even more toxic to humans and the planet than horizontal sprawl (Smart Cities Dive 2014). Freestanding high-rise towers that are objects in space make it impossible to create a hospitable public realm.

Extreme capitalism is a vicious cycle: the more high-rise we construct, the more isolated and depressed we feel, the more we consume, and the more dependent we are on the generators of GDP growth, big construction, and big energy.

Singapore and Hong Kong were encouraged by the USA to become shining models for Asia and China of what can be achieved through the capitalist system. In order to rapidly increase GDP, despite limited land area, Singapore embraced high-rise. But consumption and construction increase carbon dioxide emissions, leading WWF President Yolanda Kakabadse to observe in 2010: "Singapore…is a society that maybe is one of the best examples of what we should NOT do" (Eco-Business 2012).

Imitating Hong Kong's rapid increase in GDP, China built over 100 tall and dense uninhabited "ghost" cities simply to maximize economic growth. In the process, they discovered that rapid growth involves some health problems for humans and the earth. Their life-threatening smog is only one side effect of their breakneck economic growth. Paradoxically, this too can be counted on to increase GDP by fueling the health and pharmaceutical industry.

With high-rise construction, profit is privatized, and loss is socialized. There is a complete economic imbalance—an overinvestment in private property, and underinvestment in the public realm, and the places in the city that should belong to everyone and that represent the "common wealth." We have seen these unbalanced planning priorities at work for a long time, leading planners to create streets that are inhospitable for the pedestrian and the bicyclist.

This emphasis on privacy over community strikes hard at children who have a developmental need to grow up within a thriving community. As a result, children suffer unprecedented levels of loneliness, depression, and shyness, and they fail to develop good social skills. Bullying and violence are also the result of poor social skills. (But all these problems increase the GDP!)

Children and teenagers growing up in high-rises have an especially difficult time. Young children need almost continuous face-to-face interaction within a village or "web of sustained adult relationships" (Benson 2006). A teenager's chief developmental task is to become independently sociable within larger social circles, but high-rises do not provide a suitable environment for developing these skills.

In Tokyo, there are high levels of hikikomori—these are young people from teenagers to young adults, predominantly male, who shut themselves in their bedroom, refuse to go to school, and demand that their parents leave meals for them outside the door. There are an estimated 700,000 hikikomori in Japan and 1.55 million more on the verge of becoming hikikomori (Wikipedia n.d.-a). According to studies by Wong et al. (2014), Hong Kong is following the same path, currently with an estimated 18,500 hikikomori.

It has long been known that high-rise creates a more socially isolated living situation. While this may be welcome to some engaged in a high-stress, or socially fulfilling professional field (such as the stock market or entertainment world), for others, such as elders, small children, and mothers raising small children, studies show high-rise living can be extremely damaging to physical and mental health (Gifford 2007; Evans et al. 2003).

Cancerous Development: Konrad Lorenz first drew parallels between healthy and cancerous cells in the body and in the city: "If you look at a cancer under the microscope, a cross section with cells of healthy tissue, it looks exactly like an aerial view of a city in which the old sections are surrounded by new irregularly built regions or else by those that are monotonously geometrical—both are possible, after all. The parallels between the formation of malignant tumors and cities in a state of cultural decay are very wide ranging" (Lorenz 1990).

I explored this idea in my book *Livable Cities Observed* (Crowhurst Lennard and Lennard 1995) and in subsequent talks. But there are far more parallels than I realized at that time. Healthy cells (and a healthy urban fabric) are characterized by regular geometry, clear cellular boundaries, and slow growth and regeneration. They connect to each other, perform a clear function within their context, monitor the health of adjacent cells, and form a mutually supportive community. Malignant cells (cancerous urban developments), on the other hand, are overly large, with rapid, out of control proliferation and broken boundaries. They do not communicate with neighboring cells or their context and do their own thing irrespective of their context (Szasz 2013).

Ramray Bhat (2014) at the University of California Life Sciences Division explored similar issues: "Lastly, like the most debilitating characteristic of cancer, i.e., metastasis, ill-designed buildings and built environments cannibalize their surrounding urban landscapes by growing, dwarfing, and pushing out smaller and traditionally built structures at the interfaces… A combination of globalization, postcolonial mimicry, and aspirational urges have left burgeoning cities in India, Pakistan, Bangladesh, China and Brazil dotted with mega towers and zonings which parasitize on their surroundings through labor and energy demands creating ever widening peripheries depleted of culture, diversity, and beauty."

Mariano Bizzarri (Bizzarri and Cicina 2014), a leading researcher in carcinogenesis, who also draws parallels with cancerous urban growth, emphasizes that cancer growth depends not so much on its damaged DNA as on the cell's three-dimensional biochemical and biophysical microenvironment. In the language of the city organism, this suggests that cancerous development occurs where there is both a breakdown in the social urban fabric and the physical urban fabric.

The health of the social urban fabric lies in the degree to which community exists, or at least the degree to which people interact with each other in the public realm. The health of the built urban fabric lies in the degree to which buildings are mutually supportive (contiguous), with facades that relate to each other in scale, character, or materials, and in the degree to which they facilitate communication between people inside the building and those in the public realm.

For city-makers, the key to undoing these ills is to focus on the public realm—the "common good"—the places that connect us, and the way buildings relate to the public realm. It is the public realm that gives all of us quality of life.

Adherence to the GDP model has led us into a process that Pope Francis calls "rapidification" (Pope Francis 2015). We are following the fastest route to increasing profits (for developers and bankers), and, in so doing, we are making our cities unhealthy, inequitable, unlivable, and unsustainable.

We do not have to slavishly follow the GDP model. Today, there are many efforts around the world to develop an alternate index of success, an index that will reevaluate the costs—to human health and well-being and to the health and sustainability of the planet—created by the GDP model, and that will guide a wiser municipal, national, and global decision-making process. The economist Hazel Henderson led this effort with her Quality of Life Indicators (Ethical Markets 2017). Other indicators include the Genuine Progress Indicator (GPI) (Wikipedia n.d.-b), the Canadian Index of Wellbeing (CIW) (Wikipedia n.d.-c), the World Wildlife Fund's Living Planet Index (LPI) (Wikipedia n.d.-d), and London's Happy Planet Index (HPI)

Table 4.1 Indicators

GDP	IMCL quality of life
• Goal: increase economic growth • Focus: construction industry • Emphasizes cities as economic machines • Stresses enterprise, independence, and privacy • Favors most productive groups: neglects those less productive • Accepts suffering and marginality as the price for progress • Costs of ill health, crime, and social problems are economically valued • Segregates functions and persons • Emphasizes speed and functionality • Regulation of well-being is by technology • High rate of crime, drug, and alcohol use • Charges a fee for good experiences	• Goal: increase quality of life • Focus: health and well-being • Emphasizes humanizing and civilizing functions of cities • Stresses trust, compassion, mutual responsibility • Values wisdom; the understanding of the city as a "system" • Does not accept suffering as a price • Human processes are valued • Stresses mixed use and heterogeneity of population • Emphasizes hospitality and accessibility • Regulation of well-being is by people • Low rate of crime and drug and alcohol use • Emphasizes experiences that are free

(Happy Planet Index n.d.), as well as Bhutan's Gross National Happiness Index (Gross National Happiness n.d.).

IMCL also developed Quality of Life Indicators that state that the goal of the city is to increase quality of life, which is not measured by an economic scale but by health, happiness, and sustainability (Table 4.1). This view is founded in philosophers of the human condition; it is expressed in the work of Lewis Mumford, Jane Jacobs, and others; and it forms the foundation for the International Making Cities Livable movement.

IMCL pays special attention to children because the environment children grow up in affects all aspects of their development and can damage their physical and emotional health for the rest of their life. We propose that if we consider how all our city-making decisions affect children, we shall begin to create healthy, sustainable cities for all.

4.7 To Conclude

As I said at the beginning, if we want to make our cities healthy and livable for all, we must first make them healthy for the more vulnerable—children, elders, the disabled, and the poor. If our neighborhoods, towns, and cities do not sustain them, they are not sustainable. A city built on these principles of true urbanism provides the ideal environment for children's physical, mental, and social development and generates communities that are healthy, ecologically sustainable, and socially sustainable for all.

References

Active Living by Design. (n.d.). *About ALBD*. Retrieved June 4, 2017, from http://activelivingbydesign.org/about/.

Andersen, M. (2013). America's 10 best protected bike lanes of 2013. *People for bikes*. Retrieved June 4, 2017, from http://www.peopleforbikes.org/blog/entry/the-10-best-protected-bike-lanes-of-2013.

Baranowski, T., et al. (2000). Physical activity and nutrition in children and youth: An overview of obesity prevention. *Preventive Medicine, 31*, S1–S10.

Benson, P. (2006). *All kids are our kids* (Vol. 1, p. 104). San Francisco: Jossey-Bass.

Bhat, R. (2014). Understanding complexity through pattern languages in biological and man-made architectures: Comparisons between biological and arhitectonic patterns. *Archnet-IJAR, 8*(2), 8–19.

Bienkowski, B. (2015). More money means more trees in US cities. *The Epoch Times*. April 24. Retrieved June 29, 2017, from https://m.theepochtimes.com/more-money-means-more-trees-in-us-cities_1332965.html

Bizzarri, M., & Cicina, A. (2014). Tumor and the microenvironment: A chance to reframe the paradigm of carcinogenesis? *Hindawi Publishing Corporation BioMed Research International, 2014*, 934038., 9 pages. https://doi.org/10.1155/2014/934038.

Bristol. (2015). *European green capital. In it for good*. Retrieved June 29, 2017, from http://bristolgreencapital.org/wp-content/uploads/2014/11/bristol-2015_annual-review.pdf.

Buber, M. (1967). *A believing humanism* (p. 95). New York: Simon & Schuster.

Cohen, S. (2004). Social relationships and health. *American Psychologist, 59*(8), 676–684.

Cohen, S., Underwood, L., & Gottlieb, G. (2000). *Social support measurement and intervention*. New York: Oxford University Press.

Cooke, L. (2016). *Paris allows anyone to plant an open garden*. Retrieved June 29, 2017, from http://inhabitat.com/paris-allows-anyone-to-plant-an-urban-garden-anywhere/.

Crowhurst Lennard, S., & Lennard, H. L. (1995). *Livable cities observed*. Carmel: Gondolier Press.

Crowhurst Lennard, S., & Lennard, H. L. (2008). *Genius of the European square*. Carmel: Gondolier Press.

Dadvand, P., et al. (2012). Green space, health inequality and pregnancy. *Environment International, 40*, 110–115.

Dannenberg, A., Frumkin, H., & Jackson, R. (Eds.). (2011). *Making healthy places*. Washington: Island Press.

Day, R. (2008). Local environments and older people's health: Dimensions from a comparative qualitative study in Scotland. *Health & Place, 14*(2008), 299–312.

De Graaf, J. (2014). The greatest modern presidential speech turns 50. *Truthout*. Retrieved June 4, 2017, from http://www.truth-out.org/opinion/item/23741-the-greatest-modern-presidential-speech-turns-50.

De Vos, H. (2003). Geld en 'de rest': Over uitzwerming, teloorgang van gemeenschap en de noodzaak van gemeenschapsbeleid [Money and 'the rest': About sprawl, decline of community and the necessity of community policy]. *Sociologische gids, 50*(3), 285–311.

Donovan, G., et al. (2011). Urban trees and the risk of poor birth outcomes. *Health & Place, 17*, 390–393.

Donovan, G., et al. (2013). The relationship between trees and human health: Evidence from the spread of the emerald ash borer. *American Journal of Preventive Medicine, 44*(2), 139–145.

Eco-Business. (2012). *Singapore is top carbon culprit in Asia-Pacific*. Retrieved June 4, 2017, from http://www.eco-business.com/news/singapore-is-top-carbon-culprit-in-asia-pacific/.

Epstein, L., et al. (2006). Reducing sedentary behavior: The relationship between park area and the physical activity of youth. *Psychological Science, 17*(8), 654–659.

Ethical Markets. (2017). *Quality of life indicators in context*. Retrieved June, 2017, from http://ethicalmarketsqualityoflife.com/current-issues/.

Evans, G., Wells, N., & Moch, A. (2003). Housing and mental health: A review of the evidence and a methodological and conceptual critique. *Journal of Social Issues, 59*(3), 475–500.
Ewing, R., & Cervero, R. (2010). Travel and the built environment: A meta-analysis. *Journal of the American Planning Association, 76*(3), 265–294.
Faber, T. (2001a). Coping with ADD: The surprising connection to green play settings. *Environment and Behavior, 33*, 54.
Faber, T. (2001b). Views of nature and self discipline: Evidence from inner city children. *Journal of Environmental Psychology, 22*, 49–63.
Fjortoft, I. (2004). Landscape as playscape: The effects of natural environments on children's play and motor development. *Children, Youth and Environments, 14*(2), 21–44.
Frumkin, H., Frank, L., & Jackson, R. (2004). *Urban sprawl and public health: Designing, planning and building for healthy communities*. Washington: Island Press.
Gehl, J. (2010). *Cities for people*. Washington: Island Press.
Gele, A., & Harsløf, I. (2010). Types of social capital resources and self-rated health among the Norwegian adult population. *International Journal for Equity in Health*. Retrieved June 4, 2017, from http://www.ncbi.nlm.nih.gov/pmc/articles/PMC2848659/.
Gifford, R. (2007). The consequences of living in high-rise buildings. *Architectural Science Review, 50*, 2.
Girard, A., Self, J. L., McAuliffe, C., & Olude, O. (2012). The effects of household food production strategies on the health and nutrition outcomes of women and young children: A systematic review. *Paediatric and Perinatal Epidemiology, 26*(1), 205–222.
Grahn, P., et al. (1997). Ute på Dagis (Out in the preschool). In *Stad and land 145*. Håssleholm: Nora Skåne Offset.
Gross National Happiness. (n.d.). *Center for Bhutan Studies and Gross National Happiness Research*. Retrieved June 4, 2017, from http://www.grossnationalhappiness.com.
House, J. S., Landis, K. R., & Umberson, D. (1988). Social relationships and health. *Science, 241*(4865), 540–545.
Huynh, Q., et al. (2013). Exposure to public natural space as a protective factor for emotional well-being among young people in Canada. *BMC Public Health, 13*, 407.
International Making Cities Livable. (n.d.-a). *Mission*. Retrieved June 4, 2017, from http://www.livablecities.org/about/mission.
International Making Cities Livable. (n.d.-b). *Livable cities is gaining momentum*. Retrieved June 4, 2017, from http://www.livablecities.org/articles/livable-cities-gaining-momentum.
International Making Cities Livable. (n.d.-c). *Neighborhood squares*. Retrieved June 4, 2017, from http://www.livablecities.org/blog/neighborhood-squares.
International Making Cities Livable. (n.d.-d). *Principles for designing successful neighborhood squares*. Retrieved June 4, 2017, from http://www.livablecities.org/blog/principles-designing-successful-neighborhood-squares.
International Making Cities Livable. (n.d.-e). *Cities are killing us*. Retrieved June 4, 2017, from http://www.livablecities.org/articles/cities-are-killing-us.
Jacobs, J. (1961). *The death and life of great american cities*. New York: Random House.
Kahn, P., & Kellert, S. (Eds.). (2002). *Children and nature: Psychological, sociocultural, and evolutionary investigations*. Cambridge, MA: The MIT Press.
Kellert, S., Heerwagen, J., & Mador, M. (Eds.). (2008). *Biophilic design: The theory, science, and practice of bringing buildings to life*. Hoboken: John Wiley & Sons, Inc..
Kennedy, R. (1968). Speech at University of Kansas. transcript, March 18, 1968. *AmericanProgress.org*. Retrieved June 4, 2017, from http://images2.americanprogress.org/campus/email/RobertFKennedyUniversityofKansas.pdf.
Kirkby, M. (1989). Nature as refuge in children's environments. *Children's Environments Quarterly, 6*(1), 7–12.
Kuo, F., & Sullivan, W. (2001). Aggression and violence in the inner city. *Environment and Behavior, 33*(4), 543–571.

Kuo, F., & Taylor, F. (2004). A potential natural treatment for Attention- deficit/hyperactivity disorder: Evidence from a national study. *American Journal of Public Health, 94*(9), 1580–1586.
Lawlor, E. (n.d.). *Pedestrian pound*. Retrieved June 29, 2017, from https://www.livingstreets.org.uk/media/1391/pedestrianpound_fullreport_web.pdf.
Lennard, H. L., & Crowhurst Lennard, S. (2000). *The forgotten child*. Carmel: Gondolier Press.
Lorenz, K. (1990). *On life and living* (pp. 37–38). New York: St. Martin's Press.
Lomas, J. (1998). Social capital and health: Implications for public health and epidemiology. *Social Science & Medicine, 47*(9), 1181–1188.
Lovasi, G. S., et al. (2013, April). Urban tree canopy and asthma, wheeze, rhinitis, and allergic sensitization to tree pollen in a New York city birth cohort. *Environmental Health Perspective, 121*(4), 494–500.
Lynch, K. (1997). *Growing up in cities*. Cambridge: The MIT Press.
Malone, K. (2007). The bubble-wrap generation – Children growing up in walled gardens. *Environmental Education Research, 13*(4), 513–527.
Marks, N. (n.d.). *Happy planet index*. Retrieved June 4, 2017, from http://www.happyplanetindex.org/about.
Monheim, R. (Hrsg.). (1997). *"Autofreie" Innenstädte - Gefahr oder Chance für den Handel?* Arbeitsmaterialien zur Raumordnung und Raumplanung, H. 134, Bayreuth, Teile A und B.
Mumford, L. (1961). *The city in history*. San Diego, New York, and London: Harcourt Brace & Company.
Nestmann, F., & Hurrelmann, K. (1994). *Social networks and social support in childhood and adolescence*. Berlin and New York: Walter de Gruyter.
Nieuwenhuijsen, M., & Khreis, H. (2016). Car free cities: Pathway to healthy urban living. *Environment International, 94*, 251–262.
Nieuwenhuijsen, M, et al. (2017a). **P**ositive **H**ealth **E**ffects of the **N**atural **O**utdoor environment (PHENOTYPE). eReport #91, *#91 Healthy Communities for All Ages*. Retrieved June 29, 2017, from http://www.livablecities.org/documentationsets/91-healthy-communities-all-ages.
Nieuwenhuijsen, M. J., Khreis, H., Triguero-Mas, M., Gascon, M., & Dadvand, P. (2017b). Fifty shades of green. Pathway to healthy urban living. *Epidemiology, 28*(1), 63–71.
Pappas, M., et al. (2007). The built environment and obesity. *Epidemiologic Reviews, 29*(1), 129–143.
Pope F. (2015). *Laudato Si: Encyclical letter of the Holy Father: On care for our common home*. Retrieved June 4, 2017, from http://w2.vatican.va/content/francesco/en/encyclicals/documents/papa-francesco_20150524_enciclica-laudato-si.html, p. 15.
Pugh, T., et al. (2012). Effectiveness of green infrastructure for improvement of air quality in urban street canyons. *Environmental Science & Technology, 46*(14), 7692–7699.
Putnam, R. (2000). *Bowling alone. The collapse and revival of American community*. New York: Simon & Schuster.
Richardson, E. A., et al. (2013). Role of physical activity in the relationship between urban green space and health. *Public Health, 127*(4), 318–324.
Russell, A. (2010) Free range kids: Independence and the urban child. *PLAN7122 Planning Project*. University of New South Wales. Retrieved June 29, 2017, from https://cityfutures.be.unsw.edu.au/documents/139/Russell_thesis.pdf.
Sallis, J., et al. (1993). Correlates of physical activity at home in Mexican-American and Anglo-American preschool children. *Health Psychology, 12*, 390–398.
Schulz, W. (1994). Criteria for urban aesthetics. *Making Cities Livable Newsletter., 4*(1-2), 9–12.
Smart Cities Dive. (2014). *Ten reasons why high-rises kill livability*. Retrieved June 4, 2017, from http://www.smartcitiesdive.com/ex/sustainablecitiescollective/7-reasons-why-high-rises-kill-livability/561536/.
Swedish National Institute of Public Health. (2011). *Social health inequalities in Swedish children and adolescents*. Stockholm: Strömberg. Retrieved June 4, 2017, from https://www.folkhalsomyndigheten.se/pagefiles/12698/A2011-11-Social-health-inequalities-in-swedish-children-and-adolescents.pdf.

Szasz, O. (2013). Renewing oncological hyperthermia—Oncothermia. *Open Journal of Biophysics, 3*(4.) Retrieved June 4, 2017, from http://file.scirp.org/Html/6-1850078_38154.htm.
Topp, H. (1985). Ten simple rules of balanced urban transportation planning. *International Making Cities Livable*. Retrieved June 4, 2017, from http://www.livablecities.org/blog/ten-simple-rules-balanced-urban-transportation-planning-hartmut-topp.
US DHHS (US Department of Health and Human Services). (2008). *Physical activity guidelines for Americans*. Retrieved June 4, 2017, from http://www.health.gov/PAGuidelines/pdf/paguide.pdf.
Wang, H., Qiu, F., & Swallow, B. (2014). Can community gardens and farmers' markets relieve food desert problems: A study of Edmonton, Canada. *Applied Geography, 55*, 127–137.
Wells, N. (2000). At home with nature: Effects of 'greenness' on children's cognitive functioning. *Environment and Behavior, 32*(6), 775–795.
Wells, N., & Evans, G. (2003). Nearby nature: A buffer of life stress among rural children. *Environment and Behavior, 35*(3), 311–330.
Whyte, W. H. (1980). *The social life of small urban spaces*. New York: Project for Public Spaces.
Wikipedia. (n.d.-a). *Hikikomori*. Retrieved June 4, 2017, from https://en.wikipedia.org/wiki/Hikikomori.
Wikipedia. (n.d.-b) *Genuine progress indicator*. Retrieved June 4, 2017, from https://en.wikipedia.org/wiki/Genuine_progress_indicator.
Wikipedia. (n.d.-c) *Canadian index of wellbeing*. Retrieved June 4, 2017, from http://en.wikipedia.org/wiki/Canadian_Index_of_Wellbeing.
Wikipedia. (n.d.-d) *Living planet index*. Retrieved June 4, 2017, from http://en.wikipedia.org/wiki/Living_Planet_Index.
Wong, P., et al. (2014). The prevalence and correlates of severe social withdrawal (Hikikomori) in Hong Kong. *International Journal of Social Psychiatry, 61*(4), 330–342. https://doi.org/10.1177/0020764014543711.
World Health Organization (WHO). (n.d.) *Road safety – Speed*. Retrieved June 4, 2017, from http://www.who.int/violence_injury_prevention/publications/road_traffic/world_report/speed_en.pdf.
Young, R. (2012). *Stewardship of the built environment: Sustainability, preservation, and reuse*. Washington, Covelo, London: Island Press.

Chapter 5
Healthy Cities: A Political Movement Which Empowered Local Governments to Put Health and Equity High on Their Agenda

Agis D. Tsouros

5.1 Introduction

Healthy Cities was officially established as a strategic initiative in the European Region in 1988 to act as a vehicle of the WHO strategy Health for All at the local level (WHO 1991; Tsouros 2015). Today it is a thriving and powerful movement in most parts of the world. The aim was to put health high on the social and political agenda of the cities by promoting health, equity and sustainable development through innovation and change. Cities are societal engines for economy, human and social development. The creation of Healthy Cities was based on the recognition of the importance of action at the local and urban level and the key role of local governments. Following a decade of questioning and rethinking health and medicine and setting the values and principles of a new public health era, the 1980s provided the political legitimacy and the strategic means for taking forward an agenda for Health for All, based on powerful concepts and ideas and engaging a wide range of new actors. Most notably, the strategy Health for All (WHO 1984) and the Ottawa Charter for Health Promotion (WHO 1986) inspired new types of leadership for health that transcended traditional sectoral and professional boundaries.

The creation of the Healthy Cities project, as the WHO Regional Office's for Europe's strategic vehicle to bring Health for All (HFA) to the local level, was the result of several developments and initiatives in the early 1980s both at the local level and at WHO (Tsouros 2015). Jo Asvall, the director of the WHO Regional Office for Europe, in his speech at the European Congress on Healthy Cities in 1987

A. D. Tsouros (✉)
Global Healthy Cities, Athens, Greece

Policy and Governance for Health and Wellbeing and Healthy Cities at WHO Europe, Copenhagen, Denmark

Institute for Global Health Innovation at the Imperial College, London, UK
e-mail: agistsouros@globalhealthycities.com

M. Nieuwenhuijsen, H. Khreis (eds.), *Integrating Human Health into Urban and Transport Planning*, https://doi.org/10.1007/978-3-319-74983-9_5

(Asvall 1987) said, 'Building a healthy city becomes first and foremost a formidable challenge on how to create a movement for health where many players can be inspired and motivated for taking actions to think new and better solutions and to work together in new partnerships for health'. The following year in his speech (Asvall 1988) at a meeting to celebrate Copenhagen joining the WHO Healthy Cities project, he said, 'Why concentrate on cities? For two reasons: on the one hand, their problems are acute and rising; and on the other hand, the city level represents a particularly interesting and promising area for action in HFA. The Mayor of the city has much more power over his area than the Prime Minister has over the country; a city administration can much more easily instruct different sectors to work together in health; and … community participation is not a theoretical issue; it is daily at the finger-tips of the whole city administration'.

Jo Asvall, a truly visionary WHO regional director gave Healthy Cities huge political and strategic legitimacy from the start. He established it as a cross-cutting initiative that had the strategic mandate to actively engage local governments in the implementation of Health for All. Within a very short period the WHO Regional Office for Europe further strengthened its capacity of reaching out to new partners by establishing the health-promoting school and hospital settings networks and soon later the Regions for Health Network.

It should be stressed that Healthy Cities was launched as a political, cross-cutting and intersectoral project to be implemented through direct collaboration with cities. This was a bold and courageous move by WHO which is an organization that mainly works with and is accountable to national governments and predominantly the health sector. The importance of working at the local and community levels was reflected in many WHO resolutions since the 1970s but was not generally regarded as a green light to engaging local political leaders. Today, 29 years on Healthy Cities still represents a key strategic vehicle for implementing the new European Policy for Health and Wellbeing—Health 2020 (WHO 2013) at the local level.

The design of Healthy Cities was not meant to be static. It was launched as a value-based open system that would constantly reinvent itself and evolve, learning from practice and embracing new evidence and ideas to maintain its relevance and grounding itself on local concerns and perspectives. Healthy Cities was to be a pioneer in generating know-how for all urban communities to learn from, not an esoteric movement to benefit only its member cities.

One of the major obstacles in fully embracing the Healthy Cities concept has been the understanding of health. First, it requires understanding then adopting a meaning of health beyond the absence of disease, encompassing physical, social and mental well-being; second, it requires an appreciation of the nature and influence of the environmental, biological, social and political determinants of health; and third, it involves constantly making the case that health is important to individuals, to society and to socio-economic development.

While the overall principles underlying such an approach may appear unchanged over the years, in reality, the meaning, the content and the evidence underpinning these three requirements have vastly changed. The increasing emphasis on enduring values, such as the right to health, equity, sustainable development, and well-being, and the accumulating evidence on the social determinants of health have raised the stakes and the level of attention given to health.

The other part of the equation is dealing with change, making things happen and making arrangements that enable decision-makers, institutions, communities and citizens work together for health and well-being. Again, terms such as intersectoral action for health and community empowerment have been central in the action vocabulary of Healthy Cities' policy development. The vocabulary has expanded and conceptually evolved both in scope and depth, but the task at the core remains as challenging as ever-reaching out and engaging a wide range of agencies whose actions can impact on health. However, the world today is very different from the early phases of Healthy Cities in the late 1980s and 1990s. Several studies have drawn attention to challenges such as global interdependence and connectedness, the quickening pace of change, the added complexity of the policy environment and the increase in uncertainty.

Healthy Cities has always been driven by the enduring classic health promotion concepts which were based on the Ottawa Charter: creating supportive environments for health; making the healthy choices the easy choices; creating healthy settings, schools, workplaces, universities, health centres and neighbourhoods; and empowering individuals and communities which is a prerequisite for success.

The Ottawa Charter (WHO 1986) defined health promotion as 'the process of enabling people to increase control over, and to improve, their health'. Giving a voice to individuals and communities and creating the preconditions for empowerment and meaningful engagement are at the core of the Healthy Cities approach.

More than ever before and in the face of the fast-changing social landscapes of cities and towns, there is a need to create inclusiveness and social cohesion. Empowered communities will have the knowledge, the skills and the means to participate in decisions that affect their health and well-being and also navigate and access resources that can improve their health and quality of life.

Table 5.1 below outlines key concepts and issues that should be considered and addressed in a twenty-first century approach to Healthy Cities.

5.2 Healthy Cities Mission and Goals

The mission of Healthy Cities is to put health high on the social and political agenda of cities based on a framework of constant values and principles from its inception, namely, the right to health and well-being, equity and social justice, gender equality, solidarity and social inclusion, universal coverage and sustainable development. A Healthy City was described in terms of 11 qualities (see Table 5.2).

Healthy Cities in Europe evolved over 5-year phases. These phases allowed the regular renewal of the goals and requirements; they were long enough to see results and to evaluate progress; and participating member cities could leave the project at the end of the phase at no political cost. Every phase started afresh with a newly (re-)designated group of cities, many 'old' and several 'new blood'. The 'phase' approach proved most valuable in keeping the Healthy Cities momentum alive and strong.

Table 5.1 Modern public health concepts and issues

• Health increasingly used as a key indicator of development
• Health systems—universal coverage, patient-centred, health promotion and prevention, strong local public health infrastructures, addressing the upstream root cause of ill health
• Population-based approaches
• Whole-of-(local)-government, whole-of-society and Health in All policies approaches
• Addressing systematically social determinants (WHO 2013) of health and inequalities (WHO 2012a, b)
• Life-course approach and community resilience
• Health promotion in settings and promoting health literacy to individuals, communities and organizations
• Systematically measuring and monitoring the health of the population as well as the social, environmental and living conditions in the city
• Paying attention to the health needs of children, youth, older people, migrants and people living in poverty and the impacts of climate change

Table 5.2 Eleven qualities of Healthy City

1. A clean safe high-quality environment including affordable housing
2. A stable ecosystem
3. A strong, mutually supportive and nonexploitative community
4. Much public participation in and control over decisions affecting life, health and well-being
5. The provision of basic needs (food, water, shelter, income, safety and work) for all people
6. Access to a wide range of experiences and resources with the possibility of multiple contacts, interaction and communication
7. A diverse, vital and innovative economy
8. Encouragement of connections with the past, with the varied cultural and biological heritage and with other groups and individuals
9. A city form (design) that is compatible with and enhances the preceding characteristics
10. An optimum level of appropriate public health and care services accessible to all
11. A high health status (both a high positive health status and low disease status)

The goals and themes of every phase defined the priorities of work over the 5 years of the phase. Within the frame of the overarching goals and themes of each phase, cities had the flexibility to identify and give weight to areas that are of particular relevance to local realities. However, all cities were expected to work on the overarching and innovative themes of every phase, participating also in conceptual development and brokering the new ideas at the local level.

The agenda, themes and goals of each phase reflected WHO European priorities and strategies: global strategies and priorities and issues emerging from the urban (health, social, environmental) conditions in Europe (see the six Healthy Cities action domains (Tsouros 2017) in Table 5.3).

Looking back through the agenda and experience of Healthy Cities in Europe since its launch, one can easily trace the history of the new public health movement in the past 30 years. There is no new concept or approach in the areas of public health and sustainable development that was not embraced and tested by Healthy

Table 5.3 Healthy Cities action domains

• Political and governance
• Community level
• Policies, regulations, planning processes and city development strategies
• Services and programmes
• People and their needs; whole populations; different social groups; families; individuals
• Social, built and physical environment

Cities. Healthy Cities became on many occasions the source of innovation and leadership in areas that later gained major significance: for example, the launch of the Solid Facts (Wilkinson and Marmot 1998, 2003) publication on the Social Determinants of Health (SDH) in 1998 with professor Sir Michael Marmot led the way to the establishment of the global commission on the SDH. The strategic focus of Healthy Cities work meant a focus on upstream, high-impact approaches to health development and equity. Table 5.4 shows an overview of the themes and priorities of the Healthy Cities agenda over six phases.

Phase VI was launched as an adaptable and practical framework for delivering Health 2020 at the local level. It recognizes that each city is unique and will pursue the overarching goals and core themes of Phase VI according to its needs and processes that were sensitive and adaptable to local socio-economic, organizational and political contexts. Cities were encouraged to use different entry points and approaches but will remain united in achieving the overarching goals and core themes of the phase. Table 5.5 below shows the Phase VI priorities in more detail.

WHO Healthy Cities combined six essential features: (1) local relevance and openness to innovation and a cutting-edge public health agenda; (2) strong leadership and political commitment and a multisectoral approach to health development; (3) partnership-based management of change, transparency and democratic governance; (4) strategic thinking and planning and concrete deliverables and outcomes; (5) adaptability and receptiveness to emerging needs and ideas; and (6) commitment to solidarity and international and local networking.

Healthy Cities can exert their influence on health and equity in a wide range of mechanisms and processes including regulation (Cities are well positioned to influence land use, building standards and water and sanitation systems and enact and enforce restrictions on tobacco use and occupational health and safety regulations.), integration (Local governments have the capability of developing and implementing integrated policies and strategies for health promotion and sustainable development.), intersectoral partnerships (Cities' democratic mandate conveys authority and power to convene partnerships and encourage contributions from many sectors and stakeholders from the private and voluntary domains.), citizen engagement (Local governments have everyday contact with citizens and are closest to their concerns and priorities. They present unique opportunities for partnering with civic society and citizens' groups.) and equity focus (Local governments have the capacity to mobilize local resources and to deploy them to create more opportunities for

Table 5.4 WHO European Healthy Cities agenda—six phases

Phases	Main themes	Key strategies and political statements that defined the content of Healthy Cities: Global, WHO and Healthy Cities policies and declarations
I (1987–1992)	Creating new structures for and introducing new ways of working for health in cities. City health profiles—an essential tool	Health for All Ottawa Charter Milan Declaration of Healthy Cities
II (1993–1997)	Emphasis on intersectoral action, community participation and comprehensive city health planning	Rio Declaration on Environment and Development
III (1997–2003)	Action on health and sustainable development and healthy urban planning. Action on key NCD risk factors. Addressing the social determinants of health. City health development plans—an essential tool Partnership with other city networks in Europe	Jakarta Declaration of Health Promotion Athens Declaration of Healthy Cities Agenda 21—Rio plus 10 Health 21—Health for All in the Twenty-First Century European Sustainable Cities and Towns Campaign Millennium Development Goals
IV (2003–2007)	Increasing emphasis on partnership-based health development plans. Core themes include healthy urban planning, health impact assessment and healthy ageing	Belfast Declaration of Healthy Cities Report of the WHO Commission on the Social Determinants of Health (2008)
V (2008–2013)	Health and health equity in all local policies. Core thematic strands: caring and supportive environments, healthy living, health urban environment and design	The Tallinn Charter: Health Systems for Health and Wealth Zagreb Declaration of Healthy Cities European review of social determinants of health and the health divide Governance studies
VI (2014–2018)	Leadership for health City health diplomacy (Kickbusch and Kokeny 2017) Applying Health 2020 lens with emphasis on life-course approaches, community resilience and health literacy	European Policy and Strategy for Health and Wellbeing—Health 2020 2014 Athens Declaration of Healthy Cities (WHO 2014b) Sustainable development goals (SDGs)

poor and vulnerable population groups and to protect and promote the rights of all urban residents.).

Healthy Cities continued to expand and proved valuable during times of major changes in Europe and the world, including the fall of the Berlin Wall, the Yugoslavian wars, the expansion of the European Union, globalization, the rapid expansion of the information society and austerity waves and significant changes in the social landscape of the region.

Table 5.5 Goals and priority themes of Phase VI (2014–2018) of the European Healthy Cities

Overarching goals			
• Tackling health inequalities		• Promoting city leadership and participatory governance for health	
• Human rights and gender		• Whole-of-government and whole-of-society approaches	
		• Health and health equity in all local policies	
		• City health diplomacy	
Core themes			
Life course and empowering people	Tackling public health priorities	Strengthening people-centred health systems and public health capacity	Creating resilient communities and supportive environments
Highly priority issues			
• Early years • Older people • Vulnerability • Health literacy	• Physical activity • Nutrition and obesity • Alcohol • Tobacco • Mental well-being	• Health and social services • Other city services • Public health capacity	• Community resilience • Healthy settings • Healthy urban planning and design • Healthy transport • Climate change • Housing and regeneration

5.3 Healthy Cities Network Model

The WHO European model of Healthy Cities has two operational arms: the WHO Healthy Cities Network consisting of about 60–100 cities which are directly designated by WHO and the WHO Network of Healthy Cities Networks which brings together approximately 30 European National Healthy Cities Networks.

Table 5.6 below provides a typology of the four essential prerequisites for member cities (Tsouros 2017). Evidence of strong political commitment is fundamental requirement for the designation process. In addition to a letter from the Mayor or equivalent interested, applicant cities have to provide a city council resolution expressing support for the participation of the city in this WHO Network as well as partnership intent statements from public, private and voluntary sectors. These are crucial for the sustainability of the programme in cities.

Managing change and supporting innovation, especially when these imply new ways of working, must be supported by people who have the necessary knowledge, skills and seniority, to enable resources and mechanisms and processes to engage public sectors and agencies and civil society. The requirement for a coordinator and a project office that is strategically located within the city administration (ideally close to the Mayor's office) is of critical importance to be able to fulfil its strategic and intersectoral coordination function to the full. The profile and seniority of the coordinator have also proven crucial. Healthy Cities cannot reach its potential if it is reduced to a technical project far from the policy and strategy locus of the city. Furthermore, member cities have to establish an intersectoral steering committee

Table 5.6 Healthy Cities four action prerequisites

A	C
Explicit political commitment and partnership agreements at the highest level in the city making health, equity and sustainable development core values in the city's vision and strategies	Promoting Health in All Policies, setting common goals and priorities and developing a strategy or plan for health, equity and well-being in the city. Systematically monitoring the health of the population and the determinants of health in the city
B	**D**
Organizational structures and processes to manage, coordinate and support change and facilitate national-local cooperation, local partnerships and action across sectors, along with active citizen participation and community empowerment	Formal and informal networking and platforms for dialogue and cooperation with different partners from the public, private, voluntary and community domains

and designate a politician (the mayor or one of his deputies) to be responsible for the programme. The European experience has shown that cities can also establish project support structures outside the city organization allowing flexibility for broader collaborations with statutory and especially nonstatutory partners.

National Networks (WHO 2015) play a key strategic role in promoting the healthy Cities principles and ideas, supporting their member cities, organizing training and learning events as well as working with different ministries and participating in national programmes. National Networks also represent an important public health platform at European level. National networks are accredited by WHO at the start of each phase on the basis of explicit criteria reflecting the scope and goals of the phase as well as minimum managerial requirements similar to those applied by the WHO Network cities (WHO 2014a, b, c, d).

Unlike membership to most other international networks which is usually based on signing a statement or declaration, the members of the WHO Healthy Cities Network are required to meet political and organizational requirements and commit to addressing a set of themes and targets. This has always been a source of respectability and prestige for the membership to this WHO Network. The chapter at hand discusses the main features of Healthy Cities in the European Region and will particularly focus on a number of issues that have been and continue to be critical for its success.

Healthy Cities is a movement committed to change and innovation, and it needs to sustain its strategic course to fulfil its potential. An important attribute of Healthy Cities is the political legitimacy to address challenging issues such as equity, vulnerability, the determinants of health and sustainability. One of the greatest strengths of the Healthy Cities movement is the diversity of political, social and organizational contexts within which it is being implemented across Europe. Concepts such as healthy public policy, intersectoral action, Health in All Policies and whole-of-government and whole-of-society approaches continue to be elusive for many national governments. These concepts constitute the premise on which Healthy Cities was designed: a whole-of-local-government approach to health with strong emphasis on equity and partnerships with statutory and nonstatutory partners.

5.4 Local Leadership for Health and Sustainable Development

The well-being, health and happiness of the citizens depends on politicians' willingness to give priority to the choices that address equity and the determinants of health. Ultimately health is a political choice that should match the values and aspirations for protecting and constantly improving the health and well-being of all citizens. This means creating supportive social and physical environments and conditions for enabling all people to reach their maximum health and well-being potential. It is thus important for city leaders to visualize what society they wish to create and decide on the values that will underpin their visions for the cities.

Municipalities have evolved as key drivers of city health development, providing not only leadership but continuity and adaptability in administrative structures and processes. Leadership for health and health equity takes many forms and involves many actors, for example, international organizations setting standards, heads of governments giving priority to health and well-being, health ministers reaching out beyond their sector to ministers in other sectors, parliamentarians expressing an interest in health, business leaders seeking to reorient their business models to take health and well-being into account, civil society organizations drawing attention to shortcomings in disease prevention or in service delivery, academic institutions providing evidence on which health interventions work (and which do not) and research findings for innovation and local authorities taking on the challenge of Health in All Policies.

Such leadership for health in the twenty-first century requires new skills, often using influence, rather than direct control, to achieve results. Much of the authority of future health leaders will reside not only in their position in the health system but also in their ability to convince others that health and well-being are highly relevant in all sectors. Leadership will be not only individual but also institutional, collective, community-centered and collaborative (Kickbusch and Gleicher 2014; Kickbusch and Thorsten 2014). Such forms of leadership are already in evidence. Groups of stakeholders are coming together to address key health challenges at the global, regional, national and local levels, such as the global movement on HIV. Similar movements are emerging around noncommunicable diseases, environmental health and health promotion.

Cities have the capacity to influence the determinants of health and inequalities—'the causes of the causes' as it is commonly referred to. They can promote the health and well-being of their citizens through their influence in several domains such as health, social services, the environment, education, the economy, housing, security, transport and sport. They can do this through various policies and interventions, including those addressing social exclusion and support, healthy and active living (Edwards and Tsouros 2008a, b) (such as cycle lanes and smoke-free public areas), safety and environmental issues for children and older people, working conditions, preparedness to deal with the consequences of climate change, exposure to hazards and nuisances, healthy urban planning (Barton and Tsourou 2000) and design (neighbourhood planning, removal of architectural barriers, accessibility and proximity of services) and participatory and inclusive processes for citizens.

Intersectoral partnerships and community empowerment initiatives can be implemented more easily at the local level with the active support of local governments. Local leaders acting beyond their formal powers have the potential to make a difference to the health and well-being of local communities by harnessing the combined efforts of a multitude of actors (Kickbusch and Gleicher 2014; Kickbusch and Thorsten 2014).

In the complex world of multiple tiers of government, numerous sectors and both public and private stakeholders, local governments have the capacity to influence the determinants of health and well-being and inequities. They are well positioned to have such influence through whole-of-local-government and Health in All Policies, regulation, integrated strategies and plans and partnerships across society.

Local leadership for health and sustainable development in the twenty-first century means having a vision and an understanding of the importance of health in social, economic and sustainable development; becoming an advocate and active implementer of the health inequalities (WHO 2014a, b, c, d) and the sustainable development agenda (UN 2015); having the commitment and conviction to forge new partnerships and alliances; promoting accountability for health and sustainability by statutory and non statutory local actors; aligning local action with national policies; anticipating and planning for change; and ultimately acting as a guardian, facilitator, catalyst, advocate and defender of the right to the highest level of Health for All residents. Effective leadership for health and well-being requires strong political commitment, a vision and strategic approach, supportive institutional arrangements and networking and connecting with others who are working towards similar goals.

A manifesto for local leaders for health and well-being would read along the following lines. Local leaders should recognize that (WHO 2014a, b, c, d, 2016; PAHO 2016).

- Health is a fundamental human right, and every human being is entitled to the enjoyment of the highest attainable standard of health.
- Health should be a core value in city vision statements, policies and strategies.
- The health status of people and whole communities is profoundly affected by the conditions in which individuals are born, live and work.
- The knowledge and experience on the social, environmental, urban, cultural, commercial and political determinants of health provide the basis for how local decision-makers should understand and deal with health and well-being.
- The public health challenges of the twenty-first century to be addressed effectively require the full engagement of local governments.
- Local governments are well placed to provide effective leadership and capacity for intersectoral work for health and sustainable development, and they can promote and enable community involvement and empowerment.
- Local governments generally have primary responsibility for planning and/or delivering services critical for influencing the social determinants of health (SDH) (e.g. education, transportation, housing and urban planning, and often they have responsibility for health service delivery and public health).
- Local governments have a key and central role to play in the implementation of all the sustainable development goals (SDGs) and in particular address the strong

links between SDG 3 (good Health for All) and SDG 11 (make cities and human settlements inclusive, safe, resilient and sustainable).

Political leaders should strive to create Cities for All our citizens by:

- Fully using and integrating in their plans twenty-first-century evidence-based public health and health promotion approaches and solutions that work
- Ensuring that their policies and plans are comprehensive, systematic and strategic aiming at delivering best outcomes and maximum impact
- Integrating health and sustainable development considerations in the way their municipalities plan, design, maintain, improve and manage the build environment, infrastructure and services and by using creatively new technologies
- Valuing social diversity and investing in building trust and cohesion amongst community groups
- Employing whole-of-local-government and whole-of-society and Health in All Policies approaches in their efforts to reaching out to different partners (public and corporate) and civil society
- Focusing in engaging with other sectors on 'what they can do for health and what health can do for them' identifying win-win, synergistic and co-beneficial outcomes
- Promoting policy coherence, synergies and better coordination as well as systems enabling joint planning and accountability for health and equity
- Investing in creating adequate capacity for steering, managing and implementing our Healthy Cities initiatives and programmes
- Putting in place the resources and mechanisms for systematically assessing the health and the conditions that affect health in cities as well as for monitoring Health in All Policies and reducing health inequalities' efforts
- Publishing regularly a city health profile as a basis of identifying priorities and accountability for health in cities
- Increasing city wide policies, programmes and services for disease prevention and health promotion applying the social determinants of health (SDH), equity and economic lens and aiming at creating social and physical environments that are conducive to health and well-being as well as increasing health literacy (Kickbusch et al. 2013)
- Promoting awareness about individual responsibility and social responsibility for health through an SDH and equity perspective
- Developing strategies and plans that are framed on population-based and life-course approaches
- Developing an intersectoral integrated strategic framework and plan for health development in the city with commonly agreed (amongst different sectors and other stakeholders) goals
- Making sure that local Healthy Cities plans and activities are aligned and connected with the main city development strategies
- Developing local and national platforms, networks and fora that promote social dialogue and broad civic engagement

Fig. 5.1 Articulating and implementing a Healthy Cities vision

Figure 5.1 shows a schematic representation of key political considerations for the implementation of Healthy Cities.

There is an imperative for all truly committed Healthy Cities to work on a number of issues which are crucial for the health of urban communities. This is a minimal agenda for action (Tsouros 2017):

- To ensure that the Health in All Policies and SDG agendas are explicitly and fully integrated in cities visions and plans.
- To give high priority to community participation and empowerment and community resilience.
- To measure and systematically and comprehensively address health inequalities.
- To give all children a healthy start in life with the active involvement of different sectors (such as health, social services, education, housing and planning), families and communities.
- To create conditions for healthy and active living for all with emphasis on physical activity, healthy and sustainable nutrition, reduction of obesity and mental stress, controlling the use of alcohol and creating smoke- and drug-free cities.
- To increase health literacy amongst individuals, communities and institutions.
- To invest in healthy environments and healthy urban planning and design creating safe and clean neighbourhoods with access to greens and space for social interaction and good facilities for all and creating age and child-friendly settings.
- A key aspect of the success on the ground and sustainable evolution of Healthy Cities in Europe has been its ability to connect with other local strategies and programmes, to be a convener and facilitator of intersectoral and community dialogue and cooperation, to be continuously open to new concepts and to be sensitive to needs and emerging priorities.

One of the most promising modern ways to promoting health is the life-course approach. This means supporting good health and its social determinants throughout the life course leads to increased healthy life expectancy as well as enhanced well-being and enjoyment of life, all of which can yield important economic, societal and individual benefits. Interventions to tackle health inequities and their social determinants can be derived at key stages of the life course: maternal and child health, children and adolescents, healthy adults and healthy older people.

Without a doubt one of the most formidable goals for a Healthy City, which would require the contribution of many sectors, is giving children a healthy start in life.

A good start in life establishes the basis for a healthy life. Cities investing in high-quality early-years childcare and parenting support services can compensate for the negative effects of social disadvantage on early child development. Promoting physical, cognitive, social and emotional development is crucial for all children from the earliest years. Children born into disadvantaged home and family circumstances have a higher risk of poor growth and development.

WHO in Europe has invested in evaluating progress and achievements at the end of every phase (De Leeuw and Simos 2017; JUH 2013; HPI 2009, 2015; De Leeuw et al. 2014). This chapter has heavily drawn on the lessons learnt from almost three decades of Healthy Cities in Europe. However, a vast amount of knowledge has been generated by the European Healthy Cities movement that remains invisible, undocumented and undervalued. The stories and achievements of cities and networks need to be systematically explored and documented.

5.5 Epilogue

Mayors are emerging as powerful and influential agents for change, locally, nationally and internationally. City health diplomacy can make a true difference, but this implies coherent and strategic thinking. The local voice is essential in the decision-making governing bodies and international fora of many international organizations and can also be helpful in discussions regarding the engagement of non-state actors.

The new sustainable development agenda provides a new opportunity to strengthen health and equity in our cities and communities. Health 2020 and the sustainable development goal (SDG) agenda are mutually reinforcing and provide enormous legitimacy for strong leadership and action.

Healthy Cities can organically embrace and integrate the SDG agenda, which goes hand in hand with priority areas such as equity, vulnerability including poverty and migrants' health, community resilience, climate change and the whole determinants of health agenda. Now is the time to scale up Healthy Cities as an important global force for health and equity.

Healthy Cities is a dynamic concept which should be continuously enriched with new developments and emerging priorities and scientific evidence. It is the anticipatory quality of Healthy Cities that has made it attractive to cities in all countries, even those with very advanced public health presence. This is essential for Healthy

Cities to maintain its relevance and credibility. Its agenda, themes and goals should therefore reflect WHO regional priorities and strategies, global strategies and priorities and issues emerging from the urban (health, social, environmental) conditions in each region.

The key to using the Healthy Cities concept to its full potential is the capacity for effective leadership and intersectoral action through whole-of-government, whole-of-society and Health in All Policies approaches. Healthy Cities is a value-based movement. Equity, solidarity, sustainability and a commitment to creating Cities for All are crucial. In other words, what matters is not just the average health standard in a city, but making sure that whatever the city has to offer is apportioned to everybody.

The real danger always remains to engage in Healthy Cities using the trademark and, in reality, wasting a great concept and a great opportunity by doing fragmented low-impact ephemeral projects and not working seriously on issues such as health inequalities and healthy urban planning or addressing the special needs of disadvantaged groups.

Increasing the accessibility of the Healthy Cities movement in all regions of the world should be a priority. The time is right to create a strong global Healthy Cities movement. Health Cities provide an adaptable and practical framework for delivering Health for All at the local level. It provides an exceptional platform for joint learning and sharing of expertise and experience between cities, between national and local levels of government and between countries and regions.

Ultimately the future prosperity of urban populations depends on the willingness and ability of local decision-makers to seize new opportunities to enhance the health and well-being of present and future generations.

References

Asvall, J. E. (1987). Address at opening session, European Congress on Healthy Cities, Dusseldorf 14–18 June 1987 (unpublished original speech from Healthy Cities archives at WHO Regional Office for Europe).

Asvall, J. E. (1988). Copenhagen – A Healthy City, opening statement at a meeting to celebrate Copenhagen joining the World Health Organization project, 14 June 1988 (unpublished original speech from Healthy Cities archives at WHO Regional Office for Europe).

Barton, H., & Tsourou, C. (2000). *Healthy Urban Planning – A WHO guide to planning for people*. London and New York: Spon Press. https://www.routledge.com/Healthy-Urban-Planning/Barton-Tsourou/p/book/9780415243278.

De Leeuw, E., & Simos, J. (2017). *Healthy Cities, the theory, policy, and practice of value-based urban planning*. New York: Springer.

De Leeuw, E., Tsouros, A. D., Dyakova, M., & Green, G. (2014). *Promoting health and equity – Evidence for local policy and practice*. WHO Regional Office for Europe: Copenhagen. http://www.euro.who.int/__data/assets/pdf_file/0007/262492/Healthy-Cities-promoting-health-and-equity.pdf.

Edwards, P., & Tsouros, A. D. (2008a). *A healthy city is an active city: A physical activity planning guide*. WHO Regional Office for Europe: Copenhagen. http://www.euro.who.int/__data/assets/pdf_file/0012/99975/E91883.pdf?ua=1.

Edwards, P., & Tsouros, A. D. (2008b). *Promoting physical activity and active living in urban environments – The role of local governments: The solid facts Copenhagen*. WHO Regional Office for Europe. http://www.euro.who.int/__data/assets/pdf_file/0009/98424/E89498.pdf?ua=1.
Health Promotion International-Special Supplement on Healthy Cities. (2009, November 1). *Health Promotion International, 24*(Suppl. 1), i1–i3.
Health Promotion International-Special Supplement on Healthy Cities. (2015, June). *Health Promotion International, 30*(Suppl 1), i1–i2. https://doi.org/10.1093/heapro/dav045
Journal of Urban Health. (2013, October). Evaluating WHO Healthy Cities in Europe—issues and perspectives. *Journal of Urban Health, 90*(Suppl 1), 14–22. https://doi.org/10.1007/s11524-012-9767-6.
Kickbusch, I., & Gleicher, D. (2014). *Governance for Health in the 21st century*. WHO Regional Office for Europe: Copenhagen. http://www.euro.who.int/__data/assets/pdf_file/0019/171334/RC62BD01-Governance-for-Health-Web.pdf?ua=1.
Kickbusch, I., & Kokeny, M. (Eds.). (2017). *Health diplomacy – European perspectives*. WHO Regional Office for Europe: Copenhagen. http://www.euro.who.int/__data/assets/pdf_file/0009/347688/Health_Diplomacy_European_Perspectives.pdf?ua=1.
Kickbusch, I., & Thorsten, B. (2014). *Implementing a Health 2020 vision: Governance for health in the 21st century. Making it happen*. Copenhagen: WHO Regional Office for Europe. http://www.euro.who.int/__data/assets/pdf_file/0018/215820/Implementing-a-Health-2020-Vision-Governance-for-Health-in-the-21st-Century-Eng.pdf?ua=1.
Kickbusch, I., Pelican, J., Apfel, F., & Tsouros, A. D. (2013). *Health literacy- the solid facts*. WHO Regional Office for Europe: Copenhagen. http://www.euro.who.int/__data/assets/pdf_file/0008/190655/e96854.pdf?ua=1.
Pan-American Health Organisation. (2016). *The Santiago declaration for healthy municipalities*. Washington DC: Pan-American Health Organisation. http://www.paho.org/hq/index.php?option=com_docman&task=doc_view&gid=41598&Itemid=270&lang=en.
Tsouros, A. D. (2015). Twenty-seven years of the WHO European Healthy Cities movement: A sustainable movement for change and innovation at the local level. *Health Promotion International, 30*(S1), i3–i7.
Tsouros, A. D. (2017). *City leadership for health and sustainable development*. Published by Global Healthy Cities and the Ministry of Health of Kuwait. ISBN: 978-99966-63-20-8. http://www.kuwaithealthycities.com/city-leadership.pdf
United Nations. (2015). *Sustainable development goals, sustainable development knowledge platform*. New York: United Nations. https://sustainabledevelopment.un.org/?menu=1300.
WHO. (1986). Ottawa charter for health promotion. In *First international conference on health promotion, Ottawa* (pp. 17–21). http://www.euro.who.int/__data/assets/pdf_file/0004/129532/Ottawa_Charter.pdf?ua=1.
WHO. (2016). *Mayors consensus on healthy cities*. Geneva: WHO. http://www.who.int/healthpromotion/conferences/9gchp/9gchp-mayors-consensus-healthy-cities.pdf?ua=1.
WHO Europe: A review of progress. (1987–1990). Copenhagen: WHO Regional Office for Europe. http://www.euro.who.int/__data/assets/pdf_file/0016/101446/WA_380.pdf?ua=1.
WHO Regional Office for Europe. (1984). *Health for all targets*. WHO Regional Office for Europe: Copenhagen.
WHO Regional Office for Europe. (1991). *WHO Healthy Cities project: A project becomes a movement*.
WHO Regional Office for Europe. (2012a). *Healthy cities tackle the social determinants of inequities in health: A framework for action*. WHO Regional Office for Europe: Copenhagen. http://www.euro.who.int/__data/assets/pdf_file/0006/166137/Frameworkforaction.pdf?ua=1.
WHO Regional Office for Europe. (2012b). *Addressing the social determinants of health: The urban dimension and the role of local government*. WHO Regional Office for Europe: Copenhagen. http://www.euro.who.int/__data/assets/pdf_file/0005/166136/UrbanDimensions.pdf?ua=1.

WHO Regional Office for Europe. (2013). *Health 2020: The European policy for health and well-being*. WHO Regional Office for Europe: Copenhagen. http://www.euro.who.int/health2020.

WHO Regional Office for Europe. (2014a). *Terms of Reference and accreditation requirements for membership in the Network of European National Healthy Cities Networks Phase VI (2014–2018)*. WHO Regional Office for Europe: Copenhagen. http://www.euro.who.int/__data/assets/pdf_file/0017/244700/Terms-of-Reference-and-accreditation-requirements-for-membership-in-the-Network-of-European-National-Healthy-Cities-Networks-Phase-VI-2014-2018.pdf.

WHO Regional Office for Europe. (2014b). *Athens Healthy Cities declaration 2014*. WHO Regional Office for Europe: Copenhagen. https://etouches-appfiles.s3.amazonaws.com/html_file_uploads/a2e343ecd85e0e3f2a8d8470681d1354_AthensInternationalHealthyCitiesDeclaration.pdf?response-content-disposition=inline%3Bfilename%3D%22Declaration%22&response-content-type=application%2Fpdf&AWSAccessKeyId=AKIAJC6CRYNXDRDHQCUQ&Expires=1506796067&Signature=eY5fEIK9PGVI7W30kk3s0WohmDs%3D.

WHO Regional Office for Europe. (2014c). *Review of social determinants and the health divide in the WHO European Region. Final report*. Copenhagen: WHO Regional Office for Europe. http://www.euro.who.int/__data/assets/pdf_file/0004/251878/Review-of-social-determinants-and-the-health-divide-in-the-WHO-European-Region-FINAL-REPORT.pdf.

WHO Regional Office for Europe. (2014d). *Phase VI—The WHO European Healthy Cities goals and requirements*. WHO Regional Office for Europe: Copenhagen. http://www.euro.who.int/__data/assets/pdf_file/0017/244403/Phase-VI-20142018-of-the-WHO-European-Healthy-Cities-Network-goals-and-requirements-Eng.pdf?ua=1.

WHO Regional Office for Europe. (2015). *National healthy cities networks in the WHO European Region. Promoting health and well-being throughout Europe*. Copenhagen: WHO Regional Office for Europe. http://www.euro.who.int/en/health-topics/environment-and-health/urban-health/publications/2015/national-healthy-cities-networks-in-the-who-european-region.-promoting-health-and-well-being-throughout-europe-2015.

Wilkinson, R., & Marmot, M. (1998). *The social determinants of health – The solid facts*. WHO Regional Office for Europe: Copenhagen.

Wilkinson, R., & Marmot, M. (2003). *Social Determinants of health. The solid Facts* (2nd ed.). Copenhagen: WHO Regional Office for Europe. http://www.euro.who.int/en/health-topics/environment-and-health/urban-health/publications/2003/social-determinants-of-health.-the-solid-facts.-second-edition.

Chapter 6
Human Ecology in the Context of Urbanisation

Roderick J. Lawrence

6.1 Introduction

Humans have constructed their habitats over several millennia in each region of the world. Human habitats contain built and natural environments that provide sheltered conditions for daily life and infrastructure that maintains the supply of food, energy and water for the resident population (Boyden 1987). All the materials and processes that are necessary for the construction, functioning and maintenance of cities and large urban regions are related in some way to the availability and uses of abiotic and biological resources of ecosystems. Cities are dependent on the quantity and quality of these natural resources and the exportation of waste products in order to sustain their populations (Elmqvist et al. 2013). Energy, fuels, materials and water are transported from elsewhere and transformed into goods and services.

The high concentrations of activities, objects and people in cities and the flows between rural and urban areas mean that urbanisation is a major contributor to national economies and to environmental change at local, regional and global levels (Millennium Ecosystem Assessment 2005). One result of this urban characteristic is that cities are locations of relatively high concentrations of air, soil and water pollution, as well as all kinds of wastes. Consequently, urban populations are exposed to adverse environmental conditions in their habitat that are risks for their health (Hardoy et al. 2001).

R. J. Lawrence (✉)
Geneva School of Social Sciences (G3S), University of Geneva, Geneva, Switzerland

School of Architecture and the Built Environment, Faculty of the Professions,
University of Adelaide, Adelaide, SA, Australia

Institute for Environment and Development (LESTARI),
Universiti Kebangsaan Malaysia (UKM), Bangi, Selangor, Malaysia
e-mail: Roderick.lawrence@unige.ch

M. Nieuwenhuijsen, H. Khreis (eds.), *Integrating Human Health into Urban and Transport Planning*, https://doi.org/10.1007/978-3-319-74983-9_6

Cities and urban regions of different size provide varying numbers and kinds of community services including education, health care, leisure, tourism and welfare services (Sarkar et al. 2014). However, research confirms that residents do not have equal access to these services owing to geographical and socio-economic differences: Cities can become arenas for social differentiation, segregation and exclusion (Sarkar et al. 2014). They may also facilitate the communication of infectious disease, such as severe acute respiratory syndrome (SARS). They can be localities of social disorders (including criminality and violence), which may contribute to stress and mental illness (UN-Habitat 2010).

The ecological processes and products of cities and urban regions (such as Los Angeles, Kuala Lumpur, Paris and Tokyo) are rarely confined to administrative, geographical or political boundaries commonly associated with them. The reason for this is that cities and urban regions are complex open systems that transgress human-made borders (Dyball and Newell 2015). Hence, policies and projects that encourage land uses for either agriculture or urban development need to extend beyond traditional administrative and political boundaries as well as sector-based divisions of labour (Lawrence 2010). Therefore, coordination between geopolitical authorities within and beyond these boundaries is necessary. This chapter argues that an interdisciplinary and intersector conceptual framework based on the generic principles of human ecology should be applied in order to identify and address the diversity of public health challenges provided by urbanisation. The chapter shows that the advantage of human ecology is that its integrated, systemic framework explicitly accounts for the mutual interaction between the human and non-human components of habitats in a way that either the natural or social science disciplines have not achieved.

6.1.1 Rethinking Urbanisation

The drivers of large-scale urban development are difficult to understand (Hobbs et al. 2013). The multiple consequences of urbanisation trends are not well known and measured because they are numerous and operate at different geopolitical levels. There is a growing amount of evidence showing that modern urban and rural development programs and large housing projects have yielded many improvements to living conditions in all regions of the world. However, not all urban agglomerations or residential neighbourhoods benefit equally (UN-Habitat 2010). Some residential neighbourhoods can be characterised by relatively large numbers of migrants who are unemployed, relatively large households with low incomes and a housing stock of many non-renovated high-rise buildings constructed after 1950 (World Bank 2001). Recent events in many African, Asian and Latin American cities highlight a range of contemporary problems related to the exposure of residents to adverse environmental conditions (e.g. summer and winter smog, soil contamination and water pollution), socio-economic inequalities (that can be drivers of deprivation, delinquency, homelessness and unemployment), and political corruption (that can be a catalyst for social protests, riots and warfare).

There are no simple answers to current challenges stemming from rapid urbanisation, but it should be acknowledged that policy makers have identified and isolated problems too narrowly (Lawrence 2015). Today, there is a growing consensus that uncoordinated approaches need to be replaced by coordinated ones that account for the interrelations between ecological, economic and social dimensions of changing land uses in urban regions and how these impact on ambient living conditions and influence health and well-being. The author of this chapter argues that there is an urgent need to reconsider housing, building, transport and large-scale urban development in a broad environmental, social and political context that explicitly aims to promote public health.

The formulation and implementation of traditional sector-based contributions in housing, building, transport and urban planning should be challenged. Incremental improvements (e.g. the construction of new residential neighbourhoods) are often achieved in tandem with unintended consequences, such as direct negative impacts on environmental conditions (e.g. loss of biodiversity and public green space) and indirect impacts on the health and well-being of citizens (Mueller et al. 2017) (Fig. 6.4). These unforeseen outcomes are partly due to the number and complexity of all those factors that policy decision-makers and professional practitioners need to consider. They are also related to the recurrent lack of coordination between urban development policies, public health and other sectors including energy, housing and transportation (Khreis et al. 2016). Lack of coordination between sectors in order to promote public health has been associated by Lawrence (2010) with conceptual, institutional and social barriers including:

1. The number and the complexity of all those factors that researchers, practitioners and policy decision-makers ought to consider
2. The uncertainties and the unpredictability of the interrelations between many of these factors which are rarely admitted
3. The segmented knowledge of researchers, public administrators and practitioners who may be experts on specific subjects but they do not have an integrated perspective of what they consider
4. The lack of coordination between institutions and actors in different sectors and between people working in different geopolitical institutions
5. The lack of systematic monitoring and feedback within sectors (such as housing or transport) and especially across different local, regional and national levels.
6. The non-account of goals, priorities and values related to the ways policy decision-makers and citizens develop local economies, interpret their livelihoods and value the qualities of their habitat

The author of this chapter has argued for a fundamental rethinking of the relationships between social, economic and health inequalities and other kinds of anthropogenic problems in cities and urban regions (Lawrence 2015). The interrelations between housing markets, transport infrastructure, health systems, community services, environmental policies and land-use planning have been poorly articulated until now (Mueller et al. 2017). However, it is crucial to acknowledge the important role of cities as localities for the management of numerous resources, as places for accommodating diverse cultures and ways of life, as localities for access to medical services and health care and as significant forums for economic development at the local, national and regional levels (Kresl 2007). Although housing and urban development policies have rarely been a high priority in the manifestos of governments or political parties, there is a growing awareness led by non-governmental organisations, local government associations and research consortiums that health should be an integral component of urban planning. A rethinking of transport infrastructure as a driver of economic and urban development is feasible and necessary (Khreis et al. 2016). For example, the planning of coordinated infrastructure that facilitates cycling, walking and transit to affordable and efficient public transport can reduce exposures to the negative externalities of private car transport infrastructure and reduce sedentary lifestyles, by making active mobility attractive especially when travelling relatively short distances in urban areas.

The next section of this chapter presents the principles of human ecology and distinguishes them from those of general ecology. Then the key principles of human ecology are used to formulate and apply a conceptual framework to analyse the distinctive characteristics of urban ecosystems. This enables us to distinguish cities and large urban regions from other kinds of human habitats. These characteristics can be used by researchers, policy makers and professional practitioners to monitor and evaluate to what degree urbanisation trends impact on the health and well-being of populations in specific neighbourhoods.

6.2 Theoretical Concepts and Framework

The term 'ecology', from the ancient Greek words *oikos* and *logos*, denotes 'science of the habitat'. There is a large consensus that this term was first used by Ernst Haeckel (1834–1919), a German zoologist, in 1866 (Lawrence 2001). The word ecology commonly designates a science that studies the multiple interrelationships between organisms and their surroundings. Since the late nineteenth century, the term ecology has been interpreted in numerous ways including general and human ecology.

6.2.1 *What Is General Ecology?*

During the twentieth century, botanists and zoologists use the term 'general ecology' to refer to the interrelations between animals, fungi, plants and their immediate surroundings. The number of contributions about the science of ecology grew from the beginning of the twentieth century (Pickett et al. 2001). Animal and plant ecologists maintain that interactions between organisms and all the components of ecosystems follow principles that refer to their similarities and their differences. A community of organisms develops from simple to more complex forms through a sequence of developmental stages known as succession. This term refers to the slow progression of changes in communities of animals and plants owing to changes in ecological and climatic conditions. This evolutionary trend means that some species with a longer lifespan become dominant in a particular biotope or ecosystem for a certain time period. This trend may become a climax state: Climax is a dynamic equilibrium state that is determined by the limiting factors of the climate, soil or other ecological conditions (Pickett et al. 2001). Climax refers to the culmination of the evolution of animal and plant communities that correspond to the optimal development of the biomass with respect to specific ecological conditions. By using an analogy, some ecologists imply that human groups and communities are natural phenomena that develop by slow progression and succession processes. This interpretation means that psychological and social characteristics of human individuals and societies are analogous to biological factors, that competition between human beings is an innate biological process and that climax is the outcome. The fundamental principles of human ecology challenge this analogy by accounting for the psychological, social and cultural dimensions of human life (Boyden 1987; Lawrence 2001).

6.2.2 *What Is Human Ecology?*

In contrast to general ecology, human ecology usually refers to the study of the dynamic relationships between humans and the physical, biotic, cultural and social characteristics of their environment and the biosphere. However, this is not the original meaning of this term which was first used by Ellen Swallow Richards (1842–191). In her original contribution, she proposed human ecology in her formulation of euthenics, which she defined as a science for better living (Clarke 1973). From an institutional perspective, human ecology developed in the Department of Sociology at the University of Chicago, in the context of a rapidly urbanising city after the First World War. It was promoted by a coalition of researchers from a number of social science disciplines (including anthropology, demography, geography, psychology and sociology). These researchers shared a concern about the effects of urban living on the daily life and well-being of the residents, especially minority groups of migrants and low-income households.

Today, human ecology generally refers to the study of the reciprocal relations between people, their habitat and the environment beyond their immediate surroundings. A conceptual model of human ecology formulated in Lawrence (2001) is reproduced in Fig. 6.1. This figure is not meant to be a detailed model of people-environment relations that can provide a complete understanding of a complex and vast subject. Instead it represents an integrated model that represents the systemic interrelations between sets of biotic, abiotic and cultural factors that are combined together in any human ecosystem. Hence it does not concentrate only on specific

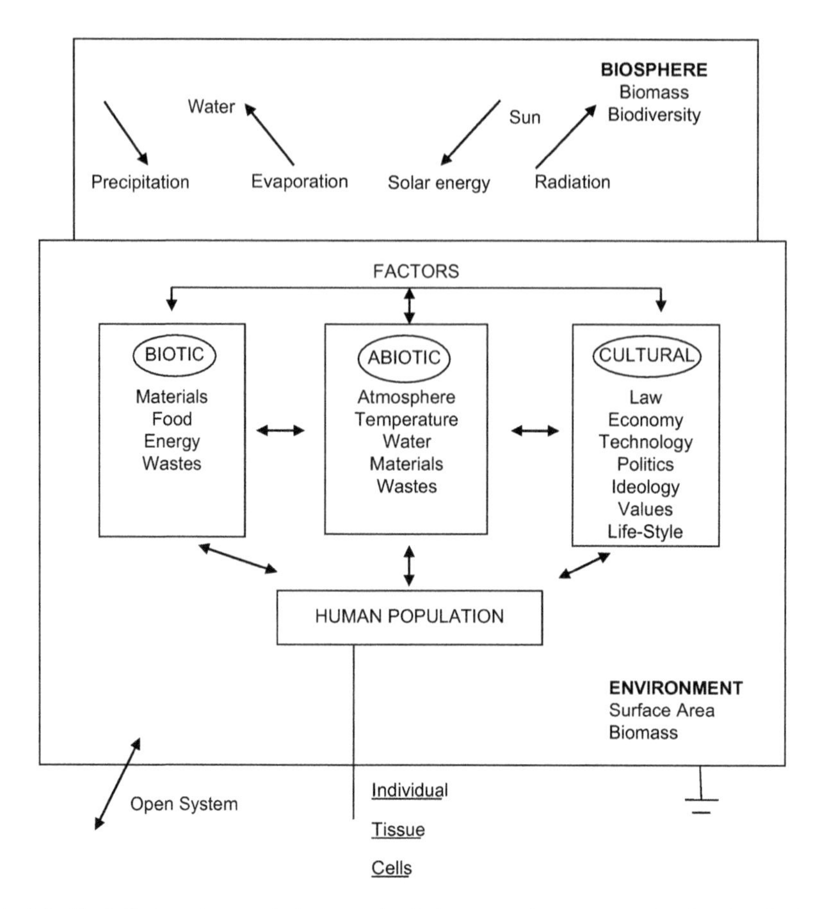

Fig. 6.1 The holistic and systemic framework of a human ecology perspective showing the interrelations between biotic factors, abiotic factors and cultural, social and individual human factors and artefacts which are delimited by situations, habitats or larger ecosystems (Source: Lawrence 2001)

components because it considers the whole system as the unit of study for people-environment relations. This integrated model can be applied to analyse different geographical areas (neighbourhoods, cities and mega-urban regions). It is a synchronic representation of a human ecosystem that is open and linked to others. The model is meant to be reapplied at different times to explicitly address both short- and long-term perspectives. This temporal perspective can identify change to any of the specific components as well as the interrelations between them.

Human ecology is explicitly interdisciplinary (Lawrence 2001). The material and non-material dimensions of human ecosystems, shown in Fig. 6.1, include genetic patrimony, especially the capacity of the human brain to interpret and transform land and other natural resources into a viable habitat; demographic characteristics such as the size and composition of human populations in mega-urban regions; the social organisation of human groups in urban neighbourhoods (including kinship relations and household structure); institutions including associations, rules and customs that regulate individual and collective behaviours; the local economy including all consumption and production processes; and, last but not least, the beliefs, knowledge religion and values of local populations (Lawrence 2001).

6.2.3 What Is Urban Ecology?

Urban ecology has been interpreted in diverse ways (Douglas et al. 2011; Young 2009). Perhaps the most common interpretation stems from the natural sciences, notably animal and plant biology. In this context, urban ecology refers to the multiple relations between animal and plant populations and the ecological conditions of their habitat, which includes significant human influences in suburban and urban areas (Pickett et al. 2001). Urban development processes can significantly modify natural habitats to the extent that some species migrate and live elsewhere, whereas other foreign species can inhabit the urban ecosystem.

A second set of interpretations of urban ecology stems from the social sciences, notably anthropology, sociology and human geography (Moran 2016). This set of interpretations is anthropocentric and deals specifically with *homo urbanus*. It analyses the mutual interaction between humans living in urban areas and the natural and human-made components of these areas. The geographical distribution of both natural and built components of urban ecosystems, as well as human populations or groups in those ecosystems, has been studied since the 1920s (Dyball and Newell 2015).

A third set of interpretations is technical and functional stemming from engineering and urban planning (Wachsmuth 2012). It considers cities and urban regions as metabolisms with the provision of infrastructure and services to supply all that is necessary to sustain human populations in them: Particular attention is given to the supply of food, energy and water by material flow analysis. Industrial ecology is one application of this kind of interpretation (Young 2009). Another application is the calculation of ecological footprints: Urban areas occupy large surfaces of land, but their ecological footprints (e.g. the quantity of resources needed to sustain them and assimilate all their wastes) exceed these surface areas many times and have significant impacts on hinterlands (Seitzinger et al. 2012). This has become a global phenomenon given that it is estimated that cities and urbanisation processes occupy only about 2–3% of the land surface of the world, whereas they need about 75% of all resources consumed globally (Harrison et al. 2000).

6.2.4 What Is Political Ecology?

Political ecology applies a different approach because it has legal and normative foundations enabling environmental problems to be addressed pragmatically by corrective measures (Lawrence 2001). These kinds of problems are meant to be overcome by legislation, technological efficiency and financial measures to reduce the impacts of human production and consumption stemming from uses of natural resources and the discharge of wastes. This interpretation has been complemented by an ethical one that has addressed property rights (including the rights of nature). Property rights are social arrangements between people that define the rights, entitlements, obligations and duties of persons, companies or an authority (the right holder) in relation to a specific entity (e.g. a component of the natural environment, such as a forest or a lake). Property rights stipulate how the right holder and other parties (non-property holders) are morally and legally required to act (Hann 1998). They create interdependence between people and natural resources as well as issues of distribution and fairness. In general, private claims, rights and responsibilities of environmental resources often fail to meet the collective or public need for environmental protection and intergenerational equity. This means that core principles of sustainable development are not met as Lawrence (2005) noted.

6.3 Ecological Public Health

Ecological Public Health posits that human health is dependent on how people live in a complex ecological system (see Chap. 33 by George Morris). Consequently, health is the outcome of the mutual interaction between humans and their immediate environment. Rayner and Lang (2012, p. 93) wrote that a key theme '… is interrelatedness, how people fit into the biosphere, how they use and care for the natural world, how all species interact, and how their interactions have consequences almost always with feedback loops'. One example of the application of this model is the growing concern about the relatedness of climate change, increasing differences in quality of life in human habitats and the health of residents in specific neighbourhoods (Whitmee et al. 2015).

The formulation of an ecological model of public health requires system thinking. Systems comprise components and subcomponents that interact directly or indirectly by two-way processes (Meadows 2009). Urban health should be interpreted as a complex system related to others (e.g. mobility and transport infrastructure) in real-world situations, as shown in Fig. 6.2. This representation shows that any urban health system comprises many proximate and distal components that interact mutually at different levels (Barton and Grant 2006). These complex systems should not be studied using linear interpretations of cause-effect models because such simplification cannot represent extant situations. Systemic models of urban health recognise that any internally or externally generated changes to one of the components of a system will impact on the other components including the initial component that was changed (Dyball and Newell 2015). Therefore, it is too restrictive to assume that improved access to green public space will change human behaviour and then improve health and well-being. It is necessary to consider how an informed public that benefits from improved health and well-being has the capacity to modify their lifestyle and influence uses of public green space. Then it is possible to monitor and evaluate what consequences occur.

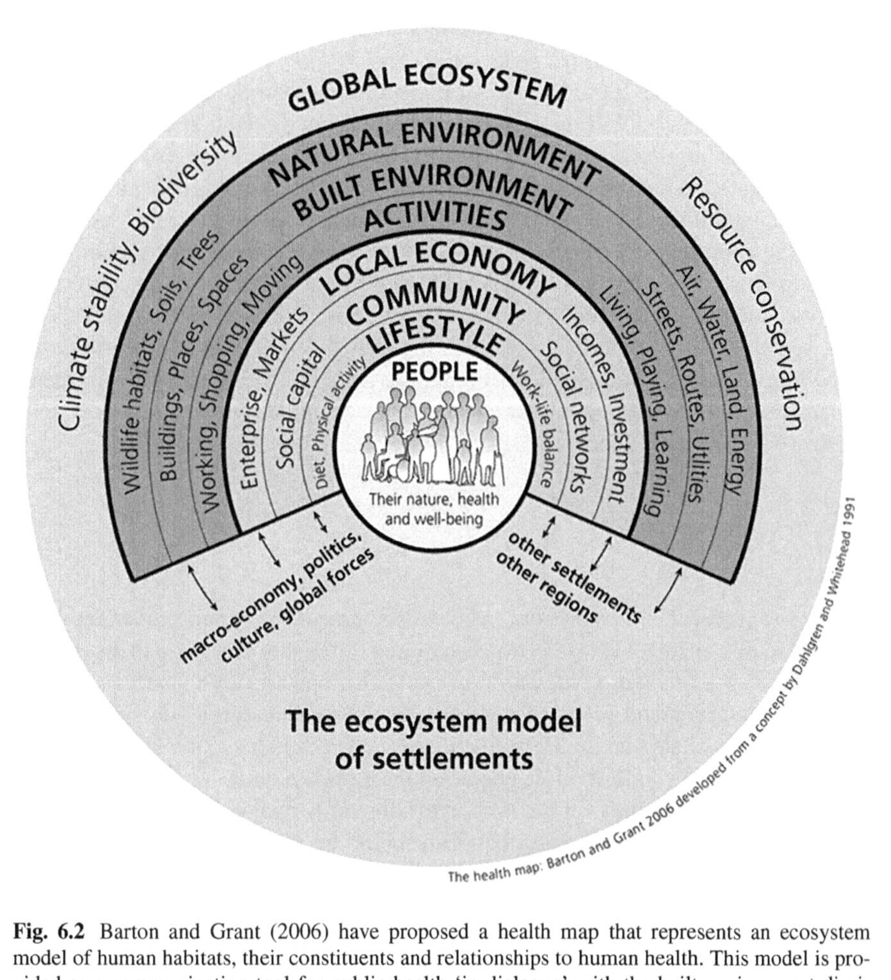

Fig. 6.2 Barton and Grant (2006) have proposed a health map that represents an ecosystem model of human habitats, their constituents and relationships to human health. This model is provided as a communication tool for public health 'in dialogue' with the built environment disciplines—(e.g. planners, architects, urban designers, landscape architects, transport planners and environmental designers). Developed from a concept by G. Dahlgren and M. Whitehead (1991)

6.4 Cities and Urban Development

Despite the global phenomenon of urbanisation, there still is no international consensus about the definition of a city, or an urban agglomeration, or a mega-urban region. Although the definition of a city varies from country to country, the United Nations uses national definitions that are commonly based on population size (Galea and Vlahov 2005). Megacities are often referred to as those with a population that exceeds ten million. Other interpretations are based on the administrative or political authority of urban areas, especially the degree of autonomy in relation to national or regional authorities. Some definitions include the socio-economic status of the resident population, especially their livelihood (e.g. the proportion of all employed people with nonagricultural occupations).

Between 1960 and 2000, the global human population doubled, and it is projected to increase to 9.7 billion in 2050 and 11.2 billion in 2100. The urban share of the global population increased from 29% in 1950 to 51% in 2011. Therefore, urban ecosystems are the habitat of more than half of the global population, and this share is predicted to increase to about 70% by 2050 (United Nations 2015). Urbanisation during the twentieth century, coupled with demographic growth, migration flows and economic development, has provided both positive and negative outcomes. The negative outcomes have been the source of numerous environmental and social concerns including loss of biodiversity following changes to land use, increasing toxic air pollution (in large cities such as Beijing, Los Angeles, Mexico and Paris), access to safe drinking water and sanitation and the accumulation of liquid and solid wastes (UN-Habitat 2010).

6.4.1 What Is Urbanisation?

In order to direct the debate between scientists, practitioners and policy decision-makers, some conceptual clarification is required. First, it is necessary to distinguish between *cities as human-made built environments* (specifically constructed buildings, public spaces and infrastructure that result from numerous decisions about how to accommodate human life) and *cities as urban processes* (including the multiple flows of energy, information, people and material resources that occur between cities and their hinterlands). It is common to adopt only one of these interpretations. The author of this chapter argues that both should be applied in a complementary way to deal with the complexity and diversity of health impacts of urbanisation. When this integrated interpretation is applied, then key principles of human ecology can be used to analyse the ordering of different kinds of natural resources, diverse groups of people and their activities as well as the goals, priorities and actions that are meant to achieve desired outcomes, especially those concerning improved health and well-being.

In essence, the construction of cities is intentional. It always occurs in a human context. Each society defines and is mutually defined by a wide range of cultural, societal and individual human factors that are implicitly or explicitly related to decisions about urbanisation. The layout, construction and intended uses of cities and urban regions involve choosing between a range of options in order to achieve objectives that may or may not give a high priority to health and quality of life. The complexity of cities and urban regions raises some critical questions such as:

1. What parameters are pertinent for a specific building task, such as the construction of a new residential neighbourhood?
2. Whose goals, intentions and values will be taken into consideration?
3. How and when will these goals and intentions be achieved?
4. What will be the monetary and nonmonetary costs and benefits of alternatives?

In order to answer these kinds of questions, it is necessary to recall the generic characteristics of cities and urban regions that have been applied over 9000 years in different regions of the world. This will be done in the next section of this chapter.

6.5 Characteristics of Cities and Urban Regions

Generic characteristics can be used to interpret differences between rural and urban areas, but this has been rare, especially in recent published research on large-scale urban development. In order to distinguish cities and urban regions from other kinds of human habitats (notably rural towns and suburban sprawl), it is important to identify their generic characteristics and then consider how they may influence health and well-being.

6.5.1 Centralisation or Decentralisation

The first characteristic of cities and urban regions is centralisation. It stems from the fact that the site of a city is chosen by humans. The choice of a specific site and the definition of the administrative and political boundaries of a city distinguish it from all other cities and their hinterlands. Studies in urban history and geography confirm that many factors have been involved in the location of cities (Bairoch 1988). For example, coastal sites for ports—for example, Cape Town, Djakarta, Hong Kong and Mumbai—can be contrasted with sites on inland trade routes such as Florence, New Delhi and Vienna. It is important to note that modern economic rationality has an interpretation of the world and human societies which has rarely accounted for the climatic, geological and biological characteristics of the location of specific cities. This has meant that urban populations in cities including Lisbon, Los Angeles and Tokyo have been confronted with unforeseen natural and human-made disasters including earthquakes, flooding and landslides over several centuries (Mitchell 1999).

During the late twentieth century, the globalisation of the public economy and private financial sectors has been increasingly concentrated in extended mega-urban regions. The era when a limited number of cities—Venice or London, for example—dominated the world economy has been superseded by networks of cities that form new polycentric world markets. This is one illustration of the principle that cities and urban regions are open rather than closed systems (Elmqvist et al. 2013). One outcome of these networks is the proliferation of invasive animal, insect and plant species via commerce and trading. These global trends have adverse impacts on natural ecosystems and their indigenous species, as well as negative impacts on human health (Sandifer et al. 2012).

In a period of accelerating change and globalisation, the growing interrelations between cities and mega-urban regions (irrespective of geographical distance) should be reconsidered in relation to their capacity to participate in and contribute to local, national and global economies in ways that support the health and well-being of residents (Kresl 2007). It should be recognised that cities and urban regions ought to have the knowledge and political commitment to deal effectively with rapid economic change including financial collapse. Macroeconomic policies and local urban development are interrelated as shown by some cities including Athens and Detroit in recent years. Consequently, the public and private sectors should form coalitions that define and implement policies that build on specific assets and potentials for community services and infrastructure that enable and sustain population health. There is also an urgent need for local and national authorities to enhance the adaptive capacity of cities to respond effectively to both predictable trends and unforeseen changes at the local and regional levels.

6.5.2 Verticality or Horizontality

The second characteristic of cities and urban regions is verticality. During the 9000-year history of cities, societies have constructed multi-storey buildings. Bairoch (1988) noted that Jericho included buildings of seven storeys. This characteristic underlies the compact or dense built environment of urban areas in contrast to the dispersed character of rural and suburban development. The height of buildings in cities increased dramatically from the late nineteenth century with the construction of skyscrapers, first in Chicago and then other cities around the world. The relations between high-rise housing conditions and health status are not easy to decipher owing to the vast number of confounding factors (Sarkar et al. 2014).

In recent decades, published research has identified and measured the relations between the specific characteristics of high-rise housing and health outcomes. Fortunately, there has been a widening of scope of scientific studies: For example, a common assumption in the 1970s that floor level above the ground of residential buildings correlated with adverse effects on mental health has been corrected and qualified by the application of explanatory factors in the field of people-environment studies, such as choice in housing markets, individual preferences, housing tenure and residential mobility. There is empirical evidence that those residents who do not choose where they live, especially households with young children who are allocated housing units in high-rise buildings, may suffer from stressors that impact negatively on their mental health (Hartig and Lawrence 2003).

6.5.3 Concentration and Density

Concentration is the third characteristic of cities and urban regions that is directly related to the two preceding ones. Urban ecosystems are dependent on the availability of natural resources and the exportation of waste products in order to sustain their populations. Cities import energy, fuels, materials and water which are

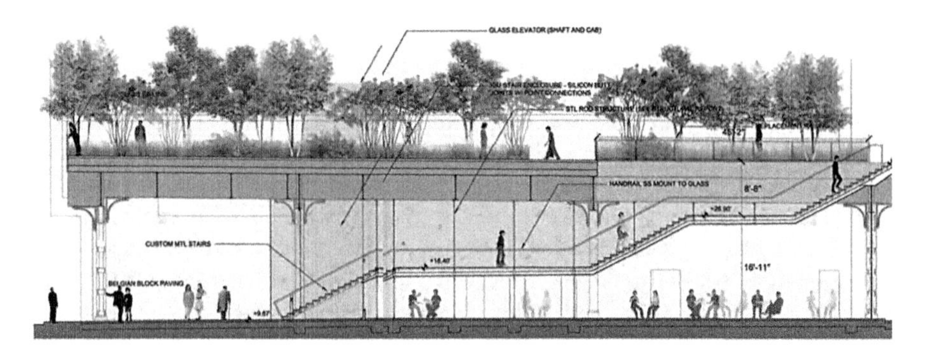

Fig. 6.3 The Highline in Manhattan, New York, is a well-known example of urban planning and design that has reconsidered promoting active living in conjunction with public green space in this dense city in terms of a people-centred not a vehicle-centred transport and mobility (Sources: New York City Department of design and construction (2010). Active design guidelines: promoting physical activity and health in design; (2010) Fit-city 5: promoting physical activity through design. Implementation of New York City's *Active design guidelines*. Source: www.thehighline.org)

transformed into goods and services. The high concentrations of activities, objects and people in cities, and the flows between rural and urban areas, mean that city authorities must manage the supply of food and water as well as the disposal of solid and liquid wastes that are risks for population health if not well managed (Hardoy et al. 2001). Urban history confirms that cities are localities that favour the rapid spread of infectious diseases, fires, social unrest and warfare (McMichael 2001).

The density of the built environment and the intensity of its activities have an influence on how cities and urban agglomerations can be sustained over time (Jencks and Burgess 2000). Many authors in recent years have argued for high-density, mixed-use urban areas (Fig. 6.3). This kind of urban ecosystem reduces the conversion of agricultural land, while it increases economic benefits related to concentrations of human activities including reduced fuel consumption for transportation, lower levels of ambient air pollution and reduced adverse exposures to noise from road traffic (Khreis et al. 2016). Some authors have claimed that urban concentration also encourages accessibility and promotes social vitality and cultural diversity while reducing social isolation and spatial segregation (Jencks et al. 2004). In contrast, other authors have suggested that there are limits to the degree of urban intensification by the containment and densification of existing or new free-standing cities; for example, when thresholds of overcrowding are surpassed leading to a lack of public open space, fewer recreation facilities and services and the absence of private gardens, then poor health may result from a lack of privacy, air pollution and noise (Jencks et al. 2004).

Too often debate about the limits to urban intensification has neglected health and other social concerns even though the promotion of quality of life has been frequently mentioned. Research shows that the social acceptability of compact urban development is a key determinant of the feasibility of this kind of development.

Social acceptability is defined by the perceptions and values of individuals and social groups. If these perceptions and values are well understood, then indicators of the 'social capacity' for urban intensification can be used in tandem with indicators of 'physical capacity' such as density thresholds (Jencks et al. 2004).

An in-depth understanding of the environmental, economic, social, behavioural, physical and geographical components of specific sites is necessary prior to the formulation of appropriate intensification projects in urban areas. This approach can be applied by local planning authorities in the framework of social and health impact assessment, as shown by recent advances in Barcelona (Mueller et al. 2017) and elsewhere, if they are willing to accept their responsibility to promote and sustain health and quality of life in their jurisdiction (Trop 2017).

6.5.4 Diversity: Cultural, Functional and Socio-economic

Diversity is a refining characteristic of cities that can be used effectively to promote ecological, economic and social well-being. Diversity is known to be an important characteristic of natural ecosystems because it enables adaptation to unforeseen (external) conditions and processes that may impact negatively and even threaten survival (Moran 2016). Likewise, lessons from history confirm that those cities with a diverse local economy have been able to cope much better with economic recessions and globalisation. This was not the case for Detroit in the late twentieth century, for example, and the consequences for the health and well-being of residents have been widely reported by mass media. Therefore, diversity—be it economic, ecological or cultural—is an important principle that enables human ecosystems to sustain health by adapting to external events or trends that negatively impact on them.

Social, economic and material diversity are inherent characteristics of cities (Holling 2001). The heterogeneity of urban populations can be considered in terms of age, ethnicity, income and socio-professional status. These kinds of distinctions are often reflected and reinforced by education, housing conditions, employment status, property ownership and material wealth. Data and statistics show that, in specific cities, different neighbourhoods are the locus of ethnic, political, monetary and professional differentiation between 'us and them' and 'here and there'. When these dimensions of human differentiation become acute, they are often reflected and reinforced by spatial segregation and social exclusion in urban agglomerations (UN-Habitat 2010). In recent decades, there have been empirical studies showing that these characteristics of urban neighbourhoods, especially acute socio-economic inequalities and lack of social cohesion, are linked to relatively high incidences of morbidity and mortality (Sarkar et al. 2014).

High levels of diversity and differentiation can be a threat to social cohesion over the long term. Viable communities in cities and urban regions can only be achieved through recognition of two key social principles of sustainable development: The first is intergenerational and intragenerational equity (Lawrence 2005).

However, it has already been noted that the principles of equity are too often omitted from ongoing debate (UN-Habitat 2010). The provision and equitable distribution of public education, vocational training, health care, social services and facilities for recreation can help address some of the root causes of delinquency and crime which can be associated with social differentiation, dependency, segregation and lack of empowerment.

The second principle is quality of life which includes issues of ownership, safety, security, aesthetics and socio-psychological dimensions of urban life. All these dimensions can impact directly or indirectly on health and well-being (Galea and Vlahov 2005). From this perspective, a broad and just interpretation of sustaining urban ecosystems extends beyond conventional contributions by natural scientists to confront a basic ethical dilemma created by some recent residential neighbourhoods including the so-called gated communities and fortress suburbs in North America (Blakely and Snyder 1997). These kinds of projects are one result of a non-willingness to address social cohesion and inclusiveness by deliberately planning for spatial segregation and social exclusion in urban and suburban areas (Fig. 6.4).

Fig. 6.4 This attractive public green space with century-old trees in the dense central business district of Sydney, Australia, is used for social contact and leisure by people who do not necessarily live or work nearby (Photo: R.J. Lawrence)

6.5.5 Information and Communication

The fifth characteristic is information and communication. Cities have always been centres for the development and exchange of ideas, information and inventions. However, during the twentieth century, some of the traditional functions of cities have been superseded or evolved with the development of information and communication technologies (ICT). The growth of new information and communication technology redefines the function of cities as centres of communication, marketing and information. In addition, the consequences for daily mobility between places of work and residence have still not been fully realised (Castells 1989). However, innovative technologies for the construction of smart cities, intelligent buildings and automated vehicles should be considered in relation to the increasing dependency of urban ecosystems on infrastructure and technologies that have no substitutes. The increasing incidence of disruptions to power supply to cities in developed countries and the hacking of communication and information systems of government agencies and public health services and hospitals in recent years should be recognised as new threats to public health and especially the health of urban populations.

Information and knowledge about urban health challenges requires commitment to systematic monitoring and evaluation of urban policies, programs and projects in relation to population health (Lawrence and Gatzweiler 2017). However, one of the anomalies of the architecture and planning professions is that monitoring and evaluation is not considered to be their responsibility. Today, too few public and private institutions are examining the range of costs and benefits of urban development and precise projects for specific communities, regional and local populations. This shortcoming can be overcome, at least partly, by health and social impact assessments and by funding for systematic monitoring and evaluation of urban ecosystems and population health.

6.5.6 Mechanisation and Metabolism

Mechanisation is the sixth characteristic. Cities have been the location of human assets, especially knowledge, technological development and complex physical infrastructures, for about 10,000 years. Cities and urban regions have depended on machinery and infrastructure to import supplies, to treat waste products and to efficiently use their increasingly complex built environments and infrastructure. Contemporary urban regions are heavily dependent on machinery for a wide range of functions and services that guarantee sanitary living conditions. Mechanical and technological characteristics of urban areas that impact directly or indirectly on health include industrial production, transportation, the processing of mass-produced foods and the increasing use of synthetic materials in the built environment. In particular, the incidence of accidents in urban areas is a major challenge for public health (UN-Habitat 2010). For example, injuries caused by motor vehicle accidents are ranked 10th among leading causes of mortality world-wide and 9th among the leading causes of disability. Today, children and young adults in all regions of the world bear a disproportionate burden of these accidents. This burden is significantly higher in urban areas compared with rural areas; it is also significantly higher in developing countries compared with developed countries (Sarkar et al. 2014).

6.5.7 Geopolitical Institutions and Actors

The seventh characteristic is political authority. The city was the *polis* in ancient Greece, meaning it had a specific political status, which is still the case today in some form of local government. Since the 1990s, much attention has been given to urban governance rather than municipal government. Governance can be defined as the sum of the ways by which individuals and institutions (public and private) plan and manage their common affairs. It is a continuing process that involves formal institutions and informal arrangements that are meant to promote mutually beneficial co-operative action. Governance is based on the effective coordination of three main components: market-based strategies for the private sector, hierarchical strategies articulated by the public sector and networking in civil society. The goal of governance in cities and urban regions should be to develop synergies between partners, so there is a better capacity to deal with the most urgent priorities (Fuchs et al. 1999 (first edition 1994)).

One example of urban governance that explicitly promotes health is the Healthy Cities project. This project was founded in 1987 by 11 European cities and the WHO Regional Office for Europe (De Leeuw and Simos 2017). Today this project is active in all regions of the world. The Health for All strategy provides the strategic framework for this project. The Healthy Cities project in the WHO European region includes four main components. First, the designated cities are committed to a comprehensive approach to achieving the goals of the project. Second, national and subnational networks work together in order to facilitate co-operation between partners. Third, multi-city action plans (MCAPs) are planned and implemented by networks of cities collaborating on specific issues of common interest.

The project involves collaboration between sectors to define a "City Health Plan" that identifies the interrelations between living conditions in urban areas and the health of residents (De Leeuw and Simos 2017). Innovative projects show that health can be improved by addressing the physical environment, as well as those social and economic factors that influence health in precise situations (such as the home, the school, the workplace). This broad interpretation means that equity and social inequalities are identified as key factors that need to be addressed in cities. In particular, the plight of vulnerable social groups (including the handicapped, homeless, unemployed, single parents and street children) can be ranked high for interventions. Hence the social and just principles of sustainable development are explicitly addressed.

6.6 Synthesis

The formulation and implementation of traditional sector-based approaches in the field of urban policies and programs does not lead to optimal results. Although there may be incremental improvements (in fields such as housing supply, public education, employment or transport infrastructure), these are often achieved in tandem with unintended consequences which may have negative impacts on local environmental conditions, the economy and the health and welfare of urban populations. In part, these outcomes are due to the number and complexity of all those factors that policy decision-makers need to consider (Lawrence and Gatzweiler 2017). They are also due to the recurrent lack of coordination in the field of public health policies and programs, which can be associated with the following factors:

1. The thematic variety and the technical complexity of specific problems related to the environment, the economy, health and well-being. Collectively, actors and institutions from different sectors need to collaborate in order to understand and address the complexity of contemporary urban health challenges.
2. The lack of consensus between specialists. There is no shared conceptual framework, methodological approaches or precise instruments for the study of population health in cities and urban regions. Moreover, there is no consensus about what instruments are most appropriate for defining, applying and monitoring urban policies and programs.
3. The lack of strategic visions and societal goals shared by politicians, professionals and the public about the definition and ordering of priorities. These visions and goals are not solely dependent on scientific knowledge. They are prescribed by human motivations, perceptions and values. Hence both qualitative and quantitative approaches are necessary, and these should be used in a complementary way.

Given the systemic nature of urban health challenges, it is necessary to consider the appropriate means and measures for the redefinition and reorientation of coordinated urban policies that are more ecologically sustainable, more socially equitable and less costly in health, monetary and ecological terms.

6.7 Conclusion

This chapter has suggested that an innovative research agenda can be based on the hypothesis that cities and urban regions have the potential to create adaptive responses rather than being considered as the overriding cause of current environmental, economic, health and other social problems. This hypothesis can be studied with respect to three basic ideas. The first is the need to examine urban lifestyles and infrastructures from three complementary perspectives—a comprehensive systemic perspective at the level of urban regions, an individual and household perspective at the level of residential neighbourhoods and a sociotechnical/institutional perspective encompassing the local, the regional and the global levels. The second idea is the need to apply a temporal framework to address these issues in the past, the present and the future. The third idea involves formulating images of the future by developing scenarios of desirable futures focused on one or more societal goals and then organising a public debate about how these goals could be achieved.

This innovative approach challenges common interpretations of urbanisation borrowed from traditional development agendas that focus narrowly on economic growth and industrialisation. In contrast, integrative principles of human ecology should be applied because they are related directly or indirectly to urban health. In essence, these principles can be used proactively in specific urban neighbourhoods to improve and sustain the health of urban populations. The importance of a radical shift in the way individuals, groups and societies think about these principles in relation to sustaining human health in cities and urban regions is a major challenge in the twenty-first century.

References

Bairoch, P. (1988). *Cities and economic development: From the dawn of history to the present.* London: Mansell.

Barton, H., & Grant, M. (2006). A health map for the local human habitat. *The Journal for the Royal Society for the Promotion of Health, 126*, 252–253.

Blakely, E., & Synder, M. (1997). *Fortress America: Gated communities in the United States.* Washington, DC: Brookings Institute Press.

Boyden, S. (1987). *Western civilisation in biological perspective: Patterns in bio-history.* Oxford: Oxford University Press.

Castells, M. (1989). *The informational city: Economic restructuring and urban development.* Oxford: Blackwell Publishing.

Clarke, R. (1973). *Ellen swallow: The woman who founded ecology.* Chicago, IL: Follet Publishing Company.

Dahlgren, G., & Whitehead, M. (1991). "The main determinants of health" model, version accessible. In G. Dahlgren, & M. Whitehead (2007) (Eds.), *European strategies for tackling social inequities in health: Levelling up Part 2.* Copenhagen: WHO Regional Office for Europe.

De Leeuw, E., & Simos, J. (Eds.). (2017). *Healthy Cities: The theory, policy, and practice of value-based urban planning.* New York: Springer.

Douglas, I., Goode, D., Houck, M., & Wang, R. (Eds.). (2011). *The Routledge handbook of urban ecology.* London: Routledge.

Dyball, R., & Newell, B. (2015). *Understanding human ecology: A systems approach to sustainability.* London: Earthscan.

Elmqvist, T., Fragkias, M., Goodness, J., et al. (Eds.). (2013). *Urbanization, biodiversity, and ecosystem services: Challenges and opportunities.* New York: Springer.

Fuchs, R., Brennan, E., Chamie, J., Lo, F.-C., & Juha, U. (Eds.). (1999). *Mega-city growth and the future.* Tokyo: United Nations University Press. (first edition 1994).

Galea, S., & Vlahov, D. (Eds.). (2005). *Handbook of urban health: Populations, methods and practice.* New York: Springer.

Hann, C. (1998). *Property relations: Renewing the anthropological tradition.* Cambridge: Cambridge University Press.

Hardoy, J., Mitlin, D., & Satterthwaite, D. (2001). *Environmental problems in an urbanizing world*. London: Earthscan.
Harrison, P., Pearce, F., & Raven, P. (2000). *AAAS atlas of population and environment*. Berkeley, CA: University of California Press.
Hartig, T., & Lawrence, R. (2003). Introduction. The residential context of health. *Journal of Social Issues, 59*, 455–473.
Hobbs, R., Higgs, E., & Hall, C. (Eds.). (2013). *Novel ecosystems: Intervening in the new ecological world order*. Oxford: Wiley-Blackwell.
Holling, C. (2001). Understanding the complexity of economic, ecological and social systems. *Ecosystems, 4*, 390–405.
Jencks, M., & Burgess, R. (Eds.). (2000). *Compact cities: Sustainable urban forms for developing countries*. London: Spon Press.
Jencks, M., Williams, K., & Burton, E. (2004). *The compact city: Sustainable urban form?* London: Taylor and Francis.
Khreis, H., Warsow, K., Verlinghieri, E., et al. (2016). The health impacts of traffic-related exposures in urban areas: Understanding real effects, underlying driving forces and co-producing future directions. *Journal of Transport and Health, 3*, 249–267.
Kresl, P. (2007). *Planning cities for the future: The successes and failures of urban economic strategies in Europe*. Cheltenham: Edward Elgar.
Lawrence, R. (2001). Human ecology. In M. K. Tolba (Ed.), *Our fragile world: Challenges and opportunities for sustainable development* (Vol. 1, pp. 675–693). Oxford: EOLSS Publishers.
Lawrence, R. (2005). Human ecology and its applications for sustainability research. In W. Leal Filho (Ed.), *Handbook of sustainability research* (pp. 121–145). Frankfurt am Main: Peter Lang.
Lawrence, R. (2010). Beyond disciplinary confinement to transdisciplinarity. In V. Brown, J. Harris, & J. Russell (Eds.), *Tackling wicked problems through the transdisciplinary imagination* (pp. 16–30). London: Earthscan.
Lawrence, R. (2015). Mind the gap: Bridging the divide between knowledge, policy and practice. In H. Barton, S. Thompson, S. Burgess, & M. Grant (Eds.), *The Routledge handbook of planning for health and well-being* (pp. 74–84). New York: Routledge.
Lawrence, R., & Gatzweiler, F. (2017). Wanted: A transdisciplinary knowledge domain for urban health. *Journal of Urban Health*. https://doi.org/10.10007/s11524-017-0182-x.
McMichael, A. (2001). *Human frontiers, environments and disease: Past patterns, uncertain futures*. Cambridge: Cambridge University Press.
Meadows, D. (2009). *Thinking in systems: A primer*. London: Earthscan.
Millennium Ecosystem Assessment. (2005). *Ecosystems and human well-being: Current state and trends* (Vol. 1). Washington, DC: Island Press.
Mitchell, J. (Ed.). (1999). *Crucibles of hazards: Mega-cities and disasters in transition*. Tokyo: United Nations University Press.
Moran, E. (2016). *Human adaptability: An introduction to ecological anthropology*. Boulder CO: Westview Press. (first edition 1982).
Mueller, N., Rojas-Rueda, D., Basagaña, X., Cirach, M., Cole-Hunter, T., Dadvand, P., & Nieuwenhuijsen, M. (2017). Urban and transport planning related exposures and mortality: A health impact assessment for cities. *Environmental Health Perspectives, 125*, 89–96. https://doi.org/10.1289/EHP220.
Pickett, A., Cadenasso, M., Grove, J., Nilson, C., Pouyat, R., Zipperer, W., & Costanza, R. (2001). Urban ecological systems: Linking terrestrial, ecological, physical and socioeconomic components of metropolitan areas. *Annual Review of Ecological Systems, 32*, 127–157.
Rayner, G., & Lang, T. (2012). *Ecological public health: Reshaping the conditions for good health*. London: Earthscan.
Sandifer, P., Sutton-Grier, A., & Ward, B. (2012). Exploring connections among nature, biodiversity, ecosystem services, and human health and well-being: Opportunities to enhance health and biodiversity conservation. *Ecosystem Services, 12*, 1–15.

Sarkar, C., Webster, C., & Gallacher, J. (2014). *Healthy cities: Public health through urban planning*. Cheltenham: Edward Elgar.
Seitzinger, S., Svedin, U., Crumley, C., Steffen, W., Abdullah, S., Alfsen, C., Broadgate, W., Biermann, F., Bondre, N., Dearing, A., et al. (2012). Planetary stewardship in an urbanizing world: Beyond city limits. *Ambio, 41*, 787–794.
Trop, T. (2017). Social impact assessment of rebuilding an urban neighborhood: A case study of a demolition and reconstruction project in Petah Tikva, Israel. *Sustainability, 9*, 1076. https://doi.org/10.3390/su9061076.
United Nations, Department of Economic and Social Affairs, Population Division. (2015). *World urbanization prospects: The 2014 revision, (ST/ESA/SER.A/366)*. New York: United Nations.
United Nations, UN-Habitat. (2010). *Hidden cities: Unmasking and overcoming inequalities in health in urban areas. UN-Habitat/WHO report*. Geneva: WHO Press.
Wachsmuth, D. (2012). Three ecologies: Urban metabolisms and the society-nature opposition. *The Sociological Quarterly, 53*, 506–523.
Whitmee, S., Haines, A., Beyrer, C., et al. (2015). Safeguarding human health in the Anthropocene epoch: Report of The Rockefeller Foundation—Lancet Commission on planetary health. *Lancet*. (published online July 16). https://doi.org/10.1016/S0140-6736(15)60901-1.
World Bank. (2001). *Attacking poverty: World development report 2000/2001*. New York: Oxford University Press.
Young, R. (2009). Interdisciplinary foundations of urban ecology. *Urban Ecosystems, 12*, 311–331.

Part II
Planning and Development

Chapter 7
The Human Habitat: My, Our, and Everyone's City

Bianca Hermansen, Bettina Werner, Hilde Evensmo, and Michela Nota

7.1 An Introduction to the Human Habitat

Cities are more than mere physical structures or modern settlement patterns. Cities are the places where we wake up, live, laugh, love, work, learn, and retire for the night. Such cities can be understood as a form of habitat and are, in fact, one of the most recent habitats on planet Earth. According to Nabhan (1997:3 cited in Steiner 2016: n/p.), a habitat is "...related to *habit*, *inhabit*, and *habitable*; it suggests a place worth dwelling in, one that has *abiding* qualities." However, in contrast to an animal or a plant habitat, the human habitat remains more or less undefined. While some think of the human habitat as related to one's home, in this chapter, we take on a broader perspective, scrutinizing the human habitat at a societal and urban scale. In doing so, we focus primarily on how cities can be planned and built in ways that foster health, quality of life, and prosperity among urban inhabitants. This entails placing emphasis on cities as *human* habitats, underlining the importance of reintroducing, or perhaps introducing, a human-centric approach to urban design. This chapter uses a working definition of health-promoting human habitats to mean well-designed, built environments that foster strong social cohesion as well as individual mental and physical well-being. Building on this definition, the chapter seeks to uncover the interconnectedness that exists between *people* and *place*. The definition of the human habitat is discussed from a Scandinavian perspective with examples of our own work, supported with relevant literature. We argue that an interdisciplinary approach to urban design is crucial to understand this relationship and, in the long term, promote quality of life in the human habitat.

Michela Nota has done all the graphic work.

B. Hermansen
CITITEK, Copenhagen, Denmark

B. Werner · H. Evensmo (✉) · M. Nota
COurban Design Collective, Copenhagen, Denmark
e-mail: hilde@courban.co; bettina@courban.co

M. Nieuwenhuijsen, H. Khreis (eds.), *Integrating Human Health into Urban and Transport Planning*, https://doi.org/10.1007/978-3-319-74983-9_7

While acknowledging that our definition of the human habitat does not cover the total complexity of what comprises urban life, we have chosen to focus on health and social factors, as we believe they constitute some of the most basic human needs and as such are of vital importance to the very existence of the human habitat. Furthermore, as the human habitat should be considered to be in a state of constant transformation, this chapter primarily focuses on and presents research and guidelines for the catalyzation of positive and health-promoting changes to the urban environment. Throughout the chapter, when using "human habitat," we will refer to this future vision of successful, livable urban areas.

7.2 MY CITY: How to Design Cities for the Individual

This section looks at the most important factors of the human habitat for the individual's mental and physical well-being. In cities, the built environment and urban form have a strong impact on the lives of individuals, their experiences, as well as their perception of their surroundings, community, and fellow citizens (Fig. 7.1). Human-centric cities are designed to facilitate positive environmental factors and human health determinants. They ensure that people feel safe and happy while, at the same time, counteracting poverty and dysfunction (Montgomery 2013). Furthermore, when planning for the well-being and quality of life of individuals, it is not only important to pay attention to how our senses come into play but also how our experiences of urban spaces are determined by our knowledge of a particular space (Gehl 1987; Holloway and Hubbard 2001). To put it simply, our sensory experience will, together with our knowledge of space, affect the way in which we perceive space, which in turn will determine our behavior in that specific spatial context. Accordingly, positive *sensory* experiences cause positive lived *experiences* that foster positive *behavior*. Imagine revisiting one of your favorite places in the city after being away for a while. The sound, smell, and sight of a familiar place and perhaps familiar faces will most likely cause you to have some form of positive

Fig. 7.1 The way we perceive, experience, and use cities depends on our individual knowledge, background, and personal traits

emotional reaction. Correspondingly, you might find yourself smiling, being at peace, and therefore engaging in specific activities such as lying or sitting down.

7.2.1 My Human Habitat: Catalyzing Safety and Trust

"Safety" encompasses many different notions, most commonly security and trust, but also absence of risk, crime, fear, and worry for oneself or for others. Safety can further refer to people's different life conditions such as financial safety and social safety. It is a multifaceted concept that can be both objective (e.g., statistical risk of crime) and subjective (e.g., individual fear for one's own safety) (Heber 2008). Security, as a component of safety, is, in public spaces, often directly related to the risk of being exposed to crime. In one way, security is something that can be bought through crime prevention measures, such as fencing, alarms and security personnel, etc. While these measures, to some extent, may prevent actual crime, the most important factor that makes public spaces safe is ensuring that people *experience* less worries about being exposed to crime. Actual safety then is always relative to people's perception of safety (ibid.). The safer an area is perceived, the safer it becomes, and vice versa. Subjective safety, *or the perception of safety*, is perhaps *the* most important success criteria of urban planning and design. As an example, if people feel that it is safe to cycle in their neighborhood, the likelihood of them choosing the bike as the mode of transportation is higher. Additionally, the more people seen cycling can further increase the perception of safety with others by making the area seem lively and safe to bike in, encouraging even more people to cycle. Particularly in residential areas, the perception of safety is crucial to residents' overall experience of space and consequently their behavior and use of space. Promoting safety as in case of the example above further improves public health by facilitating physical activity and active transportation.

Perceived safety and social interaction have the ability to enhance physical, mental, and social health (Healthy Spaces and Places 2009). In a study conducted using WHO data from three European cities, a positive correlation between people's perception of safety and the likelihood of occasional physical exercise was found (Shenassa et al. 2006). This finding was further supported by North American studies, which showed that making improvements to the residential area in terms of maintenance, rather than targeting individuals with campaigns, could achieve an increase in the perception of safety and the likelihood of physical exercise (ibid.). When managing neighborhood maintenance, community surveillance is widely regarded to be the most effective form of deterring littering and urban disorder (Wong 2012; Sundberg 2013). Community surveillance in this context should be understood as the presence of a cohesive society in which individuals look after each other. However, cohesive societies cannot be imposed from above, but rather they need to grow out of the local context and the residents themselves (Jacobs 1961a).

In addition to the presence and engagement of individuals in local community management, the design of urban spaces is an important component to catalyze safety in the human habitat. Within crime prevention literature, it has been argued that design features such as good lighting conditions, good overview of the space, places to sit, and entrance points of buildings facing the street can help to reduce crime (Loukaitou-Sideris 2006). Moreover, preventing litter and urban decay, creating safe access points to public spaces through well-designed infrastructure and connectivity, as well as facilitating flexible use for different activities to occur in the same space over the course of a day can reduce crime further (ibid.). The guiding principles presented above appear to comprise planning for safety along a physical, social, and organizational dimension. Furthermore, this research also points back to the way in which human behavior is determined by our senses and how we ascribe meaning to what we experience as safe or unsafe environments. We have, for example, been taught that littering and vandalism are signs of decay and that decay, in many instances, is associated with the presence of crime. This is where our knowledge of human behavior comes into play. In addition, the absence of lighting is connected to our senses by inhibiting visual overview of a situation, causing a feeling of lacking control and subsequently feelings of fear. This is where our senses come into play. One must understand how the interplay between physical elements, social relationships, and interpersonal sensory experiences can work together in ways that catalyze safety in the human habitat.

Digging deeper into the importance of social interaction for enhanced levels of perceived safety and quality of life of urban residents in general, it becomes apparent that research frequently connects trust—in neighbors, police, governments, and strangers—to indicators of happiness, life satisfaction, and improving urban mental and social health (Troelsen et al. 2008; Montgomery 2013). Measuring trust in a community can feel like an intangible endeavor, but trust can materialize in many concrete ways. For example, sociologists have showcased the accumulative property of trust in findings of adults that have people they trust in their lives. Their children are better equipped to handle the effects of their parents' stress; they sleep better at night, and they tackle adversity better and report being happier (Troelsen et al. 2008). Alike to designing for safety, designing for trust is based upon promoting social interaction and encounters between different people to foster tolerance and respect. Returning to the importance of social relations later, the following is a comparative example of how the interrelationships between perceptions of safety, trust, and physical design manifest differently in two parks in Scandinavia.

7.2.1.1 Trust by Design: Badeparken vs. Nørrebroparken

In 2015 CITITEK conducted a study of Badeparken in Sandefjord for the Vestfold region in Norway (Fig. 7.2). In the study, we found that the perception of safety and trust among the users of the park influenced how people utilized it. Our mappings of the use and users of Badeparken revealed that a majority of the users chose to spend time only in the north and northeast parts of the park. This user group primarily consisted of seniors, adolescents, and families with young children. Besides the fact that this part of the park had play equipment and benches, interviews also revealed that

Fig. 7.2 Badeparken is perceived as unsafe, and the design actively enforces a social and spatial division, while the design of Nørrebroparken, with its unobstructed views, does the opposite, fostering trust and a perception of safety

many people deliberately avoided other areas of the park due to feeling unsafe. The perceived lack of safety was, for most of these informants, impacted by the presence of a group of substance abusers occupying the central and southwest areas of the park. The substance abusers, on the other hand, claimed that they had few other places to stay and that the police had directed them to use this specific area of the park. Through observations and interviews, it became apparent that these two user groups did not interact in any way and in fact, to a large degree actively, avoided each other. The result was a spatial division of the park and a type of behavior that clearly indicated a lack of trust between the two user groups. Furthermore, the physical design of the park actively enforces this spatial and social division: a tall hedge located in the middle of the park effectively blocks the overview of the park in its entirety. As mentioned previously, research emphasize how lack of overview and presence of visual and physical barriers may cause discomfort in public spaces. Furthermore, it has been argued that these types of walls act to inhibit social contact and interaction (Gehl 1987:65–71), ultimately affecting opportunities for building trust between individuals. In the case of Badeparken, despite the good intentions of directing substance abusers to a designated area behind the hedge, findings from the study showed instead that this physical, visual, and social division decreased not only the overall perception of safety but the overall perception of the park as a public space.

In comparison, Nørrebroparken, a popular park in Copenhagen, is a good example of how a nondiscriminatory design approach can provide a public space for a diverse user group. In the park, a designated area was established for substance abusers, entailing benches, toilets, and semitransparent fences allowing visual overview from the outside in, as well as from the inside out (Socialministeriet 2010). This pilot project was one in a series of studies initiated by the Danish Ministry of Social Affairs, inviting underprivileged communities to partake in design processes. In a similar way to Badeparken, the substance abusers in Nørrebroparken were given a designated area to use. However, instead of "hiding away" this user group, the physical design promoted visual interaction between different user groups. Results from the study showed that the design seemed to not only cause enhanced perception of safety among all users of the park, but the substance abusers were also perceived to look after the other users (including children playing at the adjacent playground), as well as contributing to discouraging petty crimes in the area such as vandalism (ibid.). From a Scandinavian perspective and in our opinion, this is a true win-win, where the design and operation of the urban space actively contribute to the making of a good human habitat. The example also clearly illustrates the importance of involving all kinds of citizens if we are to develop context-specific inclusive design solutions.

7.2.2 My Human Habitat: Catalyzing Active Living

Promoting active living for individuals through urban design is one of the most significant methods of preventing the presence of diseases and promote a healthy human habitat. The WHO categorizes noncommunicable diseases (NCDs), which account for 63% of all annual deaths globally, as urban society's greatest public health challenge. NCDs include chronic diseases such as cancer and asthma but also lifestyle diseases such as diabetes (type 2) and cardiovascular diseases (WHO 2013). Many of these are preventable through interventions that tackle the main risk factors, namely, an unhealthy diet, harmful use of alcohol and tobacco, and minimal physical activity (ibid.). Physical activity is effectively promoted or discouraged through urban design, for example, pedestrian and bicycle infrastructure. Design cannot in itself force people to exercise, but it has the power to encourage and invite people to live actively. For example, if multiple amenities and desired destinations are close to a residence, it is more likely that a citizen will choose active transportation that harnesses the power of the human body, e.g., walking or cycling (ibid.; WHO Europe 2007; WHO 2017). Cities around the globe seeking to reduce traffic congestion and harmful emissions and improve public health advocate this model for transportation through related infrastructure, public policy, and education (WHO 2017). Providing access to this type of fast, easy, healthy, and affordable modes of transportation across all urban contexts is and will continue to be an increasingly important determinant of individual health (EEA 2006). Environments that foster physical activity typically center around parks and green spaces, playgrounds and sports facilities, as well as walkable and bikeable distances between facilities. Factors such as safe and cohesive bike and pedestrian infrastructure additionally impact the prevalence of active transport (WHO Europe 2006).

Fostering physical activity and spending time outdoors are also ways to encourage social interaction outside the home, in the streets, and in public spaces of the city. This type of urban environment also corresponds to basic social and psychological needs and mental health. Fischer (1995) refers to these needs as social interaction, privacy, stimulation, orientation, safety, and identity (Fischer 1995, cited in Troelsen et al. 2008:28–29). Providing urban residents with areas that promote both planned and spontaneous social interaction and physical activity within walking or biking distance from one's home is an effective way of catalyzing active living in the human habitat. This of course necessitates a profound understanding of the types of social, cultural, and personal factors that motivate people's choice of active living and, accordingly, the urban design and infrastructure that promote this type of behavior.

7.2.2.1 Health by Design: Activating Ørsta Municipality

Creating bikeable and walkable neighborhoods is an important step toward ensuring more active societies and a healthier population. To create safe and user-friendly design that nurtures positive changes in behavior, we need to unveil the social and physical factors that contribute to promoting walking and cycling. When cyclists' movement patterns were mapped at the intersection by the famous Queen Louise's bridge in Copenhagen, it became apparent that many cyclists chose to break the law by taking a shortcut over the pedestrian sidewalk to bypass a busy intersection. Rather than penalizing these cyclists, Copenhagen municipality choose to facilitate this behavior by formalizing the shortcut, as it was evident that the cause of this behavior was not reckless thoughtlessness, but rather traffic avoidance. The result was better utility of the space, reduced travel time for cyclists, and more space for those who bike on the surrounding streets. Pedestrians were accommodated through pedestrian crossings over the bike lane (Københavns Kommune 2013; Rasmussen 2013). This is an example of local governance that not only makes active transportation an easier choice but also facilitates urban life. With the overall purpose of facilitating these types of positive changes to the urban environment, CITITEK conducted a study on behalf of Ørsta municipality, Norway, in 2016. We mapped the movement patterns of children and youth to understand the factors that both hinder and promote their active transportation. By allowing informants to draw and talk about their own movements and experiences, it was possible to identify and create an overview of real-life, real-time vehicular traffic and infrastructure challenges, as well as concrete suggestions to improvements in infrastructure. The study from Ørsta did not only give voice to a demographic group that is often overlooked in planning processes, but the findings also provided the municipality of Ørsta with information that will allow them to more accurately plan for healthy and safe mobility among the local youth.

7.2.3 My Human Habitat: Catalyzing Social Capital

Using Bourdieu's definition, social capital can be understood as resources linked to the durable network of relationships gained from membership of a group, both individually and collectively owned (Bourdieu 1986). Although social capital is as fundamental to an individual's health and well-being as physical activity (PPS 2017), globalization has been argued to stretch social ties across time and space. Debatably, globalization has reduced the need for people to leave their homes in order to acquire social capital and increases the risk of people distancing themselves from their local human habitat (Holloway and Hubbard 2001). Despite such potential or factual challenges, as of today, face-to-face interaction continues to be the most important and fundamental form of human interaction (ibid.). Thus, for the human habitat to foster social capital among its residents, cities must offer arenas where social interaction takes place and durable relationships are established.

In his book *Happy City,* Montgomery (2013) discusses how several quality of life studies indicate that an increase in social interactions can equal or surpass the benefits of a raise in income. Consequently, the population of local neighborhoods benefit from an urban environment that encourages social interaction by inviting people to linger, converse, and live. As such, the urban environment can encourage social interaction by providing a sense of security, orientation, and opportunities for interaction through solitary and social activities (Troelsen et al. 2008). Human need for socializing covers both spontaneous and planned interactions of differing natures. One way of facilitating both these types of meetings is to create gradual transitions between private, semi private, and public spaces (Gehl 1987) as these transitions are argued to promote social appropriation, a sense of belonging, and perceived safety (Haijer and Reijndorp 2001).

Franck and Stevens (2007) draw on Lefebvre's "right to the city" and the "right to habit and inhabit" when they argue for what they call "loose space." Loose space is public space that facilitates activities that they are not intended for. In this way, they allow for people themselves to appropriate the space for their own uses and unexpected and unintended activities (Franck and Stevens 2007; Haijer and Reijndorp 2001). One example of this type of appropriation is Queen Louise's Bridge in Copenhagen, which after a renovation process aimed at improving conditions for cyclists and pedestrians unintentionally turned it into one of the most popular hangout spots in Copenhagen. Even if the bridge had not been designed to facilitate staying activities except for a few benches here and there, the citizens began to appropriate the space by sitting down and spending time on the railings of the bridge and on the pavement. The bridge had suddenly become a space where friends and strangers meet, sit down, eat, and play music from one's cargo bike or even dance—without blocking the way for pedestrians or cyclists. Thus, the importance of loose spaces lies in the anonymity of a city, the "open-endedness" if you will. On one hand, the anonymity of public urban space can break down social roles and make people engage in unexpected activities that they might not otherwise engage with (ibid.) On the other hand, due to the same anonymity and strangers' relationship with each other, people require a *reason* to engage in socialization (Franck and Stevens 2007). The idea of loose space and the transitions between private and public spaces can create a sense of familiarity and belonging in how the spaces are easy for the individual to appropriate. Over time, these experiences can create opportunities for socialization as the perception of these factors encourage people to use and stay in these spaces (Gehl 1987). The following example describes a design solution to facilitate social interactions in Cairo, Egypt.

7.2.3.1 Social Capital by Design: Cairo Passageways

In 2014, the city of Cairo was marred by financial and political instability as a consequence of the Egyptian Revolution (Fantz 2016). During the Spring of 2014, Bianca Hermansen, together with a group of Danish and Egyptian architects and artists, was invited by Cairo Lab for Urban Studies, Training and Environmental

Fig. 7.3 While living up to the Egyptian government's strict requirements for public spaces, the parklet design of the Kodak passageways effectively fosters the rebuilding of community, neighbor-to-neighbor trust, increasing social capital and subsequently contributing to the human habitat of Cairo

Research (CLUSTER) to participate in the development of a democratic urban plan for downtown Cairo, starting with two pilot projects, the Kodak and the Philips Passageways (Fig. 7.3). The project aimed to "...develop an urban design and art project...highlighting existing and emerging initiatives activating underutilized public spaces..." (CLUSTER 2017). While emphasizing diversity, inclusivity, safety, and positive sensory experiences for people using the passageways both indoor and outdoor were activated with different types of cultural programs and designs to revitalize the public spaces (ibid.). As a consequence of the many riots and demonstrations in the aftermath of the Arab Spring, the Egyptian government had in 2014 enforced a law sanctioning gatherings of more than ten people (Human Rights Watch 2013). In CITITEK's winning concept for the pilot project, this prohibition was embraced by removing one third of the paving in the passageway to satisfy city officials. Now, in theory, the space would appear to become less public and effectively discourage too large gatherings. However, the socio-spatial consequence would be the opposite. While discouraging many people to gather at the same place, the Kodak passageway monofunctional stone paving was replaced by a lush green parklet, where narrow paths connecting small subspaces were designed to foster the rebuilding of community and neighbor-to-neighbor trust. Subsequently, this would increase social capital, which is the main driver of rebuilding and reinventing the human habitat of Cairo. Observations after the implementation of the new passageway design showed not only an increase in the number of people using the space but first and foremost an increase in the diversity of activities at different times of the day. People would no longer just walk or stand in the area, but staying activities such as cultural events and even a wedding took place.

7.3 OUR CITY: How to Design Cities for Community

While the previous subchapter—MY CITY—focused on individual's sensory and personal experiences of the human habitat at a local scale, this subchapter addresses the human habitat at a societal scale (Fig. 7.4). However, this division does not mean that the two scales should be understood as independent of one another.

Fig. 7.4 A city designed for everyone can encompass all different user groups with minimal adaption. The well-designed built environment fosters strong social cohesion as well as individual mental and physical well-being, and a human habitat is present

Rather, the relationships that manifest themselves between people and place at one scale will always influence and be influenced by human-environment interactions at other scales (Holloway and Hubbard 2001). Just as for the individual, the health and well-being of social groups are also affected by the urban design of the city. Thus, the inevitable presence of a variety of social populations in a city, including both socioeconomically privileged and underprivileged groups, children and seniors, etc., means that equity in access to health needs to be emphasized within the urban and transport planning agenda. The perspectives of sustainable development and the significant influence of social inequality on future generations mean that it is imperative to assess, address, and mitigate the impact of inequality-related action and policies for future generations (WHO 2014a).

7.3.1 Our Human Habitat: Catalyzing Proximity

Urban sprawl, "...a form of urbanization distinguished by leapfrog patterns of development, commercial strips, low density, separated land uses, automobile dominance, and a minimum of public open space," as defined by Gillham (2002:383), is one of the greatest challenges our contemporary cities are trying to overcome after previous "unplanned" or "ad hoc" urban planning (ibid.). Urban sprawl is indisputably the least environmentally, socially, and economically sustainable form of urban development, and yet it has made its mark on cities, small and large, across the world (Ewing et al. 2003; Gargiulo et al. 2012). From a human habitat perspective, urban sprawl should be considered a major threat to the public health and well-being of urban citizens. First, sprawl facilitates car dependency causing not only sedentary and physical inactivity but also enhanced levels of air pollution and climate change, which both directly and indirectly affect the presence of diseases and epidemics (Frumkin 2002). The second, but perhaps more debated, argument criticizes suburbanization and urban sprawl for deteriorating social ties, trust, and civic engagement (ibid.). Putnam (cited in Frumkin 2002:209) has, for example, estimated a 10% decline in social activities for every 10 min driving time. Considering the destructive and harmful consequences for urban life and livability, in our opinion, urban sprawl has no place in the human habitat.

The model of the compact city was developed to combat the consequences of sprawl in urban development (Dieleman and Wegener 2004). According to its advocates, the compact city promises a sustainable and health-promoting urban design strategy, which will bring life and activity back to urban centers and prevent further sprawl. Yet, the health benefits of the compact city are dependent not only on the density of people but also the degree of mixed use and proximity. For this paper, there is an important distinction to make between the two. Here, *density* considers numeral content, e.g., number of people, businesses, services, etc. within an area, while *proximity* is concerned with the access to and distance between people, businesses, services, etc. in an area (see Fig. 7.5). By considering proximity to, and between, facilities and services such as employment and income, schools, retail,

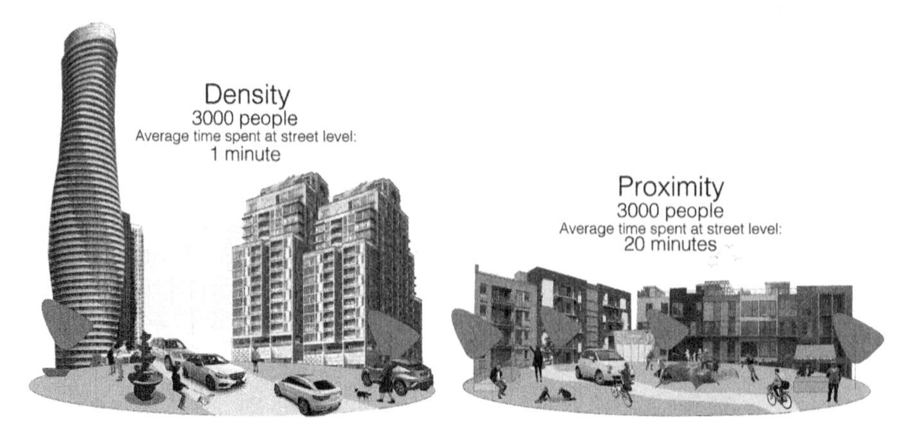

Fig. 7.5 While *density* considers numeral content within an area, *proximity* is concerned with the access to and relationship and distance between people, businesses, services, etc. in an area

and public spaces including squares and green spaces, decision-makers can ensure a human habitat when planning for and designing neighborhoods. Below follow two examples of the importance of planning for proximity to green space and healthy food.

7.3.1.1 Green Space Proximity

Access to natural environments should be considered one of the fundamental pillars of the human habitat and a catalyst for improved health and well-being. However, people's access to these types of environments has changed with rapid urbanization that limits opportunities to utilize green spaces for many urban residents. This is especially true in cities where planning for the compact city has been prioritized at the expense of the preservation of urban green structures (Jansson 2014). There is a need to acknowledge both the public health, and environmental and biodiversity benefits green spaces can provide (ibid.). Research shows that green spaces promote physical activity, improve the mental well-being of individuals, and provide important arenas for social interaction (Lee et al. 2015; Troelsen et al. 2008). Reflecting upon these benefits, designing for proximity to green spaces is essential as studies show that the closer we live to green spaces, the more we use them (Schipperijn et al. 2010; Sotoudehnia and Comber 2011). In fact, if green spaces are located more than 300–400 m from one's residence, they are used significantly less (Grahn and Stigsdotter 2003). In this regard, it is important to closely consider both actual and perceived distance in planning processes, as both may influence people's use of urban green spaces. In addition to proximity, use of green spaces is further influenced by both their size and the facilities they offer (Schipperijn et al. 2010; Van

Herzele and Wiedemann 2003; Bedimo-Rung et al. 2005). People's use of green spaces depends on both personal and demographic factors. This underpins the importance of understanding which type of facilities and activities need to be in place to promote increased use among different population groups (Lee et al. 2015; Payne et al. 2002). Understanding how these factors affect the interrelationship between people and place is consequently important to ensure democratic and health-promoting urban design practices (Lee et al. 2015). At the same time, uncovering this interrelationship will mitigate the risk of wrongful investments in facilities that do not live up to the full potential that green spaces can bring to public health and the human habitat.

7.3.1.2 Food Proximity

Access to food is vital for the very existence of the human species, and as such nutritious foods are a determining factor for the health and well-being of people at all stages of life (Azétsop and Joy 2013). Malnutrition, micronutrient deficiencies, overweight, and obesity have high social and economic costs for individuals, families, communities, and governments (WHO 2014b). Ensuring food security, defined as "...a situation that exists when all people, at all times, have physical, social and economic access to sufficient, safe and nutritious food […]" (Azétsop and Joy 2013:1), should thus be considered an essential component of the human habitat. Research from Europe and the United States shows that urban food security is closely connected to both social and environmental determinants, as low-income households have the lowest consumption of fruit and vegetables (WHO 2014a; Hilmers et al. 2012). Furthermore, neighborhoods with lower socioeconomic levels often have a higher density of fast food outlets, less physical access to healthy food options, and higher rates of obesity among adolescents than socioeconomically privileged neighborhoods (Hilmers et al. 2012; Baker et al. 2006). Local governments wanting to secure a healthy population and reduce social inequity in health need to simultaneously address spatial, economical, and knowledge barriers that make the healthy food choice a difficult choice to make. To plan for the human habitat, it is therefore crucial that healthy and nutritious food is available and affordable in all types of residential areas (WHO Europe 2007). This entails not only local policy-making that encourages supermarkets to provide a wide variety of healthy food choices but also the type of food offered in public institutions such as schools, hospitals, senior homes, public offices, libraries, etc. Through targeted partnerships, policy-making, and set standards, governments should also strive to make healthy food a main priority among privately owned businesses, organizations, and other actors (Hilmers et al. 2012). This can, for example, be done through means such as subsidies, tax deduction programs, or other economic incentives that will pay off in the long run in the form of public health-care savings (ibid.).

7.3.2 *Our Human Habitat: Catalyzing Diversity*

On a societal scale, the human habitat consists of the total sum of all its individuals from all population groups. In the field of natural sciences, diversity is acknowledged as promoting rich and healthy habitats, enabling systems to be resilient and allowing all organisms to adapt to change (Steiner 2016). Correspondingly, it is inherently important that the human habitat also encompass diversity. Thus, the spatial layout and design of urban spaces need to cater to different groups of people across geographical, demographic, and socioeconomic divides. According to WHO (2014a), socioeconomic status, both on a national and regional level, is proportional to factors such as life expectancy, healthy life years, and child mortality rate, where higher socioeconomic status increases the likelihood of higher life expectancy, etc. Evidence increasingly suggests that socially underprivileged people and those who live in neighborhoods of lower socioeconomic status have limited opportunities for outdoor activity (WHO Europe 2007). Furthermore, several studies conducted by WHO Europe have shown that in underprivileged neighborhoods, people are less likely to go outside for physical activity or socializing. Correspondingly, recreational or sports facilities are also less likely to be present in these areas (ibid.). The less people can achieve in terms of individual resources, the more important it is that they be able to draw on collective resources (WHO 2014a). This requires planning for diversity along two parallel lines. First, planning for diversity in the human habitat entails creating equal access to amenities, recreational facilities, and opportunities for active living across all types of urban neighborhoods. Second, to achieve community building, social cohesion, and diversity, the human habitat also needs to facilitate the bringing together of different types of people at the local level. Thus, at the core of planning for diversity is the idea of a nondiscriminatory, inclusive approach to the way in which we plan and design the human habitat.

One measure that can help combat health inequality and allow for all citizens' equal access to health-promoting resources and facilities is *universal design*. As defined by the United Nations (2017), article 2, "Universal design means the design of products, environments, programs and services to be usable for all people, to the greatest extent possible, without the need for adaptation or specialized design." The definition builds on a democratic principle where all citizens have the same rights and equal opportunities for participating in society. In this way, universal design is concerned with collective benefits and recognizing diversity rather than providing homogenous solutions (United Nations 1999). We see plenty of opportunities for creating an inclusive human habitat derived from a universal design approach. However, universal design should not only be concerned with physical access but also include considerations of how social and cultural barriers may come into play. As an example, a study from Copenhagen showed that even though approximately the same amount of men and women were passing by three selected public spaces (Charlotte Ammundsens plads, Prags Boulevard, and Multipladsen), there was a clear male dominance among the users staying in these spaces. The male users were also observed to have a higher physical activity level than the female users

(Copenhagen Municipality 2011). This example shows that other factors apart from physical access can act as determinants for the use and users of a public space. If the goal is to create an inclusive and diverse human habitat, universal design should also be concerned with uncovering how factors such as identity, ownerships, and sense of belonging are interrelated to the physical design of spaces. In this regard co-creation and citizen participation are important tools for achieving universal design and democratic human habitats.

7.3.2.1 Making People Visible Through Data: Sandefjord Municipality Merger

In 2015 CITITEK conducted a study in Sandefjord, Norway, on behalf of Vestfold region, as they were facing the challenge of merging three municipalities and centralizing public cultural facilities and services. Through gathering data about users and user patterns in current cultural facilities, the findings from the study underlined the importance of these facilities as arenas for social interactions. Furthermore, the study revealed that cultural facilities were experienced by their users as important contributors to their health, well-being, and quality of life. In particular, one of the local libraries was found to be an important meeting place for seniors and youth. The library thus acted as a catalyst for intergenerational meetings and catered to a diverse user group. However, interviews elucidated that the importance of the library as a meeting place seemed to be determined by its spatial location and users' proximity to the facility. This finding was further supported through observational mapping of the users' choice of transportation, which showed that a substantial proportion of the people using the library arrived by foot or by bike. Many of the senior users were also arriving with walkers, clearly indicating physical proximity to the library. The importance of the library as a contributor to enhanced public health was thus a combination of the activities it offered, as well as its local geographical positioning. Accordingly, our recommendation to the region was not to centralize the library services, as it would potentially harm the current public health benefits by inferring with users' current use and access through proximity.

7.3.3 Our Human Habitat: Catalyzing Democratic Change

As Jacobs (1961b:238) stated, "Cities have the capability of providing something for everybody, only because, and only when, they are created by everybody." As relevant today as in 1961, the quote underlines the importance of introducing a human-centric, inclusive, and local approach to the way in which we design and organize our cities. While many features of cities may be comparable across space, the human habitat should not be understood as a blueprint that can be copied and pasted from city to city. Rather, planning for the human habitat requires careful consideration and the tailoring of solutions to the local context to which they are intended. This, again, requires tapping into the local knowledge, resources, and experiences of the individuals who reside in the human habitat. In our own work, we have found placemaking to be a foundational tool in co-creating the democratic city. Placemaking maximizes shared values in the public realm by allowing the physical, social, and cultural identities that define a place to shape the collaborative process of strengthening the connection between a place and its people. Placemaking processes attempt to utilize community assets and potentials that can contribute to health and happiness when creating public spaces (PPS 2017).

7.3.3.1 Co-creation: The New Nordic of Urban Design

Participatory methods have become a norm when incorporating citizen perspectives into development projects. Citizen involvement through participation in local and regional policy is already a practice that is required by law for public development projects in Scandinavia (Mulder 2012). However, depending on the degree of participation, citizens often have little involvement in actual spatial design and visible change. Klausen et al. (2013) use Arnstein's "participation staircase" to illustrate citizen participation on the political agenda. The degree of citizen participation ranges from the lowest form of involvement, *information*, to inform citizens of the agenda, through *consultation*, *dialogue*, and *decision-making*, to the highest degree of participation, *co-governance* or *co-creation* (ibid.). Co-creation in this sense allows all participants and actors in a design process, whether it is a product, service, or policy, the same role. There is an emphasis in co-creation processes that the participation of all actors is *meaningful* and that there is an "equal base for participation" where all participants are considered partners in the process (Axelsen et al. 2014:3).

In Scandinavia, we are today witnessing the beginning of a paradigm shift whereby citizens, who previously have been on the lower end of the participation staircase, as passive recipients of public services, are transforming into active and informed *co-creators,* interested in public service creation and problem solving (Umeå Municipality 2009). In our experience, the same paradigm shift is true for urban design. The emerging methodological approach of co-creation blurs the disciplines of research and design, shifting the focus from how and what is designed to *whom the design is for.* There is a much larger emphasis on future social and political change compared to previous design research (Sanders and Stappers 2008). Co-creation strives not only for a redistribution of power and benefits but brings focus to identity policies and diversity so that matters of identity construction and recognition can become more prominent arguments for participation (Hansson et al. 2013). It is therefore an effective tool to account for different population groups' needs and untapped potentials in urban development, design, and placemaking processes.

To ensure successful urban design that caters for a human habitat and quality of life, a high degree of citizen involvement and co-creation is required. In this process, however, representation is as important as the participation itself since representation increases the chances for a successful solution that will cater to the majority of the population. In our work, we first and foremost ensure representation through mapping and interviewing the actual users of a space while simultaneously seeking out those who do not use the space. In this way, all users and nonusers can share their experiences with a higher rate of representation than a voluntary public hearing. In our experience, a public hearing is not representative. It requires that people actively seek to participate and have access to do so, which is often not the case in all population groups. Without full community representation, we cannot retrieve accurate information about what the community represents nor a consensus to inform design solutions, ultimately undermining the development of a democratic human habitat.

7.4 EVERYONE'S CITY: Catalyzing the Human Habitat

This chapter has discussed the reciprocal relationship between people and place through the lens of the human habitat. We have argued that the way people experience and subsequently use an urban space is, to a large degree, shaped by the design of that specific space. Consequently, urban design is to be understood as an important tool to promote and foster positive human behavior and experiences as well as public health and well-being. As we have discussed throughout this chapter, a human-centric approach to urban design is crucial to prevent and counteract the negative consequences of health-compromising urbanization and to promote quality of life in the human habitat. Put simply: in order to develop a truly inclusive and healthy human habitat, we need to put people first.

If we are to create cities for everyone, we need to understand all the factors at play when analyzing the way people perceive and use their surrounding environment. This should first and foremost be done through in-depth research and data collection that will allow for an empirically grounded understanding of people-place interactions at a local level. In our experience, using empirical data as the foundation for urban design makes it difficult for any policy maker, municipality, or practitioner to ignore the needs, desires, and voices of the citizens. Data about people and urban life can thus be a powerful force to push forward a human-centric approach in urban development. Moreover, the complexity of the relationship between people and place underlines the importance of a multidisciplinary approach to assess challenges and opportunities in urban design. Working together across disciplines to promote state, professional, and citizen collaboration will allow for a better understanding of current urban systems, as well as the processes needed for effective urban interventions. Furthermore, the cases and literature presented in this chapter are primarily from a Scandinavian perspective. This is important to point out, not only because there is no blueprint in urban design, it is context dependent, but also because we argue that this Scandinavian approach is at the forefront of urban design that can promote the human habitat. The guidelines and catalyzers presented in this chapter are to be seen as tools to push the urban design, planning, and transport agenda forward globally to develop and co-create our cities for the

Fig. 7.6 The guiding principles for designing for the human habitat

human habitat. By following these guidelines, we hope that we, together, can accelerate the process of designing cities for everyone (Fig. 7.6).

References

Axelsen, L. V., Mygind, L., & Bentsen, P. (2014). Designing with children: A participatory design framework for developing interactive exhibitions. *The International Journal of the Inclusive Museum, 7*, 1–17.

Azétsop, J., & Joy, T. R. (2013). Access to nutritious food, socioeconomic individualism and public health in the USA: A common good approach. *Philosophy, Ethics and Humanities in Medicine, 8*, 16.

Baker, E. A., Schootman, M., Barnidge, E., & Kelly, C. (2006). The role of race and poverty in access to foods that enable individuals to adhere to dietary guidelines. *Preventing Chronic Disease, 3*(3), A76. The National Center for Biotechnology Information.

Bedimo-Rung, A. L., Mowen, A. J., & Cohen, D. A. (2005). The significance of parks to physical activity and public health: A conceptual model. *American Journal of Preventive Medicine, 28*(2005), 159–168.

Bourdieu, P. (2008 [1986]). The forms of capital. In Readings in economic sociology. Biggart, N.. Oxford, UK: Blackwell, pp. 280–291.

CLUSTER. (2017) Cairo downtown passages—Kodak passage. Available from http://clustercairo.org/cluster/design/cairo-downtown-passages-kodak-passage

Copenhagen Municipality. (2011). Byens Bevægelsesrum—Et studie af byrums evner til at fremmer fysisk aktivitet og møde mellem mennesker. *Ministeriet for bolig, by og landdistrikter,* Copenhagen.

Dieleman, F., & Wegener, M. (2004). Compact city and urban sprawl. *Built Environment, 30*(4), 308–323.

EEA: European Environment Agency. (2006). Urban sprawl in Europe: The ignored challenge. *EEA Reports*, No. 10.

Ewing, R., Schmid, T., Killingsworth, A., Zlot, A., & Raudenbush, S. (2003). Relationship between urban sprawl and physical activity, obesity, and morbidity. *American Journal of Health Promotion, 18*(1), 47–57. Springer.

Fantz, A. (2016). Egypt's long, bloody road from Arab spring hope to chaos. *CNN,* published 27 Apr 2016. Available from http://edition.cnn.com/2016/04/27/middleeast/egypt-how-we-got-here/.

Franck, K., & Stevens, Q. (2007). *Loose space: Possibility and diversity in urban life*. London and New York: Routledge.

Frumkin, H. (2002). Urban sprawl and public health. *Public Health Reports, 117*, 201–217.

Gargiulo, V., Sateriano, A., Bartolomei, R. D., & Salvati, L. (2012). Urban sprawl and the environment. Retrieved from http://eprints.uni-kiel.de/20777/1/gi412.pdf#page=46.

Gehl, J. (1987). *Life between buildings: Using public space*. New York: Van Nostrand reinhold.

Gillham, O. (2002). What is sprawl? In M. Larice & E. Macdonald (Eds.), *The Urban Design reader (2013)*. London and New York: Routledge.

Grahn, & Stigsdotter. (2003). Landscape planning and stress. *Urban Forestry & Urban Greening., 2*, 1–18.

Hajer, M., & Reijndorp, A. (2001). *In search of new public domain, analysis and strategy*. Rotterdam: NAi.

Hansson, K., Cars, G., Ekenberg, L., & Danielsson, M. (2013). The importance of recognition for equal representation in participatory processes: Lessons from Husby. In M. Krivý & T. Kaminer (Eds.), *The participatory turn in urbanism* (Vol. 7, No. 2, pp. 81–98). Footprint.

Heber, A. (2008). *En guide till trygghetsundersökningar om brott och trygghet*. Elanders, Göteborg: Tryggare och mänskligare Göteborg.

Healthy Spaces and Places. (2009). Design principles safety and surveillance. *Healthy Spaces & Places*. Available from www.healthyplaces.org.au.

Hilmers, A., Hilmers, D. C., & Dave, J. (2012). Neighborhood disparities in access to healthy foods and their effects on environmental justice. *American Journal of Public Health, 102*(9), 1644–1654. The National Center for Biotechnology Information.

Holloway, L., & Hubbard, P. (2001). *People and place: The extraordinary geographies of everyday life*. Harlow, England: Pearson Education.

Human Rights Watch. (2013). Egypt: Draft law would effectively ban protests—Amend repressive draft assembly law. *Human Rights Watch*, Published 30 Oct 2013. Available from https://www.hrw.org/news/2013/10/30/egypt-draft-law-would-effectively-ban-protests.

Jacobs, J. (1961a). The uses of sidewalks: Contact. In M. Larice & E. Macdonald (Eds.), *The Urban Design reader (2013)*. London and New York: Routledge.

Jacobs, J. (1961b). *The death and life of great American cities*. New York: Vintage.

Jansson, M. (2014). Green space in compact cities: The benefits and values of urban ecosystem services in planning. *The Nordic Journal of Architectural Research, 26*(2), 139–160.

Klausen, J. E., Arnesen, S., Christensen, D. A., Folkestad, B., Hansen, G. S., Winsvold, M., & Aars, J. (2013). Medvirkning *med virkning*: Innbyggermedvirkning i den kommunale beslutningsprosessen. *NIBR/Uni Rokkansenteret*. Oslo: Nordberg trykk.
Københavns Kommune. (2013). *Cykelregnskabet 2012*. Copenhagen: Miljø og teknikforvaltningen.
Lee, A. C. K., Jordan, H. C., & Horsley, J. (2015). Value of urban green spaces in promoting healthy living and wellbeing: Prospects for planning. *Risk Management and Healthcare Policy, 8*, 131–137.
Loukaitou-Sideris, A. (2006). Is it safe to walk? 1. Neighborhood safety and security considerations and their effects on walking. *Journal of Planning Literature, 20*, 219–232.
Montgomery, C. (2013). *Happy city*. Canada: Doubleday.
Mulder, I. (2012). Living labbing the Rotterdam way: Co-creation as an enabler for urban innovation. Available from http://timreview.ca/article/607.
Payne, L. L., Mowen, A. J., & Orsega-Smith, E. (2002). An examination of park preferences and behaviours among urban residents: The role of residential location, race, and age. *Leisure Sciences, 24*(2002), 181–198.
PPS. (2017). The Power of 10: Applying placemaking at every scale. *Project for Public Spaces*. Retrieved from http://www.pps.org/reference/the-power-of-10/.
Rasmussen, A. H. (2013). Lovbrudd for det fælles bedste. *Information,* published 19 Oct 2013. Available from https://www.information.dk/moti/2013/10/lovbrud-faelles-bedste.
Sanders, L., & Stappers, P. (2008). Co-creation and the new landscapes of design. *CoDesign, 4*(1), 5–18.
Schipperijn, J., Ekholm, O., Stigsdotter, U. K., Toftager, M., Bentsen, P., Kamper-Jørgensen, F., & Randrup, T. B. (2010). Factors influencing the use of green space: Results from a Danish national representative survey. *Landscape and Urban Planning, 95*(2010), 130–137.
Shenassa, E. D., Liebhaber, A., & Ezeamama, A. (2006). Perceived safety of area of residence and exercise: A pan-European study. *American Journal of Epidemiology, 163*(11), 1012–1017.
Socialministeriet. (2010). In M. Stender, S. Mertner Vind, K. Hauxner, J. Raun Nielsen, & S. Willems (Eds.), *Byen som dagligstue? Byfornyelse med plads til socialt udsatte*. Copenhagen: Socialministeriet.
Sotoudehnia, F., & Comber, L. (2011). Measuring perceived accessibility to urban green space: An integration of GIS and participatory map. Available from http://www.agile-online.org/Conference_Paper/CDs/agile_2011/contents/pdf/shortpapers/sp_148.pdf.
Steiner, F. (2016). *Human ecology: How nature and culture shape our world*. Washington: Island Press.
Sundberg, K. W. (2013). Preventing crime through informed urban design. *Security Solutions Magazine*. Retrieved December, 2013, from http://www.safedesigncouncil.org/preventing-crime-through-informed-urban-design/.
Troelsen, J., Toftager, M., Nielsen, G., & Kaya Roessler, K. (2008). De bolignære områders betydning for sundhed. *Movements*. Syddansk Universitet, Institut for Idræt og Biomekanik.
Umeå Municipality. (2009). *Curiosity and passion—The art of co-creation*. Umeå Municipality.
United Nations. (2017). Universal design. *Article 2—Definitions,* Division for Social Policy and Development Disability.
United Nations. (1999). Summary of the "International seminar on environmental accessibility; planning and design of accessible urban development in developing countries" held in Beirut 30th of November–3rd of December 1999. Available from http://www.un.org/esa/socdev/enable/disisea.htm.
Van Herzele, A., & Wiedemann, T. (2003). A monitoring tool for the provision of accessible and attractive urban green spaces. *Landscape and Urban Planning, 63*, 109–126.
Wong, S. (2012). *What have been the impacts of World Bank Community-driven development programs?* Washington, DC: The World Bank.
World Health Organization—WHO. (2013). Fact sheet: 10 facts on noncommunicable diseases. Available from http://www.who.int/features/factfiles/noncommunicable_diseases/en/.

World Health Organization—WHO. (2014a). In M. Marmot (Ed.), Review *of social determinants and the health divide in the WHO European Region: Final report*. UCL Institute for Health Equity.
World Health Organization—WHO. (2014b). European food and nutrition action plan 2015–2020. Available from http://www.euro.who.int/__data/assets/pdf_file/0008/253727/64wd14e_FoodNutAP_140426.pdfhttp://www.euro.who.int/__data/assets/pdf_file/0008/253727/64wd14e_FoodNutAP_140426.pdf.
World Health Organization—WHO. (2017). Fact sheet: Physical activity. Available from http://www.who.int/mediacentre/factsheets/fs385/en/.
WHO Europe. (2006). In M. Braubach & S. Schoeppe (Eds.), *The solid facts: Promoting physical activity and active living in urban environments*. Copenhagen: WHO Europe.
WHO Europe. (2007). In P. Edwards & A. Tsouros (Eds.), *Tackling obesity by creating healthy residential environments*. Copenhagen: WHO Europe.

Chapter 8
Superblocks for the Design of New Cities and Renovation of Existing Ones: Barcelona's Case

Salvador Rueda

8.1 An Urban Model (Rueda 1995) to Cope with the Challenges of the Beginning of the Century

We are in the midst of an era change with a new paradigm and new rules. We are transforming from an industrial age to the age of information and knowledge. The industrial society was characterized by its consumption of resources and its "independence" of the laws of nature. The competitive strategy of cities was based on the consumption of resources: soil, materials, and energy. In general, the city that was better organized in the consumption of more natural resources took a competitive advantage. However, the false belief about the "independence" of the laws of nature led to the excessive use of energy and technology with a great capacity of transformation. The result has been an unsustainable impact on the systems and ecosystems of Earth. The uncertainties about the future are so grand that they oblige us to change the rules of the game and create a new paradigm that increases our capacity of resilience and preparedness. The foundations for the new paradigm are based on:

1. A change of strategy of competition among cities based on information and knowledge which involves, at the same time, the dematerialization of economy.
2. A change of metabolic regime. Industrial society was based, mainly, on the consumption of fossil fuels, as if there were no limits. The excessive consumption of fossil energy and technology was the cause of the simplification of ecosystems on all scales, that is to say, in the generation of generalized entropy with irreversible effects on much of the impacted ecosystems. The new metabolic regime has to be based on entropy.

S. Rueda (✉)
BCNEcologia, Barcelona, Spain
e-mail: rueda@bcnecologia.net

M. Nieuwenhuijsen, H. Khreis (eds.), *Integrating Human Health into Urban and Transport Planning*, https://doi.org/10.1007/978-3-319-74983-9_8

3. An accommodation of the laws of nature, where the exploitation of supporting systems does not exceed its carrying capacity and regeneration. In urban systems, the recycling and regeneration of the current fabrics take precedence over the production of new cities. In both cases, the accommodation of the laws of nature forces us to formulate new tools, among them, the formulation of a new urban model and a new urbanism: Ecosystemic Urbanism.

The battle of sustainability and execution of the new paradigm will take place in the cities, in the design of new cities, and, among all, in the regeneration and recycling of current ones. The current urbanism does not accommodate the new challenges of the beginning of the century. To deal with them, we need to create new conceptual models and different instruments, starting by considering the city as a complex ecosystem, the most complex that humanity has created. Ecologists usually deal with the complexity of ecosystems with the construction of models. Based on analyses of a lot of urban systems, an urban model emerges which is compact in its morphology, complex (mixed in uses and biodiversity) in its organization, metabolically efficient, and socially cohesive. There are four axes closely related that interact synergistically to give integrated answers to urban realities in the process of rehabilitation and regeneration and, also, to accompany planners of new urban developments. The model goes from a city scale to a neighborhood scale.

8.2 The City as a System of Proportions

The set of elements that constitute a city are a result of a system of proportions with its relations and restrictions (Rueda et al. 2012). These proportions, which are the result of multiple factors and interests, can generate dysfunctions of diverse nature. In favor of an easy comprehension and bridging the gap, a "paella" (typical Spanish dish) is a system of proportions. Even if you put the best ingredients, if you don't add salt, the paella will be bland, and if you add too much, it will be unedible.

The same occurs in the city. If the creation of a city has an excessive extension of suburbs, it is very much probable that social segregation takes place by incomes and/or cultures and/or ethnics. There will be inequality. In the suburb, which usually has the residence as its almost exclusive function, the lack of basic services and public facilities will force residents to use the car to access to them. However, most of the population does not have autonomous access because they don't have driver licenses, because they are too young or too old or because they simply don't own a car.

If the modal distribution is excessively tilted toward the private vehicle, the massive occupation of public space by car, congestion and air quality, etc. will result in poor liveability and poor urban quality. The energy consumption and the emission of greenhouse gases will increase, and the air quality will become worse with great impact on human health. The dysfunctions generated by the wrong way of building cities generate a good number of proportions that create dysfunctions. This analysis could be extended to other aspects linked to mobility, infrastructures, equipments, or economic promotion. These variables need to be linked and accommodated by a system of proportions that avoids dysfunctions. The impacts have to be linked necessarily to an intentional urban model. It is necessary, then, to find the system of proportions that allows us to achieve the appropriate model.

8.3 The Principles of Ecosystemic Urbanism: The Basis of the System of Proportions of a More Ecologist City

If the city is a system of proportions, it is necessary to know what are the underlying principles in order to produce a compact, complex, efficient, and socially cohesive city.

The bases of the new Ecosystemic Urbanism (Rueda et al. 2014) are:

8.3.1 The Context of the Urban Action

Dealing with the transformation of cities forces us to take into account the environment (system of support) from all sides: environmental, economic, and social. The adopted solutions can neither create more dysfunctions in the context nor in the secondary variables that accompany it. In any case, they should mitigate them and have a maximum dialog with the environment in a way that the actions support a factor of improvement of the conditions and of the uncertainties at global scale. This obliges us to think about the challenges, vulnerabilities of every emplacement (whether physical, social, or cultural), availability and usage of local resources, habits and lifestyles, etc. that characterize the urban systems, with the objective of transforming the cities into a more efficient and habitable way.

8.3.2 Land Occupation and Morphology of Cities

Urban morphology refers to the form and distribution of the built space and public space. The compactness or dispersion of urban fabrics determines the proximity between the urban uses and functions. The way of occupying the city and its level of usage allows us to foster an urban space which is a social integrator, to develop with efficiency those urban functions linked to the sustainable mobility and the provision of services and basic public facilities and to promote interchange and interactions among complementary ones. The density of population and activities gives a certain critical mass that generates public space, that makes viable public transportation, that gives sense to the existence of public facilities, that generates the necessary diversity of legal entities to make a city, etc. At this point, the objectives and criteria of the model of occupation as well as the relation between areas must be dealt with in the most efficient way, to achieve more sustainable urban fabrics and at the same time to reduce the pressure on the support systems.

8.3.3 Urban Functionality: Model of Mobility and Public Space

Urban functionality, defined by the mobility and services patterns of each city, determines, largely, the quality and liveability of public space. There is a need to develop a model of mobility and a more sustainable public space, in order to guarantee a more accessible, comfortable, safe, and multifunctional public space where people can be citizens and exercise in the public space the rights to interchange, culture, leisure, expression, and demonstration, besides the right to move. With the current model of mobility, cities dedicate most of the public space to mobility, and in these conditions, the ultimate aspiration is to be a pedestrian, a mean of transportation. At least 75% of public space should be dedicated to citizen rights. Public space should acquire the maximum liveability but being, at the same time, comfortable (without noise, without air pollution, and with the greatest thermal comfort), attractive (with a high diversity of activities, with attractive activities, and with the greatest biodiversity), and ergonomic (accessible, with liberated space to exercise all the rights, and with a good relation between built heights and street widths).

8.3.4 Urban Complexity

Urban complexity refers to the degree of urban organization of a city. Both in natural and urban systems, the increase of complexity results in a growth of the organization contributing to stability and continuity of the system itself. Diversity and abundance of legal entities (in the urban systems) or live organisms (natural systems) are the key conditions to increase organized information. Diversity of legal entities (economic activities, associations, and institutions) must be the highest in order to increase the complexity of economic and social capital. The proliferation of activities and its diversity must pay attention and give service to residents and also to the model of city of knowledge. That's why the conditions of land and services must be created to increase the activities dense in knowledge (@) and its networks, which are the basis of the "smart" city, since they are the ones that have the most useful information. Among the activities dense in knowledge are the ones that propose an increase of self-production. Furthermore, it is necessary to define the adequate proportion of building for residence and for legal entities (between 25% and 30%) which will mainly occupy the front of the façade.

8.3.5 Urban Green and Biodiversity

Biodiversity is the wealth of living forms of a city. The urban environment is an artificialization, with impermeabilization of much of the soil and deep alteration of the embossed, air quality, soil and water, climate, and hydrological regime, facts that lead to the loss of habitats and/or accommodation of the urban species to the specific conditions of the city. The objectives of sustainability of biodiversity cannot be separated from other areas such as the construction, the urbanism, or the mobility.

8.3.6 Metabolic Efficiency

Efficiency is a concept related to urban metabolism, which are the fluxes of materials, water, and energy that constitute the support to any urban system to maintain its organization and avoid contamination or simplification of its organization. The management of natural resources must achieve the highest efficiency in its use with the least perturbation of ecosystems.

8.3.7 Social Cohesion

Social cohesion refers to people and social relations in the urban system. Social cohesion in the urban context refers to the degree of coexistence between groups of people with different cultures, ages, incomes, or professions. The increase of social cohesion is deeply linked to accessibility to housing, in a certain area, for people with different incomes, culture, and ethnicities in “adequate” proportions and also to the provision of public facilities that guarantee the best liveability, located at a distance that can be covered by foot in 5–10 min, depending on the facility.

8.3.8 Management and Governance

Every objective of the model or models of more sustainable cities and metropolis requires an adequate organization to achieve it. At this point, we should ask ourselves if current techniques and technologies are enough to reduce the uncertainties derived from the current model of the occupation of the territory and urban model. We are forced to rethink the management mechanisms to achieve them. How do we accommodate organizations in new challenges? Given the complexity that characterizes the urban systems, it is essential to review the current approach of predominant policies and management models, based on a vertical, segmented, and partial structuring.

8.4 Instruments of Ecosystemic Urbanism

8.4.1 Ecosystemic Urbanism Has Three Levels

The current urban planning, with two dimensions, is unable to include the set of variables that incorporate the principles/objectives mentioned above. Ecosystemic urbanism draws on three dimensions, in height, on the surface, and in the subsurface, to integrate the set of variables and principles which must be addressed by the challenges previously cited (Fig. 8.1).

8.4.2 A New Urban Cell: The Superblock (Rueda et al. 2017)

Besides this, Ecosystemic Urbanism proposes a city through a new model, an urban cell of 16/20 ha that is called superblock that has the adequate dimensions to develop and integrate the set of principles and objectives and which arises as the basis of a new urban and functional model in the cities. Like Cerdà (father of the word urbanism and designer of the Eixample in Barcelona), one could consider that the key element in the city is not the housing but the intervia, the cell that becomes a piece of the mosaic of a road network, where the continuity of movement obliges the occupation of roads in its totality and not one by one. The new cell defines an intervia of 400 × 400 m which will be the place to apply to the Ecosystemic

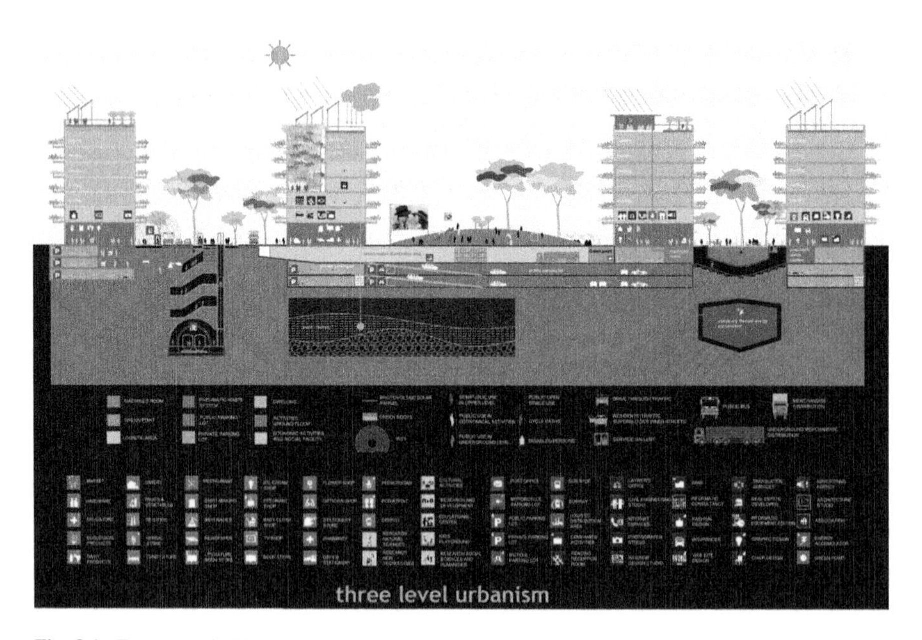

Fig. 8.1 Ecosystemic Urbanism's section. Source: BCNecologia

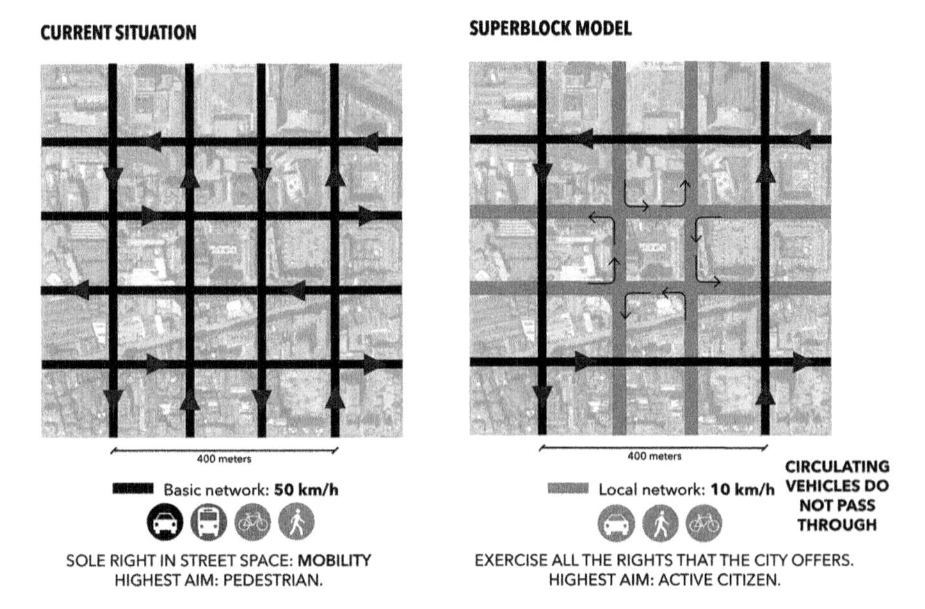

Fig. 8.2 Networks scheme, current and future, based on superblocks. Source: BCNecologia

Urbanism and develop, at the same time, the new model of mobility and public space that we will mainly consider in this chapter below. To make it more applicable, we will focus on the city of Barcelona, specifically Eixample, the more complex part of the city.

The superblock is a cell of nine blocks in the case of the Eixample in Barcelona, defined by a network of basic roads that connects origins and destinations of the whole city (Fig. 8.2). When the cell is being reproduced along the urban system, the size accommodates the morphological and functional characteristics of the current city (it must be highlighted that the superblocks project is a urban recycling project), liberating the maximum surface of public space, which is nowadays occupied by traffic and, at the same time, guarantee the functionality and organization of the system. The new cell is defined by the perimeter basic roads, where through and connecting traffic circulate at a maximum speed of 50 km/h. Interior roads (intervias) of superblock constitute a local network with limited speed of 10 or 20 km/h, speeds which allow shared urban uses. A superblock cannot be crossed, which means that the movements in its interior only make sense if their origin or destination is in the intervias. Therefore the neighborhood streets will be with little or no noise or pollution, etc. and allow more than the 70% of space that is currently occupied by the through traffic for the movements by foot or bicycle.

The reasons to choose the dimensions of 3 × 3 for the superblock are based on the characteristics of cars which, at a speed of 20 km/h (which is the average urban speed in Barcelona), will spend a similar amount of time to go around the superblock as a person walking around the block at about 4 km/h. With the presence of main crossings every 400 m, the traffic lights synchronization is much more efficient (with these distances one can think of prioritization of the traffic lights for public transport) and avoids disruption of the main flux because of turns (two out of three turns are avoided).

8.5 Superblock, Base for a Functional and Urbanism Model: The Case of Barcelona

Superblocks aim to be the basis of a functional model of any city, but they can also become the basis of a new urbanism model. The average number of people in a superblock in Barcelona is higher than 6.000 inhabitants. More than three quarters of Catalan municipalities have less than 6.000 inhabitants, and a county town like Viella (in the North of Catalonia) has 5.500 inhabitants. These urban systems have more public facilities than a much larger city. It only seems reasonable that the superblock gets the attention that deserves an entity with such population. Urban superblocks, as one of the tools of Ecosystemic Urbanism, allow for the application of the set of principles that were referred to above. Every superblock emerges as a little city. In this section, we will focus on the superblock as the basis of a new functional model and the consequences for public space.

The defining roads of superblocks (in red) comprise a network of basic roads where the urban transport circulates: collective transport, private vehicle, emergencies, services and, if the section allows it, bicycles. This network of basic roads, which results in the greatest orthogonality, allows access to the city at the highest speed admitted by law (50 km/h) (Fig. 8.3).

The basic network of superblocks reduces the length of the total of roads that nowadays is used by through traffic by 61% (see Fig. 8.3). This dramatic reduction does not lead to a proportional drop-off in the circulation of vehicles for the same level of service (the same speed of the vehicles circulating). In Barcelona, with a reduction of vehicles of about 13%, we can achieve a level of service similar to the current one. It maintains, thus, the functionality and organization of the system. Figure 8.4 shows that the space dedicated nowadays to traffic is near 15 million of squared meters and the length of the roads dedicated to mobility nearly 912 km, or 85% of the total roads of Barcelona.

8.5.1 The Impacts of the Current Mobility Model

Mobility currently brings the biggest dysfunctions to the city. The use of public space is restricted to mobility, and Barcelona dedicates more than 60% of public space and 85% of roads to traffic. Air pollution emitted by motorized traffic has an

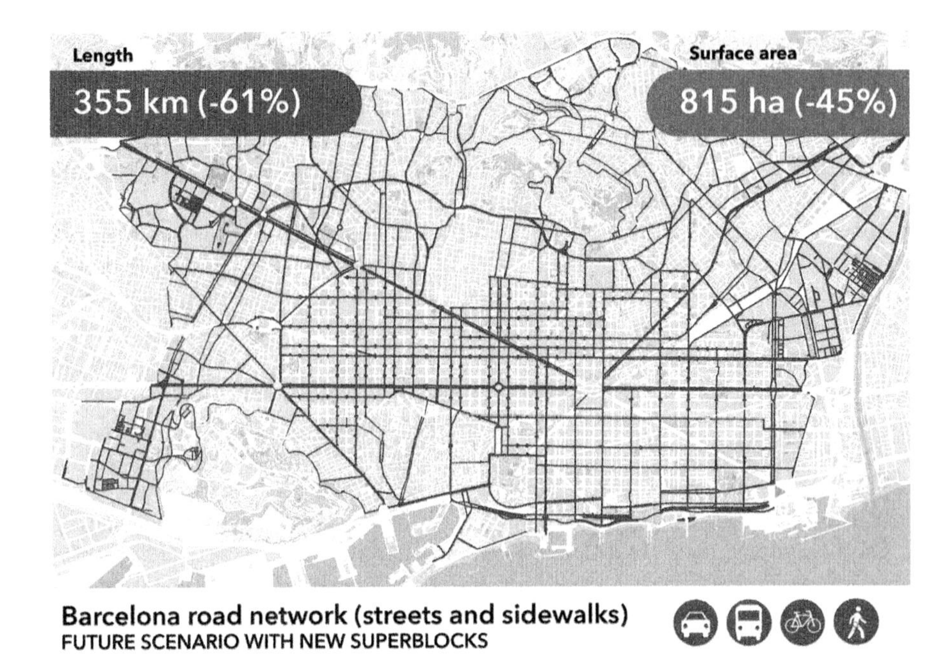

Fig. 8.3 Map of superblocks in Barcelona. Public space (in red) dedicated to mobility. Source: BCNecologia

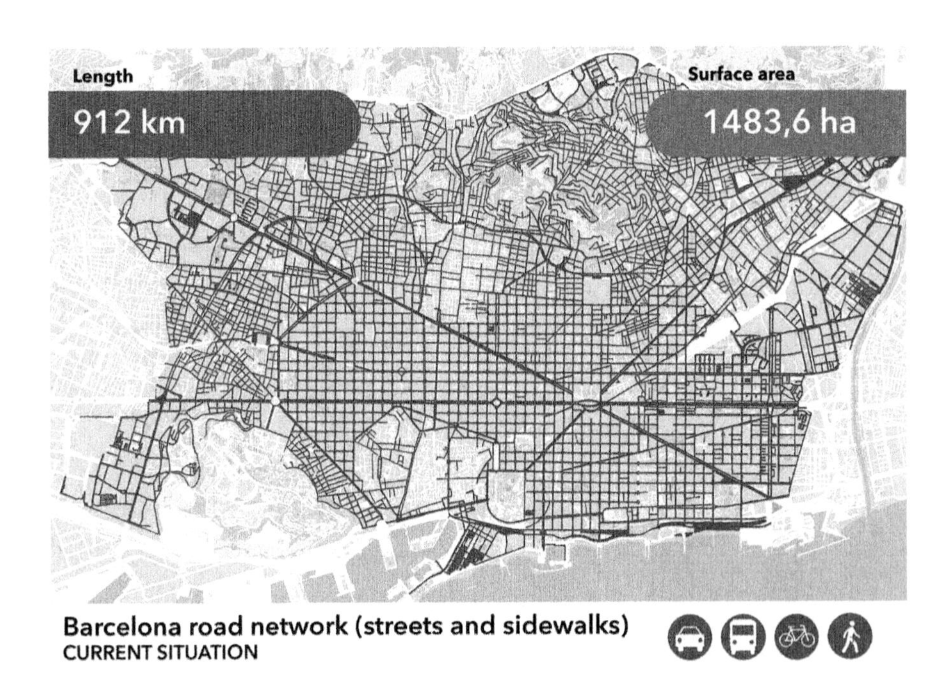

Fig. 8.4 Public space of Barcelona dedicated to through traffic in the current situation. Source: BCNecologia

inacceptable impact on the health of population in the Metropolitan Area of Barcelona. In a study conducted by ISGlobal CREAL in an area of 56 municipalities, including Barcelona, it was estimated that air pollution causes 3.500 premature deaths per year, 1.800 hospitalizations by cardiovascular reasons, 5.100 cases of chronic bronchitis in adults, 31.100 cases of children bronchitis, and 54.000 asthma attacks among children and adults (Künzli and Pérez 2007).

The impacts of air pollution on health are today the main problem to solve out of all the problems caused by the current model of mobility. The liveability index in most of the city is below minimal. The green surface in central Eixample is 1.85 m^2 per inhabitant when the WHO recommends 9 m^2/inhab. Eixample is also the district with the most traffic; almost 50% of the population is exposed to inadmissible levels of noise (diurnal values >65 dbA).The negative economic impact runs in millions of Euros per year (according to the World Bank, for Spain it was 45.000 million Euros/year in 2013, considering only the impact on the health). Black asphalt and the emissions of cars are responsible for the most important part of the urban heat island. This increase of more than 2° of average temperature (at summer nights can surpass the 5° of difference compared to periphery) is especially painful and in some cases mortal, for the most vulnerable people, elderly, children, and sick people, particularly when the heat waves occur because of climate change. Traffic accidents cause 30 deaths per year in Barcelona and more than 30 injuries per km/year in Eixample. Visual intrusion and landscape deterioration (landscape as expression of the integration of different variables) convert Barcelona in a "pressure cooker" affecting 85% of the length of streets in the city.

The results of a study conducted by ISGlobal (Mueller et al. 2017) for Barcelona and its Metropolitan Area show clearly the impact that some of the exposures have on morbidity of the citizens of Barcelona (Fig. 8.5).

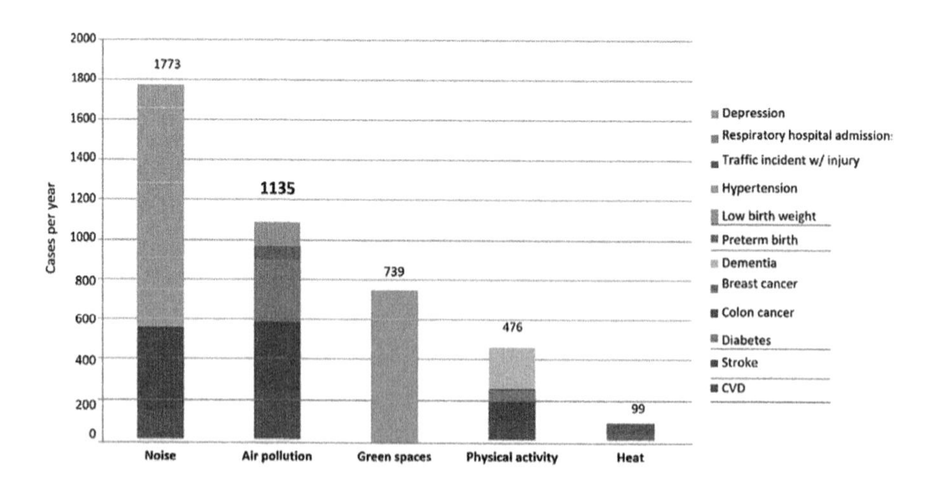

Fig. 8.5 Morbidity in Barcelona by different reasons. Source: ISGlobal

The result is a city that is not ready to address the great challenges of the beginning of century: sustainability in the information age. There is need for a new ecosystemic model with its corresponding system of proportions that includes, at the same time, the reduction of: polluting emissions, noise, energy... and the increase of: green, staying spaces; legal entities diversity, and also dense in knowlegde... An urban model that extents along the city and takes into account the current modes of mobility.

To address these serious problems, the Sustainable Urban Mobility Plan of Barcelona approved by the city council (March 2015) proposes to extend the superblocks throughout the city and reduce 21% of circulating vehicles. With this reduction, it is estimated that the levels of air pollution in all the measurement stations will be below the guideline values. With a reduction of 21% of circulating vehicles, the level of service of traffic and the environmental conditions of basic roads in a superblocks scenario will be much better than now.

With this reduction of vehicles, the percentage of people exposed to levels below the guidelines for air pollution will be 94% (today it is 56%) and for noise 73.5% (nowadays it is 54%). The liveability index will increase significantly in all the neighborhoods of the city (Fig. 8.6).

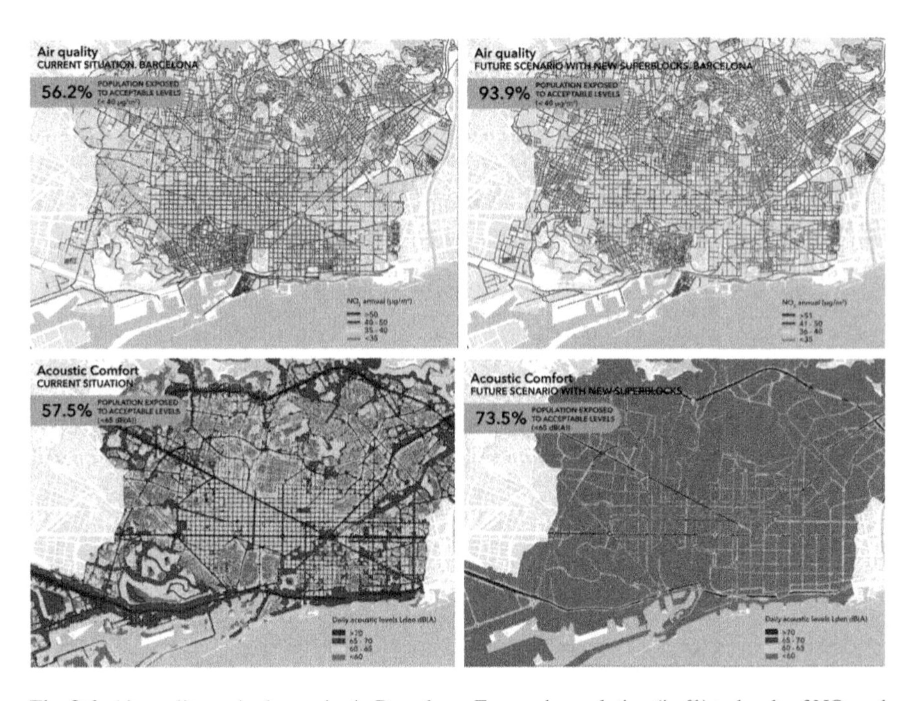

Fig. 8.6 Air quality and urban noise in Barcelona. Exposed population (in %) to levels of NO_2 and decibels legally admissible. Current scenario and with superblocks. Source: BCNecologia

8.5.2 The Bus Network

Superblocks allow the integration of networks of through traffic (cars, buses and bicycles) in its periphery, and priority of movements by foot and bicycle on the inside. Orthogonal networks are the most efficient in urban systems. With the new orthogonal network of busses, the topology of the new network, combined with the distance of the stops every 400 m, contributes to the increase of commercial speed of busses, more so than the traffic lights prioritization or the construction of bus lanes (Fig. 8.7). With the same number of busses that have a frequency of coming every 14–15 min nowadays, there will be a move to a frequency of busses coming around every 5 min in the future (it will be like a surface underground). The service will be the same in the center and in the periphery of the city. The design of the network allows an average waiting time at each stop of around 2 min, which for our mental clock does not feel like waiting. It is a network that connects any origin with any destination with only one transfer in 95% of cases. It will be an intelligible network like the underground, and the estimated number of transfers will be similar to the underground. In the Eixample fabric, the perpendicularity of the lines of the network allows the presence of a unique stop for the horizontal and vertical lines in all the intersections (in the octagonal crossroads).

The bicycle network will also be adjusted to the superblocks structure (Fig. 8.7). Superblock's periphery hosts the transport network of bicycles, with exclusive lanes, and shares the street section with busses and cars. The interior of the superblocks, at 10 or 20 km/h, allows bicycles in both directions, to cross the superblock. Their speed must be adjusted to the speed of pedestrians and the uses that are in place at that moment. If necessary, the rider needs to step off the bicycle. The conditions of intervias allow children go to school by bicycle or by foot without being accompanied by an adult.

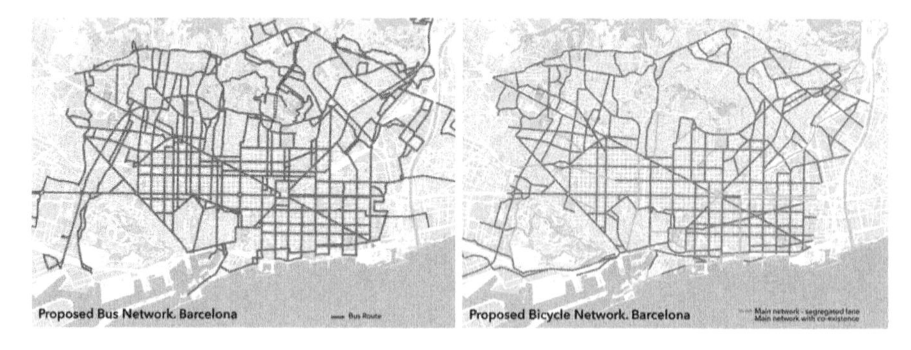

Fig. 8.7 Bus network and bicycle network in Barcelona in a superblocks scenario. Source: BCNecologia

The integration of electric engines for the transport is on the agenda of many cities. There is no doubt than that the electric bicycle is the electric vehicle to promote. It does not pollute, it does not make noise, and it practically does not consume energy (the consumed energy for a journey by electric bicycle, adding the used metabolic energy and the consumed electricity, is less than the energy metabolically consumed doing the same journey by foot); it allows an average person to overcome high slopes of up to 20%, and it is healthy and adjusts the effort to the context. During the summer, it even refrigerates—allowing its use in the most severe season without sweating. The homologated motor works up to 25 km/h reducing the severity of accidents. The average distance of the classical bicycle is about 5 km, while electric bicycle increases it up to 10 km, which is the distance from one end to the other in the municipality of Barcelona.

The electric bicycle is, for a distance of 10.5 km and a speed 30% higher than the speed of a classic bicycle, the most competitive form of mobility.

The isoform distribution of the networks along the city gives a level of service which is equitable for the bus and bicycle networks, something previously only the car had.

8.5.3 *The Pedestrian Network*

At present, Barcelona has 230 ha of streets with specific space for pedestrians or with speeds limited to 20 km/h. If we add the surface used by the pedestrians in big avenues, the total is about 15.8% of the public space of roads. With the implementation of superblocks, 6.22 million of square meters are liberated. This will make it the most important recycling proposal in the world, without demolishing a single building (Fig. 8.8).

Fig. 8.8 Map for pedestrians and other uses. Current scenario and with superblocks. Source: BCNecologia

8.5.4 Uses in Public Space and Citizen Rights: From Pedestrians to Citizens

Maybe the most radical aspect of the proposal is the reconversion of most of the urban space to multiple uses and rights. Radical because it goes to the roots of the meaning of public space. A city exists when, first, there is public space and, second, when they are reunited in a limited space by a determined number of legal entities that are complementary and "work" synergistically. We can find ourselves in an urbanization of undetached houses with space for the car to get to the garage. In this case, we talk about urbanized space, but not public space. In urbanization it makes little sense to have a market and a cultural event or, even, find children playing with balls in the middle of the street.

A city starts to become a city when there is public space, since it is the "house of everybody," the meeting place for interchange, leisure and staying, culture, expression and democracy, and, also, the movement. Public space makes us citizens and we are so when we have the possibility of occupying it for the exercise of all the rights mentioned above. Today, the limited possibility of exercising the citizen rights relegates us as pedestrians. Giving citizens public space that was lost due to the current model of mobility is key to the new model of mobility and public space based on superblocks. Electric vehicles can reduce part of the noise (but not the noise that at some speeds comes from the friction between the tires and the bearing surface and not from the engine) and part of the air pollution (as almost half of the pollution caused by particles come from the "dust" lifted by the wheels, that comes from tires particles, brakes, bearing lubricant oils, etc., that, as it is known, contain heavy metals and components of high toxicity). What they cannot reduce is the space that they occupy, which, in a compact city in general and particularly in Barcelona, which is the most scarce good.

Superblocks provide citizens with nature in almost 70% of the space of the city. Superblocks allow not only for the integration of transportation networks but also green networks. Furthermore, spaces are not crossed by any network of mobility: cars, busses, and bicycles. Rights are obtained through speeds that are compatible with the use of the space by the most vulnerable people (for instance, the pass of blind people, children playing). If a superblock allows a bus network, a car network, or a bicycle network with signalized lane, then it is no longer a superblock because it is not compatible with ALL the rights.

8.5.5 The Green Network that Appears with the Implementation of Superblocks

Many urban planners consider that the permeability of the soil is a good indicator of naturalization of an urban fabric. The presence of permeable soils rebalances the water cycle: it favors the infiltration of rainwaters and retains water through the

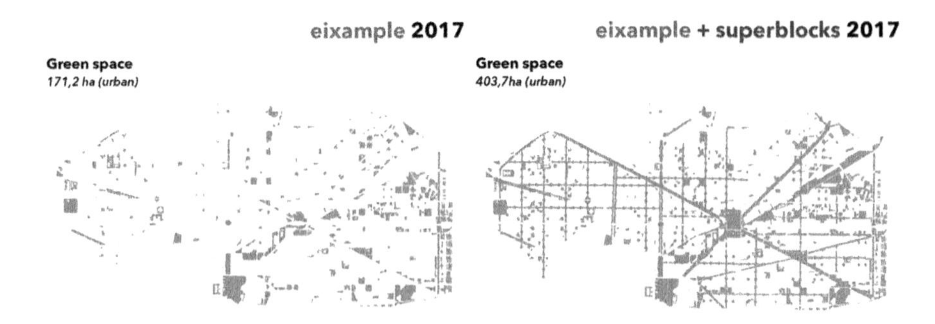

Fig. 8.9 Green space in current situation and with superblocks. Source: BCNecologia

different vegetable surfaces. The vegetation protects the soil from excessive insulation and protects it from the compaction that provokes the direct impact of raindrops on the soil. By enabling the water to stay longer on surface, it increases the possibility that it infiltrates into the phreatic layers and reduces the risk of floods. It boosts the closing of the cycle of organic matter, by providing the urban soil with surfaces of compost generated by organic waste self-composting. Green spaces and land reservation for urban gardens constitute spaces for generating community among the inhabitants of a neighborhood or territorial unit. Surfaces with vegetable cover help to mitigate the emissions of CO_2, by fixing this gas through the photosynthetic process. Vegetable surfaces are, furthermore, potential captors of polluting particles and help to propitiate thermal comfort, minimizing the heat island effect. Besides, surfaces with trees proportionate acoustic and mechanic comfort, reducing the effect of noise and wind in the urban environment.

There are few areas with permeable soils in Barcelona (Fig. 8.9). The current green surface in the area of Eixample is only 171.2 ha. The number of square meters per inhabitant is 2.7 m^2, very far below the 9 m^2/inhab the WHO recommends. In a city like Barcelona, with such a lack of free spaces, superblocks allow us to obtain better values of corrected compactness (balance between urban compression and decompression). The release of only the interiors of blocks in Eixample is clearly insufficient although necessary. In Eixample, greed occupied green spaces. Superblocks will allow us to restore part of the green space that is so needed.

With superblocks, green surfaces increase significantly to 403.7 ha of potential green while maintaining city's functionality. It will increase from 2.7 m^2/inhab to 6.3 m^2/inhab for the whole area of Cerdà's plan. In the Sant Martí area, it increases up to 7.6 m^2/inhab. Street transformation and substituting cars with green space allow us to obtain urban landscapes like the ones that are shown in Fig. 8.10. It shows the project presented by the city council to the neighbors of the pilot superblock of Poblenou, for the section of Sancho d'Àvila between Llacuna and Roc Boronat streets (Fig. 8.10). As Oriol Bohigas said: "A street will have for Latins an infinity of values that a garden won't" (Bohigas 1958).

Una calle tendrá para los latinos una infinidad de valores que no tendrá nunca un jardín. (Bohigas, O. 1958)

Fig. 8.10 Sections of a street with single platform in the interior of superblock and proposal of transformation of Sancho d'Àvila street between Llacuna and Roc Boronat streets. Sources: BCNecologia and Barcelona City Council

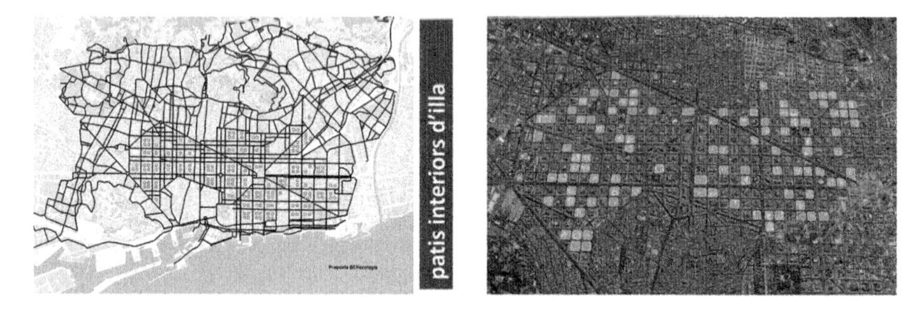

Fig. 8.11 Junctions that become new squares in the superblocks model and block's interiors released and potentially releasable in central Eixample. Sources: BCNecologia and Barcelona City Council

A square has been and is the very place for public space. In the case of Eixample of Barcelona, often, there are sidewalks of only 5 m of width with little green space (1.85 m^2/inhabitants). With the superblocks project, the number and surface areas of new squares that are in the junctions of the Eixample fabric will multiply. In a superblock type of 3 × 3 blocks, four new squares of about 1.900 m^2 will be created (Fig. 8.11). One-hundred thirty nodes will become complete squares of 1.900 m^2, adding around 24.7 ha, while there will be 20 new squares with a surface of 2/3rd of a complete surface, adding 3 more ha. In total there will be about 150 new squares with a surface of 27.7 ha.

Green from the interiors of block and green covers can also be added. Environmental benefits increase with the increase of urban green surface in height and surface. When you add the public space's green surface and green covers (with

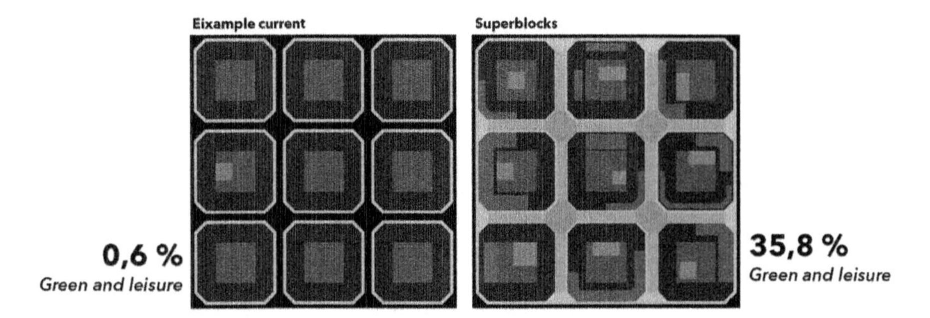

Fig. 8.12 Percentage of current green surface and potential green spaces (in % of total surface) in an area equivalent to a superblock type which include: public space green, blocks' interior patio's green and green covers. Source: BCNecologia

an estimated an occupation of 30%) and the green surface in the interior of blocks (counting 1.500 m^2 per block), green surface per inhabitant increases up to 9.6 m^2/inhabitants (Fig. 8.12).

8.5.6 Superblocks: A Tool that Helps in Mitigation and Adaptation to Climate Change (Climate Change Assessment)

Urban heat islands form due to the emissions of heat by part of the series of surfaces that configure physical parts of the city. The emission of heat to the atmosphere is a consequence of different factors, for example, climate conditions of the place, urban morphology, distribution of materials of the different urban surfaces (pavements, covers, and façades), and also by the heat produced by the activity of the city (traffic, building's conditioning, etc.) (Martín-Vide 2015).

Currently, cities are characterized by the predominance of impermeable surfaces. This aspect produces an important concentration of heat, especially in the asphalt pavements of urban roads and the buildings. One of the areas of discussion of climate change is precisely the mitigation of urban heat island because its impact has repercussions both for energy consumption and people's well-being, reducing the comfort levels during the day and above all during night. Heat absorbed during the day is freed during night, increasing notably the temperatures of summer nights. That is why it is important to implement measures that allow mitigating urban heat generation and reducing urban centers temperature.

To reduce the current heat island effects, it is necessary to reduce the number of vehicles circulating and the consumption of fossil fuel energy, as well as to extend a green carpet that allows us to increase our resilience toward heat waves that will become more frequent with climate change. Superblocks will help in the mitigation and in the adaptation of the problem.

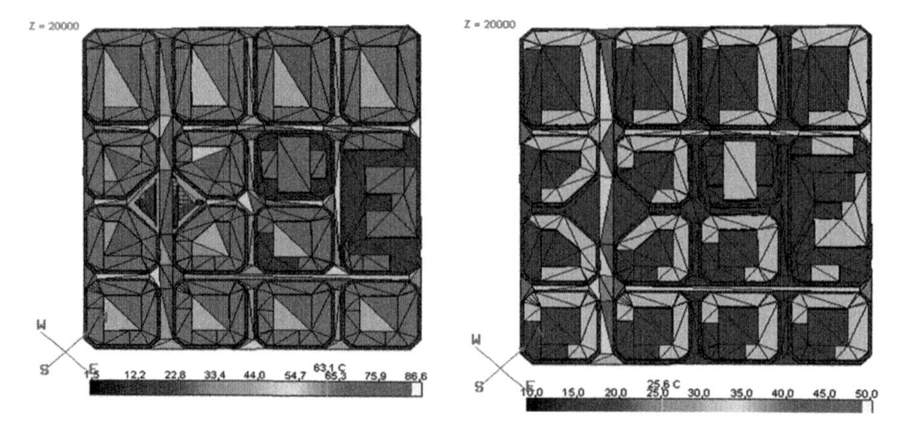

Fig. 8.13 Thermal simulation comparing Eixample in current situation and with superblocks. Source: BCNecologia

Through simulations of heat transfer for a summer day in Barcelona, one can demonstrate the repercussion of the current urban morphology and with superblocks. The superblock scenario will substitute the asphalted pavements in intervias with semipermeable pavements and the greatest cover of road trees. It will dedicate 30% of cover surfaces to green covers and release 1.500 m^2 of interior patios of each of the blocks to green spaces. This has been analyzed in two areas of superblocks in the left part of Eixample (Fig. 8.13).

Results clearly indicate how the current situation of urban fabric in Eixample is warmer than the superblocks proposal. In terms of heat balance,[1] while the current fabric is *72.1 kW/m²*, the superblocks proposal obtains a balance of *46.3 kW/m²*, *35.9%* less than the current situation. This energy translates into lower surface temperatures of pavements, buildings, and green spaces.

References

Bohigas, O. (1958). En el centenario del Plan Cerdà. Cuadernos de arquitectura.

Climate Change. *The IPCC Scientific Assessment WMO/IPCC/UNEP*. Cambridge: Cambridge University Press.

Künzli, N., & Pérez, L. (2007). Els beneficis per a la salut pública de la reducció de la contaminació atmosfèrica a l'àrea metropolitana de Barcelona. CREAL.

[1] The heat balance by radiation refers to the sum of heat gained by radiation both in short and long wave length. In incident solar radiation (short wave), a part is absorbed, and another proportion is reflected according to material's albedo. Absorbed radiation accorded to the thermal capacity, and emissivity of materials is equally emitted as long wave to the rest of materials. From the albedo characteristics and emissivity of materials in pavements, façades, and covers, it is calculated the total of heat produced during the day in energy units of kW/m^2.

Martín-Vide, J. (2015). *La isla de calor en el Área Metropolitana de Barcelona y la adaptación al cambio climático.* Proyecto METROBS 2015. Ed. AMB.

Mueller, N., Rojas-Rueda, D., Basagaña, X., Cirach, M., Cole-Hunter, T., Dadvand, P., Donaire-Gonzalez, D., Foraster, M., Gascon, M., Martinez, D., Tonne, C., Triguero-Mas, M., Valentín, A., & Nieuwenhuijsen, M. (2017). Health impacts related to urban and transport planning: A burden of disease assessment. *Environment Int, 107*, 243–257.

Rueda, S. (1995). Ecologia Urbana: Barcelona i la seva Regió Metropolitana com a referents. Ed. Beta Editorial.

Rueda, S., et al. (2012). *Ecosystemic Urbanism Certification.* Ed. BCNecologia.

Rueda, S., et al. (2014): *Ecological Urbanism: its application to the design of an eco-neighborhood in Figueres.* Ed. Agencia de Ecología Urbana de Barcelona.

Rueda, S. et al (2017). *Les superilles per al disseny de noves ciutats i la renovació de les existents. El cas de Barcelona.* Papers-59: Nous reptes en la mobilitat quotidiana. Politiques publiques per a un model més equitatiu i sostenible. Publicacions IERMB.

Chapter 9
Informal Settlements and Human Health

Jason Corburn and Alice Sverdlik

9.1 Introduction

Urban informal settlements are communities that are highly diverse and have a range of locally specific names, such as *barrios*, *bustees*, *mjondolo* or *favelas* (Gilbert 2007). According to UN-Habitat, 'slums' are human settlements that have one or more of the following deprivations: (1) inadequate access to safe water, (2) inadequate access to sanitation and other infrastructures, (3) poor structural quality of housing, (4) overcrowding and (5) insecure residential status (UN-Habitat 2016b). Yet, as we emphasize throughout this chapter, there is no single definition that adequately captures the characteristics of 'informal settlements' that contribute to poor health or well-being. For instance, the UN also defines 'informal settlements' as unplanned squatter areas that lack street grids and basic infrastructure, with makeshift shacks erected on unsanctioned subdivisions of land. As we suggest, Engaging with the health issues facing residents of self-built communities will require exploring not just how physical deprivations might influence health outcomes but also how economic poverty, social inequalities and political disenfranchisement may act to stymie well-being or support resilience.

Here we use the terms 'slums' and 'informal settlements' to denote largely self-built urban communities, which are rarely recognized officially and typically are denied life-supporting services and infrastructure. Holistic urban development

J. Corburn (✉)
Department of City & Regional Planning, School of Public Health, UC Berkeley, Berkeley, CA, USA

Institute of Urban & Regional Development, Center for Global Healthy Cities, UC Berkeley, Berkeley, CA, USA
e-mail: jcorburn@berkeley.edu

A. Sverdlik
International Institute for Environment and Development (IIED), London, UK
e-mail: alice.sverdlik@iied.org

M. Nieuwenhuijsen, H. Khreis (eds.), *Integrating Human Health into Urban and Transport Planning*, https://doi.org/10.1007/978-3-319-74983-9_9

strategies are urgently needed to overcome vast health inequalities and to support well-being in 'slums' or 'informal settlements', which offer precarious shelter to over 880 m city dwellers globally (UN-Habitat 2016a, b). Over half of the world's population already live in urban areas, and future growth will be concentrated in cities of the Global South that often experience significant housing deficits and profound health challenges, particularly in informal settlements (WHO and UN-Habitat 2016). These areas vary widely in their living conditions and associated environmental health risks, but residents may experience several overlapping challenges including elevated poverty levels, low-quality shelter, food insecurity and political exclusion.

Many health outcomes are worse in urban slums in the Global South than in neighbouring urban areas or even rural areas (Ezeh et al. 2016; Harpham 2009). Moreover, the formal health sector typically encounters slum residents only when they develop complications of preventable chronic diseases. This takes a costly toll on these neglected communities and already limited healthcare resources. Indeed, the urban poor living in informal settlements face a 'triple threat' of injuries, infectious diseases and non-communicable conditions (NCDs) like diabetes and heart disease (WHO and UN-Habitat 2016). Residents of informal settlements typically face multiple risks due to (1) hazardous shelter and local environmental conditions; (2) limited or non-existent access to safe water, sanitation, public transport and clean energy; (3) tenure insecurity; (4) exclusion from affordable, high-quality healthcare, education, refuse collection and other vital services; (5) spatial segregation; (6) violence and insecurity; and (7) political marginalization (Corburn and Riley 2016).

The Sustainable Development Goals (SDGs) include a target for upgrading 'slums'[1] and recognise that reducing inequality will require attention to the environmental and social conditions that keep the urban poor unhealthy, disabled and subject to premature mortality (WHO and UN-Habitat 2016). The social determinants of health (SDOH) are factors outside medical care that shape health outcomes, such as safe housing, food access, political and gender rights, education and employment status (De Snyder et al. 2011). In informal settlements, residents are often burdened with multiple and overlapping challenges that can undermine the SDOH, from spatial segregation and entrenched poverty to overcrowded shelter inadequate infrastructure and tenure insecurity. These factors typically combine and heighten the risk of exposures to environmental pathogens and limit access to life-supporting services, resulting in greater prevalence of infectious, and NCDs in poor urban areas (Sverdlik 2011). Slum-dwellers also experience health inequities, which are the result of spatial, political and economic exclusion. As the UN-Habitat and the World Health Organization (WHO) stated in their 2010 report, *Hidden Cities: Unmasking and Overcoming Health Inequities in Urban Settings*:

---

[1] SDG 11 includes the target 'By 2030, ensure access for all to adequate, safe and affordable housing and basic services and upgrade slums' (http://www.undp.org/content/undp/en/home/sustainable-development-goals/goal-11-sustainable-cities-and-communities/targets/).

> Health inequities are the result of the circumstances in which people grow, live, work and age, and the health systems they can access, which in turn are shaped by broader political, social and economic forces. They are not distributed randomly but rather show a consistent pattern across the population, often by socioeconomic status or geographical location. No city—large or small, rich or poor, east or west, north or south—has been shown to be immune to the problem of health inequity.

9.2 Informal Settlements and Urban Health

By 2016, close to 900 million people were living in informal settlements or almost one in eight people globally (UN-Habitat 2016a). While the proportion of city dwellers living in slums decreased from 46% to 29% in the Global South between 2009 and 2014, the absolute numbers living in slums increased from 689 to 881 million over the same period (UN-Habitat 2016a). However, the percentage of urban slum-dwellers varies across regions, ranging from 21% in Latin American cities to 28% in Southeast Asia, 31% in South Asia and 56% in sub-Saharan African cities (UN-Habitat 2016a).

While city living can be healthy for most residents, *where you live* in that city plays a critical role in determining well-being and life chances (Dye 2008; Ezeh et al. 2016). For instance, data from Kenya and the City of Nairobi suggest that the health benefits from urbanization are not uniform (Fig. 9.1). In Nairobi's informal settlements, child mortality rates reached 151 per 1000 births in 2012, versus just 62 in Nairobi overall, 84 in other urban areas and 113 in rural areas (APHRC 2014). In every region of the world, the wealthiest urban households maintain a health advantage over the poorest, but this is magnified in urban informal settlements (Lilford et al. 2016).

The life-threatening and disease-inducing aspects of slum living include pervasive socio-spatial exclusion, constant discrimination and lack of recognition by government (WHO and UN-Habitat 2016). Further, the UN highlights that there is no one single defining set of health risks in informal settlements and different ethnic, religious, age and other groups living in slums experience a multitude of contrasting social, economic and environmental health threats, as noted below:

> Different vulnerable groups living in slums are particularly affected: women are more likely to have lower education levels and face high rates of teen pregnancies, children are constantly exposed to a whole range of impacts, unskilled youth are excluded from economic and employment opportunities, people with disabilities suffer due to the slums' dilapidated infrastructure and migrants, refugees and internally displaced persons affected by conflict and economic crisis also face additional levels of vulnerability and marginalisation through their uncertain status and lack of resources. (UN-Habitat 2016b)

Informal settlements are usually created via land invasions and squatting or informal subdivision, sale and development of vacant land. Residents are forced to use low-cost building materials to construct shelter and often build community assets incrementally. Thus, urban informal settlements are very dynamic communities with a complex combination of assets and risks, making conventional epidemiologic approaches to studying population health and 'neighbourhood effects' on health difficult for measuring and monitoring health status (Lilford et al. 2016).

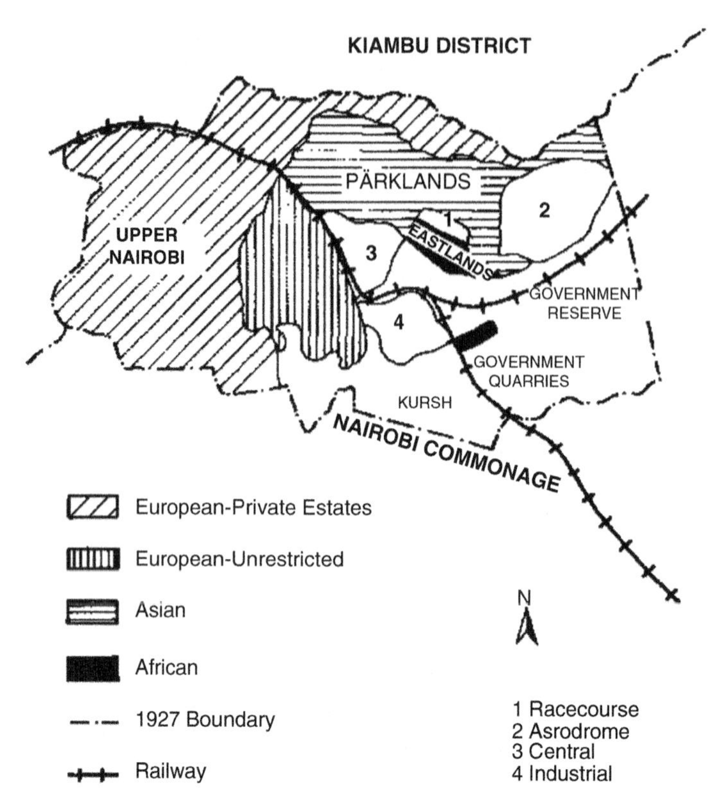

Fig. 9.1 The segregation of residential areas in Nairobi, 1909. Source: Obudho (1997)

Table 9.1 highlights the major physical and socioeconomic dimensions of informal settlements, as well as their links to community health. While drawing upon UN-Habitat's definition, Table 9.1 incorporates additional aspects such as hazardous locations, violence and insecurity and political disempowerment in informal settlements. We offer more detail on the categories and findings from Table 9.1 in the discussion below.

9.3 Key Drivers of Health and Well-Being in Urban Informal Settlements

9.3.1 Spatial Segregation

Contemporary informal settlements are frequently located on dangerous sites, such as areas at high risk of flooding or landslides, which partly reflect legacies of colonial-era planning (Fig. 9.1). In many colonial cities, the indigenous populations were not allowed by the occupying power to live in urban areas (Home 2013).

Table 9.1 Informal settlements and selected health risks

Example slum characteristics	Definition and indicators (examples)	Community health risks (select)
Overcrowding	>2 persons/room or <5 m^2 per person	Spread of TB, influenza, meningitis, skin infections and rheumatic heart disease
Low-quality housing structure	Inferior building materials dirt floors and substandard construction	Vulnerability to floods, extreme heat/cold, burns and falling injuries
Hazardous housing sites	Geological and site hazards (e.g. industrial waste sites, garbage dumps, railways, wetlands, steep slopes, etc.)	Acute poisoning; unintentional injuries, landslides, flooding, toxic contamination, environmental pollutants, leptospirosis, cholera, malaria, dengue, hepatitis, drowning
Inadequate water access	<50% of households have affordable, 24/7 access to piped water/public standpipe	Malaria, dengue and diarrhocal diseases, cholera, typhoid, hepatitis; increased HIV/AIDS vulnerability
Inadequate sanitation access	<50% of households with sewer, septic tank, pour-flush or ventilated improved latrine	Faecal-oral diseases, hookworms, roundworm; missed schooldays during girls' menstruation; malnutrition and children's stunting; safety/sexual violence for women from unsafe toilets
Limited services and infrastructure	Inadequate healthcare, drainage, roads, energy, transport, schools and/or refuse collection	Traffic injuries; lack of emergency provision; fires; flooding/drowning; waste burning and air pollution; respiratory diseases and cancer
Tenure insecurity	Lack of formal title deeds to land and/or structure	Fear; increased hypertension; diabetes; low birthweight newborns
Poverty and informal livelihoods	Low incomes, few assets and access to credit; lack of social protection	Increased occupational hazards; maternal health complications; vaccine-preventable diseases; perinatal diseases; drug-resistant infections
Violence and insecurity	Elevated crime, including domestic- and gender-based violence	Homicides; hypertension; obesity; sexual violence; vulnerability to STIs, especially for young people forced into sex work
Political disempowerment	Low or no governmental responsiveness to needs and services	Lack of health services; poor education; preventable hospitalizations; typhus, leptospirosis, cholera, chronic respiratory diseases, growth retardation

Instead, the indigenous population was either only permitted to labour in the city or to settle on land designated for them, which was frequently the most risky, flood prone and/or adjacent to growing industries (ibid., also Fox 2014). Informal settlements not only reflect household poverty and urban population growth but also represent a legacy of discriminatory, segregationist planning, national policies that have compounded economic exclusion and municipal governments that are incapable or unwilling to serve the urban poor.

9.3.2 *Insecure Residential Status*

The lack of secure land or housing tenure forces the urban poor to settle on undesirable, often risky land (e.g. flood-prone areas, waste dumps and steep slopes subject to landslides) where they are often at elevated risk of forcible eviction and natural disasters. Shelter tenure security is widely acknowledged as a pivotal aspect of human health for the urban poor (Haines et al. 2013). Lack of housing tenure may be associated with health risks such as homelessness, increased poverty and exposure to cold and environmental toxins, leading to infectious and non-communicable diseases. Eviction threats may contribute to constant stress that can compromise the immune system and cause hypertension and cardiovascular disease, glucose intolerance and insulin resistance, increased susceptibility to infection and inflammation and the death of neurons in the hippocampus and prefrontal cortex (Shonkoff et al. 2012). Social and legal exclusion can become embodied in poor health outcomes, taking a short- and long-term toll on the well-being of the urban poor.

However, when residents of informal settlements gain secure housing, their well-being seems to improve. For example, in Buenos Aires, children living in informal settlements whose families received land titles had better weight-for-height scores compared to households that did not receive titles (Galiani and Schargrodsky 2004). A longitudinal follow-up study revealed that titled families also had reduced household size, lowered poverty and improved children's education (Galiani and Schargrodsky 2010). In Indian cities, lack of formal recognition and residential insecurity often deepens poverty, vulnerability and health risks when it inhibits the urban poor from such things as a ration card, which can provide subsidized food, kerosene for cooking, support formal employment and access to other basic services (Agarwal and Srivastava 2009; Subbaraman et al. 2012).

9.3.3 *Poverty and Employment*

Residents of informal settlements are often among the most impoverished urban residents. Lack of economic resources for slum-dwellers harms human health in a number of ways, including:

- Limiting access to food, water, energy and other essential services
- Increasing vulnerability to eviction and homelessness due to inability to pay rent
- Limiting access to expensive healthcare and medications
- Redirecting resources from education to basic needs
- Forcing the poor into risky informal labour and/or sex work and exposing the poor to toxins and STIs

For instance, many women in informal settlements are home-based workers (HBW) and face acute occupational health hazards due to inadequate seating or work-tables, poor ventilation and exposure to toxic substances that can result in respiratory problems, eye strain, body aches and injuries (Chen and Sinha 2016; Lund et al. 2016). However, employment training with and for the urban poor living in slums can transform health outcomes. For example, the Juventud y Empleo (JE) youth training programme in the Dominican Republic provided life skill and technical training, along with an internship at an approved private sector firm (Ibarrarán et al. 2015). Two years after the project ended, the 2008 initial cohort was significantly more likely than a control group to report having very good health, and there was a 45% drop in female participants' pregnancy rates, as compared to a control group (ibid.).

9.3.4 *Poor Structural Quality of Housing and Overcrowding*

Slum housing is often extremely dense and poorly built with substandard or flammable materials (Haines et al. 2013). Houses built against hillsides are subject to landslides during heavy rain, and inferior building standards cause many thousands of deaths from earthquakes, especially where urbanization and poverty collide (Manda and Wanda 2017). Informal housing is often constructed with low-quality materials, such as scrap-metal doors or roofs that can be easily lifted and removed, while residents may lack adequate locks, gates and other ways of deterring criminal entry (Meth 2017). Dense, often inflammable housing is at high risk of arson, and settlements' lack of access by police and emergency services only compounds local vulnerabilities to crime (Satterthwaite and Bartlett 2017). With cooking, sleeping and living in one unventilated room common in slum housing, residents risk increased respiratory infections, meningitis and asthma and are more likely to contract tuberculosis. Overcrowded housing is also associated with rheumatic heart disease, a chronic and debilitating condition facilitated by increased transmission of group A *Streptococcus pyogenes* infections and lack of early treatment (Riley et al. 2007).

Poor planning and lack of open, recreation spaces are again common features in informal settlements that can adversely influence health. Parks and other green spaces can promote physical activity, thereby helping to prevent cardiovascular disease and other NCDs (Prüss-Ustün et al. 2016). Globally, an estimated 23% of adults and 81% of adolescents are insufficiently active (i.e. engage in less than 150 min of moderate-intensity activity each week), but parks, active transport and higher-density urban planning strategies can promote physical activity and improve quality of life for the urban poor (Cervero 2013).

9.3.5 *Water, Sanitation and Food Security*

Poor water quality and lack of sanitation are leading causes of morbidity and mortality worldwide (JMP 2017), particularly in urban slums. Diarrhoeal diseases are a leading contributor to global child mortality, causing an estimated 20% of all under-5 deaths globally, and in the Global South, 58% of all cases of diarrhoea are attributed to inadequate WASH (water, sanitation and hygiene) (Prüss-Ustün et al. 2016). Many slum-dwellers still struggle with intermittent, low-quality or poorly maintained WASH.

In Nairobi's Mukuru informal settlement, home to over 300,000 people, only 3.6% of households have access to adequate toilets, and 29% had access to adequate water (Corburn et al. 2017). Diarrhoeal diseases caused by inadequate sanitation put children at multiple risks leading to vitamin and mineral deficiencies, malnutrition and stunting (Bartram and Cairncross 2010). Sustained or long-term exposure to excreta-related pathogens—including helminthes or worms—in early life limits cognitive or brain development and lowers long-term disease immunity. As inadequate slum sanitation contributes to the cascading impacts on children of waterborne illness, malnutrition and, in turn, stunting, this can result in poorer cognitive development and performance in school for young people in slums (Niehaus et al. 2002). Restricted toilet opportunities for women have been shown to increase the chance of urinary tract infections (UTIs) and chronic constipation by 80% (Cheng et al. 2012). Poor access to water and sanitation in slums can also contribute to the high prevalence of undernutrition in childhood (especially stunting) alongside elevated levels of overweight/obesity during adulthood, particularly among women (Kimani-Murage et al. 2015).

Many life-threatening infectious diseases are associated with contaminated water in slums, such as cholera and hepatitis. Lack of access to water also restricts water intake and can inhibit cooking, bathing and personal hygiene. Infrequent bathing is associated with scabies and bacterial skin infections, a subset of which (i.e. group A *Streptococcus*) can lead to acute glomerulonephritis.

Women and girls are especially burdened by water collection, as well as being disproportionately affected by inadequate sanitation due to their greater needs for privacy, particularly during menstruation. Women and girls in urban slums face greater risk of sexual assault when walking to use public toilets especially at night (Sommer et al. 2015). Stemming from these gender-inequitable time burdens of collecting water or seeking a private site for open defecation, women and girls may experience long-term declines in human and financial capital if they forego schooling or income-generating activities (Chant and McIlwaine 2016). Inadequate WASH can also imperil women and girls' dignity and self-respect while increasing their feelings of shame and humiliation (Massey 2011).

The lack of adequate toilets also disproportionately impacts women living with HIV. People living with HIV/AIDS are particularly vulnerable to intestinal parasites, since they tend to suffer from more frequent diarrhoeal episodes than those with stronger immune systems. When frequent diarrhoea leads to insufficient nutrient absorption and weight loss, intestinal parasites can be lethal for people living with HIV (West et al. 2012). Frequent diarrhoea can limit the efficacy of antiretroviral (ARV) drugs that can reduce mortality from HIV.

9.3.6 Energy and Air Pollution

Low-income urban households often struggle to access safe, reliable energy sources due to the high costs of electricity, gas and secure household connections and/or storage (Singh et al. 2015). Slum-dwellers frequently rely on informal or makeshift electricity connections and a mix of unclean solid fuels, known as 'fuel stacking' rather than a linear transition up the 'energy ladder' towards clean energy (Sovacool 2011). Globally, exposure to household air pollution (including from unclean cooking fuels) is responsible for 33% of the total disease burden from respiratory infections, and in total, exposure to household air pollution from solid fuels causes an estimated 4.3 million deaths annually (Prüss-Ustün et al. 2016). Furthermore, an estimated 268,000 deaths occur annually as a result of burns (from exposure to fire, heat or hot substances) and the overwhelming majority is in the Global South (ibid.). Indeed, providing access to clean energy (SDG 7) can support respiratory health and reduce fire risks or burns and other accidental injuries (SDG 3), as well as helping to eliminate poverty (SDG 1) and supporting inclusive economic growth (SDG 8). Clean energy can also create significant climate and health co-benefits (SDG 13) and can support gender equality (SDG 5) because women disproportionately utilize unclean fuels for cooking (WHO 2016).

The Global Burden of Disease 2013 study estimated that small particulate matter (PM2.5) was associated with 670,000 premature deaths annually in Africa (globally 5.5 million, of which approximately 90% occurred in low- and middle-income countries), divided roughly equally between indoor and outdoor exposures. The urban poor are more likely to live closer to roadways and polluting industrial facilities thus exposing them to greater levels of pollutants (Kinney et al. 2011). Chronic exposure to elevated levels of toxic air pollution negatively influences not only respiratory and pulmonary health but also contributes to premature births and low birth weights, cancers and premature death (Prüss-Ustün et al. 2016).

PM concentrations are particularly high in Africa's informal settlements (Egondi et al. 2016). Sources of outdoor air pollution in urban slums are mainly dust, burning trash and vehicle and industrial emissions. Due to poor ventilation in these settings, outdoor air pollutants regularly infiltrate into dwellings, raising levels of indoor air pollution. This combination of indoor and outdoor air pollution increases the burden of air pollution in deprived urban areas. For instance, in Nairobi's informal settlements, respiratory illness, asthma and acute respiratory infections are the leading contributors to the mortality burden among children (Egondi et al. 2013).

9.3.7 Transport and Injuries

Road traffic injuries cause an estimated 1.25 million fatalities each year—pedestrians, cyclists and motorcyclists represent almost half of those killed (Prüss-Ustün et al. 2016). Urban transport plans often privilege cars over non-motorized transport (NMT) or public transport. Yet, the urban poor travel on foot or utilize informal para-transit options such as minibuses or rickshaws that vary widely in coverage, safety and performance. Improved planning for NMT can reduce injuries and provide safe spaces for street-level vendors (Khayesi et al. 2010). Public transport and NMT can also enhance air quality, foster neighbourhood vitality and promote physical activity (both for cycling and walking), all of which contribute to reducing non-communicable diseases (WHO and UN-Habitat 2016). Safe paths and roads can foster a sense of place and enhance the social, economic and spatial integration of slums into the entire fabric of the city, all of which are recognized as key social determinants of urban health (Ezeh et al. 2016).

Improved paths or roads can also support access by emergency vehicles and enhance the livelihoods of vendors, manufacturing workshops and others who rely upon streets for income-generating activities (UN-Habitat 2014). Examples from Latin America indicate that improving streets, pavement or public transport can generate multiple health benefits for informal areas. In Acayucan, Mexico, street asphalting in an informal settlement increased residents' wealth and contributed to increased appliance ownership and home improvements (Gonzalez-Navarro and Quintana-Domeque 2016). In Medellín, Colombia, cable cars ('Metrocables') constructed in the poorest Comunas (districts) linked previously segregated and poor areas with the entire city. The results were improved economic status for the poor residents, significant reductions in gun homicides and improvements in social cohesion (Cerda et al. 2012).

9.3.8 Safety and Security

Along with the threat of forcible eviction, many residents of informal settlements face elevated risks of crime or interpersonal violence (IPV). Many informal settlements are excluded from the benefits of formal policing or are subject to harsh military-style occupation. Young men in the favelas of Brazil are up to five times more likely to die from homicide than their urban counterparts (Barcellos and Zaluar 2014). Violence towards women has been associated with the absence of basic services like safe toilets and street lighting (Sommer et al. 2015). Homicide victims are overwhelmingly poor, urban male youths (Muggah 2014). Many informal settlements are excluded from the benefits of formal policing or are subject to harsh military-style occupation by the army.

Women's safety audits in Indian, Latin American and African cities have identified poorly lit areas, dilapidated bus stops or other hotspots that have placed women at elevated risk of assault or crime (Trujillo et al. 2015). These participatory approaches have led to subsequent interventions such as enhanced lighting, paved footpaths and other infrastructures, as well as raising awareness of the need to combat violence against women and other vulnerable groups (ibid.). Additionally, there are recent experiments with utilizing mobile phones to report crimes, prevent violence and improve relations between police and the urban poor. A Smart Policing

Initiative, spearheaded by the Igarapé Institute, created a mobile phone application to promote police transparency and accountability in low-income areas of Brazil and South Africa.[2]

During the Violence Prevention through Urban Upgrading (VPUU) programme in Cape Town's Khayletisha informal settlement, interventions prioritized youth employment, vigorous community participation and social inclusion. The programme used urban design strategies to promote passive surveillance and create integrated community centres called 'Active Boxes' to provide services and overall resulted in a 33% reduction in the murder rate in less than 4 years (VPUU 2011).

9.3.9 *Climate Change Vulnerability*

Climate change will likely exacerbate the health and socioeconomic disparities in cities, with slum-dwellers particularly vulnerable to extreme weather and other projected impacts (Bartlett and Satterthwaite 2016). Slum-dwellers face a set of interrelated health risks from climate change, including increased susceptibility to infectious diseases, food-borne diseases, flooding, sea-level rise and landslides, heat events and drought-induced food insecurity (Scrovronick et al. 2015). Heat stress contributes to cardiorespiratory disease and death due to heat waves. Climate change-induced weather can influence the adverse health impacts of aeroallergens and air pollution. Mosquito and tick-borne diseases, such as malaria and dengue, are known to increase during floods, and higher temperatures reduce the development time of pathogens in vectors and increase potential transmission to humans. Survival of bacterial pathogens is related to temperature. Extreme rainfall can affect the transport of disease organisms into the water supply.

9.3.10 *Available Health Services*

Informal settlements often have an acute dearth of affordable, high-quality healthcare (Shetty 2011). Slums tend to have an inconsistent patchwork of public-, private- and charity-based providers, with many 'informal' healthcare workers. Inadequate or inappropriate care permits the progression of preventable diseases, such as hypertension and diabetes, and increases the risk of drug-resistant infections, such as multidrug-resistant TB. Vaccination coverage for children aged 12–23 months in India's slums was only 23% compared to 60% of the same aged children living in urban India, largely due to inadequate infrastructure and a lack of community awareness and mobilization (Agarwal et al. 2005).

[2] See https://igarape.org.br/en/smart-policing/.

9.4 Improving Health and Equity Through Slum Upgrading

Improving the well-being of those living in informal settlements will require integrated projects that utilize multipronged strategies and strengthen communities, while they remain in place. This is what is commonly referred to as 'slum upgrading', which can combine shelter improvements (adequate housing, water, roads and other infrastructures) with tenure security, political recognition, support for livelihoods and enhanced social services. In situ upgrading (where residents of informal settlements have the right to remain in place) recognize that despite the environmental health risks in informal settlements, these areas can still support well-being, and residents usually benefit from remaining in place. For instance, informal settlements often provide much-needed affordable shelter and proximity to jobs; they may also have dense social networks and vibrant cultural, economic or political institutions. Meanwhile, when areas are declared 'slums', communities may become highly stigmatized, and government responses are typically exclusionary policies or even demolition campaigns that may only entrench residents' poverty. Disparaging generalizations about 'slums' thus downplay the considerable social, economic and cultural opportunities created in most informal settlements, but we argue that these benefits can be strengthened via holistic, in situ upgrading projects. Although governments and aid agencies have supported upgrading projects for decades, these initiatives' major possibilities for health improvement have been ignored (Corburn and Sverdlik 2017; Turley et al. 2013) (Fig. 9.2).

Some national governments have embraced upgrading, as in Thailand's Baan Mankong (Boonyabancha 2009), while other initiatives are spearheaded by local governments such as in Cape Town (VPUU 2011). Available funding is often insufficient to support large-scale upgrading interventions, leading to a plethora of 'boutique' projects that cannot reach all of a city's informal settlements. However, regardless of scale, upgrading projects can 'build citizenship' for marginalized populations and develop innovative co-produced solutions that can improve health Jaitman and Brakarz 2013).

For example, cases from Pakistan and India indicate that community-led sanitation partnerships not only have improved health but also catalysed local empowerment and enhanced governance. Since 1982, the Orangi Pilot Project (OPP) in Pakistani cities has spearheaded low-cost sewerage combining internal initiatives in informal settlements' lanes with external trunk provision by the state (Hasan 2010). In Mumbai and Pune, residents developed a low-cost sanitation model, designed and managed public toilet blocks that have improved multiple determinants of health (SHARE 2016).

Importantly, slum upgrading can be an inclusive process generating vital shelter improvements as well as far-reaching benefits for health, poverty reduction and other Sustainable Development Goals (SDGs). Upgrading not only improves living conditions (housing, infrastructure, services, tenure security, etc.), but it may also influence municipal governance (e.g. by forging participatory partnerships and dialogue with local officials), reduce levels of social stratification and promote resilience to climate change.

Fig. 9.2 Upgrading in Nairobi, Kenya. Source: J. Corburn, used with permission

To help realize these benefits and guide key urban stakeholders (including elected and public health officials, aid agencies, residents and NGOs), we recommend the following principles:

1. Reframe upgrading as a health promotion intervention and regularly collect data to track the health and socioeconomic impacts.
2. Codesign interventions with residents, utilizing participatory processes.
3. Develop inter-sectoral partnerships between elected officials, public health officers, city planners, medical practitioners, shelter and utilities agencies, law enforcement and residents.
4. Promote multisectoral solutions during upgrading.
5. Implement complementary policies, such as revised building standards, greater recognition for informal workers and alternative tenure regularization strategies.
6. Minimize residential displacement, so as to strengthen social capital and local livelihoods.
7. Bolster social and economic assets, including community-based organizations (CBOs), youth groups and other grassroots institutions.
8. Support economic development via improved shelter quality, infrastructure construction and maintenance, trainings and access to credit and other support for local enterprises.
9. Recognize upgrading as a strategic entry point to improve security and promote inclusive urban development.
10. Incorporate for the urban poor and all city dwellers.

9.5 Conclusions and Future Research Needs

Slum health is a highly complex issue, and interventions are often hampered by a lack of reliable population and place-based exposure data in informal settlements (Corburn and Riley 2016). We have summarized the key health issues facing many urban slum-dwellers and emphasized that their root causes are in spatial and material deprivation and pervasive discrimination as well as inadequate or inappropriate urban and transport planning. Poor health in urban informal settlements is not due to the behaviours or lifestyles of the urban poor. Responses to inequitable health outcomes and living conditions in urban slums must not just treat people and send them back into the living and working conditions that are making them sick in the first place. Integrated slum upgrading, when it is a participatory and inclusive process, can be one strategy that addresses immediate population needs, substandard built and social environments as well as deep-seated structural inequities that keep the poor unhealthy (Corburn and Sverdlik 2017). A sector-by-sector or disease-by-disease approach will be inadequate to reduce the unnecessary suffering and premature mortality experienced in urban informal settlements today.

References

African Population and Health Research Centre (APHRC). (2014). *Population and health dynamics in Nairobi's informal settlements: Report of the Nairobi cross-sectional slums survey*. Nairobi: APHRC.

Agarwal, S., & Srivastava, A. (2009). Social determinants of children's health in urban areas in India. *Journal of Health Care for the Poor and Underserved, 20*(Suppl. 4), 68–89.

Agarwal, S., Bhanot, A., & Goindi, G. (2005). Understanding and addressing childhood immunization coverage in urban slums. *Indian Pediatrics, 42*, 653–663.

Barcellos, C., & Zaluar, A. (2014). Homicides and territorial struggles in Rio de Janeiro favelas. *Revista de Saúde Pública, 48*(1), 94–102. https://doi.org/10.1590/S0034-8910.2014048004822.

Bartlett, S., & Satterthwaite, D. (Eds.). (2016). *Cities on a finite planet: Towards transformative responses to climate change*. London: Routledge.

Bartram, J., & Cairncross, S. (2010). Hygiene, sanitation, and water: Forgotten foundations of health. *PLoS Medicine, 7*(11), e1000367.

Boonyabancha, S. (2009). Land for housing the poor—By the poor: Experiences from the Baan Mankong nationwide slum upgrading programme in Thailand. *Environment and Urbanization, 21*(2), 309–329.

Cerda, M., et al. (2012). Reducing violence by transforming neighborhoods: A natural experiment in Medellín, Colombia. *American Journal of Epidemiology, 175*(10), 1045–1053. https://doi.org/10.1093/aje/kwr428.

Cervero, R. (2013). Linking urban transport and land use in developing countries. *Journal of Transport and Land Use, 6*(1), 7–24.

Chant, S., & McIlwaine, C. (2016). *Cities, slums and gender in the global south: Towards a feminised urban future*. London: Routledge. 9780415721646.

Chen, M. A., & Sinha, S. (2016). Home-based workers and cities. *Environment and Urbanization, 28*(2), 343–358.

Cheng, J. J., Schuster-Wallace, C. J., Watt, S., Newbold, B. K., & Mente, A. (2012). An ecological quantification of the relationships between water, sanitation and infant, child, and maternal mortality. *Environmental Health, 11*. http://www.ehjournal.net/content/11/1/4.

Corburn, J., & Riley, L. (2016). *Slum health: From the cell to the street*. Berkeley: University of California Press. ISBN: 9780520281073.

Corburn, J., & Sverdlik, A. (2017). Slum upgrading and health equity. *International Journal of Environmental Research and Public Health, 14*(4), 342. https://doi.org/10.3390/ijerph14040342.

Corburn, J., Asari, M., Makua, J., et al. (2017). *Mukuru situational analysis: Existing conditions in Nairobi's informal settlement*. Berkeley: University of California.

De Snyder, V. N., Friel, S., Fotso, J. C., Khadr, Z., Meresman, S., Monge, P., & Patil-Deshmukh, A. (2011). Social conditions and urban health inequities: Realities, challenges and opportunities to transform the urban landscape through research and action. *Journal of Urban Health, 88*, 1183–1193.

Dye, C. (2008). Health and urban living. *Science, 319*(5864), 766–769.

Egondi, T., Kyobutungi, C., Ng, N., Muindi, K., Oti, S., van de Vijver, S., Ettarh, R., & Rocklöv, J. (2013). Community perceptions of air pollution and related health risks in Nairobi slums. *International Journal of Environmental Research and Public Health, 10*(10), 4851–4868. https://doi.org/10.3390/ijerph10104851.

Egondi, T., Muindi, K., Kyobutungi, C., Gatari, M., & Rocklöv, J. (2016). Measuring exposure levels of inhalable airborne particles (PM2.5) in two socially deprived areas of Nairobi, Kenya. *Environmental Research, 148*, 500–506. https://doi.org/10.1016/j.envres.2016.03.018.

Ezeh, A., Oyebode, O., Satterthwaite, D., Chen, Y. F., Ndugwa, R., Sartori, J., Mberu, B., Melendez-Torres, G. J., Haregu, T., Watson, S. I., et al. (2016). The history, geography, and sociology of slums and the health problems of the people who live in slums. *Lancet*. https://doi.org/10.1016/S0140-6736(16)31650-6.

Fox, S. (2014). The political economy of slums: Theory and evidence from Sub-Saharan Africa. *World Development, 54*(2), 191–203.
Galiani, S., & Schargrodsky, E. (2004, July). *Effects of land titling on child health*. IDB Working Paper No. 194. Available at SSRN: https://ssrn.com/abstract=1814746 or https://doi.org/10.2139/ssrn.1814746
Galiani, S., & Schargrodsky, E. (2010). Property rights for the poor: Effects of land titling. *Journal of Public Economics, 94*, 700–729.
Gilbert, A. (2007). The return of the slum: Does language matter? *International Journal of Urban and Regional Research, 31*, 697–713.
Haines, A., Bruce, N., Cairncross, S., Davies, M., Greenland, K., Hiscox, A., Lindsay, S., Satterthwaite, D., & Wilkinson, P. (2013). Promoting health and advancing development through improved housing in low-income settings. *Journal of Urban Health, 90*, 810–831.
Harpham, T. (2009). Urban health in developing countries: What do we know and where do we go? *Health & Place, 15*, 107–116.
Hasan, A. (2010). *Participatory development: The story of the Orangi Pilot Project-Research and Training Institute and the Urban Resource Centre, Karachi, Pakistan*. Oxford: Oxford University Press.
Home, R. (2013). *Of planting and planning: The making of British colonial cities* (2nd ed.). Abingdon: Routledge. 272 pages.
Ibarrarán, P., Kluve, J., Ripani, L., & Rosas Shady, D. (2015). *Experimental evidence on the long term impacts of a youth training program*. Inter-American Development Bank.
Jaitman, L., & Brakarz, J. (2013). Evaluation of slum upgrading programs: Literature review and methodological approaches. *IDB Technical Note, 60*. https://publications.iadb.org/bitstream/handle/11319/6021/Evaluation%20of%20Slum%20Upgrading%20Programs.pdf?sequence=1.
JMP. (2017). *Progress on drinking water, sanitation and hygiene: Joint Monitoring Programme 2017 update and SDG baselines*. Geneva: WHO and UNICEF.
Khayesi, M., Monheim, H., & Nebe, J. M. (2010). Negotiating "streets for all" in urban transport planning: The case for pedestrians, cyclists and street vendors in Nairobi, Kenya. *Antipode, 42*(1), 103–126.
Kimani-Murage, E. W., Muthuri, S. K., Oti, S. O., Mutua, M. K., van de Vijver, S., & Kyobutungi, C. (2015). Evidence of a double burden of malnutrition in urban poor settings in Nairobi, Kenya. *PLoS ONE, 10*(6), e0129943. https://doi.org/10.1371/journal.pone.0129943.
Kinney, P., et al. (2011). Traffic impacts on PM2.5 air quality in Nairobi, Kenya. *Environmental Science & Policy, 14*(4), 369–378.
Lilford, R. J., Oyebode, O., Satterthwaite, D., et al. (2016). Improving the health and welfare of people who live in slums. *Lancet*. https://doi.org/10.1016/S0140-6736(16)31848-7.
Lund, F., Alfers, L., & Santana, V. (2016). Towards an inclusive occupational health and safety for informal workers. *New Solutions: A Journal of Environmental and Occupational Health Policy, 26*(2), 190–207.
Manda, M., & Wanda, E. (2017). Understanding the nature and scale of risks in Karonga, Malawi. *Environment and Urbanization, 29*(1), 15–32.
Marco Gonzalez-Navarro & Climent Quintana-Domeque. (2016, May). "Paving streets for the poor: Experimental analysis of infrastructure effects," The Review of Economics and Statistics, MIT Press, vol. *98*(2), 254–267.
Massey, K. (2011). Insecurity and shame: Exploration of the impact of the lack of sanitation on women in the slums of Kampala Uganda, LSHTM and SHARE Briefing Note, 11 pages.
Meth, P. (2017). Informal housing, gender, crime and violence: The role of design in urban South Africa. *The British Journal of Criminology, 57*(2), 402–421.
Muggah, R. (2014). Deconstructing the fragile city: Exploring insecurity, violence and resilience. *Environment & Urbanisation, 26*(2), 345–358. https://doi.org/10.1177/0956247814533627.
Niehaus, M., Moore, S., Patrick, P., Derr, L. L., Lorntz, B., Lima, A. A., et al. (2002). Early childhood diarrhoea is associated with di- minished cognitive function 4 to 7 years later in children in a northeast Brazilian shantytown. *American Journal of Tropical Hygiene and Medicine, 66*, 590–593.

Obduho, R. A. (1997). Nairobi: National capital and regional hub. In C. Rakodi (Ed.), *The urban challenge in Africa: Growth and management of its large cities*. Tokyo: UN University Press.
Prüss-Ustün, A., Wolf, J., Corvalán, C., Bos, R., & Neira, M. (2016). *Preventing disease through healthy environments: A global assessment of the burden of disease from environmental risks*. WHO.
Riley, L. W., Ko, A. I., Unger, A., & Reis, M. G. (2007). Slum health: Diseases of neglected populations. *BMC International Health and Human Rights, 7*. https://doi.org/10.1186/1472-698X-7-2.
Satterthwaite, D., & Bartlett, S. (2017). The full spectrum of risk in urban centres: Changing perceptions, changing priorities. *Environment and Urbanization, 29*(1), 3–14.
Scovronick, N., Lloyd, S. J., & Kovats, R. S. (2015). Climate and health in informal urban settlements. *Environment and Urbanization, 27*(2), 657–678.
SHARE (Sanitation and Hygiene Applied Research for Equity (SHARE). (2016). Emergence of community toilets as a public good: The sanitation work of Mahila Milan, NSDF and SPARC in India. http://www.shareresearch.org/research/emergence-community-toilets-public-good
Shetty, P. (2011). Health care for urban poor falls through the gap. *The Lancet, 377*(9766), 627–628.
Shonkoff, J., et al. (2012). The lifelong effects of early childhood adversity and toxic stress. *Pediatrics, 129*(1), e232–e246. https://doi.org/10.1542/peds.2011-2663.
Singh, R., Wang, X., Mendoza, J. C., & Ackom, E. K. (2015). Electricity (in) accessibility to the urban poor in developing countries. *Wiley Interdisciplinary Reviews: Energy and Environment, 4*(4), 339–353.
Sommer, M., Hirsch, J. S., Nathanson, C., & Parker, R. G. (2015). Comfortably, safely, and without shame: Defining menstrual hygiene management as a public health issue. *American Journal of Public Health, 105*, 1302–1311. pmid:25973831.
Sovacool, B. K. (2011). Conceptualizing urban household energy use: Climbing the "Energy Services Ladder". *Energy Policy, 39*(3), 1659–1668.
Subbaraman, R., O'Brien, J., Shitole, T., Sawant, K., Bloom, D. E., & Patil-Deshmukh, A. (2012). Off the map: The health and social implications of being a non-notified slum in India. *Environment and Urbanization, 24*, 643–663.
Sverdlik, A. (2011). Ill health and poverty: A literature review on health in informal settlements. *Environment and Urbanization, 23*, 123–155.
Trujillo, H. R., Siegel, E., Clayton, M., Shapiro, G., & Elam, D. (2015). Securing cities: Innovations for the prevention of civic violence. In R. Ahn, T. F. Burke, & A. M. McGahan (Eds.), *Innovating for healthy urbanization* (pp. 97–122). USA: Springer.
Turley, R., Saith, R., Bhan, N., Rehfuess, E., & Carter, B. (2013). Slum upgrading strategies involving physical environment and infrastructure interventions and their effects on health and socio-economic outcomes. *The Cochrane Database of Systematic Reviews*. https://doi.org/10.1002/14651858.CD010067.pub2.
UN-Habitat. (2014). *Streets as tools for urban transformation in slums: A street-led approach to citywide slum upgrading*, Nairobi.
UN-Habitat. (2016a). *World cities report 2016: Urbanization and development emerging futures*. Nairobi, Kenya: UN-Habitat. http://wcr.unhabitat.org.
UN-Habitat. (2016b). *Slum almanac 2015–2016: Tracking improvement in the lives of slum dwellers*. https://unhabitat.org/slum-almanac-2015-2016/
Violence Prevention Through Urban Upgrading (VPUU). (2011). *A manual for safety as a public good*. VPUU. http://vpuu.org.za
West, B. S., Hirsch, J. S., & El-Sadr, W. (2012). HIV and H2O: Tracing the connections between gender, water and HIV. *AIDS Behavior, 17*, 1675–1682.
WHO. (2016). *Burning opportunity: Clean household energy for health, sustainable development, and wellbeing of women and children*. Geneva: WHO.
WHO and UN-Habitat. (2016). *Global report on urban health: Equitable, healthier cities for sustainable development*. http://www.who.int/kobe_centre/measuring/urban-global-report/en/

Chapter 10
Complex Systems Modeling of Urban Development: Understanding the Determinants of Health and Growth Across Scales

Luís M. A. Bettencourt

10.1 Introduction

Cities and associated processes of growth and change are becoming an increasingly important focus of many convergent scientific traditions, models, and policy (UN-Habitat 2012; Bettencourt 2014).

This surge of interest is driven to a large extent by the fact that urbanization is arguably the most important global event in recent Earth's history. Worldwide urbanization is the defining culmination of the Anthropocene, with human populations increasingly controlling global resource flows from cities (Ellis 2015; Crutzen 2006). This transformation is accompanied by humanity's reaching peak population around the mid-twenty-first century, and its spatial concentration in cities larger than ever before, with Mumbai, Delhi, and other Asian and African cities, is expected to surpass 50 million people over the next few decades (Hoornweg and Pope 2017). At the same time, a necessary transition to sustainable development and conscious planetary stewardship must take place in order to render this process ultimately beneficial for humanity and for nature (Bolay 2012; Brelsford et al. 2017).

For all their ubiquity and exuberance, cities and processes of urbanization have remained poorly understood scientifically. A few scientific traditions, such as sociology, economics, and geography, have developed some scholarship dedicated to cities, but this theme has remained mostly off the mainstream of these and any other disciplines.

A similar situation applies to policy and practice. Cities—for all their importance as engines civilization, innovation, and wealth creation—have remained at the

L. M. A. Bettencourt (✉)
Mansueto Institute for Urban Innovation and Department of Ecology and Evolution, University of Chicago, Chicago, IL, USA

Santa Fe Institute, Santa Fe, NM, USA
e-mail: bettencourt@uchicago.edu

M. Nieuwenhuijsen, H. Khreis (eds.), *Integrating Human Health into Urban and Transport Planning*, https://doi.org/10.1007/978-3-319-74983-9_10

mercy of improvised policy and of utopian or partisan planning practices, which have often generated tragic unintended consequences. Well-informed practices—knowledgeable of the processes taking place in cities and how to best affect them positively—have been rare and usually the province of inspired urbanists and insightful local politicians, not the result of applied scholarship (Jacobs 1992; Lynch 1984).

Besides the increasingly global character of the urban phenomenon, which makes obvious many of the communalities of cities and urbanization across space (Bettencourt et al. 2007), the advent of ubiquitous information technologies are making it possible to observe and measure many more quantities relevant to understanding cities across scales, from individuals to nations. The result of this convergence has been a surge of interest in cities by scientists and technologists as well as by businesses and nonprofit organizations, often under rubrics such as *smarter cities* (Batty et al. 2012; Bettencourt 2014), *resilience* (Bush and Grayson 2009), or *sustainability* (Keivani 2010). However, for these goals to be achieved and for cities to fulfill their promise for human development and sustainable economic growth quickly, much more needs to be understood scientifically.

The obvious question when attempting to model cities is where to start plugging in and connecting the avalanche of current data to models and theory. Cities are extremely complex interconnected environments (Bettencourt and West 2010; Bettencourt et al. 2007), so that any model that focuses on only some aspects of the city—for example, its economy or its built environment—tends to remain descriptive and fail to engage with the true nature of the phenomenon.

Thus, in my view, most scientific opportunity for theory of cities lies at the interconnections between current models. Connecting and mutually constraining current frameworks will in turn necessarily change some of their assumptions and building blocks, creating new and more powerful syntheses. It should also be pointed out that this type of interdisciplinary articulation and synthesis is relatively unusual in the social sciences, which is another reason why urban theory as the nexus of perspectives spanning cognitive, social, economic, political, and cultural aspects of human societies is so transformational and important.

This chapter is dedicated to a brief review of modeling approaches of processes of urban development and of cities more generally. There are, of course, many such approaches, across a range of traditional disciplines. In this light, my main objective is to establish a general road map and emphasize connections between different frameworks and perspectives, rather than provide a deep technical dive into each set of methods. I will emphasize the nature of variables and spatial and temporal scales at which various approaches work well and those at which they are inadequate. I will also attempt to identify where different modeling traditions meet and must better constrain each other.

Health occupies a central role in any attempt at synthesis of urban models. Historically, public health has always been the pioneering field in urban issues, driving much of our understanding of cities and associated policy. Issues of epidemiology, especially related to cholera outbreaks in the nineteenth century (Colwell 1996; Rosenberg 2009), resulted in some of the first quantitative analyses of urban social data. The resulting findings revolutionized medicine by establishing the basis for a science of contagious diseases, associated with pathogen spread mediated by the structure of human social contacts and by urban services and infrastructure (or lack thereof). Issues related to asthma and other respiratory diseases and to conditions associated with poverty and lifestyles, such as crime, diabetes, obesity, or mental health, are at the core of the most difficult urban issues today. Thus, I will illustrate throughout the manuscript how different approaches to modeling cities relate to human public health and how the current frontier in health studies, dedicated to issues of personalized medicine and of lifestyle and exposure over extended time scales, continues to drive at the heart of outstanding question for urban theory and modeling.

I will end with a challenge for the type of theory and models needed to describe urban development fully and the kind of data necessary to assist their construction and testing along the way.

10.2 Models of Cities and Urban Development

There are clearly many ways to navigate the many existing approaches to modeling cities and urbanization, so that any choice is likely to reflect the author's worldview and be necessarily incomplete.

One of my objectives in this piece is to attempt to erase modeling choices based on different disciplines—sociology versus economics, for example—which are artificial in light of the phenomena at hand. Adopting this problem-based approach, I find temporal and spatial scale considerations a useful set of dimensions to organize different kinds of urban approaches; see Table 10.1. Most current models describe cities well in situations involving short-time horizons and/or steady-state spatial equilibria. However, most models fail over the long term, when issues of growth, human development, personal histories, and environmental change come into play.

Table 10.1 The five modeling approaches to cities and urbanization discussed in the paper and their main areas of focus and time scales of applicability

Modeltype	Time scale	Quantities predicted	Results
Agent-based models	Minutes-days	Transportation, city logistics, urban services	Strategic optimization
Spatial equilibrium models	Day-week	Economic output, infrastructure networks, built environment, density, social interactions	Scaling relations (agglomeration effects) for city averages
Contagion models	Days-years	Outbreaks of spreading processes	Critical thresholds, interaction + recovery rates
Life-course models	Human life time	Risk, health, crime	Critical events, trauma, cumulative (dis)advantage
Growth models	Years-decades	Economic, demographic growth	Growth rates, size distributions

Similarly, most models struggle with the immense heterogeneity of cities, in terms of different characteristics of their populations (income, ethnicity, behavior) as well as of places (different quality of services, including environmental quality and access to justice). I start by discussing a number of recent computational models—typically *agent-based*—describing cities over the short term. These are especially useful in considerations of efficiency and optimization, e.g., tied to *smarter cities* policies, as well as traffic management and emergency evacuation situations. Next, I describe *spatial equilibrium* models, which have a long tradition in geography and economics and more recently in complex systems. The contrast between these two types of models provides a few useful lessons about issues of emergence that are not always well codified in agent-based models. I then briefly review the uses of *contagion models*, well known in public health policy but also useful for accounting for some processes of cultural change. I also describe *life-course models*, which take the long view and a cumulative perspective on an agent's behavior and his/her choices and therefore deal with important processes of learning and adaptation. Finally, I briefly discuss the current state of *growth models*, especially for population and economic quantities and contrast these typical large-scale population approaches to the rich individual perspectives captured by life-course models. I close with a discussion of the shortcomings as well as the opportunities ahead regarding the integration and synthesis of these various modeling approaches and progress in underlying urban theory toward creating predictive frameworks for cities and urbanization across scales.

10.2.1 Agent-Based Models Over the Built Environment

Agent-based models (Gilbert 2008; de Marchi and Page 2014) have become increasingly common as a means to model cities (Batty 2007) and other socioeconomic systems (Fig. 10.1). Their use is predicated on the idea that agents (typically people, but in other relevant cases land parcels or institutions) can be endowed with a range of decision rules over a space of actions. For example, an agent may choose different paths to work depending on time of day and current congestion over the road network.

These decisions and actions are in turn functions of those of other agents, so that models are iterated computationally over time to produce collective behavior. The most important feature of these models is the property of *emergence*, that is, that collective behavior may manifest complex patterns not predictable from the simple rules at play at the agent level (Batty 2007; Epstein 1999; Schelling 2006). This idea is well known in complex systems modeling, but it is also a fundamental strategy for a number of traditions in the social sciences, such as economics and sociology, which seek to derive "macrobehavior" from "micromotives" (Schelling 2006).

In the past, agent-based models have been used to model many aspects of cities, from land use expansion to residential choice and from health to traffic. Earlier models used cellular automata (spatial local rule sets) and other simple rules

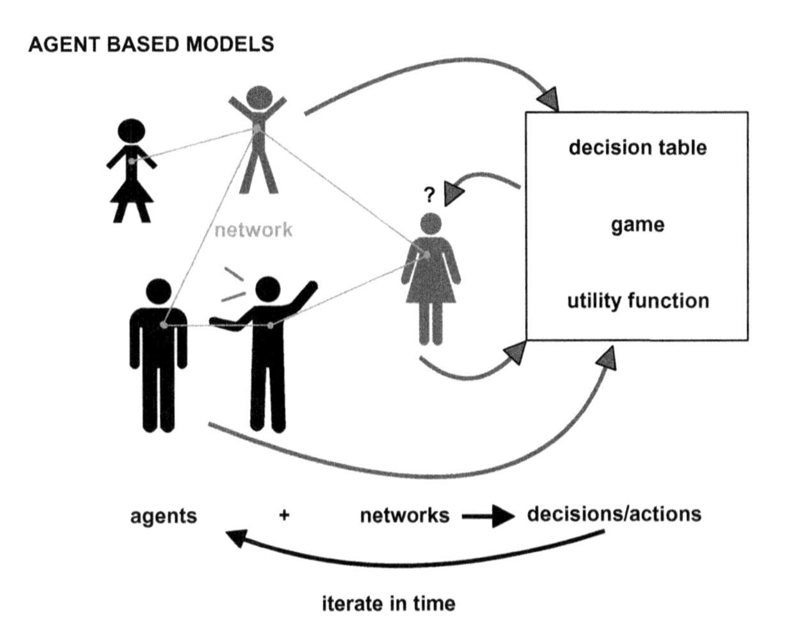

Fig. 10.1 Agent-based models. In these models, agents with a number of given internal states, often connected to each other in structured networks (blue lines) and embedded in space, are updated based on a given set of rules. Update rules vary considerably from model to model and may be based on classical decision or game theory approaches, utility functions, contagion probabilistic rules (see Fig. 10.4), or other models of choice and behavior. In some cases, agents may not be people at all, for example, for models describing changing land uses or urban expansion, where "agents" become land parcels. Agent-based models can be computationally intensive for large systems in structured environments. The essence of these models is the idea of *emergence*, that simple rules at the individual level can generate complex collective behavior as they are iterated over time

(seemingly dissociated from decisions made by people and institutions) to try to understand issues related to city form, such as the spatial growth of cities, the fractal dimension of the built environment (how much of a city's land is paved over), and thus questions of urban form (Batty 2007; Benguigui et al. 2000). While a number of interesting findings expressing the spatial expansion of cities and their densification of land uses with population size were observed and modeled in this way, resulting patterns remained mostly context specific (Benguigui et al. 2000; Shen 2002), and this type of approach eventually lost some steam. For example, the fractal dimension of cities in terms of area filling by built surfaces turns out to be quite variable and dependent of city size and other factors. As a consequence, cities have been found not to be simple spatial fractals. These models remain interesting for exploring the development of spatial patterns of urban expansion from potentially simple local rules, though (Benguigui et al. 2000; Batty 2007; Shen 2002).

Another strand of models important in sociology and urban economics deals with residential choice and neighborhood structure. Much of the modeling history of this field started with the Schelling model of residential choice (Schelling 2006;

Bruch and Mare 2006). Its objective was to attempt to explain why cities are often so strongly spatially segregated by race, especially in the USA. The model proposes a very simple local rule by which individuals with a given trait (race) choose to stay or move neighborhoods so as not to be in a local (neighborhood level) racial minority. Since then the assumptions of the model have been contested and compared with actual choices made by urbanites (Bruch and Mare 2006; Bruch 2014; Browning et al. 2016), showing that transitions to segregation are less abrupt than the model predicts. Nevertheless the Schelling model provided a very simple instance of both emergence (neighborhood segregation) and macrobehavior resulting from very simple micromotives. It is in part responsible for a rich literature on *neighborhood effects* in cities in sociology (Sampson 2012; Ioannides and Topa 2010; Sampson 1997) and remains relevant in light of strong spatial segregation at the neighborhood level in cities around the world not just in terms of race but also related to income, public health outcomes, educational attainment, and other important variables implicated in human development (Bailey et al. 2017; Bruch 2014; Lichter et al. 2012; Brelsford et al. 2017).

A larger range of agent-based models has been associated with human mobility in cities and with traffic in particular (Helbing 2012; Balmer et al. 2006; Batty et al. 2012). These models are typically large computer simulations that take the built space of cities (all roads and buildings) as their "stage" and move people and vehicles in terms of simple local rules. These models often implement and extend older models of traffic in cities, which rely of matching trip originations and destinations with specific flows over the street network. The most important of these models historically was arguably TRANSIMS (Barrett et al. 1995; Nagel et al. 1999; Smith et al. 1995), which replicated the entire population and built environment of several cities, including Portland and Los Angeles, and simulated their daily movement using novel computational algorithms for statistical extrapolation from census data and scheduling in high-performance computing environments (Nagel and Rickert 2001).

Besides traffic simulations, agent-based models have become important tools for public health and emergency management (Epstein 1999). In terms of the latter, traffic related to an evacuation could be simulated and strategies could in this way be tested and to some extent optimized. Such solutions are used in practice for emergency planning by many local and federal agencies in the USA and elsewhere.

One of the most compelling applications of these models has arguably been in health. A version of these models, based on TRANSIMS, called EpiSimS, and many other variants by other mathematical epidemiology groups have become common as a means to include structured social networks based on spatial coincidence over the built environment that may promote effective contagion transmission events (Mniszewski et al. 2008a, b). Conversely, these models were also used to test social distancing, vaccination strategies, and other potential public health interventions and policies. As such, agent-based models are popular in scenario planning and decision support by policymakers (Mniszewski et al. 2008a, b). Although I am less aware of other uses, models of population exposure to localized sources, such as those related to air or water pollution, or with risk associated with buildings can in similar ways be simulated and tested in these type of models.

Some criticisms of these approaches are that (1) they are computationally intensive (early models ran more slowly than real time); (2) they do not tend to produce significantly better predictions, at least in the aggregate, than population level models (see below); and (3) they lack adaptive human behavior and especially behavior change or strategic decision-making.

Thus, these models, despite immense detail, matching each individual and each aspect of the built environment of a given city, typically still lack some of the fundamental features of realistic agent behavior in urban environments, at least in situations where learning, adaptation, and strategic decision-making must come into play.

The future of these models is likely tied to the ability to assimilate real-time data and thus make short-term predictions, updated by observations, rather than being used to predict long-term growth and change. There are many interesting precedents for this sort of model "refresh" via data assimilation, for example, in weather modeling, via Bayesian methods. The enormous and detailed corpus of data available on mobility, traffic, health, and other relevant variables may therefore become the greatest opportunity for the next generation of agent-based models, which can in this way remain more accurate and predictive of real behavior in very specific local contexts.

10.2.2 Spatial Equilibrium Network Models

These are arguably the most successful and mature approaches to quantifying the interconnections between many of the properties of cities (Fig. 10.2). Their principal merit is that they allow us to calculate relationships between various properties of cities and thus predict, for example, economic performance, energy use, or traffic congestion from the population size or area extent of an urban area (Bettencourt and West 2010; Bettencourt 2013).

These models stand in contrast to most computational agent-based models in that they start by expressing certain key macroscopic constraints that lead to (short-term) spatial equilibria. The earliest version of these ideas, to the best of my knowledge, is von Thünen's model of a central market, sometimes known as the *Isolated State* (Fujita et al. 2001). As the name indicates, this is not a model of a city at all, but of a market where agricultural products are sold. His question was how to calculate land rents in the area adjacent to the market, knowing the price of products brought to market and their associated transportation costs. A simple analysis shows that products that are more valuable and that have higher transportation costs (such as fruit) should be grown closer to market, while those that can be more easily transported and fetch lower prices per weight (grain) should be grown further away.

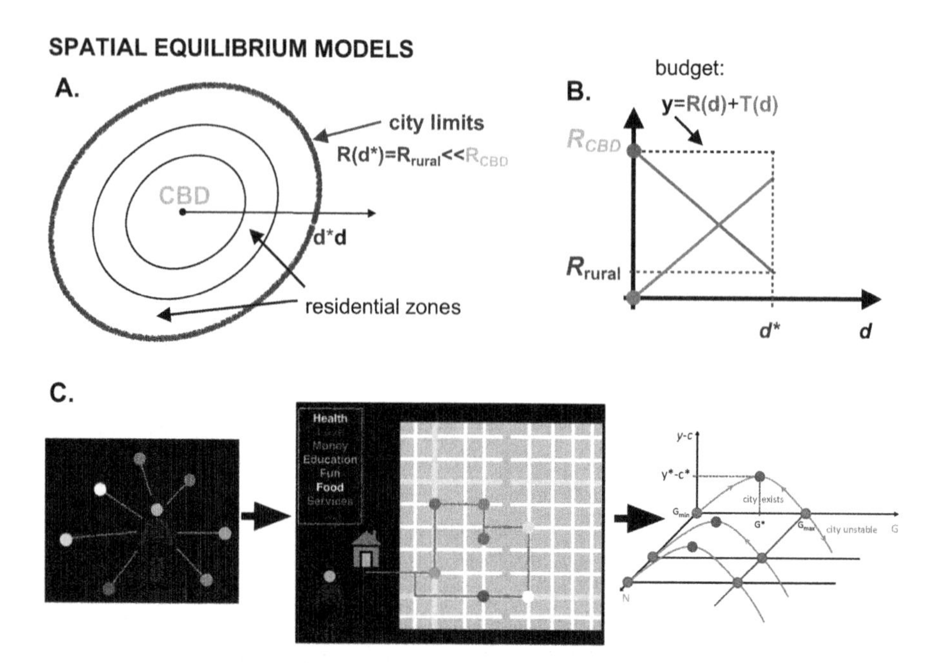

Fig. 10.2 Spatial equilibrium models. Spatial equilibrium models go back a long way, to the first instances of quantitative modeling in urban geography and economics. These models today still describe the most developed socioeconomic urban theory, associated with urban economics and complex systems approaches to cities. Panels **a** and **b** show the essential early ideas (as in the Alonso model; see text) of a city as a spatial area with a dense population concentration. Over this area, individuals earn urban incomes, *y*, which are assumed to be higher (in nominal terms) than rural incomes. They must, in turn, spend part of their budget, *y*, on land rents *R* and transportation *T*, which are functions of distance, *d*, to the central business district (CBD). In the simplest (very unrealistic) models, the city is thus radially symmetric: all people work at the CBD and locate their residences at some distance *d*. City limits are defined as the places were urban land rents equal rural land rents, R_{rural}. In more modern and realistic models, Panel **c**, a person's multidimensional socioeconomic "income" is the result of a set of social exchanges over time embedded in the built space of cities, resulting in a population scale-independent spatial equilibrium (right panel) that can be independent of city size. Such models describe many of the observed aggregate properties of cities accurately, such as a variety of scaling (or agglomeration) relations expressing the size of the economy, the extent of a city's infrastructure, land rents, average building heights, the average speed of walking, and many other quantities as functions of city size

In this way von Thünen created the first model of heterogeneous land use, tied to economic market value and transportation costs (Fujita et al. 2001; Fujita 1990).

In the 1960s, William Alonso inverted this logic and applied essentially the same model to cities (Alonso 1977). He was interested in predicting urban land rents (land value) and in socioeconomic patterns where different kinds of people (rich or poor) may live closer to the center of cities. It is said that he found it paradoxical that in Europe and Latin America richer households live in city centers, while in the USA, they often prefer suburbs and inner cities tend to be poor.

Alonso was able to use von Thünen's logic to characterize a city as an agglomeration of people in space, where the equilibrium between urban and rural land uses determines the edge of the city. This model also made clear that there is a trade-off between higher land rents and higher transportation costs. So, for example, on the same budget, a household working in the city center may prefer to pay higher rents and lower transportation costs and live close to work or vice versa (Alonso 1977). Many elaborations of this model, including preferences (via utilities) that break this symmetry, as well as population heterogeneity and the location of businesses still constitute today the core of most models of urban economics (Glaeser 2008).

These models tend however to have too many parameters, resulting from the introduction in the model specification of utility functions, production functions, transport costs, and other ingredients (Glaeser 2008; Henderson 1991). Most of their predictions are only "qualitatively quantitative," in that they may predict, for example, that lower transportation costs will lead to lower population densities but that the relationship between these urban characteristics remains a function of a (large) number of unspecified input parameters and modeling choices, such as the choice of utility function, which has a status analogous to a force in physics. So, in my view, while Alonso-type models are core to any understanding of cities, much more work needs to be done to understand model ingredients in terms of choices that are both simpler and truer to the character of cities.

Other important equilibrium models deal with spatial agglomeration at the regional level and less with the internal structure of a city. The most famous of these models is the core-periphery model (Fujita et al. 2001), for which Paul Krugman won the Economics Nobel Prize in 2008. To date this remains the only Nobel Prize related to spatial economics or with cities, an issue that boggles the mind given the critical role of cities in shaping economic and social structure as well as problems of development. Krugman's core-periphery model consists of three basic ingredients: (1) a two-sector economy made up of primary producers (farmers) and manufacturers; (2) the existence of transportation costs, which reduce the value of products when moved between point of production and point of consumption; and (3) a "taste for variety" in that consumers (who are also producers) prefer to consume different products. The model has been elaborated by many authors and is the core of modern regional economics and economic geography. Below I give the reader only the essence of the original model (Krugman 1991).

The three main ingredients of the core-periphery model, at least at their most basic, are as follows. First, imagine a very simple economy with only two sectors: food producers tied to the land (spatially immobile) and manufacturers who can locate anywhere and thus can be agglomerated in what would be a city or dispersed to match the density of food producers. Both farmers and manufacturers are also consumers. The question then is how should manufacturers arrange themselves spatially?

The other two ingredients supply the opposing forces that determine such an arrangement. First, Krugman introduced into the model a utility function for consumers that contains a certain kind of increasing returns to scale. In detail, he used a Dixit-Stiglitz model (a common type of constant elasticity of substitution (CES) function). This choice inputs into the model a "force" that is maximized when all products (maximum variety) can be consumed by all people. Note that this is an input to the model and not a derivation: its form is artificial, though a common ingredient of standard economics modeling.

The final ingredient is the inclusion of transportation costs. In the original model, these were included as "iceberg costs" meaning that the value of a product decreases ("melts away") proportionally to distance traveled from production to consumption points. The model is then solved for maximum overall utility of consumption subject to the budgets derived from production and transportation costs, for various spatial configurations of manufacturers.

In this way, Krugman shows that for sufficiently low transportation costs and for a large manufacturing sector relative to farmers, the formation of cities becomes inevitable (Krugman 1991; Fujita et al. 2001). In other words, when transportation costs are sufficiently low, manufacturing can be spatially concentrated minimizing the costs of supplying manufacturing consumers and servicing the periphery from a central point. Conversely, when transportation costs are high, or the sector is small, manufacturers should be distributed to match farmers and their consumption.

The main virtue of the model is to provide a very simple mechanism—grounded on standard ingredients of economic theory—for the formation of cities within a background of dispersed agricultural production, tied to land. In other words, the model gives an economics-based argument for the formation of a differentiated spatial core and periphery within a region that may have started off with a spatially homogeneous distribution of population, production, and consumption.

Most of the assumptions of the model are very simplistic, and many known functions of real cities, such as their capacity for innovation and growth and as nexus of information in human societies, are entirely missing. The model also has little to say on any other urban phenomena, besides the spatial allocation of economic production, viz., consumption, e.g., to do with health, civics, politics, or any other dimensions of human and social development. Important socioeconomic forces, such as land rents, congestion, violence, and challenges of public health, which tend to diminish the advantages of cities are also missing.

For these reasons, the core-periphery model remains a model of regional agglomeration and in some specific ways at odds with the Alonso model of land use in cities, which, for example, emphasizes the role of lower transportation costs (relative to wages) in making cities less dense, not more concentrated. In the Alonso model, this is due to the easing of land rents when more space becomes available and possibly also to a preference for the consumption of more land by each household.

To make sense of this apparent contradiction, it is necessary to go beyond current standard approaches of economic geography and consider the more fundamental role of socioeconomic interactions in human societies as the mediators of many phenomena, economic and otherwise, and to develop more realistic transportation costs, associated with the nature of urban built spaces. These considerations can to a large extent be borrowed with modifications from other modeling and conceptual traditions in sociology and epidemiology (social networks, influence, and contagion), social psychology (trade-offs in behavior between density and opportunity), geography (spatial networks and the built environment), and engineering and physics (transportation planning, efficiency of flows of people, goods, and information).

Recently, these ideas have been converging in the field of complex systems, in interaction with growing amounts of empirical evidence, which has helped inspire, test, and refine quantitative models of cities (Bettencourt 2013; Ortman et al. 2015). The last few years has seen a much more realistic approach to these foundational ideas, using much more data from actual urban systems around the world and modeling that includes more realistic and empirically sound ingredients.

In this work, cities have continued to be conceptualized as steady-state equilibria. However, both sides of the balancing forces have been elaborated and changed (Bettencourt 2013). On the one hand, the idea of a simple given utility function common to all agents baking in increasing returns to scale has been replaced by a more general socioeconomic network expressing the various contacts that each individual has with others over time. These networks can be quite different for different people and express in general the exchange of any socioeconomic quantity, from a pathogen to money and from violence to pro-social civic behavior. The critical insight here is that increasing returns to scale effects observed in economics (e.g., higher average profits for companies in larger cities) are likely due to *network effects*. Network effects are well known in information and communication networks and express the fact that the value of each node's quantities is proportional not to the number of other nodes in the network but to connections, which naively scale with the square of the number of nodes. This *superlinear* effect, when computed via the interactions an individual has, on average, over the more realistic built environment of cities then provides a general approach to understanding the variation of many urban properties with city population size (Bettencourt 2013; Bettencourt et al. 2007; Ortman et al. 2015).

A self-consistent computation of transportation costs over the transportation networks of cities provides us with the expectation that overall transportation costs and socioeconomic outputs vary with city size with the same superlinear dependence on population, and thus the latter remain a fixed fraction of the former, as assumed typically in the Alonso model.

This work continues to expand and improve, based on the more detailed study of urban built spaces, including transportation networks (Barthélemy 2011; Batty et al. 2012; Brelsford et al. 2017), building forms, etc., and resulting from the comparative analysis of economic performance, innovation systems, health outcomes, and many other fundamental quantities characterizing cities, which can now be studied in urban systems worldwide. Such studies have also been systematically applied to urban systems throughout history (Ortman et al. 2015, 2014, 2016; Cesaretti et al. 2016), which provide a unique and sometimes simpler window into many urban phenomena.

The main advantage of these spatial equilibrium network models is that they produce predictable, empirically observed relationships for many important urban quantities, including number and intensity of social interactions, mobility, and qualities of the built environment, all of which are implicated in health outcomes, as well as all other socioeconomic outputs. Related to contagion models, see below, the predicted increase in socioeconomic interactions per capita in larger cities predicted by these models, and observed, e.g., in cell phone networks (Schlapfer et al. 2014), implies that disease outbreaks may typically have larger reproductive numbers, leading to faster spread in larger urban populations.

Going forward, spatial equilibrium network models need to be able to increasingly deal with quantities beyond averages and provide a true statistical picture of cities, including issues of heterogeneity and inequality and how they play across different scales, from neighborhoods to cities and urban systems. Fast progress in the last few years in this direction bodes well for these much needed developments.

10.2.3 Contagion Models

Perhaps the best-known type of models among those described here, especially to readers from public health, are contagion models (Fig. 10.3).

These models typically describe a population of individuals who are spatially bound and in interaction with each other. These interactions may be "well-mixed" or structured in social networks. Individuals belong typically to classes that describe their condition, viz., a particular issue, for example, a communicable disease such as influenza or HIV. They may be susceptible, exposed, infected, or recovered relative to this condition and may progress in this order from class to class as a result of contacts with other infected individuals and due to the interaction of the pathogen with the host's immune system. This type of models, with appropriate modifications, has also been commonly been used to account for the spread of information in human societies, such as ideas, fashions, panics, and so on.

In the last few years, due to increasing interest in the network structure of human societies and built spaces (transportation networks), there has been a shift toward the use of probabilistic computational models of contagion over structured networks, over more aggregate "compartment" models, which assume that populations can in principle interact all-to-all.

This spatialization and personalization of contagion models also allows for mode detailed studies of groups at risk and factors associated with space, such as air pollution, lead poisoning, socioeconomic factors, heterogeneous services, and so on. Most such models are thus a type of agent-based model, where updates depend on the agent's state, viz., spreading quantities in the population. The simplest version of such models is the Reed-Frost model (Britton 2010; Newman 2002). In this implementation, the state of an agent in a network of other agents is updated probabilistically based on the current state of the agent and its neighbors over the network. So, for example, a susceptible individual will become exposed and may become infected if any of its neighbors are infectious, with a given probability. Updating the state of all agents in this way yields a population-level dynamics. Structured populations have some interesting properties, in that parameters predicting the expected number of neighbors being infected by an infectious agent (the reproductive number or

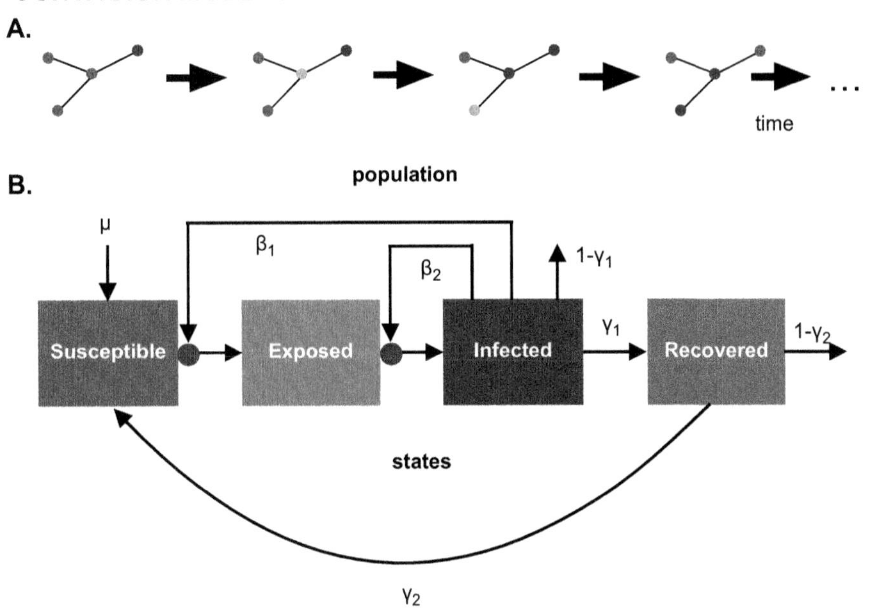

Fig. 10.3 Contagion models. Contagion models are an important set of tools to describe the spread of communicable diseases in human populations. They have also been used extensively to model the spread of information, at least in relatively simple cases. These models can be seen as a particular instance of agent-based models, where individuals are represented by their epidemic state (susceptible, exposed, infected, recovered, etc.) and transition between these states given a set of rules iterated over time. These models may deal with a population divided into epidemic states and assumed to be well mixed (all interactions between individuals are equally probable, panel **b**), or they may be described by probabilistic rules over structured networks (panel **a**), embedded in physical space. Rates describing the progression of individuals through epidemic states (shown as Greek letters) are usually given as inputs to models. These models can also be used as ways to measure parameters, given data, using standard regression or more sophisticated Bayesian inference methods

branching ratio for the process) depend not only on the average number of neighbors (or degree) of the network but also on its variance (Castellano and Pastor-Satorras 2010) and possibly on higher moments. This is a general characteristic of multiplicative random processes but has special importance for public health in situations when populations may be very heterogeneous in terms of risks and contacts. Successful interventions at bringing the process of spread below threshold can then be achieved by vaccinating or isolating super-spreaders (individuals with the largest number of contacts) (Castellano and Pastor-Satorras 2010).

Despite the growing importance of computational, probabilistic models of contagion, simpler population models still have a role to play and permit in general much simpler, analytic predictions for important quantities such as the rate of spread of a contagious agent and the estimation of thresholds for contagion and the effect of any policy intervention.

Like agent-based models, the greatest criticism of contagion models, networked or not, is the rigidity of some of their parameters in situations when, either due to policy interventions or to behavior change, their values is expected to change. Examples are "distancing" public policies aiming at reducing effective contacts between individuals, such as quarantines or "stay at home" curfews as those implemented in Mexico City recently, during the outbreak of swine flu in 2009. Similarly, it is to be expected that individuals aware of upcoming outbreaks will change behavior spontaneously, by seeking vaccination, wearing masks, and other precautions that reduce their risk of contracting or spreading the disease.

As in other agent-based models of human environments, a general trend has been to try to mitigate these issues by modeling contagion parameters over time or by estimating them using real-time data. Bayesian methods for data assimilation that, for example, adapt contact rates in compartment models have been developed and demonstrated to be useful in this context (Bettencourt and Ribeiro 2008). Nevertheless, the ultimate solution to these issues and to better models of contagion hinges on developing more sophisticated social and cognitive modeling approaches that include social behavioral changes in the presence of outbreaks of both biological and informational quantities.

10.2.4 Life-Course Models

Life-course models are approaches to modeling the state of agents in a society (usually a specific person) based on their internal states, viz., their environment and a set of discrete events over time (Willekens 1999; Elder 1998). The central idea of these models is that the state of an agent at the present time and its future behavior and decisions are a function of its personal history. So, for example, an individual may become richer by keeping her job, being occasionally promoted, and winning the lottery. All of these events reflect discrete choices and chance events that build up an individual's characteristic (wealth) over the life-course (Fig. 10.4).

For this reason, life-course models are typically very different in character from other modeling approaches discussed in this paper. They emphasize the uniqueness and contingency of an individual's personal history, and are consequently aimed at accounting for and explaining diverging outcomes for different individuals over time, even as they share the same broad environment, such a living in the same neighborhood or the same city (Elder 1998; Murray et al. 2011; Sampson and Laub 1990).

Most life-coarse models are not especially quantitative. Their emphasis is typically on either special events that may trigger negative outcomes for particular individuals (Murray et al. 2011; Warr 1998)—such as youth incarceration—or on the (sometimes slow) accumulation of events (e.g., lead exposure) that may lead to a later condition. For this reason, life-course models have been found so far to be

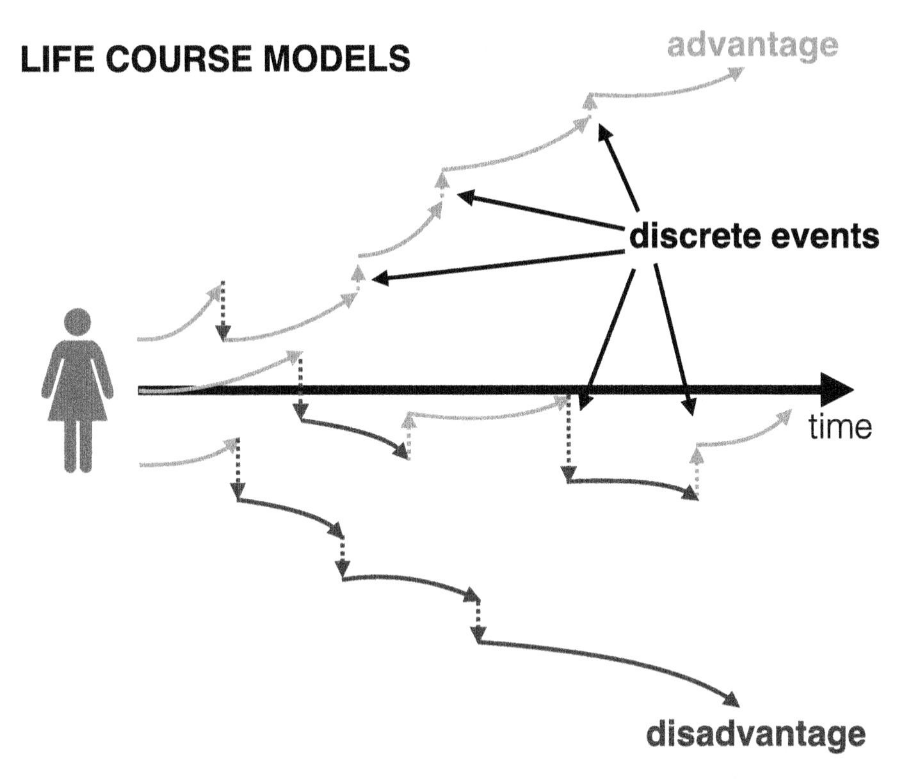

Fig. 10.4 Life-course models. Life-course models represent an agent's life as a series of events that affect their internal states and accumulate over time. Events typically are seen as triggering positive (green) or negative (red) outcomes that are statistically dependent on each other over time, possibly generating trajectories characterized by cumulative advantage or disadvantage. Most interesting models are multidimensional and describe how, say, progress in education and training may lead to improved socioeconomic status or how delinquency and interactions with the criminal justice system may lead to cumulative socioeconomic or health degradation. Thus, life-course models can be seen as another type of agent-based model but one in which the accumulation of events is *statistically dependent* over time. For these reasons life-course models bridge short-term agent-based models (Figs. 10.1 and 10.3) and models of growth and development at the population level (see Fig. 10.5)

useful in studies of criminology, especially cohort analysis—and in personalized medicine, associated with environmental or lifestyle conditions.

One of the most productive ideas arising from these models is that of *cumulative (dis)advantage* (Hannon 2003; O'Rand 1996; Shuey and Willson 2008), where vicious or virtuous cycles of decisions and events that feedback on each other tilt someone's life course in ever more negative or positive outcomes. A familiar example may be in the area of youth delinquency, where instability at home may lead to school absences, and petty criminal behavior, which may then escalate as an individual faces challenges catching up at school and acquires a bad reputation for trouble.

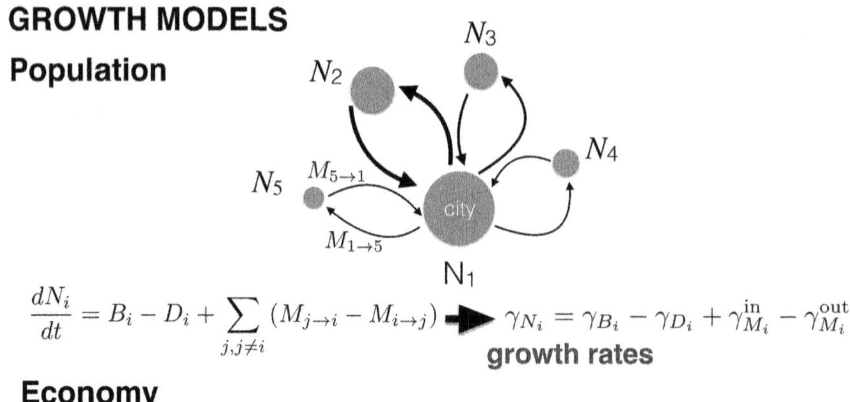

Output = Information X Labor X Capital

$$Y(I, N, K) = A(I) N^{\alpha} K^{1-\alpha} \quad \rightarrow \quad \gamma_Y = \gamma_A + \alpha \gamma_N + (1-\alpha)\gamma_K$$

growth rates

Fig. 10.5 Growth models. Growth models describe how populations (such as a city) grow both demographically (upper panel) and in other ways, for example, economically (lower panel). Demographic models rely on strict accounting and conservation laws, so that the number of people in a given city is simply the result of processes of birth, death, and migration. Economic models invoke, in addition, concepts that rely on a deeper understanding of how value is created. Economic growth models place an especially important emphasis on knowledge or information, I, as the driver of increases in economic productivity. This is often written in terms of a prefactor (total factor productivity, $A(I)$, a function of knowledge) that multiplies traditional production factors—labor and capital—in a given production function such as a Cobb-Douglas form, shown. All models of growth express processes of change in terms of growth rates, shown here as γ, with subscripts corresponding to each quantity. Computing these growth rates from more fundamental and predictive theory remains the greatest challenge for these models. Substantial progress in measuring growth rates empirically—at different scales and in a variety of situations—is, however, creating new opportunities for modeling and theory describing growth in cities

Another set of issues are health challenges related to poverty, where someone's physical environment may create health outcomes that are debilitating of their economic performance or acquisition of human capital. This, in turn, tends to throw the person into more unhealthy situations that compound their sinking socioeconomic status and so on. Examples of people doing well (virtuous cycles) in all these fronts are also easy to imagine and document, but are less often analyzed.

Life-course models are critically important in explaining heterogeneity of outcomes in human societies (Shuey and Willson 2008; Elder 1998; Murray et al. 2011). I expect that they will gain much greater prominence in the near future as issues of inequality in health, wealth, and opportunity receive more attention and become more amenable to quantitative statistical analysis. The challenge for these models is to square what is specific to each personal history with the identification of broad patterns of risk and prediction in order to nudge behavior and design environments that produce positive and holistic long-term outcomes for every individual. Data on health, education, and wealth over the life course of individuals is becoming increasingly more available, leading to many opportunities to also make these models more mathematical and quantitative.

10.2.5 Growth Models

Urban models that involve cumulative change in time, over the long term, assume a very different character depending on the quantity at play and the time scales involved (Fig. 10.5).

This is true even within health, where a fast contagious disease such as influenza can be well described quantitatively by standard epidemiological contact models over a few days, as we discussed above. Contrast that to a condition such as diabetes or heart disease, which requires taking into account many accumulating factors through a person's life course over many decades. Or indeed to how do households or businesses manage their cash flow every few weeks, while modern urban economies double in size every a few decades?

Bridging the gap between short-term and long-term change and development in cities remains a general challenge. As, we have seen, models of short-term dynamics are typically more developed because the dynamics they describe can be observed more easily in practice and even tested by formal experiments and data. Long-term models, however, necessarily involve myriads of events that sometimes accumulate change and sometimes not, both at the individual level and for the entire population.

Most growth models to date have focused on quantities that are measured consistently and/or are of particular importance to human societies. Two salient examples are population and the size of the economy (measured as gross domestic product).

Population growth models illustrate in a straightforward way both what is well known and what remains open in modeling processes of growth. Imagine the population of a given city (or any other entity with clear boundaries). Given its population today, we can readily compute its population at a future date by adding up births with in-migration and subtracting deaths and out-migration. This is pure accounting and is practiced extensively in population biology and human formal demography.

What remains difficult is the modeling of the various rates associated with change. Indeed, births, deaths, and migration can be defined on a per person basis, which makes modeling their overall accounting very simple. But how should we compute birth rates? Or death rates? Or indeed rates of migration from one city or town to another?

A gigantic, mostly empirical, literature exists on each of these topics, though not always an urban-rural transition or at the level of different cities (Dye 2008; Coale 1989; Montgomery 2008). Changes in fertility are known, for example, to be strongly dependent on urban to rural transitions, human development, and especially the status of women (Montgomery 2008; Coale 1989; Adsera 2005; D'Addio and d'Ercole 2005; McDonald 2000). It is widely observed that urbanization, development, and gender equity (usually measured simply as women's labor participation) are associated with lower birth rates per capita and with greater investment in the health and education of each child. However, these findings are more qualitative than quantitative (Becker and Barro 1988; D'Addio and d'Ercole 2005). While structural factors are certainly correlated to fertility choices, they don't give a recipe for reversing population trends, which are systemic: many societies experiencing very low birth rates struggle to enact policies that promote increases in fertility, once their demographic transition has set in (Frejka et al. 2008; Mcdonald 2006).

Death rates are generally much more predictable, and a result principally of public health policies, broadly defined. This may be simply the result that most people would rather not die and that it is biology and health that primarily influences death rates (Coale 1989). For example, over the time span of a single century, progress in epidemiology has all but eliminated deaths from contagious diseases in developed cities (Armstrong et al. 1999). Progress in the treatment of cardiovascular diseases has also recently reduced related deaths (Mathers and Loncar 2006). The expansion of these developments to entire populations including poorer segments of the population and those most at risk is also typical of the recent history of most cities and nations. Its results are substantially lifting longevity by as much as 20 years in some cases, from 1960s to 1980s (Oeppen and Vaupel 2002). In this way, death rates are knowable, and well modeled by current empirical trends, given data and a sufficiently strong and effective public health system (Lichtenberg 2002).

Finally, migration rates remain hard to predict in detail, while broad empirical trends have been known for a long time (Greenwood 1997; Molho 1986). Migration flows between cities tend to be reasonable well modeled by gravity-type models, where population flows are proportional to the population of each city divided by a factor that grows with increasing distance between the two places (Buch et al. 2004). It has become increasingly clear though that people migrate for many interconnected reasons, from tangible jobs and opportunities to general aspirations, which make most current models of migration decisions—as of other decisions reviewed here—quite inadequate (Greenwood and Hunt 2003; Molho 1986; Zhao 2003). Nevertheless, aggregate migration flows between cities are quite predictive of their future values, with slow decadal trends modulating these values, so that reasonable practical estimations of migration at least at the level of entire cities are possible.

What these considerations imply is that the prediction of growth rates remains in general a challenge for models of urban change and development. Deep dives into the processes involved are necessary and require knowledge of context, life history, and a sense of strategic behavior on the part of agents and populations. This problem becomes much worse when quantities other than population are considered.

For example, current models of economic growth emphasize innovation and knowledge in setting up growth rates (Romer 1986, 1989; Jones 2005). The central idea of these so-called models of *endogenous economic growth* is that, controlling for flows of labor and capital (i.e., essentially at fixed human population), it is the expansion of knowledge that can add economic value to existing production (Acemoglu 2009; Jones 2005). In other words, it is the improvement in the recipe for the use of labor and capital that creates additional value. While cities and urbanization are the context in which economic growth always takes place, cities are a necessary but not sufficient condition for economic expansion (Glaeser et al. 1995; de Briggs et al. 2015; UN-Habitat 2003). Clear examples abound as specific cities cease sometimes to grow economically if their dominant economic sectors become uncompetitive, such as those dedicated to large-scale manufacturing of cars, steel, coal, etc. Entire nations, such as Brazil and to a lesser extent the USA, have experienced periods in time with no expansion of GDP despite the population growth of cities.

Current models to calculate economic growth rates, usually at the national level, are rather simple in their attempts to incorporate knowledge (Acemoglu 2009; Barro and Sala-i-Martin 2003). This leads to a number of counterfactual properties, such as strong scale dependencies on population size (or innovator population size), which are not observed in practice (Jones 1999; Jones and Romer 2009). A recent focus of interest has been on the role of institutions and culture in propitiating economic growth (Acemoglu 2009; Storper 2013), but again there are some obvious caveats to these ideas when we consider which cities or nations are growing the fastest at the moment worldwide and their institutions. It is also very difficult to translate these complicated issues into a calculus of a single growth rate for the economy.

Another associated issue is that growth rates are stochastic quantities, so that we would like to have a statistical theory that predicts not only their expected value over some time period but also its higher moments, starting with their variance (known as volatility) (Peters and Gell-Mann 2016). Such a theory would need to express many of the properties of the various models discussed above and play out at different scales from individuals to cities and urban systems. All this remains a challenge for socioeconomic theory, though many glimmers of hope and much empirical evidence exist now that may produce better theory in the near future.

It is perhaps important to clarify why—despite apparently strong growth rates—models of stationary equilibrium and other approaches appropriate for short-term modeling have remained relevant. A growth rate for population or an urban economy of, say, 3% sets the typical temporal scale for the system to change as a whole. A 1% growth rate means that a system's size doubles in 72 years, so 3% would imply a doubling in 24 years. Equilibrium models—and indeed epidemic and agent-based approaches—aim to describe a city over characteristic time periods of days or weeks and thus are perfectly appropriate in the short term. It is this strong separation of time scales, between daily hustle and change over a lifetime, that allows us to use many of the models discussed here in practice, even while theory and modeling developments on the time scales of decades remain underdeveloped.

10.3 Outlook

In this manuscript, I have discussed the state of the art of leading approaches to modeling cities and urban development. My objective was to introduce the reader to a number of very diverse approaches, which originate in very distinct intellectual traditions and that aren't typically compared and contrasted against each other. Each of these traditions has a large and detailed literature, which is not reviewed here in any detail. I hope the curious reader finds this paper a good departing point to further reading, though.

The picture emerging from these diverse traditions shows how we have *good short-term models* that capture the essential character and dynamics of socioeconomic processes, social interactions, and epidemiology in cities, over their built environment. Much of the progress in the last few years in this front has been achieved through the uses of more data and computation, which can render these models more detailed and testable, as well as applicable to their context of interest.

Over the medium term and over the scale of entire cities, more complex interdisciplinary models of cities as self-consistent social and infrastructural networks over built spaces are also emerging, thanks to comparative evidence from urban systems throughout the world and across history. Critical to these developments has been the perspective from complex systems that social and economic processes must be consistent with each other and constrained realistically by the characteristics of actual urban built environments. This has allowed ideas of sociology, economics, social psychology, engineering, physics, and health to converge into a predictive quantitative scaling theory of cities, which derives the average dependence of urban quantities on city size.

Challenges of making these ideas and models more realistic and predictive remain on at least three fronts, specifically (1) accounting for heterogeneity and inequality among agents across scales, (2) deriving rates of change and growth for urban quantities, and (3) wrapping these objectives into a statistical theory capable of predicting not only average dynamical outcomes but also associated higher moments.

These challenges are well illustrated by the current state of life-course models as well as theories of urban growth. Bridging these two approaches—the rich path-dependent perspective of events and agency of real people in cities to aggregate and predictable outcomes for the population—encapsulates the central challenge ahead.

For public health this convergence of a dynamical population perspective with individuals histories is not only natural but the focus of most compelling current advances, tied to issues of violence, poverty, lifestyles, and environmental factors that bridge together aspects of urban social and physical environments over the life course of individuals (Dye 2008; Corburn 2004; Galea and Vlahov 2005).

References

Acemoglu, D. (2009). *Introduction to modern economic growth*. Princeton: Princeton University Press.

Adsera, A. (2005). Vanishing children: From high unemployment to low fertility in developed countries. *The American Economic Review, 95*(2), 189–193.

Alonso, W. (1977). *Location and land use: Toward a general theory of land rent. 6. Printing*. Cambridge, MA: Harvard University Press.

Armstrong, G. L., Conn, L. A., & Pinner, R. W. (1999). Trends in infectious disease mortality in the United States during the 20th century. *JAMA, 281*(1), 61–66. https://doi.org/10.1001/jama.281.1.61.

Bailey, N., van Gent, W. P. C., & Musterd, S. (2017). Remaking urban segregation: Processes of income sorting and neighbourhood change: Remaking urban segregation. *Population, Space and Place, 23*(3), e2013. https://doi.org/10.1002/psp.2013.

Balmer, M., Axhausen, K., & Nagel, K. (2006). Agent-based demand-modeling framework for large-scale microsimulations. *Transportation Research Record: Journal of the Transportation Research Board, 1985*, 125–134. https://doi.org/10.3141/1985-14.

Barrett, C., Berkbigler K., Smith L., Loose V., Beckman R., Davis J., Roberts D., & Williams M. (1995). *An operational description of transims*, June. https://trid.trb.org/view.aspx?id=460374.

Barro, R. J., & Sala-i-Martin, X. I. (2003). *Economic growth* (2nd ed.). Cambridge, MA: The MIT Press.

Barthélemy, M. (2011). Spatial networks. *Physics Reports, 499*(1–3), 1–101. https://doi.org/10.1016/j.physrep.2010.11.002.

Batty, M. (2007). *Cities and complexity: Understanding cities with cellular automata, agent-based models, and fractals*. Cambridge, MA: MIT Press.

Batty, M., Axhausen, K. W., Giannotti, F., Pozdnoukhov, A., Bazzani, A., Wachowicz, M., Ouzounis, G., & Portugali, Y. (2012). Smart cities of the future. *The European Physical Journal Special Topics, 214*(1), 481–518. https://doi.org/10.1140/epjst/e2012-01703-3.

Becker, G. S., & Barro, R. J. (1988). A reformulation of the economic theory of fertility. *The Quarterly Journal of Economics, 103*(1), 1–25. https://doi.org/10.2307/1882640.

Benguigui, L., Czamanski, D., Marinov, M., & Portugali, Y. (2000). When and where is a city fractal? *Environment and Planning B: Planning and Design, 27*(4), 507–519. https://doi.org/10.1068/b2617.

Bettencourt, L. M. A. (2013). The origins of scaling in cities. *Science, 340*(6139), 1438–1441. https://doi.org/10.1126/science.1235823.

Bettencourt, L. M. A. (2014). The uses of big data in cities. *Big Data, 2*(1), 12–22. https://doi.org/10.1089/big.2013.0042.

Bettencourt, L. M. A., & Ribeiro, R. M. (2008). Real time Bayesian estimation of the epidemic potential of emerging infectious diseases. *PLoS One, 3*(5), e2185. https://doi.org/10.1371/journal.pone.0002185.

Bettencourt, L., & West, G. (2010). A unified theory of urban living. *Nature, 467*(7318), 912–913. https://doi.org/10.1038/467912a.

Bettencourt, L. M. A., Lobo, J., Helbing, D., Kuhnert, C., & West, G. B. (2007). Growth, innovation, scaling, and the pace of life in cities. *Proceedings of the National Academy of Sciences, 104*(17), 7301–7306. https://doi.org/10.1073/pnas.0610172104.

Bolay, J.-C. (2012). What sustainable development for the cities of the south? Urban issues for a third millennium. *International Journal of Urban Sustainable Development, 4*(1), 76–93. https://doi.org/10.1080/19463138.2011.626170.

Brelsford, C., Lobo, J., Hand, J., & Bettencourt, L. M. A. (2017). Heterogeneity and scale of sustainable development in cities." *Proceedings of the National Academy of Sciences*, May, 201606033. https://doi.org/10.1073/pnas.1606033114.

Briggs, de S., Xavier, R. P., & Rubin, V. (2015). *Inclusive economic growth in America's cities: What's the playbook and the score?*. SSRN Scholarly Paper ID 2621876. Rochester, NY: Social Science Research Network. https://papers.ssrn.com/abstract=2621876.

Britton, T. (2010). Stochastic epidemic models: A survey. *Mathematical Biosciences, 225*(1), 24–35. https://doi.org/10.1016/j.mbs.2010.01.006.

Browning, C. R., Cagney, K. A., & Boettner, B. (2016). Neighborhood, place, and the life course. In M. J. Shanahan, J. T. Mortimer, & M. K. Johnson (Eds.), *Handbook of the life course, Handbooks of sociology and social research* (pp. 597–620). New York: Springer International Publishing. http://link.springer.com/chapter/10.1007/978-3-319-20880-0_26.

Bruch, E. E. (2014). How population structure shapes neighborhood segregation. *American Journal of Sociology, 119*(5), 1221–1278. https://doi.org/10.1086/675411.

Bruch, E. E., & Mare, R. D. (2006). Neighborhood choice and neighborhood change. *American Journal of Sociology, 112*(3), 667–709. https://doi.org/10.1086/507856.

Buch, C. M., Kleinert, J., & Toubal, F. (2004). The distance puzzle: On the interpretation of the distance coefficient in gravity equations. *Economics Letters, 83*(3), 293–298. https://doi.org/10.1016/j.econlet.2003.10.022.

Bush, W., & Grayson, M.. (2009). *Critical infrastructure resilience final report and recommendations*. .National Infrastructure Advisory Council. http://www.dhs.gov/xlibrary/assets/niac/niac_critical_infrastructure_resilience.pdf.

Castellano, C., & Pastor-Satorras, R. (2010). Thresholds for epidemic spreading in networks. *Physical Review Letters, 105*(21), 218701. https://doi.org/10.1103/PhysRevLett.105.218701.

Cesaretti, R., Lobo, J., Bettencourt, L. M. A., Ortman, S., Smith, M. (2016). *Population-area relationship in medieval European cities*. Santa Fe Institute. Retrieved October 20, 2015, from http://www.santafe.edu/research/working-papers/abstract/6c15cd3a937adc784c0ed8385e1be3ad/.

Coale, A. J. (1989). Demographic transition. In *Social economics* (pp. 16–23). London: The New Palgrave. Palgrave Macmillan. https://link.springer.com/chapter/10.1007/978-1-349-19806-1_4.

Colwell, R. R. (1996). Global climate and infectious disease: The cholera paradigm. *Science, 274*(5295), 2025–2031.

Corburn, J. (2004). Confronting the challenges in reconnecting urban planning and public health. *American Journal of Public Health, 94*(4), 541–546. https://doi.org/10.2105/AJPH.94.4.541.

Crutzen, P. J. (2006). The 'Anthropocene'. In *Earth system science in the anthropocene* (pp. 13–18). Berlin, Heidelberg: Springer. https://doi.org/10.1007/3-540-26590-2_3.

D'Addio, A. C., & d'Ercole, M. M. (2005). *Trends and determinants of fertility rates: The role of policies*. 27. OECD social, employment and migration working papers. OECD Publishing. https://ideas.repec.org/p/oec/elsaab/27-en.html.

Dye, C. (2008). Health and urban living. *Science, 319*(5864), 766–769. https://doi.org/10.1126/science.1150198.

Elder, G. H. (1998). The life course as developmental theory. *Child Development, 69*(1), 1–12. https://doi.org/10.1111/j.1467-8624.1998.tb06128.x.

Ellis, E. C. (2015). Ecology in an anthropogenic biosphere. *Ecological Monographs, 85*(3), 287–331. https://doi.org/10.1890/14-2274.1.

Epstein, J. M. (1999). Agent-based computational models and generative social science. *Complexity, 4*(5), 41–60. https://doi.org/10.1002/(SICI)1099-0526(199905/06)4:5<41::AID-CPLX9>3.0.CO;2-F.

Frejka, T., Hoem, J., & Sobotka, T.. (2008). *Childbearing trends and policies in Europe*. BoD – Books on Demand.

Fujita, M.. (1990). *Urban economic theory: Land use and city size*. 1, paperback ed. Cambridge: Cambridge University Press.

Fujita, M., Krugman, P., & Venables, A. J. (2001). *The spatial economy: Cities, regions, and international trade*. Cambridge, MA: The MIT Press.

Galea, S., & Vlahov, D. (2005). URBAN HEALTH: Evidence, challenges, and directions. *Annual Review of Public Health, 26*(1), 341–365. https://doi.org/10.1146/annurev.publhealth.26.021304.144708.

Gilbert, N.. (2008). *Agent-based models*. SAGE.

Glaeser, E. L. (2008). *Cities, agglomeration, and spatial equilibrium. Lindahl lectures*. Oxford: Oxford University Press.

Glaeser, E. L., Scheinkman, J. A., & Shleifer, A. (1995). Economic growth in a cross-section of cities. *Journal of Monetary Economics, 36*(1), 117–143. https://doi.org/10.1016/0304-3932(95)01206-2.

Greenwood, M. J. (1997). Chapter 12: Internal migration in developed countries. In M. R. Rosenzweig & O. Stark (Eds.), *Handbook of population and family economics. 1. Part B* (pp. 647–720). Amsterdam: Elsevier. http://www.sciencedirect.com/science/article/pii/S1574003X97800049.

Greenwood, M. J., & Hunt, G. L. (2003). The early history of migration research. *International Regional Science Review, 26*(1), 3–37. https://doi.org/10.1177/0160017602238983.

Hannon, L. (2003). Poverty, delinquency, and educational attainment: Cumulative disadvantage or disadvantage saturation? *Sociological Inquiry, 73*(4), 575–594. https://doi.org/10.1111/1475-682X.00072.

Helbing, D. (2012). Agent-based modeling. In *Social self-organization, Understanding complex systems* (pp. 25–70). Berlin, Heidelberg: Springer. https://link.springer.com/chapter/10.1007/978-3-642-24004-1_2.

Henderson, J. V. (1991). *Urban development: Theory, fact, and illusion*. New York: Oxford Universiyt Press.

Hoornweg, D., & Pope, K. (2017). Population predictions for the world's largest cities in the 21st century. *Environment and Urbanization, 29*(1), 195–216. https://doi.org/10.1177/0956247816663557.

Ioannides, Y. M., & Topa, G. (2010). Neighborhood effects: Accomplishments and looking beyond them. *Journal of Regional Science, 50*(1), 343–362. https://doi.org/10.1111/j.1467-9787.2009.00638.x.

Jacobs, J.. (1992). *The death and life of great American cities*. Vintage, Books ed. New York: Vintage Books.

Jones, C. I. (1999). Growth: With or without scale effects? *The American Economic Review, 89*(2), 139–144.

Jones, C. I. (2005). Chapter 16: Growth and ideas. In P. Aghion & S. N. Durlauf (Eds.), *Handbook of economic growth. 1. Part B* (pp. 1063–1111). Amsterdam: Elsevier. http://www.sciencedirect.com/science/article/pii/S1574068405010166.

Jones, C., & Romer, P. (2009). *The new kaldor facts: Ideas, institutions, population, and human capital. w15094*. Cambridge, MA: National Bureau of Economic Research. http://www.nber.org/papers/w15094.pdf.

Keivani, R. (2010). A review of the main challenges to urban sustainability. *International Journal of Urban Sustainable Development, 1*(1–2), 5–16. https://doi.org/10.1080/19463131003704213.

Krugman, P. (1991). Increasing returns and economic geography. *Journal of Political Economy, 99*(3), 483–499. https://doi.org/10.1086/261763.

Lichtenberg, F. R. (2002). *Sources of U.S. longevity increase, 1960–1997*. Working Paper 8755. National Bureau of Economic Research. http://www.nber.org/papers/w8755.

Lichter, D. T., Parisi, D., & Taquino, M. C. (2012). The geography of exclusion: Race, segregation, and concentrated poverty. *Social Problems, 59*(3), 364–388. https://doi.org/10.1525/sp.2012.59.3.364.

Lynch, K. (1984). *Good city form*. Cambridge, MA: MIT Press.

de Marchi, S., & Page, S. E. (2014). Agent-based models. *Annual Review of Political Science, 17*(1), 1–20. https://doi.org/10.1146/annurev-polisci-080812-191558.

Mathers, C. D., & Loncar, D. (2006). Projections of global mortality and burden of disease from 2002 to 2030. *PLoS Medicine, 3*(11), e442. https://doi.org/10.1371/journal.pmed.0030442.

McDonald, P. (2000). Gender equity in theories of fertility transition. *Population and Development Review, 26*(3), 427–439. https://doi.org/10.1111/j.1728-4457.2000.00427.x.

Mcdonald, P. (2006). Low fertility and the state: The efficacy of policy. *Population and Development Review, 32*(3), 485–510. https://doi.org/10.1111/j.1728-4457.2006.00134.x.

Mniszewski, S. M., Del Valle, S. Y., Stroud, P. D., Riese, J. M., & Sydoriak, S. J. (2008a). Pandemic simulation of antivirals + school closures: Buying time until strain-specific vaccine is available. *Computational and Mathematical Organization Theory, 14*(3), 209–221. https://doi.org/10.1007/s10588-008-9027-1.

Mniszewski, S. M., Del Valle, S. Y., Stroud, P. D., Riese, J. M., & Sydoriak, S. J. (2008b). EpiSimS simulation of a multi-component strategy for pandemic influenza. In *Proceedings of the 2008 spring simulation multiconference* (pp. 556–563). SpringSim '08. San Diego, CA, USA: Society for Computer Simulation International. http://dl.acm.org/citation.cfm?id=1400549.1400636.

Molho, I. (1986). Theories of migration: A review. *Scottish Journal of Political Economy, 33*(4), 396–419. https://doi.org/10.1111/j.1467-9485.1986.tb00901.x.

Montgomery, M. R. (2008). The urban transformation of the developing world. *Science, 319*(5864), 761–764. https://doi.org/10.1126/science.1153012.

Murray, E. T., Mishra, G. D., Kuh, D., Guralnik, J., Black, S., & Hardy, R. (2011). Life course models of socioeconomic position and cardiovascular risk factors: 1946 birth cohort. *Annals of Epidemiology, 21*(8), 589–597. https://doi.org/10.1016/j.annepidem.2011.04.005.

Nagel, K., & Rickert, M. (2001). Parallel implementation of the TRANSIMS micro-simulation. *Parallel Computing, Applications of Parallel Computing in Transportation, 27*(12), 1611–1639. https://doi.org/10.1016/S0167-8191(01)00106-5.

Nagel, K., Beckman, R., & Barrett, C.. (1999). *TRANSIMS for urban planning*.

Newman, M. E. J. (2002). Spread of epidemic disease on networks. *Physical Review E, 66*(1), 016128. https://doi.org/10.1103/PhysRevE.66.016128.

O'Rand, A. M. (1996). The precious and the precocious: Understanding cumulative disadvantage and cumulative advantage over the life course. *The Gerontologist, 36*(2), 230–238. https://doi.org/10.1093/geront/36.2.230.

Oeppen, J., & Vaupel, J. W. (2002). Broken limits to life expectancy. *Science, 296*(5570), 1029–1031. https://doi.org/10.1126/science.1069675.

Ortman, S. G., Cabaniss, A. H. F., Sturm, J. O., & Bettencourt, L. M. A. (2014). The pre-history of urban scaling. *PLoS One, 9*(2), e87902. https://doi.org/10.1371/journal.pone.0087902.

Ortman, S. G., Cabaniss, A. H. F., Sturm, J. O., & Bettencourt, L. M. A. (2015). Settlement scaling and increasing returns in an ancient society. *Science Advances, 1*(1), e1400066. https://doi.org/10.1126/sciadv.1400066.

Ortman, S. G., Davis, K. E., Lobo, J., Smith, M. E., Bettencourt, L. M. A., & Trumbo, A. (2016). Settlement scaling and economic change in the Central Andes. *Journal of Archaeological Science, 73*, 94–106. https://doi.org/10.1016/j.jas.2016.07.012.

Peters, O., & Gell-Mann, M. (2016). Evaluating gambles using dynamics. *Chaos: An Interdisciplinary Journal of Nonlinear Science, 26*(2), 023103. https://doi.org/10.1063/1.4940236.

Romer, P. M. (1986). Increasing returns and long-run growth. *Journal of Political Economy, 94*(5), 1002–1037. https://doi.org/10.1086/261420.

Romer, P. (1989). *Endogenous technological change, w3210*. Cambridge, MA: National Bureau of Economic Research. http://www.nber.org/papers/w3210.pdf.

Rosenberg, C. E. (2009). *The cholera years: The United States in 1832, 1849, and 1866*. Chicago, IL: University of Chicago Press.

Sampson, R. J. (1997). Neighborhoods and violent crime: A multilevel study of collective efficacy. *Science, 277*(5328), 918–924. https://doi.org/10.1126/science.277.5328.918.

Sampson, R. J. (2012). *Great American City: Chicago and the enduring neighborhood effect*. Chicago, IL: University of Chicago Press.

Sampson, R. J., & Laub, J. H. (1990). Crime and deviance over the life course: The salience of adult social bonds. *American Sociological Review, 55*(5), 609–627. https://doi.org/10.2307/2095859.

Schelling, T. C. (2006). *Micromotives and macrobehavior (new ed.) with a new preface and the Nobel lecture. Fels lectures on public policy analysis*. New York: Norton.
Schlapfer, M., Bettencourt, L. M. A., Grauwin, S., Raschke, M., Claxton, R., Smoreda, Z., West, G. B., & Ratti, C. (2014). The scaling of human interactions with city size. *Journal of the Royal Society Interface, 11*(98), –20130789. https://doi.org/10.1098/rsif.2013.0789.
Shen, G. (2002). Fractal dimension and fractal growth of urbanized areas. *International Journal of Geographical Information Science, 16*(5), 419–437. https://doi.org/10.1080/13658810210137013.
Shuey, K. M., & Willson, A. E. (2008). Cumulative disadvantage and black-white disparities in life-course health trajectories. *Research on Aging, 30*(2), 200–225. https://doi.org/10.1177/0164027507311151.
Smith, L., Beckman, R., Anson, D., Nagel, K., & Williams, M.. (1995). *Transims: Transportation analysis and simulation system*. LA-UR-95-1664; CONF-9504197-1. Los Alamos National Lab., NM (United States). https://www.osti.gov/scitech/biblio/111917.
Storper, M. (2013). *Keys to the city: How economics, institutions, social interactions, and politics shape development*. Princeton: Princeton University Press.
UN-Habitat. (2003). *The challenge of slums: Global report on human settlements*. UN-HABITAT. http://mirror.unhabitat.org/pmss/listItemDetails.aspx?publicationID=1156.
UN-Habitat. (2012). *State of the world's cities 2012/2013 , prosperity of cities. State of the world's cities*. UN-Habitat. http://mirror.unhabitat.org/pmss/listItemDetails.aspx?publicationID=3387.
Warr, M. (1998). Life-course transitions and desistance from crime. *Criminology, 36*(2), 183–216. https://doi.org/10.1111/j.1745-9125.1998.tb01246.x.
Willekens, F. J. (1999). The life course: Models and analysis. In *Population issues, The plenum series on demographic methods and population analysis* (pp. 23–51). Dordrecht: Springer. https://link.springer.com/chapter/10.1007/978-94-011-4389-9_2.
Zhao, Y. (2003). The role of migrant networks in labor migration: The case of China. *Contemporary Economic Policy, 21*(4), 500–511. https://doi.org/10.1093/cep/byg028.

Chapter 11
Implementing Car-Free Cities: Rationale, Requirements, Barriers and Facilitators

Mark Nieuwenhuijsen, Jeroen Bastiaanssen, Stephanie Sersli, E. Owen D. Waygood, and Haneen Khreis

11.1 Introduction

The private car has dominated the second half of the twentieth century, which some have described as "the century of the car" (Gilroy 2000). The car has become an essential part of our contemporary societies and daily lives, in which it plays an important role in the economic functioning of cities, regions and countries as it facilitates the movement of some people and goods and allows some people to participate in essential out-of-home activities such as employment, accessing health

M. Nieuwenhuijsen (✉)
ISGlobal, Centre for Research in Environmental Epidemiology (CREAL), Barcelona, Spain

Universitat Pompeu Fabra (UPF), Barcelona, Spain

CIBER Epidemiologia y Salud Publica (CIBERESP), Madrid, Spain
e-mail: mark.nieuwenhuijsen@isglobal.org

J. Bastiaanssen
Institute for Transport Studies (ITS), University of Leeds, Leeds, UK
e-mail: tsjb@leeds.ac.uk

S. Sersli
Simon Fraser University, Burnaby, Vancouver, Canada
e-mail: smsersli@sfu.ca

E. O. D. Waygood
Université Laval, Quebec City, QC, Canada
e-mail: Owen.Waygood@esad.ulaval.ca

H. Khreis
ISGlobal, Centre for Research in Environmental Epidemiology (CREAL), Barcelona, Spain

Universitat Pompeu Fabra (UPF), Barcelona, Spain

CIBER Epidemiologia y Salud Publica (CIBERESP), Madrid, Spain

Institute for Transport Studies (ITS), University of Leeds, Leeds, UK
e-mail: haneen.khreis@isglobal.org

M. Nieuwenhuijsen, H. Khreis (eds.), *Integrating Human Health into Urban and Transport Planning*, https://doi.org/10.1007/978-3-319-74983-9_11

services and interacting with social contacts. However, these advantages have come at the expense of the environment and society with great adverse environmental, social and public health impacts attributable to increasing car travel and usage (Lucas 2012; Khreis et al. 2016; Gärling and Steg 2007). Further, the gains by those who can profit from the infrastructure for cars come at the expense of those who do not use them through increased danger, barrier effects and reduced accessibility as education, jobs, retail and services agglomerate and locate in places with poor to no access by means other than by car (e.g. Delbosc and Currie 2011; Lucas and Jones 2012; Lucas et al. 2016). This has resulted in many cities having a (real or perceived) dependence on cars.

Some cities, recognizing these impacts, are aiming to reduce car usage and shift towards active and more sustainable transport modes such as walking and cycling and public transport. Hamburg, Helsinki, Madrid and Oslo have recently announced their plans to become partly private car-free cities (Cathkart-Keays 2015). Many cities have introduced different policies that aim to reduce motorized traffic including implementing car-free days, investing in cycling infrastructure and pedestrianization, restricting parking supply and increasing public transport provision (Cathkart-Keays 2015; Nieuwenhuijsen et al. 2016). One of the main cited drivers of such transitions is to mitigate climate change effects through reduction of greenhouse gases. Yet, these measures have many other benefits (Woodcock et al. 2009).

Here we consider a car-free city as a city without private cars but one that may still have buses, lorries, taxis, emergency vehicles, motor bikes or even shared cars as necessary to move goods and people. The characteristics we envisage are that the largest mode share is taken by public and active transport and that these are also the modes at the top of the hierarchy for transport planning and engineering. Furthermore, the motor vehicles remaining on the roads should be as sustainable and healthy as possible—e.g. by being electric, having speed restrictions as well as other restrictions in terms of time and areas of the city they can be.

Nieuwenhuijsen et al. (2016) recently described the potential health impacts of car-free cities including significant reductions in traffic-related air pollution, noise and temperature in the areas freed from cars. These reductions are likely to lead to a significant reduction in premature mortality and morbidity (Nieuwenhuijsen et al. 2016; Mueller et al. 2017). Car-free initiatives, if undertaken at a sufficiently large scale, can also result in positive distal effects and climate change mitigation through greenhouse gas reductions (Nieuwenhuijsen et al. 2016), which in turn have their own positive impacts on public health (Khreis et al. 2017). Furthermore, traffic and parked vehicles often dominate large areas, which reduces the viability of other means of transport and different land usage and affects the visual quality of the physical environment. The reduction of the number of cars, and their infrastructure and parking spaces, provides previously non-existing opportunities to increase green space and green networks in cities, which in turn can lead to many beneficial health effects (Nieuwenhuijsen et al. 2016, 2017). All these measures are likely to lead to higher levels of active mobility and physical activity, which may be the pathway providing the largest public health benefits (Mueller et al. 2017) and provide more opportunities for people to interact with each other in public and green space. As a form of regular physical activity, walking and cycling, for instance, can have positive impacts on health including reduced morbidity burden and all-cause mortality risk (Arsenio and Ribeiro 2015).

Furthermore, the transition to car-free cities may substantially improve the livability of neighbourhoods—especially in those neighbourhoods that bear disproportional burdens of pollution, social disadvantage, crashes and public transport disinvestment (Bullard 2003). Inadequate public transport services and lack of access to a private car due to legal, financial or physical reasons have been found to significantly decrease people's access to essential out-of-home activities such as employment opportunities, health services and social contacts which, in turn, can severely increase levels of unemployment, deteriorate health and/or lead to social isolation (SEU 2003; Lucas 2012). Children have become a marginalized group in the planning for cars with numerous negative outcomes on their well-being (e.g. Waygood et al. 2017) and likely greater parental time consumption due to chauffeuring (e.g. Mattsson 2002; Waygood 2009). A transition from car-dominated urban landscapes and transport policies towards car-free cities that are considering the mobility needs of all people to access key destinations, regardless of their access to private cars, would, therefore, constitute an important step towards a more inclusive and just urban environment.

Despite emerging initiatives and a growing awareness of the environmental, health and social benefits of car-free cities, the academic literature on how to make this transition, including prerequisites, potential barriers and facilitators for car-free cities, remains scarce. In this chapter, we aim to describe the rationale behind car-free cities and early developments, strategies for the implementation of car-free policies and prerequisites, barriers and facilitators. We also highlight research gaps and provide suggestions for future work.

11.2 Car-Free Cities

11.2.1 Early Steps Towards a Car-Free City

Towards the end of the twentieth century, many cities started to pedestrianize streets of their centres (Topp and Pharoah 1994). Although the main reason was to make city centres more welcoming to people and provide an economic boost to rundown and deflated city centres (e.g. Hass-Klau 2014), we consider the experiences of these car-free areas as the first steps towards restricting the use of cars. These car-free areas may allow vehicles to enter at night, and access is allowed (sometimes only at certain times of day) for deliveries and servicing. In many cases, it seems that overall traffic diminished, and there were not significant spillover effects to surrounding areas (e.g. Goodwin 2001). Other cities that have achieved reductions in car ownership and use have experienced a rise in taxis and other service and delivery vehicles, underscoring the potential to replace private cars with public or shared alternatives (Cervero and Tsai 2004; Metz 2013; Nijland and van Meerkerk 2017; Buehler et al. 2017).

11.2.2 Rationale for the Car-Free City

Car-free cities are still perceived as a radical idea, and policy-makers need a strong rationale to make the shift. Yet, the reasons for car-free cities are more pressing than ever. A strong argument, which is often mentioned by cities considering this transition, is the need for low-carbon transport given the imminent threat of climate change (Garfield 2017). Furthermore, there is an urgent need to reduce pollution to create healthier environments (Nieuwenhuijsen et al. 2016). The health burden of urban and transport-related exposures is particularly high, e.g. causing up to 20% of premature mortality in cities (Mueller et al. 2017), and is more immediate than climate change effects. There is also the need to create socially inclusive (urban) environments (Litman 2003) with evidence showing that car use has negative impacts on community social interactions (Appleyard 1980) or children's community connections (Waygood and Friman 2015), while active transport means like walking are associated with greater community networks (Grannis 2011) and children's social interactions (Waygood et al. in press). These issues are not mutually exclusive, for example, low-carbon transport has the potential to produce healthier cities when there is a shift from private car use to clean public and active transportation (Gouldson et al. 2017), and the reduction of traffic leads to lower levels of air pollution and noise which promotes and protects public health (Nieuwenhuijsen 2016), as well as reduces barrier effects and enhances community cohesion (Khreis et al. 2017).

11.2.3 Early Developments Towards Car-Free City Advocacy

A theoretical design for a car-free city of one million people was first proposed by J.H. Crawford in 1996 and further refined in his books, *Car-free Cities (Crawford* 2000*)* and *Car-free Design Manual* (Crawford 2009). In 1997, the Lyon Protocol was published for the design and implementation of car-free districts in cities (http://www.carfree.com/lyon.html). Created during a weeklong conference in Lyon, participants envisioned what would be needed to implement car-free districts. This included (1) identifying interested parties; (2) gathering necessary data, e.g. on land use, traffic and pollution; (3) conceptualizing preliminary concepts and proposals; (4) facilitating dialogue through media attention; (5) engaging the political process; (6) gradually implementing car reduction measures; (7) engaging with, consulting and listening to local communities; and (8) city planners developing final plans for full implementation. Although not specifically mentioning car-free cities, work by Newman and Kenworthy highlighted many of the problems with car dependence and the benefits of moving away from such development patterns (Newman and Kenworthy 1999, 2015).

To be clear, our concept of a car-free city is not a city free of motorized vehicles, and this vision is agreed on by other authors (Loo 2017). There are significant advantages related to motorized vehicles such as movement of heavy goods, aiding immobile people or traveling long distances. The key point is effectively eliminating the need and use of *private* motorized vehicles, especially in cities. To reduce or even eliminate the need for private vehicles in cities, numerous steps are necessary. Clearly this will involve improving alternatives and on the long term, adjusting the city form and land-use to support active travel and (private) car-free living. However, to make such "hard" changes, numerous "soft" processes (see, e.g. Loo 2017) are required as the attitudes, beliefs and knowledge of decision-makers and the general population have for decades been sold on the advantages of private cars with the negative implications not sufficiently considered. The next section will describe the prerequisites to move towards car-free cities. The final sections discuss strategies, barriers, facilitators and the general changes needed.

11.3 Prerequisites to Move Towards Car-Free Cities

We suggest nine essential prerequisites underpinning the transition to car-free cities, illustrated in Fig. 11.1.

The first prerequisite is political vision and leadership.

Cities have become the power house of change and innovation and are leaders in new environmental initiatives such as C40 Cities Climate Leadership Group, Local Governments for Sustainability (ICLEI) and United Cities and Local Governments (UCLG), where mayors of the cities take an active role (Gouldson et al. 2015). An essential ingredient for the planning of car-free cities is the leadership of city mayors and political support across the political spectrum with a shared vision across the majority, recognition and acknowledgement of the key challenges associated with the use and overreliance on cars and as such dedicating funding for solutions.

Cities like New York and Paris have been able to make progress in paving the way for a renewed view on sustainable mobility through clear and determined leadership of their mayors who have implemented a wide range of measures to restrict the use of cars. These include speed restrictions and shared street events (Kutsch 2016), temporarily closing several streets and offering free public transport rides (Fitzsimmons and Wolfe 2016), city centre car-free days and speed restrictions in other districts (Chrisafis 2015; Anzilotti 2016).

The second prerequisite towards car-free cities is a paradigm shift from mobility centre transport planning and policies towards one of accessibility.

A paradigm shift from mobility centre transport planning to one of accessibility has been suggested by various authors (e.g. Banister 2008). This includes changes in typical transport investment appraisal methods such as cost-benefit analyses (Khreis et al. 2016) which provide no insight into latent travel demand (Martens and Hurvitz 2011) and in which cyclists and pedestrians are typically not incorporated (de Graaf et al. 2015). Where the conventional method of predicting travel demand and providing the required means was thought to reduce congestion and therewith improve the economic performance of cities and regions, it paradoxically leads to increased traffic volumes (e.g. Goodwin and Noland 2003) as it enhanced car dependency (Levine and Garb 2002) and unleashed latent demand, ultimately

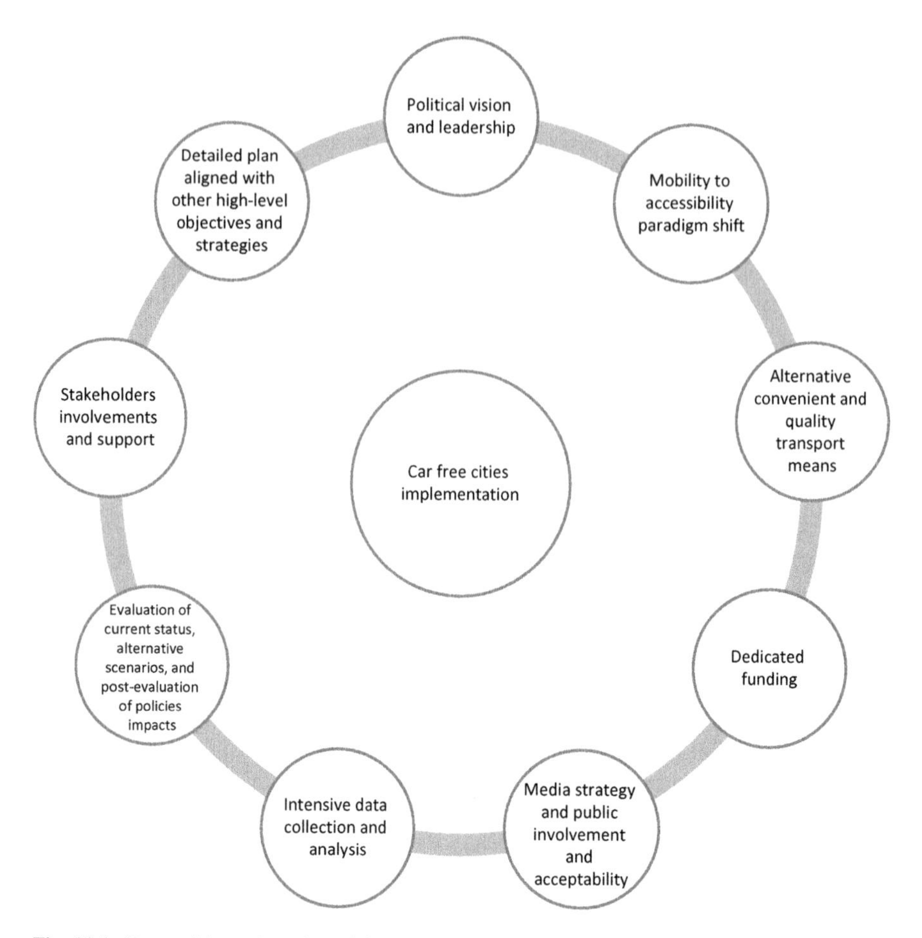

Fig. 11.1 Prerequisites of car-free cities

increasing related levels of pollution, road casualties, community severance, spatial dispersal of activities and social inequalities (Khreis et al. 2017). Some would argue that planning for an elimination of congestion is even disingenuous as it cannot be eliminated (Goodwin 2004). As everyone needs access to essential services (jobs, health, education, food), some authors have shown that such an approach requires that accessibility is the measure of most importance and that reducing congestion for those with already high levels of accessibility should not be the priority (Martens 2016).

Despite the embrace of accessibility planning by some urban and transport planners from the 1990s onwards and the emergence of "predict-and-prevent" regulations and restrictive car-related transport policies (Owens 1995), a fully fledged transition never took place. Conversely, car-related travel and infrastructure continued to increase in most cities, as transport planning remained demand-driven: predict and provide (car) mobility and facilitate this with the belief that it boosts the economy, although the role that further road construction plays in the economy is not so clear (Banister 2001; Button and Hensher 2001). The dominant role of the car and the consequences for people that lack access to it has been linked to social exclusion (SEU 2003; Lucas 2012), as lack of mobility can severely hamper peoples' access to key out-of-home activities, resulting in unemployment, deterioration of health or social isolation. For children, this focus on ablest adults (i.e. those who can afford and are physically and mentally able to use a car) has manifested in reduced independence (e.g. Shaw et al. 2015) which is unfortunate, as active and independent travel is positively associated to many well-being measures (Waygood et al. 2017).

An important condition for the transition towards car-free cities is therefore the recognition of mobility as a precondition for full participation in society (Martens and Bastiaanssen 2014). Hence, transport policies should shift from facilitating mobility as a purpose in itself to ensuring sufficient levels of accessibility, which provide people with access to key destinations necessary for their full participation. In practice, this implies that (car) travel demand should no longer guide transport interventions. At places where people cannot make necessary trips with transport means other than the private car, the transport system is of insufficient quality and needs to be enhanced (ibid).

This leads to our third and fourth prerequisites: alternative convenient and quality transportation means and dedicated funding.

Since many key destinations already have high accessibility by car, a significant policy and budget shift is required to invest in high-quality public transport linkages, as well as bicycle and pedestrian facilities that meet the needs of all people, e.g. public transport services that also run during off-peak hours, are universally accessible and provide near-complete coverage. Such enhancements need to meet the needs of shift and weekend workers and those whose jobs are in non-central locations. Further than serving just workers, the transport system should also cater for the needs of children and their caregivers, the elderly, and people with disabilities to be fully inclusive (Greed 2006). A more complete coverage should not solely focus on accessing job locations, but the services and amenities that the entire population needs. This will require integrated land-use, housing, essential services and transport planning, which are all key elements of planning for accessibility.

By establishing the missing transport links and services, sufficient levels of accessibility can be ensured for all while also offering the potential for a modal shift by current car drivers. These physical barriers to modal shift must be low before psychological barriers (e.g. attitudes, beliefs) will be effective (e.g. Swim et al. 2009). It is more difficult to have low car use in car-dependent areas than to have low car use in more active travel-friendly locations (e.g. Schwanen and Mokhtarian 2005). Active travel-friendly locations have been shown to maintain low car use over decades even in economies that showed growth (e.g. Sun et al. 2009).

Our fifth prerequisite for planning for car-free cities is intensive data collection and analyses.

Creating car-free cities will require intensive data collection and analyses in terms of land use (e.g. use, densities of population/services/amenities/jobs, diversity), mobility patterns (e.g. distance travelled, mode, accessibility, parking demand), demographics (e.g. age, gender, income, employment status and occupation, physical and mental ability, vulnerable groups), environmental pollutants (e.g. air pollution, noise and heat islands), traffic crashes, social preferences (e.g. cultures, believes, attitudes), health (e.g. health status, physical activity levels) and economics (economic activity, employment, job creation, etc.). This is necessary to demonstrate and develop an understanding of the current status, the interrelationship between the different factors and the impact of current practices, allowing the potential impacts of alternative scenarios (e.g. by using health impact assessments) to be estimated and documented (Nieuwenhuijsen et al. 2017). Further, such data collection can be beneficial to rigorously evaluate the impact of car-free policies (by comparing pre- and post-situations) as they are implemented.

Evaluation of pre- and post-implementation impacts forms the core of our sixth prerequisite.

Cities are becoming more interested in post-evaluations, something currently lacking in research and practice, but that is becoming increasingly important as more radical policies, which may be politically unpopular, are adopted (Sadik-Khan and Solomonow 2016). The collection and analysis of pre- and post-data can be particularly powerful in the call for and defensibility of changing practices (Whitfield et al. 2017). For example, in some regions, data shows that almost half of car journeys are less than 5 km (Xia et al. 2013) with some cities aiming to eliminate such "ridiculous" car trips (e.g. "No ridiculous car trips", Malmö, Sweden[1]). A reasonable argument can be made that these could be feasibly substituted by active transport modes, which are both healthier and environmentally friendlier, but actual evidence is missing.

In many regions, there are gaps in relevant datasets, which limit research and policy-making. There is a need for further and new datasets linking transport and health, while some gaps can be filled by applying data fusion to existing datasets and/or the synthesis of datasets from different fields.

Strong citizen and business support is crucial in the creation of car-free cities and forms the seventh prerequisite.

Citizen and business participation is essential to obtain commitment for the proposed changes and vision. Citizens' needs and convenience are tightly linked to public acceptability, calling for more public participation in the planning and policy-making process, which needs to become more transparent to those affected first-hand. Public acceptability and citizens' movements are core to successful implementation and radical change (Banister 2008). The process may not be straightforward and linear, and it may contain a multitude of policy measures to achieve the overarching goal. Indeed, people need to realize the benefits of car-free policies, and their accessibility needs (currently catered for by cars) must be catered for by alternative convenient and quality transport means (e.g. Loo 2017). The success of transitioning to car-free cities will largely depend on baseline car dependency and the extent to which other modes can replace this in a convenient and

[1] http://www.eltis.org/resources/videos/no-ridiculous-car-trips-malmo-sweden

triple bottom line efficient[2] manner (including economic, environmental and social/health measures). Further, business support is essential, and leaving them out of municipalities, decision-making can lead to backlashes and slow down progress, e.g. car-free-related restrictions in Oslo have been opposed and hampered by trade associations and shopkeepers (Cathcart-Keays 2017).

The eighth prerequisite is the development of a detailed plan aligned with other strategies.

The development of a detailed plan must be aligned with other strategies and context-specific priorities such as climate change, economic development (more jobs, more homes, more businesses and higher productivity), regeneration and urban air quality. A detailed plan is essential for consultation purposes and implementation of the car-free cities' strategy. It should also include indicators for measuring progress including, e.g. mode share, public transport punctuality, traffic injuries, CO_2 emission, air quality and transport-related health impacts. The alignment with other strategies and priorities is also key, and the climate change co-benefits discussion can be an important resource in this regard (Gouldson et al. 2017).

Our ninth prerequisite is the development of a timely media plan to explain the purpose and process of the proposed changes.

It is essential to get the media on board as the agents between decision-makers and those affected (Sadik-Khan and Solomonow 2016). The purpose of a media plan is to determine the best way to convey messages to target audiences. It must identify desired reach (amount of people), frequency (number of times the message reaches people), cost and penetration (audience reached). The messages must also consider audience stage of change towards low or no car use (e.g. Bamberg 2014; Waygood et al. 2012), social norms (e.g. De Groot et al. 2008) and values (e.g. Graham et al. 2009). Media campaigns are sometimes argued to be ineffective at creating behaviour change. Media campaigns may target the reasons why the change is needed, which is required to stimulate consideration of behaviour change, but is not in itself likely to create behaviour change (e.g. Waygood et al. 2012; Bamberg 2014). If people do not consider car use to be a problem, then they will not contemplate change. Providing rational and emotional evidence will be necessary to convince individuals due to the dual processing nature of human decision-making (e.g. Evans 2003). This must take into consideration the role of personal and social norms with respect to culpability (e.g. de Groot et al. 2008). It is only at this point that people will consider change and look for options (Waygood et al. 2012; Bamberg 2014), which our third prerequisite aims to provide. Lastly, many people who work to reduce the societal burden of excessive car use will use equity and harm/care arguments. This may seem self-evident to them, but it is related to their moral foundations. There are key moral foundations that are influential to people of different political leanings, and these must be understood and messages that speak to those morals developed. Using a limited number of moral-type arguments will limit the

[2]The use of the word efficient can also be a barrier to communication as without defining it, it may refer to traffic flow, government expenses, emissions/pollutants per trip, personal cost per trip, health benefits per trip, etc.

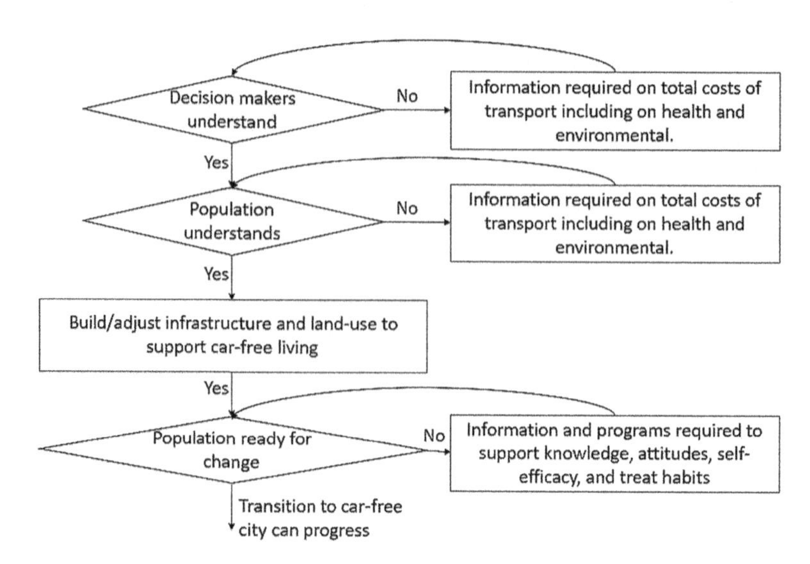

Fig. 11.2 Information cycle related to support for change towards car-free cities

message to only those individuals with the same moral foundations. Broadening the horizon of moral arguments for the transition towards car-free cities allows for substantiated decisions and incorporation of a wider variety of views in transport policy and projects (Graham et al. 2009).

This requires a long-term strategy targeting different target groups holding various perceptions, attitudes and behaviours towards the implementation of car-free cities (Fig. 11.2). At the same time, media can play an important role in agenda setting, which would be essential for prioritizing car-free cities in transport policies and within politics (McCombs and Shaw 1972).

Having described the nine prerequisites to implementing car-free cities, we turn our attention to the various considerations such as urban form, infrastructure and lifestyles that influence travel choices. We end with strategies that could be or have been used to move towards car-free cities.

11.4 Considerations, Facilitators/Changes Needed and Strategies

11.4.1 Urban Form

Urban form is an important aspect to consider for the implementation of car-free cites. One of the key urban form characteristics is density or compactness: a compact city such as Amsterdam, when compared to a sprawling city such as Atlanta, can reduce travel distances and the need to travel (Khreis et al. 2017). Studies repeatedly find that travel distance is the strongest correlate of walking and cycling

(Gouldson et al. 2017). Factors related to urban form that are often found to influence travel behaviour have been previously labelled as the five Ds and include Density, Diversity, Design, Destination accessibility and Distance to transit (Ewing and Cervero 2010). These are all important considerations in the design of car-free cities: density refers to population density or residential units, diversity refers to the land-use mix (e.g. mix of commercial and residential land use), design refers to the street network characteristics which can impact travel behaviour, destination accessibility refers to the ease by which destinations can be reached, and distance refers to the minimum distance to public transport (Van Wee 2018 forthcoming). Density and diversity can reduce travel distances and influence mode choice promoting greater active travel and public transport use no matter a family's life cycle stage (e.g. Waygood et al. 2015) and can maintain this for decades (Sun et al. 2009). Therefore, compact cities are more suited and easier to convert to car-free cities than sprawling cities (e.g. in real-world car-free neighbourhood; see Loo 2017).

Sprawling cities would need stronger urban and transport planning policies to increase density and diversity and decrease travel distance. Urban form can also influence the attractiveness and amount of active travel in different ways. Attractive scenery, green space and attractive and safe infrastructure (segregated cycle lanes, wide pavements, street furniture, green and blue space) may increase the appeal and the levels of walking and cycling. However, attention must be paid to not negatively affect distances to locations or create areas where individuals would feel unsafe due to a lack of transparency of green spaces.

11.4.2 Changes to Infrastructure and Public Space

Transport infrastructure often covers a vast amount of the available public space in cities, with cars being the most space-intensive form of urban transport during use (e.g. Héran and Ravalet 2008) and when not in use, i.e. when parking (McCahill and Garrick 2012). When moving at 50 km/h, the dynamic space (i.e. the vehicle plus the stopping distance) consumed per passenger by cars is estimated to be 20 times larger than by bus (Héran and Ravalet 2008). Where most city centres are historically organized around slower modes with key destinations often within walking distance, the creation of suburbs is strongly reliant on the private car (Frumkin 2002), and a great deal of space and infrastructure has been dedicated to roads and parking with the creation of new cities and the expansion of old ones. Therefore, one of the key considerations in the transition towards car-free cities will be how to change infrastructure that was mainly designed for cars, to infrastructure that caters to active and public transportation. Changing the road capacity/infrastructure to reduce car traffic (e.g. road diets and surrounding greening) can also potentially provide more public and green/blue spaces, which are beneficial for health and well-being (Nieuwenhuijsen et al. 2017; Gascon et al. 2015).

Over the years, some cities have transformed large highways into urban parks, green space and smaller, less congested streets (Martinez 2016). The city of Seoul was able to remove a central city elevated highway and restore the river that had been hidden by it. That space was transformed into an attractive, more natural walking boulevard that has measurably reduced the heat island effect (Kang and Cervero 2009; Kim et al. 2009) and improved traffic flow (Lee and Anderson 2013). Another example is Quito's United Nations boulevard where parking space was reallocated to pedestrian traffic making the streets a safer place for pedestrians' daily activities (Habitat 2016).

The organization of the street network is also important. The street network design in many cities of Japan uses arterial streets which are spaced at 500–700 m combined with internal local streets which are "shared space" (i.e. no sidewalks, all users negotiate the space) and narrow. These conditions naturally restrict vehicle speeds. Car use is low (<20% of household trips) in such areas (Sun et al. 2009), and active travel is over 50% (Waygood et al. 2015). Recently, Barcelona has begun introducing Superblocks (Superilles) by making relatively small changes to their existing street network to create them. A superblock will consist of nine existing blocks of the grid with internal travel by motorized modes restricted. Car, scooter, lorry and bus traffic will then be restricted to just the roads in the superblock perimeters. They will only be allowed in the within-perimeter streets if they are residents or providing local businesses, and at a greatly reduced speed of 10 km/h (typically the speed limit across the city is 30–50 km/h in specific areas) (Bausells 2016). The (elements of) superblocks' design can be introduced elsewhere. Superblocks will be complemented by the introduction of 300 km of new cycling lanes (from currently around 100 km), as well as an orthogonal bus network that has already been put in place, whereby buses navigate a series of main thoroughfares (Bausells 2016).

11.4.3 Changes in Lifestyles

Another big challenge is how to cater for changes in societal lifestyles. Simply put, this pertains to changing expectations away from using the private car with all its conveniences, direct accessibility and flexibility to using public and active transportation which at times requires more effort and may take longer. People often choose to use the car because it appears to be the easiest and fastest option available and that is also often considered a cheap one as people will often only consider running costs such as fuel consumption and not the sunk costs (ownership, insurance, etc.). Indeed, the literature lists determinates for transport mode choice including cost, time and access factors (e.g. Collins and Chambers 2005), the built environment (e.g. Ewing and Cervero 2010), cohort effects (e.g. Sun et al. 2012) and, especially for active transportation, distance (e.g. Morency et al. 2011); topography (Parkin et al. 2008); actual or perceived safety, beliefs about the environmental threat of cars and pro-environmental behaviours (Collins and Chambers 2005; Shen et al. 2008; Anable 2005); and their social responsibility (Loo 2017), the latter two relating to Fig. 11.2.

Unfortunately, most cities have been planned around the private car, making the car the convenient, cost-effective and primary mode of transport while decreasing the appeal, safety and convenience of active travel means (Khreis et al. 2016). However, understanding the factors that influence mode choice highlights the potential for transport planners and policy-makers to make active and public transport convenient and accessible options for citizens. For example, bus-dedicated lanes and solutions such as bus rapid transit can make public transport highly competitive with the car making trips almost as fast and potentially cheaper; complemented by green walkways or segregated cycling lanes from and to public transport hubs, the modal shift and positive impacts may be further reinforced, and access needs for other segments of the population that have no access to the car also met. Another important consideration is the transition into different stages of life as these change household transport needs, but given supportive built environments, car use can be kept minimal (e.g. Sun et al. 2009). The difference in terms of parents' time can be considerable as car-dependent areas require chauffeuring as opposed to environments built for all users (e.g. Waygood 2009).

The presence and proximity of destinations and sidewalks, connectivity, aesthetics, traffic on walking routes and their safety can all influence people's preferences and encourage alternative modes (Gouldson et al. 2017). People's beliefs about the use of the car and their environmental consciousness are also modifiable, and these issues are tightly linked with the need for targeted media strategies, strong leadership and knowledge transfer. The emergence of E-bikes can be a particularly important technology that can overcome the inconvenience of hilliness and to some extent trip length. Other soft measures that can be implemented include cycling sharing schemes, providing cycling parking, showering facilities in workplaces and better integration between cycling and public transport.

11.4.4 Barriers

Various barriers exist, and not all of them can be mentioned or addressed here. Beyond the barriers of infrastructure, car habits and various attitudes and beliefs is also the concern of a reducing car access on retailers. However, experience from numerous countries would suggest that this is not the case. As Germany pedestrianized its central areas, there was initially the belief that retail would negatively be affected, yet this was not found to be the case (Hass-Klau 1993). Two decades later, a similar pattern was seen in the UK where Lawlor (2014) reviewed the effect of a reduction of cars on existing business performance (footfall and retail), urban regeneration (new business, rental income, employment, social exclusion, etc.), improved consumer and business perceptions and business diversity. Case study evidence suggests that well-planned improvements to these public spaces can boost footfall and trading by up to 40% and that investing in better streets and spaces for walking can provide a competitive return compared to other transport projects. People who walked to the shops spent up to six times more than people who arrived by car. They also suggested that better streets and places may create a virtuous circle by raising self-esteem for residents and promoting investor confidence in an area. Studies have linked the quality of public spaces to people's perceptions of attractiveness of an area, contributing towards their quality of life and influencing where they shop. This was also reflected in the findings of New York City as they reduced car access and improved conditions for pedestrians (Sadik-Khan and Solomonow 2016). However further research is needed as the evidence is still scarce and not available for full car-free cities.

A further important consideration is that a more democratic (as in all residents, not just those who are physically, mentally and financially capable of owning and operating a private motorized vehicle) transportation system would likely address problems such as food deserts (where grocery stores are accessible only by private vehicle) and the movement towards larger retail centres that are again mainly accessible by private vehicle and reduce the vitality of cities. That reduced vitality can have important impacts on city tax revenues while at the same time increasing demands on those revenues through increased infrastructure requirements due to urban sprawl. A perception of time and money savings is typically based on ignoring the personal transport costs, the city infrastructure costs and the actual time required by the entire population to reach such dispersed locations.

Other concerns are related to important issues around inequalities. Car-free cities might mean increased difficulties of accessing vibrant centres by the portion of the population who do not live in the vicinity of these centres and who have relied on their private cars to access goods, services or opportunities. It is important that this portion of the population are given sufficient consideration as not to deter their opportunities or exacerbate existing inequalities posed by transport disadvantage (though typically the transport disadvantaged are those without cars living in car-oriented development (e.g. Currie et al. 2009)). Park and ride schemes may provide some solutions to this potential issue. Ideally, development and opportunities should not be restricted to certain centres of agglomeration, and part of development budgets should be put into ensuring that other towns and centres have their needs and opportunities accessible too.

One problem is that people often consider a car-free city from the view point of their current situation. An individual who has chosen a location that is car dependent cannot see how a car-free city would be an improvement to their current situation. They have established their lifestyle patterns around an assumption of high mobility. A strong car habit is likely to be hard to break (Verplanken and Wood 2006; Verplanken et al. 1997). As such, changes will require time as people adjust their travel patterns. Programs that directly address changes in travel habits due to life changes, such as moving, starting a job, having a baby, retiring, will be key in allowing a transition to travel patterns and transport systems that are not car-centric. Such programs can be seen in other domains and are referred to as upstream (i.e. context change) and downstream (i.e. information) (Verplanken and Wood 2006).

There are strong political and economic forces such as the car industry and lobby groups that may hinder the realization of car-free cities, but which can be tackled with strong leadership and city governance. The motor lobby in particular is a powerful and diffuse force (Irwin 1987, Douglas et al. 2011) and has previously successfully opposed traffic restriction measures, from restrictions on parking space (Higginson 2013) to scrutinizing and opposition of proposals for car-free zones in some German cities (Hajdu 1988), leading municipal authorities to delay plans or scrap them altogether. It has also opposed other health-benefiting interventions related to improving traffic safety (Sfetcu 2014) and successfully weakened and delayed car emission regulations in Europe (The Greens, European Free Alliance Group 2015, Neslen 2015). We acknowledge that enormous benefits have been accrued by global economies from car manufacturing and car mobility, and that these will continue, but there are also significant costs that society and governments are paying in terms of increasing inequalities, premature mortality and increased morbidity, which also takes its toll on the workforce, productivity and health services.

11.4.5 Strategies

General policy measures and strategies that can be considered in cities which aim at going (partly) car-free include reducing car use gradually through slow incremental changes to restrict traffic in certain areas/roads, or a big bang approach (with sudden large changes or restricting or banning car use in large parts of the city). Copenhagen followed the former approach by taking away a small percentage of road space each year, while Oslo was planning to do the latter with the centre of the city. Slow incremental changes may have an advantage of creating awareness and familiarity with the concept of car-free planning, as well as allowing residents to experience the benefits of being car-free. Generally, it may be easier to make gradual changes as opposed to sudden large changes due to psychological aspects (e.g. Kahneman and Tversky 1979). It is important to note that transitions are chaotic and will always, at first, encounter resistance. Both short-term (e.g. experiments with closing off streets, bike-sharing) and long-term actions (e.g. system/institutional changes such as paradigm shift and legislation) need to be taken.

There are already a wide range of policy measures that can reduce traffic (e.g. Goodwin 2001; Khreis et al. 2017), and the number of available measures continues to expand. Detailed information on the performance of some of policy measures can be found from the Victoria Transport Policy Institute (VTPI) Transportation Demand Management (TDM) encyclopaedia (www.vtpi.org/tdm) (VTPI 2014) and the Knowledgebase on Sustainable Urban Land use and Transport (KonSULT) knowledgebase (www.konsult.leeds.ac.uk). Other sources such as ELTIS provide case studies of successful policy interventions (www.eltis.org). For example, several car use reducing policy measures are indexed (and described) in KonSULT including road user charging, regulatory restrictions, land use to support public transport, parking charges, development density and mix, pedestrian areas and routes, cycle networks, private parking charges, company travel plans, vehicle ownership taxes and others. An overview on these measures, performance and their likely impacts on public health can be found in Khreis et al. (2017). These measures are likely to perform best when packaged in calculated strategies aiming at reinforcing the facilitators and weakening the barriers (e.g. Cairns et al. 2004). Lessons on behavioural change can also come from outside of the transport field (e.g. Avineri and Goodwin 2010).

11.5 In Conclusion

Transport planning and policy-making cater too much for private motorized travel and too little for public and active transport modes that could benefit public health, mitigate the effects of traffic-related environmental exposures and lead to more inclusive and just cities. Decades of planning and investments in car infrastructure attracted cars to the cities, and it will take decades to overturn this. Large car-oriented infrastructures continue to dominate with relatively small proportions of the budget allocated to and little work done for quality active transport provision across most regions. There is an urgent need to rebalance and provide better and safer infrastructures and policy support for active and public transport modes. A car-free city would provide a catalyst for better town planning by removing the need to facilitate car mobility and ensuring that urban areas are planned around people, functionality and better built environments instead. In this paper, we have given the rationale and identified a number of prerequisites for a car-free city and further described potential strategies and barriers for such implementation. This paper is not meant to be the final word, but a starting point to discuss a vision that may seem to be far in the distance but desirable, needed and possible for many cities. Changes require planning and discussion and a good theoretical and evidence base, which we have contributed to in this paper.

Furthermore, there are many other ongoing changes that may facilitate or hinder the process. Technological advancements such as electric cars and autonomous cars have been proposed as the solution to current issues, but they have only a limited number of advantages and potentially many disadvantages. The use of electric cars may reduce emissions (or at least displace them outside of the city), but it does not reduce the number of cars in the city, and they do not help to reduce other health problems such as deaths and injuries due to crashes or a lack of daily physical activity. Autonomous vehicles may reduce the number of cars parked in the city, and to some extent the number of cars circulating in the city, but it has also the potential to increase car use in certain population groups and may increase actual kilometres travelled as vehicles would need to travel between locations of disparate users, thereby increasing the problems related to car use. If using autonomous vehicles is as easy as waiting for an elevator (an autonomous vertical vehicle) to travel up a few floors, many people will choose to wait rather than making the small effort to travel short distances. These changes, which seem to have become a focal point of recent transport policy, need more investigation and should be integrated within the detailed plans or proposals for car-free cities. It is not that such technologies won't play a role in future transport services, but they should not be considered a silver bullet to all current transport problems.

There are yet many uncertainties in terms of acceptability and behaviour change when introducing the car-free city and also the likely changes in terms of air pollution, noise, temperature, social cohesion and physical activity. This will entail evaluations of health impacts of (planned) interventions in cities, including changes in perceptions and attitudes and health impact modelling of future scenarios. Further research and research synthesis are needed to build a good evidence base, which is currently nonexistent. It is also vital to better understand how mobility patterns change with the introduction of car-free measures.

Acknowledgement The review was conducted without any specific funding source. We thank James Woodcock for some helpful comments and input into the paper.

References

Anable, J. (2005). 'Complacent car addicts' or 'aspiring environmentalists'? Identifying travel behaviour segments using attitude theory. *Transport Policy, 12*, 65–78.

Anzilotti E. (2016). *Paris introduces car-free Sundays, City lab*. Retrieved from https://www.citylab.com/transportation/2016/04/paris-introduces-car-free-sundays/480609/.

Appleyard, D. (1980). Livable streets: Protected Neighborhoods? *The Annals of the American Academy of Political and Social Science, 451*, 106–117.

Arsenio, E., & Ribeiro, P. (2015). The economic assessment of health benefits of active transport (Chapter 2). In M. Attard & Y. Shiftan (Eds.), *Sustainable urban transport*. UK: Emerald Publishing. https://doi.org/10.1108/S2044-994120150000007011.

Avineri, E., & Goodwin, P. (2010). *Individual behaviour change: Evidence in transport and public health*. London: The Department for Transport.

Bamberg, S. (2014). *Psychological contributions to the development of car use reduction interventions, handbook of sustainable travel* (pp. 131–149). Dordrecht: Springer.

Banister, D. (2008). The sustainable mobility paradigm. *Transport Policy, 15*(2), 73–80.

Banister, D. (2001). *Transport planning, handbook of transport systems and traffic control* (pp. 9–19). Bingley: Emerald Group Publishing Limited.

Bausells M. (2016). Retrieved from https://www.theguardian.com/cities/2016/may/17/superblocks-rescue-barcelona-spain-plan-give-streets-back-residents.

Buehler, R., Pucher, J., Gerike, R., & Götschi, T. (2017). Reducing car dependence in the heart of Europe: Lessons from Germany, Austria, and Switzerland. *Transport Reviews, 37*(1), 4–28.

Bullard, R. (2003). Addressing urban transportation equity in the United States. *Fordham Urban Law Journal, 31*, 1183–1209.

Button, K. J., & Hensher, D. A. (2001). *Introduction handbook of transport systems and traffic control* (pp. 9–19). Bingley, UK: Emerald Group Publishing Limited.

Cairns, S., Sloman, L., Newson, C., Anable, J., Kirkbride, A., & Goodwin, P. (2004). *Smarter choices—Changing the way we travel*. London: Department for Transport.

Cathcart-Keays, A. (2017). Retrieved from https://www.theguardian.com/cities/2017/jun/13/oslo-ban-cars-backlash-parking.

Cathkart-Keays, A. (2015). Retrieved from http://www.theguardian.com/cities/2015/dec/09/car-free-city-oslo-helsinki-copenhagen.

Cervero, R., & Tsai, Y. (2004). City CarShare in San Francisco, California: Second-year travel demand and car ownership impacts. *Transportation Research Record: Transportation Research Board, 1887*, 117–127.

Cheng, J., Xu, Z., Zhu, R., Wang, X., Jin, L., Song, J., & Su, H. (2014). Impact of diurnal temperature range on human health: A systematic review. *International Journal of Biometeorology, 58*(9), 2011–2024.

Chrisafis A. (2015). *All-blue skies in Paris as city centre goes car-free for first time, The Guardian Cities*. Retrieved from https://www.theguardian.com/cities/2015/sep/27/all-blue-skies-in-paris-as-city-centre-goes-car-free-for-first-time.

Collins, C. M., & Chambers, S. M. (2005). Psychological and situational influences on commuter-transport-mode choice. *Environment and Behavior, 37*(5), 640–661.

Crawford, J. H. (2000). *Carfree cities*. International Books. isbn:978-90-5727-037-6.

Crawford, J. H. (2009). *Carfree design manual*. International Books. isbn:978-90-5727-060-4.

Currie, G., Richardson, T., Smyth, P., Vella-Brodrick, D., Hine, J., Lucas, K., Stanley, J., Morris, J., Kinnear, R., & Stanley, J. (2009). Investigating links between transport disadvantage, social exclusion and well-being in Melbourne—Preliminary results. *Transport Policy, 16*, 97–105.

De Groot, J. I., Steg, L., & Dicke, M. (2008). Transportation trends from a moral perspective: Value orientations, norms and reducing car use. In F. N. Gustavsson (Ed.), *New transportation research progress* (pp. 67–91). New York: Nova Science.

Delbosc, A., & Currie, G. (2011). Exploring the relative influences of transport disadvantage and social exclusion on well-being. *Transport Policy, 18*(4), 555–562.

Douglas, M. J., Watkins, S. J., Gorman, D. R., & Higgins, M. (2011). Are cars the new tobacco? *Journal of Public Health, 33*(2), 160–169.

Evans, J. S. B. T. (2003). In two minds: Dual-process accounts of reasoning. *Trends in Cognitive Sciences, 7*, 454–459.

Ewing, R., & Cervero, R. (2010). Travel and the built environment. *Journal of the American Planning Association, 76*, 265–294.

Fitzsimmons E. and Wolfe J. 2016, New York today: A car-free day. *New York Times*. Retrieved from https://www.nytimes.com/2016/04/22/nyregion/new-york-today-a-car-free-day.html?_r=0.

Frumkin, H. (2002). Urban sprawl and public health. *Public Health Reports, 117*(3), 201.

Garfield, Leanne 2017. *12 major cities that are starting to go car-free, Business Insider*. Retrieved from http://www.businessinsider.com/cities-going-car-free-2017-2/#oslo-will-implement-its-car-ban-by-2019-1.

Gärling, T., & Steg, L. (2007). *Threats from car traffic to the quality of urban life: Problems, causes and solutions*. Bingley: Emerald Group Publishing Limited.

Gascon, M., Triguero-Mas, M., Martínez, D., Dadvand, P., Forns, J., Plasència, A., & Nieuwenhuijsen, M. J. (2015). Mental health benefits of long-term exposure to residential green and blue spaces: A systematic review. *International Journal of Environmental Research and Public Health, 12*, 4354–4379.

Gilroy, P. (2000). Driving while black. In D. Miller (Ed.), *Car cultures*. New York: Berg Publishers.

Goodwin, P. (2004). *The economic costs of road traffic congestion*. London, UK: ESRC Transport Studies Unit University College London.

Goodwin, P., & Noland, R. B. (2003). Building new roads really does create extra traffic: A response to Prakash et al. *Applied Economics, 35*, 1451–1457.

Goodwin, P. B. (2001). *Traffic reduction, handbook of transport systems and traffic control* (pp. 21–32). Bingley, UK: Emerald Group Publishing Limited.

Gouldson, A. P., S. Colenbrander, A. Sudmant, N. Godfrey, J. Millward-Hopkins, W. Fang, & X. Zhao. (2015). *Accelerating low carbon development in the World's cities*.

Gouldson, A., Sudmant, A., Khreis, H., Papargyropoulou, E. (2017) The wider economic and social benefits low carbon development in cities: A systematic review of the state of the evidence base. In *The New Climate Economy 2017*. In preparation.

de Graaf, S., Hoogendoorn, S., & Barmentlo, H. (2015). De opleving van modellering van langzaam verkeer. *NM Magazine, 3*, 18–19.

Graham, J., Haidt, J., & Nosek, B. A. (2009). Liberals and conservatives rely on different sets of moral foundations. *Journal of Personality and Social Psychology, 96*, 1029.

Grannis, R. (2011). *From the ground up: Translating geography into community through neighbor networks*. Princeton, NJ: Princeton University Press.

Greed, C. (2006). Making the divided city whole: Mainstreaming gender into planning in the United Kingdom. *Tijdschrift voor Economische en sociale geografie, 97*, 267–280. https://doi.org/10.1111/j.1467-9663.2006.00519.x.

Habitat III 2016. *Turning roads into streets—Road space allocation and public space resilience, Quito*. Retrieved from https://habitat3.org/programme/turning-roads-into-streets-road-space-allocation-and-public-space-resilience/.

Hajdu, J. C. (1988). Pedestrian malls in West Germany: Perceptions of their role and stages in their development. *Journal of the American Planning Association, 54*(3), 325–335.

Hass-Klau, C. (1993). Impact of pedestrianization and traffic calming on retailing: A review of evidence from Germany and the UK. *Transport Policy, 1*, 11.

Hass-Klau, C. (2014). *The pedestrian and the city*. New York: Routledge.

Héran, F., & Ravalet, E.P.J. (2008). *La consommation d'espace-temps des divers modes de déplacement en milieu urbain, Application au cas de l'île de France*. PREDIT.

Higginson, M. (2013). Workplace parking levies as an instrument of transport policy. *Public Transport International, 62*(3).

Irwin, A. (1987). *Risk and the control of technology: Public policies for road traffic safety in Britain and the United States*. Dover, NH: Manchester University Press.

Kahneman, D., & Tversky, A. (1979). Prospect theory: An analysis of decision under risk. *Econometrica, 47*, 263–291.

Kang, C. D., & Cervero, R. (2009). From elevated freeway to urban greenway: Land value impacts of the CGC project in Seoul, Korea. *Urban Studies, 46*, 2771–2794.

Khreis, H., May, A. D., & Nieuwenhuijsen, M. J. (2017). Health impacts of urban transport policy measures: A guidance note for practice. *Journal of Transport & Health, 6*, 209–217.

Khreis, H., Warsow, K. M., Verlinghieri, E., Guzman, A., Pellecuer, L., Ferreira, A., Jones, I., et al. (2016). The health impacts of traffic-related exposures in urban areas: Understanding real effects, underlying driving forces and co-producing future directions. *Journal of Transport & Health, 3*(3), 249–267.

Kim, K. R., Kwon, T. H., Kim, Y.-H., Koo, H.-J., Choi, B.-C., & Choi, C.-Y. (2009). Restoration of an inner-city stream and its impact on air temperature and humidity based on long-term monitoring data. *Advances in Atmospheric Sciences, 26*, 283–292.

Kutsch, T. (2016) *A new New York? Manhattan's oldest neighbourhood goes car-free, kind of …, The Guardian Cities*, Retrieved from https://www.theguardian.com/cities/2016/aug/15/new-york-manhattan-car-free-shared-streets-financial-district.

Lawlor, E. 2014. *The pedestrian pound*. The business case for better streets and places. Retrieved from http://www.livingstreets.org.uk/sites/default/files/content/library/Reports/PedestrianPound_fullreport_web.pdf.

Lee, J. Y., & Anderson, C. D. (2013). The restored Cheonggyecheon and the quality of life in Seoul. *Journal of Urban Technology, 20*, 3–22.

Levine, J., & Garb, Y. (2002). Congestion pricing's conditional promise: Promotion of accessibility or mobility? *Transport Policy, 9*(3), 179–188.

Litman, T. (2003). *Social inclusion as a transport planning issue in Canada*. Canada: Victoria Transport Policy Institute.

Litman, T. (2014). Retrieved March 2, 2016, from http://www.vtpi.org/wwclimate.pdf.

Loo, B. P. Y. (2017). Realising car-free developments within compact cities. In *Proceedings of the Institution of Civil Engineers-Municipal Engineer*. Great Britain: Thomas Telford Ltd.

Lucas, K. (2012). Transport and social exclusion: Where are we now? *Transport Policy, 20*, 105–113.

Lucas, K., Bates, J., Moore, J., & Carrasco, J. A. (2016). Modelling the relationship between travel behaviours and social disadvantage. *Transportation Research Part A: Policy and Practice, 85*, 157–173.

Lucas, K., & Jones, P. (2012). Social impacts and equity issues in transport: An introduction. *Journal of Transport Geography, 21*, 1–3.

Martens, K., & Hurvitz, E. (2011). Distributive impacts of demand based modelling. *Transportmetrica*, 181–200.

Martens, K. & Bastiaanssen, J. (2014). *An index to Measure Accessibility Poverty Risk*. Paper presented at the Colloquium Vervoersplanologisch Speurwerk [Transport conference], November 20–21, 2014, Eindhoven.

Martens, K. (2016). *Transport justice: Designing fair transportation systems*. New York: Routledge.

Martinez (2016) Retrieved from http://www.archdaily.com/800155/6-cities-that-have-transformed-their-highways-into-urban-parks?platform=hootsuite.

Mattsson, K. T. (2002). Children's (in)dependent mobility and Parents' chauffeuring in the town and the countryside. *Tijdschrift voor Economische en Sociale Geografie, 93*, 443–453.

McCahill, C., & Garrick, N. (2012). Automobile use and land consumption: Empirical evidence from 12 cities. *Urban Design International, 17*, 221–227.

McCombs, M., & Shaw, D. (1972). The agenda-setting function of mass media. *Public Opinion Quarterly, 36*(2), 176.

Morency, C., Paez, A., Roorda, M. J., Mercado, R., & Farber, S. (2011). Distance traveled in three Canadian cities: Spatial analysis from the perspective of vulnerable population segments. *Journal of Transport Geography, 19*, 39–50.

Metz, D. (2013). Peak car and beyond: The fourth era of travel. *Transport Reviews, 33*(3), 255–270.

Neslen, Arthur (2015). *EU caves in to auto industry pressure for weak emissions limits*, The Guardian, Retrieved from http://www.theguardian.com/environment/2015/oct/28/eu-emissions-limits-nox-car-manufacturers.

Newman, P., & Kenworthy, J. (1999). *Sustainability and cities: Overcoming automobile dependence*. Washington, DC: Island Press.

Newman, P., & Kenworthy, J. (2015). *The end of automobile dependence* (pp. 201–226). Washington, DC: Island Press.

Nieuwenhuijsen, M. (2016). *Urban and transport planning, environmental exposures and health-new concepts, methods and tools to improve health in cities*. http://ehjournal.biomedcentral.com/articles/supplements/volume-15-supplement-1.

Nieuwenhuijsen, M. J., Khreis, H., Verlinghieri, E., Mueller, N., & Rojas-Rueda, D. (2017). Participatory quantitative health impact assessment of urban and transport planning in cities: A review and research needs. *Environment International, 103*, 61–72.

Nieuwenhuijsen, M. J., Khreis, H., Verlinghier, E., & Rojas-Rueda, D. (2016). Transport and health: A marriage of convenience or an absolute necessity. *Environment International, 88*, 150–152.

Nijland, H., & van Meerkerk, J. (2017). Mobility and environmental impacts of car sharing in the Netherlands. *Environmental Innovation and Societal Transitions, 23*, 84–91.

Owens, S. (1995). From `predict and provide' to 'predict and prevent'? Pricing and planning in transport policy. *Transport Policy, 2*(1), 43–49.

Parkin, J., Wardman, M., & Page, M. (2008). Estimation of the determinants of bicycle mode share for the journey to work using census data. *Transportation, 35*(1), 93–109.

Sadik-Khan, J., & Solomonow, S. (2016). *Street fight: Handbook for an urban revolution*. New York: Viking.

Santos, G., Behrendt, H., Maconi, L., Shirvani, T., & Teytelboym, A. (2010). Part I: Externalities and economic policies in road transport. *Research in Transportation Economics, 28*(1), 2–45.

Sfetcu, Nicolae (2014). *The Car Show*. Retrieved from https://books.google.co.uk/books?id=Zy-luAwAAQBAJ&pg=PT516&lpg=PT516&dq=car+manufacturer+oppose+traffic+restrictions&source=bl&ots=8RiZzrT1mb&sig=xra7J2gF0d75oAXGHLW8wMJfeIg&hl=en&sa=X&ved=0ahUKEwi16fXi5PLMAhUDJMAKHbpaBogQ6AEIMzAE#v=onepage&q=car%20manufacturer%20oppose%20traffic%20restrictions&f=false.

Shaw, B., Bicket, M., & Elliott, B. (2015). *Children's independent mobility: An international comparison and recommendations for action*. London: Policy Studies Institute.

Shen, J., Sakata, Y., & Hashimoto, Y. (2008). Is individual environmental consciousness one of the determinants in transport mode choice? *Applied Economics, 40*(10), 1229–1239.

Social Exclusion Unit. (2003). *Making the connections: Final report on transport and social exclusion*. UK, London: Office of the Deputy Prime Minister.

Sun, Y., Waygood, E., Fukui, K., & Kitamura, R. (2009). Built environment or household life-cycle stages: Which explains sustainable travel more? *Transportation Research Record: Journal of the Transportation Research Board, 2135*, 123–129.

Sun, Y., Waygood, E. O. D., & Huang, Z. (2012). Do automobility cohorts exist in urban travel? *Transportation Research Record: Journal of the Transportation Research Board, 2323*, 18–24.

Swim, J., Clayton, S., Doherty, T., Gifford, R., Howard, G., Reser, J., Stern, P., & Weber, E. (2009). *Psychology and global climate change: Addressing a multi-faceted phenomenon and set of challenges*. A report by the American Psychological Association's task force on the interface between psychology and global climate change. Retrieved April 15, 2010.

The Greens, European Free Alliance Group 2015, *InfluenceMap EU NOx emissions: How the automotive industry shaped policy: A report prepared for the European Parliament greens/ EFA Group*. Retrieved December 2015, from http://www.Sven-giegold.De/wp-content/uploads/2016/01/Greens_EP_Automotive_Jan-2016_report.pdf.

Topp, H., & Pharoah, T. (1994). Car-free city centres. *Transportation, 21*, 231–247.

Wee, V. (2018). Land-use policy, travel behavior and health. In M. Nieuwenhuijsen & H. Khreis (Eds.), *Urban development, environment exposures, and health: A framework for integrating human health into urban and transport planning*. New York: Springer.

Verplanken, B., & Wood, W. (2006). Interventions to break and create consumer habits. *Journal of Public Policy & Marketing, 25*, 90–103.

Verplanken, V., Aarts, H., & van Knippenberg, A. (1997). Habit, information acquisition, and the process of making travel mode choices. *European Journal of Social Psychology, 27*, 539–560.

Waygood, E., Friman, M., Olsson, L. E., & Taniguchi, A. (2017). Transport and child well-being: An integrative review. *Travel Behaviour and Society, 9*, 32–39.

Waygood, E. O. D. (2009). *What is the role of mothers in transit-oriented development? The case of Osaka–Kyoto–Kobe, Japan, Women's issues in transportation 4th international conference* (pp. 163–178). Irvine, CA: Transportation Research Board.

Waygood, E. O. D., & Friman, M. (2015). Children's travel and incidental community connections. *Travel Behaviour and Society, 2*, 174–181.

Waygood, E. O. D., Friman, M., Olsson, L. E., & Taniguchi, A. (in press). Children's transportation and incidental social interaction: Evidence from Canada, Sweden, and Japan. *Journal of Transport and Geography*. (in press).

Waygood, E. O. D., Sun, Y., & Letarte, L. (2015). Active travel by built environment and lifecycle stage: Case study of Osaka metropolitan area. *International Journal of Environmental Research and Public Health, 12*, 15900–15924.

Waygood, O., Avineri, E., & Lyons, G. (2012). The role of information in reducing the impacts of climate change for transport applications. In T. Ryley & L. Chapman (Eds.), *Transport and climate change* (pp. 313–340). UK: Emerald.

Whitfield, G. P., Meehan, L. A., Maizlish, N., & Wendel, A. M. (2017). The integrated transport and health impact modeling tool in Nashville, Tennessee, USA: Implementation steps and lessons learned. *Journal of Transport & Health, 5*, 172–181.

Woodcock, J., Edwards, P., Tonne, C., Armstrong, B. G., Ashiru, O., Banister, D., Beevers, S., Chalabi, Z., Chowdhury, Z., Cohen, A., Franco, O. H., Haines, A., Hickman, R., Lindsay, G., Mittal, I., Mohan, D., Tiwari, G., Woodward, A., & Roberts, I. (2009). Public health benefits of strategies to reduce greenhouse-gas emissions: Urban land transport. *The Lancet, 374*, 1930–1943.

Xia, T., Zhang, Y., Crabb, S., & Shah, P. (2013). Cobenefits of replacing car trips with alternative transportation: A review of evidence and methodological issues. *Journal of Environmental and Public Health, 2013*, 1–14.

Chapter 12
Planning for Healthy Cities

Marcus Grant

12.1 Planning for Healthy Cities

A document submitted by the WHO to the United Nations Conference on Housing and Sustainable Urban Development, Habitat III, in Quito states that 'Healthy urban policies can significantly reduce infectious and non-communicable diseases and enhance wellbeing' (WHO 2016, p. 4). I take this as an indication that the role of the built environment in addressing the wider determinants of health, in addition to its accepted role on addressing infectious disease, is now well evidenced and is starting to be accepted. Cities are being pushed to the front of the call for action on health (WHO-UN 2016). Despite this, good practice exemplars of cities *deliberately* using a comprehensive approach for policy and practice to provide healthier urban environments are not common.

The focus of this chapter is urban form, the processes that control it and its influence on supporting health and health equity for non-communicable diseases and enhanced wellbeing. Planning for healthy cities needs to cover two public health objectives;

- firstly, the control and reduction of 'what is bad for people's health', such as exposures to air pollution, noise, heat; and
- secondly, providing support for the creation of, and access to, 'what is good for people's health' such as good quality green space, inclusive communities, healthy food and options for everyday activity.

The overall goal being to support long healthy lives for all. But to what degree does urban form matter? A recent study covered five international exposure recommendations across these two objectives. It looked at compliance with recommenda-

M. Grant (✉)
Environmental Stewardship for Health, Bristol, UK
e-mail: marcusxgrant@citieshealth.world

M. Nieuwenhuijsen, H. Khreis (eds.), *Integrating Human Health into Urban and Transport Planning*, https://doi.org/10.1007/978-3-319-74983-9_12

tions for physical activity, air pollution, noise, heat, and access to green spaces. The study quantified the associations between exposures and mortality to estimate preventable premature deaths. It is estimated that annually almost 20% of premature mortality could be prevented if international recommendations were complied with (Mueller et al. 2017).

With such a complex subject, this chapter can only aspire to give an introduction. My intention in writing the chapter is to provide an orientation, key concepts, some useful frameworks and tools, a scattering of examples, references to useful texts and examples of relevant research and activity.

The chapter starts with an overview as if planning for healthier cities was a game. What is the goal and what are the rules of this game, what does the playing field look like and who are the players? There is next an outline of planning for healthy cities in the WHO European Healthy City Network, a look at the relevance of the Sustainable Development Goals and a review of health equity and city planning. In the next section, some of the main elements and drivers of city development are outlined, including patterns of development and the importance of the planning system, concluding with a logic framework for an integrated city planning process for health. The next section provides an overview of how city form influences risks and challenges to health.

The following two sections look in more detail at actions that can be taken at two scales, strategic city planning and place-making. Strategic planning includes discussions of growth patterns and sprawl, strategic processes and climate change. The local planning section looks at place-making, the neighbourhood scale and the importance of involving communities. The chapter concludes by drawing together some of the main threads, in particular spatial leadership for health, and co-generating new knowledge and healthier places for more effective action.

12.1.1 Playing the Game: Rules, Tactics and Strategy

The nature and strength of the rules will vary according to the local context. Many policies influence city planning in addition to spatial planning policy per se. In response, the strategy, tactics and game play will need to be determined by local stakeholder players. In many cases, city planners may see their role as umpires. The best of these umpires will need to see rules adopted that to allow for a high level of consensus building, transparency and participation, so that everyone wins (Grant and Barton 2013).

All elements of planning for health need to be covered as a complete package. Such an approach was adopted in a WHO document setting out the 12 principles of Healthy Urban Planning in 2000 (Barton and Tsourou 2000). Table 12.1 shows the suggested schema; this has been adapted to better cover some important determinants of health that were not covered in the original, namely, biodiversity, food issues, and energy and waste (as resources).

Table 12.1 The 12 objectives of Healthy Urban Planning

Do planning policies and proposals promote and encourage health through:	
Supporting healthy personal lifestyles?	Providing safety and the feeling of safety?
Promoting social cohesion and social capital?	Supporting equity?
Providing quality housing?	Ensuring good air quality and a high-quality visual environment?
Access to work?	Adopting a sustainable approach to water, sanitation and drainage?
Access to local facilities and services?	Wise use of land and resources and support for biodiversity?
Promoting access to local, sustainable food and food production?	Addressing climate mitigation and adaptation issues?

Source: Adapted from Barton and Tsourou (2000)

12.1.2 The Playing Field

This game of planning for health is played on the playing pitch of people's lives in urban areas. At whatever scale you set up the game, city region, district or neighbourhood, the pitch has a fuzzy boundary; it is affected by nearby games in other urban areas and by other far away games that influences its resources, economics and politics. The 'HealthMap' (Fig. 12.1) is a useful metaphor for playing field and indicates that there are consequences for planetary and individual health of any intervention.

The HealthMap places people and their health at the centre with planetary health as the global context. In six concentric arcs, from individual lifestyle to the natural urban environment, it captures urban components of the determinants of health. The three outer arcs are rendered in a darker colour to highlight the core locus of operation for those actors who intervene directly in the built environment. The HealthMap is a systemic tool. Treat any definitions of components as loose and all implied relationships as fluid; in each application stakeholders need to reassess the relevance of the map to their local situation. For planning a healthy city, each of the arcs can be translated into specific and measureable objectives (Table 12.2).

Knowledge in this field is rapidly developing. Research points to the potential of using multilevel ecological models of public health, to better understand physical environment effects on health and health equity (McLeroy et al. 1988; Sallis and Owen 2002; Hanlon et al. 2011; Vardoulakis et al. 2016). The validity of using complementary scientific approaches needs to be recognised; and the understanding within cities of what works and why it is an essential ingredient (Pawson and Tilley 1997). Disparate stakeholders have found that the HealthMap can be a useful device for negotiating, understanding and developing a consensus for action.

There has recently been further discussion of the wider societal determinants of health in reaction to a narrow 'personal lifestyles are to blame' approach to solving public health problems. An integrative review of frameworks that focus on social determinants and lifestyles as 'bridging' concepts between the fields of public health and environmental sustainability is useful in this context (Graham and White 2016). The review found that frameworks which included human health as an

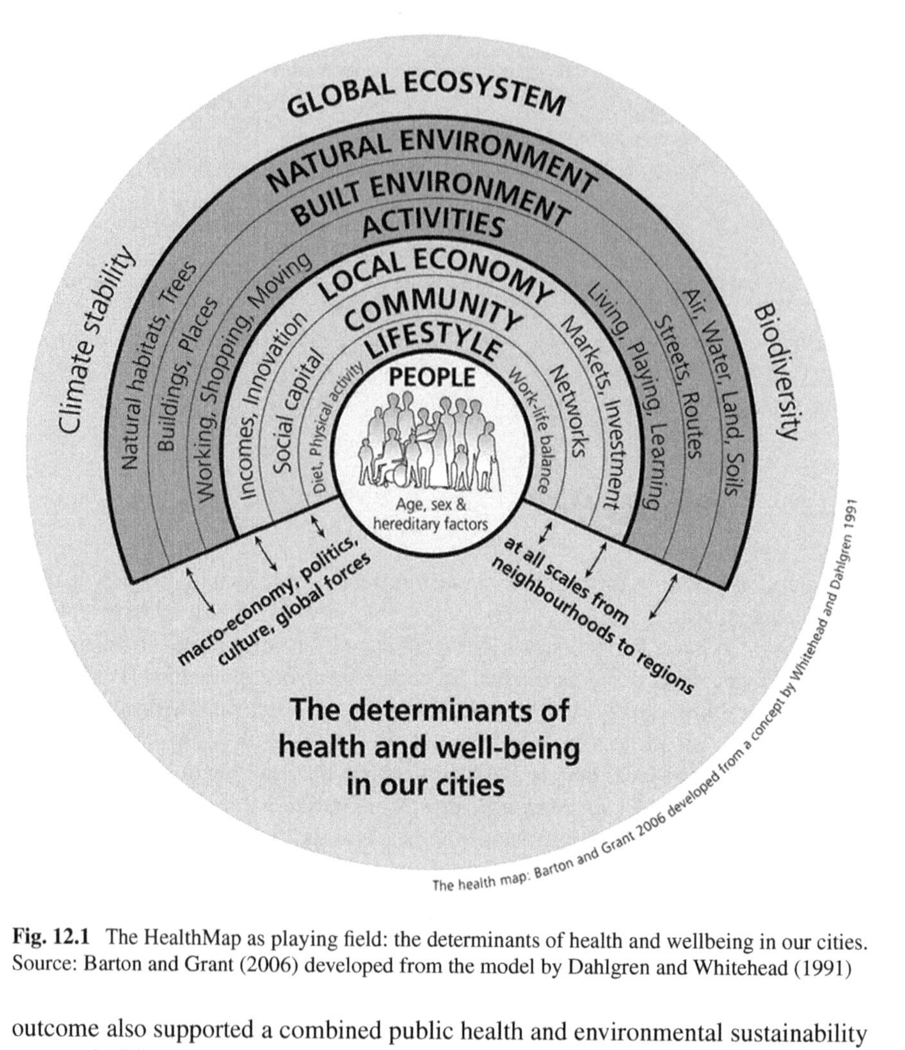

Fig. 12.1 The HealthMap as playing field: the determinants of health and wellbeing in our cities. Source: Barton and Grant (2006) developed from the model by Dahlgren and Whitehead (1991)

outcome also supported a combined public health and environmental sustainability approach (Fig. 12.2). This logic model provides another useful way to understand the route from social determinants to human health.

12.1.3 The Players

Using planning to achieve healthier cities occurs at the intersection of two disciplines. Although having a history of collaboration, and having common roots, the two disciplines have built-up working practices that make joint working difficult (Arthurson et al. 2016). A simple caricature of the two disciplines would be:

Public health: Population level interventions focused on people.

Planning: Population level interventions focused on place.

Table 12.2 Linking the HealthMap to urban planning objectives

Arcs of the HealthMap	Objectives for Healthy Urban Planning
1. People	• Providing for the needs of all groups in the population • Reducing health inequalities • Involving people in supporting a healthier local environment
2. Lifestyle	• Promoting active travel • Promoting physically active recreation • Facilitating healthy food choices
3. Community	• Facilitating social networks, social cohesion and inclusion • Supporting a sense of local pride and cultural identity • Promoting a safe environment
4. Economy	• Promoting accessible job opportunities for all sections of the population • Encouraging a resilient and buoyant local economy
5. Activities	• Ensuring retail, educational, leisure, cultural and health facilities are located to be accessible to all • Providing good-quality facilities, responsive to local needs
6. Built environment	• Ensuring good quality, variety and supply of housing • Promoting a green urban environment supporting mental wellbeing and contact with nature • Planning an aesthetically rewarding environment, with acceptable noise levels
7. Natural environment	• Promoting good air quality • Ensuring security and quality of water supply and sanitation • Ensuring soil conservation and quality • Reducing risk of environmental disaster
8. Global ecosystems	• Reducing transport- and building-related greenhouse gas emissions • Improving city and city region support for global biodiversity

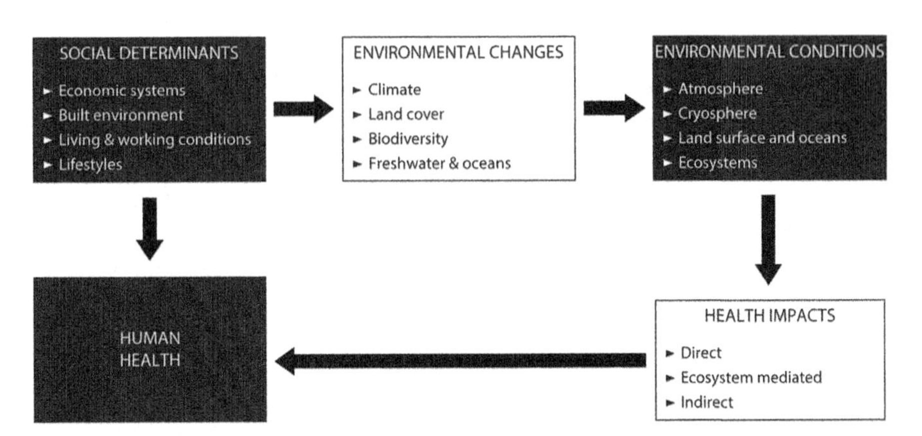

Fig. 12.2 Social determinants: environmental and health outcomes. Source: Graham and White (2016)

Of course in reality, public health can be concerned with place and environment, but in other instances having just a narrow focus on the individual and planning in many quarters see its purpose as places for people, but in other quarter it focuses entirely on serving economic market-led goals. However, 'planning for healthy cities' can allow each of these traditions to play to their strengths, reaffirm their roots and at the same time respond to current and future challenges (Barton 2009). In public health, at its best this agenda calls for the expression of a 'Health in All Policies' approach (Leppo et al. 2013) and also supports a call for 'Asset-Based Community Development' (GCPH 2011, 2012). Planning for healthy cities has a strong part to play in integrating the four levels of health promotion: environmental, social, organisational and individual (Kelly et al. 1993). It also can be a building block in the development of the fifth wave of public health (Hanlon et al. 2011). The fifth wave of public health sees emergent qualities that relate well to whole city planning such as recognising the need to deal with complex adaptive systems; maintaining a supportive, creative and cooperative mind-set; and a rebalancing of orientation to integrate objective science with 'the subjective' (lived experience, inner transformation) and 'intersubjective' (shared symbols, meanings, values, beliefs and aspirations) (Kelly et al. 1993).

In terms of the culture of planning, 'planning for healthy cities' strengthens collaborative planning (Healey 1996, 2003), placing process at the centre, with 'public goods' for both people and the environment as an objective. It can help to rebalance spatial planning, moving it away from the recent dominance of a market orientation and placing people and communities at the heart of planning again. There is a history of value propositions sitting at the heart of both these realms of human activity: public health and planning. In both too, there is an opposing force that claims of 'objectivity' and value free action. Choosing to adopt a healthy city approach is unashamedly value based (de Leeuw 2017).

12.1.4 Planning for Healthy Cities: The World Health Organisation European Healthy Cities Network

An approach for using the physical planning and design of cities was specifically developed within the WHO European Network of Healthy Cities. This initiative began in 1988 and is still continuing today, with Phase VII launched in 2018. Each phase lasts about 5 years, with cities across the WHO European region eligible to apply phase by phase. At the end of every phase, an evaluation helps to guide the development of the themes and develop working practices for the next phase (Belfast Healthy Cities 2014).

Phase I (1988–1992) involved 34 cities with the goal of introducing new ways of working for urban health based on a whole city approach. Phase II (1993–1997) included 38 cities. It was more action-oriented, with a strong emphasis on healthy public policy and comprehensive city public health planning. Phase III (1998–2002) saw talk of a 'European Healthy Cities movement' with 55 WHO designated cities and also national networks in several countries for non-designated cities. Moving towards a more strategic health development approach, it focused on constructing intersectoral city health development plans. This was intended to promote strategic planning for health across different sectors at city level. The Healthy Urban Planning approach emerged in one of the four themes: transport, environment, planning and housing. The other three themes were integrated information, lifelong learning and mental health and young people. The 12 objectives of Healthy Urban Planning were first used within this phase. Cities who wanted to focus on healthy urban planning formed a subnetwork and evaluated their experiences of the approach (Barton et al. 2003). This helped embed 'integration' and 'co-operation' as a key principles within the next phase.

Phase IV (2003–2008) saw consolidation of this spatial approach, and 'Healthy Urban Planning' became the name of an identified theme, with a thematic subnetwork of cities meeting regularly. Almost 100 designated cities joined this phase. The other three themes were healthy ageing, health impact assessment and physical activity and active living. Across all themes was an emphasis on equity, tackling the determinants of health, sustainable development and participatory and democratic governance. Work continued under a 'Healthy urban environment and design' theme in Phase V (2009–2013) (WHO EURO 2009). Phase VI (2014–2018) encompassed two strategic goals: improving health for all and reducing health inequities and improving leadership and participatory governance for health. A spatial approach was nested within 'creating resilient communities and supportive environments' one of the four core themes.

The evaluation of Phase V showed that well-designed interventions in the field of urban planning, transport or housing could all be very effective at meeting multiple objectives, addressing several of the Healthy Urban Planning objectives in a single intervention (Grant 2015). It also showed that the cities involved were using both strategic Healthy Urban Planning and local participatory neighbourhood design initiatives to tackle the health equity issues they had identified.

12.1.5 Sustainable Development Goals and Healthy Urban Planning

On 25 September 2015, the United Nations General Assembly formally adopted the 2030 Agenda for Sustainable Development, with a set of 17 Sustainable Development Goals (SDGs) and associated targets. A cursory glance at the 17 SDGs shows health to be the concern of SDG3 'good health and wellbeing'. Health advocacy in the global processes leading to the development of the SDG goals (from the Millennium Development Goals) focused. On 'maximising healthy lives' through reducing the burden of major non-communicable diseases, ensuring universal health coverage and access and accelerating previous commitments (Anon

2013). Concern has been expressed that through this process and under the weight of many pressing global concerns, health objectives will be weakened. Remedies suggested by Hill et al. (2014) include reframing health in terms of social sustainability and relating health to the whole sustainable development agenda; recognition that action for health cannot be limited to SDG3 is spreading (WHO 2017) with SDG11, cities, and SDG17, partnerships, just two from many others that are relevant.

The message of this chapter is the influence of the built and natural urban environment on how people choose *or are obliged* to live, to move and to consume in the places they live (i.e. people's lifestyles). Through this route, urban environments affect personal health and well-being. In addition, these environments we create and the lifestyles they promote, both affect the wider ecological processes and systems on which everyone relies for health and wellbeing (i.e. many other aspects of sustainable development); refer again to Fig. 12.2. Using urban planning to address health provides an approach that can link the two agendas: health and sustainability. Implementation and processes for Healthy Urban Planning are very close to those required for sustainable cities and towns. As a consequence, urban planning for health resonates across several other SDGs in addition to SDG3 'good health and wellbeing', including:

SDG8: Promote sustained, inclusive and sustainable economic growth, full and productive employment and decent work for all

SDG7: Ensure access to affordable, reliable, sustainable and modern energy for all

SDG11: Make cities and human settlements inclusive, safe, resilient and sustainable

SDG12: Ensure sustainable consumption and production patterns

Sustainable Development Goals can and should be used as a policy spur for healthier planning.

12.1.6 Health Equity and City Planning

Spatial planning and design has a role to play in reducing health inequities. SDG10,'reduce inequality within and amongst countries', contains seven targets for reducing inequalities, and health equity targets are often embedded into national health policy. The capacity of local environments to support health is of particularly importance for population groups disadvantaged by relative poverty, unemployment, low status and disability. Those who, for financial, physical or cultural/racial reasons, are more vulnerable, and have fewer choices open to them, find themselves typically in locations and settings that are less conducive to good health with little ability to move away or gain respite from unhealthy working and living environments. Many features of the built environment that detract from health are more likely to be experienced in areas of socio-economic disadvantage (Jones and Yates 2013). An outline of spatial factors that can exacerbate health inequity is found below.

12.1.6.1 Transport

In children aged 5–19 years, unintentional injuries are the leading cause of death, the majority of these being a result of road traffic collisions (Peden 2008).

There are very wide socioeconomic differentials in the levels of death and serious injury from road traffic. A study in England showed that children in the most deprived 10% of areas are four times more likely to be hit by a car as children in the least deprived 10%. Traffic-related issues affecting health such as poor air quality and noise also disproportionally affect those living in disadvantaged neighbourhoods (WHO 2007, 2011b). There is also now sufficient evidence to support an association between the exposure to traffic-related air pollution and the development of childhood asthma (Khreis et al. 2017).

12.1.6.2 Neighbourhoods and Facilities

Access to local facilities such as shops, schools, health centres and places of informal recreation is important for health and well-being both for the exercise taken in getting there (generally) on foot and the social interaction en route or at the facilities (Croucher et al. 2007). This is particularly important for lower-income groups who get much of their physical activity from active travel rather than leisure time recreation. Deteriorating features of an urban environment such as dilapidation, vandalism, graffiti and litter are disproportionately found in disadvantaged areas with a corresponding loss of amenity for walking to be pleasurable. In such areas children are less likely to be let out to play, leading to reduced physical activity and exacerbating health problems such as obesity which is more prevalent in lower-income groups. A cross-sectional survey of 12 European cities found that, compared to respondents from areas with low levels of litter and graffiti, those from areas with higher levels were 50% less likely to be physically active and 50% more likely to be overweight (Ellaway et al. 2005).

12.1.6.3 Local Shopping Streets

A high density of cheap, fast-food outlets can contribute to health problems, particularly those associated with obesity, for example, type 2 diabetes, hypertension and coronary heart disease (Robinson 2004). In the United States, it has been estimated that up to 10% of children's total energy intake now comes from fast foods, compared to 2% in the late 1970s. This increase in fast-food intake amongst children is significantly contributing to the rising childhood obesity in high-income countries (Bowman et al. 2004). Townshend (2017) talks of 'toxic high streets' found especially in low-income neighbourhoods with a concentration of retail outlets that could present a risk to health such as betting shops, discount alcohol, vaping emporia, pawn shops and pay day lending, take-away food and tanning salons.

12.1.6.4 Green Space

Inequality in mortality is lower in populations living in the greenest areas. Evidence shows that populations that are exposed to the greenest urban environments also have lowest levels of health inequality, accounting for income deprivation (Mitchell and Popham 2008). However, green space is not equally available to all of the population, with poorer neighbourhoods often lacking green space or with poorly maintained or vandalised green areas. Benefits of increases in physical activity and improved mental health only arise where the green space is high quality, accessible and safe (Ward Thomson et al. 2016; WHO EURO 2016).

12.1.6.5 Inequity of Climate Change Impacts

There are two particular aspects of climate change, influenced by the built environment, which are likely to impact disproportionately on the disadvantaged:

Increasing temperatures: exposure to heat is a cause of morbidity and mortality in the urban environment with lower socio-economic and ethnic minority groups more likely to live in warmer neighbourhoods and suffer greater exposure to heat stress (Harlan et al. 2006). High residential densities, sparse vegetation and having little or no open space, all often found in poorer neighbourhoods, have been correlated with higher temperatures.

Flooding: urban flooding from sea level rise and fluvial inundation presents an increasing risk to health through drowning, injuries, infectious diseases, stress and loss of essential services. The effects of flooding can be particularly devastating to already vulnerable populations, such as children, older people, the disabled, ethnic minorities and those with low incomes (WHO 2003). The disadvantaged may also live in more vulnerable areas; a study in England found there were eight times more people in the most deprived decile living in the tidal floodplain compared to the least deprived (Walker et al. 2003).

12.1.6.6 Developing a Strategic Focus on Health Equity

An evaluation of Phase V of the European Healthy Cities Network demonstrated that health equity could be addressed project by project or could be absorbed into the way a city conducts its spatial and transport policy (Grant 2015). The main subgroups targeted were older people, children, residents living in disadvantaged neighbourhoods and cultural minorities (Grant and Lease 2014). Health equity checks can be made part of policy and project impact assessment. In Whitechapel, London, the local masterplan's Sustainability Assessment included an Equality Impact Assessment which identified lower life expectancy and vulnerable groups, including minority ethnic groups and those with ill-health, as a key inequality issue and recommended actions to reduce inequalities (LBTH 2006). These included improving outdoor space and indoor leisure facilities to make them inclusive. In

Ljubljana, Slovenia, a member of the European Healthy Cities Network, they have developed a comprehensive and systematic city-wide approach to improve accessibility for disabled citizens. Measures include changes to the physical environment, as well as to public services, public transport, information, communication and the cultural and recreation sectors (Grant and Lease 2014).

12.2 City Development and Its Drivers

Rapid, unplanned, unsustainable patterns of urban development combined with continuing population urbanisation have always made cities in the developing nations focal points for environmental and health hazards (WHO 2002). Public health issues of waste disposal, provision of safe water and sanitation and injury prevention are familiar issues at the interface between urban inequalities, local environments and health.

Economic development has often been hailed as essential in a package of solutions to these problems. However, a rise in non-communicable disease and weakening of the determinants of health is emerging in the urban settlements in high-income countries. Diseases associated with sedentary lifestyles or poor-quality food intake on the one hand and increasing gaps in health inequalities, loss of social cohesion and support networks on the other are gaining public health focus in the developed cities in high-income countries. Worryingly some commentators are finding that a rise in these non-communicable health risks is now compounding the more frequently encountered communicable disease risks found in the informal of many low- and middle-income countries (Oni and Unwin 2015; Oni et al. 2016).

12.2.1 City Development

The idea of a rationally designed master plan for a city or urban area being delivered on the ground is a notion that non-built environment professionals sometimes hold in their mind as to how towns and cities develop. Although this is not unknown, it is a rarity, and such as the spatial process of development does not in itself even predict a successful place to live. City development patterns, as found and lived 'on the ground', result from an interplay of many drivers. Distribution of land use function (such as housing, employment, retail or mixed-use) plus the separation or mixing of these functions is a strong determinant of health. Three key influences on spatial patterns are:

Territorial context: Place; geography and geology, natural features, scale of development, patterns of ownership of land

Processes and drivers: Market and economic forces, land values, urban population growth, public investment, community interests, public involvement

Spatial policy context: Local, regional and national planning policy. Other sectoral plans (such as protected areas, catchment management and transport) that can influence development, the nature of the spatial planning system itself

These influences are non-exclusive and are interdependent. They can give rise to a plethora of spatial development patterns even within the same city. An examination of the features and driving forces of development in Hangzhou, China, found three patterns of development within the category of urban sprawl alone: infilling, edge development and leapfrog growth (Yue et al. 2013).

12.2.2 Spatial Planning Systems

The nature of the planning policy context provides important context. Different countries have very different spatial planning systems, stemming from different traditions influencing how planning happens. These styles of planning are deeply embedded in the complex historical conditions and cultures of particular places (Nadin and Stead 2008). The EU Compendium of Spatial Planning Systems and Policies (1997) used a set of criteria to discern idealised types or traditions of spatial planning. These criteria were:

- Legal family context
- Scope of the system in terms of policy topics covered
- Extent of national and regional planning
- Locus of power or relative competences between central and local government
- Relative roles of public and private sectors
- Maturity of the system or how well it is established in government and public life
- Apparent distance between expressed goals for spatial development and outcomes

Variations in these criteria give rise to four 'idealised' planning types: comprehensive integrated, land use regulation, regional economic and urbanism. However, Nadin and Stead acknowledged that each country will display its own unique variations on these idealised types. In addition to influencing that way that health can be incorporated into planning, these differences need to be examined if attempting any kind of learning or comparisons using international case studies (Grant 2015).

12.2.3 An Integrated City Planning Process for Health

So if the form and nature of the urban environment are critical to urban population health, how do we attempt to modify existing places or create new ones? Certainly we need to join forces with local communities and use local data. Using spatial planning and urban design to achieve health and health equity outcomes can be enacted as one or more isolated activities or projects or can be nested within a holistic healthy cities approach (Grant 2015). This agenda can be addressed at any

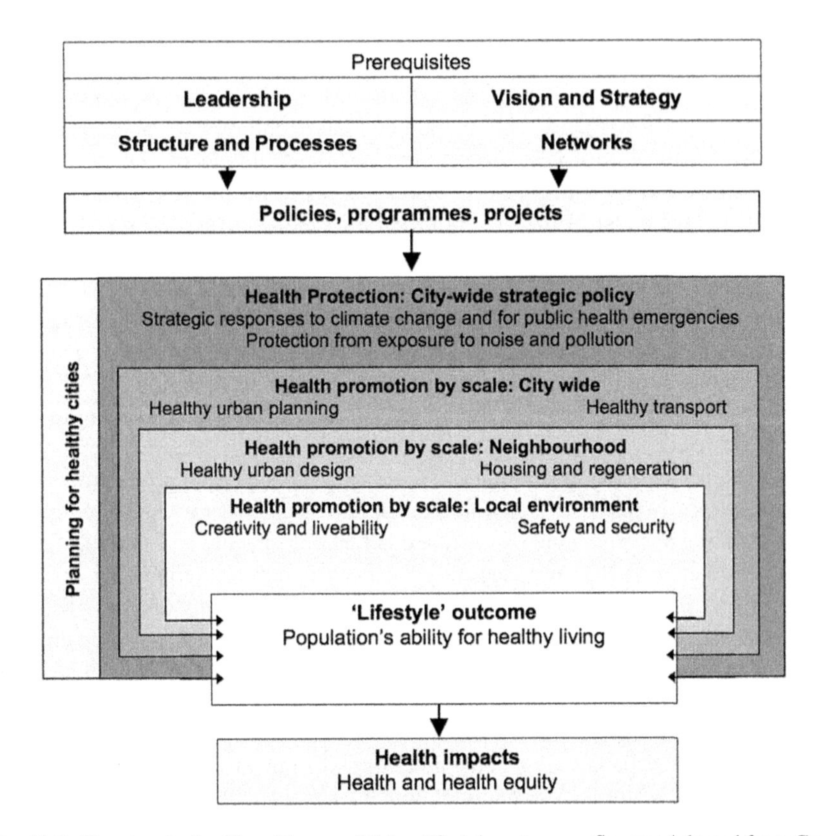

Fig. 12.3 Planning for healthy cities: modifying lifestyle outcomes. Source: Adapted from Grant (2015)

scale, from building and street, home-patch and neighbourhood to district, city and city region. Pivotal to success is an understanding of the multilevel context; prerequisites such as leadership and vision; different modes of working such as through policies, programmes or projects; and the different scales of working that can be employed (Fig. 12.3). Scale as an important factor in planning for health is slowly being recognised (Garfinkel-Castro et al. 2017). Planning for healthy cities is best addressed at all scales simultaneously, potentially triggering a health in all policies approach.

Planning for healthy cities must comprise of both health protection, usually city-wide policy through licensing, regulation and standards, and health promotion. Health promotion itself can be broken down into city-wide spatial policy for Healthy Urban Planning and transport, neighbourhoods scale activity such as urban design and housing and promoting innovation, liveability, safety and security at the local level. All of this should be directed towards an outcome that supports people's ability for healthy living across different sub-groups, with the goal of health and health equity.

12.3 City Form and Non-communicable Disease

The fabric of our towns and cities is the outcome of the way we plan, design and manage the territory of places, spaces, facilities and buildings of our urban habitat impacts on health from both a positive and negative perspective (RCEP 2007; Rydin et al. 2012). This physical form of an urban area includes the different land uses, where they are located in relation to each other and how they are distributed. This encompasses size, configuration and the nature of both human-built environments and natural features. However, the form is not immutable. Even in the most developed cities, there is room for manipulation and change to benefit health, as experience has shown through decades of remodelling to reduce the dominance of private motorised transport and increase the modal share of cycling and walking in many European cities.

The evidence base demonstrating associations between spatial urban form and the health outcomes is disparate and voluminous. Due to different urban patterning and spatial planning systems, much knowledge is location specific and not transferable. For some urban characteristics, causality has been found or inferred. But it is those characteristics of urban form that can be influenced by planning and design, which need to be the focus. Evidence of risks and challengers to health in urban environments from a spatial planning perspective (Grant et al. 2009) is presented here as a matrix linking parameters of urban environment to five risks to health (Table 12.3).

Risks to health such as air pollution, noise exposure, physical activity levels, social impacts and pathologies and unintentional injuries are all associated with urban elements that are open to manipulation (Grant et al. 2009; WHO 2016; Ewing and Cervero 2010; Barton 2009).

Four urban elements have been selected as they represent broad areas for urban policy and spatial intervention. At the strategic scale, we have what is termed 'land use pattern', straddling strategic to local we have 'transport' and 'green space, and at local scale we have 'urban design' (Fig. 12.4).

Transport infrastructure and the land use pattern are interdependent. The existence of transport networks affects the pattern of accessibility which helps deter-

Table 12.3 Urban elements as spatial determinants of health

Risk and challenges to health	Urban elements to manipulate			
	Land use pattern	Transport	Green space	Urban design
Physical activity	**X**	**X**	**X**	**X**
Social impacts	**X**	**X**	**X**	**X**
Air quality	**X**	**X**	**X**	**X**
Noise exposure	**X**	**X**	X	X
Adverse microclimate	X	X	**X**	**X**
Unintentional injuries	X	**X**	O	X

Main interactions between the urban components and risks and challenges to health. Source: Adapted from Grant et al. 2009

X = major interaction, X = minor interaction, O = very little interaction

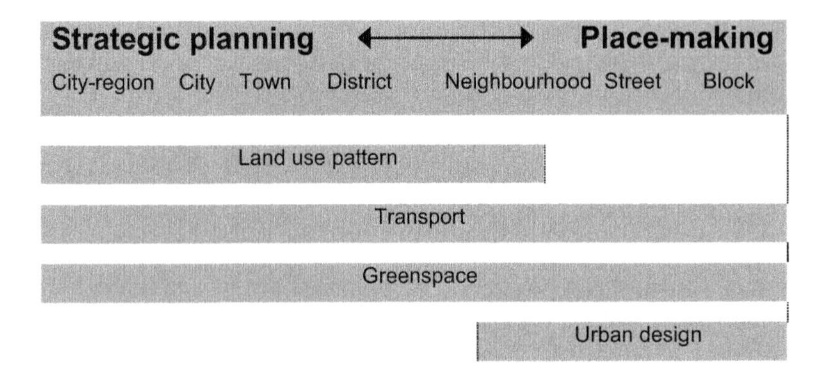

Fig. 12.4 Coverage of urban scale by components of the urban environment. Source Grant et al. (2009)

mine where land use development occurs. The pattern of use determines movement patterns, which in turn triggers demand for extra transport provision.

12.3.1 Land Use Pattern

In most settlements, the land use pattern results from the decisions of a myriad of land owners, in a context of market forces, mediated or not and to varying degrees by a spatial planning system. Spatial planning policies separating uses into different areas (zoning) or large area spatial designs (masterplanning) can have a profound effect on the land use pattern.

Land use pattern comprises the nature, disposition and density of land use. Issues of layout, networks, connectivity, accessibility, distribution and availability of facilities and functions are influenced. In terms of impacts on health, we can discern impacts at a series of distinct but nested scales. These are the region, city, town, district and neighbourhood scales. Several commentators discuss the general evidence of the impact of land use patterns on health (Lavin et al. 2006; Rao et al. 2007; Barton 2009). There is also evidence of significant health impacts at the smaller spatial scales of the street, the block; these are outlined in the section on 'urban design'.

12.3.2 Transport

This refers to the transport infrastructure, public and private, for all modes of movement. At the larger scales, this encompasses rail, road and water connections between a city and its hinterland and between a city and other cities. At the smallest scale, this may be the nature of how side streets meet main roads, of residential pavement management and of domestic cycle parking. Each of these has its own impact on the wider determinants of health. Transport policy and infrastructure investment is usually determined by transport specialists, often all too weakly co-ordinated with other aspects of strategic spatial planning.

12.3.3 Green Space

Green space confers many benefits for both physical and mental health and well-being (Mass et al. 2006; Ward Thompson et al. 2012) and is a critical component for urban health (Nieuwenhuijsen et al. 2017; Brown and Grant 2005; Coutts and Hahn 2015). Urban green space includes a huge variety of land from country parks and river corridors running through cities to residential gardens and pocket parks. It includes land in public, commercial and individual ownerships. It includes a wide range of uses including public and private gardens and squares, amenity and sports open space (often associated with mown grass), play space (often associated with shrubberies and mown grass), river and canal corridors and movement greenways and other functional green space such as allotments, churchyards and cemeteries. It comprises natural and seminatural habitats (including derelict and previously developed land) and managed verges, parks and gardens. In some settlements, this broad category could also include remnant countryside now within urban boundary such as woodlands, cliff ridges and coastlines. Green space is also taken to include elements of urban nature, not necessarily connected with a designated land use, such as street trees, vegetated walls and green roofs. As such responsibility for urban green space is usually spread across several departments in a municipality, including those responsible for civic amenity and parks, biodiversity and nature conservation, public housing, street trees and allotments, ownership and management may be in private hands.

Increasing the urban green space is an adaptive and also a mitigating response to climate change, reducing heat island effects. Presence or absence of trees and vegetation which cool areas through shading and evapotranspiration is important (Gartland 2012). Climate change mitigation also occurs through urban tree planting in colder climates due to increased wind friction with the potential to reduce heat loss from residential areas in winter. Urban heat island reduction policies should specifically target vulnerable residential areas and take into account equitable distribution and preservation of environmental resources (RCEP 2007). The progressive greening of cities is critical across a number of health objectives.

12.3.4 Urban Design

The way we experience and use our immediate environments in towns and cities is determined at the smaller spatial scale of the street, the public square, the development block and individual buildings (Rao et al. 2007). This includes much of what is referred to as 'the public realm', the work of landscape architects and urban designers, but strongly influenced by the form and facades of architecturally designed buildings and the decisions of those designing roads. Urban design can also determine the degree of social mixing or segregation of communities through the locational control of social housing in new build and neighbourhood regeneration.

Urban areas will also have to adapt to climate change. When designed correctly buildings have a reduced need for mechanical air conditioning in areas with lower external ambient temperatures. In more dense urban areas, decentralised energy distribution systems, such as combined heat and power and community heating networks, which can contribute to climate change mitigation (Barton et al. 2010) become more viable. The nature of the urban materials contributes to urban heat islands (Gartland 2012). Buildings, tarmac and paving absorb and store heat, increasing air temperature, particularly noticeable at night. To moderate temperature then the essential parameters are surface roughness, colour and porosity (all affecting albedo characteristics).

12.4 Strategic Planning: Working the Whole City

Strategic planning may be formulated through plans setting out a vision or objectives for the next 10, 20, 25 or more years for a city or city region. Work at this scale is often an attempt to determine broad future spatial patterns for investment including for employment, housing and transport. Plans may also include resource strategies including for energy, waste and water.

12.4.1 Growth Patterns and Urban Sprawl

Development pressures in urban areas are often acute, with demand for housing and employment outstripping supply. At a strategic planning level, the resolution, in terms of settlement growth, can been categorised into a number of archetypal spatial patterns relating to strategic policy. Different strategic policy results in different spatial patterns of development. Some idealised types of strategic policy and their resultant spatial development patterns are given in Table 12.4.

The impact on health cannot be wholly predicated at this strategic scale. Even the outcomes of the compact city option, often advocated for sustainability and quality of life, may be adverse for health and health equity if urban design and governance at the micro-scale are poor. Outcomes are always dependent on the local context. However low-density 'urban sprawl', which is market driven and often occurs if strategic policies are not implemented or are weak, has been shown to damage health (Lavin et al. 2006; Frank et al. 2012).

Table 12.4 Strategic urban growth policy and resultant spatial patterns

Strategic urban growth policy	Resulting spatial pattern
Intensification: infill and build within urban existing boundaries	Compact city
Edge expansion: allow connected growth on any boundary	Planned extensions
Linear or corridor expansion: allow growth connected or not only along transport corridors	Planned extensions and satellite towns
Exurbs: built new suburbs dependent on the exiting city	Satellite towns
New settlement: built new self-supporting independent settlements	Satellite towns
Free market: allow development to without controls over location and form	Dispersal

Sprawl is a possibility across all these strategic approaches, except intensification; and it is actually built into the free market policy. Urban sprawl can be said to occur where residential densities are not high enough to support a good range of local amenities and services, leading to the creation of environments where travelling by car is necessary for daily living, with a detrimental impact on walkability in the public realm and on equity. Urban sprawl now occurs at the periphery of most urban centres. It results in increased consumption of energy, resources, transport and land, raising levels of greenhouse gas emissions and air and noise pollution (EEA 2009a).

12.4.2 Strategic Planning Processes

Three strategic planning processes need to be manipulated to set the stage for health and healthy place-making at the local level.

Urban planning: Healthy Urban Planning sees the integration of health considerations into urban planning processes, programmes and projects. This involves establishing the necessary capacity and political and institutional commitment to achieve this goal. City-wide plans such as for housing, employment, retail or education need to be scrutinised. Health impact assessment can be a useful tool, but it is important that its methodology aligns with iterative and creative planning and design processes (Barton and Grant 2008; Grant and Barton 2013).

Transport planning: This needs to be viewed as movement planning not traffic planning. People need to move between home and other locations to access work, shops, education and leisure. Health is facilitated by building everyday physical activity into people lives—relevant to their ability. Planning for walking (often overlooked) and public transport needs careful attention. Access needs to be promoted, facilitating everyone, including the very young and people with limited mobility, to reach their required destination without having to use a car. The emergence of car-free and restricted car developments points to one way forward for health and sustainable development (Nieuwenhuijsen and Khreis 2016).

Green and blue infrastructure planning: Green infrastructure needs planning at the strategic scale and detailed design at the local level. The research into the relationship between green infrastructure and human health is well advanced (WHO 2017b). Well-planned strategic green and blue infrastructure can have multiple benefits for health, providing populations with access to nature and amenity; providing opportunities for urban farming, food growing and silviculture, and providing active long-distance transport routes and microclimate amelioration. This is in addition to a number of ecosystem service functions that support planetary health such as support for biodiversity through provision of species migration corridors and habitat reserves.

12.4.3 Climate Change

Spatial planning at the strategic scale has a major role to play in enabling cities to cope with extreme weather events. This includes the proper planning of the water cycle, including rain water capture, grey water reuse, flood risk management and sustainable urban drainage. There is an extensive literature on risks of climate change for health (WHO 2014). Urban planning needs to address building and transport energy use, two sources of climate emissions. In Europe, urban populations account for 69% of all energy use (EEA 2009b). Transport accounts for 21% of the climate gas emissions levels determined by both mode of travel used and spatial land use distribution (EC 2008). Two responses to climate change have been established, mitigation and adaptation. Mitigation is technological change and substitution to reduce resource inputs and emissions. Adaptation refers to measures to reduce the vulnerability of natural and human systems to climate change effects (Verbruggen 2007). With both responses, co-benefits should be sought which also increase the likelihood of healthier urban lifestyles.

12.5 Place-Making: Creating Human Habitat

Places are the everyday and intimate living environments that promote or detract from population health. What we have called strategic planning sets the scene for local environments. However design and planning decisions at the local level and how local people are involved (or not) create places. The idea of place exists at several scales, from individual rooms and apartments to neighbourhoods. When thinking of place in the context of health, we include physical characteristics: local buildings, streets and what people view as their own neighbourhood. However, the concept of place also has social dimensions: the relationships, support networks, social contacts and other aspects of a community. Urban design can support, detract from or even destroy this social dimension. Place must be seen as human habitat whose diverse characteristics combine to create, or undermine, health and well-being. The characteristics of a place have a bearing on behaviour and lifestyle choices open to individuals who live there. Importantly, what constitutes a healthy place can vary for different groups within society. The elderly, young, disadvantaged and infirm, for example, all have particular requirements of a place if it is to support their health and well-being.

Although individual studies often lack the methodology to show causality (a consistent methodological problem when researching real urban situations), there is a wealth of guidance providing design characteristics that support population health in a local environment (see, e.g. Barton et al. 2010). There is also a growing experience amongst practitioners about the kinds of processes that are successful in creating healthier places. People matter, and perceptions of neighbourhood are strongly associated with health and well-being (Croucher et al. 2007).

There are two distinct ways that cities can tackle place-making. First, through neighbourhood-based projects, these usually focus on specific locations. Projects may improve existing neighbourhoods or lay the foundations for new neighbourhoods. Two projects from the WHO Healthy Cities will illustrate this. Firstly, an urban design project in Kirikkale, Turkey, to transform a successful local street market whose growth has compounded local problems of high traffic volumes and air pollution (Grant 2015). Secondly, in Helsingborg, where the focus was a residential problem in four apartment blocks with high levels of vandalism; this was linked to a lack of local pride, lack of natural surveillance and no sense of community. Following a pilot study, an urban design-led gardening project saw rich health outcomes for physical activity, well-being and social capital (Grant and Lease 2014). Secondly, some projects at this scale are not based on a specific location but cover a specific type of local place found repeatedly across a city, such as the 'Healthy Streets' programme in London (TfL 2017) or the greening of transport corridors in Barcelona (Ajuntament de Barcelona 2013).

12.5.1 Neighbourhood Projects as Place-Based Health Promotion

Neighbourhood environment are places of particular importance; at this scale evidence consistently indicates that there is an association between the built environment, health and well-being and levels of physical activity (Saelens et al. 2003; Croucher et al. 2007). Research has often focussed on the measurement of specific environmental attributes that may determine health, such as population density, land use diversity, street network design, accessibility to transit and greenery and aesthetics; but causality remains illusive in such studies (McCormack and Shiell 2011; Feng et al. 2010). Neighbourhoods that support walking are associated with higher levels of physical activity and lower levels of obesity. Accessible neighbourhood resources are strongly associated with levels of physical activity. Local green space plays an important role in facilitating exercise and promoting health and well-being (WHO 2016).

One example of integrated neighbourhood work can be found in Gyor, Hungary, a WHO Europe designated Healthy City. Here, a physical regeneration programme combined with social interventions aimed to reduce health inequalities in a multicultural neighbourhood of the city, where many of the social and health problems were associated with a run-down urban environment. They began with assessments of physical interventions, which included renovations of the physical fabric, creating public areas for leisure, sport and play and general greening. To complement these physical improvements, the city introduced several social interventions with the main aims of community building and health and lifestyle improvements. These were delivered through programmes focusing on young people, children, families, mothers-to-be, mothers with young children and the Roma community (Grant and Lease 2014).

12.5.2 Involving Communities in Neighbourhoods for Health

The aspiration should be to engage local communities and stakeholders in designing and managing places, settings and communities consistent with the needs of users throughout the life course. And the good news is that 'communities get it!'. In Bristol and elsewhere, the HealthMap (Fig. 12.1) has been used as a 'health lens' through which local communities could understand their locality in terms of what is keeping them healthy and what is undermining their health. This approach can enable people's lived experiences to feed into neighbourhood planning processes, such as community health impact assessments of a regeneration plan (Hewitt and Grant 2010). In Belfast, UK, school children have participated, through a range of innovative range of engagement methods, in voicing their needs for an inclusive, resilient and child-friendly city. This approach has been particularly effective in bringing the poor quality of their walking routes to school to the attention of local politicians (BHC 2016).

The Scottish National Health Service has developed a tool to support healthier place-making (Scottish Government 2015). Called the 'Place Standard', the tool provides a method for communities and other stakeholders to come together to assess their local environments against 14 parameters, each with a physical manifestation that is known to influence health. The tool helps to combine social, economic, physical, cultural and historical characteristics of a location. Sixty-five separate instances of Place Standard use were recorded between December 2015 and February 2017. An evaluation report indicated how useful it has been in raising awareness and developing conversations leading to action about health and place-making (Scottish Government 2017).

12.5.3 Recognising Co-benefits

Planning for health is an approach that is strengthened, in its execution and outcomes, if the potential for co-benefits is recognised. In theory and in practice, a single intervention can be supported by multiple stakeholders, from several budgets if it delivers benefits for several city objectives. A good example of this is a programme in the Indian state of Bihar aimed at reducing the gender gap in secondary school enrolment by providing girls who continued to secondary school with a bicycle to improve access to school. Analysis found that this was responsible for a 32% increase in girls' enrolment in secondary school and a corresponding reduction in gender gap by 40% (Muralidharan and Prakash 2017). The initiative also proved good value for money compared with other options. Framing this in terms of Sustainable Development Goals (SDGs), it supported gender equality (SDG5). But this is the kind of initiative that would also support targets for good health and well-being (SDG3), quality education (SDG4) and sustainable cities and communities (SDG11) amongst others.

A global study of premature mortality due to non-compliance with international directives for the urban environment gives another example of the potential for co-benefits. Its findings emphasised the need for the reduction of motorised traffic through the promotion of active and public transport and the provision of green infrastructure, both of which could also provide opportunities for physical activity and mitigation of air pollution, noise and heat (Mueller et al. 2017). A third example can be found in Cape Town, South Africa. A public-private partnership formed to launch a housing retrofit programme in a low-income neighbourhood of Khayelitsha Township. Housing units were upgraded with insulation, low-energy lighting and solar water heaters with consequent savings in carbon dioxide emission and energy cost for each household. These measures lowered the risk of tuberculosis by reducing dampness in dwellings and improved hygiene by encouraging washing with warm water. The associated reductions in air pollution also lowered the risk of pneumonia and other respiratory illnesses (WHO 2011a). Focussing on co-benefits allows a stronger coalition of stakeholders to come together and support these more complex interventions.

12.6 Setting the Scene for Effective Action

Planning for healthy cities creates the potential for new city level collaboration for urban health. This is now supported globally. The Shanghai Consensus for Healthy Cities 2016 (Mayors Forum 2016) identified priority action areas for cities to achieve SDG3 'ensure healthy lives and promote well-being for all at all ages' and SDG11 'make cities and human settlements inclusive, safe, resilient and sustainable'. The Quito Implementation Plan, stemming from Habitat III, the United Nations Conference on Housing and Sustainable Urban Development, provides a platform for national action (WHO 2016). These lay the basis for national and city level action to put the wider determinants of health and health equity into city and transport planning.

12.6.1 City Leadership for Healthy People

In the quest for 'the creation of health', the causes of the wider determinants of health (and ill-health) have been the subject of continual dialogue (such as Hippocrates 1849; Hancock 1985; Dahlgren and Whitehead 2007; Burns 2014; de Leeuw and Simos 2017). The refrain often heard in the context of national policy and echoed in city leadership is that economic growth is the main determinant of health (Povall et al. 2007) and that economic growth as an objective in itself will promote human development and improve health as a matter of course. However it is not as simple as that. Economic development can bring with it widening inequalities that frustrate health objectives (Pickett and Wilkinson 2015) and even the way that economic development is measured is subject to challenge on the grounds of being insensitive or even blind to human and planetary well-being (Jackson and Senker 2011). Moreover, the association at national or city level between average wealth and health can often mask a deepening inequality. After a basic level of economic development, further improvements often skew benefit to those in higher socio-economic groups (CSDH 2008). As part of what looks like a circular argument, health may need to constantly reposition itself in relation to more dominant political objectives (Kickbusch 2012). For example, the European Union set an additional 2 years of healthy life expectancy as a key determinant of its economic growth policy through expecting a longevity dividend in return. This same type of reasoning was been applied when negotiating the Ministerial Declaration on NCDs at the United Nations in 2011, by calculating the significant loss to national economies that emerging powers will experience if they do not address the health challenges at hand (Kickbusch 2012).

With the drive for economic prowess in the minds of many city leaders, there also has been recent discussion of the commercial determinants of non-communicable disease (Kickbusch et al. 2016; Buse et al. 2017). In Bristol, the public health section of the local authority has published a number of working papers of the negative impact on health of job creation. Even though employment is a key determinant of health, we cannot say that job creation is always pro-health especially where the work is insecure, low paid and shift work or involves long commuting times. City leaders will need to carefully assess political and economic drivers locally in order to clarify, strengthen and position a healthy city or Healthy Urban Planning approach. However, with several competing drivers of city development, many conflicting with health goals, it can be an uphill struggle to attempt to use planning and urban design for such value-based activity without strong political support (Tsouros 2017).

In terms of action, city leaders will need to:

1. Prioritise city policies that create co-benefits between health and well-being and other social, economic and environmental goods.
2. Engage with public agencies, spatial planners, voluntary bodies, businesses and industry and all other actors whose activities influence the ability of places to support and create health.
3. Engage local communities in identifying the physical characteristics of place that most support the health and well-being.
4. Promote development that is supportive to groups of all ages and levels of ability.
5. Ensure sustainable use and access to natural resources and reduce vulnerabilities to climate risks for communities.

12.6.2 Coalitions for Healthy Planning

Policy makers and researchers need to join forces with city leaders and help them steer investments in physical change to better support health, together with their communities. Pivotal to this success is the contribution of spatial planning, through Healthy Urban Planning and healthy transport; it offers the opportunity to shape the form and function of cities and the neighbourhoods where people live:

- Through planning, development and design
- Across all sectors; city governance, urban management, community development, landscape, architecture, urban planning and transport
- By partnerships; in policy, in practice and in research
- For initial build, for regeneration and for retrofit
- At all scales; from building and street, home-patch and neighbourhood to district, city and city region

12.6.3 Changing the Game

Revisiting the analogy of a game. The ever-growing community of those interested in healthier cities now needs to take charge of how this game is being played. We need to find common cause between action on the ground in cities and the research into that action. So we should not forget that one goal of planning for healthy cities should be to obtain better knowledge of how to improve urban health through planning, urban design and transport. We need research methods that can support and critically analyse multidisciplinary groups, fully including the communities involved, as we all develop our understanding of neighbourhood characteristics that support health.

There are currently two significant areas where more knowledge is required:

- Firstly, the economic case still needs proving and articulating in a way that politicians will find meaningful. There is an implicit hunch that 'building-in' health to cities and their neighbourhoods makes long-term financial sense. Can we now develop the research to prove it?
- Secondly, in this multidisciplinary, multi-professional and multi-stakeholder domain, evidence is being 'created', and evidence is being 'consumed' across quite different traditions of validity and relevance. However, we have yet to develop a common language. Without this how can we identify the misconnections and outright gaps that still exist within our ever-growing landscape of evidence and interventions?

I would suggest the following key questions for practitioners, researchers and city leaders with their communities, to keep asking of their own local context:

- How can I develop and spread an understanding of health in relation to urban form?
- What urban components and determinants of health are locally most dominant?
- What urban planning solutions have been successful and how can we adapt and implement them more widely?

References

Ajuntament de Barcelona. (2013). *Barcelona green infrastructure and biodiversity plan 2020. Barcelona, Spain (in Spanish, English Summary).* Retrieved from http://scholar.google.com/scholar?hl=en&btnG=Search&q=intitle:Barcelona+green+infrastructure+and+biodiversity+plan+2020.#0

Anon 2013. Global thematic consultation on health: Health in the post-2015 agenda. *Report of the Global Thematic Consultation on Health*. New York: The World We Want.

Arthurson, K., Lawless, A., & Hammet, K. (2016). Urban planning and health: Revitalising the alliance. *Urban Policy and Research, 34*(1), 4–16.

Barton, H. (2009). Land use planning and health and well-being. *Land Use Policy, 26*, S115–S123.

Barton, H., & Grant, M. (2006). A health map for the local human habitat. *The Journal for the Royal Society for the Promotion of Health, 126*(6), 252–253. ISSN 1466-4240 developed from the model by Dahlgren and Whitehead, 1991.

Barton, H., & Grant, M. (2008). Testing time for sustainability and health: Striving for inclusive rationality in project appraisal. *The Journal of the Royal Society for the Promotion of Health, 128*(3), 130–139.

Barton, H., & Tsourou, C. (2000). *Healthy urban planning*. Copenhagen/London: World Health Organisation/SPON.

Barton, H., Tsourou, C., & Mitcham, C. (2003). *Healthy urban planning in practice: Experience of European cities*. Copenhagen: WHO Regional Office for Europe.

Barton, H., Grant, M., & Guise, R. (2010). *Shaping neighbourhoods: For local health and global sustainability*. Oxford: Routledge.

Belfast Healthy Cities. (2014). *Belfast: A WHO Healthy City through 25 years . . .* Belfast: Belfast Healthy Cities.

BHC. (2016). *Taking action for child friendly places: First steps – strategic approach and action plan for Belfast*. Belfast: Belfast Healthy Cities. Retrieved June 20, 2017, from http://www.belfasthealthycities.com/sites/default/files/publications/TakingAction.pdf.

Bowman, S., Gortmaker, S., Ebbeling, C., Pereira, M., & Ludwig, D. (2004). Effects of fast-food consumption on energy intake and diet quality among children in a national household survey. *Journal of the American Academy of Pediatrics, 113*, 112–118.

Brown, C., & Grant, M. (2005). Biodiversity and human health: What role for nature in healthy urban planning. *Built Environment, 31*, 326–338. http://doi.org/10.2148/benv.2005.31.4.326.

Burns, H. (2014). What causes health? *Journal of the Royal College of Physicians Edinburgh, 44*, 103–105.

Buse, K., Tanaka, S., & Hawkes, S. (2017). Healthy people and healthy profits? Elaborating a conceptual framework for governing the commercial determinants of non-communicable. *Globalization and Health*. http://doi.org/10.1186/s12992-017-0255-3.

Coutts, C., & Hahn, M. (2015). Green infrastructure, Ecosystem services, and human health. *International Journal of Environmental Research and Public Health, 12*(8), 9768–9798. http://doi.org/10.3390/ijerph120809768.

Croucher, K., Policy, H., & Beck, S. (2007). *Health and the physical characteristics of urban neighbourhoods: A critical literature review final report*. Glasgow: Glasgow Centre for Population Health.

CSDH. (2008). *Closing the gap in a generation: Health equity through action on the social determinants of health. Final report of the Commission on Social Determinants of Health*. Geneva: World Health Organization.

Dahlgren, G., & Whitehead, M. (1991). "The main determinants of health" model, version accessible. In G. Dahlgren & M. Whitehead (Eds.), (2007). *European strategies for tackling social inequities in health: Levelling up Part 2*. Copenhagen: WHO Regional Office for Europe. http://www.euro.who.int/__data/assets/pdf_file/0018/103824/E89384.pdf

Dahlgren, G., & Whitehead, M. (2007). *European strategies for tackling social inequities in health: Levelling up Part 2*. Copenhagen: WHO Regional Office for Europe.

EC. (2008). *Progress towards achieving the Kyoto objectives, Communication from the Commission*. Brussels, 19.11.2008, Com(2008) 651 Final/2, Commission of the European Communities.

EEA. (2009a). *About the urban environment, European Environment Agency website*. Retrieved June 7, 2009, from http://www.eea.europa.eu/themes/urban/about-the-urban-environment.

EEA. (2009b). *Ensuring quality of life in Europe's cities and towns*. EEA Report No. 5/2009. Luxembourg: Office for Official Publications of the European Communities.

Ellaway, A., Macintyre, S., & Bonnefoy, X. (2005). Graffiti, greenery, and obesity in adults: Secondary analysis of European cross sectional survey. *British Medical Journal, 331*, 611–612.

Ewing, R., & Cervero, R. (2010). Travel and the built environment: A meta-analysis. *Journal of the American Planning Association, 76*(3), 265–294.

Feng, J., Glass, T. A., Curriero, F. C., Stewart, W. F., & Schwartz, B. S. (2010). The built environment and obesity: A systematic review of the epidemiologic evidence. *Health & Place, 16*(2), 175–190.

Frank, L., Kavage, S., & Devlin, A. (2012). *Health and the built environment: A review*. The Canadian Medical Association: Quebec.

Garfinkel-Castro, A., Kim, K., Hamidi, S., & Ewing, R. (2017). Obesity and the built environment at different urban scales: Examining the literature. *Nutrition Reviews, 75*(Suppl. 1), 51–61. http://doi.org/10.1093/nutrit/nuw037.

Gartland, L. M. (2012). *Heat islands: Understanding and mitigating heat in urban areas*. Oxford: Routledge.

GCPH. (2011). *Asset based approaches for health improvement: Redressing the balance*. Glasgow Centre for Population Health. Retrieved from http://www.gcph.co.uk/assets/0000/2627/GCPH_Briefing_Paper_CS9web.pdf

GCPH. (2012). *Putting asset based approaches into practice: Identification, mobilisation and measurement of assets*. Glasgow Centre for Population Health Retrieved from http://www.gcph.co.uk/assets/0000/3433/GCPHCS10forweb_1_.pdf

Graham, H., & White, P. C. L. (2016). Social determinants and lifestyles: Integrating environmental and public health perspectives. *Public Health, 141*, 270–278.

Grant, M. (2015). European Healthy City Network Phase V: Patterns emerging for healthy urban planning. *Health Promotion International, 30*(Suppl. 1), i54–i70.

Grant, M., & Barton, H. (2013). No weighting for healthy sustainable local planning: Evaluation of a participatory appraisal tool for rationality and inclusivity. *Journal of Environmental Planning and Management, 56*(9), 1267–1289. ISSN 0964-0568.

Grant, M., & Lease, H. (2014). *MOTHER REPORT: WHO EURO Healthy Cities Phase V Evaluation: Healthy urban environment and design*. Bristol: University of the West of England.

Grant, M., Coghill, N., Barton, H., & Bird, C. (2009) .*Evidence review on environmental health challenges and risks in urban settings*. For WHO European Centre for Environment and Health. Technical Report. WHO Collaborating Centre for Healthy Cities and Urban Policy, University of the West of England (UWE), Bristol. Available from http://eprints.uwe.ac.uk/10384

Hancock, T. (1985). The mandala of health: A model of the human ecosystem. *Family & Community Health, 8*(3), 1–10.

Hanlon, P., Carlisle, S., Hannah, M., Reilly, D., & Lyon, A. (2011). Making the case for a "fifth wave" in public Health. *Public Health, 125*(1), 30–36.

Harlan, S. L., Brazela, A. J., Prashada, L., Stefanovb, W., & Larsenc, L. (2006). Neighborhood microclimates and vulnerability to heat stress. *Social Science & Medicine, 63*(11), 2847–2863.

Healey, P. (1996). The communicative turn in planning theory and its implications for spatial strategy formation. *Environment and Planning B: Planning and design, 23*(2), 217–234.

Healey, P. (2003). Collaborative planning in perspective. *Planning Theory, 2*(2), 101–123.

Hewitt, S., Grant, M., & Bristol Partnership. (2010). *Building health into our plans from the start: Report and review of the health impact assessment workshop on the Knowle West Regeneration Strategy*. Bristol: NHS Bristol/Bristol City Council/University of the West of England. Available from http://eprints.uwe.ac.uk/15235.

Hill, P. S., Buse, K., Brolan, C. E., & Ooms, G. (2014). How can health remain central post-2015 in a sustainable development paradigm? *Globalization and Health, 10*(1), 18.

Hippocrates. (1849). On air, waters, and places. In *The genuine works of Hippocrates*. New York: W. Wood. Translated with a commentary by Francis Adams. London: The Sydenham Society.

Jackson, T., & Senker, P. (2011). Prosperity without growth: Economics for a finite planet. *Energy & Environment, 22*(7), 1013–1016.

Jones, R., & Yates, G. (2013). *The built environment and health: An evidence review*. Briefing paper 11. Glasgow Centre for Population Health.

Kelly, M. P., Charlton, B. G., & Hanlon, P. (1993). The four levels of health promotion: An integrated approach. *Public Health, 107*(5), 319–326.

Khreis, H., Kelly, C., Tate, J., Parslow, R., Lucas, K., & Nieuwenhuijsen, M. (2017). Exposure to traffic-related air pollution and risk of development of childhood asthma: A systematic review and meta-analysis. *Environment International, 100*, 1–31.

Kickbusch, I. (2012). Addressing the interface of the political and commercial determinants of health. *Health Promotion International, 27*(4), 427–428. http://doi.org/10.1093/heapro/das057.

Kickbusch, I., Allen, L., & Christian, F. (2016). The commercial determinants of health. *Lancet, 4*(2016), e895–e896. http://doi.org/10.1016/S2214-109X(16)30217-0.

Lavin, T., Higgins, C., Metcalfe, O., & Jordan, A. (2006). *Health impacts of the built environment: A review*. Dublin and Belfast: The Institute of Public Health in Ireland.

LBTH. (2006). *Equality impact assessment: Aldgate masterplan supplementary planning document*. London Borough of Tower Hamlets. Retrieved August 6, 2017, from http://democracy.towerhamlets.gov.uk/documents/s4098/Aldgate%20Masterplan%20draft%20EqIA.pdf.

de Leeuw, E. (2017). Cities and health from the neolithic to the anthropocene. In E. de Leeuw & J. Simos (Eds.), *Healthy cities: The theory, policy, and practice of value-based urban planning* (pp. 3–30). New York: Springer.
de Leeuw, E., & Simos, J. (Eds.). (2017). *Healthy cities: The theory, policy, and practice of value-based urban planning*. New York: Springer.
Leppo, K., Ollila, E., Pena, S., Wismar, M., & Cook, S. (2013). *Health in all policies-seizing opportunities, implementing policies*. Finland: Ministry of Social Affairs and Health.
Maas, J., Verheij, R. A., Groenewegen, P. P., De Vries, S., & Spreeuwenberg, P. (2006). Green space, urbanity, and health: How strong is the relation? *Journal of Epidemiology & Community Health, 60*(7), 587–592.
Mayors Forum. (2016). Shanghai consensus on healthy cities 2016. In *9th global conference on health promotion*. Retrieved August 6, 2017, from http://www.who.int/healthpromotion/conferences/9gchp/9gchp-mayors-consensus-healthy-cities.pdf?ua=1.
McCormack, G. R., & Shiell, A. (2011). In search of causality: A systematic review of the relationship between the built environment and physical activity among adults. *International Journal of Behavioral Nutrition and Physical Activity, 8*(1), 125.
McLeroy, K. R., Bibeau, D., Steckler, A., & Glanz, K. (1988). An ecological perspective on health promotion programs. *Health Education Quarterly, 15*, 351–377.
Mitchell, R., & Popham, F. (2008). Effect of exposure to natural environment on health inequalities: An observational population study. *Lancet, 2008*(372), 1655–1660.
Mueller, N., Rojas-Rueda, D., Basagaña, X., Cirach, M., Cole-Hunter, T., Dadvand, P., et al. (2017). Urban and transport planning related exposures and mortality: A health impact assessment for cities. *Environmental Health Perspectives, 125*, 1–36. http://doi.org/10.1289/EHP220.
Muralidharan, K., & Prakash, N. (2017). Cycling to school: Increasing secondary school enrollment for girls in India. *American Economic Journal: Applied Economics, 9*(3), 321–350.
Nadin, V., & Stead, D. (2008). European spatial planning systems, social models and learning. *disP – The Planning Review, 44*(172), 35–47.
Nieuwenhuijsen, M. J., & Khreis, H. (2016). Car free cities: Pathway to healthy urban living. *Environment International, 94*, 251–262. http://doi.org/10.1016/j.envint.2016.05.032.
Nieuwenhuijsen, M. J., Khreis, H., Triguero-Mas, M., Gascon, M., & Dadvand, P. (2017). Fifty shades of green. *Epidemiology, 28*(1), 63–71.
Oni, T., & Unwin, N. (2015). Chronic non-communicable diseases and infectious diseases. In A. D. G. Aikins and C. Agyemang (Eds.), *Chronic non-communicable diseases in low and middle-income countries* (pp. 30–49). cabi.
Oni, T., Smit, W., Matzopoulos, R., Adams, J. H., Pentecost, M., Rother, H. A., Albertyn, Z., Behroozi, F., Alaba, O., Kaba, M., & van der Westhuizen, C. (2016). Urban health research in Africa: Themes and priority research questions. *Journal of Urban Health, 93*(4), 722–730.
Pawson, R., & Tilley, N. (1997). *Realistic evaluation*. Sage.
Peden, M. M. (Ed.). (2008). *World report on child injury prevention*. Copenhagen: World Health Organization.
Pickett, K. E., & Wilkinson, R. G. (2015). Income inequality and health: A causal review. *Social Science & Medicine, 128*, 316–326.
Povall, S., Whitehead, M., Gosling, R., & Barr, B. (Eds.). (2007). *Focusing the equity lens: Arguments and actions on health inequalities*. Liverpool: World Health Organization Collaborating Centre for Policy Research on Social Determinants of Health.
Rao, M., Prasad, S., Adshead, F., & Tissera, H. (2007). The built environment and health. *The Lancet, 370*(9593), 1111–1113.
RCEP. (2007). *The urban environment, royal commission on environmental pollution*. London: The Stationery Office.
Robinson, V. (2004). Fast food and obesity—Is there a connection? *Health and Healthcare in Schools, 4*, 12.

Rydin, Y., Bleahu, A., Davies, M., Dávila, J. D., Friel, S., De Grandis, G., Groce, N., Hallal, P. C., Hamilton, I., Howden-Chapman, P., & Lai, K. M. (2012). Shaping cities for health: Complexity and the planning of urban environments in the 21st century. *Lancet, 379*(9831), 2079.

Saelens, B. E., Sallis, J. F., Black, J. B., & Chen, D. (2003). Neighborhood-based differences in physical activity: An environment scale evaluation. *American Journal of Public Health, 93*, 1552–1558.

Sallis, J. F., & Owen, N. (2002). Ecological models of health behavior. In K. Glanz, B. K. Rimer, & F. M. Lewis (Eds.), *Health behavior and health education: Theory, research and practice* (3rd ed., pp. 462–484). San Francisco, CA: Jossey-Bass.

Scottish Government, NHS Health Scotland, Architecture and Design Scotland. (2015). Place Standard. Retrieved August 14, 2017, from http://www.gov.scot/Resource/Doc/229649/0062206.pdf.

Scottish Government, NHS Health Scotland, Architecture and Design Scotland. (2017). Place Standard process evaluation: Learning from case studies in year one. Retrieved August 14, 2017, from http://www.healthscotland.scot/media/1394/place-standard-process-evaluation_may2017_english.pdf.

Tfl. (2017). *Healthy streets for London*. The Mayor's Office: Transport for London Retrieved from http://content.tfl.gov.uk/healthy-streets-for-london.pdf

Thompson, C. W., Roe, J., Aspinall, P., Mitchell, R., Clow, A., & Miller, D. (2012). More green space is linked to less stress in deprived communities: Evidence from salivary cortisol patterns. *Landscape and Urban Planning, 105*(3), 221–229.

Townshend, T. G. (2017). Toxic high streets. *Journal of Urban Design, 22*(2), 167–186. http://doi.org/10.1080/13574809.2015.1106916.

Tsouros, A. D. (2017). Healthy cities: A political project designed to change how cities understand and deal with health. In E. de Leeuw & J. Simos (Eds.), *Healthy cities: The theory, policy, and practice of value-based urban planning*(pp. 489–504). New York: Springer.

Vardoulakis, S., Wilkinson, P., & Dear, K. (Eds.). (2016). Healthy polis: Challenges and opportunities for urban environmental health and sustainability. *Environmental Health, 15*(Suppl. 1), S30.

Verbruggen, A. (Ed.). (2007). Glossary. In B. Bert Metz, O. Davidson, P. Bosch, R. Dave, & L. Meyer (Eds.), *IPCC fourth assessment report*. Cambridge: Cambridge University Press.

Walker, G., Fairburn, J., Smith, G., & Gordon, M. (2003). *Environmental quality and social deprivation*. R&D Technical Report E2-067/1/TR. Bristol: Environment Agency.

Ward Thompson, C., Aspinall, P., Roe, J., Robertson, L., & Miller, D. (2016). Mitigating stress and supporting health in deprived urban communities: The importance of green space and the social environment. *International Journal of Environmental Research and Public Health, 13*(4), 440.

WHO. (2002). *The World Health Report 2002 – Reducing risks, promoting healthy life*. Geneva: World Health Organization.

WHO. (2003). *Extreme weather events: Health effects and public health measures*. Fact Sheet EURO/04/03, Copenhagen, Rome 29 September 2003. Retrieved June 7, 2009, from http://www.euro.who.int/document/mediacentre/fs0403e.pdf.

WHO. (2007). Quantifying burden of disease from environmental noise: Second Technical Meeting Report, Bern, Switzerland.

WHO. (2011a). *Health in the green economy: health co-benefits of climate change mitigation – Housing sector*. Geneva: World Health Organization.

WHO. (2011b). *Burden of disease from environmental noise: Quantification of healthy life years lost in Europe*. Copenhagen: World Health Organization Regional Office for Europe.

WHO. (2014). Promoting health while mitigating climate change. In *Technical briefing for the World Health Organization conference on health and climate discussion draft* (pp. 1–33). Retrieved from http://www.who.int/phe/climate/conference_briefing_2_promotinghealth_27aug.pdf

WHO. (2016). *Health as the pulse of the new urban agenda: United Nations conference on housing and sustainable urban development, Quito, October 2016*. Geneva: World Health Organization.

WHO. (2017a). *EUR/RC67/9: Roadmap to implement the 2030 Agenda for Sustainable Development, building on Health 2020, the European policy for health and well-being*. World Health Organisation.

WHO. (2017b). Urban green space interventions and health: A review of impacts and effectiveness. *Full report*. World Health Organization Regional Office for Europe. Retrieved from http://www.euro.who.int/en/health-topics/environment-and-health/urbanhealth/publications/2017/urban-green-space-interventions-and-health-a-review-of-impacts-and-effectiveness.-full-report-2017

WHO EURO. (2009). *Phase V (2009–2013) of the WHO European Healthy Cities Network: Goals and requirements*. Copenhagen: WHO Regional Office for Europe.

WHO EURO. (2016). Urban green spaces and health: A review of the evidence. Copenhagen: WHO Regional Office for Europe.

WHO-UN. (2016). *Global report on urban health: Equitable, healthier cities for sustainable development*. Geneva: World Health Organization and UN-Habitat.

Yue, W., Liu, Y., & Fan, P. (2013). Measuring urban sprawl and its drivers in large Chinese cities: The case of Hangzhou. *Land Use Policy, 31*, 358–370.

Part III
Accessibility and Mobility

Chapter 13
Land Use Policy, Travel Behavior, and Health

Bert van Wee

13.1 Introduction

The transport system, together with the land use system, allows people to participate in activities at different places and transport goods between different locations. In case of people: it allows us to travel to work, friends, and relatives, amenities, health-care services, shops, schools, and many more destinations. This results in important economic and other accessibility-related benefits for society.

But these benefits come at considerable costs for the users, and (via the tax payer), for governments (e.g., costs of infrastructure, subsidies for public transport), and for society (environmental impacts, safety impacts).

An important effect category that is largely (but not only) of a non monetary nature is health. Health effects of the transport system can be both positive and negative. This chapter's focus is on travel-related health effects. The four dominant categories of such health effects for the traveler are accidents, exposure to pollutants, exercise, and well-being, and in addition, there are health effects for others, such as exposure to air pollution and noise for people living near heavily trafficked roads. Health is an upcoming theme in the field of transport, both in policy making and in research. Since 2013 there is even a journal titled *Journal of Transport and Health*. The growing interest in transport and health is partly related to the increasing awareness of the health risks of transport policies and the health benefits of active modes (walking and cycling), and more specifically because cycling is becoming more popular in many cities and regions across the world (Pucher and Buehler 2012), and receives a lot of attention in academic research anyway. Of the ten most downloaded papers of the journal *Transport Reviews*, six were about cycling (assessed 2-12-2016), and all these papers were published recently, since 2008. Health impacts of travel for people other than the user, mainly exposure to air

B. van Wee (✉)
Transport and Logistics Group, Delft University of Technology, Delft, The Netherlands
e-mail: G.P.vanWee@tudelft.nl

M. Nieuwenhuijsen, H. Khreis (eds.), *Integrating Human Health into Urban and Transport Planning*, https://doi.org/10.1007/978-3-319-74983-9_13

pollutants and noise, have been addressed in the literature and policy making for decades. The same applies for accident risks of travel. The *increasing* attention paid to transport and health is related to all four health-related categories of effects for the traveler as presented above: there is an increasing awareness of cycling being a form of exercise (and thus improves health), and an activity that might increase well-being, but cycling also results in an increased intake of pollutants and is often risky (at least expressed as risks per kilometer).

But the relationships between transport and health include more than those related to active modes. People traveling by car, public transport, and aircraft are also exposed to risks and often to pollutants and experience varying levels of well-being.

Several policy categories can influence the impact of the transport system on health. Regulations for new road vehicles have an impact on emission and exposure levels, and on road safety of both people using these vehicles, and people experiencing the risk of being hit by these vehicles. Pricing policies (e.g., subsidies on public transport, levies on fuels, taxes on cars) and parking policies influence mode choice and therefore exposures and health effects. Infrastructure policies influence the (un) attractiveness to travel to distinguished destinations, via influencing travel times, travel costs, and effort. Specific public transport policies (such as those having an impact on the services offered and tariffs) influence mode choice and the intensity of using public transport (number of trips, distances traveled). Land use policies influence which activities are located were, and next in multiple ways influence travel behavior and next health, but also the health impacts of the transport system in other ways (see Sects. 13.4 and 13.5).

Despite the increasing awareness of the relationships between transport and health, to the best of my knowledge, there is no systematic overview of policies to influence the impact of the transport system on health. This chapter aims to reduce this knowledge gap. It is beyond the aims of this chapter to discuss all policies but focuses on one important and less researched category: land use policies. Land use policies are the most relevant to the scope of the book (which is on urban development) in which this chapter is included. More specifically, this chapter aims to answer the question:

How can land use policies influence the impact of travel behavior on health?

This general question is answered by answering next sub-questions:

1. In which ways does travel behavior influence health?
2. How can the impact of travel behavior on health be conceptualized?
3. How does land use influence the transport system, travel behavior, and next health?
4. Which land use policies can influence the impact of land use via the transport system on health?

Note that land use policies have way more effects than those related to health only and can influence the environmental pressure of the transport system on ecosystems (e.g., by changing mode choice) or levels of accessibility. These effects are excluded from the current chapter (see Van Wee 2011, for a discussion on the environmental and accessibility benefits of land use policies).

Before I explain the impacts of travel behavior and the wider transport system on health, it is important to make explicit how health is defined. The World Health Organization (WHO) defines health as "a state of complete physical, mental and social well-being and not merely the absence of disease or infirmity" (http://www.who.int/about/definition/en/print.html). Following earlier work (van Wee and Ettema 2016), in this chapter I also consider health to be broader than the absence of disease or infirmity, but I adopt a less broad approach than the WHO, by excluding the social dimension. This is because the social dimension is only indirectly related to the links between travel behavior and health.

Section 13.2 presents two conceptual models expressing the relationships between transport and health, answering the first two sub-questions. Next, Sect. 13.3 explains the relationships between land use and the transport system and travel behavior as far as relevant for the health impacts of travel, answering sub-question 3, and Sect. 13.4 answers sub-question 4 presenting options for land use policies to influence the transport system and next health. Section 13.5 discusses some key topics for the impact of land use via travel behavior and the transport system, on health. Section 13.6 finally summarizes the main conclusions of this chapter.

13.2 Transport and Health: A Conceptualization

I first discuss health impacts related to people's traveling. Van Wee and Ettema (2016) propose a conceptual model for the relationships between travel behavior and health. I take this model, in a slightly revised form, as a point of departure. Below I present the model and summarize the underpinnings. For a more elaborate discussion of the model and sources used, the reader can refer to Van Wee and Ettema (2016) (Fig. 13.1).

Figure 13.1 makes clear that health of people traveling is primarily determined by the following components:

- Level of physical activity (Block A)
- Air pollution intake (Block B)
- Casualties/accidents (Block C)
- (Subjective) well-being (Block D)

These four main factors are interrelated, a first example being that the use of active modes may result in an increase of subjective well-being (Olsson et al. 2013), but on the other hand, accidents (crashes/falls) can decrease the use of active modes because people become disabled or because they become scared to use these modes (see Lee et al. 2015). Another example: high concentrations of air pollutants are unhealthy and can also reduce the willingness of people to walk or cycling, reducing the health benefits of physical activity.

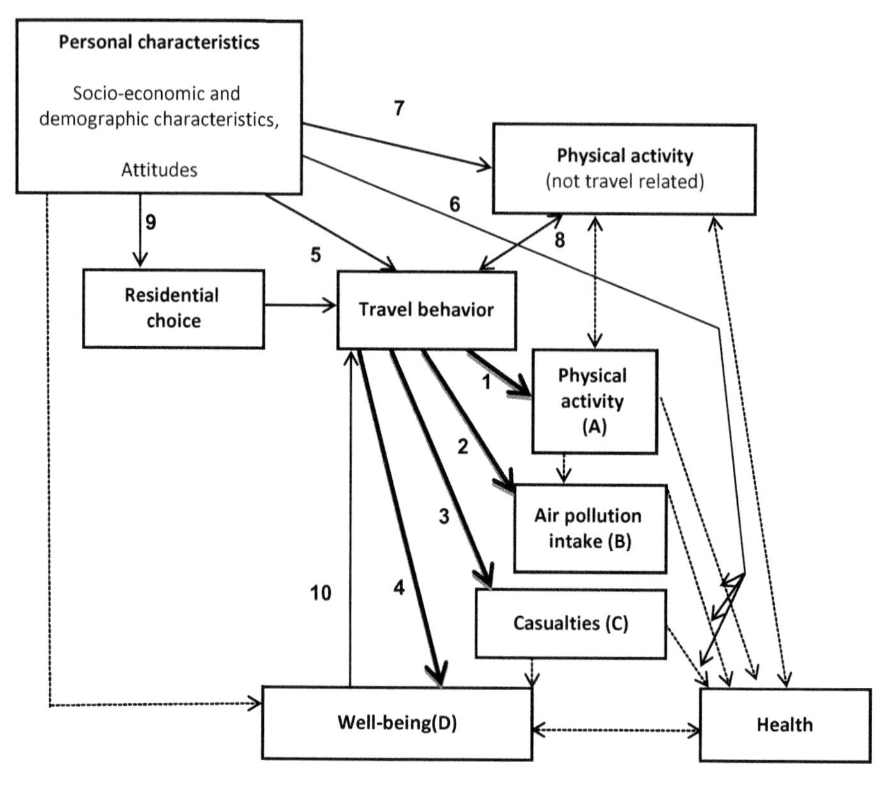

Fig. 13.1 Conceptual model for the dominant relationship between travel behavior and health of people traveling (source: Van Wee and Ettema 2016)

Arrows 1–4 express the direct impact of travel behavior on these four factors. Distances traveled influence mode choice and vice versa: the bike is not an option if a person needs to travel for 100 km, and a person preferring to cycle will chose less remote destinations. Consequently, travel behavior influences the level of travel-related physical activity of people (arrow 1—e.g., Handy 2014). If people walk or cycle, their intake of pollutants per unit of time can be higher compared to when they would drive, especially when they walk or cycle close to heavily trafficked roads (arrow 2—e.g., Nyhan et al. 2014). But if they travel in areas with lower concentrations of pollutants, the intake can also be lower compared to when they would drive. People traveling by underground are exposed to relatively high concentrations of particulate matter (PM) originating from mechanical friction processes (e.g., Şahin et al. 2012). Therefore, distance traveled and mode choice influence the levels of air pollution exposures. Other risk factors are also mode-dependent (Wegman 2013), and in case of road traffic, these vary between road types (Amoros et al. 2003) (arrow 3). In Swedish cities (momentary), well-being is highest for people who commute by active modes, followed by traveling by car and finally public transport (e.g., Olsson et al. 2013). The authors hypothesize that desirable physical exercise might explain the high level of well-being for active modes, as well as the relatively short travel distances—long commuting distances are less appreciated. They do not explain the difference between driving and public transport. Next, travel influences (subjective) well-being because people can reach locations of activities and services (e.g., De Vos et al. 2013) (arrow 4).

In addition, several second-order relationships exist, as expressed in Fig. 13.1:

- Socioeconomic and demographic characteristics and travel behavior (arrows 5–7)
- Physical activity: walking and cycling versus wider activity patterns (arrow 8)
- Subjective well-being and the use of active modes (arrow 10)
- Self-selection effects (arrows 5 and 9)

13.2.1 Socioeconomic and Demographic Variables (Arrows 5–7)

The importance of socioeconomic and demographic variables (such as age, gender, education level, and household characteristics) for travel behavior (arrow 5) is confirmed by many studies (e.g., Stipdonk et al. 2013). In addition, these variables can mediate the impact of physical activity, air pollution intake, and crashes/falls on health (arrow 6). For example, falling from a bicycle in general will have more impact on an 80-year-old person than on a 15 years old, and obese people will benefit more from an increase of physical activity (Bauman 2004). Comparably, personal characteristics may also have an impact on non-travel-related physical activity (arrow 7), and next its impact on health, as well as on the impact of well-being on health.

13.2.2 Interaction of Travel-Related Physical Activity and Other Physical Activity (Arrow 8)

Walking and cycling levels can be related to other forms of physical activity. People may substitute these two forms. It could be that a person does not go to the gym because she already walks or cycles frequently. On the other hand, it is also possible that people who walk or cycle feel fitter and therefore also engage more in other forms of physical activity. For example, they might take the stairs and not the elevator. Such relationships are hardly studied in the literature, and the results are inconclusive.

13.2.3 Causality of Subjective Well-Being and the Use of Active Modes (Arrow 10)

It is possible that people with a high level of subjective well-being walk and cycle (arrow 10) more than average, but to the best of my knowledge, there is hardly any literature on this topic, an exception being Baruth et al. (2011) who conclude that people with higher levels of subjective well-being achieved more than the average increase in physical activity levels during a physical activity intervention program. It could be that a higher level of subjective well-being increases the willingness to change behavior, but this is rather speculative. The literature generally studies the reverse causality: the impact of walking and cycling on mental health and mood (see above).

13.2.4 Self-Selection Effects (Arrows 5 and 9)

A potential important phenomenon relevant for the relationships between travel behavior and health is self-selection. People self-select in many ways. For example, people with higher incomes generally live in neighborhoods with more expensive houses. This form of self-selection is generally included in research by including socioeconomic and demographic variables, and therefore I do not further discuss it. But people can self-select in many other respects, the most often studied form being residential self-selection based on preferences for modes or travel attitudes more generally (e.g., Cao et al. 2009). The impact of attitudes on travel behavior is conceptualized via arrow 5 in Fig. 13.1 and the impact of attitudes on residential self-selection by arrow 9. Residential self-selection can also be influenced by health considerations (not conceptualized in Fig. 13.1). For instance, people who think exercise is important may chose a residential location that encourages walking or cycling.

As expressed by the heading of Fig. 13.1, this figure conceptualizes the dominant relationships related to travel behavior for the people traveling. In addition, there are effects on other people. Figure 13.2 conceptualizes these impacts.

Figure 13.2 shows that people traveling affect the health of others, nearby or "local" effects being a first category (noise, air pollution, risks, and "barrier effects" (e.g., crossing ability) being dominant effects). People exposed to these effects are other road users, and people staying near roads, such as residents, and children at schools located near heavily trafficked roads. Traffic also contributes to larger-scale air pollutions in the form of smog. To keep Fig. 13.2 as simple as possible, these are not explicitly included in another box but assumed to be included in the "nearby" box. In addition, people traveling and the infrastructure they use result in barrier effects: people cannot easily cross streets because of traffic, or there are no nearby physical options to cross motorways, other main roads, or railways. Secondly, transport contributes to climate change, mainly due to CO_2 emissions, and climate change will have a range of health-related effects (e.g., Patz et al. 2005), such as exposure to flood risks, extremely hot temperatures, and the spread of diseases. Next, travel behavior in the long run will induce land use changes, as often expressed in the land use and transport interaction literature (e.g., Wegener and Fürst 1999). For example, if more people travel by car, companies, shops, and services value car accessibility higher and might prefer to be located at locations well accessible by car. And such land use changes influence accessibility levels. For example, a shift of activities to locations well accessible by car might result in social exclusion of those

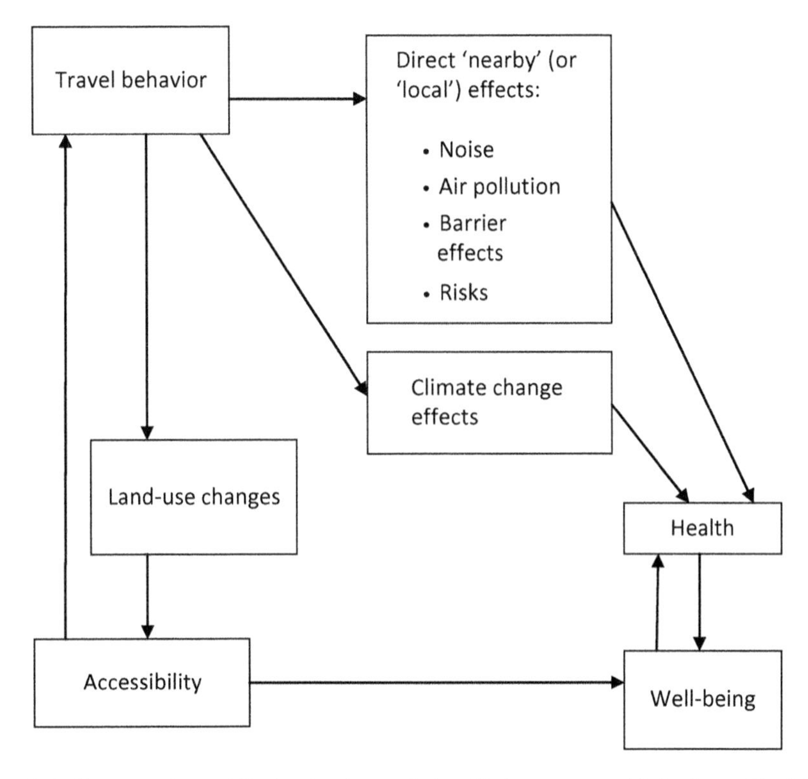

Fig. 13.2 Conceptual model for the dominant relationship between travel behavior and health of others than the traveler

not having a car available (e.g., Lucas 2004), decreasing the well-being of people and next their health. And changing land use has an impact on travel behavior. Land use changes also influence the nearness of green space and health effects due to exposure of green space.

13.3 Land Use, the Transport System, and Health

I now move to the impact of land use on the transport system and next on health.

Figure 13.3 conceptualizes the ways in which land use can influence the four blocks that influence health of people traveling (physical activity, air pollution intake, casualties/accidents, and well-being) as presented above. Note that the figure is not limited to the direct effects of travel behavior via these four blocks but takes a broader perspective also focusing on the locations of travel, which comprise both the locations of infrastructure, as well as the use of infrastructure by people traveling.

Figure 13.3 shows that land use (the locations of activities) can influence the four blocks relevant for health in multiple ways. First, land use can influence travel

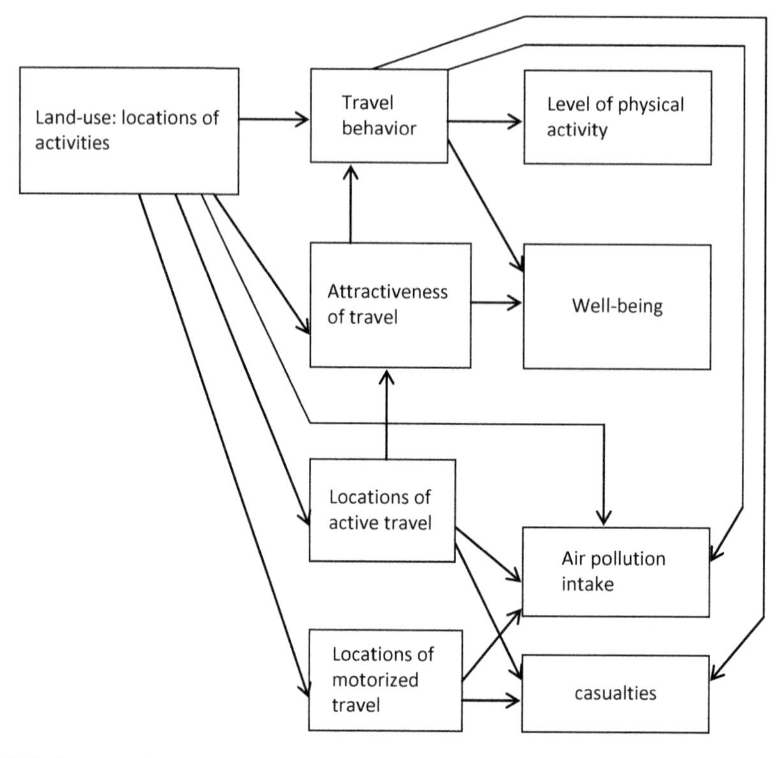

Fig. 13.3 Impacts of land use, via the transport system, on physical activity, air pollution intake, casualties, and well-being of people traveling

behavior and next the levels of physical activity and well-being (and next health—see above). The land use factors that are often found to have an impact on travel behavior are sometimes labeled as the five Ds and include density, diversity, design, destination accessibility, and distance to transit (Ewing and Cervero 2010). Density is expressed in units of a variable per unit of surface, e.g., population size, number of jobs, or houses per square kilometer. Higher densities, at least theoretically, allow for shorter trip distances because destinations are nearer. Shorter distances increase the attractiveness and convenience of walking and cycling and increases the share of these modes. And an increase in the levels of walking and cycling can increase both physical activity and well-being, as explained above. Diversity expresses the level to which different land use categories (dwellings, shops, medical services, schools, jobs, etc.) are mixed. Higher level of mixed land use can reduce travel distances. For example, if all shops and services would be concentrated in the center of a town, people on average would have to travel longer compared to when shops and services would be distributed over neighborhood centers (and the town center). Consequently, mixed land use also influences mode choice because, as explained above, active modes are relatively attractive for shorter distances. Design expresses street network characteristics and can influence travel behavior in many ways. For example, a grit-based street pattern can reduce travel distances compared to other street patterns that force people to take longer routes. And next, as explained above, it can influence the share of active modes. Destination accessibility expresses how (un) easy it is to reach the locations of destinations. It is often expressed in travel times or distances to (potential) destinations. The nearer destinations, the shorter travel distances and the higher the share of active modes. Distance to transit expresses the shortest distance (or sometimes time) to travel to a bus stop or railway station and influences mode choice, in particular the share of transit. Results of many studies reveal that land use does influence travel behavior, after controlling for socioeconomic and demographic variables and even after controlling for (attitudes based) residential self-selection. But the influence is not very strong and limited compared to socioeconomic and demographic variables (Ewing and Cervero 2010, and see many references in that study for further underpinnings of this general conclusion). But there still is discussion about the quantitative results, the interpretation, and the desirability of related land use policies (e.g., Stevens 2017).

Note that design can also relate to the characteristics of infrastructure, not only to land use, and infrastructure characteristics also influence travel behavior. And there is a gray area, related to reduced car access, or even car-free zones. Some researchers interpret this as infrastructure planning, but one can also see this as a form of land use planning. In case of this latter interpretation, it is important to note that zones with little or no motorized traffic, often central urban areas, will have better air quality and will be more attractive for pedestrians and cyclists—see, for example, Nieuwenhuijsen and Khreis (2016) who conclude that such areas have direct and indirect health benefits, but that the size and conflicts between different effects are yet unclear.

Secondly, land use can influence the attractiveness of travel. Nice scenery and attractive infrastructure (cycle lanes, wide pavements) increase levels of walking and cycling (Meurs and Haaijer 2001) and can improve the well-being of travel directly, e.g., due to enjoying the scenery (see Gatersleben and Uzzell 2007).

Thirdly, the locations of active travel and motorized travel infrastructure matter. The locations of active travel infrastructure matter, at least because of the attractiveness of the specific route taken (see previous point) and next because the specific routes taken have an impact on risks and the intake of pollutants. This is because risk factors vary between road types (see above), and the concentrations of pollutants also vary between roads. The magnitude of these impacts to some extent not only relates to the characteristics of the infrastructure and the direct environment of infrastructure but also to the locations of motorized transport (where and when do which motorized vehicles drive), as conceptualized by the arrow from "locations of motorized travel" to "air pollution intake" and "causalities". If, for example, roads for motorized road traffic are adjacent to cycle lanes, the intake of pollutants will be higher compared to cases in which the distance to the nearest road with motorized traffic is larger. And also for people traveling by motorized modes, the intake depends on the concentrations of pollutants which vary by road, and the risks vary by road type (and even road).

The impact of land use on travel behavior also influences the health of others, as conceptualized by Fig. 13.4.

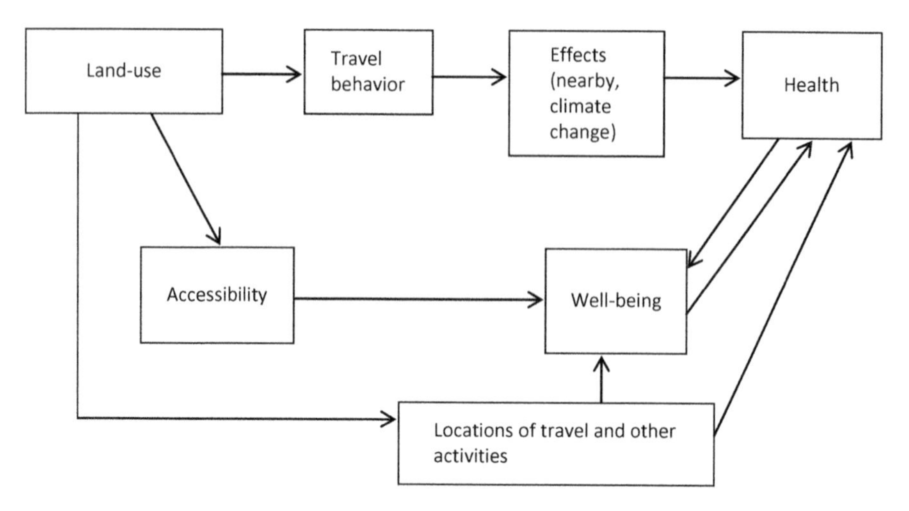

Fig. 13.4 Impacts of land use, via the transport system, on health of others than the traveler

Land use influences travel behavior and accessibility and next health in the way as conceptualized in Fig. 13.2. In addition, it has an impact on where people drive, and where other people stay (travel, live, work, shop, etc.), and therefore on their exposure to the negative impacts of people traveling (air pollution, noise, barrier effects, long-term climate change impacts). It may even influence well-being because of other negative environmental implications of travel, such as the impact of parked and driving vehicles, regardless of risks, pollution, and noise. For example, if many cars are parked in streets in residential areas, it may prevent children from playing on the street.

13.4 Land Use Policies

The next question is: How can land use policies influence travel behavior, the attractiveness of travel, the locations of active and motorized travel, and the locations of other activities (working, living, shopping, etc.)? I first discuss the determinants that land use can influence, followed by a discussion on specific policy instruments. Table 13.1 presents the main determinants for land use policies.

Firstly, land use policies can influence all Ds as presented above and next travel behavior. But the options for this influence differ across regions and countries/states, depending on the policy instruments available and the planning culture at stake. In several European countries, it is much more common for policy makers to develop land use policies than in the USA, although planning concepts like *Transit Oriented Development* and *New Urbanism* have gained popularity during the past decades (Cervero and Radisch 1996; Handy 1996, 2005).

Land use policies can, among others, include policies encouraging building in high densities and policies that stimulate mixed land use (diversity) and building

Table 13.1 Determinants influenced by land use policies and their impact on travel-related health

Determinants for land use policies	Impact on:
Density	Travel behavior, locations of active and motorized travel
Diversity	Travel behavior, locations of active and motorized travel
Design	Travel behavior, locations of active and motorized travel
Destination accessibility	Travel behavior, locations of active and motorized travel
Distance to transit	Travel behavior, locations of active and motorized travel
Attractiveness of infrastructure	Quality of the environment/attractiveness, route choice
Attractiveness of the areas near infrastructure	Quality of the environment/attractiveness, route choice
Any form of land use in general	Locations of other activities

near railway stations. Such policies often have synergetic effects. For example, not only building near stations will increase the share of the train in travel, but building in high densities near stations will further increase the potential of rail. See Van Wee (2002) for a more elaborate discussion of how land use policies can influence travel behavior.

Secondly, land use policies influence the attractiveness of travel, by improving the attractiveness of both the infrastructure and the areas adjacent to infrastructure. Infrastructure can be made attractive, for example, by constructing attractive noise barriers and using asphalt with nice colors. The area adjacent to infrastructure can be made attractive by vegetation or water areas and nice buildings near infrastructure. Infrastructure can be planned making use of already nice areas. The attractiveness of infrastructure and the adjacent areas influences the attractiveness of a given route and can also influence route choice.

Thirdly, all journeys have an origin, a destination, and a route connecting both, and land use planning can influence the locations of activities and consequently the locations of active and other forms of travel: where do people travel? The determinants influenced by land use policies are the same five Ds as discussed above. In addition, the routes that people chose between given origins and destinations can be influenced by land use planning, as explained above, by influencing the attractiveness of infrastructure and the adjacent areas.

Fourthly, land use policies can influence all activities other than travel and consequently the levels of exposure to pollution, noise, and barrier effects and well-being-related effects.

Which specific policy instruments do authorities have available as far as land use policies are concerned? The way in which land use policies are implemented varies between countries. I distinguish between direct and indirect policies. Direct policies directly determine which land use categories are (not) allowed at which locations, zoning being a dominant instrument type. Indirect policies can be manifold. For example, several policies can influence land values and prices, and these values influence land use as expressed by bid-rent theory (e.g., Alonso 1964), examples being restrictions on urbanization and anti-speculation policies. Regulations, e.g., with respect to maximum speeds, can influence the negative impacts of motorways on the environment and indirectly the (un)attractiveness of the surroundings for specific land use categories. It is beyond the scope of this chapter to discuss all indirect policies. I limit myself to policies related to the level of service of the transport system, distinguishing infrastructure policies and public transport policies.

Infrastructure influences land use, as recognized by so-called Land Use—Transport Interaction models (LUTI models) (e.g., Wegener and Fürst 1999). Areas around railway, metro, or tram stations are attractive for some use categories (e.g., companies with many office jobs) because of the high accessibility by public transport; areas near motorway that exists may be attractive for other use categories (e.g., distribution centers for goods). On the other hand, areas close to motorways or rail infrastructure may be less attractive because of noise or air pollution.

Parking policies are a next category of infrastructure policies that may influence land use. Abundant and free parking may have a positive influence on the attractiveness for cars but is likely to negatively influence the quality of the urban environment, making some areas less attractive. Especially central urban areas suffer from driving and parked cars, and many cities have introduced restrictions on parking (see Mingardo et al. 2015 for a conceptualization of the development parking policies over time).

In addition to infrastructure policies, specific public transport policies can influence land use indirectly. Not only do the locations of stations have an impact on land use, as explained above, but so do bus stops. And for all forms of public transport, the services offered (as expressed by time tables) matter: the "better" the services, the higher the likeliness that stations and stops influence land use.

13.5 Discussion

In this section, I discuss the content of this chapter from the perspective of the relevance for research and (land use) policy making.

13.5.1 Lack of Integrative Approaches

This chapter made explicit that the relationships between travel behavior and health are manifold and complex. In the debates and research papers on land use and travel behavior, health is only seldom addressed. The framing of why land use could matter is much more related to the environmental impacts of travel behavior, due to mode choice and distanced traveled (mainly by car). But this chapter has made it clear that land use and land use policies have a much broader link with health than related to environmental impacts only. I think the general theme of land use, travel, and health therefore is poorly studied.

A key element in this general theme is the complexity of the many relationships as presented in this paper, and this complexity is hardly addressed in the academic literature and in (land use) policy making, although there are studies addressing parts of these complex relationship (e.g., Nieuwenhuijsen 2016), and studies discussing qualitatively the complex relationships (e.g., Khreis et al. 2016). There certainly is a gap in the academic literature with respect to quantifying the complex relationships as discussed in this chapter. The relationships between land use policies and health via travel behavior are poorly addressed in policy documents in general, and to the best of my knowledge, the complex relationships are about absent. This is understandable, both in the case of research and policy making. Focusing on research, I think to fully study the complex relationships is about undoable, at least if these need to be studied simultaneously and quantitatively, with one large dataset of multiple combined datasets. Data collection then would be very complex, and respondents need to provide a lot of data, probably leading to low response rates. I think the best way forward is to split the full picture in parts and study these. Combining those parts, preferably quantitatively, probably leads to a better understanding of the complex relationships.

More specifically, I next discuss some specific topics that are poorly understood, the first one being related to Fig. 13.1: we do not know the interactions between travel as a form of physical activity and other types of physical activity. And we poorly understand the combined effect of all four blocks on health. There are a few attempts to at least include multiple effects. For example, De Hartog et al. (2010) studied the combined effect of cycling on physical activity, the intake of air pollutants, and accidents and concluded that a shift from driving to cycling increases expected life years. But the study did not include the (other) complex relationships relevant for the health benefits of the assumed substitution from driving to cycling, as conceptualized in Fig. 13.1. Note that this discussion is not only related to the impact of land use on travel and next health but is of a more general nature, though also relevant for the link with land use.

The relationships as conceptualized in Fig. 13.2 between land use, accessibility, and travel behavior are much better understood. The main challenge is to link these relationships to the right part of the figure: the impacts on health via intermediate effects ("nearby," climate change, well-being). A lot of literature focuses on the impact of land use and land use policies on travel behavior, ignoring emissions, exposure to emissions, risks, well-being, and health.

Focusing on Fig. 13.3, the impact of land use on travel behavior has been studied frequently, although there still are important debates, as addressed above. And the locations of travel are also well addressed. A more or less separate strand of literature studies the impact of emissions via dispersion to exposure, and these relationships are also relatively well known. The same applies to the impact of this spatial distribution of travel (mainly by road type) on causalities. All other relationships as conceptualized in Fig. 13.3 are poorly understood, and these are promising challenges for future research.

Most relationships conceptualized in Fig. 13.4 are addressed in the preceding figures, the exception being the impact of the spatial distribution of activities on well-being. To the best of my knowledge, this is also an understudied topic and thus an interesting topic for future research.

I now switch to Table 13.1. As mentioned above, the impact of the 5Ds on travel behavior is relatively well understood, although important debates remain. The impacts of land use on the attractiveness of infrastructure and the surrounding areas have received way less attention of literature, and this is a promising area of future research. For policy making it is even more understandable that the complexities as presented in this chapter are poorly recognized. It is about impossible to "sell" policy measures to decision makers, if the effects are communicated in a complex way. A way out could be to summarize the health effects of candidate land use policies in terms of differences in expected life years, quality-adjusted life years or comparable indicators, probably added with a brief description addressing who are affected (e.g., categories of travelers, neighborhoods) and in which way (e.g., due to changes in exercise or exposure to risk or pollutants).

13.5.2 The Evaluation of Land Use Policies

A next topic is the question of how to evaluate land use policies that aim to improve health via travel behavior. Let us assume land use policies influence health in a positive way, in any of the ways conceptualized in Fig. 13.3. Does this mean that these policies should thus be implemented? The answer to this question is not necessarily "yes". This is because such policies have many more impacts and these can all be relevant to social welfare. In Van Wee (2002), I give an overview of relevant effects:

- Accessibility effects: how (un)easy can people reach destinations, and can companies transport goods between destinations?
- The option value: how do people value options to travel, even if they do not use these (see Geurs et al. 2006)?
- The consumers' surplus of travel: of how much more value is traveling for a traveler than it costs?
- Safety effects.
- The valuations of dwellings and the residential area, regardless of travel implications.
- Financial aspects: land use policies can influence costs. For example, building within the existing urban area is generally more expensive than building adjacent to the current urban area.
- Robustness: how robust is the land use and transport system for trend breaks, like disruptions due to climate change policies, strong changes in energy prices (up or down), or trend breaks in mobility behavior? Will it fulfill its role in societies under such changing conditions?

So, for final decisions, it is important to at least take into consideration the most important effects of candidate policy options. If this is not done, policy makers and next decision makers are poorly informed about the pros and cons of these options, while the role of policy-related research is to inform decision makers.

13.6 Conclusions

This section summarizes the most important conclusions that follow from this chapter. A first conclusion is that travel behavior can influence health via (1) level of physical activity, (2) air pollution intake, (3) casualties/accidents, and (4) (subjective) well-being.

A second conclusion is that the impacts of travel on health depend on personal characteristics, other forms of physical activity, and residential choice, and the interrelationships between these factors, and the impacts of travel behavior on health, are rather complex and under researched.

Third, these complex interrelationships are only partly understood. Consequently, several research challenges remain.

Fourth, land use can influence health via the transport system in multiple ways. It influences travel behavior, the attractiveness of travel, and the locations of active and motorized modes.

Fifth, travel behavior can be influenced by land use policies via the five Ds: density, diversity, design, destination accessibility, and distance to transit.

Sixth, land use can also influence the attractiveness of travel, and the locations of origins and destinations of trips, and route choice.

Seventh, decision making with respect to land use policies should not only be based on health impacts but should include many other aspects, at least accessibility effects, the consumers' surplus of travel, safety effects, the valuations of dwellings, and the residential area, regardless of travel implications, financial aspects, and the robustness of the land use and transport system.

References

Alonso, W. (1964). *Location and land-use. Toward a general theory of land rent.* Cambridge: Harvard University Press.

Amoros, E., Martin, J. L., & Laumon, B. (2003). Comparison of road crashes incidence and severity between some French counties. *Accident Analysis & Prevention, 35*(4), 537–547.

Baruth, M., Lee, D. C., Sui, X., Church, T. S., Marcus, B. H., Wilcox, S., & Blair, S. N. (2011). Emotional outlook on life predicts increases in physical activity among initially inactive men. *Health Education & Behavior, 38*(2), 150–158.

Bauman, A. E. (2004). Updating the evidence that physical activity is good for health: an epidemiological review 2000–2003. *Journal of Science and Medicine in Sport, 7*(1), 6–19.

Cao, X., Mokhtarian, P. L., & Handy, S. L. (2009). Examining the impacts of residential self-selection on travel behaviour: A focus on empirical findings. *Transport Reviews, 29*(3), 359–395.

Cervero, R., & Radisch, C. (1996). Travel choices in pedestrian versus automobile oriented neighborhoods. *Transport Policy, 3*(3), 127–141.

De Vos, J., Schwanen, T., van Acker, V., & Witlox, F. (2013). Travel and subjective well-being: A focus on findings, methods and future research needs. *Transport Reviews, 3*(4), 421–442.

De Hartog, J. J., Boogaard, H., Nijland, H., & Hoek, G. (2010). Do the health benefits of cycling outweigh the risks? *Environmental Health Perspectives, 118*(8), 1109–1116.

Ewing, R., & Cervero, R. (2010). Travel and the built environment. A meta analysis. *Journal of the American Planning Association, 76*(3), 265–294.

Gatersleben, B., & Uzzell, D. (2007). Affective appraisals of the daily commute: comparing perceptions of drivers, cyclists, walkers and users of public transport. *Environment and Behavior, 39*(3), 416–431.

Gatersleben, B., & Uzzell, D. (2007). Affective appraisals of the daily commute: comparing perceptions of drivers, cyclists, walkers and users of public transport. *Environment and Behavior, 39*(3), 416–431.

Geurs, K., Haaijer, R., & van Wee, B. (2006). Option value of public transport: Methodology for measurement and case study for regional rail links in the Netherlands. *Transport Reviews, 26*(5), 613–643.

Handy, S. (1996). Methodologies for exploring the link between urban form and travel behavior. Transportation Research Part D: Transport and Environment, *1*(2), 151–165.

Handy, S. (2005). Smart growth and the transportation-land-use connection: What does the research tell us? *International Regional Science Review, 28*(2), 146–167.

Handy, S. (2014). Health and travel. In T. Gärling, D. Ettema, & M. Friman (Eds.), *Handbook of sustainable travel*. Dordrecht/Heidelberg/New York/London: Springer.

Khreis, H., Warsow, K. M., Verlinghieri, E., Guzman, A., Pellecuer, L., Ferreira, A., Jones, I., Heinen, E., Rojas-Rueda, D., Mueller, N., Schepers, P., Lucas, K., & Nieuwenhuijsen, M. (2016). The health impacts of traffic-related exposures in urban areas: Understanding real effects, underlying driving forces and co-producing future directions. *Journal of Transport and Health, 3*(3), 249–267.

Lee, A. E., Underwood, S., & Handy, S. (2015). Crashes and other safety-related incidents in the formation of attitudes toward bicycling. *Transportation Research Part F, 28*, 14–24.

Lucas, K. (Ed.). (2004). *Running on empty: Transport, social exclusion and environmental justice*. Bristol: Policy Press.

Meurs, H., & Haaijer, R. (2001). Spatial structure and mobility. *Transportation Research Part D, 6*(6), 429–446.

Mingardo, G., Van Wee, B., & Rye, T. (2015). Urban parking policy in Europe: A conceptualization of past and possible future trends. *Transportation Research Part A, 74*, 268–281.

Nieuwenhuijsen, M. J. (2016). Urban and transport planning, environmental exposures and health-new concepts, methods and tools to improve health in cities. *Environmental Health, 15*(Suppl 1), 38.

Nieuwenhuijsen, M. J., & Khreis, H. (2016). Car free cities: Pathway to healthy urban living. *Environment International, 94*(2016), 251–262.

Nyhan, M., McNabola, A., & Misstear, B. (2014). Evaluating artificial neural networks for predicting minute ventilation and lung deposited dose in commuting cyclists. *Journal of Transport and Health, 1*(4), 305–315.

Olsson, L. E., Gärling, T., Ettema, D., Friman, M., & Fujii, S. (2013). Happiness and satisfaction with work commute. *Social Indicators Research, 111*(1), 255–263.

Patz, J. A., Campbell-Lendrum, D., Holloway, T., & Foley, J. A. (2005). Impact of regional climate change on human health. *Nature, 438*, 310–317.

Pucher, J., & Buehler, R. (Eds.). (2012). *City cycling*. Cambridge/London: MIT Press.

Şahin, Ü. A., Onat, B., Stakeeva, B., Ceran, T., & Karim, P. (2012). PM10 concentrations and the size distribution of Cu and Fe-containing particles in Istanbul's subway system. *Transportation Research Part D, 17*(1), 48–53.

Stevens, M. R. (2017). Does compact development make people drive less? *Journal of the American Planning Association, 83*(1), 7–18.

Stipdonk, H., Bijleveld, F., Van Norden, Y., & Commandeur, J. (2013). Analysing the development of road safety using demographic data. *Accident Analysis and Prevention, 60*, 435–444.

Van Wee, B. (2002). Land-use and transport: research and policy challenges. *Journal of Transport Geography, 10*(2002), 259–271.

Van Wee, B. (2011). Evaluating the impact of land use on travel behaviour: The environment versus accessibility. *Journal of Transport Geography, 19*(6), 1530–1533.

Van Wee, B., & Ettema, D. (2016). Travel behaviour and health: A conceptual model and research agenda. *Journal of Transport and Health, 3*(3), 240–248.

Wegener, M., & Fürst, F. (1999). *Land-use transport interaction: State of the art*. Dortmund: Universität Dortmund, Insititut für Raumplanung.

Wegman, F. (2013). Road Safety. In B. Van Wee, J. A. Annema, & D. Banister (Eds.), *The transport system and transport policy. An introduction*. Cheltenham: Edward Elgar.

Chapter 14
Complex Urban Systems: Compact Cities, Transport and Health

Mark Stevenson and Brendan Gleeson

14.1 The Urban Age

The urban age has been declared. A new conversation welcomes the fact that humanity is now preponderantly an urban species, *Homo urbanis*. The major transnational institutions bestow great significance to urbanisation as a force shaping human fortunes (OECD 2010; UN-Habitat 2009, 2012; UNICEF 2012; World Bank 2010). For the past half-decade, the United Nations has broadcast the message of a new urban ascendancy. UN-Habitat enthuses, 'A fresh future is taking shape, with urban areas around the world becoming not just the dominant form of habitat for humankind, but also the engine-rooms of human development as a whole' (2012:v).

For the first time, the centre stage of human contest is urban. Much commentary would have us believe that cities are more than human stages: infernal machines at the heart of the crisis (e.g. Miles and Miles 2004). Grandly flagged statistics report their overwhelming contribution to global consumption and despoliation. Urban landscapes are said to consume around three-quarters of the world's energy and generate the same proportion of its greenhouse gas emissions (Urry 2011). Dobbs et al. (2012) estimate that global urban growth until 2025 will drive an 80 million cubic metre increase in water demand and necessitate newly built floor space equivalent to an area the size of Austria. The hard spectre of the 'consumptive city' looms over the global environmental consciousness. This truism of the age neglects the

M. Stevenson (✉)
Transport, Health and Urban Design Research Hub, Melbourne School of Design and Melbourne School of Engineering, The University of Melbourne, Parkville, VIC, Australia
e-mail: mark.stevenson@unimelb.edu.au

B. Gleeson
Melbourne Sustainable Society Institute, The University of Melbourne, Parkville, VIC, Australia
e-mail: brendan.gleeson@unimelb.edu.au

M. Nieuwenhuijsen, H. Khreis (eds.), *Integrating Human Health into Urban and Transport Planning*, https://doi.org/10.1007/978-3-319-74983-9_14

dialectic of urbanisation, in which cities are *simultaneously* engines and artefacts of the underlying process of accumulation, of money, matter, bodies and ambition.

Our long love affair with the city has reached new heights. It will only continue and intensify. By 2050, it is expected that three out of every four humans will live in an urban setting. The world population will have grown by around a third to number ten billion (Urry 2011). In the developed world, nearly nine in ten people will be urban denizens, even as some national populations decline.

Suburbanisation was the key form of urban growth in Western cities during the twentieth century. It continues apace, now accommodating the desires of aspirant middle classes in the developing world. Hamlin reports, 'What the US did in the 1950s with 160 million people, China is now doing with more than a billion – moving to suburbia' (2013:36). In the fractious 'developing city', the suburban shift must largely take the forms of 'imprisoned freedom', gated villa estates for the elite and isolated tower dormitories for the middle classes.

The progressive abolition of slums and the growth of lower-density suburbs were undeniably linked to, if not solely a cause of, a vast increase in the health and well-being of urban populations. It was, however, also a model of human growth freighted with self-endangerment, but this was not to become clear until late into the twentieth century. In 2011, Urry reflects on the 'high carbon lives' that were born and ordained in its car-dependent fabric (2011:64–65). Recognition of the limits, contradictions and dangers of continued urban dispersal led to the formulation of a new urban policy idea, the compact city.

14.2 The Compact City

The compact city has dominated the international urban imagery since the 1990s, most particularly amongst the urban professions, planning, architecture and design (Breheny 1995; Neuman 2005; OECD 2010). The ideal has been disputed in scholarship (e.g. Gordon and Richardson 1997; Troy 1996) and by industry lobbies associated with traditional suburban development (e.g. Moran 2006). In the United States, often fractious and heated debates have counter-posed 'sprawl' with 'smart growth', traditional development with 'New Urbanism' (see Chap. 4), car-based development with 'transit-oriented development' and so forth. The contest has been sharply drawn, producing highly polarised debates in North America and Australia in particular. In the United States, Hirschhorn's 2005 book reports that *sprawl* kills and annihilates comprehensively by stealing 'your time, health and money'.

Whilst professional and academic planning historically identified itself with, and facilitated, suburbanisation (Hall 1992), its mindset and policy outlook has increasingly fixed on the compact city as a leading policy goal. A high degree of international professional consensus has settled on guiding morphological principles, which Forster summarises as 'containment, consolidation and centres' (2006:178). In Australia, the compact city construct is central to all contemporary metropolitan planning policy. Forster portrays the compact vision splendidly:

> 'In 20 to 30 years-time, if the plans come to fruition, our major cities will be characterised by limited suburban expansion, a strong multi-nuclear structure with high density housing around centres and transport corridors, and infill and densification throughout the current inner and middle suburbs. Residents will live closer to their work in largely self-contained suburban labour sheds, and will inhabit smaller, more energy efficient houses. The percentage of trips using public transport, walking or cycling will have doubled'. (Forster 2006:179)

In the United States, Neuman wrote, 'The compact city, we are told, is more energy efficient and less polluting because compact city dwellers can live closer to shops and work and walk, bike or take transit' (2005:12).

Although freighted with structural principles—notably dispersed concentration and multi-nucleation—the construct has been most powerfully associated with its qualifier, 'compact', engendering a strong emphasis on density and thereby the containment of urban development. Not all scholars associated with pro-compaction arguments support a simple generalisable relationship between density and energy use or other urban consumption forms and values. Newman and Kenworthy (1999), for example, advocate a compact city model with strongly defined structural features, especially polycentrism and the spatial focusing of densification. Neither does 'compact' necessarily equate to high density in all analysis nor advocacy (though it does in many). Some compact models and planning frameworks explicitly favour spatially targeted medium-density development (Searle and Filion 2011).

What is undeniably true, however, is that institutionally and materially, planning response to the compact city ideal has focused on densification. Forster explained the assumption that urban compaction would achieve sustainability by '...producing higher densities more favourable to public transport than the private car, through developing smaller dwellings and blocks that use less water and power, and through reducing the impact of urban expansion on surrounding ecosystems' (2006:178–179). The new urban commentaries have tended to effuse about the many putative remunerations of density, which include sustainability, but also economic benefits, such as innovation and agglomerative synergies. As Christensen et al. explain, the technocratic enthusiasm for density as a 'magic lever' for urban improvement: '... urbanists, like Glaeser, dismiss many ecological concerns out of hand as impediments to efficient economies of scale in cities, believing that the efficiency of cities is sufficient on its own to solve our ecological problems' (2012:B10). This postulate is broadly shared by the new urban physics that has asserted 'scaling effects' from population size and urban density (Gleeson 2015).

The tendency towards exclusive focus on density remains to be explained fully but could reflect the constrained institutional possibilities of planning intervention: specifically, the difficulties of broader intervention in global urbanisation processes (including economic drivers) which have tended to drive physical dispersion of activity and development (Troy 1996). In the neo-liberal city, it may be easier to pursue densification, usually through relaxation of planning controls, than to actively guide the structure of metropolitan development towards less wasteful ends (e.g. transport and land use integration, creation of subregional centres). And yet, some compact urbanisms are likely only to arouse social ire and resistance, further constraining planning possibility.

At the institutional level and within major policy domains, the goal of urban containment and compaction has been strongly linked to reduction in energy use and mitigation of greenhouse gas (GHG) emissions. The UN believes that 'spatially compact and mixed-use urban developments have generally significant benefits in terms of GHG emissions' (UN-Habitat 2011:4). The World Bank agrees: 'Increasing density could significantly reduce energy consumption in urban areas… cities that are denser produce less emissions' (2010:19, 28). The principal global climate science institution, the Intergovernmental Panel on Climate Change, identifies urban low density as a barrier to GHG emission reduction (Mees 2010:53).

The lineage of the density argument lies in scholarship gaining voice in the 1990s that asserted a strong observed relationship between urban form and transport use, especially at the metropolitan scale (especially, Newman and Kenworthy 1989, 1999; Newman et al. 1992). Mees (2010) traced an earlier airing of the argument to the *1960 Chicago Area Transport Study*. Lower-density cities were held to engender car dependence and to render public transport unviable or unattractive. Automobile dependence was associated with relatively higher metropolitan levels of fuel consumption and GHG emissions. Reflecting the broad contemporary acceptance of the argument, the World Bank summarises the case for compaction: 'As density increases, people use more public transportation and non-motorized forms of transport, lowering transportation energy use per capita' (2010:27). The argument for compaction has now extended beyond the sequelae associated with car dependence to include health impacts of transport such as motor vehicle crashes, traffic-related air pollution, noise, heat island effect, lack of green space, physical inactivity, climate change and social exclusion and community severance (Khreis et al. 2017).

Whilst the original emphasis on GHG reduction through densification was on the transport sector, the analytical and professional ambit of compaction was progressively extended in many quarters to include other forms of energy use, especially household stationary energy. UN-Habitat, typifying this broader association, remarks that 'Density may also affect household energy consumption, as more compact housing uses less energy for heating' (2011:17). The Australian Government's (2011) National Urban Policy Statement reflects this generalised association of density and energy use: 'How cities are planned, their density and spread, and the buildings and infrastructure within them, provide an enormous opportunity to reduce greenhouse gas emissions' (Australian Government 2011:42).

In recent years, an accumulating evidence base, including data on revealed consumption, has thrown shadows of doubt over the bright generalisations linking density and sustainability (Neuman 2005). The many strands of critique emerging from this widening empirical base address both conceptual and methodological issues (Gray et al. 2010). What is rarely if ever acknowledged is that the scientific failure to link density changes with sustainability outcomes has a deeper history of epistemological discussion and refinement, especially in geographical debates that took issue with positivism (law bound social science) from the 1970s. Sifting the growing counter-evidence on density reveals serious scientific problems, such as physical determinism and ecological fallacy (inferring individual qualities from group attributes), which were well rehearsed by geographers in earlier decades.

14.2.1 Geography Matters

The early analyses in favour of the compact city form drew sustained scholarly criticism. A central object of contest were the much reproduced cross-metropolitan comparisons of density and transport use forwarded by Newman and Kenworthy (Mees 2010). Critical response alleged empirical and methodological errors, especially incorrect measurement (of urban densities), variable constraint and causal inference on the basis of broad, bi-variate correlation (Breheny 1995; Gomez-Ibanez 1991). Re-examination of empirical sets was held to reveal weak or highly variable relationships between densities and transport behaviour at the metropolitan scale (Mees 2010). Troy (1996) and Breheny (1995) countered that urban structure (fundamental land use patterns) was a greater influence on energy use and sustainability than form (density and aesthetics).

In the past decade, the accumulation of sub-metropolitan scale information about density and energy use, in tandem with new modelling approaches (e.g. ACF 2007), has strengthened earlier scientific doubts about the compact city generalisations. Evaluations of urban density and household consumption differ widely in what forms of energy they are measuring: usually all, or components of, stationary and transport energy consumption. Terminology and framing vary (operational-embodied, direct-indirect, stationary-transport-fugitive, etc.), as do point of analysis (inputs-end use), objects of analysis (building forms and social correlates), and data types and sources. Density itself can be measured in widely different ways (Mees 2010). Some studies find evidence of negative relationships between urban density and forms of energy use, especially transport (see Mees 2010). Taken together, however, their findings disrupt and confound the proposition that density determines energy consumption in widely predictable ways. Wright's recent review of Australian evidence concludes that the relationship between housing density and energy use '… is not collinear and is impacted by other variables' (2010:7). Consider a brief sample of the evidence base.

An extensive study of revealed domestic energy use in Sydney, Australia, by Myors et al. (2005) showed that, in this setting, higher-density forms were more not less consumptive than medium- and lower-density development. Another Australian residential study by Perkins et al. (2009) in Adelaide, showed that whilst compaction had the potential to reduce transport emissions, high-density small dwellings were associated with higher per capita energy use than suburban dwellings. In Oslo, Norway, Holden and Norland's (2005) survey work on household emissions considered wider travel behaviour, including air travel, and found that per capita emissions declined with rising density to a certain point, but then rose progressively in the most heavily populated neighbourhoods. In particular, '… extreme densities statistically correlate with high energy use for long-distance leisure-time travel' (2005:2163). The authors concluded that structural change—decentralised concentration—offered the best prospects for reducing urban energy use. In France, Gaigné and Thisse (2010) found that as density increases, so do the negative consequences of intensification, namely, traffic congestion and thereby emissions. That is, without countervailing changes (e.g. pricing, demand management), intensification may generate rising congestion and longer work-trips.

These counterfactual findings and analytical doubts break new empirical ground but struggle to gain attention in an urban conversation enthralled with power of density to enhance human well-being. The newer waves of studies at the sub-metropolitan scale suggests strongly that density does condition human behaviour, including energy use, but that its influence is strongly over-determined by multi-causation, context dependent and thus subject inevitably to a high degree of spatial variability. The evidence base thus underlines the distinction between idealised and material urban realities, between the compact city ideal and actually existing consolidation. The findings do not logically foreclose on the mitigating potential of intensification, but they do point to the difficulty of its material realisation. Harvey (1997) believes that the 'New Urbanism' guiding many compact city visions is, amongst other things, deeply utopian.

A frictional urban reality obviously challenges the compact ideal. And what of science? The more simplistic claims about density are clearly marked by ecological fallacy. Large, bivariate analyses of energy use at the metropolitan scale do not support a fixed or non-scalar relationship between density and urban behaviour. In dense cities, there are high-density neighbourhoods that consume more than less populated areas. In geography, the earlier identification of this fallacy showed the importance of careful, multi-scaled spatial analysis that is sensitive to context and acknowledges the multi-causal—viz. socio-spatial—conditioning of human behaviour. Determination is both multi-scalar and variegated by scale. Thus, cross-level fallacy is a possibility at any empirical scale (Johnston et al. 2001:191). There appears to be a weak relationship between density and energy consumption at the metropolitan scale (Mees 2010). At the sub-metropolitan and local scales, however, the connection is even weaker, foreclosing on any meaningful generalisation.

The progressive unearthing of this fallacy lends weight to earlier claims (e.g. Troy 1996) that compact city arguments were marked by environmental determinism. This feature of early twentieth-century geography '… stimulated a focus on the role of physical features – site and situation – as determinants of urban foundations and growth…' (Johnston et al. 2001:876). The tenacious grip of deterministic thinking was greatly loosened if not completely released by the break with positivism in the social sciences, including human geography, from the 1960s. Importantly, for compact city debates and planning, its transcendence imposes a double burden, viz. to ensure that all morphological constructs—of form and structure—do not assume the status of determining forces in urban analysis.

As Mees (2010) established, the density thresholds usually asserted for public transport reliability have poor, even bogus, empirical foundations. They also, however, consistently undermine the institutional possibility of public transport provision in suburban areas which can never be expected to redevelop to the densities demanded by the thresholds:

> 'Environmentalists who argue in this manner can unintentionally provide support for the continuation of unsustainable transport policies...many British advocates of sustainable transport seem more interested in higher-density housing than in fixing public transport'. (Mees 2010:51).

14.2.2 The 'Parallel Universe' Problem

Compact city advocates frequently fail to consider the concrete forms and directions of urbanisation which are largely at counter-purpose to the vision. The Australian geographer Forster (2006) termed this a 'parallel universe' problem. For Forster, the bright diorama of the compact city is a world away from the murkier reality of Australian metropolitan urbanisation, marked by the rapid and powerful dispersion which is common to many western cities:

> '[T]here is a mismatch between the strategies' consensus view of desirable future urban structure, based on containment, consolidation and centres, and the complex realities of the evolving urban structures. In particular, the current metropolitan strategies do not come to terms with the dispersed, suburbanised nature of much economic activity and employment and the environmental and social issues that flow from that, and they are unconvincing in their approaches to the emerging issues of housing affordability and new, finer-grained patterns of suburban inequality and disadvantage' (2006:173).

The compact city ideal thus '...sits dangerously at odds with the picture of increasing geographical complexity that emerges clearly from recent research on the changing internal structure of Australian cities since the early 1990s' (2006:180). Breheny earlier aired analogous concerns for the United Kingdom, pointing to the considerable momentum behind urban decentralisation in that country and doubting that this 'powerful force' could be contained 'beyond existing levels of constraint' (1995:99). None of this, of course, is to regard dispersion as desirable.

14.3 Compact Cities, Transport and Health

The relationship between density and particularly compact cities and urban form appears to be far more contingent and context based than commonly supposed in many compact city policies. Density of course is an important dimension of urban form and structure, but it acts in complex ways with other social and morphological features to shape cities, human behaviour and population health.

It is not surprising that both the urban age and the compact city ideal will contribute in numerous ways to our health: from where we live, how we travel, the food that is available, how clean the air and water is and much more. Rapid urbanisation and a policy environment that incentivise a city of short distances in which there are

Table 14.1 Land use elements associated with transport mode choice

Land use element	Relationship with transport
Density	
Definition: Population density, residential unit density, intersection density or recreation space density	Associated with transport mode choice even when accounting for socio-demographic elements of city residents
Diversity	
Definition: Alternately named mixed land use	Associated with travel behaviour and transport choice especially the tendency for increased walking, cycling and public transit use
Distance	
Definition: The average shortest street routes from a place of residence or workplace in an area to the nearest public transport option	Strongly associated with transport mode particularly if the distances between home, workplace or public transit are reduced
Design	
Definition: Refers to the characteristics and layout of land including streets, building setbacks, intersection connectivity, aesthetics, footpaths and other physical infrastructure	Associated with reductions in motor vehicle distances travelled and increased active transport such as walking and cycling along with increased access to parks, recreational facilities and improved aesthetics

higher residential and population densities, greater mixed land use and urban design amenable to walking and cycling (Gordon and Richardson 1997) can be both beneficial and detrimental to the health of city residents: from providing access to health and social services (Sørgaard et al. 2003) to the pollution from car emissions and industrial plants situated closer to residential properties. This dichotomy highlights the complex urban age and points to what Forster described earlier in this chapter as the 'parallel universe' (Forster 2006).

Acknowledging the complexity of the urban age and the potential co-benefits associated with compact cities, recent research has begun to unravel elements of this complexity by assessing the health co-benefits of a compact city albeit from the perspective of changes in key elements of land use and subsequent transport mode choice (Stevenson et al. 2016) (Table 14.1). The approach, namely, a health impact assessment framework (Dannenberg et al. 2008), integrated findings from an extensive meta-analysis in which Ewing and Cervero (2010) identified key land use elements that influence transport modal choice (see Table 14.1). The influence of these land use elements on transport mode choices, namely, public transit, walking, cycling and private motor vehicle use, was assessed with respect to their influence on three risk exposures: physical inactivity, road deaths and injury and air pollution (see Fig. 14.1 infographic).

There is considerable evidence highlighting the increased risks associated with physical inactivity and many non-communicable diseases (Lee et al. 2012), and the health impact assessment adopted by Stevenson et al. (2016) considered the varying transport modes on physical activity or, more appropriately, physical inactivity and its effect on overweightedness and obesity and cardiovascular disease.

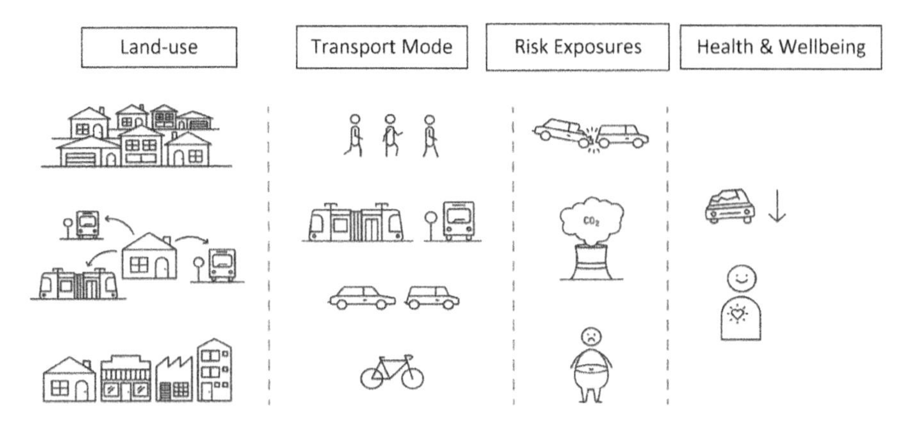

Fig. 14.1 Compact city—land use, transport and health model

The impact of land use on transport mode choice also goes beyond rates of physical inactivity with evidence highlighting that increased exposure to motorised vehicles heightens a city resident's risk of death and physical injury (De Hartog et al. 2010). For example, the relative risk of death and injury associated with transport varies by mode, by the interaction of the transport mode and location and by proportional traffic volumes within the transport system (Bhalla et al. 2007; Elvik 2009; World Health Organization 2013). The risk of death and injury, therefore, is greatest amongst vulnerable road users or the most unprotected users of the road system, namely, pedestrians, bicyclists and motorcyclists, whilst users of public transport have the lowest risk per kilometre travelled.

An increasing reliance on private motor vehicle use as a consequence of sprawling suburbs (as observed across many North American and Australasian cities) has also resulted in a heightened exposure to vehicle emissions and air pollution and consequent adverse health outcomes (Briggs et al. 2008; Brunekreef and Holgate 2002; Chen et al. 2008). There is considerable evidence highlighting that exposure to fine particulate matter (PM_{10} and $PM_{2.5}$) associated with emissions from road transport (Ministry of the Environment Ontario 2010) contributes significantly to the burden of disease, specifically an increased incidence of cardiovascular disease (due to elevated blood pressure) (Lee et al. 2014) and respiratory disease (Janssen et al. 2011; Karakatsani et al. 2012; Tamura et al. 2003). For further discussion on the significance of this burden, we refer you to Chap. 21 on air pollution.

Taking account of the land use elements, transport mode choices and the varying risk exposures in relation to health and well-being, a compact city scenario was modelled across six selected cities (Melbourne, Australia; London, United Kingdom; Copenhagen, Denmark; Delhi, India; Sao Paulo, Brazil; and Boston, United States of America). To reflect a 'compact city', the land use densities in each city were increased by 30% (from city densities described in 2014), the diversity of land use was also enhanced by 30% (from mixed-use ratios based on 2014 estimates), and average distances to public transit were also reduced by 30%. As a consequence of increasing the compactness of the cities, opportunities for changing transport mode choices were assessed such as supporting a 10% modal shift away from private motor vehicle use to active transport modes, namely, walking and cycling (Fig. 14.1).

The approach to increasing these land use elements to reflect a compact city resulted in considerable reductions (between 13 and 19% in cardiovascular disease alone) in the burden of disease (Stevenson et al. 2016), highlighting what could be achieved if cities were able to plan or enhance current land use such that it reflected a city of short distances, greater mixed use and at the same time, increased public transit and presided over a population transition to active transport using safe infrastructure. This study, the first to highlight that a compact city influences the health of its citizens, points to the need for increased efforts to understand other transport exposures such as noise and exposure to green and blue spaces (Nutsford et al. 2016) which is reported to effect mental health. Further it is necessary to understand the effects of compact cities not only it relates to health but other urban systems including, housing, transport, energy and water.

14.3.1 Complex Cities: The Need for Policy Decision Support

Cities comprise multiple, constantly changing, urban systems, and the transport system is an excellent illustration of such change. Globally, the transport system is undergoing not only a digital revolution but a significant paradigm shift involving increasing electrification of motor vehicles and automation of motor vehicles along with on-demand shared mobility (Fulton). Each of these changes will alter the way cities operate, bringing with them potential health co-benefits. Some of the likely health benefits from these changes will be reduced rates of respiratory and cardiovascular disease due to enhanced air quality, fewer road deaths as the human error in the driving task is minimised and greater access to health and social services for marginalised populations poorly served by public transit. However, the changing transport system is not all-positive. The advent of the combustion engine in the early twentieth century heralded a new era of mobility, not dissimilar to the current digital mobility era. Elements of the early transport system failed, resulting in new and necessary services that continue to operate today, services such as insurance schemes to compensate road users injured whilst using the transport system. Similarly, delivering a transport system in the digital mobility era will be associated with considerable risk. Duchemin and Marembaud (2015) highlight a number of these risks, namely, (1) increased inequities due to disparities in service provision and cost, (2) job losses due to automation, (3) unfair competition by industry players, (4) reduced tax revenues and (5) system failures due to a lack of reliability. The implication of such disruptors to the current transport system is significant with a far-reaching effect across numerous urban systems including employment and productivity, health, social services and more; delivery of safe and sustainable urban systems is overwhelmingly complex.

Given the complexity of our cities, it is not sufficient to merely compare urban and non-urban areas or compare the varied urban systems with respect to key health outcome measures. Instead, the relationship between urbanisation and health needs to be observed across multiple levels (such as the levels of government) and multiple influences including global influences. Seldom are the associations between key urban systems and health simple linear relationships but rather relationships that are affected by an array of inputs and feedback across multiple levels.

Recently, the complex relationship between cities, transport and health has been assessed using health impact assessment frameworks. Woodcock et al. (2009) examined transport emissions associated with large cities and the likely health consequences using such an approach and found that reductions in CO_2 emissions associated with active transport (walking and cycling) delivered significant health benefit than would be expected from increased use of low-emission motor vehicles. Although the health impact assessment approach is delivering valuable insights with respect to key risk exposures and their relationship to health, they are limited in that they cannot assess the city as a dynamic system with multiple interactions, relationships and consequences.

Cities are adaptive complex systems which, by definition, means '…they are systems made up of many individual, self-organising elements capable of responding to others and to their environment' (Glouberman et al. 2006:328). Urban systems in a city comprise an array of elements with varying relationships and interactions, and a change in any part of the system will result in changes in associated elements in the broader environment. Therefore, cities are considered complex because various components of the system interact and change, and the change is not always observable by assessing the characteristics of an individual component, but rather by considering multiple levels and interactions.

As Vlahov and Galea (2003) and Galea et al. (2005) purport, it is important to embrace the complexity of cities, and this can be achieved by changing the focus of researchers, planners and policymakers from disease outcome to one where the attention is on understanding the multi-level risk exposures in cities (exposures such as the risk per vehicle kilometre travelled, air pollution and other elements) that influence the health and well-being of city residents.

The emergence of systems dynamic modelling in the 1950s (Forrester 1971), a mathematical modelling approach used to understand complex issues, has been valuable in providing important insights to the complexity of cities. A recent application of a systems dynamic approach has also concurred that compact cities that influence transport mode choice can deliver health gains (McClure et al. 2015). The systems dynamic modelling approach provided considerable insight with respect to the delivery of the compact city. For example, McClure et al. (2015) research specifically outlined elements of compactness that were relevant to enhanced health outcomes, namely, '…desirable land use changes are ones that increase the density of urban dwelling space so residential, work and leisure environments are co-located within geographically defined nodes and where nodes are connected by safe, clean, and rapid mass transport options' (McClure et al. 2015:S229).

Although the effects of the transport system (as highlighted here) on health are becoming well described, other urban systems such as housing or employment are not well described in part, because they occur against a backdrop of dynamic environmental, technological, population and economic conditions that evolve over years that makes attempts to predict key health outcomes challenging. As highlighted, there are now a number of methods that can be used to understand the varying inputs and outputs and health-related outcomes of any policy choice.

Such complexity is best explored using policy simulation derived from the integrated models. The development of a *policy decision platform* that investigates the processes that will lead to healthier and more sustainable cities is of paramount importance. So what does such a platform look like? A policy decision platform would comprise a collection of data from varied sources, standardised in a way and modelled such that relationships between key urban systems can be explored with respect to the resilience of a city (Ilmola 2016). The platform would be capable of assessing the varied effects of compact city scenarios along with the utility to explore altered city attributes such as land use mix, transport systems or changes to housing affordability and the resultant change in population health outcomes.

14.4 Conclusions

Cities around the world are dealing with transport challenges as a consequence of changing population demographics. In 2007, 51% of the world's population lived in cities, and it is estimated that this will increase to 70% by 2050 (United Nations 2004, 2011). These projections are reflected in population growth estimates that will see the world's population likely increase by 66% from seven billion people in 2013 to 10.5 billion people over the next 40 years (United Nations 2011). There is no doubt that the twenty-first century is a century characterised by urbanisation: the urban age.

Cities are complex systems, and to meet the major challenges of this century, cities will need to respond rapidly to accommodate growing populations, the growing inequities particularly on the urban fringe, the ongoing challenges to mitigate climate change and the increasing rates of chronic disease and road trauma. Building cities that are sustainable, productive and capable of delivering favourable health outcomes is an urgent priority. As highlighted in this chapter, a city of short distances, greater mixed land use and increased public transit, namely, the compact city, is capable of delivering important health outcomes. Despite this, the notion of a compact city is disputed in scholarship (Gordon and Richardson 1997; Troy 1996) and by industry (e.g. Moran 2006).

There are numerous policies focused on achieving enhanced health outcomes for residents of cities including the World Health Organization's 'Healthy Cities', 'Safe Cities' and 'Child Friendly Cities'. However, there is a paucity of information on how cities can achieve the goals of such policies, including those related to compact cities. There is an urgent need for robust decision tools and platforms for policymakers and key city-stakeholders. Such tools would enable stakeholders to consider a variety of policies and interventions that relate to the delivery of key urban systems including transport and housing and thereby ensure efficacious interventions are implemented and thereby have a profound impact on the health and well-being of growing urban populations.

References

ACF. (2007). Consuming Australia: Main findings. Retrieved from www.acfonline.org.au/consumptionatlas.

Australian Government. (2011). *Our cities, our future. A national urban policy for a productive, sustainable and liveable future*. Canberra.

Bhalla, K., Ezzati, M., Mahal, A., Salomon, J., & Reich, M. (2007). A risk-based method for modeling traffic fatalities. *Risk Analysis, 27*(1), 125–136.

Breheny, M. (1995). The compact city and transport energy consumption. *Transactions of the Institute of British Geographers, 20*, 81–101.

Briggs, D. J., de Hoogh, K., Morris, C., & Gulliver, J. (2008). Effects of travel mode on exposures to particulate air pollution. *Environment International, 34*(1), 12–22. https://doi.org/10.1016/j.envint.2007.06.011.

Brunekreef, B., & Holgate, S. T. (2002). Air pollution and health. *Lancet, 360*(9341), 1233–1242. https://doi.org/10.1016/s0140-6736(02)11274-8.

Chen, H., Namdeo, A., & Bell, M. (2008). Classification of road traffic and roadside pollution concentrations for assessment of personal exposure. *Environmental Modelling & Software, 23*(3), 282–287. https://doi.org/10.1016/j.envsoft.2007.04.006.

Christensen, J., McDonald, R., & Denning, C. (2012). Ecological urbanism for the 21st century. *The Chronicle Review, 58*(21), B9–B11.

Dannenberg, A. L., Bhatia, R., Cole, B. L., Heaton, S. K., Feldman, J. D., & Rutt, C. D. (2008). Use of health impact assessment in the US: 27 case studies, 1999–2007. *American Journal of Preventive Medicine, 34*(3), 241–256.

De Hartog, J., Boogaard, H., Nijland, H., & Hoek, G. (2010). Do the health benefits of cycling outweigh the risks? *Environmental Health Perspectives, 118*, 1109–1116.

Dobbs, R., Remes, J., Manyika, J., Roxburgh, C., Smit, S., & Schaer, F. (2012). Urban world: Cities and the rise of the consuming class. Retrieved from Washington, DC.

Duchemin, B., & Marembaud, O. (2015). *The digital revolution and changes to individual and collective mobility*. Paris: Economic, Social and Environmental Council.

Elvik, R. (2009). The non-linearity of risk and the promotion of environmentally sustainable transport. *Accident Analysis & Prevention, 41*(4), 849–855. https://doi.org/10.1016/j.aap.2009.04.009.

Ewing, R., & Cervero, R. (2010). Travel and the built environment: A meta-analysis. *Journal of the American Planning Association, 76*(3), 265–294.

Forrester, J. W. (1971). Counterintuitive behavior of social systems. *Technological Forecasting and Social Change, 3*, 1–22.

Forster, C. (2006). The challenge of change: Australian cities and urban planning in the new millennium. *Geographical Research, 44*(2), 173–182.

Gaigné, C. R. S., & Thisse, J. (2010). *Are compact cities environmentally friendly?* Lyon – St Étienne: Groupe d'Analyse et de Théorie Économique.

Galea, S., Freudenberg, N., & Vlahov, D. (2005). Cities and population health. *Social Science & Medicine, 60*(5), 1017–1033.

Gleeson, B. (2015). *The urban condition*. London: Routledge.

Glouberman, S., Gemar, M., Campsie, P., Miller, G., Armstrong, J., Newman, C., Siotis, A., & Groff, P. (2006). A framework for improving health in cities: A discussion paper. *Journal of Urban Health, 83*(2), 325–338.
Gomez-Ibanez, J. A. (1991). A global view of automobile dependence. *Journal of the American Planning Association, 57*(3), 376–379.
Gordon, P., & Richardson, H. W. (1997). Are compact cities a desirable planning goal? *Journal of the American Planning Association, 63*(1), 95–106.
Gray, R., Gleeson, B., & Burke, M. (2010). Urban consolidation, household greenhouse emissions and the role of planning. *Urban Policy & Research, 28*(3), 335–346.
Hall, P. (1992). *Urban and regional planning* (3rd ed.). London: Routledge.
Hamlin, K. (2013). Soviet-style suburbia heralds environmental disaster. *Australian Financial Review*. 8(November):36.
Harvey, D. (1997). The new urbanism and the communitarian trap. *Harvard Design Magazine, 1*(2), 68–69. Winter/Spring.
Hirschhorn, J. (2005). *Sprawl kills*. New York: Sterling & Ross.
Holden, E., & Norland, I. T. (2005). Three challenges for the compact city as a sustainable urban form: Household consumption of energy and transport in eight residential areas in the greater Oslo region. *Urban Studies, 42*(12), 2145–2166.
Ilmola, L. (2016). Approaches to measurement of urban resilience. In *Urban resilience* (pp. 207–237). Cham: Springer.
Janssen, N. A., Hoek, G., Simic-Lawson, M., Fischer, P., Van Bree, L., Ten Brink, H., Keuken, M., Atkinson, R. W., Anderson, H. R., & Brunekreef, B. (2011). Black carbon as an additional indicator of the adverse health effects of airborne particles compared with PM^ sub 10^ and PM^ sub 2.5. *Environmental Health Perspectives, 119*(12), 1691.
Johnston, R., Gregory, D., Pratt, G., & Watts, M. (2001). *The dictionary of human geography* (4th ed.). Oxford: Blackwell.
Karakatsani, A., Analitis, A., Perifanou, D., Ayres, J. G., Harrison, R. M., Kotronarou, A., Kavouras, I. G., Pekkanen, J., Hämeri, K., & Kos, G. P. (2012). Particulate matter air pollution and respiratory symptoms in individuals having either asthma or chronic obstructive pulmonary disease: A European multicentre panel study. *Environmental Health, 11*(1), 75.
Khreis, H., May, A. D., & Nieuwenhuijsen, M. J. (2017). Health impacts of urban transport policy measures: A guidance note for practice. *Journal of Transport & Health, 6*, 209–227.
Lee, I.-M., Shiroma, E. J., Lobelo, F., Puska, P., Blair, S. N., Katzmarzyk, P. T., & Group, L. P. A. S. W. (2012). Effect of physical inactivity on major non-communicable diseases worldwide: An analysis of burden of disease and life expectancy. *The Lancet, 380*(9838), 219–229.
Lee, B. J., Kim, B., & Lee, K. (2014). Air pollution exposure and cardiovascular disease. *Toxicology Research, 30*(2), 71–75.
McClure, R. J., Adriazola-Steil, C., Mulvihill, C., Fitzharris, M., Salmon, P., Bonnington, C. P., & Stevenson, M. (2015). Simulating the dynamic effect of land use and transport policies on the health of populations. *American Journal of Public Health, 105*(S2), S223–S229.
Mees, P. (2010). *Transport for suburbia: Beyond the automobile age*. London: Earthscan.
Miles, S., & Miles, M. (2004). *Consuming cities*. New York: Palgrave Macmillan.
Ministry of the Environment Ontario. (2010). *Fine particulate matter*. Retrieved from http://www.airqualityontario.com/science/pollutants/particulates.php
Moran, A. (2006). *The tragedy of planning*. Retrieved from Melbourne
Myors, P., O'Leary, R., & Helstroom, R. (2005). Multi unit residential buildings energy & peak demand study. *Energy News, 23*(4), 113–116.
Neuman, M. (2005). The compact city fallacy. *Journal of Planning Education and Research, 25*(1), 11–26.
Newman, P. G., & Kenworthy, J. R. (1989). *Cities and automobile dependence: An international sourcebook*. Aldershot: Gower.
Newman, P., & Kenworthy, J. (1999). *Sustainability and cities: Overcoming automobile dependence*. Washington, DC: Island Press.

Newman, P., Kenworthy, J., & Robinson, L. (1992). *Winning back the cities*. Sydney: Pluto Press.
Nutsford, D., Pearson, A., Kingham, S., & Reitsma, F. (2016). Residential exposure to visible blue space (but not green space) associated with lower psychological distress in a capital city. *Health and Place, 39*, 70–78.
OECD. (2010). *Cities and climate change*. Paris: OECD Publishing.
Perkins, A., Hamnett, S., Pullen, S., Zito, R., & Trebilcock, D. (2009). Transport, housing and urban form: The life cycle energy consumption and emissions of city centre apartments compared with suburban dwellings. *Urban Policy and Research, 27*(4), 377–396.
Searle, G., & Filion, P. (2011). Planning context and urban intensification outcomes: Sydney versus Toronto. *Urban Studies, 48*(7), 1419–1438.
Sørgaard, K. W., Sandlund, M., Heikkilä, J., Hansson, L., Vinding, H. R., Bjarnason, O., Bengtsson-Tops, A., Merinder, L., Nilsson, L.-l., & Middelboe, T. (2003). Schizophrenia and contact with health and social services: A Nordic multi-centre study. *Nordic Journal of Psychiatry, 57*(4), 253–261.
Stevenson, M., Thompson, J., de Sá, T. H., Ewing, R., Mohan, D., McClure, R., Roberts, I., Tiwari, G., Giles-Corti, B., & Sun, X. (2016). Land use, transport, and population health: Estimating the health benefits of compact cities. *The Lancet, 388*(10062), 2925–2935.
Tamura, K., Jinsart, W., Yano, E., Karita, K., & Boudoung, D. (2003). Particulate air pollution and chronic respiratory symptoms among traffic policemen in Bangkok. *Archives of Environmental Health: An International Journal, 58*(4), 201–207.
Troy, P. N. (1996). *The perils of urban consolidation*. Sydney: Federation Press.
UN-Habitat. (2009). *Global report on human settlements: Planning sustainable cities*. London: UN Habitat/Earthscan.
UN-Habitat. (2011). Global report on human settlements 2011, Abridged edition. London: United Nations Human Settlement Programme/Earthscan.
UN-Habitat. (2012). *State of the World's cities 2012/2013*. Nairobi, Kenya: UN Human Settlements Programme.
UNICEF. (2012). *Children in an an urban world*. New York.
United Nations. (2004). *World population to 2300*. New York: United Nations Publications.
United Nations. (2011). *World population prospects: The 2010 revision*.
Urry, J. (2011). *Climate change and society*. Cambridge: Polity.
Vlahov, D., & Galea, S. (2003). Urban health: A new discipline. *The Lancet, 362*(9390), 1091–1092.
Woodcock, J., Edwards, P., Tonne, C., Armstrong, B. G., Ashiru, O., Banister, D., Beevers, S., Chalabi, Z., Chowdhury, Z., & Cohen, A. (2009). Public health benefits of strategies to reduce greenhouse-gas emissions: Urban land transport. *The Lancet, 374*(9705), 1930–1943.
World Bank. (2010). *Cities and climate change: An urgent agenda*. The World Bank: Washington, DC.
World Health Organisation. (2013). *Global status report on road safety*. Luxembourg: WHO Press.
Wright, K. (2010). The relationship between housing density and built-form energy use. *Environment Design Guide, 65*, 1.

Chapter 15
Advancing Health Considerations Within a Sustainable Transportation Agenda: Using Indicators and Decision-Making

Josias Zietsman and Tara Ramani

15.1 Introduction

Transportation is an integral part of human life, providing access to goods, services, and economic opportunities. Transportation is also a major consumer of fossil fuel energy and is similarly responsible for a large share of global greenhouse gas emissions. The transport sector also has several other negative externalities, ranging from safety impacts to pollutant emissions and other environmental disturbances. In this context, the concept of sustainable transportation has emerged as a common means of framing issues related to transportation. In recent years, public health has also emerged as a driver or consideration in the transportation planning arena. Several aspects related to health and transportation (such as safety, exposure to pollutant emissions and noise, opportunities for active living, access to healthy food, etc.) also relate directly to sustainability considerations. Promoting the use of indicators and decision-making aligned toward health is therefore potentially compatible with a sustainable transportation agenda. As discussed later in the chapter, the exact definition of indicators may vary, but they can broadly be viewed as qualitative or quantitative criteria that can be used to assess progress toward a specific goal or objective.

This chapter discusses indicators and decision-making for sustainable transportation planning and how health considerations can be addressed in this process. Therefore, it discusses the advancement of public health considerations within a sustainable transportation agenda. Specifically, the chapter covers the following:

- Sustainable transportation as a framework for transportation planning
- Framing health as a sustainability issue
- Indicators for sustainable transportation and health

T. Ramani · J. Zietsman (✉)
Texas A&M Transportation Institute, College Station, TX, USA
e-mail: zietsman@tamu.edu

M. Nieuwenhuijsen, H. Khreis (eds.), *Integrating Human Health into Urban and Transport Planning*, https://doi.org/10.1007/978-3-319-74983-9_15

- Framework to advance health considerations in a sustainable transportation paradigm
- A case study application of health-oriented indicators versus sustainability indicators

In the authors' view, it is very important for transportation practitioners to understand the interrelationships between approaches to promote "sustainable transportation" and those that promote "health and transportation." Given that the concepts of health and sustainability are broad, and often extend beyond the transport sector, they individually pose challenges to practitioners tasked with advancing these considerations. Further, sustainability or sustainable transportation already has considerable buy-in as a framing device for transportation planning and policy globally. Thus, viewing health considerations within a sustainable transportation paradigm can help leverage existing indicators, tools, and resources to address health considerations, without detracting from broader sustainability issues.

15.2 Sustainable Transportation

Sustainability is recognized today as a broad concept that can hold several meanings (Beatley 1995), and the most widely cited definition of "sustainable development" remains that of the Brundtland Commission. In a report titled *Our Common Future* (WCED 1987), the commission defined sustainable development as "development which meets the needs of current generations without compromising the ability of future generations to meet their own needs." Kidd traced the evolution of sustainability from these environmentally/ecologically focused roots, crediting biologists, ecologists, and environmental planners for taking leadership in the area. He also acknowledged that the concept now signifies something much broader, "encompassing a wide range of economic, political and social goals" (Kidd 1992). Current discussion of sustainability, in line with this evolution, often refers to the three "dimensions" or "pillars" of sustainability or the "triple bottom line"—namely, the environment, the economy, and society/social equity.

There are differing perspectives on what sustainable transportation means. Similar to discussions of sustainability in general, literature on sustainable transportation mostly discusses sustainability along environmental, economic, and social dimensions—with some viewing equity as separate from the social dimension and instead as an overarching inter- and intra-generational issue (Gudmundsson et al. 2015; Holden et al. 2013; Zietsman et al. 2011). Several authors have put forward frameworks and approaches for sustainable transportation, including policy and planning frameworks (Hall 2006), performance measure-based approaches and frameworks (Litman 2010; Ramani et al. 2011a; Zietsman et al. 2011), as well as the application of popular generalized ecological and sustainability frameworks to transportation (Jeon and Amekudzi 2005; Pei et al. 2010). A comprehensive view of sustainability principles in the transportation context is as follows and is shown in Fig. 15.1.

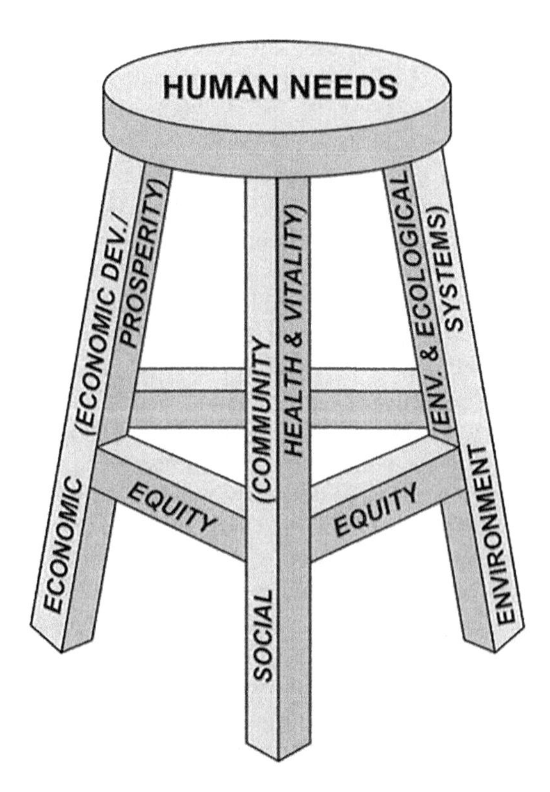

Fig. 15.1 Principles of sustainability and the significance of equity. Source Zietsman et al. (2011)

Sustainability entails meeting human needs for the present and future while:

- *Preserving and restoring environmental and ecological systems,*
- *Fostering community health and vitality,*
- *Promoting economic development and prosperity, and*
- *Ensuring equity between and among population groups and over generations.* (Zietsman et al. 2011)

Notable in this definition is the lack of a specific mention of transportation. The intent is to emphasize that the principles of sustainability are not sector-specific, and that sector-specific goals can be derived from the same. Additionally, we see the emphasis on equity or the distributional element as a reinforcing aspect of all principles (as shown in Fig. 15.1), covering both inter- and intra-generational equity aspects.

15.3 Health and Transportation as a Sustainability Issue

Health has emerged as an area of emphasis in transportation planning in recent years, with increasing collaboration between public health and transportation agencies, as they recognize linkages between the two areas (Ramani 2017a). Early discussions of health as it relates to transportation covered aspects such as active living, the built environment, air quality, and traffic safety (Frank 2000; Frank et al. 2006; Killingsworth et al. 2003; Lee and Moudon 2004; Litman 2003). Environmental justice, health disparities, and the need for collaborative research between the transportation and health disciplines were also noted (Sallis et al. 2004). In the USA, the focus on active living as the primary linkage between health and transportation was also reflected in case studies of collaborative planning efforts among health and transportation professionals (Burbidge 2010; Lyons et al. 2012). However, the health in transportation planning framework from the Federal Highway Administration (FHWA) also introduced a concept for a holistic approach to health and transportation that covered four main elements (1) active transportation, (2) safety, (3) air pollution, and (4) access to opportunities for healthy lifestyles (Lyons et al. 2012, 2014). The equity element was also addressed in terms of impacts on vulnerable populations. Another view of the connections between transportation and health is to look at direct effects (such as physical activity, air pollution, safety, etc.) and indirect effects (including access to jobs, services, medical care, etc.) (CDPH 2013). The Transportation and Health tool developed by the US Department of Transportation (USDOT) and the Centers for Disease Control and Prevention (CDC) similarly covers a broad range of areas for health and transportation interlinkages, from alcohol-impaired fatalities to commute mode shares, land-use mix, seat belt use, and housing and transportation affordability (USDOT 2015). Similarly, as discussed further in other chapters, the pathways of action linking transportation to health can be viewed as covering motor vehicle crashes, air pollution exposure, noise exposure, increased urban temperature exposure, lack of green space, physical inactivity, climate change effects, social exclusion, and community severance (Khreis et al. 2017).

If we consider the abovementioned goals and issues to be reflective of priorities in terms of health and transportation, we see that they overlap with areas that are commonly discussed in the sustainable transportation area, covering aspects like safety, mode shares and nonmotorized transportation, emissions, and accessibility issues. This leads us to emphasize that *health and transportation is a sustainability issue*, touching on several of the social, economic, and environmental facets important to sustainability. There is limited literature exploring parallels and conflicts between health and sustainability in the transportation context. However, there are examples of more general discussions on human health/public health in relation to sustainability/sustainable development as a whole. McMichael discusses these issues in two papers—one that addresses a new vision for sustainability and another on population health as a bottom line for sustainability (McMichael 2006; McMichael et al. 2003). As stated, "…public health researchers have a responsibility to ensure that their societies understand that, in the final analysis, sustainability is about ensuring positive (and equitable) human experience – of which health is fundamental…"(McMichael 2006). King made the case for health as a sustainable

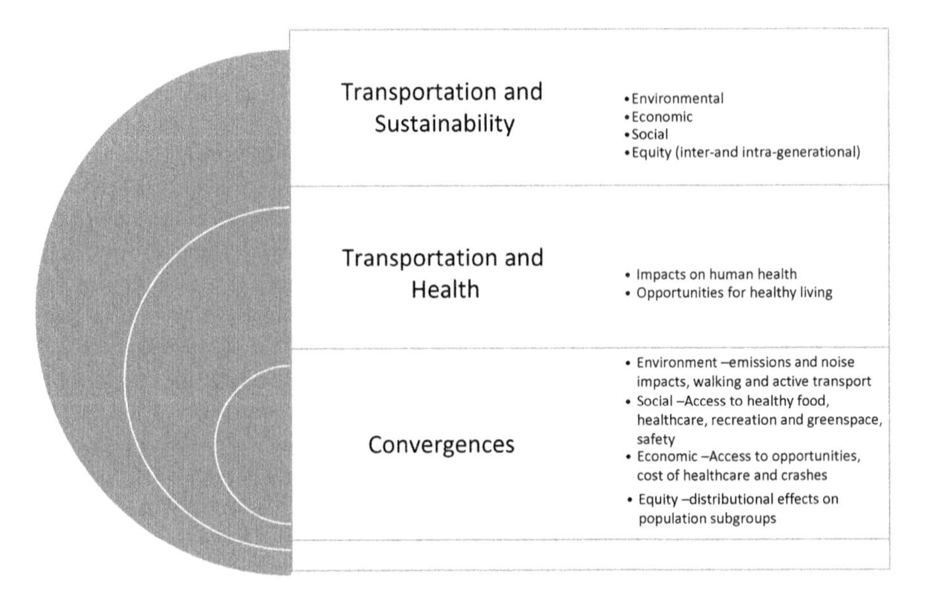

Fig. 15.2 Health in a sustainability framework—convergences

state, tracing how health broadened from being an individual concern to a family concern, and finally a social issue (King 1990). Additionally, in the context of climate change as a sustainability issue, a vast body of literature has also emerged on the effects of climate change on human health, which range from heat- and cold-related illnesses and deaths, air pollution effects, spread of infectious diseases, flood- and drought-related death and diseases, etc. (Haines et al. 2006; McMichael et al. 2006; Patz et al. 2005).

An examination of the issues that come to the surface in terms of health and transportation show a clear overlap with the environmental-, economic-, social-, and equity-related considerations of sustainability. The realm of health and transportation can be viewed as having two main aspects:

1. Transportation's impact on human health
2. Transportation as a provider of opportunities for healthy living

In each of these aspects, there are subareas that converge with sustainability considerations, as shown in Fig. 15.2. It is important to understand these convergences and commonalities to see how health fits into a sustainable transportation agenda. As detailed in a mapping exercise conducted to assess sustainability-related discourse, including the concepts of livability, resilience, and health (Ramani 2017b), it is important to note that health does not equally emphasize all aspects of holistic sustainability (i.e., environmental, economic, and social dimensions, and inter- and intra-generational equity).

15.4 Transportation Planning and Decision-Making in the Context of Sustainability

Sustainability considerations are not necessarily a driver of transportation planning everywhere. However, there is increased movement toward a more sustainability-oriented approach to planning. For example, Hall contrasted a "sustainability approach" with the current approach to planning (Hall 2006), and similarly, Zietsman and Rilett (2002) discussed a paradigm shift needed in assessing sustainable transportation. We also see that the principles of sustainability are often implicitly driving transportation planning, even if there is no explicit mandate to do so, as seen in the US context (Ramani et al. 2015). This is echoed by the Sustainable Urban Mobility Planning (SUMP) guidelines that distinguishes SUMP from traditional transport planning by its multimodal focus, integrated planning approaches, focus on people and equality of life, engagement of stakeholders, and comprehensive impact assessments (Wefering et al. 2013). Keeping this in mind, we propose viewing sustainable transportation planning as the "baseline" from which the health context can be discussed and introduced through overlapping priorities and synergies.

In this chapter, we use the term transportation planning in the general sense to describe activities related to the provision of transportation infrastructure and services to the public. Different nations and areas have different approaches, institutional structures, governance models, and local priorities that drive the exact transportation planning processes. A discussion of this is beyond the scope of this chapter, but it is important to note that transportation planning as defined here actually encompasses a set of domains—long-range/strategic planning, short-range programming, project development, construction, operations, and maintenance (Gudmundsson et al. 2015). However, we emphasize the "planning" aspects since generally, long-range planning drives the decisions made in terms of specific project selection and implementation.

In general, as noted by Amekudzi and Meyer "Transportation planning occurs within a societal and legal context that largely influences both the substance and approach toward planning ….(and)… historically, transportation planning has reflected the policy concerns and issues of the times in which it was occurring" (Amekudzi and Meyer 2006). Several authors have tackled some formalization to the approaches to decision-making in transportation (Gudmundsson et al. 2015), and a useful means on understanding the prevailing approaches to planning is to consider the classification posited by Emberger et al., who discuss *vision-led, plan-led*, and *consensus-led* approaches, while recognizing that in practice, a combination or hybrid approach prevails, with different aspects being given different levels of emphasis in various contexts (Emberger et al. 2008).

For purposes of discussing decision-making and indicators for transportation, we make the following generic observations regarding transportation planning processes:

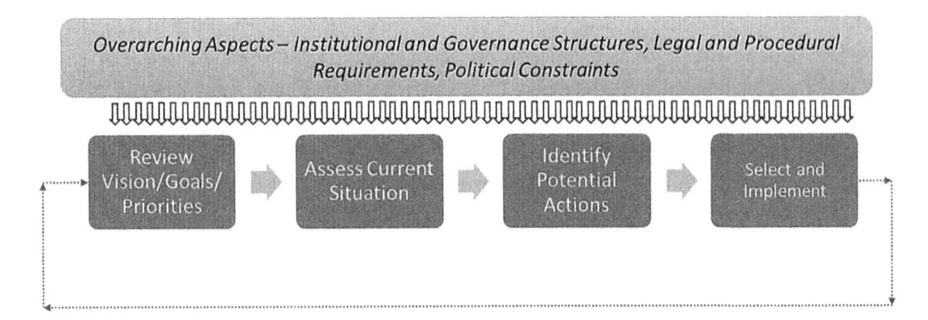

Fig. 15.3 Generic, simplified transportation planning process

- Transportation planning is a complex set of activities, involving multiple stakeholders and institutions operating under a range of legislative environments.
- Overall, transportation planning reflects local priorities and higher-level national policy goals—the extent of emphasis on each may vary.
- There are variations to the extent to which decision-making processes are formalized (e.g., through the use of multi-criteria decision-making (MCDM) processes) and whether these types of processes are used in a *decision-making* versus *decision-support* role.
- There are variations to how aspects such as multimodalism, public-private partnerships, collaboration with other agencies, stakeholder engagement processes, engagement of private sector, etc. are handled.

Keeping this in mind, we developed a generic, simplified transportation planning process for purposes of discussion. It can be viewed as a cyclical process as shown in Fig. 15.3. It involves (1) identification of goals (also termed as objectives or priorities), (2) assessment of current situation, (3) identification of potential actions, and (4) selection and implementation of these actions. Monitoring and review of the impacts of actions taken can serve as a feedback mechanism to adjust goals, priorities, and actions. These key steps are undertaken in a context-specific environment, under specific institutional, governance, political, and legal contexts. In this context, indicators (or performance measures) can serve a valuable purpose in different aspects of the process, ranging from tracking progress toward goals to supporting decisions or assessing benefits of different courses of action. This is discussed further in the next section.

15.5 Indicators and Decision-Making for Sustainable Transportation

As mentioned previously, exact mechanisms, processes, and frameworks for transportation planning vary. At the very basic level however, most planning is done in a way that aligns decisions or investments with higher-level goals or priorities. In this context, *indicators* become very relevant as a means of evaluating or assessing progress toward these goals or priorities. Indicators are also sometimes termed as performance measures, and may also be known as related terms such as parameter, measure, proxy, value, variable, etc. (Gallopín 1996; Gudmundsson et al. 2015). Indicators have different definitions and applications (Gilbert et al. 2003; Joumard and Gudmundsson 2010), but here we use the definition "a variable, or a combination of variables, selected to represent a certain wider issue or characteristic of interest" (Gudmundsson et al. 2015). Indicators are a very useful tool for the understanding and assessment of broad, multi dimensional concepts, such as sustainability, and can range from simple values to complex indices/functions and can be nonnumerical. A common approach to using indicators or performance measures is to view them as part of a hierarchy in which they derive from goals or subgoals, with the term "performance measure" used to define indicators used in a goal-oriented setting (Gudmundsson et al. 2015; Zietsman et al. 2011).

Our previous work assessed and defined the uses of indicators for decision-making, as follows: (1) description, (2) evaluation, (3) accountability, (4) decision support, and (5) communication. These indicator applications often blend into each other, for example, the communication function is implied as a part of any indicator-based assessment, while the evaluation, accountability, and decision-support applications can derive from indicators used for descriptive purposes alone. However, it is important to keep in mind that certain indicators tend to be more suited for certain types of functions (Zietsman et al. 2011).

A common indicator typology relevant to the context of sustainable transportation is to classify indicators as process (or input), output, and outcome measures, based on what they signify from an organizational perspective, scale, and time frame (Gudmundsson et al. 2015; Zietsman et al. 2011). Generally, process or input indicators represent actions or inputs into a system (e.g., investment levels in a particular program to increase transit usage); output indicators or measures can be viewed as the results as manifesting in the transportation system (e.g., increased transit usage as a result of the program), while outcomes may represent progress toward broader or longer-term goals (e.g., reduction in percentage of people driving as a result of a program). In the context of sustainability (and also health), it is important to keep in mind an emphasis on outcomes, while also acknowledging that it is often easier to exert control over process or output indicators in the transport sector, while effecting meaningful change in outcomes is more challenging given broader issues beyond transportation that may be in play.

Since indicators have a wide range of uses outside the context of sustainability, a question arises about what makes an indicator an indicator of sustainability. Maclaren characterizes sustainability indicator as being (1) integrating, (2) forward -looking, (3) distributional, and (4) developed with input from multiple stakeholders in the community (Maclaren 1996). Others emphasize the need to focus on outcomes (Marsden et al. 2005; Zietsman et al. 2011), while other compilations of sustainable transportation indicators identify and classify indicators based on specific goals or areas of interest relevant to sustainability (Castillo and Pitfield 2010; Jeon and Amekudzi 2005; Lautso and Toivanen 1999; Litman 2007; Zietsman et al. 2011).

Others also focus on the urban scale, quantifying sustainable transportation indicators for the comparison of cities (Haghshenas and Vaziri 2012; WBCSD 2015).

Often, indicators are also combined or used to develop composite indicators or scoring systems, for applications such as decision support. These generally follow approaches based in the field of multi-criteria decision-making (MCDM) or multi-criteria decision analysis (MCDA), including the use of fuzzy-logic-based approaches or Delphi methods to elicit priorities and values and to score indicators (Accorsi et al. 1999; Castillo and Pitfield 2010; Jeon et al. 2010; Ramani et al. 2009). Other approaches to assess sustainability in the transportation context that use some form of indicator-based analyses include sustainability rating systems (Muench et al. 2010; Oswald and McNeil 2009), or modified versions of a benefit-cost analysis (McVoy et al. 2013).

15.6 Adapted Indicator-Based Framework for Sustainable Transportation and Health

Previous research by the authors developed a generally applicable sustainability framework based on a hierarchy of goals, objectives, and performance measures (indicators) to help transportation agencies understand and apply sustainability concepts (Ramani et al. 2011b; Zietsman et al. 2011). In this section, we present a modified framework that can be used, along with a discussion of sustainable transportation goals and indicators relevant in the context of health. A simplified overview of the framework is shown in Fig. 15.4.

As discussed, the overall context or high-level vision will be driven in this case by the convergences between health and sustainability in transportation. This context guides the development of goals and subgoals (objectives), which in turn can be used to identify appropriate indicators, which can be applied for purposes such as description, evaluation, ensuring accountability, for decision support, or communication. Every step of the process will involve feedback and continuous adjustments, to ensure that the indicator applications allow for progress to be made toward health and sustainability goals.

In implementing this framework, it is useful to understand where the convergences between health and sustainable transportation lie in terms of operation goals, objectives, and indicators relevant to transportation agencies or to the transport sector. In the development of the generally applicable sustainability framework, we also developed a set of goals for sustainable transportation, as listed below (Zietsman et al. 2011):

1. Safety—Provide a safe transportation system for users and the general public.
2. Basic accessibility—Provide a transportation system that offers accessibility that allows people to fulfill at least their basic needs.
3. Equity/equal mobility—Provide options that allow affordable and equitable transportation opportunities for all sections of society.

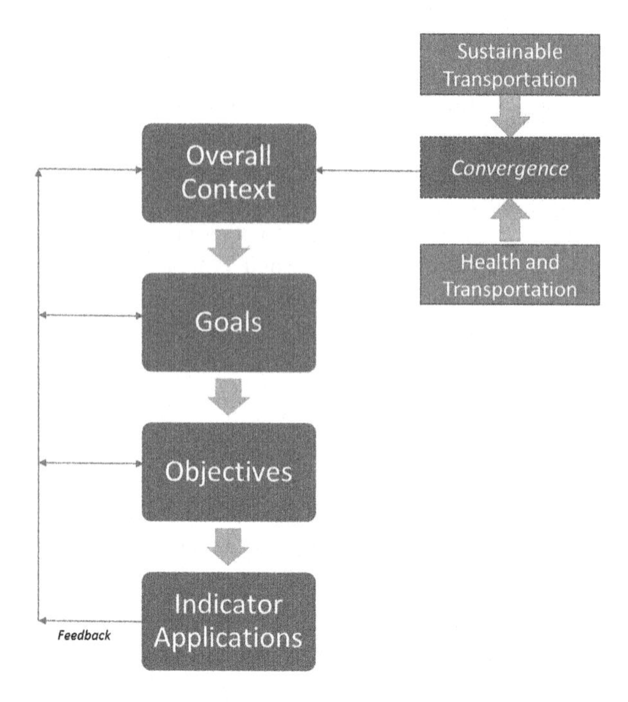

Fig. 15.4 Simplified framework for addressing health in a sustainability framework. Adapted from Zietsman et al. (2011)

4. System efficiency—Ensure that the transportation system's functionality and efficiency are maintained and enhanced.
5. Security—Ensure that the transportation system is secure from, ready for, and resilient to threats from all hazards.
6. Prosperity—Ensure that the transportation system's development and operation support economic development and prosperity.
7. Economic viability—Ensure the economic feasibility of transportation investments over time.
8. Ecosystems—Protect and enhance environmental and ecological systems while developing and operating transportation systems.
9. Waste generation—Reduce waste generated by transportation-related activities.
10. Resource consumption—Reduce the use of nonrenewable resources and promote the use of renewable replacements.
11. Emissions and air quality—Reduce transportation-related emissions of air pollutants and greenhouse gases.

These goals formed the basis for a compendium of objectives and indicators that can be used as a reference to identify relevant indicators and performance measures. From an examination of the set of comprehensive sustainable transportation goals, we see that the following sustainable transportation goals are most relevant to health and transportation issues:

- Safety—in terms of injuries and deaths attributable to motorized and other transportation
- Accessibility—in terms of access to destinations that promote health and access in terms of alternative modes
- Equity/equal mobility—in terms of mobility by nonmotorized and alternative modes from an equity perspective and emissions reduction perspective
- Emissions and air quality—in terms of health impacts of air pollution

Table 15.1 shows a set of sample objectives and measures from the compendium that are relevant from a health perspective. It should be noted that not all relevant health-related exposures are covered in this list, but the list demonstrates that several common sustainable transportation indicators can also function as health indicators, and vice-versa.

The indicators listed in the table are only a sample of sustainable transportation -related measures relevant to health. There are several other indicators and assessment methods for health-related to sustainable transportation. This includes tools and criteria that can be used to evaluate healthy places in relation to transportation and land use—including walkability audits, health impact assessments (HIAs), and certifications such as the Leadership in Energy and Environmental Design (LEED) for neighborhood development (Dannenberg and Wendel 2011). Built environment-related health measures include metrics such as population density, land-use mix, access to recreational facilities, street pattern, sidewalk coverage, vehicular traffic, and crime (Brownson et al. 2009). In a study of opinions on health and transportation among practitioners in the UK, the three most important health-related transportation issues were traffic casualties, air quality, and walking/cycling, followed by social inclusion (Davis 2005). As discussed further in the case study in the next section, identifying appropriate indicators also depends on obtaining appropriate data at the relevant scale of analysis.

15.7 Case Study: Contrasting Health Indicators and Sustainability Indicators

This chapter posits that health considerations can be incorporated into a sustainable transportation agenda and that there are indicators for sustainable transportation that can appropriately reflect converging considerations. It is also important, however, to understand the potential implications of a "health-oriented" versus "sustainability-oriented" perspective in the use of indicators and decision-making. That is, even when there are theoretical and conceptual overlaps between health and sustainability, the choice of indicators to reflect each concept may vary and result in differences in measured outcomes. This case study presents the findings from an indicator-based analysis conducted using a "health-oriented" versus "sustainability-oriented" perspective for the El Paso metropolitan area in the USA. The study is part

Table 15.1 Example sustainable transportation indicators relevant to health considerations

Goal/sample objective	Sample indicator(s)
Safety/reduce the number and severity of crashes	Change in the number and severity of crashes
	Change in the number of crashes by crash type and contributing factor
	Change in the number and severity of truck crashes
Safety/ensure safety considerations are addressed for all modes	Change in the number and severity of crashes by user type (e.g., pedestrian, bicycle, transit user, freight)
Accessibility/ensure accessibility to essential destinations	Change in travel time (by mode) to schools, health services, grocery stores, civic and public spaces, recreation
Equity/ensure comparable transportation system performance for all communities	Change in level of transportation service for disadvantaged and non-disadvantaged neighborhoods
Equity/ensure reasonable transportation options for all communities	Change in the percentage of disadvantaged population with convenient access to high quality transit service
Equity/ensure accessibility to jobs and essential destinations for all communities	Change in the level of access for disadvantaged populations to jobs, schools, health services, grocery stores, civic and public spaces, recreation
Air quality/reduce activity that generates pollutant emissions (travel, trip length, mode split, emissions)	Change in trips, vehicle trips, VMT, percent nondriver, tons of emissions per day
Air quality/reduce polluting exhaust emissions (criteria pollutants and GHGs)	Change in emissions by criteria pollutant, total and mode/ton mile
Air quality/increase land use compactness, density, and balance of interacting uses (compactness, density, balance)	Change in jobs/housing balance
Air quality/increase the use of nonmotorized modes	Change in planned miles of transit routes, pedestrian facilities, designated bike facilities, population within one mile of transit, connectivity index (pedestrian facilities, bike facilities, transit routes
Air quality/reduce proximity of air pollution-sensitive land uses to major pollution sources (high volume highways)	Change in sensitive receptors within close proximity, residential population within critical distance, percentage of ethnic/racial population groups within critical distance

From Zietsman et al. (2011)

of a broader investigation of sustainability-related discourse and potential impacts in the context of transportation planning in the USA (Ramani 2017b).

El Paso is located in West Texas, at the U-Mexico border, across from the city of Ciudad Juarez in Mexico. The analysis presented here compares the results from a GIS-MCDA-based analysis in which indicators were developed at the transportation planning level to contrast concepts of "health" and "sustainability" as commonly conceptualized in literature and practice in the USA. The process involved

(1) identification of representative indicators, (2) quantification of these indicators using spatially disaggregated data, and (3) combining the indicators into composite sustainability and health indices using a simplified aggregation method based on MCDA processes. Data from the El Paso Metropolitan Planning Organization's travel demand model and other sources were used to quantify a set of indicators and develop the sustainability and health indices.

The approach to indicator selection was to rely on examples from the literature and the stated goals of a health- and a sustainability-oriented perspective. It therefore is representative of a good-faith attempt by a transportation practitioner to quantify the concepts, taking into account constraints such as the need for a uniform analysis and data availability. As seen from discussions earlier in the chapter, there are several convergences between the areas of sustainability and health. At the same time, there are differences in emphasis in the discourse surrounding these concepts. This analysis intends to represent these nuances through the choice of indicators. Keeping this in mind, the indicators selected were as follows:

Sustainability-oriented indicator set: This set of indicators were selected reflecting environmental, economic, and social factors that tended to prevail in discussions of sustainable transportation planning. It included the following:

- Mode split
- Greenhouse gas emissions
- Criteria pollutant emissions
- Land use type
- Safety

The greenhouse gas and criteria pollutant emissions cover the environmental aspects; the safety measures address social considerations, while the land use measure and mode split measure can be viewed as addressing both the social and economic dimensions. This is consistent with several sources that discuss sustainability indicators (Haghshenas and Vaziri 2012; Jeon and Amekudzi 2005; Litman 2007; WBCSD 2015; Zietsman et al. 2011).

Health-oriented indicator set: This set of indicators were selected reflecting transportation's relationship to human health, especially in relation to four key elements that prevail in discussions of health and transportation in the USA—safety, air quality, active living opportunities, and access to critical destinations. The selection was intended to represent commonalities as well as distinctions from a "sustainable transportation" perspective and included the following:

- Traffic density
- Safety
- Proximity to clinics and hospitals
- Proximity to parks and recreation

The selection of indicators is consistent with literature on the subject (CDPH 2013; Lyons et al. 2012, 2014; USDOT 2015). This included a measure of traffic intensity used as a proxy for exposure to traffic-related emissions (Rioux et al. 2010; Rowangould 2013), which also accounts for mode split/use of nonmotorized modes by taking into account vehicle miles of travel. The safety indicator, proximity to parks and recreational facilities (active living opportunities and access to green space) and to clinics and hospitals (critical destinations), represents the other considerations. As discussed previously, these indicators are not pure "health indicators" but rather represent a health-oriented slant on sustainable transportation indicators. The two indicator sets also show an overlap in terms of the safety measure, which is understandable given the convergences between the two concepts. Further, the traffic intensity measure was selected for the health-oriented indicator set to represent an "exposure-type" planning metric for traffic and emissions, taking into account traffic levels (indirectly addressing mode split) and also serving as a proxy for the health impacts of vehicular emissions. A case can be made for the mode split and emissions metrics to also serve directly as health-related indicators instead. However, the intent of this analysis was to develop two representative indicator sets that, taken together, best represented the concepts of "health" and "sustainability."

The indicators were quantified at the level of a traffic analysis zone (TAZ) for the study region, consistent with transportation-planning level analyses and the level of data disaggregation available for the indicators. The indicators were scaled and aggregated into a health index and sustainability index, expressed on a 0–1 scale, with a high number representing better performance relative to other zones (Ramani 2017b). The results of the contrast for the 2010 and 2040 conditions are as shown in Fig. 15.5. The findings indicate that there isn't a very strong correlation between the health index and sustainability index values for a particular zone (correlation coefficient of 0.48 for 2010 and 0.45 in 2040). That is, areas that do well (darker colors on the map) in terms of sustainability do not necessarily do well from a health index perspective. This indicates that focusing solely on health-oriented indicators could potentially result in certain aspects of sustainability being neglected. The results are presented in further detail in the analysis documentation (Ramani 2017b). Other observations of relevance include the high correlations of the health index values over the years and the sustainability index values over the years, i.e., demonstrating the relatively low levels of change in all indicators over time. This shows that it is challenging to effect change through the transportation system alone, in the absence of changes addressing land use, socio economic, and other factors. The findings of this case study should not be taken as a concrete indication of differences between the concepts of "health" and "sustainability," but rather as a reflection of the importance of indicator selection on measured outcomes. For example, different indicator choices in this context may produce different results; similarly, different priority weights assigned to indicators may also have an impact. Thus, any indicator-based analysis exercise should attempt to understand the broader context and avoid "cherry-picking" of convenient indicators.

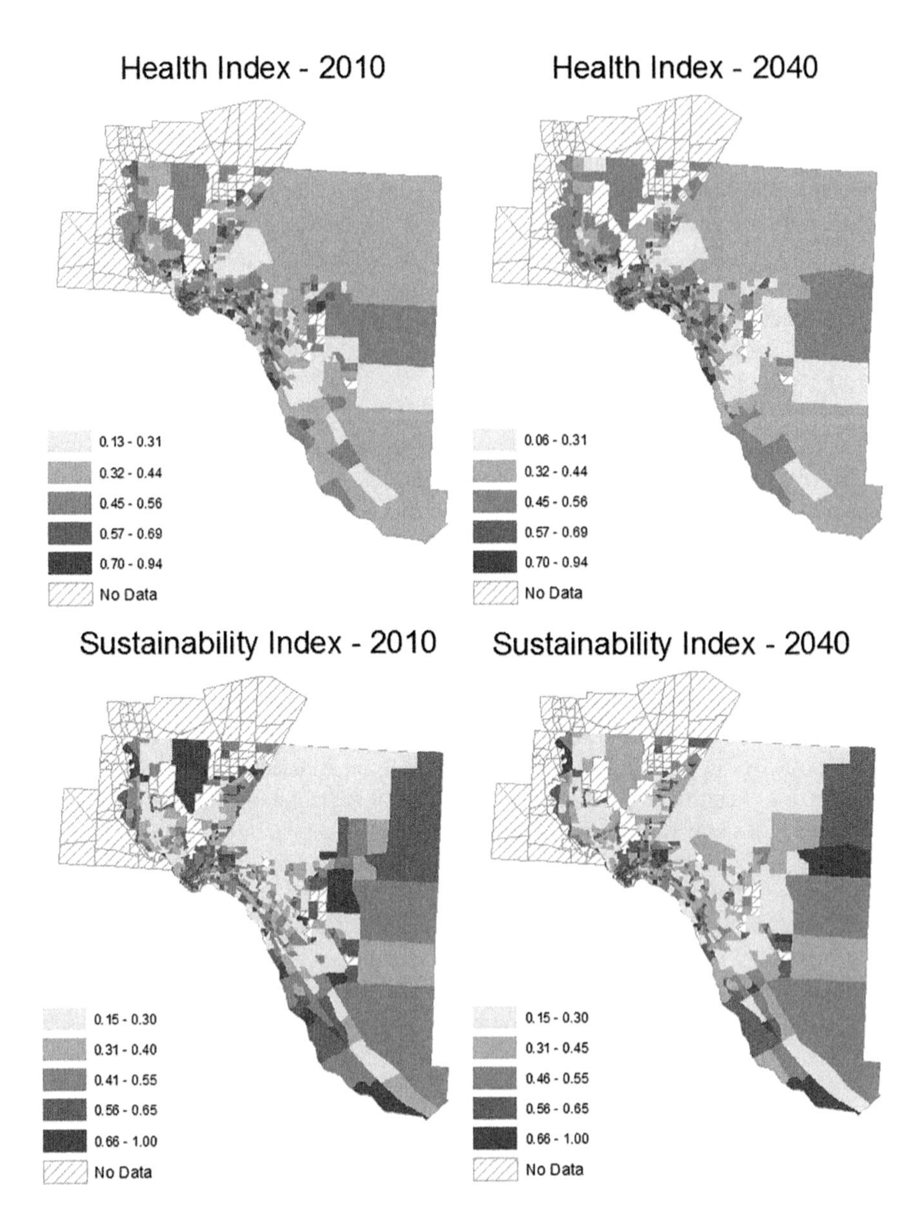

Fig. 15.5 Case study results

15.8 Summary and Conclusions

This chapter provides our perspective and analysis of indicators and decision-making for sustainable transportation, in relation to the emerging discipline of public health. We believe that it is important to advance health within a sustainable transportation agenda. Sustainability is a broad and overarching concept, and there is recognition of the need for sustainability to be a driver of transportation planning activities. Sustainability has evolved from being an environmentally focused concept to function as a "catch-all" for economic, environmental, and social goals that we strive for; while this can be a detriment, it also provides a useful framework for setting a vision and goals for transportation.

Health has similarly emerged as a broad and holistic concept, and we see that it has several overlaps and convergences with the concept of sustainability. We can therefore make the case that *health is a sustainability issue*, and an adapted framework to show how addressing sustainability can also address health issues through the use of appropriate goals, objectives, and indicators. We adapted a framework and indicator-based approach to sustainable transportation to pursue a joint health-sustainability agenda. As practitioners in the disciplines of health and transportation come together, it is important for them to acknowledge synergies between the two concepts. However, as the indicator-based case study showed, despite convergences between the two areas, health and sustainability are not necessarily proxies for each other. It also demonstrated the importance of indicator selection and the inherent complexities of developing universally agreed upon indicator sets to reflect concepts such as health and sustainability. The use of indicators geared solely toward health priorities may run the risk of certain aspects of sustainability being neglected. Similarly, a focus on sustainability may potentially reduce emphasis on health-related issues that do not serve higher-level sustainability goals.

In conclusion, the established field of sustainable transportation provides a useful jumping-off point for practitioners to understand and apply concepts of health. Decision-making process and indicators geared toward common goals can help practitioners advance health considerations in the transportation agenda.

References

Accorsi, R., Zio, E., & Apostolakis, G. E. (1999). Developing utility functions for environmental decision making. *Progress in Nuclear Energy, 34*, 387–411.

Amekudzi, A., & Meyer, M. D. (2006). Considering the environment in transportation planning: Review of emerging paradigms and practice in the United States. *Journal of Urban Planning and Development, 132*, 42–52.

Beatley, T. (1995). The many meanings of sustainability: Introduction to a special issue of JPL. *Journal of Planning Literature, 9*, 339–342. https://doi.org/10.1177/088541229500900401.

Brownson, R., Hoehner, C., Day, K., Forsyth, A., & Sallis, J. (2009). Measuring the built environment for physical activity. *American Journal of Preventive Medicine, 36*, S99–S123.e112.

Burbidge, S. K. (2010). Merging long range transportation planning with public health: A case study from Utah's Wasatch Front. *Preventive Medicine, 50*(Suppl 1), S6–S8. https://doi.org/10.1016/j.ypmed.2009.07.024.

Castillo, H., & Pitfield, D. E. (2010). ELASTIC—A methodological framework for identifying and selecting sustainable transport indicators. *Transportation Research Part D: Transport and Environment, 15*, 179–188. https://doi.org/10.1016/j.trd.2009.09.002.

CDPH. (2013). *A primer for California's public health community on regional transportation plans and sustainable communities strategies*. Sacramento: California Department of Public Health.

Dannenberg, A. L., & Wendel, A. M. (2011). Measuring, assessing, and certifying healthy places. In *Making healthy places* (pp. 303–318). New York: Springer.

Davis, A. (2005). Transport and health—What is the connection? An exploration of concepts of health held by highways committee chairs in England. *Transport Policy, 12*, 324–333.

Emberger, G., Pfaffenbichler, P., Jaensirisak, S., & Timms, P. (2008). "Ideal" decision-making processes for transport planning: A comparison between Europe and South East Asia. *Transport Policy, 15*, 341–349. https://doi.org/10.1016/j.tranpol.2008.12.009.

Frank, L. D. (2000). Land use and transportation interaction: Implications on public health and quality of life. *Journal of Planning Education and Research, 20*, 6–22. https://doi.org/10.1177/073945600128992564.

Frank, L. D., Sallis, J. F., Conway, T. L., Chapman, J. E., Saelens, B. E., & Bachman, W. (2006). Many pathways from land use to health: Associations between neighborhood walkability and active transportation, body mass index, and air quality. *Journal of the American Planning Association, 72*, 75–87. https://doi.org/10.1080/01944360608976725.

Gallopín, G. C. (1996). Environmental and sustainability indicators and the concept of situational indicators. A systems approach. *Environmental Modeling & Assessment, 1*, 101–117.

Gilbert, R., Irwin, N., Hollingworth, B., & Blais, P. (2003). Sustainable transportation performance indicators (STPI). Transportation Research Board (TRB), CD ROM.

Gudmundsson, H., Hall, R. P., Marsden, G., & Zietsman, J. (2015). *Sustainable transportation: Indicators, frameworks, and performance management. Springer texts in business and economics*. Berlin, Heidelberg: Springer-Verlag. https://doi.org/10.1007/978-3-662-46924-8.

Haghshenas, H., & Vaziri, M. (2012). Urban sustainable transportation indicators for global comparison. *Ecological Indicators, 15*, 115–121. https://doi.org/10.1016/j.ecolind.2011.09.010.

Haines, A., Kovats, R. S., Campbell-Lendrum, D., & Corvalan, C. (2006). Climate change and human health: Impacts, vulnerability and public health. *Public Health, 120*, 585–596. https://doi.org/10.1016/j.puhe.2006.01.002.

Hall, R. P. (2006). *Understanding and applying the concept of sustainable development to transportation planning and decision-making in the US*. Cambridge: Massachusetts Institute of Technology.

Holden, E., Linnerud, K., & Banister, D. (2013). Sustainable passenger transport: Back to Brundtland. *Transportation Research Part A: Policy and Practice, 54*, 67–77. https://doi.org/10.1016/j.tra.2013.07.012.

Jeon, C. M., & Amekudzi, A. (2005). Addressing sustainability in transportation systems: Definitions, indicators, and metrics. *Journal of Infrastructure Systems, 11*, 31–50. https://doi.org/10.1061/(Asce)1076-0342(2005)11:1(31).

Jeon, C. M., Amekudzi, A. A., & Guensler, R. L. (2010). Evaluating plan alternatives for transportation system sustainability: Atlanta metropolitan region. *International Journal of Sustainable Transportation, 4*, 227–247. https://doi.org/10.1080/15568310902940209.

Joumard, R., & Gudmundsson H. (2010). *Indicators of environmental sustainability in transport: An interdisciplinary approach to methods*. INRETS report, Recherches R282, Bron, France, 422 p.

Khreis, H., May, A. D., & Nieuwenhuijsen, M. J. (2017). Health impacts of urban transport policy measures: A guidance note for practice. *Journal of Transport & Health*. https://doi.org/10.1016/j.jth.2017.06.003.

Kidd, C. V. (1992). The evolution of sustainability. *Journal of Agricultural and Environmental Ethics, 5*, 1–26. https://doi.org/10.1007/bf01965413.

Killingsworth, R. E., De Nazelle, A., & Bell, R. H. (2003). Building a new paradigm: Improving public health through transportation. *ITE Journal, 73*, 28–32.

King, M. (1990). Health is a sustainable state. *The Lancet, 336*, 664–667. https://doi.org/10.1016/0140-6736(90)92156-C.

Lautso, K., & Toivanen, S. (1999). SPARTACUS system for analyzing urban sustainability. *Transportation Research Record: Journal of the Transportation Research Board, 1670*, 35–46. https://doi.org/10.3141/1670-06.

Lee, C., & Moudon, A. V. (2004). Physical activity and environment research in the health field: Implications for urban and transportation planning practice and research. *Journal of Planning Literature, 19*, 147–181. https://doi.org/10.1177/0885412204267680.

Litman, T. (2003). Integrating public health objectives in transportation decision-making. *American Journal of Health Promotion, 18*(1), 103–108.

Litman, T. (2007). Developing indicators for comprehensive and sustainable transport planning. *Transportation Research Record: Journal of the Transportation Research Board, 2017*, 10–15. https://doi.org/10.3141/2017-02.

Litman, T. (2010). *Sustainability and livability: Summary of definitions, goals, objectives and performance indicators.*

Lyons, W., Peckett, H., Morse, L., Khurana, M., & Nash, L. (2012). *Metropolitan area transportation planning for healthy communities* (No. DOT-VNTSC-FHWA-13-01). John A. Volpe National Transportation Systems Center (US).

Lyons, W., Morse, L., Nash, L., & Strauss, R. (2014). *Statewide transportation planning for healthy communities* (No. DOT-VNTSC-FHWA-14-01). United States. Federal Highway Administration.

Maclaren, V. (1996). Urban sustainability reporting American Planning Association. *Journal of the American Planning Association, 62*, 184–202.

Marsden, G., Kelly, C., Snell, C., & Forrester, J. (2005). *Sustainable transport indicators: Selection and use.* DISTILLATE Deliverable C1. Institute for Transport Studies, University of Leeds, UK.

McMichael, A. J. (2006). Population health as the 'bottom line' of sustainability: A contemporary challenge for public health researchers. *The European Journal of Public Health, 16*, 579–581. https://doi.org/10.1093/eurpub/ckl102.

McMichael, A. J., Butler, C. D., & Folke, C. (2003). New visions for addressing sustainability. *Science, 302*, 1919–1920. https://doi.org/10.1126/science.1090001.

McMichael, A. J., Woodruff, R. E., & Hales, S. (2006). Climate change and human health: Present and future risks. *The Lancet, 367*, 859–869. https://doi.org/10.1016/S0140-6736(06)68079-3.

McVoy, G., Gunasekera, K., Sousa, L. R., & Schaffner, P. (2013). An analytical framework for sustainability analysis of transportation investments across the triple bottom line using a common metric. In *2013 international conference on ecology and transportation (ICOET 2013)*, 2013.

Muench, S. T., Anderson, J., & Bevan, T. (2010). Greenroads: A sustainability rating system for roadways. *International Journal of Pavement Research and Technology, 3*, 270–279.

Oswald, M., & McNeil, S. (2009). Rating sustainability: Transportation investments in urban corridors as a case study. *Journal of Urban Planning and Development, 136*, 177–185. https://doi.org/10.1061/(ASCE)UP.1943-5444.0000016.

Patz, J. A., Campbell-Lendrum, D., Holloway, T., & Foley, J. A. (2005). Impact of regional climate change on human health. *Nature, 438*, 310–317. https://doi.org/10.1038/nature04188.

Pei, Y., Amekudzi, A., Meyer, M., Barrella, E., & Ross, C. (2010). Performance measurement frameworks and development of effective sustainable transport strategies and indicators. *Transportation Research Record: Journal of the Transportation Research Board, 2163*, 73–80.

Ramani, T. (2017a). *Assessing alternative conceptualizations of sustainable transportation.* Paper presented at the Transportation Research Board 96th Annual Meeting, Washington, DC.

Ramani, T. (2017b). *Validating alternative conceptualizations of sustainable transportation.* College Station: Texas A&M University.

Ramani, T., Zietsman, J., Eisele, W., Rosa, D., Spillane, D., & Bochner, B. (2009) *Developing sustainable transportation performance measures for TX DOT'S strategic plan.* Texas Transportation Institute Technical Report 0-5541-1.

Ramani, T., Zietsman, J., Gudmundsson, H., Hall, R., & Marsden, G. (2011a). Framework for sustainability assessment by transportation agencies. *Transportation Research Record, 2242,* 9–18.

Ramani, T., Zietsman, J., Gudmundsson, H., Hall, R., & Marsden, G. (2011b). Framework for sustainability assessment by transportation agencies. *Transportation Research Record: Journal of the Transportation Research Board, 2242,* 9–18.

Ramani, T., Zietsman, J., & Pryn, M. R. (2015). Towards sustainable transport planning in the United States working paper prepared for SUSTAIN Research Network, Danish Research Council.

Rioux, C. L., Gute, D. M., Brugge, D., Peterson, S., & Parmenter, B. (2010). Characterizing urban traffic exposures using transportation planning tools: An illustrated methodology for health researchers. *Journal of Urban Health, 87,* 167–188. https://doi.org/10.1007/s11524-009-9419-7.

Rowangould, G. M. (2013). A census of the US near-roadway population: Public health and environmental justice considerations. *Transportation Research Part D: Transport and Environment, 25,* 59–67. https://doi.org/10.1016/j.trd.2013.08.003.

Sallis, J. F., Frank, L. D., Saelens, B. E., & Kraft, M. K. (2004). Active transportation and physical activity: Opportunities for collaboration on transportation and public health research. *Transportation Research Part A: Policy and Practice, 38,* 249–268. https://doi.org/10.1016/j.tra.2003.11.003.

USDOT. (2015). *Transportation and health tool.* United States Department of Transportation. Retrieved November 5, 2015 from http://www.transportation.gov/transportation-health-tool.

WBCSD. (2015). *Sustainable urban mobility – Methodology and indicator calculation method world business council on sustainable development*

WCED. (1987). *Our common future. Oxford paperbacks.* Oxford; New York: Oxford University Press.

Wefering, F., Rupprecht, S., Bührmann, S., Böhler-Baedeker, S., & GmbH, R. C. F. B. (2013). *Guidelines: Developing and implementing a sustainable urban mobility plan.* Brussels: European Commission.

Zietsman, J., & Rilett, L. R. (2002). *Sustainable transportation: Conceptualization and performance measures.* College Station, TX: Southwest Region University Transportation Center.

Zietsman, J., Ramani, T., Reeder, V., Potter, J., & DeFlorio, J. (2011). *A guidebook for sustainability performance measurement for transportation agencies* (Vol. 708). Transportation Research Board.

Chapter 16
Measure Selection for Urban Transport Policy

Anthony D. May

16.1 Introduction

Measure selection is the process of identifying the most suitable and cost-effective policy measures to achieve a city's vision and objectives and overcome its transport-related problems. Logically, therefore, it comes after the process of developing a vision, objectives and targets, which are discussed in previous chapters.

Measure selection is a challenge for five principal reasons. Firstly, cities have a very wide range of measures available to them; these include building new road and rail infrastructure, providing new public transport services, managing the road network more effectively, encouraging behavioural change, providing improved information, charging for use of the transport system and modifying development patterns to reduce travel demands. They can be applied to the private car but also to buses, trains, trams, cycling and walking and to freight vehicles. It is all too easy to overlook solutions which would be more effective. Secondly, many stakeholders and politicians will have preconceived ideas as to what should be done, and evidence suggests that these solutions are often not the most cost-effective. Thirdly, the most cost-effective measures are often not the most easily implemented; split responsibilities, lack of funding and public opposition can limit what is done. Fourthly, a transport plan is likely to draw on several measures, but its performance and ease of implementation will depend on how these measures are packaged. Finally, a transport plan needs to be more than a wish list of measures; prior to implementation each measure needs to be defined in detail, assessed in terms of its likely impact, and appraised in terms of its potential contribution.

A. D. May (✉)
University of Leeds, Leeds, UK
e-mail: a.d.may@its.leeds.ac.uk

M. Nieuwenhuijsen, H. Khreis (eds.), *Integrating Human Health into Urban and Transport Planning*, https://doi.org/10.1007/978-3-319-74983-9_16

Measure selection, sometimes referred to as option generation, has been highlighted as one of the weaknesses of urban transport policy formulation (Eddington 2006). A failure to consider the full range of possible measures can lead to:

- An over-reliance on preconceived ideas
- A tendency to focus on supply-side measures rather than demand-side measures
- Lack of experience of the wider range of measures available
- Lack of evidence of the performance of those measures in other contexts

By definition, a measure which more effectively meets a city's objectives will be more beneficial. One that is more acceptable will stand a greater chance of being implemented and thus actually producing benefits. One which offers greater value for money will be able to realise those benefits while making less demand on limited budgets.

An effective package can combine those measures which are themselves most effective in ways which achieve synergy—by making the whole more effective than the sum of the parts—and which overcome barriers to implementing them, such as lack of acceptability. Examples of both these concepts are described more fully in May et al. (2012).

To assist in this process, the European Commission has recently published guidance on measure selection (May 2016). In this chapter, we summarise key elements of that guidance and look more specifically at ways in which specific measures can contribute to protecting and improving public health. In Sect. 16.2 we consider the range of measures and the evidence on them. In Sect. 16.3 we outline the principles of packaging. In Sect. 16.4 we discuss the concept of strategies. In Sect. 16.5 we consider the barriers to using individual measures. Section 16.6 presents a Measure Option Generator developed to support the guidance. Section 16.7 considers the potential of the range of measures to contribute specifically to improved public health. Section 16.8 concludes with a series of recommendations for good practice.

16.2 The Range of Policy Measures

A "measure" is an action which can be taken to contribute to one or more policy objectives or to overcome one or more identified problems. Examples range from building new transport infrastructure to managing the way in which that infrastructure is used and from service provision to regulation and pricing. There is a growing range of measures available to transport professionals. A total of 64 measures are included in the Measure Option Generator (see Sect. 16.6). Some of these, such as low emission zones, bike sharing and crowd sourcing, are relatively new. In all, planners and policymakers have access to around twice as many measures as they did 30 years ago.

There are several ways of categorising these measures. One is the distinction between "supply side" and "demand side". On the supply side are measures which add to the capacity of the transport system to move people and freight. On the demand side are measures which affect how people and freight operators use the transport system. Demand-side measures are often grouped under the title Transport Demand Management. Another categorisation considers the type of impact which the measure has. This is the approach which is adopted in the Measure Option Generator, which distinguishes between land-use measures, infrastructure measures, management and service measures, attitudinal and behavioural measures, information provision and pricing measures.

Unfortunately, evidence on the performance of many of these measures is incomplete. Cities and governments rarely set aside a budget for ex post evaluation of a measure's impact. All too often, the view is taken that the final step is to implement a measure, rather than to learn from that experience. This will particularly be the case where a measure has not been as effective, as hoped, since those involved will not want their decisions questioned publicly. It is particularly important for cities to take the opportunity to measure and evaluate the impacts of new measures and to make that information available to others. In particular, information on measures which have been less successful than planned can help others avoid making the same mistakes.

Even where experience is available, it may not be directly relevant in another context. Light rail will work better in larger cities than in smaller ones. Walking and cycling provision are more effective in high-density areas than in lower-density ones. Parking controls are likely to be more effective in city centres than elsewhere. Regulatory controls will be more acceptable in some cultures than in others. For all of these reasons, it can be difficult to judge how transferable experience with successful policy measures will be. This is a further reason for encouraging as much experience as possible to be recorded.

In understanding the performance of individual measures, it is important to consider their impacts in:

- Changing the demand for travel
- Changing the supply of transport facilities
- Changing the cost of provision and operation of the transport system

Initial responses (e.g. changes in mode) may lead to secondary ones (e.g. increases in overcrowding). Each of these types of change will in turn affect performance against the objectives and hence reduce (or increase) problems. Tracing all these impacts can be difficult, and causal chain diagrams (Fig. 16.1) can help to understand them. A first principles assessment of this kind can help to assess the potential contribution of a measure, and our Measure Option Generator is structured on this basis.

Changes in demand result from a myriad of decisions by individual travellers. When faced with a new policy measure, or with a change in an existing one, such as a fare increase, the individual traveller has a number of options, including changing the number of journeys made, combining journeys, changing destination, changing departure time, changing mode and changing route. The scale of response will depend on the circumstances. Those who are directly exposed to a change will respond more strongly than those for whom the impact is indirect. Those who have

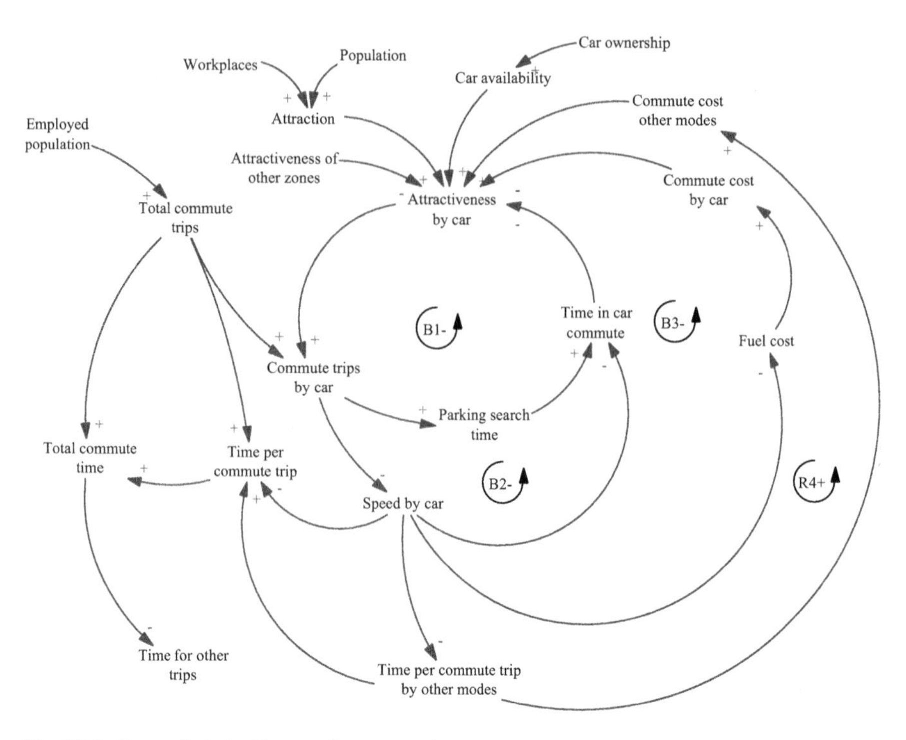

Fig. 16.1 A causal chain diagram for commuting

fewer alternatives will be more reluctant to change. Longer-term responses may well be stronger, as people have more time to respond. People are more likely to change when they experience life cycle changes, such as having a baby or changing jobs. Elasticities of demand are often used to understand the scale of such responses (e.g. Paulley et al. 2006).

Changes in the supply of transport can take a number of forms, including changes in capacity, the allocation of capacity, permitted speeds, access time, costs of use and information. Some of these will have a direct influence on travellers, while others will only affect them if they become aware of them. For most policy measures, it will be clear how they affect supply, but the scale of this impact may be difficult to assess.

The principal types of financial cost are capital costs of new infrastructure, operating and enforcement costs and costs of maintenance and replacement. Pricing measures such as congestion charging will, in addition, generate a revenue stream which will both reduce the net cost of the measure and influence demand. Changes in these costs and revenues are crucial in determining whether an individual policy measure, or the overall strategy, provides value for money. Low-cost measures typically offer greater value for money than major infrastructure projects.

There are two main sources of information on these effects: empirical evidence and predictive computer models. Each has its limitations. Empirical evidence can be collected from before and after studies of the implementation of a particular measure or a package. Ideally such studies will be carried out whenever a new type of measure is implemented or when a measure is implemented in a different context. However, as noted above, cities are often reluctant to spend scarce resources on such studies, and national governments rarely invest in them. Even where studies are conducted, they are often less than comprehensive. Information is needed on resulting changes in demand, supply and costs of travel, as well as on changes in outcome indicators for each of the objectives of interest to other cities. Moreover, performance will be affected by context and may not be transferable. Predictive models can in principle overcome these constraints by enabling impacts on demand, and hence on outcome indicators, to be predicted in a number of different contexts. But models themselves have limitations, as discussed in more detail in Sect. 3.5.2 of May (2016).

There are several useful sources of evidence from those studies which have been conducted, including CiViTAS (2016), ELTIS (2015) and Evidence (2016) projects. ELTIS case studies are added at the rate of perhaps ten per month and cover both policy measures and planning practice. A large amount of evidence has been collected for the Measure Option Generator, which now covers 64 measures and over 200 case studies (see Sect. 16.6).

16.3 The Use of Packages

No one measure on its own will be sufficient to achieve a city's objectives or overcome its problems. Most cities will include several policy measures in their transport plans and need to think about how these different measures might interact. This is the concept behind creating a policy package.

The key to developing a package is to identify which policy measures will work well together and which measures may be needed to make other measures viable. Thus, within a policy package, policy measures can interact in one of two different ways:

- They can achieve more together than either would on its own; this is the principle of synergy.
- They can facilitate other measures in the package by overcoming the barriers to their implementation (which are considered more fully in Sect. 16.5).

London offers a good example of both principles of packaging. Congestion charging was unpopular and was expected to affect lower-income car users. However, its revenue was used to finance an increase in bus services in inner London. These bus services helped overcome the barriers to congestion charging, while congestion charging overcame the financial barrier to increasing bus services. Moreover, the two together achieved a greater switch away from car use than either would have done on its own (TfL 2007).

It is difficult to find empirical evidence on the performance of packages given the problems of needing to implement several measures together and of isolating their effects from external changes. An alternative approach is to use predictive models to assess how measures might operate together. As an example, the PROPOLIS study (Lautso et al. 2004) tested a common set of policy measures in seven European cities, using each city's own model to produce a standard set of sustainability indicators. The two most effective policy measures were improvements to public transport, through faster, more reliable services and lower fares, and charges for car use, through road pricing or higher parking charges. Land use measures, tested on their own, had little impact but helped to intensify the effect of the transport measures and to reduce their potential contributions to urban sprawl. Few of the infrastructure projects being planned by the cities were as cost-effective as these public transport, pricing and land use measures.

16.4 Strategies

Cities may find it easier to think about the overall strategy which they wish to pursue than to list the measures which they want to use. At its simplest, a strategy is a combination of measures to address a city's objectives. More specifically, a strategy can be a direction of change which a city wants to achieve in the transport system. Such strategies are not objectives in their own right but changes which should contribute to the city's chosen objectives.

For example, a city might wish to reduce car use. Presenting this as an objective is likely to attract criticism that the city is "anticar". But demonstrating that reducing car use should help to improve the environment, liveability, health and safety links the strategy directly to objectives and hence helps justify it.

The KonSULT knowledgebase (Sect. 16.6) distinguishes between six strategies:

- Reducing the need to travel
- Reducing car use
- Improving public transport
- Improving the use of the road network
- Improving walking and cycling
- Improving freight operations

Cities may wish to pursue many or all of these, and the Measure Option Generator (Sect. 16.6) allows the user to do so and indicate priorities among them. For each strategy, it is then possible to identify the measures which potentially contribute to it. Some measures will contribute directly to one strategy (e.g. bike sharing is clearly linked to improving walking and cycling). Others may contribute to several; for example, mixed development will reduce the need to travel but may also make it easier to provide for public transport, walking and cycling. Measures appropriate for one or more of these strategies can then be combined into a package of measures designed to achieve a given change in the direction of the transport system, following the principles outlined in Sect. 16.3.

16.5 The Principal Barriers to Measure Selection

A barrier is an obstacle which prevents a given policy measure being implemented or limits the way in which it can be implemented. As a result, some measures may be rejected, making the transport plan less effective. For example, demand management measures in larger cities can control congestion and improve the environment. But cities may be tempted to reject them simply because they will be unpopular. The emphasis should therefore be on how to overcome these barriers, rather than simply how to avoid them. The Measure Option Generator (Sect. 16.6) identifies six types of barrier:

- Legal and regulatory
- Financial
- Governance and institutional
- Political acceptability
- Public acceptability
- Technical

Legal barriers include lack of legal powers to implement a particular measure, legal responsibilities which are split between agencies and regulations which require involvement of the private sector. A survey of European cities in PROSPECTS (May and Matthews 2007) indicates that land use, road building and pricing are the policy areas most commonly subject to legal and institutional constraints.

Finance for implementing transport plans will typically come from five sources: national and regional government, local taxation, transport users, developers and other sources such as bonds, bank loans and private investment. Some of these sources of funding will be assigned to certain types of project; for example, the French *versement transport* charge on local firms can only be used to improve public transport (Cerema 2015). Some will be for infrastructure (capital funding) rather than management measures (revenue funding). Both of these constraints are likely to lead to less cost-effective strategies. As shown in Fig. 16.2 (Goodwin 2010), infrastructure projects typically have much lower benefit/cost ratios than management projects. The guidance on measure selection offers fuller advice on how best to overcome financial barriers.

Governance issues are considered more fully in other chapters. As they and the EU guidance (May 2016) emphasise, lack of direct control, intervention by other tiers of government and involvement of private sector stakeholders may limit a city's ability to adopt the full range of policy measures. The PROSPECTS project (May and Matthews 2007) found that it is typically medium-sized cities which suffer most from such governance barriers; smaller cities often have more freedom, while larger ones often have more power.

Political acceptability barriers arise where politicians fear lack of public acceptance, when different political parties hold opposing views or where pressure groups or the media oppose a measure. Public acceptability may differ from political acceptability if politicians have not kept in touch with changes in the public's views.

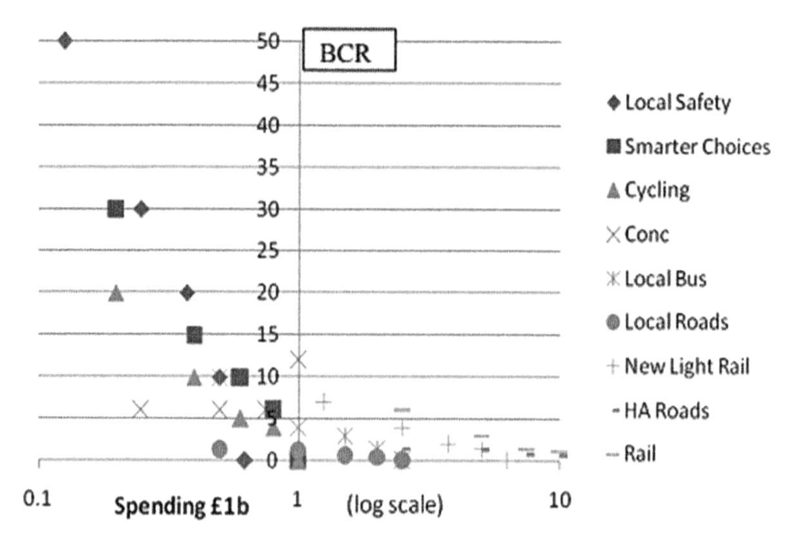

Fig. 16.2 Capital costs and benefit/cost ratios for different types of policy measure (Source: Goodwin (2010))

It may well differ by socio-economic group and can be influenced by cultural attributes, such as attitudes to enforcement. The surveys in PROSPECTS (May and Matthews 2007) found that road building and pricing are the two policy areas which are most commonly subject to constraints on political acceptability. Public transport operations and information provision are generally less affected by acceptability constraints.

Technical barriers are more obvious. For land use and infrastructure, these may well include difficulties in land acquisition. For management and pricing, enforcement and administration are key issues. For infrastructure, management and information systems, engineering design and availability of technology may limit progress. Generally, lack of key skills and expertise can be a significant barrier to progress and will be aggravated by rapid changes in the types of policy being considered and the emergence of new technologies.

Acceptability and governance barriers can be reduced by effective participation and cooperation, as considered in Chap. 26. Effective packaging, as outlined in Sect. 16.3 above, can reduce acceptability and financial and governance barriers. Legal and technical barriers are harder to overcome in the short term.

16.6 The Measure Option Generator

The Measure Option Generator developed for the EC guidance on measure selection (May et al. 2016) has been incorporated into the Knowledgebase on Sustainable Urban Land use and Transport (www.konsult.leeds.ac.uk). KonSULT itself was developed, with support from the UK government and the EC, with the aim of assisting policymakers, professionals and interest groups to understand the challenges of achieving sustainability in urban transport and to identify appropriate policy measures and packages for their specific contexts. It consists of three elements: a Measure Option Generator, a Policy Guidebook, which contains the information on each of the policy measures in the knowledgebase, and a Decision-Makers' Guidebook. We describe the first two in this section.

In the Policy Guidebook, policy measures are grouped into six high-level categories of land use interventions, infrastructure projects, management and service measures, attitudinal and behavioural measures, information provision and pricing interventions. Each measure is described following a standard structure:

- Summary: a one-page summary of the description and findings
- Taxonomy and description, which describes what the measure is, how it works, what it tries to do and how it contributes to different strategies
- First principles assessment, which assesses from first principles how it affects demand, supply and finance; how, through these impacts, it might contribute to policy objectives and the resolution of policy problems; and what the barriers are to its implementation
- Evidence on performance, which summarises a series of case studies which provide empirical evidence on their contribution to policy objectives and problem resolution
- Policy contribution, which combines the findings of the previous two sections to summarise the measure's contribution to policy objectives and to the resolution of policy problems and identifies the areas of a city in which it might most usefully operate
- References

To ensure consistency of treatment, a standard eleven-point scoring method is applied, ranging from +5 (a highly positive contribution) to −5 (a highly negative contribution) throughout the knowledgebase. These scores underpin the operation of the Measure Option Generator. Each of the concepts used, including objectives, problems, strategies and barriers, is more fully described in the Decision-Makers' Guidebook.

The Measure Option Generator allows cities quickly to identify those policy measures which may be of particular value in their context. Users specify their context, including their objectives and strategy, and the Measure Option Generator provides an ordered list of the 64 measures contained in the Policy Guidebook. From the Measure Option Generator screen, the user begins this process by specifying the type of area they are concerned with (corridor, town centre, outer suburb, etc.).

The next screen then prompts the user to decide whether to base their search on objectives or problems or indicators. An objective-led search and a problem-oriented one should lead to the same overall strategy, provided that the problems identified are consistent with the objectives set. The user is thus required to adopt one of these approaches, to avoid double counting. The user can also assign weights ranging from 0 to 5 to the each of the chosen objectives (or problems) to indicate their relative importance in the user's local context. This addresses the concern that objectives may be in conflict and that it may help to specify a hierarchy of objectives (or problems). This stage is one to which stakeholders might usefully contribute, and the Measure Option Generator is designed to be used interactively.

The third screen prompts the user to select the strategies they envisage adopting. As outlined in Sect. 16.4, the strategies included in the Measure Option Generator describe broad directions of policy, such as reducing the need to travel or improving walking and cycling. Users can reflect a mixed approach by assigning weights from 0 to 5 to indicate the relative importance of each selected strategy.

Based on these input values which specify the context of interest to the user, KonSULT's Measure Option Generator produces a list of the 64 available policy measures ranked according to their potential relevance and ability to contribute to the specified context. The example shown (Fig. 16.3) emphasises environmental, social and economic objectives, but does not specify any particular strategy. In this case, the first six ranked measures include pedestrian areas and routes (under the "infrastructure" category), land use to support public transport (under the "land use measures" category), cycle networks (under the "infrastructure" category), accident remedial measures (under the "management and service measures" category), road user charging (under the "pricing" category) and intelligent transport systems (under the "management and service measures" category).

A different specification of context will generate a different ranking, and this can be used to check on the robustness of any given policy measure. The output in Fig. 16.3 also provides a broad indication of the cost for each measure and the timescale for implementation. Users can thus limit their search to low-cost or rapidly implemented measures.

The output as in Fig. 16.3 is not intended to be prescriptive but to prompt the user to investigate measures which might not previously have been considered. Once again, this feature can be used interactively with stakeholders, who may be prompted to debate the relative merits of the more highly ranked measures. At any stage, the user can click on any of the measures listed and transfer immediately to the fuller information on that measure in the KonSULT Policy Guidebook.

As a next step, the output in Fig. 16.3 can be used to develop packages of measures. The KonSULT Package Option Generator allows the user to consider packaging in one of two ways. The first involves taking a preferred policy measure (such as bus rapid transit) and identifying other measures which might support it. These are referred to as complementary measures. The second involves true packaging, where several measures are chosen which work well together. Computationally, assessing packages of several measures from a long list can rapidly become complex, so the packaging option is limited to packages of up to five measures chosen from a list of up to ten measures.

As explained in Sect. 16.3, measures can work together in one of two ways, by achieving synergy or by helping to overcome barriers. Users can choose either of these approaches in searching for complementary measures or packages. The calculation of synergy is based on detailed research using predictive models to assess the interaction of different pairs, and sets, of policy measures (May et al. 2016).

Measure Option Generator

The list below shows all the policy measures within KonSULT in rank order based on their ability to contribute to the context which you have specified.

The absolute scores are arbitrary, but by comparing them you can judge the relative contribution of different measures.

To find out more about any of the measures listed, simply click on it.

By clicking on the Package Option Generator button you can investigate how these policy measures can combine with one another. The process is explained in subsequent screens.

Previous Screen
Package Option Generator...
Save results
Start again

rank	measure	category	cost	timescale	score
1	Pedestrian areas & routes	Infrastructure	medium	medium	83
2	Land use to support public transport	Land Use Measures	neutral	long	60
3	Cycle networks	Infrastructure	medium	medium	52
4	Accident remedial measures	Management and service measures	medium	short	51
5	Road user charging	Pricing	neutral	medium	45
6	Intelligent transport systems	Management and service measures	medium	medium	45
7	Regulatory restrictions	Management and service measures	low	short	44
8	Parking standards	Land Use Measures	low	long	43
9	Development density and mix	Land Use Measures	high	long	43
10	Trams and light rail	Infrastructure	high	long	40

Fig. 16.3 Ranking of policy measures in KonSULT (Source: www.konsult.leeds.ac.uk)

The assessment of the overcoming of barriers is based on the scores for each measure against the barriers of governance, political acceptability, public acceptability and finance in the Policy Guidebook.

As an example, starting with the output in Fig. 16.3, the user could choose the "packages" tool, select the synergy method, set the size of the desired package to five measures (the maximum) and select the first ten ranked measures from Fig. 16.3 for possible packaging. The package option generator produces 252 ranked packages of size 5 in this case, the first five of which are shown in Fig. 16.4.

The first package, with cycle networks, intelligent transport systems, road user charging, pedestrian areas and land use to support public transport, is a logical combination, given the selected objectives. Road user charging reduces car use and hence supports the environment; measures to support walking and cycling reinforce this and provide alternatives available to all; intelligent transport systems improve the efficiency of the transport system; and, as outlined in the PROPOLIS study (Lautso et al. 2004), land use measures help avoid road user charging leading to relocation of activities.

Presentation Options

Number of packages policy measures: 100
Minimum score: -100
Apply Changes
Save results

Rank	Measure1	Measure2	Measure3	Measure4	Measure5	Total
1	Cycle networks	Intelligent transport systems	Road user charging	Pedestrian areas & routes	Land use to support public transport	61
2	Cycle networks	Road user charging	Pedestrian areas & routes	Land use to support public transport	Bike sharing	59
3	Intelligent transport systems	Road user charging	Pedestrian areas & routes	Land use to support public transport	Bike sharing	59
4	Cycle networks	Intelligent transport systems	Road user charging	Pedestrian areas & routes	Bike sharing	59
5	Cycle networks	Intelligent transport systems	Pedestrian areas & routes	Land use to support public transport	Bike sharing	59

Fig. 16.4 Ranking of packages in KonSULT (Source: www.konsult.leeds.ac.uk)

16.7 Addressing the Public Health Objective

As initially designed, KonSULT did not directly address public health as an objective, since it had not then (in 2002) emerged as a public policy objective for transport in its own right. However, KonSULT does consider specific objectives of liveable streets, environmental enhancement and safety, all of which have implications for health. More recently, an attempt has been made to assess the implications for public health of each of the 64 policy measures currently included in KonSULT, with the intention in due course of including public health as an additional objective (Khreis et al. 2017). In this section, we summarise the main findings of that analysis.

An initial literature review was used to identify the range of potential health impacts of transport policy measures. These included motor vehicle crashes, air pollution exposure, noise exposure, increased urban temperature, loss of green space and biodiversity, physical inactivity, climate change and social exclusion.

The assessment involved considering the potential health impact in each of these categories for each of the 64 measures. This assessment was based on the existing text in the "summary", "first principles assessment" and "evidence on performance" sections of the Policy Guidebook, using professional judgement and the principles identified from an initial literature review. In addition, where the impacts were unclear or contested (e.g. in the case of low emission zones and electric vehicles), further literature search was carried out to establish the current evidence, and the following studies on intervention effects were consulted (Holman et al. 2015; Morfeld et al. 2014; Ji et al. 2012; Timmers and Achten 2016).

The resulting assessment describes the direction of the expected health impacts (positive or negative), but does not attempt to quantify the scale of each impact. Such assessment is difficult to make given the current limited evidence base. Figure 16.5 is an example of the mental models that governed the impacts,

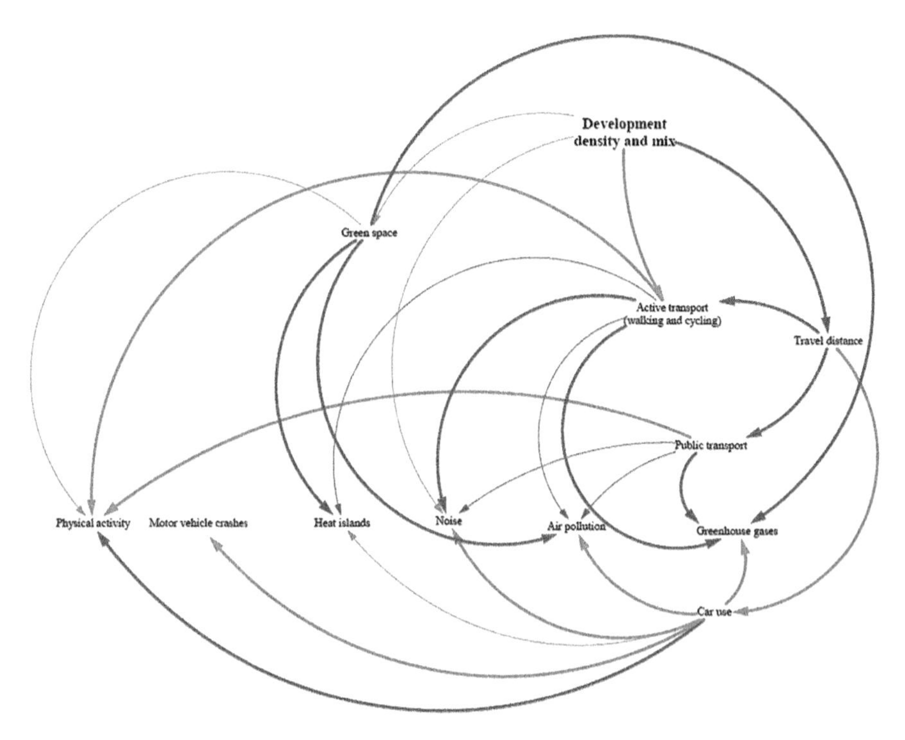

Fig. 16.5 Mental model for the interactions between development density and mix and specific health impacts, showing the pathways leading to premature mortality and morbidity. Source: Khreis et al. 2017

assignment, as applied to the first policy measure in KonSULT: "development density and mix".

Note: The green arrows indicate a positive impact (increase associated with increase). The red arrows indicate a negative impact (increase associated with decrease). The thicker arrows indicate effects for which the evidence is stronger than for the thinner arrows.

The first category of interventions in KonSULT, *land use*, includes four individual policy measures. Land use policy measures such as development density and mix and land use to support public transport can have health impacts both by affecting the level of travel and by affecting the overall travel patterns. Higher densities of activities can improve accessibility, reduce the need for motorised travel and encourage shorter journeys and increased levels of walking, cycling and physical activity. This can result in reductions in air pollution, noise and climate change effects and possibly reduce local heat islands and motor vehicle crashes as a result of reductions in road traffic levels. Dense and mixed developments can help make public transport provision more viable. Encouraging public transport use through land use planning can have positive health impacts by increasing the accessibility of urban areas and the convenience of public transport use and hence encourage a mode shift away from private car use to more active travel and physical activity. Further positive health impacts are possible if there is an increase in green space provision and a decrease in inequalities by supporting the mobility needs of vulnerable groups by transport means other than the private car. The health impacts of parking standards policies vary, depending on the direction of these policies. If the amount of parking required, or permitted, for new developments is reduced, then developers might rethink where to position their developments to provide access for their target customers by transport means other than the private car. This can have positive health impacts through the same pathways above if the development is positioned in dense and mixed urban space and/or near public transport hubs. On the contrary, the generous provision of parking for new developments can reinforce the use of the private car for travel from and to the development and can increase local air and noise pollution as a result of the induced travel demand associated with the new development. Further negative impacts are possible if the resulting development leads to a decrease in green space or biodiversity. The impact of developers' contributions depends on the infrastructure they support.

The second category of interventions, *infrastructure*, includes nine individual policy measures. Many infrastructure policy measures including trams and light rail, new rail stations and lines, bus rapid transit, park and ride, terminals and interchanges, cycle networks and pedestrian areas and routes can have positive health impacts by increasing active travel and physical activity and reducing traffic levels and consequently traffic-related air pollution, noise, heat island effect and climate change effects. If there is an increase in green space, more health benefits are expected. Furthermore, as green space may improve pedestrians' and cyclists' experience, such green space provision may also reinforce a shift from the private car to using these active travel modes. Further positive impacts are expected if there is a reduction in inequalities, for example, by supporting the travel of vulnerable groups by transport means other than the private car. As some of these infrastructure policy measures also tend to increase the geographical accessibility of urban space, integrating these interventions within a wider land use framework is desirable and can help to better realise potential positive health impacts. On the other hand, measures like new rail stations and lines may encourage urban sprawl, new low-density development, longer distance travel and higher associated emissions. Further negative impacts are possible if there is a decrease in exposure to green space and an increase in inequalities by unaffordable fares, land acquisition or displacement of vulnerable groups. From the infrastructure category, new road construction and off-street parking can have negative health impacts by increasing car use and therefore reducing active travel and physical activity, increasing air pollution, heat, noise and climate change effects and possibly motor vehicle crashes. Further negative health impacts will occur if the land uptake for the new infrastructure leads to a decrease in green space or biodiversity loss or an increase in severance.

The third category of interventions, *management and service*, includes 23 individual policy measures. Many of the management and service policy measures can have a positive health impact through the reduction of motor vehicle crashes (e.g. road maintenance, conventional traffic management, intelligent transport systems, accident remedial measures, traffic calming measures, physical restrictions, regulatory restrictions, bus services and priorities, cycling promotion measures, pedestrian crossing facilities, lorry routes and bans, road freight fleet management systems and new rail services). Several will also reduce air pollution, noise and heat island effects (e.g. road maintenance using materials to reduce air pollution and noise, traffic management and urban traffic control, intelligent transport systems, high-occupancy vehicle lanes, physical restrictions, regulatory restrictions, low emission zones, parking controls, bus service and priorities, cycling promotion measures, lorry routes and bans and new rail services). Some of these measures such as physical road restrictions and parking controls may free up urban space that could be utilised for green or public space. With the exception of improvements in cycle and pedestrian facilities, measures in this category often do little to increase levels of active travel and physical activity and may have negative impacts through increase in inequalities.

The fourth category of interventions, *attitudinal and behavioral*, includes ten individual policy measures. The health impacts of attitudinal and behavioral policy measures such as promotional activities, personalised journey planning and company or school travel plans are harder to predict and depend on the direction and content of the measures, but in general are likely to result in positive health impacts through increased levels of active travel and physical activity, and reduction in air pollution, noise and climate change effects and possibly motor vehicle crashes. Similarly, ride and bike sharing, car clubs, flexible working hours and telecommunications are likely to result in positive health impacts and higher flexibility in mobility patterns. Promoting low-carbon vehicles is a controversial measure, and lessons learnt from the European diesel car boom indicate that this measure can negatively impact air quality and health through the increased exposure to nitrogen oxides and particulate matter. On the other hand, electric cars should have a positive contribution to air quality and health through reductions in tailpipe emissions, provided that a target for clean electricity generation is also implemented. However, pollution from tyres, brakes, road surface wear and corrosion will remain.

The fifth category of interventions, *information provision*, includes nine individual policy measures. For some of the information provision policy measures, the health impacts are unclear (e.g. crowd sourcing), while for others there may be positive health impacts through a reduction in motor vehicle crashes (e.g. conventional signs and marking, variable message signs, barrier-free mobility) and air pollution and climate change effects via reducing stop-start driving and idling and encouraging and facilitating the use of public transport (e.g. conventional time tables and service information, trip planning systems).

The final category of interventions, *pricing*, includes nine individual policy measures. Pricing policy measures are often likely to have positive health impacts through general reductions in car use and traffic levels, taxing the most polluting fuels, regulating the age of the vehicle stock, reducing the convenience of motoring and parking, decreasing public transport fares to increase patronage and providing integrated ticketing that allows passengers to transfer within or between different public transport modes with ease and convenience. These measures can possibly slightly increase active travel and physical activity and reduce levels of air pollution, noise, heat island effect, climate change and motor vehicle crashes.

The brief assessment above suggests that there are measures in each category which are likely to improve public health, with the most common pathways arising from reduced car use and its positive effects on air and noise pollution, crashes and heat island effects. Many measures also serve to encourage active travel, either directly or by increasing the opportunities for adopting it. Impacts on climate change and on green space and biodiversity are more limited. Climate change reduction is assisted mainly by land use measures which reduce the need to travel and pricing measures targeted on car use. Social exclusion can be reduced by land use measures and those which directly promote public transport, walking and cycling. However, many infrastructure projects will have a negative impact by encouraging the separation of wealthy and poorer neighbourhoods.

Overall, land use measures appear to offer the most significant benefits, by reducing the need to travel, enhancing green space and facilitating shorter distance travel by active modes. The only measures in doubt in this category are parking standards and developer contributions, where the impacts will be depend critically on how these standards and contributions are used. The second most effective category appears to be pricing, particularly in the case of low and integrated fares which facilitate greater public transport use and help reduce social exclusion, and congestion and parking charges, which can help reduce car use. The categories of management and services, awareness and information all contain measures which can be effective provided that they are appropriately designed. The category of infrastructure appears to be the least likely to be effective and the most likely to aggravate problems of climate change, loss of green space and social exclusion.

16.8 Conclusions and Recommendations

Option generation is a key input to any transport plan and is often one of the weakest links. The guidance on Measure Selection (May 2016) contains a series of recommendations which are based on the approaches outlined in Sects. 16.2–16.6 and are worth repeating here:

1. Avoid thinking about solutions before you have agreed on your vision and objectives. These will help you to understand what problems you face. Measures can then be thought of as ways of overcoming those problems (Sect. 16.2).
2. In looking at possible measures, cast your net as widely as possible. Look at the different types of measure and the information on them. Try to understand how each works and can thus contribute to your objectives (Sect. 16.2).
3. Think about the principles of packaging the measures that you are interested in; packaging can help in achieving enhanced performance, but it can also help to overcome barriers to implementation (Sect. 16.3).

4. Decide whether there are particular strategies that you want to pursue (like reducing the need to travel) (Sect. 16.4).
5. Be clear as to the constraints that you face. Who is responsible for each of the types of measure that you are considering? What level of funding is available? How acceptable are different measures likely to be? But don't take these constraints as reasons for not pursuing a given measure; you can use packaging and careful design to overcome them (Sects. 16.4 and 16.5).
6. Involve your stakeholders and public in selecting the measures and packages which you might adopt. But also consider using the Measure Option Generator, which may offer a tool for stakeholder and public involvement (Sect. 16.6 and Chap. 26).
7. Ensure that each shortlisted measure is designed in sufficient detail to ensure that it can be implemented and that stakeholders and the public know what to expect (See Sect. 3.5 of May 2016).
8. Assess the likely impacts (on objectives and problems) of each of these detailed designs. This will require an ability to predict what might happen and can be assisted by predictive models and appraisal methodologies (See Sect. 3.5 of May 2016).
9. Use these predictions to appraise each detailed measure and package against your objectives. This will help you to prioritise the measures which you adopt and may suggest ways in which individual designs can be enhanced (See Sect. 3.5 of May 2016).

To these should of course be added the recommendation above to carry out a full ex post evaluation of the impacts of any newly implemented measure and to disseminate that evaluation so that others may benefit from the experience gained.

It is planned to add public health as an objective in the KonSULT knowledgebase in the near future. In the meantime, it appears that land use and pricing measures offer the greatest promise for enhancing public health. The management and service, awareness and behavioural and information categories all include measures which are likely to contribute positively to health enhancement. On balance, infrastructure measures appear the least likely to assist in a public health campaign.

References

Cerema. (2015). *Le versement transport: une contribution essentielle au financement des transports urbains, Collection Essentiel, "Le point sur" – Mobilité et transports*. Cerema: Lyon.

CiViTAS. (2016). Retrieved from www.civitas.eu.

Eddington, R. (2006). *The Eddington Transport Study. Main report: Transport's role in sustaining the UK's productivity and competitiveness*. London: The Stationery Office.

ELTIS. (2015). Retrieved from www.eltis.org.

Evidence. (2016). Retrieved from www.evidence-project.eu.

Goodwin, P. B. (2010). *Transport and the economy: Evidence to the House of Commons Transport Committee. Memorandum TE4*. London: The Stationery Office.

Holman, C., Harrison, R., & Querol, X. (2015). Review of the efficacy of low emission zones to improve urban air quality in European cities. *Atmospheric Environment, 111*, 161–169.

Ji, S., Cherry, C. R., Bechle, M., Wu, Y., & Marshall, J. D. (2012). Electric vehicles in China: Emissions and health impacts. *Environmental Science & Technology, 46*, 2018–2024.

Khreis, H., May, A. D., & Nieuwenhuijsen, M. J. (2017). Health impacts of urban transport policy measures: A guidance note for practice. *Journal of Transport & Health, 6*, 209–227.

Lautso, K., et al. (2004). *Planning and research of policies for land use and transport for increasing urban sustainability (PROPOLIS). Final report*. Brussels: European Commission.

May, A. D., & Matthews, B. (2007). Decision making processes. In S. Marshall & D. Banister (Eds.), *Land use and transport planning: European perspectives on integrated policies*. Oxford: Elsevier.

May, A. D., Kelly, C., Shepherd, S., & Jopson, A. (2012). An option generation tool for potential urban transport policy packages. *Transport Policy, 20*, 162–173.

May, A. D., Khreis, H., & Mullen, C. A. (2016). *Option generation for policy measures and packages: the role of the KonSULT knowledgebase. Proceedings of the 14th World Conference on Transport Research. Shanghai, July 2016*. Amsterdam: Elsevier.

May, A. D., (2016). Measure selection: selecting the most effective packages of measures for sustainable urban mobility plans. Retrieved from www.eltis.eu and www.sump-challenges.eu/kits.

Morfeld, P., Groneberg, D. A., & Spallek, M. F. (2014). Effectiveness of low emission zones: Large scale analysis of changes in environmental NO 2, NO and NO x concentrations in 17 German cities. *PLoS ONE, 9*, e102999.

Paulley, N., et al. (2006). *The demand for public transport*. Crowthorne: TRL.

Timmers, V. R., & Achten, P. A. (2016). Non-exhaust PM emissions from electric vehicles. *Atmospheric Environment, 134*, 10–17.

Transport for London. (2007). *Central London Congestion Charging: impacts monitoring; fifth annual report*. London: TfL.

Chapter 17
(Un)healthy Bodies and the Transport Planning Profession: The (Im)mobile Social Construction of Reality and Its Consequences

António Ferreira

17.1 Introduction

> The ways along which we walk are those along which we live
> (Ingold and Vergunst 2008, p. 1)

Transport planning is an activity with key consequences for our individual and collective futures. As the quotation from Ingold and Vergunst's suggests, our lives are greatly influenced by the ways along which we *walk*. However, we move more and more seated inside motorised transport means and not due to our own physical abilities as when we walk (or run or cycle). Motorised transport is becoming increasingly relevant in determining how we live our lives, use and shape our bodies, understand our identities and conceptualise the world we inhabit (Urry 2004). In others words, our understanding of reality is becoming increasingly influenced by transport technologies and lifestyles that not only create but actually presuppose ill and feeble bodies with a single outstanding ability: the capacity to remain seated for endless hours in closed spaces such as cars and planes, offices and living rooms.

The ability to stay immobile on a chair for extensive periods of time should not be seen as trivial. As Tofler (1980) notes, during the Industrial Revolution, it was very difficult to discipline peasants coming from the countryside to work in factories where the obedient capacity to remain at their workstations doing the same task for endless hours was required (see as well McNally 2012). Their bodies were used to freedom, physically mobile daily lives and the great outdoors and were therefore not prepared to accept this level of monotony and restraint. The primary schools

A. Ferreira (✉)
CITTA–Research Centre for Territory, Transports and Environment, University of Porto, Porto, Portugal
e-mail: acf@fe.up.pt; http://citta.fe.up.pt/about-us/staff/acf2018

M. Nieuwenhuijsen, H. Khreis (eds.), *Integrating Human Health into Urban and Transport Planning*, https://doi.org/10.1007/978-3-319-74983-9_17

where the children of factory workers were forced to stay indoors for long periods of time, perfectly immobile at their desks and following orders from their teachers all day, were therefore of critical importance to prepare future generations of workers. First generation peasant workers were very difficult to deal with.

Since the Industrial Revolution, we have achieved a remarkable accomplishment as a global species: never had we before so many humans perfectly disciplined and physically able (and actually wanting) to remain immobile on a chair all day. The world is now populated by what Bowman calls "sitting ninjas" (2014). Our ability to stay seated for so long is an extraordinary adaptation to the demands of contemporary society. Today we geographically move a lot while we physically move very little. We are becoming a deeply physically immobile and geographically mobile species. In other words, *immobile mobility* dominates a good part of our reality.

This book chapter has three key goals. *First*, it aims at objectively defining and highlighting the importance of the process of (im)mobile construction of reality presented above. It will allow showing the extent to which the dominance of immobile mobility is problematic and particularly when this is experienced by transport planners in a way that frames their understanding of personal life and work. *Second*, and following the previous point, it aims at alerting the reader to the importance of the paradox of transport planners—whose societal role is making decisions about (geographical) mobility—having a job that is fundamentally (physically) immobile, as it is characterised by large amounts of desk and computer work and long periods of time held in meeting rooms. The emergence of very powerful digital technologies has further exacerbated this situation, making it possible for transport planners to plan both major transport systems and street-level infrastructures without ever leaving their desks. Worryingly, but naturally, to this sedentary life is associated a "flatland worldview"—a key Wilberian concept that will be described in detail later (Wilber 2000b). *Third*, and finally, this paper aims at proposing how to end this paradox so that transport planners can better contribute to a physically mobile society where health and physical mobility are prioritised. This is an important and timely subject because human health is experiencing a crisis that is to a large extent caused by our physical immobility to which we are not at all adapted to as a biological species (Bowman 2014; Forencich 2006). This crisis is so serious that it has been considered a global pandemic responsible for numerous deaths every year (Lee et al. 2012).

The present chapter is structured as follows. I will start by presenting the different forms of mobility and immobility the chapter considers. Then I will argue that any profession has a set of restrictions to the range of ideas, forms of knowledge and behaviours that are considered acceptable for those practicing it. When professionals cross those boundaries, they are incurring in a form of career-specific deviance likely to have negative consequences for their professional success. However, it is their responsibility as well to critically reflect on the extent to which those boundaries are excessively restrictive and limiting too much their understanding of reality. If the boundaries are problematic, they need to negotiate their size and shape so that what is considered deviant is reassessed. When the boundaries of deviance are out of place, what happens is that professionals will be working in a destructive or, at very best, counterproductive form. After this, I will remember that transport planning is a profession that is all about mobility. A paradox emerges as a result of that: how can professionals that are leaving increasingly less their desks and meeting rooms, and therefore are experiencing an increasingly more physically immobile existence due to the boundaries of their profession, be the same people that are responsible for planning geographical mobility? This is very negative and can only lead to one consequence: there are growing numbers of people who are very geographically mobile and very physically immobile, that is, immobile-mobility is growing in frequency, volume and intensity. This happens because planners (as any human being) shape reality according to their own understanding and experience of it. As more and more people experience this immobile mobility co-created by techno-bureaucratic planners, there will be more and more people who find it natural that transport planners are just techno-bureaucrats whose work is desk-based. This is a self-reinforcing feedback cycle, and the consequence of this cycle is more and more physical immobility, obesity and health problems.

A specific form of rationality is behind this process: I shall call it the "Brain in the Jar" monorational form of thinking. This is the specific rationality of people with feeble and inactive bodies whose work is all about deep techno-bureaucracy. I propose an alternative to this. This alternative is based on a much more comprehensive and open approach to transport planning, one that includes and honours the subjective and embodied experience of reality. This transcends the realms of interdisciplinary thinking, which in my view is becoming increasingly less relevant as it was framed by the logic of the Brain in the Jar paradigm. It requires "polyrational" (a theoretical concept adapted to planning theory by Davy 2008) and "all quadrants" thinking (sometimes addressed to in its simplified form as "I, We, It" thinking by Wilber 1998), where body and mind, science and subjectivity all have a critical role to play. I end this chapter explaining these concepts and their deep meaningfulness. I strongly believe that doing that has the potential to reshape the transport planning profession towards a situation where health and well-being become central and where the practice of the profession will be more productive and emotionally rewarding.

17.2 Multiple Forms of (Im)mobility

The knowledge about how to promote active mobility through land use and transport planning measures is abundant and consistent (see, e.g. Alfonzo 2005, Christiansen et al. 2016, Saelens and Handy 2008, Frank et al. 2007). Basically, there is consensus that promoting high-density mixed land uses with high street connectivity and plenty of green spaces is very effective to increase physical mobility. This body of knowledge has been considered robust enough to be applicable across diverse demographic, geographical and cultural settings. This realisation has led Reis and associates to invite researchers to stop focusing their key efforts on deepening this body of knowledge and instead thinking more about how to scale up interventions aimed at promoting active travelling and disseminating the existing knowledge worldwide (Reis et al. 2016).

The benefits of promoting active mobility are very likely to be huge, as its health benefits have been measured in economic terms as extremely high (Cavill et al. 2008) and its synergies in the fight against climate change have been understood as very substantial (Laverty et al. 2012; Woodcock et al. 2009). As Giles-Corti and associates put it, amid "growing concerns about the impact of rising obesity and physical inactivity levels, climate change, population growth, increasing traffic congestion and declining oil supplies, multiple sectors are now promoting active transportation as an alternative to driving" (2010, p. 122). This raises the important question of why promoting active mobility is not the top priority for the vast majority of transport planners today. I propose a possible (and necessarily partial) answer to this intriguing question: transport planners tend to be physically inactive people due to the techno-bureaucratic nature of their profession. This makes it difficult for them to be effective in promoting active mobility. I will do my best to explain the ramifications of this argument throughout this book chapter.

At the core of the argument is the following idea: individuals' geographical and physical mobilities strongly influence their daily experiences, ideological understandings and identity (Ferreira et al. 2012; Pred 1981b, a). Both forms of mobility have a role to play in the kind of daily practices that we take for granted and that we accept as standard because we have frequent, and easy, access to them. This will contribute to our perceptions of what should be seen as familiar and in the long term to the definition of our ideological perspectives. Essentially, mobility plays a key role in the *social construction of our collective reality* (a concept initially explored by Berger and Luckmann 1966) because of the people, things and experiences that it exposes us to and that it hides from us. This is a very deep and meaningful process particularly because different types of physical and geographical mobilities are associated with different urban forms, economic arrangements and political ideologies (Walks 2008). At last but not the least, the combination of all these factors requires and eventually determines very different uses and shapes for the human body. For example, people living in low-density areas with low street connectivity and that use cars often have a greater probability of being obese and have poorer health (for further insights see Badland and Schofield 2005, Christiansen et al. 2016, Falconer et al. 2015, Frank et al. 2007, Saelens and Handy 2008, Sallis et al. 2016).

It is important to distinguish *physical mobility* from *active mobility* because the later refers to geographical mobility that is achieved by means of physical activity (as in walking or cycling). The former refers to a more general category that includes all forms of physical activity, which might lead or not to geographical mobility. For example, a manual worker might be very physically mobile (because he lifts objects and performs multiple whole-body movements throughout his working day) and not actively mobile (as he might stay inside the same workshop all day and drive to work). A fitness enthusiast might train intensively in a gym to which he goes by car. This distinction is important because it is yet to be fully clarified whether and under which circumstances active mobility promotes or reduces other forms of physical mobility, and vice versa. Indeed, *compensation* of one for the other might (or might not) take place, as already discussed in some studies (e.g. Falconer et al. 2015; Panter et al. 2011).

The process through which patterns of physical and geographical immobility and mobility shape our bodies, ideologies, social practices, values and cognition can be named as the *(im)mobile construction of reality*. It is important to realise that it is difficult to be fully aware of the extent to which this process is powerful and affects our beings. It operates in a very subtle but pervasive way. In order to bring more awareness about it, an *ecological approach* to self, transport technology and culture is probably the best. This is an approach where the "user and the machine are [understood as] intimately related, emotionally and socially, and that these relationships are part of processes that create what is considered to be valuable and morally relevant in a given technological practice or culture" (Kronlid 2008, p. 258). This ontological approach, the same author highlights, is very different from the instrumental approach typically adopted in transport studies.

The instrumental approach assumes a linear and positivistic logic where the user of transport technologies is conceptualised as a detached creator, master and user of transport technologies, and also as an objective assessor of their merits and drawbacks. The problem of the instrumental logic is that it fails to take into consideration the extent to which humans are deeply shaped, conditioned and perhaps even dominated by the (transport) technologies they use and create—see the classical paper on the *system of automobility* by Urry (2004) for further insights. Following the same ecological vein, Kesselring (2008) alerts us to the emergence and importance of *motile hybrids*: highly complex and intertwined assemblages of human beings, technologies, architectural forms and economic and social relations. In these hybrid forms, the human element is unlikely to play the creator-master-user role typically assumed in traditional transport studies.

Decision makers working in the transport profession—in common with all human beings—are individuals exposed to this process of (im)mobile construction of reality I have done my best to explain above. However, decision makers working in the transport domain have a special characteristic, which is that they are at the very core of the process of creation, assessment and implementation of future transport developments. This means that there is a particularly strong and direct feedback cycle between the (im)mobile construction of reality that transport decision makers were subject to, the use of the body they accept as normal, the ideological understandings and cognitive predispositions they have and the kind of (im)mobile construction of reality they will help to create in their professional lives.

17.3 The Taken-for-Granted Wisdom and the Importance of Deviance

It is important to critically reflect on what kind of research transport planning academics are willing to undertake in their professional activity and what kind of knowledge transport planners accept as relevant and valid. Simon and Dippo, two critical ethnographers, consider "all modes of knowing and all particular knowledge forms as ideological, hence the issue is not whether one is 'biased'; but rather, whose interests are served by one's work" (1986, p. 196). The same authors state that academics must become aware that the knowledge they produce is constrained by their own histories and hosting institutions (1986, p. 200). On this see as well the introduction of *Friends of Friends* by Boissevain (1974). Something similar could be said about transport planners: most certainly the work that they undertake as professionals is also constrained by their own histories and their hosting institutions. This psychodynamic view of planning has been explored in great depth by authors such as Baum (1983, 1987). As stated by Tewdwr-Jones (2002, p. 70):

> I firmly believe that (planning) individuals can hold preferences, gathered independently from experiences and influences, not only from relations with other contacts but through varying sources, including media, culture, education, and environment, and bring these to bear on their professional planning activities.

In line with these considerations, we can address feminist theory. Feminism has shown that the forms of knowledge which become legitimated and accepted are dependent on who has generated that knowledge and how the knowledge was generated (see, e.g. Monk and Hanson 1982). There is also, potentially, a self-perpetuating power dimension in the social construction of knowledge. The reason for this is that "those who have the power to validate their own models of the world can validate their own power in the process" (Spender 1981, p. 1). Feminism focuses on the role of gender in this process. Here the focus is on the relationship between physical immobility and geographical mobility. The reason why this is a very important issue is that what urban and transport planners are willing to do in their professional practice is necessarily related to their ideologies and the functioning and shape of their bodies. However, the development of planning ideologies is a result of the opportunities planners have to learn about alternative practices, body uses and ideas and their ability to deviate from mainstream tendencies. Mobility presents them to (or hides from them) many learning opportunities. When people are presented to a place, practice or idea that challenge what they accept as a given, it becomes necessary to reconsider to what extent they are right to do what they do and to think what they think. Planners are no exception to this. The insights provided by Alan Pred are particularly relevant to understand this process (1981a, p. 38):

> When an individual's life path, or biography, becomes associated with a given family role or committed to a specialized role with some other institution she must as a consequence intermittently steer her daily path to activity bundles belonging to specific routine or non-routine projects.

In other words, the adoption of a life project is associated to specific patterns of mobility and immobility which allow the individual to be in the places and do the actions that materialise the chosen life project. Pred uses as an example of a life project a family role, but we could alternatively mention a professional career in planning. Conversely, the possibility of adopting a specific life project is constrained by the individual's capacity to have frequent access to the places where the project can become a reality through the corresponding practices.

It should not be forgotten that the adoption of a certain life project means the creation of new mobility constraints to numerous places and people, opportunities and ideologies. As Pred also explains, when people commit to the (1981a, p. 46):

> Discipline of an institutional project, joining activity bundles at specific times and places, they cannot (owing to their indivisibility) directly know, or be aware, of events and phenomena occurring elsewhere at the same time. Nor, because of their individually finite time resources and the time demands of all travel, can they come to directly know of later happenings and created information that are placed outside their 'reach' by virtue of such commitment. Moreover, the frame of meaning provided by project-based language acquisition may place certain knowledge, or understanding, of events and information beyond the mental access of an individual or an entire group.

We can say then that to work as a planner (or as a medical doctor, or as an astronaut—this applies to all professions) means to create career-specific constraints and boundaries to oneself that will impede the person to have access to several places, practices, knowledge, ideologies and uses of their own body. A partial way out of this condition is to realise that it is indeed impossible to know or experience everything, and so planners (as any professional) benefit from dedicating some energy to analyse which forms of "unknowing" (Thrift 1979; Thrift and Pred 1981) they are creating for themselves through the daily practices and embedded/embodied ideologies that they adopt. This relates to the concept of deviance.

The practices that we undertake in our daily lives are influenced by those around us because people react negatively when we deviate. This is the reason why DeLamater (1968) stressed the importance of geographical mobility when an individual wants to undertake deviant behaviour or to learn how to perform it. DeLamater also stressed the importance of "learning and opportunity structures" (1968). These have a social and a geographical dimension. We have to meet people who will provide us with the knowledge, sometimes the means and frequently the moral support, to undertake deviant behaviour. To find these people, if they do not form part of our primary socialisation circles, we need to move to where they can be found. It is also important to acknowledge that different people judge the same behaviour differently. We all know that what is deviant in some contexts might be mainstream in others. It is therefore pertinent to reflect on the drivers and consequences of transport planning being an increasingly more techno-bureaucratic and physically immobile profession essentially concerned with geographical mobility.

17.4 The Brain in the Jar Metaphor

According to Zygmunt Bauman, metaphors can describe particularly well the effects of the mobile construction of reality experienced both by individuals and society at large (Bauman 1995). He dedicates long and very creative passages to describe the nomad, the tourist and the vagabond, which are for him key metaphors to understand what are the consequences of high levels of mobility today. Following his style, I would like to propose that the best metaphor to think about the

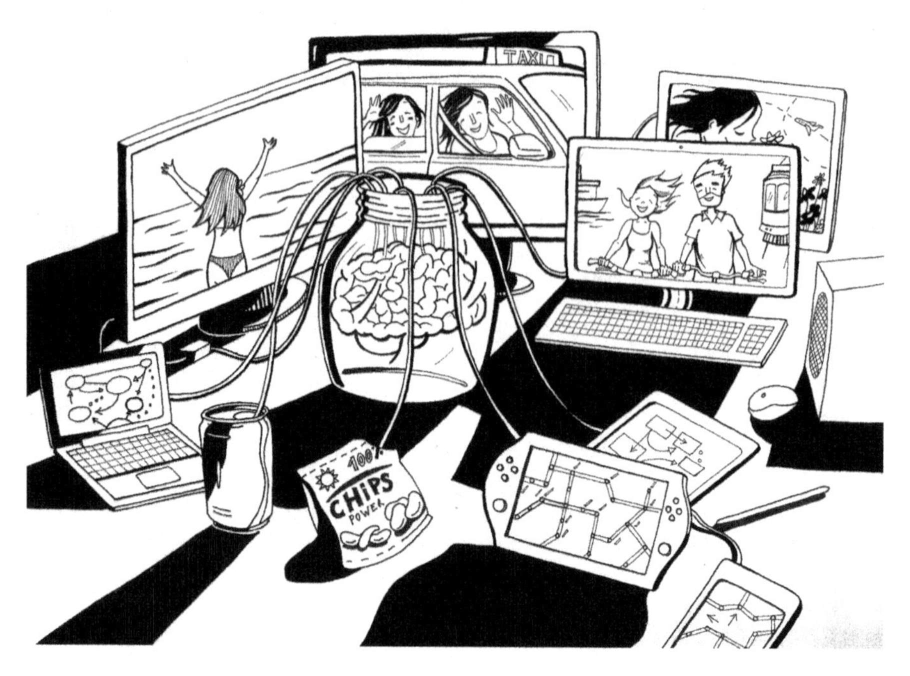

Fig. 17.1 Brains in the Jar live a hard life, particularly when their work is aimed at promoting health and well-being

condition of many planners today is the *Brain in the Jar*. A depiction of this is shown in Fig. 17.1.

I find this a sufficiently entertaining and benevolent metaphor to make us all smile a little while identifying where the problem lies. Note that the use of monsters could also be made, consider, for example, the extent to which zombies are part of popular culture today as they represent much of the fears we encounter in contemporary daily life: being trapped in an existence characterised by scarcity (the zombie is always hungry), lack of purpose (the zombie is typically depicted roaming aimlessly through streets and corridors) and boredom (the zombie is frequently found repeating the same action without any result, e.g. banging with the head on a glass wall). This uncomfortable metaphor for people working without any motivation or engagement (a condition experienced by alarming numbers of people, as explained by Laloux 2014, p. 62) while living in a harsh capitalist world has not escaped very serious academic scrutiny (McNally 2012). However, I prefer to keep things a little bit lighter on this occasion. So I would describe the Brain in the Jar working in transport planning this way (I will call this character Ben Jar):

> Ben Jar wakes up with effort and sits on the bed to dress up. The serious lower back pain that torments him strikes immediately, and dressing up proves to be a mission against evil forces. Coffee in abundant quantities is needed to raise awareness to this sleep-deprived individual to levels where driving is possible. Sleeping is very difficult as the back and neck hurt in most lying positions. The house is abandoned while eating a sugar-intensive donut

while sipping extra-strong capsule coffee from a widely advertised brand. The car trip to work is a living nightmare as back pain strikes every time the clutch is pressed or Ben has to turn his neck to see things around. Leaving the car takes almost one minute as pain is now almost at killing level.

Ben Jar spends all morning in meetings where other Brains in the Jar decide about future mobility schemes seated on ergonomic chairs that alleviate some of their back pain (some have knee and hip pain instead, or as well). A new highway will be built soon and this requires from these Brains intense desk studies and endless PowerPoint presentations. Ben considered visiting the area where the highway is going to pass through (a natural reserve they have just downgraded so they can build the highway), but there are not roads there (just walking paths) and therefore it is impossible for the team to go there without taking pain killers for the back, hips or knees. As the stomach ulcer cannot accept those anymore, Ben Jar uses Google Earth and the local authority digital databases to check the place out from his desktop computer.

Ben checks quickly what kind of media coverage this project is getting at the moment. For that, he likes to use Facebook and a couple of websites from local newspapers. It is clear that the project is terribly unpopular among local people. They consider that building a road through the reserve is an environmental and cultural crime. There are rare birds in the area and also some sacred rocks that are important for these people, who like to walk in the middle of the bush to go and visit the rocks and probably perform ridiculous rituals there. Ben chuckles. "Sacred rocks..." he says to himself. "How can people believe in this kind of thing?" he asks with a frown and eats some crisps while checking again the excellent results the cost-benefit assessment has awarded to the project. He then feels sad and tired again and so he sips from his energy drink. At least tonight there will be some fun, as he is going to watch the new episode of his favourite series about zombies on his new widescreen television.

17.5 The Problem of Monorational Thinking

As we say goodbye to Ben Jar, I would like to mention the work of Benjamin Davy (2008). He explains that, as the urban dweller learns to develop a protection against the stimuli of the city in order to survive—this was Georg Simmel's widely known insight: the "flâneur" and his "blasé" attitude (Simmel 1971)—planners sometimes do the same when they perform their professional practice. The protection of the planner is a simplified mode of thinking which Davy calls "monorational". This is a mode of thinking in which the planner just pays attention to a specific, narrow, rationality, in opposition to the mode of thinking that Davy recommends: the "polyrational" (on complex modes of thinking associated with planning, see as well Ferreira et al. 2009, 2012).

The polyrational mode of thinking is one that simultaneously welcomes different concepts and approaches, rationalities and paradigms (some of which might in fact be incompatible) as part of the process of understanding a given subject in a complex way. Becoming polyrational is today a form of deviance for many planners, as it negates and is negated by the design of the urban environment, the dominant mobility practices and the professional standards of planning, which are becoming increasingly more techno-bureaucratic, as I already mentioned. On this topic, Davy says that people (2008, p. 303):

> …who associate hairdressers with beauty parlours or a bakery with a butcher shop, are thinking monorationally. We owe to monorational thinking that this street continues around the corner because the local highway engineer did not think of anything but constructing more roads. Monorationality explains why subdivisions of terrace houses follow upon subdivisions of terrace houses; the developer only thought of young middle-class families, who wish to live next door to other young middle-class families (and can afford it). Monorationality establishes order. This order arranges conforming uses and developments into a pattern of spatial proximity. Anything that does not fit does not count, and remains neglected.

This author provides then a theory for why the spatial segregation of urban functions is so common in planning. As a result of this, residential areas become very distinct and separated from industrial areas—which in some cases might be a good thing due to the pollutants and other negative externalities associated with some industrial activities; but also separated from green areas and other amenities—and this is not so good for health and mental well-being, as a growing body of research shows (Nieuwenhuijsen 2016; Nieuwenhuijsen et al. 2017; Triguero-Mas et al. 2015).

Plans based on spatial segregation of functions and people are easy to conceptualise as they are generated by the simpler (and unfortunately the dominant) mode of thinking: the monorational. They are also the plans that many people desire so that modes of behaviour perceived by them as deviant do not take place in their neighbourhoods. Mobility and transport networks become consequently very important. Indeed, there is consistent evidence that urban environments with mixed land uses, high population densities and high street connectivity promote walking and cycling (Sallis et al. 2016; Badland and Schofield 2005). There are also good reasons to believe that active modes of transport are the best to promote connections among people *as they travel* (Brömmelstroet et al. 2017)—a good example of a polyrational way of conceptualising travelling, in opposition to the dominant monorational approach where travelling is perceived just as a means (or as a disutility) to reach a given destination. As places and transport systems are purposefully (even though possibly unconsciously) made monofunctional, extensive geographical movement becomes needed for having access to diverse urban land uses and people connections. The greater the distances one needs to overcome, the less physical mobility is adequate to access what is needed and more likely is that people become Brains in Jar travelling by car everywhere. As an example of this, evidence was found that when the land use mix is higher the probability of obesity is lower among local residents. It was also found that when time spent in cars is higher the probability of obesity is also higher (Frank et al. 2004). The more one becomes an unfit Brain in the Jar used to travel by car everywhere, which is not stimulating at all in terms of social interaction or body preparedness for physical mobility, the more acceptable is that monorational landscapes are considered the best choice. It is then that planning processes that are monorational and result in monorational landscapes and monorational travelling can be perceived as successful. This is a snake with the tail in its mouth. We can therefore convincingly argue that monorational thoughts among planners who are used to travel by car tend to reinforce the presence of urban environments and societies where physical immobility is common and well-accepted, and vice versa.

It should be noted however that Davy, or myself, is not the first to present a theory of this kind. Before ourselves Christopher Alexander presented ideas in a somewhat comparable style (Alexander 1965a, b). Another possible reference is the classic work of Jane Jacobs (1970). The relevance that these key theorists attribute to the way planners think is meaningful. But how can we challenge dominant ways of thinking?

17.6 Challenging the Brain in the Jar Paradigm

There is a long-established tradition in transport planning to argue that interdisciplinary thinking is most needed for this profession to develop itself and expand its horizons; I have been myself an advocator of this idea (Ferreira et al. 2009, 2012, 2013). I have in fact developed at a certain stage the quite firm belief that interdisciplinary thinking would be the single best approach to combat the tendency in transport planning to think monorationally, to combat the dominance of the private car (Ferreira and Batey 2010), to stop ignoring health issues in transport studies (Khreis et al. 2016) and to stop putting so much faith on technological solutions for highly complex social and environmental problems as those we face today (for an overview of this topic, see Morozov 2014).

To my own surprise, I have become in the last couple of years increasingly more sceptical about the power of interdisciplinary thinking to change anything for the better. This scepticism has appeared in my mind because most contemporary academic disciplines and universities have become dominated by what Ken Wilber describes as the *flatland worldview* (Wilber 2000a, b). This means that, as long as the academic disciplines contributing to transport planning share this worldview, the ability to think beyond monorationalism stays limited, no matter how much interdisciplinary thinking is done. In other words, both profoundly disciplinary and completely interdisciplinary thought can be equally monorational and equally *flat*.

Using the Brain in the Jar metaphor is useful to explain what I mean above. If we have a gathering of three Brains in the Jar, it makes little difference that Brain A is a human geographer, Brain B is an artist and Brain C is a transport planner. They are all in a jar, and therefore, in that way, their cognition is fundamentally monorational and conditioned by their disembodied state. It is then natural that they fundamentally rely on computer-based analysis, techno-bureaucratic procedures and abstract thinking. It is also natural that they will have little capacity to promote physical mobility and health, as their condition is fundamentally disconnected from and in fact terribly ignorant of those subjects. Even if one of them is a health expert, this health expertise will be also framed by the Brain in the Jar paradigm. It is not experienced from "the inside". However, they will believe that their knowledge is comprehensive and their ideas creative because they form an interdisciplinary team.

In order to explore how this situation can be challenged, I would like to focus on the work of Ken Wilber, a key proponent of the flatland concept. For a description and a critical analysis of the concept, see Slaughter (1998). Perhaps the key criticism

Ken Wilber has been exposed to might be the tendency he has to base arguments on his own meditative and personal development practices, which are not fully accounted on in his own writings—see again Slaughter (1998). His willingness to mix spirituality with social science and history and philosophy with deep metaphysics is also seen with great scepticism by some readers. Being these criticisms important to acknowledge, it is relevant to mention that they are in themselves quite well-aligned with the flatland worldview core values and methodological preferences. In flatland, writing like Ken Wilber does is a form of deviance as his logic is not monorational.

The flatland worldview can be described as a way of interpreting reality where only concepts and processes that can be clearly and precisely mapped out are considered valid; the rest is perceived as subjective and biased and therefore must be supressed from serious professional or academic environments. This means that it is considered perfectly fine to base decisions exclusively on performance indicators, on supposedly objective conclusions from scientific research and on econometric tools such as cost-benefit analyses. For a critique of these analyses, see, for example, Naess (2006). It is therefore considered as perfectly appropriate that transport planners determine which future developments should take place fundamentally using the outputs of transport and econometric models and without leaving their offices and meeting rooms. The flatland worldview is that of the Brain in the Jar.

A fully embodied and physically active person who works as a planner, experiencing the city using his or her own physical means and senses as a way to gather information to support professional decision-making, is increasingly less an image to be taken into consideration with any seriousness in the Brain in the Jar flatland paradigm. All the focus should be on guaranteeing techno-bureaucratic accountability and engaging in procedural activities. As a result, diseased human beings talking about databases in front of computer screens transmit a powerful image of professionalism in Brain in the Jar flatland. It is upsetting to see that the lack of attention that office workers in general and many transport planners in particular pay to their sore bodies and minds entrapped all day in air-conditioned cars, offices and meeting rooms is so well-mirrored by the lack of attention paid to all the non-measured negative impacts created by their work.

According to the flatland worldview, the inner experience of things is always considered of secondary importance. There is no recognised depth and *insideness* to experiences. What matters is their surface and their exterior value, as measured by scientifically validated performance indicators or by peer validation. This lack of inner depth has led many of us to a point where there is a troublesome feeling of lack of purpose in life beyond achieving some sort of socially recognised goal, again and again. According to Ken Wilber, this is the reason why we have become focused on expanding the *surface realm*, in pursuit of external goals that succeed each other to infinity upon completion. This is the reason why flatland is focused on economic growth, achieving more (of the same, e.g. never ending economic growth or more career advancements) and therefore fierce competition. This is also the reason why flatland is concerned with fitness (the external and easily measurable appearance of health, as perceived according to a Brain in the Jar logic) and not health as something that is lived internally and in the intimacy of the individual human experience.

17.7 The Four Quadrant Model Applied to Transport Planning

In order to better explain Wilber's flatland concept and its applicability to transport planning, it is constructive to introduce the four quadrant model he proposed (see Fig. 17.2). For further insights on the model see, for example, Wilber (2003). This model proposes that there are four types of analysis that should be applied to any given topic and that all these forms of analysis must be made in order to fully grasp the topic. These include the exteriorcollective quadrant, which is characterised by attention being paid to large-scale phenomena and processes using objective/scientific means of analysis. This quadrant, when applied to transport planning, is typically concerned with aggregate indicators of performance. Radical Brain in the Jar flatland-type transport planning accepts only this quadrant as valid. This is the transport planning of the hardcore techno-bureaucrat focused on levels of service observed in the transport networks. Note that the problem is not that this is considered (in fact, this *must* be considered). The problem is when this type of analysis is considered *enough*. Some insights about the shortcomings of doing so were provided by Timms et al. (2014).

There is also the exterior-individual quadrant, which focuses on the objective analysis of particular and small-scale phenomena. It is interested in analysing

Fig. 17.2 Ken Wilber's four quadrant model

	Interior	Exterior
Individual	• Personal beliefs and mindsets • The experience of being healthy • Personal views of the world and self	• Individual behaviour • Stated or manifested preferences • Fitness status
Collective	• Culture • Collective values and worldviews • Popular understandings • Tacit agreements and preferences	• Material objects, infrastructures and technologies • Institutional structures • Social practices • Law and protocol

people individually, as famously recommended by Hagerstrand (1970). All the research in transport studies that is concerned with stated preferences of travellers belongs to this quadrant. Unfortunately, some transport researchers from this quadrant became too enthusiastic and started arguing that the researchers from the other quadrants were wrong or that their knowledge is irrelevant. That is the flatland worldview in operation: when ideas from one quadrant are presented as better, and not as complementary, to those coming from other quadrants.

Ken Wilber explains that the dominance of the two exterior quadrants in contemporary society is a result of the enlightenment and is active since the Industrial Revolution. Nevertheless, a number of authors have proposed that planning would benefit from considering the interior quadrants too. These are all about subjectivity and perception. This is obviously a brave choice as this type of research is not valued as equally meaningful in flatland as that coming from the exterior quadrants. A fine example of interior quadrants work is that of Tewdwr-Jones, who claimed that the *personal dynamics* of planners need to be taken into consideration when thinking about planning practice (Tewdwr-Jones 2002). Consider as well the contributions from Baum (1980, 1983, 1987) or Gunder and Hillier (2004). I have myself made some attempts to contribute to enrich these quadrants (Ferreira 2013; Ferreira et al. 2009).

Regarding work from the interior quadrants applied to the experience of mobility itself (and not the experience of the planners as professionals), one can mention as an example the contributions provided by Sheller (2004) who has explored how automobility *feels*. See as well my own contribution where we simulate through a gamified academic paper the inner experiences of a commuter walking back home (Ferreira et al. 2012). This kind of work must not be seen as a substitute for that coming from the exterior quadrants, but as complementary. Seeing it as a substitute would bring a new pathological understanding of reality that is dominated by the interior quadrants. Subjectivity would be gained, but objectivity would then be lost. Ken Wilber's key message is that when knowledge that belongs to one quadrant is reduced to knowledge from other quadrant, the result is indeed that a reduction. The inner experience of the subjective needs to be accepted as valid as the scientific measurement of the objective. Likewise, the realities of individuals and particulars (those covered by the individual quadrants) are as important as the large-scale realities of social groups, countries, organisations and ecosystems when studied in macroscopic terms (those covered by the collective quadrants). One reality cannot and should not be reduced to any other. Gross simplification and bias are the result of those reductions. In my view, this is precisely what we need to start changing to improve transport planning. Transport planning needs indeed to be interdisciplinary, but it needs to be all quadrants too as that is what will help solving the problem of the Brain in the Jar flatland where monorational thinking dominates.

The abovementioned means that we need to be able to inform our professional activities by means of engaging with them physically, emotionally and even spiritually. We need to be able to talk about these topics and express those realities with much greater freedom. We need our bodies to be engaged in the activities we contribute to intellectually, and we need to accept the importance of spirituality, health, meaning of life, happiness, purpose and what we feel when we feel it. Recent research shows that much can be gained from this in terms of productivity of organisations, not to mention in terms of well-being in the workplace (Laloux 2014). We need to be accepted as people with an interior dimension that is intrinsically subjective and emotional; planners are definitely no exception to this (Ferreira 2013). That is not our professional weakness and our bias, but our most precious quality and source of wisdom as human beings. Ignoring this can only lead to frustration and resentment, ultimately to poor health. This is becoming increasingly more accepted in a number of highly effective and sophisticated organisations, as described by Laloux (2014). This author shows the extent to which companies that allow their employees to leave the Brain in the Jar predicament (as I call it) and instead ask them to express themselves fully in the workplace have remarkable success. One of the characteristics of these organisations is the complete absence of techno-bureaucratic structures. For this, however, their workers need to be capable of operating at very sophisticated levels of human and professional development where integrative and polyrational thinking becomes not only possible but natural, as explored by Beck and Cowan (2006). This requires a complete rethink of organisations and education of professionals so that the problem of the Brain in the Jar can be solved by means of human/individual and organisational/collective evolution. This is a truthfully exciting future for transport planning in particular and for many other professions in general. I believe that the positive consequences of this in the social realm and the natural environment will be massive. Are we then ready to stop living as Brains in the Jar and evolve to something more productive and integrative, healthy and rewarding?

17.8 Conclusion: A Utopian Vision for the Future of Transport Planning

When he wakes up, he feels refreshed and happy. Another beautiful day is waiting for him. Without effort, he gets out of bed and dresses himself while he sings quietly not to wake up his partner. After some brief exercises, muscle stretches and a brief meditation (interior individual quadrant work), he takes a healthy breakfast sitting outside of his house with a couple of neighbours. Even though they don't talk much, they like doing this as it feels good having friends and company. They bond through this time they spend together in silence (interior collective quadrant work). He leaves the house on his bike some time afterwards, after wishing them a good day and being nurtured by their mindful presence and smiles. Yesterday he walked; tomorrow he might use his roller blades. Keeping things varied is nice and helps him to understand better the journeys to work that different people choose or have to do (interior individual quadrant work). On his way to the office, he enjoys the ride as many people recognise him and say hi.

With concern, he realises that in one of the crossroads he passes by regularly there are issues with cyclists again. It seems that there are conflicts with car drivers happening there all the time; a cyclist he knows confirms this when he asks her opinion (exterior individual quadrant work). He buys a piece of fruit and sits outside the teashop. He uses his mobile device to communicate that he will be doing field work before heading to the office. No problem doing this; checking things immediately is standard practice among transport planners.

He watches the traffic for a while with his trained eye, and he starts to formulate a hypothesis about what the problem is with the traffic flow (external collective quadrant work). Then he cycles back and forth in the area, and he is now confident that he understands what the problem is, as he can feel it while he cycles (internal individual quadrant work): poor street design. He remembers that this looked good on paper, but cycling through it is indeed a bit counter-intuitive. He calls his colleague; she is already in the office. She appears 20 minutes later on her sports wheel chair as she is disabled. This is urgent business as there is safety issues at stake. They discuss and explore alternatives. Then they go to the office and ask for a meeting with other people to decide what to do. They start preparing the meeting using a mix of technological devices and social interaction approaches (external collective quadrant work).

A couple of days later, they have the meeting at work, where they present the problem to their colleagues. They all check the digital databases and use the integrated transport model. It becomes clear that the spot is indeed prone to accidents and has some issues going on at the social level (external collective quadrant work). This has already been identified, but other priorities have been placed ahead of this one. No problem, they conclude, and they will do it now.

The team of planners goes to the crossroads with their bikes and cycles and walks around, talks with people individually, particularly those who know the area well (External Individual Quadrant work), and uses the opportunity to identify other issues that might exist in the area and have not been identified yet. They use both pen and paper and portable computer devices connected to their office databases and the integrated transport model. They start sketching alternatives using various digital and nondigital techniques and discuss them with the people that start gathering around spontaneously. All know that when planners show up, it is time to stop everything and participate in what they are doing (internal and external collective quadrant work).

After some time, the teashop area is filled with people who have gathered to discuss the crossroads and how to make the area more liveable, safe and enjoyable. The planners take notes, inform people, ask things, make jokes, laugh together with the locals, hear complaints and file them directly in the right databases, run simulations in the transport model that can be activated real time at long distance and show the simulations to people. People see the results, talk about them, bond with the planners and understand better what they do and why they do it (internal individual quadrant and external collective quadrant work). More people who were just passing by also take part in this; it becomes a proper social gathering.

At the end of the day, the community has a reasonably clear idea about what they will do about the crossroads in particular and the area in general. People volunteer to help in the reconstruction of the site; this is standard practice: when there are construction works in public space, the public helps, and at the end of the day, there is a street party offered by the local authority. This is the way the local authority has found to promote a sense of belonging and ownership of public space while reducing costs. Then the planners return home to their families and friends, and they feel good. People say bye to them and to each other as they leave. Next day, the planners will have another fantastic day at work!

Our planner is the last one to leave the area. He sits for a while watching people passing by. He is tranquil and breathes deeply. For him, making the city enjoyable and its people healthy and happy is more than a job. It is his life purpose (internal individual quadrant).

Acknowledgements I would like to show my gratitude to Haneen Khreis and Mark Nieuwenhuijsen for their dedicated help and constructive comments during the writing of this text. The picture of the Brain in the Jar was produced by Catarina França, and it was used with her permission (catarinafrancaillustrations.com).

References[1]

Alexander, C. (1965a). A city is not a tree, Part 1. *Architectural Forum, 122*, 58–62.

Alexander, C. (1965b). A city is not a tree, Part 2. *Architectural Forum, 122*, 58–61.

Alfonzo, M. A. (2005). To walk or not to walk? The hierarchy of walking needs. *Environment and Behavior, 37*, 808–836.

Badland, H., & Schofield, G. (2005). Transport, urban design, and physical activity: an evidence-based update. *Transportation Research Part D: Transport and Environment, 10*, 177–196.

Baum, H. (1980). Sensitizing planner to organization. In P. Clavel, J. Forester, & W. Goldsmith (Eds.), *Urban and regional planning in an age of austerity*. New York: Pergamon.

Baum, H. (1983). *Planners and public expectations*. Cambridge: Schenkman Publishing Company.

Baum, H. (1987). *The invisible bureaucracy: The unconscious in organizational problem solving*. Oxford: Oxford University Press.

Bauman, Z. (1995). *Postmodern ethics*. Oxford: Basil Blackwell.

Beck, D., & Cowan, C. (2006). *Spiral dynamics: Mastering values, leadership and change*. Oxford: Blackwell Publishing.

Berger, P. L., & Luckmann, T. (1966). *The social construction of reality*. New York: Doubleday.

Boissevain, J. (1974). *Friends of friends: Networks, manipulators and coalitions*. Oxford: Basil Blackwell.

Bowman, K. (2014). *Move your DNA*. Washington: Propriometrics Press.

Brömmelstroet, M. T., Nikolaeva, A., Glaser, M., Nicolaisen, M., & Chan, C. (2017). Travelling together alone and alone together: mobility and potential exposure to diversity. *Applied Mobilities, 2*, 1–15.

[1] Mapping a relationship among planning practice, theory and education.

Cavill, N., Kahlmeier, S., Rutter, H., Racioppi, F., & Oja, P. (2008). Economic analyses of transport infrastructure and policies including health effects related to cycling and walking: A systematic review. *Transport Policy, 15*, 291–304.
Christiansen, L. B., Cerin, E., Badland, H., Kerr, J., Davey, R., Troelsen, J., Dyck, D. V., Mitáš, J., Schofield, G., Sugiyama, T., & Salvo, D. (2016). International comparisons of the associations between objective measures of the built environment and transport-related walking and cycling: IPEN adult study. *Journal of Transport and Health, 3*, 467–478.
Davy, B. (2008). Plan it without a condom! *Planning Theory, 7*, 301–317.
Delamater, J. (1968). On the nature of deviance. *Social Forces, 46*, 445–455.
Falconer, C., Leary, S., Page, A., & Cooper, A. (2015). The tracking of active travel and its relationship with body composition in UK adolescents. *Journal of Transport and Health, 2*, 483–489.
Ferreira, A. (2013). Emotions in planning practice: a critical review and a suggestion for future developments based on mindfulness. *Town Planning Review, 84*, 703–719.
Ferreira, A., & Batey, P. (2010). University towns and the multi-layer transport model: Excavating the connections between academic ethos and transport problems. *Planning Theory and Practice, 11*, 573–592.
Ferreira, A., Batey, P., te Brömmelstroet, M., & Bertolini, L. (2012). Beyond the dilemma of mobility: Exploring new ways of matching intellectual and physical mobility. *Environment and Planning A, 44*, 688–704.
Ferreira, A., Marsden, G., & Te Brommelstroet, M. (2013). What curriculum for mobility and transport studies? A critical exploration. *Transport Reviews, 33*, 501–525.
Ferreira, A., Sykes, O., & Batey, P. (2009). Planning theory or planning theories? The hydra model and its implications for planning education. *Journal for Education in the Built Environment, 4*, 29–54.
Forencich, F. (2006). *Exuberant animal: The power of health, play and joyful movement*. Bloomington: Author House.
Frank, L., Andresen, M., & Schmid, T. (2004). Obesity relationships with community design, physical activity, and time spent in cars. *American Journal of Preventive Medicine, 27*, 87–96.
Frank, L. D., Saelens, B., Powell, K., & Chapman, J. E. (2007). Stepping towards causation: Do built environments or neighborhood and travel preferences explain physical activity, driving, and obesity? *Social Science and Medicine, 65*, 1898–1914.
Giles-Corti, B., Foster, S., Shilton, T., & Falconer, R. (2010). The co-benefits for health of investing in active transportation. *NSW Public Health Bulletin, 21*, 122–127.
Gunder, M., & Hillier, J. (2004). Conforming to the expectations of the profession: A Lacanian perspective on planning practice, norms and values. *Planning Theory and Practice, 5*, 217–235.
Hagerstrand, T. (1970). What about people in regional science? *Regional Science Association Papers, 24*, 6–21.
Ingold, T., & Vergunst, J. L. (2008). Introduction. In T. Ingold & J. L. Vergunst (Eds.), *Ways of walking: Ethnography and practice on foot*. Surrey: Ashgate.
Jacobs, J. (1970). *The economy of cities*. New York: Vintage Books.
Kesselring, S. (2008). The mobile risk society. In W. Canzler, V. Kaufmann, & S. Kesselring (Eds.), *Tracing mobilities: Towards a cosmopolitan perspective*. London: Ashgate.
Khreis, H., Warsow, K., Verlinghieri, E., Guzman, A., Pellecuer, L., Ferreira, A., Jones, I., Heinen, E., Rojas-Rueda, D., Mueller, N., Schepers, P., Lucas, K., & Nieuwenhuijsen, M. (2016). The health impacts of traffic-related exposures in urban areas: Understanding real effects, underlying driving forces and co-producing future directions. *Journal of Transport & Health, 3*, 249–267.
Kronlid, D. (2008). Ecological approaches to mobile machines and environmental ethics. In S. Bergmann & T. Sager (Eds.), *The ethics of mobilities: Rethinking place, exclusion, freedom and environment*. Aldershot: Ashgate.
Laloux, F. (2014). *Reinventing organisations: A guide to creating organisations inspired by the next stage of human consciousness*. Nelson Parker: Brussels.

Laverty, A., Webb, E., Mindell, J., & Millett, C. (2012). Is being concerned about the environment good for your health? *The Lancet, 380*, S56.

Lee, M., Shiroma, E., Lobelo, F., Puska, P., Blair, S., & Katzmarzyk, P. T. (2012). Effect of physical inactivity on major non-communicable diseases worldwide: An analysis of burden of disease and life expectancy. *The Lancet, 380*, 219–229.

Mcnally, D. (2012). *Monsters of the market: Zombies, vampires and global capitalism*. Chicago: Haymarket Books.

Monk, J., & Hanson, S. (1982). On not excluding half of the human in human geography. *Professional Geographer, 34*, 11–23.

Morozov, E. (2014). *To save everything, click here: Technology, solutionism, and the urge to fix problems that don't exist*. London: Penguin Books.

Naess, P. (2006). Cost-benefit analyses of transportation investments: Neither critical nor realistic. *Journal of Critical Realism, 5*, 32–60.

Nieuwenhuijsen, M. (2016). Urban and transport planning, environmental exposures and health-new concepts, methods and tools to improve health in cities. *Environmental Health, 15*(suppl 1), 38.

Nieuwenhuijsen, M., Khreis, H., Triguero-Mas, M., Gascon, M., & Dadvand, P. (2017). Fifty shades of green: Pathway to healthy urban living. *Epidemiology, 28*, 63–71.

Panter, J., Jones, A., van Sluijs, E., & Griffin, S. (2011). The influence of distance to school on the associations between active commuting and physical activity. *Pediatric Exercise Science, 23*, 72–86.

Pred, A. (1981a). Power, everyday practice and the discipline of human geography. In A. Pred (Ed.), *Space and time in geography – Essays dedicated to Torsten Hagerstrand*. CWK Gleerup: Lund.

Pred, A. (1981b). Social reproduction and the time-geography of everyday life. *Geografiska Annaler. Series B, Human Geography, 63*, 5–22.

Reis, R., Salvo, D., Ogilvie, D., Lambert, E., Goenka, S., & Brownson, R. C. (2016). Scaling up physical activity interventions worldwide: Stepping up to larger and smarter approaches to get people moving. *The Lancet, 388*, 1337–1348.

Saelens, B. E., & Handy, S. (2008). Built environment correlates of walking: A review. *Medicine and Science in Sports and Exercise, 40*, S550–S566.

Sallis, J., Cerin, E., Conway, T., Adams, M., Frank, L., Pratt, M., & Davey, R. (2016). Physical activity in relation to urban environments in 14 cities worldwide: A cross-sectional study. *The Lancet, 387*, 2207–2217.

Sheller, M. (2004). Automotive emotions: Feeling the car. *Theory, Culture & Society, 21*, 221–242.

Simmel, G. (1971). The metropolis and mental life. In D. N. Lievine (Ed.), *Georg Simmel: On individuality and social forms*. Glencoe: The University of Chicago Press.

Simon, R. I., & Dippo, D. (1986). On critical ethnographic work. *Anthropology & Education Quarterly, 17*, 195–202.

Slaughter, R. (1998). Transcending flatland: Implications of Ken Wilber's meta-narrative for futures studies. *Futures, 30*, 519–533.

Spender, D. (1981). Introduction. In D. Spender (Ed.), *Men's studies modified*. Oxford: Pergamon Press.

Tewdwr-Jones, M. (2002). Personal dynamics, distinctive frames and communicative planning. In P. Allmendinger & M. Tewdwr-Jones (Eds.), *Planning futures: New directions for planning theory*. London: Routledge.

Thrift, N. (1979). *The limits to knowledge in social theory: Towards a theory of practice*. Canberra: Department of Human Geography, Australian National University.

Thrift, N., & Pred, A. (1981). Time-geography: A new beginning. *Progress in Human Geography, 5*, 277–286.

Timms, P., Tight, M., & Watling, D. (2014). Imagineering mobility: Constructing utopias for future urban transport. *Environment and Planning A, 46*, 78–93.

Tofler, A. (1980). *The third wave*. New York: Bantan Books.

Triguero-Mas, M., Dadvand, P., Cirach, M., Martínez, D., Medina, A., Mompart, A., Basagaña, X., Gražulevičienė, R., & Nieuwenhuijsen, M. (2015). Natural outdoor environments and mental and physical health: Relationships and mechanisms. *Environment International, 77*, 35–41.
Urry, J. (2004). The 'System' of automobility. *Theory, Culture & Society, 21*, 25–39.
Walks, R. A. (2008). Urban form, everyday life, and ideology: Support for privatization in three Toronto neighbourhoods. *Environment and Planning A, 40*, 258–282.
Wilber, K. (1998). *The essential Ken Wilber*. Boston: Shambhala Publications.
Wilber, K. (2000a). *Integral psychology: Consciousness, spirit, psychology, therapy*. Boston: Shambhala Publications.
Wilber, K. (2000b). *A theory of everything: An integral vision for business, politics, science and spirituality*. Boston: Shambhala.
Wilber, K. (2003). *Kosmic consciousness*. Sounds True: Boulder.
Woodcock, J., Edwards, P., Tonne, C., Armstrong, B., Ashiru, O., Banister, D., Beevers, S., & Chalabi, Z. (2009). Public health benefits of strategies to reduce greenhouse-gas emissions: urban land transport. *The Lancet, 374*, 1930–1943.

Part IV
Environmental Exposures, Physical Activity and Health

Chapter 18
Built Environment and Physical Activity

Billie Giles-Corti, Lucy Gunn, Paula Hooper, Claire Boulange, Belén Zapata Diomedi, Chris Pettit, and Sarah Foster

18.1 Introduction

There is a growing interest globally on how city planning affects health, particularly physical activity given its important role in preventing major chronic diseases (Giles-Corti et al. 2016). City planning can improve or harm human health through the opportunities created for health-promoting (or health-damaging) lifestyles.

B. Giles-Corti (✉) · L. Gunn · C. Boulange
NHMRC Centre for Research Excellence in Healthy Liveable Communities, Melbourne, VIC, Australia

Centre for Urban Research, RMIT University, Melbourne, VIC, Australia
e-mail: billie.giles-corti@rmit.edu.au; lucy.gunn@rmit.edu.au; claire.boulange@rmit.edu.au

P. Hooper
NHMRC Centre for Research Excellence in Healthy Liveable Communities, Melbourne, VIC, Australia

Centre for the Built Environment and Health, The University of Western Australia, Perth, WA, Australia
e-mail: paula.hooper@uwa.edu.au

B. Z. Diomedi
NHMRC Centre for Research Excellence in Healthy Liveable Communities, Melbourne, VIC, Australia

School of Civil Engineering, University of Queensland, St Lucia, Australia
e-mail: b.zapatadiomedi@uq.edu.au

C. Pettit
Faculty of Built Environment, University of New South Wales, Kensington, NSW, Australia
e-mail: c.pettit@unsw.edu.au

S. Foster
Centre for Urban Research, RMIT University, Melbourne, VIC, Australia
e-mail: sarah.foster@rmit.edu.au

M. Nieuwenhuijsen, H. Khreis (eds.), *Integrating Human Health into Urban and Transport Planning*, https://doi.org/10.1007/978-3-319-74983-9_18

The World Health Organization (WHO 2016a) reaffirmed this view in its 2016 Shanghai Declaration, declaring that "Health is created … in the neighbourhoods and communities where people live, love, work, shop and play" (World Health Organization 2016b). WHO also recognised that a healthy city is a sustainable city, pronouncing that "Health is one of the most effective markers of any city's successful sustainable development …" and pledged that it would "accelerate the implementation" of the UN's Sustainable Development Goals by investing politically and financially in health promotion (United Nations General Assembly 2015), including an emphasis on creating healthy cities (World Health Organization 2016a).

The WHO's commitment reflects growing public, policy and scientific interest over the last decade in the effects of the built environment on the health and wellbeing of urban dwellers and health equity (WHO and UN Habitat 2016). This has been fuelled by multi-sector concerns about the effects of city planning associated with rising levels of chronic disease and obesity, low-density car-dependent suburbs on the fringes of cities as well as rapid urbanisation, population growth, transport-related air pollution, greenhouse gas emissions and climate change (Giles-Corti et al. 2016). In 2011, the UN acknowledged that multi-sector, whole-of-society action was required to curb chronic disease (United Nations 2011), and there is evidence that healthy, active lifestyles support both individual and planetary health (Watts et al. 2015). To this end, the UN's Sustainable Development Goals include Goal 11, to create more sustainable, resilient, inclusive human settings and cities, and Goal 3, to create health and wellbeing for all (United Nations General Assembly 2015).

Scientific, community and policy concerns are based in part on a growing body of evidence on the effects of the built environment on physical activity (Althoff et al. 2017; Reis et al. 2016; Sallis et al. 2016a). A simple Web of Science search using the keywords "environment" AND "walk*" OR "physical activity" in human populations up to the year 2000 identifies only 17 relevant journal articles. However, changing the date range to 2001–2010, the same keywords reveal 570 articles and 1286 article for 2011–2018. This exponential growth reflects the emergence of a new field of active-living research focused on the built environment, with researchers from many disciplines—including public health, transport, planning, engineering and ecology—bringing their unique perspectives to answer related research questions.

With the proliferation of articles exploring the relationship between the built environment and physical activity, there is a genuine need to assess the state of current active-living research and, after almost two decades of studies, to identify what is needed to advance the field. This chapter begins identifying which physical-activity behaviours are affected by different aspects of the built environment and then reflects on factors that might be contributing to inconsistencies in the evidence. It then considers the gaps in the literature, before suggesting opportunities for new collaborative research that might change policy and practice.

18.2 Which Physical-Activity Behaviours Are Affected by Which Built-Environment Features?

Physical activity is a complex behaviour. It covers different purposes (recreational, transport, work-related), types (formal and informal sport, walking and cycling for different purposes, formal and informal exercise such as jogging and gym, domestic chores such as gardening) and intensity levels (moderate, vigorous, light). Activity occurs in different settings (home, work, neighbourhoods) and is likely to be influenced by different environmental attributes. Hence, more than a decade ago, systematic reviewers began recognising that different built-environment features were associated with different types of behaviour, prompting calls for behaviour-specific environmental exposure measures to be studied and context-specific environments (Humpel et al. 2002; Giles-Corti et al. 2005b). As active-living built-environment research has evolved, two different types of physical-activity behaviours have attracted the most attention: walking for transport and walking for recreation—particularly in residential neighbourhoods.

18.2.1 Walking for Transport

In the last 5 years, numerous systematic reviews of (mainly) cross-sectional evidence have confirmed earlier reviews that physical activity—principally walking for transport—is associated with three main built-environment features: higher residential density, mixed land use or access to local destinations required for daily living and connected street networks (either measured individually or combined into a composite "walkability" index). These associations persevere irrespective of how physical activity is measured (self-reported behaviour, accelerometry, steps) and the age of the adult population (adults or older adults) (Cerin et al. 2017; Sugiyama et al. 2012; McCormack and Shiell 2011; Van Holle et al. 2012).

However, there is a complex relationship between these variables. For example, density *alone* is unlikely to encourage physical activity. Higher-density development with few local destinations or little public transport—as is now being built in some cities—is simply high-rise sprawl and continues to foster motor-vehicle dependency and traffic congestion. Rather, the relationship between higher-density development and walking is apparent because density is generally a proxy for other environmental characteristics (including demographics; car ownership; access to local destinations, employment, shops and services; frequent public transport; connected street networks that make destinations more proximate) that directly influence choice of transport mode and hence levels of physical activity (Boarnet and Crane 2001; Transportation Research Board 2005). Numerous studies report that when both density and accessibility are included in the same model, the effects of density attenuate, suggesting that the accessibility of shops, services and public transport is more important than density per se (Transportation Research Board 2005).

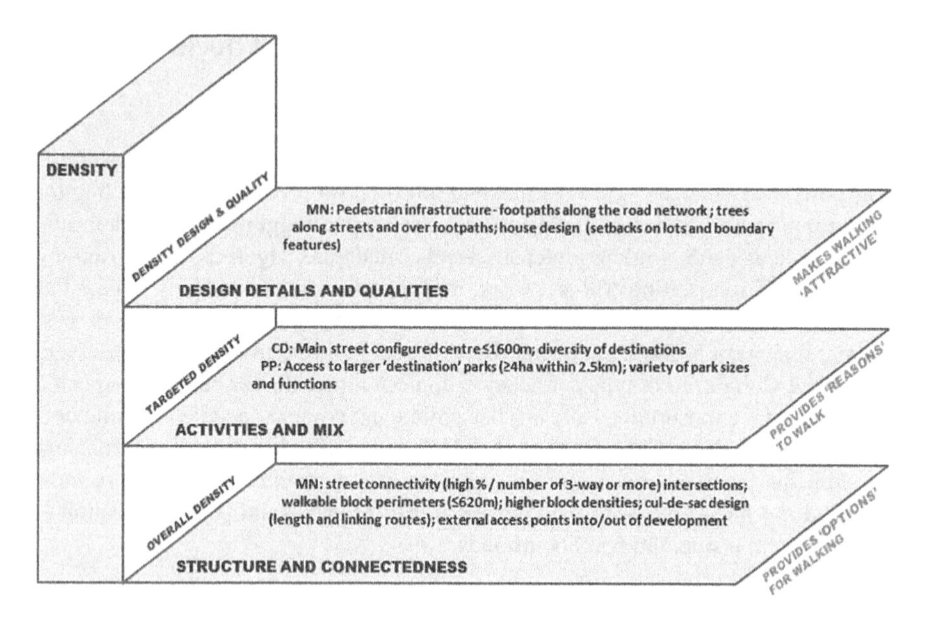

Fig. 18.1 The building blocks of a healthy, liveable neighbourhood (source: Hooper et al. 2015)

Nevertheless, increased residential density is a crucial building block in a healthy and liveable community that encourages active lifestyles. Recognising policy-makers' need for guidance on how to create such communities, Hooper and colleagues explored the interacting effects of different urban design features (Hooper et al. 2015). As shown in Fig. 18.1, the neighbourhood's structure and connectedness (for instance, street connectivity) facilitate walking, the activities and mix of destinations provide reasons to walk and the design details and qualities of the neighbourhoods make walking attractive. Higher density helps make all the other building blocks more efficient and effective: the denser the population, the greater the likelihood an area will have shops, services and accessible and frequent public transport. Nevertheless, more research is required on the levels of density that will maximise health benefits and minimise potential harm (Giles-Corti et al. 2012, 2014).

Although not widely explored in the health literature, the importance of environmental characteristics at both trip origin and destination is studied by transport academics in North America. This suggests that higher dwelling densities at both origin *and* destination increase walking and public transport use while discouraging private motor-vehicle use (Transportation Research Board 2005). The importance of the environment at both ends of the journey was highlighted in a study that found that using public transport was 16 times more likely when residents in a suburban development lived and worked within 400 m of a public transport stop, compared with those without transport access at either end of their journey (Badland et al. 2014a). The importance of environmental characteristics at both origin and destination warrants further research.

18.2.2 Walking for Recreation

Evidence of associations between the built environment and recreational walking (walking for leisure, recreation or exercise) is less consistent than for transport walking (McCormack and Shiell 2011; Sugiyama et al. 2012; Saelens and Handy 2008). In an attempt to assist policy-makers, Sugiyama and colleagues extended previous reviews by considering both access to destinations (such as parks, sports fields and playgrounds) and route attributes (sidewalks, street connectivity, aesthetics, traffic, safety) associated with walking, including recreational walking (Sugiyama et al. 2012). They confirmed that "no dominant environmental attribute" consistently predicted recreational walking. Nevertheless, they found some evidence that recreational walking was associated with access to recreational destinations such as parks and the aesthetic appeal of routes. Moreover, *all* studies reviewed that assessed the *quality* of recreational destinations revealed associations with recreational walking. So, for volitional behaviours such as recreational walking, destination quality may be an important—albeit often ignored or under-explored—element of the built environment that warrants further investigation (Kaczynski et al. 2014; Koohsari et al. 2013b; Taylor et al. 2011; Giles-Corti et al. 2005a; Sugiyama et al. 2010, 2015).

A review by Bancroft et al. (2015) focussed on the proximity and density of parks associated with objectively measured physical activity in the United States. They found "no consistent pattern of results" relating park exposure to physical activity yet observed "stronger park–physical activity associations for analyses with smaller buffer sizes" (p. 280). A strength of this review was that it comprehensively summarised findings from papers reviewed. This enabled our team to delve into the finding to better understand inconsistencies in findings. An analysis of the result summaries revealed a lack of agreement between exposure measures and standards for measuring environmental exposures that may be contributing to measurement error. This may have accounted for inconsistent findings (Koohsari et al. 2015b). Our analysis suggested that both researchers and reviewers may be neglecting to apply the same level of rigour to measuring and critiquing environmental exposures, as they do to outcome measures such as physical activity. Some of these methodological issues are now considered.

18.3 Methodological Problems Contributing to Inconsistencies in the Literature

18.3.1 Buffer Size

A major consideration for built-environment and physical-activity studies is the measurement and reporting of exposure, including whether there is sufficient variation to observe an effect. Indeed, close examination of exposure measures reveals why there may be inconsistencies in the literature (Bancroft et al. 2015). Bancroft's review observed considerable diversity in the exposure measures used. Buffers, for instance, ranged in size from 1.6 km or more down to 400 or 200 m, 50 m from a GPS/accelerometer point or the block where a child lived. Unspecified buffer distances such as "within walking distance" were defined as 10 or 20 min or distance to the nearest park. Aside from some mixed findings in a study applying two buffers (400 m and 1.6 km), no other results were significant. In contrast, in five studies incorporating smaller buffers of 400–800 m (three with children, one with adults and one with seniors), all but the seniors' study reported significant correlation with physical activity.

Bancroft and colleagues suggested that "exposure reporting bias" may explain the findings, "in which authors may have coded exposures in multiple ways and then presented only the findings most consistent with their hypotheses" (p. 27). An alternative explanation might be that, in this emerging field, investigators are (appropriately) using inductive research to explore which (if any) buffers are associated with outcomes of interest. Nevertheless, Bancroft and colleagues may be correct, and, apart from "fishing" for findings, it is also plausible that neither investigators nor reviewers pay sufficient attention to the buffer sizes that suit studying (in this case) recreational walking and, indeed, ignore the way communities are designed by urban designers. In other words, they are failing to consider what constitutes an appropriate environmental exposure for the built-environment feature or the recommended or even optimal "dose" of an environmental feature that could produce an effect.

To explain the significance of these omissions, consider applying the same approach to a hypothetical systematic review of medication use. Manufacturers of hypertension drugs, for instance, generally specify a recommended dose of their drug. If medical practitioners prescribed patients a quarter or twice the dose, what effects would we expect to observe? Moreover, if a systematic review of hypertension treatments failed to differentiate between doctors who complied with recommended doses and those who ignored them, or if the review did not consider dose at all, and simply reported inconsistent positive, negative and non-significant findings, what would we conclude? Perhaps that the review had not adequately assessed the quality of the evidence and that we should not trust its conclusions. We might also question why it got published in the first place.

The same applies to studies of environmental correlates. The importance of considering the "dose" of the intervention and the way communities are planned is highlighted in a study in Perth, Western Australia. It found that 99.1% of respondents had a park within 1.5 km of their home and 22.2% within 400 m (see Table 18.1) (McCormack et al. 2008; Sallis 2008). This is because Perth's planners and urban designers have guidelines ensuring that communities have accessible public open space. We do not know how alike or dissimilar Perth is to other cities in this regard. However, it is plausible that studies of recreational walking using large buffer sizes (1.6 km) are producing non-significant findings because there is insufficient variation in the exposure measure. Conversely, depending on the sample size, very small buffers (less than 200 m, or block size) may result in insufficient statistical

Table 18.1 Descriptive statistics for destination variables (Perth, Western Australia, 1995)

		Respondents with destination within 400 m	Respondents with destination within 1500 m
Destination	*n*	%	%
Beach	1394	0.4	7.5
Park	1394	22.2	99.1
River	1391	0.6	7.8
School	1391	7.7	61.3
Post box	1380	41.7	99.1
Bus stop	1284	79.3	100.0
Transit station	1391	1.5	30.8
Convenience store	1391	22.7	74.4
Newsagent	1391	13.2	63.8
Shopping mall	1394	8.6	82.6

Source: McCormack et al. (2008)

power: that is, insufficient numbers of people have parks in such close proximity. Moreover, if parks are too close, one might expect *lower* levels of physical activity, not only because the parks are likely to be small and may not invite physical activity but because very little recreational walking is done *en route*. Researchers need to think carefully about the exposure measures they are adopting. A good starting point might be to consider buffer sizes found in the literature to be associated with different types of physical activity but also to review local planning policies that are governing the built forms observed in different cities (Giles-Corti et al. 2015). This would be more useful than "fishing" for appropriate exposures and would provide a rationale for selecting different-sized buffers, which are likely to vary for different exposure measures.

18.3.2 Buffer Type

Another factor leading to measurement error is the buffer type used to capture environmental exposure data, particularly radial buffers (as the crow flies), road network buffers or administrative boundaries. Radial buffers are often used due to data availability. Road network buffers are generally considered the gold standard, but if "cleaned" road network data are not available, significant work is required to topologically prepare the data in order to perform a systematic networked buffer analysis across a city. Failure to prepare data adequately increases measurement error.

In areas with cul-sacs or curvilinear street networks, radial buffers may significantly overestimate access to local destinations compared with grid or connected street networks. Figure 18.2 shows 800 m radial buffers in neighbourhoods with grid-pattern (left) and curvilinear (right) street networks, with the corresponding

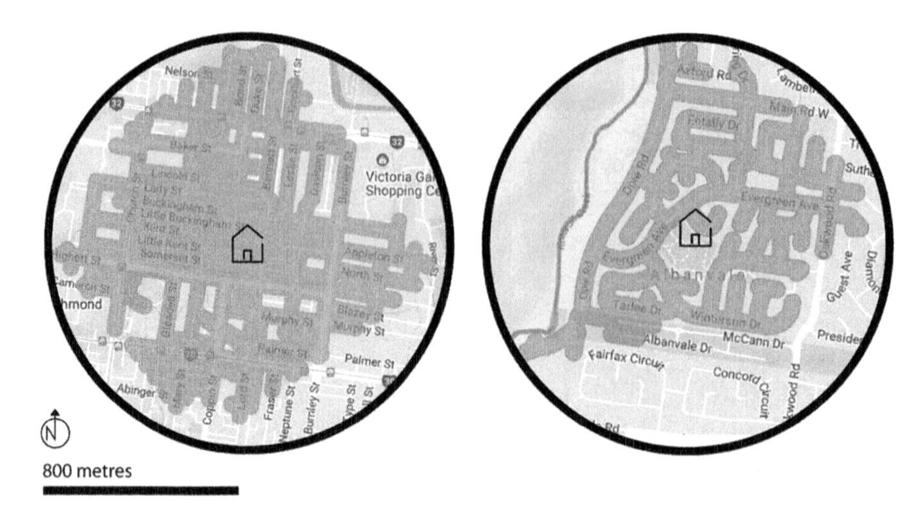

Fig. 18.2 Radial buffer (circle) and road network buffer (shaded) on grid-patterned street network (left) and curvilinear-patterned street network (right)

road network buffers (shaded) based on travelling along the road network. While *radial* buffers are of equal size irrespective of the street pattern, the area of the *road network* buffer is considerably smaller in neighbourhoods with curvilinear street networks (0.6 km^2) than grid pattern (1.4 km^2). Radial buffers therefore *overestimate* the environments to which residents living in areas with curvilinear street networks are exposed. Yet, systematic reviews rarely report study results by the types of buffers used or consider the buffer types used at all.

18.3.3 Boundary Type and Size

Similarly, many studies use administrative boundaries (e.g. census tract) to measure their environmental exposures. The size of these artificial boundaries is typically based on number of households, so in higher-density neighbourhoods (inner city), the area of the administrative boundary will be substantially smaller than in lower-density (outer suburban) areas (see Fig. 18.3), further contributing to error when measuring exposure. An alternative approach to area-level measures was that adopted by Hooper (2014). Rather than using generic environmental exposure measures, she developed policy-relevant area-level measures for all dwellings within specific housing developments and found significant associations with walking. Notably, she overcame "edge" effects for study participants living on the outer edge of housing developments, by adding an 800-m buffer to the housing development to capture the environments of neighbouring housing developments. This approach better reflects how communities are planned and developed in practice

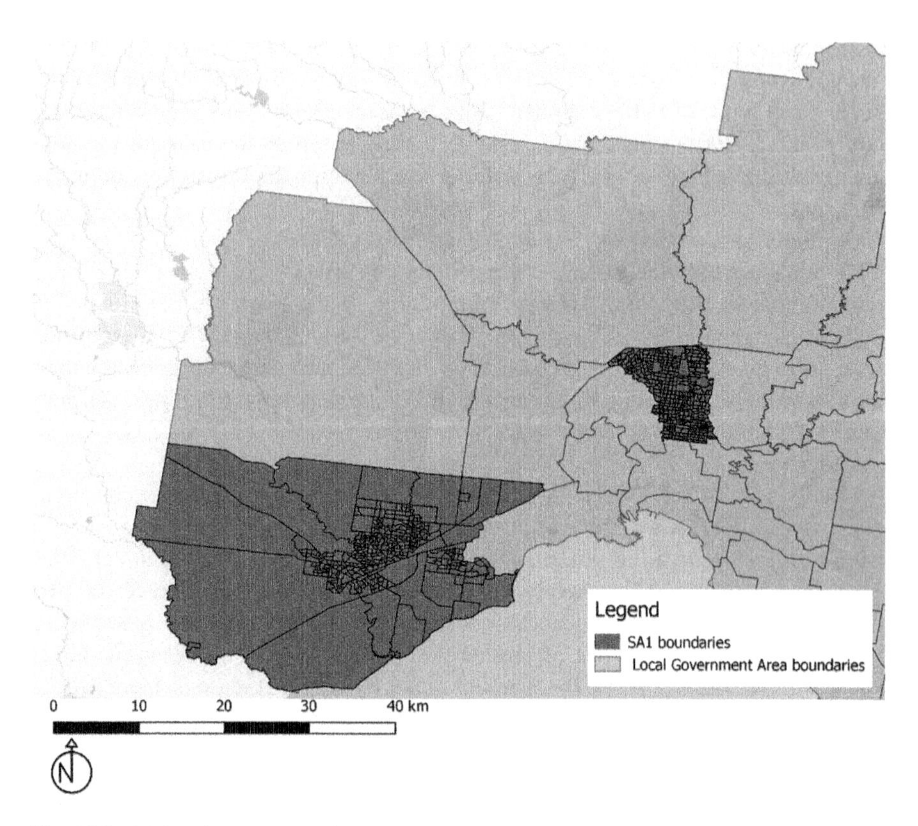

Fig. 18.3 A Victorian example of a local government and SA 1 (400 dwellings) boundaries in high-density and low-density areas

(rather than at the individual household level or over administrative boundaries) and, if housing development boundaries are available, may offer an alternative approach to using administrative boundaries.

18.3.4 Implications for Further Research and Systematic Reviews

Despite the maturity of the built environment and health field, many researchers and systematic reviewers ignore the importance of the scale, size and type of buffers used for different built-environment exposure measures. Although some reviews are now reporting whether or not administrative boundaries are used (Cerin et al. 2017), most fail to assess the quality or type (radial, road network, administrative boundary) and/or size of the buffer, nor do they consider the appropriateness of the buffer size for different types of destinations (e.g. public open space compared with a shop). This is important, because buffer size influences the variability observed in environmental exposures. Reviewers continue to combine results—irrespective of the quality of the environmental exposure measure—without differentiating between the size and types of buffers applied. With few exceptions (Ding and Gebel 2012; Panter et al. 2008), systematic reviewers simply report inconsistent or non-significant findings without any thoughtful consideration about why this might be occurring.

Similarly, as with Bancroft and colleagues' review (Bancroft et al. 2015), reviewers continue to combine findings from studies of different population age groups (adults, children, older adults), rather than reporting results separately. Environmental factors associated with walking by each of these groups may differ. It would be more instructive if results were reported separately, to make these differences apparent, rather than combining and reporting mixed or inconsistent findings. Journals could play an important role by requesting this as part of their quality assessment procedures.

Finally, some reviews combine studies in ways that do not make sense conceptually. For example, walkability is generally an urban concept, but some reviews include studies comparing urban and rural communities and small regional cities (Hajna et al. 2015). Others combine results of studies measuring perceived and objective built-environment features, despite now well-established mismatches between these different types of measures (Ball et al. 2008; Gebel et al. 2011; Koohsari et al. 2015a). While both are important for different reasons, there is little to be gained by combining studies with measures that measure different things.

Tighter guidelines are required for journals about the reporting of studies and systematic reviews that include built-environment measures. For example, future systematic reviews should report separately on studies using road network or radial buffers, and administrative boundaries of different sizes, as well as studies of different population groups. As in any research area, understanding the variables is critical, and a comprehensive review of the existing literature and its variables is required by those new to the area. As research advances and technology improves, journals should consider not publishing studies with poorly conceived exposure measures, unless they are part of a methodological study comparing measures of different types. The latter have considerable value, as they help identify and refine better measures (Brown et al. 2009; Christian et al. 2011b; Learnihan et al. 2011; Chin et al. 2008).

Some studies may continue to incorporate exposure measures known to have limitations (e.g. significant measurement error associated with large administrative boundaries or radial rather than road network buffers). This often reflects data inequities, particularly in lower-income countries where high-quality geographic information system (GIS) data are not readily available, or there are insufficient resources to clean data. However, the measurement errors and inconsistencies this is likely to produce (and the potential to attenuate findings) need to be considered when these studies are incorporated into reviews. Data providers and national statistics agencies therefore have a critical role to play in making GIS data more accessible *and* in improving data quality. Ideally, these organisations should work together to prepare and clean data to create high-value data sets that can be used for a variety of core purposes but which can also be incorporated into walkability studies, to help reduce measurement error (Chin et al. 2008). Until high-quality data are more readily available, particularly in the developing world, research funders need to fully fund the expensive process of cleaning GIS data (Jenks and Malecki 2004), and data providers and researchers should explore how they can overcome licencing barriers to enable cleaned GIS data to be shared between groups.

18.4 Where Are the Gaps?

Interventions in the built environment that are designed to reduce chronic diseases and health inequity complement urban planning efforts focused on creating cities that are more "liveable", compact and pedestrian-friendly, less automobile-dependent and more socially inclusive (Major Cities Unit 2010; Department of Infrastructure and Transport 2011; Western Australian Planning Commission 2007). Despite a rapid growth of research in this field, there is still a gap between the concept and actual creation of healthy, liveable communities (Hooper et al. 2013; Whitzman 2007; Curtis and Punter 2004). Transport and planning professionals have dismissed much public health evidence as merely "accepted wisdom", retorting: "Tell us something we don't already know" (Allender et al. 2009). While they acknowledge that health evidence adds credibility to their endeavours to create better communities, it often lacks the specificity they need to change practice—a major challenge to public health academics. The sections that follow reflect on gaps in the evidence. Further research here, if policy-relevant and codesigned with policy-makers and practitioners, could inform policy development and practice (Giles-Corti et al. 2015), perhaps providing the basis for health-promoting urban design and the "smart cities" of the future.

18.4.1 Beyond Measuring Associations, Towards Defining Thresholds

Establishing the existence and magnitude of any association between the built environment and human health has been a crucial first step in active-living research. Understanding this relationship is important, partly because changing the built environment can be costly (Cao 2010; Boarnet et al. 2008).

The next important step is to provide evidence to planners, policy-makers and designers about *how* to design walkable environments that promote healthy, active lifestyles. Attention has turned towards setting standards for *how much* of a built-environment feature leads to healthy behaviours (the "dose") and identifying any dose–response relationships. In seeking to identify thresholds for interventions, researchers must study metrics of the built environment, to inform urban design standards that can be used to develop built-environment interventions and to design communities that improve people's health.

18.4.2 Non-linear Methods

Models exploring these relationships need to consider non-linear forms that allow for a threshold to be established. Non-linear forms can be introduced in regression models via variable definitions such as polynomials, but the preferred method to date has relied on using smoothing or cubic splines in the Generalized Additive Mixed Model (GAMM) (Wood 2009). This allows for flexible, non-linear forms that are useful for uncovering dose–response, linear, non-linear and threshold relationships. To date, research using these methods has primarily been conducted by researchers in the International Physical Activity and the Environment Network (IPEN), combining comparable data from several countries.

IPEN's early studies have examined non-linear relationships between perceived built-environment measures derived from the Neighbourhood Environment Walkability Scale (NEWS) and a variety of health measures (Kerr et al. 2013). For example, a three-country study found that minutes spent in motorised transport started to decline when the perceived number of destinations exceeded 10, or the destination was perceived to be no further than 20 mins' walk away (Van Dyck et al. 2012). Other IPEN papers have found that higher levels of perceived aesthetics led to increased leisure-time walking and that higher levels of perceived walking and cycling infrastructure led to increases in moderate to vigorous leisure-time activity (Van Dyck et al. 2013, 2014). Similarly, there were non-linear relationships between various neighbourhood perceptions and walking and cycling for transport (Kerr et al. 2016). Specifically, transport walking increased for higher perceived residential density, before plateauing, yet the results for transport cycling suggested a negative overall relationship with perceived residential density. Similarly, an IPEN paper found a curvilinear relationship between objectively measured accelerometer data of daily minutes of moderate to vigorous activity, and a measure of the perceived land-use mix access, whereby only higher scores perceiving greater quantities of places and transit stops within walking distance of home led to an increase in daily minutes of moderate to vigorous activity (Cerin et al. 2014).

These IPEN papers are important, because they have tested new statistical methods and found evidence of non-linear dose–response relationships with perceived built-environment measures. However, studies of perceptions are of limited practical use to planners, policy-makers and designers. The case for designing healthy cities and for developing evidence to inform standards for built-environment interventions must incorporate *objective* measures. For example, using a series of indicator variables, Eom and Cho explored dose–response relationships between land-use mix, population and street densities and walking and private-vehicle trips (Eom and Cho 2015). They found that, in Seoul, walking increased once population densities reached between 9132 and 16,101 persons per km^2, but beyond this level of density walking decreased. Such evidence is clearly important, suggesting that once cities become too densely populated, they may discourage walking. Such "diminishing returns" also have implications for planning and investment: they suggest an optimal point for certain built-environment features for supporting healthy and active behaviours, beyond which there is limited benefit.

To this end, potentially important IPEN data from 14 cities around the world was recently published in *The Lancet*. Using GAMMs, Sallis and colleagues explored non-linear relationships between accelerometry physical-activity data and objective measures of the built environment (Sallis et al. 2016b). No dose–response or threshold relationships were found between objectively measured accelerometry data and residential, street, park or public transport density. However, another IPEN study—using similar built-environment measures and studying self-reported transport walking—found curvilinear relationships with objectively measured residential density, street connectivity and park density exposures: the residential density threshold for any transport walking was 12,000 dwellings per km^2, for street connectivity 200–250 intersections per km^2, and participants with up to 10 parks within 1 km of their home were between 2.5 and 3.5 times more likely to do any self-reported walking. In this study, only linear relationships were observed for residential density, street connectivity and land-use mix exposures for self-reported transport cycling suggesting that there are no specific threshold quantities for these exposures (Christiansen et al. 2016).

18.4.3 Lack of Specificity

One reason why the IPEN paper could not identify thresholds using objective measures of total physical activity and transport cycling is the lack of specificity between environmental exposures and behaviour. For example, total physical activity includes many types of activity, but the objective measures of the built environment studied are relevant only to transport walking. As discussed above, behaviour-specific and environment-specific models are needed (Giles-Corti et al. 2005b). Such approaches should also be applied to threshold analyses, with exposure type chosen according to the outcome variable being investigated. By way of example, Kerr et al. (2016) found that "walkable environments" did not appear to be associated with cycling. However, a person can cycle further than they can walk in a given time, so different thresholds may exist for cycling compared with walking. For instance, while many transport walking studies use buffer sizes of up to 1.6 km, a much larger buffer (such as 3.2 km) may be more relevant for cycling. Also, cycling is generally under-studied and requires further investigation.

As research in active living evolves, it remains to be seen what advice researchers will offer to policy-makers, should several different or conflicting thresholds be found for different health behaviours or health benefits, or indeed what should be done if none are found at all. For example, what will our advice be if linear relationships suggest increasing improvements in physical activity for any improvement in the built environment (Sallis et al. 2016b). We would argue that urban design should aim to maximise the number of residents able to achieve (at the very least) the recommended levels of physical activity in their local communities.

18.4.4 Comprehensive Ecological Models Incorporating Individual, Social and Environmental Factors

Examining multiple levels of influence reflects an ecological approach to research and interventions (McLeroy et al. 1988), yet the adoption of ecological models is relatively new in active-living research. Despite these models' potential usefulness in conceptualising research, Ball et al. (2006) have argued that many studies lack strong and well-conceptualised theoretical models for testing the interactions and pathways among personal, social and environmental factors.

Hence, in the last two decades, social-ecological frameworks (Barton and Tsouros 2000; Sallis et al. 2006a) have been proposed, developed and used to examine the effects of the built environment on physical activity (Giles-Corti and Donovan 2002). These models have mapped the interrelationships between individual, social, physical and policy determinants and have identified a range of urban design attributes related to physical activity—primarily for recreation and transportation purposes (Sallis et al. 2012).

Nevertheless, a major gap in the literature is studies that simultaneously incorporate multiple levels of influence: individual, social environment and built environment (Brownson et al. 2009; Lee and Moudon 2004; Panter and Jones 2010; McCormack and Shiell 2011; Mokhtarian and Cao 2008). This is becoming a particular challenge for secondary analyses of existing studies, as few studies have sought to simultaneously examine the relative influence on behaviour of these three factors (Giles-Corti and Donovan 2002, 2003). This is a prime opportunity for future research to inform comprehensive interventions.

18.4.5 Natural Experiments and Longitudinal Studies

A major criticism of built-environment research is its reliance on cross-sectional studies, which limit our ability to determine cause and effect. The main question is self-selection: does neighbourhood design change residents' behaviour, or do residents who prefer more active lifestyles simply choose to live in neighbourhoods where it is easier or more pleasant to walk? In other words, the relationship between the built environment and walking behaviour is confounded by walking preference (Cao 2010).

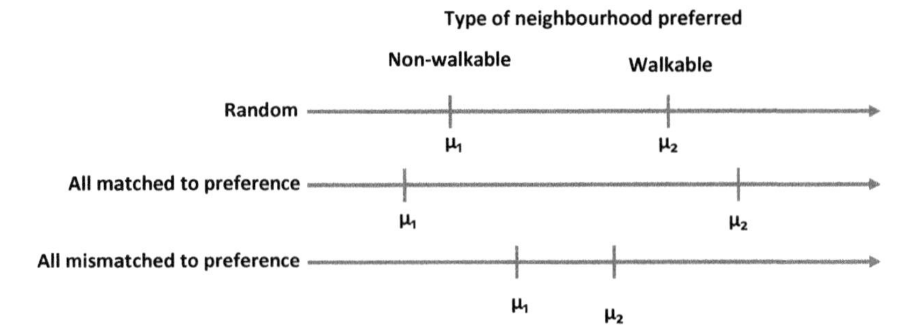

Fig. 18.4 The relationship between self-selection and misestimation (Modified from: Cao 2010)

Nevertheless, from a public health perspective, self-selection may be a moot point. In most cities, most people—irrespective of their preferences—do not live in walkable neighbourhoods. Indeed, at least in Australia, there is evidence that more than 60% of residents of low-walkable neighbourhoods (Bull et al. 2015) or non-transit-oriented developments (Kamruzzaman et al. 2016) would prefer to live in a more walkable neighbourhood or in a transit-oriented development, while fewer residents (just over 30%) of walkable neighbourhoods or transit-oriented developments would prefer to live elsewhere.

Cao's hypothesis on the effects of self-selection and misestimating the effects of neighbourhoods on walking illustrates this point (Cao 2010). Figure 18.4 shows the hypothesised mean walking levels of people based on their preference to live in non-walkable or walkable neighbourhoods: if people could be randomised to live in varying types of neighbourhoods, those who prefer non-walkable neighbourhoods are hypothesised to walk less than those who prefer walkable neighbourhoods. If, on the other hand, people could live in their preferred type of neighbourhood, then those preferring a non-walkable neighbourhood would walk less, and those preferring a walkable neighbourhood would walk more. Conversely, if people were mismatched (those who preferred a non-walkable neighbourhood were allocated to live in a walkable neighbourhood, and vice versa), Cao hypothesises that those who prefer a non-walkable neighbourhood but who live in a walkable area would walk more than they would have done if left to their own preference, and those who prefer a walkable one would walk less than they would otherwise. This is important from a public health perspective. It suggests that if more walkable communities became the norm, rather than the exception—as is currently the case in most cities—then everyone would walk more: those who prefer to walk and even those who don't, particularly if walkable communities implemented strategies to discourage driving. Given the current lack of physical activity among a large proportion of the population, the health benefits of even a modest increase in the numbers of people who are moderately physically active could be large, producing significant annual cost savings to health systems (Stephenson and Bauman 2000).

Nevertheless, systematic reviewers repeatedly call for more natural experiments and longitudinal studies as an alternative to randomised controlled trials (Merom et al. 2003; Ogilvie et al. 2007; Petticrew et al. 2005; Kaczynski and Henderson 2007; Kerr et al. 2012; McCormack and Shiell 2011; Sugiyama et al. 2012; Pearce and Maddison 2011; Lee and Moudon 2004; Panter and Jones 2010; Mokhtarian and Cao 2008; Van Cauwenberg et al. 2011; Ding and Gebel 2012; Durand et al. 2011; Saelens and Handy 2008; Cao et al. 2009; Renalds et al. 2010; Ewing et al. 2003), since—with limited exceptions (Wells and Yang 2008)—it is not feasible to randomise study participants to live in certain homes or environments. The aim of these studies is to satisfy causal inference concerns (Mokhtarian and Cao 2008), and they include panel studies that comprise participants who relocate ("movers").

Other longitudinal studies have monitored health changes over time in participants living in high- or low-walkable neighbourhoods. For example, based on ecological data, living in higher-walkable neighbourhoods appeared to protect people's health and was associated with decreased prevalence of overweight, obesity and diabetes between 2001 and 2012 (Creatore et al. 2016). At each time point, living in highly walkable areas was associated with significantly higher walking, cycling and public transit use, and lower car use, than living in the least-walkable areas. However, there was little change in daily walking and cycling frequencies; these "increased only modestly" over time in highly walkable areas. But perhaps more importantly (although not considered by the authors), as participants aged, active behaviours did not *decline* in those living in higher-walkable areas.

"Movers" studies include those that follow participants in longitudinal or panel studies who relocate to new neighbourhoods (Krizek 2003; Hirsch et al. 2014a, b, c). These have shown varying results, mainly because people who relocate often move to neighbourhoods with similar characteristics (Krizek 2003). Hence, although people relocate, environmental exposures that encourage more walking may not have changed. To overcome this problem, Hirsch and colleagues studied movers who relocated to neighbourhoods with a 10-point higher "walk score". This was associated with 16 min more weekly walking, an 11% higher chance of meeting goals for transport walking and a modest reduction in body mass index. However—not surprisingly given the study's exposure measure—there was no association with recreational walking (Hirsch et al. 2014a).

Other types of studies that are informative for built-environment and health research are natural experiment studies of the effects of policy changes on physical activity. These might involve significant changes to the built environment following transport interventions, such as congestion charging or new infrastructure for public transport or cycling (Goodman et al. 2013, 2014; Sahlqvist et al. 2013; Ogilvie et al. 2011, 2012; Panter et al. 2016, b; Heinen et al. 2015a; Martin et al. 2015; Sahlqvist et al. 2015; Merom et al. 2003 #921; Cohen et al. 2008), implementation of new urban design policies (Egan et al. 2003, 2007; Sallis 2009; Story et al. 2009) or upgrades to public open space (Veitch et al. 2014, 2017; Cohen et al. 2009a, b). In many cases, the built environment is changed *around* study participants; the question is whether the "dose" of the intervention is sufficient to change behaviour or whether people substitute one behaviour for another. In other cases, natural experiments involve monitoring people who relocate to new types of housing (Wells and Yang 2008) or to new suburbs designed according to new subdivision design codes (Giles-Corti et al. 2008). In the latter case, a unique, large-scale study conducted by our team in Perth, Western Australia, is described in Sect. 18.5.

18.4.6 Policy-Relevant and Practice-Relevant Natural Experiments

Academic research—including active-living research—is often criticised for being irrelevant or poorly linked to policy and practice (Oliver et al. 2014), even though influencing urban planning policy and practice is often a stated goal of built-environment and active-living research (Goldstein 2009). This shortcoming has led to calls for research to be better aligned with current and future policy environments and planning practices (Orton et al. 2011).

Notably, Allender and colleagues found that much of the built-environment and health evidence lacks the specificity needed by government planners to inform planning policy (Allender et al. 2009). Evidence-based active-living or public health recommendations are often provided to planners and policy-makers without any obvious links to existing policies or legislation or guidance on how to achieve change.

Planning professionals and policy-makers have said that to help translate health research into planning policy and practice, more research is required that assesses the effectiveness of planning regulations and policies in changing public health behaviours, through the evaluation and documentation of innovative communities and environmental or planning policies, programmes or codes that promote active living (Allender et al. 2009; Koohsari et al. 2013a; Durand et al. 2011; Talen 1996). Planning academics have even criticised their own field for neglecting to determine the degree to which plans (e.g. planning policies or guides for future urban development) have been implemented and adhered to, with few studies quantitatively assessing the implementation processes and success (Talen 1996).

While few studies have rigorously examined the role of specific urban planning policies in producing built environments that contribute to healthy and liveable neighbourhoods, cities are increasingly being viewed as "urban laboratories". This reflects growing recognition of the valuable knowledge that can be created concurrently with urban development, when new ideas are implemented and their effects measured (Karvonen and Van Heur 2014).

This is also important given the rise in popularity of planning movements, such as New Urbanism and Smart Growth, that have emerged in an attempt to counter planning codes thought to be responsible for suburban sprawl. Proponents of these codes focus on recalibrating and reconfiguring the design of suburbs, based on traditional neighbourhood designs that create places for living, working and recreation (Audirac 1999) and incorporating higher-density, mixed-use, pedestrian-oriented, walkable communities (CNU 1997; Duany et al. 2000; Calthorpe 1993; Kim 2000). As more cities adopt the rhetoric of (and often explicitly refer to) these codes in their planning work, there has been interest in assessing their claims and how well they improve health and wellbeing. While a growing number of case studies and research papers measure built-environment concepts that can be linked to New Urbanism or Smart Growth principles (such as walkability and land-use mix), an explicit connection between these principles and better health has not yet been demonstrated (Durand et al. 2011). Although the principles of New Urbanism or Smart Growth make theoretical sense for creating healthier communities, there is virtually no empirical work directly assessing and linking the implementation of specific planning policies based on these principles to actual health (Talen 1996). This is a serious gap in the literature: as more communities implement planning policies based on these principles, there is a need for greater certainty about whether, and under what conditions, policy change might improve overall community health and wellbeing (Durand et al. 2011).

As new or modified planning policies, codes and movements are adopted, tested or implemented, opportunities emerge for natural experiments to monitor the effects of these interventions on active-living and health behaviours and to provide more rigorous evidence than is contained in the cross-sectional studies that dominate the literature. Evaluations of urban planning policies and codes that promote active living are important, both for providing prescriptive and policy-specific evidence to inform urban policy and practice and for strengthening academic evidence (Allender et al. 2009; Durand et al. 2011; Koohsari et al. 2013a; Taylor et al. 2011; Brownson et al. 2009). In case studies or natural experiment evaluations of planning policies and practices, results that are directly relatable to the policy and its implementation are essential. We need this type of evidence in order to know which aspects of the policy are (or are not) being implemented and to identify those that may (or may not) improve people's health (Hooper et al. n.d.; Brownson et al. 2009).

18.5 A Western Australian Example

A unique opportunity to evaluate the health benefits of urban policy in situ arose in 1998, when the Western Australian government tested its "Liveable Neighbourhoods Community Design Guidelines" (LN): a new planning policy aiming to create more sustainable, liveable and healthy communities (Western Australian Planning Commission 1997). Based on a local interpretation of New Urbanism (CNU 1997), the trial of LN, which commenced in 1997, was designed to eventually replace conventional development controls and design codes that had led to car dependence and suburban sprawl with an alternative approach to suburban neighbourhood design. LN aimed to create "more compact, pedestrian-friendly neighbourhoods, with good links to public transport services" (Western Australian Planning Commission 2000). Important aims were to reduce dependence on private motor vehicles, encourage more active forms of transport (walking, cycling and public transport) and foster a sense of community.

The RESIDE study was designed as a natural experiment in 74 new greenfield developments (19 liveable, 11 hybrid and 44 conventional) under construction across the Perth metropolitan region in 2003 (Giles-Corti et al. 2008). All people who purchased house-and-land packages in these developments (n = 10,193) were invited to participate in the study by the state water authority, which is notified of new owners after land title is transferred. Eligible participants were those currently building a home in the selected development, who intended to relocate to that home, and who were willing to complete three surveys and wear a pedometer for 1 week (on three occasions) over a 5-year period. (In 2010, additional funding was secured to conduct a fourth follow-up.) Participants completed questionnaires on four occasions: Time 1, baseline, 2003–2005, during construction of their new home and before relocation (n = 1813; 33.4% response rate); Time 2, 2004–2006, ≈1 year after relocation to their new home (n = 1465); Time 3, 2006–2008, ≈3 years after relocating (n = 1229); and Time 4, 2011–2012, ≈6–9 years after relocating (n = 565). Using a GIS, objective built-environment measures were generated for all participants at each time point, to provide a set of consistent and comparable (i.e., longitudinal) measures across the study period. These measures quantified the built-environment characteristics of participants' neighbourhoods (defined as the area accessible along the street network within 1600 m of participants' homes), including access to different types of public open space, shops and public transport; provision of footpaths, street connectivity, land-use mix, residential density and (standardised) neighbourhood walkability measures.

The intention of RESIDE was to undertake policy-relevant research and, in particular, to evaluate current state government urban design policy. Hence, findings were published not only in the academic literature (Hooper et al. n.d.) but also in a user-friendly summary for policy-makers (Bull et al. 2015). Briefly, RESIDE's longitudinal findings supported cross-sectional evidence that neighbourhood walkability (especially land-use mix and street connectivity), local access to public transport and access to a mix of different types of local destinations are important determinants of transport walking (Knuiman et al. 2014), but not of body mass index (Christian et al. 2011a). Moreover (after full adjustment for individual, social-environment and built environment factors and neighbourhood preference), RESIDE found that, following relocation, among those people who now had greater access to transport-related destinations, transport walking increased by 5.8 min per week for each type of destination gained and recreational walking by 17.6 min per week for each type of recreational destination gained (Giles-Corti et al. 2013).

However, the LN policy itself did not appear to be effective. We found no significant differences in the transport or recreational behaviours of those who relocated to neighbourhoods designed according to the LN policy, compared with others. While the policy helped to create more supportive environments, the dose of these interventions was insufficient to encourage more walking (Christian et al. 2013). We proposed a longer-term follow-up, but our findings also raised the question of whether there had been a policy failure or a failure to implement the policy as intended (Hooper et al. 2014).

Hence, a unique aspect of RESIDE was the quantification of the *actual* implementation of the policy on the ground, to better understand whether the observed behaviours were products of under-delivery of the policy or of fundamental shortcomings in the policy itself (Hooper et al. 2014). Using spatial measures tailored to the urban design features required by the policy, the evaluation identified which, and how much, of the design features had been implemented. None of the developments (including those classified as LN developments by the Department of Planning) had implemented the full suite of requirements as intended by the policy. Indeed, overall across all new developments, the LN policy was only around 50% implemented. Nevertheless, for every 10% increase in levels of overall policy compliance, the odds of participants walking for transport in the neighbourhood increased by 53% (Hooper et al. 2014), the odds of sense of community increased by 22%, the odds of good mental health increased by 8% (Hooper et al. 2014) and the odds of being a victim of crime decreased by 40% (Foster et al. 2015). These process evaluation findings were of considerable interest to the Department of Planning and led to the RESIDE findings being translated into a recent policy review. The process evaluation was also able to show which, if any, aspects of the policy were producing the desired physical-activity increases and health benefits. Moreover, it allowed the study team to communicate to the Department of Planning that the policy levers were effective but that there needed to be greater focus on achieving higher levels of implementation. RESIDE revealed the need to measure and assess actual *implementation* of planning policies. Without such evidence, natural experiments are simply academic, offering policy-makers no new information that they "don't already know" (Allender et al. 2009).

18.6 Where to Next?

Despite the wealth of knowledge garnered in the last two decades about the influence of the built environment on physical activity, there is still much to be done. First, most of the evidence to date is from developed countries, with research in developing countries only beginning to emerge very recently (Giles-Corti et al. 2016). While some correlates appear to be universal (Sallis et al. 2016b), there is still much to learn about the extent, type and quantity of interventions needed in different countries, and the conditions under which supportive built environments will be effective (Giles-Corti et al. 2016). For example, do the same interventions work (and for whom) in very hot climates or in cities that are actually, or are perceived to be, unsafe?

But even in developed countries, important questions remain about the interplay between urban design attributes at the local level and the contribution of regional planning to the overall system and whether there is an economic argument for building walkable neighbourhoods that produce co-benefits across multiple sectors.

18.6.1 Complex Systems Modelling

As discussed earlier, social-ecological models (Barton and Tsouros 2000; Sallis et al. 2006b) consider multiple levels of influence on behaviour and have been used to map the interrelationships between individuals and social, physical and policy determinants, enabling the identification of a range of local urban design attributes related to physical activity, primarily for recreation and transportation purposes (Sallis et al. 2012).

However, in the recent *Urban Design, Transport, and Health* series published in *The Lancet* (Giles-Corti et al. 2016), we argued that opportunities for active living arise out of synergies between the "Eight Ds": five local urban design and transport factors (density, design, distance to transit, destinations and desirability) and three regional factors (destination accessibility, distribution of employment and demand management). It is now clear that simply designing local neighbourhoods that encourage walking and cycling is not enough; communities need metropolitan regional planning interventions that ensure public transport access to jobs that are (typically) inequitably distributed across metropolitan regions. Furthermore, without demand management strategies to reduce the attractiveness of driving (e.g. congestion charging, manipulating the amount and cost of parking), it is unlikely that higher use of active transport will ever be achieved.

The question of how to plan and design cities that encourage active living and physical activity is multifaceted and complex. In established areas, the general conditions that support active living evolve and transform over time as the local urban fabric changes (e.g. when land undergoes zoning changes or local destinations are replaced with other types). At the same time, broader metropolitan infrastructure projects (new suburbs, road expansions, public transport upgrades, relocation of government agencies) affect established areas also. Assessing the effects of all these changes requires a complex systems approach.

In complex systems, analysis focuses on the behaviour of the system, which emerges from the interactions between the components (Von Bertalanffy 1969). This mode of thinking involves the use of either analytical deterministic models, where the model output is dependent on the parameter values and the initial conditions, or stochastic probabilistic models, where the model allows for random variation in one or more inputs over time (Ross 2014). Complex systems can be analysed using causal loop diagrams and can be operationalised using digital planning tools that support urban planning and policy-making (Boulange et al. 2017). Digital tools and planning support systems also enable end users to understand, explore and interact with these models via easy-to-use visual interfaces. Digital planning tools enable planners to simulate possible interventions before they finalise their plans and to explore myriad connected factors via an interactive mapping interface made available through a desktop GIS (Pelzer et al. 2013; Arciniegas and Janssen 2012; Boulange et al. 2017).

Research into the relationship between the built environment and physical activity could benefit in several ways from complex system modelling:

18.6.1.1 Bringing Together Researchers, Planners, Policy-Makers and Practitioners

Planning support systems based on complex systems help planners and decision-makers understand—from a variety of perspectives—how planning will affect the design and functioning of the city, depending on the outcomes of interest (e.g. health or transport). One example is the Walkability Planning Support System (Boulange et al. 2015, 2017), built in Community Viz, a now well-established ArcGIS extension (Walker and Daniels 2011). Ideally, planning support system models are designed collaboratively with a range of participants (including planners and policy-makers), to stimulate communication about how a complex system works. By participating in the design, all parties can develop a shared understanding of the main relationships and dynamics of the system and identify vulnerabilities and leverage points for changing the system structure. They can also provide valuable feedback on the outputs of the model as it is being built and refined (Newell and Proust 2009; Newell et al. 2008). Techniques such as face validity and collaborative modelling are important to build into the complex model workflow, to ensure that the end users trust the structure and behaviour of the model.

Planning support systems and interactive mapping tools such as the Australian Urban Research Infrastructure Network portal (AURIN) offer datasets from custodians such as health, housing and transport departments, to model the city as a complex system revealing previously unexplored spatio-temporal relationships (Delaney and Pettit 2014; Pettit et al. 2015). Such analysis helps us understand the various dimensions of a city, further supported by urban "big data" such as spatio-temporal housing data at the property level (Pettit et al. 2017), or effects on physical activity measured using smartphones with built-in accelerometry (Althoff et al. 2017).

Collaborative processes aimed at mapping the complex relationship between the built environment and physical activity are a novel way to communicate built environment research and increase awareness among policy-makers of the interdisciplinary nature of urban planning and research. Furthermore, the potential of urban big data becomes evident when using new, interactive mapping environments such as CityViz, which allows bicycle and other data to be used in city planning and decision-making (Goodspeed et al. in press).

For example, Fig. 18.5 shows rider travel movements in Melbourne derived from the Bicycle Network's Riderlog smartphone application. These data can be used by city planners to understand cycling patterns and behaviour across the city, such as the proportion of female versus male cyclists and their different cycling patterns. They can also be mapped and analysed to determine hot spots of increased infrastructure demand or danger for cyclists (and therefore where to invest in upgrades). Understanding patterns and demographics in cycling behaviour is important in rapidly urbanising cities (like Melbourne), particularly cities promoting more sustainable, healthy and safe transport modes.

Fig. 18.5 Urban big data: rider travel movements in Melbourne, from the Bicycle Network's Riderlog smartphone application

18.6.1.2 Validating Urban Health Indicators

CityViz and Community Viz (the Walkability Planning Support System) demonstrate that, with appropriate data, complex systems models can be built to explore how multifaceted associations between urban design features and transport systems result in health risks among population groups (Badham 2010). However, some complex systems allow for dynamic modelling; this in turn allows researchers to explore causal relationships that help us understand how cities affect people's physical and mental health (Gatzweiler et al. 2017).

These types of analyses are also useful when developing and validating *indicators* of the condition of the urban health system and of the built environment itself, offering a convenient way to store information from scenarios tested in the complex system model (Badland et al. 2017a; Murphy et al. 2017). Indicators can support good urban management and urban policy (Badland et al. 2014b; Lowe et al. 2015). In particular, they can help decision-makers measure and monitor cities' performance against population health or environmental protection standards and trigger alerts when unwanted results occur.

18.6.1.3 Simulating Urban Future Scenarios

Complex models could be built to examine the interplay of urban design attributes at the local level while accounting for the effect of regional planning on people's physical activity. This would allow researchers to work with policy-makers and practitioners to determine optimum levels of density, street connectivity and destination access to increase physical activity. At the same time, one could model commuting patterns that are active transport-friendly, for example; after a model has been validated, projection and "what if?" analyses can be undertaken. By varying input parameters, analysts can simulate interventions and estimate the repercussions of changing one or more areas of the system. Although model simulations cannot be relied upon to give accurate predictions of what will happen following interventions, they are a useful tool to describe the likely results and identify potential unintended consequences (such as harm to health) (Urban et al. 2011; Lich et al. 2013).

Simulation may be particularly useful when estimating the likely effect of urban planning policies or urban design requirements on levels of physical activity over time. Simulations could be run using agent-based modelling (ABMs), which model synthetic populations and the ways in which agents (e.g. residents in a local neighbourhood) might respond to changes in the built environment under various hypothetical changes to the transport network (Badland et al. 2017b).

In modelling urban futures and health, we need to understand the likely effects on our cities of disruptive technologies. Uber and Airbnb are two notable examples that are already challenging planners and policy-makers, and they have the potential to fundamentally change the way our cities function. Dockless bike-sharing schemes are also being implemented in many cities; these, along with other disruptive technologies and the data they can provide, need to be considered when understanding and modelling the ever-evolving complexity of the city. Finally, autonomous vehicles are being trialled in some cities, with dissemination on the horizon: what changes will they bring, especially to the physical-activity patterns of urban dwellers? Measuring and monitoring to create early warning systems of the effects of these major changes on cities and on the health and wellbeing of residents is a priority for future research.

18.6.2 Economic Evaluations

Low levels of physical activity worldwide represent a health and economic burden for individuals and societies (Ding et al. 2016; Kyu et al. 2016). But the health benefits of physical activity are rarely considered in economic evaluations of environmental interventions and policies (Mulley et al. 2013). One reason may be that studies examining the economic merit of environmental initiatives are in their infancy, with no agreement on appropriate methods (Brown et al. 2016). Lack of full consideration of the health benefits of environmental initiatives that foster physical activity may underestimate their value to society. This might lead to the misallocation of resources towards car-oriented infrastructure.

Three recent systematic reviews that include economic evaluations of active-transport initiatives (walking and cycling) indicate significant heterogeneity in approaches for the inclusion of physical-activity-related health benefits (Brown et al. 2016; Doorley et al. 2015; Mueller et al. 2015). Methods to estimate demands on infrastructure or the effectiveness of interventions are diverse and focused on transport walking and cycling, without offsetting the substitution effect from other forms of physical activity such as recreational walking and cycling (Brown et al. 2016). Likewise, quasi-experiments suggest that new transport infrastructure may not influence total physical activity (Panter et al. 2016), due to a substitution effect: people who use this infrastructure may substitute this activity for participation in other forms of physical activity, with no overall gain. This may lead economic evaluations to overestimate the health benefits.

Most evaluations to date have used a cost–benefit analysis (CBA) framework (Brown et al. 2016; Doorley et al. 2015; Mueller et al. 2015), which is consistent with methods used when appraising government initiatives (Mulley et al. 2013). But CBAs present a problem, as a monetary value needs to be placed on goods without a market value, such as health (Drummond et al. 2005). Nevertheless, in the transport field, agreed methods have been developed to evaluate non-market goods: for example, travel-time savings are now included in routine evaluations of transport projects, despite initial controversy (Mulley et al. 2013).

The World Health Organization has developed the Health Economic Assessment Tool (HEAT) as a homogeneous framework to evaluate the health benefits of interventions that promote walking and cycling (Kahlmeier et al. 2014). Although a promising effort, it quantifies the effect of walking and cycling on premature mortality only, ignoring any improvement to quality of life. Users of the tool still need to produce demand estimates from the proposed initiative, which, as highlighted, has been generally weak. In addition, HEAT (like most of the literature to date) has focused on the transport sector. The built environment is broader than this and includes the planning of cities (for instance, based on the Eight Ds). We need standardised, transparent methods for quantifying the health benefits of physical activity. This could be achieved through collaborative research between urban and transport planners and health economists. Tools such as the Walking Planning Support System model, in combination with established methods used in health economics that incorporate mortality and quality-of-life measures (Vos et al. 2010), may help us evaluate the economic merit of building neighbourhoods that promote active living. However, public health researchers also need to improve their understanding of the policy-making process, if they hope to generate health-improving change in non-health sectors such as transport and planning (Carey and Crammond 2015).

18.7 Conclusion

Research on the built environment and physical activity is maturing, and it is important that we learn from past lessons. While many built-environment features appear to influence walking for transport or recreation, there are many inconsistencies in the literature, which could in part be due to the way exposure measures are conceptualised and measured. Assumptions that researchers have access to high-quality data across cities are unfounded, particularly in many developing countries. There is a critical need to create and provide access to clean, reliable and rich data for statistical analysis and data modelling. In developed countries, researchers have access to many good urban datasets, but there are significant gaps (e.g. footpath data), and quality and availability vary between cities. Higher-quality, more accessible data will strengthen our understanding of the built environment and physical activity.

A number of promising areas of research might also support the translation of findings into policy and practice: identifying thresholds for built-environment interventions; assessing the relative influence of individual, social and built-environment features; using complex system modelling to simulate interventions; and evaluating the economics of built-environment interventions. To avoid simply confirming conventional wisdom in the planning and transport professions, researchers are encouraged to codesign research with policy-makers, practitioners and disciplines outside health. Through interdisciplinary multi-sector research, there is potential to help shape the cities in which future generations will live, work and play.

References

Allender, S., Cavill, N., Parker, M., & Foster, C. (2009). Tell us something we don't already know or do!' – The response of planning and transport professionals to public health guidance on the built environment and physical activity. *Journal of Public Health Policy, 30*, 102–116.

Althoff, T., Sosic, R., Hicks, J. L., King, A. C., Delp, S. L., & Leskovec, J. (2017). Large-scale physical activity data reveal worldwide activity inequality. *Nature, 547*, 336–339.

Arciniegas, G., & Janssen, R. (2012). Spatial decision support for collaborative land use planning workshops. *Landscape and Urban Planning, 107*, 332–342.

Audirac, I. (1999). Stated preference for pedestrian proximity: An assessment of new urbanist sense of community. *Journal of Planning Education and Research, 19*, 53–66.

Badham, J. (2010). *A compendium of modelling techniques*. Canberra: The Australian National University.

Badland, H., Foster, S., Bentley, R., Higgs, C., Roberts, R., Pettit, C., & Giles-Corti, B. (2017a). Examining associations between area-level spatial measures of housing with selected health and wellbeing behaviours and outcomes in an urban context. *Health & Place, 43*, 17–24.

Badland, H., Hickey, S., Bull, F., & Giles-Cort, B. (2014a). Public transport access and availability in the RESIDE study: Is it taking us where we need to go to? *Journal of Transport & Health, 1*, 45–49.

Badland, H., Whitzman, C., Lowe, M., Davern, M., Aye, L., Butterworth, I., Hes, D., & Giles-Corti, B. (2014b). Urban liveability: Emerging lessons from Australia for exploring the potential for indicators to measure the social determinants of health. *Social Science & Medicine, 111C*, 64–73.

Badland, H. M., Rachele, J. N., Roberts, R., & Giles-Corti, B. (2017b). Creating and applying public transport indicators to test pathways of behaviours and health through an urban transport framework. *Journal of Transport & Health, 4*, 208–215.

Ball, K., Jeffery, R., Crawford, D., Roberts, R., Salmon, J., & Timperio, A. (2008). Mismatch between perceived and objective measures of physical activity environments. *Preventive Medicine, 47*, 294–298.

Ball, K., Timperio, A. F., & Crawford, D. (2006). Understanding environmental influences on nutrition and physical activity behaviours: Where should we look and what should we count? *International Journal of Behavioral Nutrition and Physical Activity, 3*, 33.
Bancroft, C., Joshi, S., Rundle, A., Hutson, M., Chong, C., Weiss, C. C., Genkinger, J., Neckerman, K., & Lovasi, G. (2015). Association of proximity and density of parks and objectively measured physical activity in the United States: A systematic review. *Social Science & Medicine, 138*, 22–30.
Barton, H., Tsouros, C., & World Health Organization Regional Office for Europe. (2000). *Healthy urban planning: a WHO guide to planning for people*. Abingdon: Taylor & Francis.
Boarnet, M., & Crane, R. (2001). The influence of land use on travel behavior: specification and estimation strategies. *Transportation Research Part A Policy and Practice, 35*, 823–845.
Boarnet, M. G., Greenwald, M., & McMillan, T. E. (2008). Walking, urban design, and health – Toward a cost-benefit analysis framework. *Journal of Planning Education and Research, 27*, 341–358.
Boulange, C., Pettit, C., & Giles-Corti, B. (2017). The walkability planning support system: An evidence-based tool to design healthy communities. In S. Geertman, A. Allan, C. Pettit, & J. Stillwell (Eds.), *Planning support science for smarter urban futures*. Cham: Springer International Publishing.
Boulange, C., Pettit, C. J., Arciniegas, G., Badland, H., & Giles-Corti, B. (2015). A simulation-based planning support system for creating walkable neighbourhoods. In J. Ferreira & R. Goodspeed (Eds.), *The 14th international conference on computers in urban planning and urban management – planning support systems and smart cities, July 7–10th 2015* (pp. 1–22). Boston: MIT.
Brown, B. B., Yamada, I., Smith, K. R., Zick, C. D., Kowaleski-Jones, L., & Fan, J. X. (2009). Mixed land use and walkability: Variations in land use measures and relationships with BMI, overweight, and obesity. *Health & Place, 15*, 1130–1141.
Brown, V., Zapata-Diomedi, B., Moodie, M., Veerman, J. L., & Carter, R. (2016). A systematic review of economic analyses of active transport interventions that include physical activity benefits. *Transport Policy, 45*, 190–208.
Brownson, R., Hoehner, C., Day, K., Forsyth, A., & Sallis, J. (2009). Measuring the built environment for physical activity: State of the science. *American Journal of Preventive Medicine, 36*, S99–123.
Bull, F., Hooper, P., Foster, S., & Giles-Corti, B. (2015). *Living liveable: The impact of the Liveable Neighbourhoods Policy on the health and wellbeing of Perth residents*. Perth: The University of Western Australia.
Calthorpe, P. (1993). *The next American metropolis: Ecology and urban form*. New Jersey: Princeton Architectural Press.
Cao, X. (2010). Exploring causal effects of neighborhood type on walking behavor using stratification on the propensity score. *Environment and Planning A, 42*(2), 487–504.
Cao, X., Mokhtarian, P., & Handy, S. (2009). Examining the impacts of residential self-selection on travel behaviour: A focus on empirical findings. *Transport Reviews, 29*, 359–395.
Carey, G., & Crammond, B. (2015). Action on the social determinants of health: Views from inside the policy process. *Social Science & Medicine, 128*, 134–141.
Cerin, E., Cain, K. L., Conway, T. L., Van Dyck, D., Hinckson, E., Schipperijn, J., De Bourdeaudhuij, I., Owen, N., Davey, R. C., Hino, A. A., Mitas, J., Orzanco-Garralda, R., Salvo, D., Sarmiento, O. L., Christiansen, L. B., Macfarlane, D. J., Schofield, G., & Sallis, J. F. (2014). Neighborhood environments and objectively measured physical activity in 11 countries. *Med Sci Sports Exerc, 46*(12), 2253–2264.
Cerin, E., Nathan, A., van Cauwenberg, J., Barnett, D. W., Barnett, A., & Council on, E. & Physical Activity - Older Adults working, g. (2017). The neighbourhood physical environment and active travel in older adults: a systematic review and meta-analysis. *International Journal of Behavioral Nutrition and Physical Activity, 14*, 15.

Chin, G. K., Van Niel, K. P., Giles-Corti, B., & Knuiman, M. (2008). Accessibility and connectivity in physical activity studies: The impact of missing pedestrian data. *Preventive Medicine, 46*, 41–45.
Christian, H., Giles-Corti, B., Knuiman, M., Timperio, A., & Foster, S. (2011a). The influence of the built environment, social environment and health behaviors on body mass index. Results from RESIDE. *Preventive Medicine, 53*, 57–60.
Christian, H., Knuiman, M., Bull, F., Timperio, A., Foster, S., Divitini, M., Middleton, N., & Giles-Corti, B. (2013). A new urban planning code's impact on walking: The residential environments project. *American Journal of Public Health, 103*, 1219–1228.
Christian, H. E., Bull, F. C., Middleton, N. J., Knuiman, M. W., Divitini, M. L., Hooper, P., Amarasinghe, A., & Giles-Corti, B. (2011b). How important is the land use mix measure in understanding walking behaviour? Results from the RESIDE study. *International Journal of Behavioral Nutrition and Physical Activity, 8*, 55.
Christiansen, L. B., Cerin, E., Badland, H., Kerr, J., Davey, R., Troelsen, J., van Dyck, D., Mitas, J., Schofield, G., Sugiyama, T., Salvo, D., Sarmiento, O. L., Reis, R., Adams, M., Frank, L., & Sallis, J. F. (2016). International comparisons of the associations between objective measures of the built environment and transport-related walking and cycling: IPEN adult study. *J Transp Health, 3*, 467–478.
CNU. 1997. Congress for the new urbanism [online]. Congress for the new urbanism. Retrieved Sept 24 2010, from, http://www.cnu.org/.
Cohen, D., Sehgal, A., Williamson, S., Golinelli, D., McKenzie, T. L., Capone-Newton, P., & Lurie, N. (2008). Impact of a new bicycle path on physical activity. *Preventive Medicine, 46*, 80–81.
Cohen, D. A., Golinelli, D., Williamson, S., Sehgal, A., Marsh, T., & McKenzie, T. L. (2009a). Effects of park improvements on park use and physical activity: policy and programming implications. *Am J Prev Med, 37*, 475–480.
Cohen, D. A., Sehgal, A., Williamson, S., Marsh, T., Golinelli, D., & McKenzie, T. L. (2009b). New recreational facilities for the young and the old in Los Angeles: Policy and programming implications. *Journal of Public Health Policy, 30*(Suppl 1), S248–S263.
Creatore, M. I., Glazier, R. H., Moineddin, R., Fazli, G. S., Johns, A., Gozdyra, P., Matheson, F. I., Kaufman-Shriqui, V., Rosella, L. C., Manuel, D. G., & Booth, G. L. (2016). Association of neighborhood walkability with change in overweight, obesity, and diabetes. *JAMA, 315*(20), 2211.
Curtis, C., & Punter, J. (2004). Design led sustainable development: The liveable neighbourhood experiment. *The Town Planning Review, 75*, 31–65.
Delaney, P., & Pettit, C. J. (2014). Urban data hubs supporting smart cities. In S. Winter & C. Rizos (Eds.), *CEUR workshop proceedings, 07–09 April 2014* (pp. 13–25). Canberra: Research@ Locate'14.
Department of Infrastructure and Transport. (2011). *Our cities, our future. A national urban policy for a productive, sustainable and liveable future*. Canberra: Department of Infrastructure and Transport.
Ding, D., & Gebel, K. (2012). Built environment, physical activity, and obesity: What have we learned from reviewing the literature? *Health & Place, 18*, 100–105.
Ding, D., Lawson, K. D., Kolbe-Alexander, T. L., Finkelstein, E. A., Katzmarzyk, P. T., van Mechelen, W., & Pratt, M. (2016). The economic burden of physical inactivity: A global analysis of major non-communicable diseases. *The Lancet, 388*, 1311–1324.
Doorley, R., Pakrashi, V., & Ghosh, B. (2015). Quantifying the health impacts of active travel: Assessment of methodologies. *Transport Reviews, 35*, 559–582.
Drummond, M. F., Schulpher, M. J., Torrance, G. W., O'Brien, B. J., & Stoddart, G. L. (2005). *Methods for the economic evaluation of health care programmes*. Oxford: Oxford University Press.
Duany, A., Plater-Zyberk, E., & Speck, J. (2000). *Suburban nation: The rise of sprawl and the decline of the American dream*. New York: North Point Press.

Durand, C., Andalib, M., Dunton, G., Wolch, J., & Pentz, M. (2011). A systematic review of built environment factors related to physical activity and obesity risk: Implications for smart growth urban planning. *Obesity Reviews, 12*, e173–ee82.
Egan, M., Petticrew, M., Ogilvie, D., & Hamilton, V. (2003). New roads and human health: A systematic review. *American Journal of Public Health, 93*, 1463–1471.
Egan, M., Petticrew, M., Ogilvie, D., Hamilton, V., & Drever, F. (2007). "Profits before people"? A systematic review of the health and safety impacts of privatising public utilities and industries in developed countries. *Journal of Epidemiology and Community Health, 61*, 862–870.
Eom, H.-J., & Cho, G.-H. (2015). Exploring thresholds of built environment characteristics for walkable communities: Empirical evidence from the Seoul Metropolitan area. *Transportation Research Part D, 40*, 76–86.
Ewing, R., Schmid, T., Killingsworth, R., Zlot, A., & Raudenbush, S. (2003). Relationship between urban sprawl and physical activity, obesity and morbidity. *American Journal of Health Promotion, 18*, 47–57.
Foster, S., Hooper, P., Knuiman, M., Bull, F., & Giles-Corti, B. (2015). Are liveable neighbourhoods safer neighbourhoods? Testing the rhetoric on new urbanism and safety from crime in Perth, Western Australia. *Social Science & Medicine, 164*, 150–157.
Gatzweiler, F. W., Zhu, Y.-G., Roux, A. V. D., Capon, A., Donnelly, C., Salem, G., Ayad, H. M., Speizer, I., Nath, I., & Boufford, J. I. (2017). *Advancing health and wellbeing in the changing urban environment: Implementing a systems approach*. New York: Springer.
Gebel, K., Bauman, A. E., Sugiyama, T., & Owen, N. (2011). Mismatch between perceived and objectively assessed neighborhood walkability attributes: Prospective relationships with walking and weight gain. *Health & Place, 17*, 519–524.
Giles-Corti, B., Broomhall, M. H., Knuiman, M., Collins, C., Douglas, K., Ng, K., Lange, A., & Donovan, R. J. (2005a). Increasing walking: How important is distance to, attractiveness, and size of public open space? *American Journal of Preventive Medicine, 28*, 169–176.
Giles-Corti, B., Bull, F., Knuiman, M., McCormack, G., Van Niel, K., Timperio, A., Christian, H., Foster, S., Divitini, M., Middleton, N., & Boruff, B. (2013). The influence of urban design on neighbourhood walking following residential relocation: Longitudinal results from the RESIDE study. *Social Science & Medicine, 77*, 20–30.
Giles-Corti, B., & Donovan, R. J. (2002). The relative influence of individual, social and physical environmental determinants of physical activity. *Social Science and Medicine, 54*, 1793–1812.
Giles-Corti, B., & Donovan, R. J. (2003). Relative influences of individual, social environmental, and physical environmental correlates of walking. *American Journal of Public Health, 93*(9), 1583.
Giles-Corti, B., Hooper, P., Foster, S., Koohsari, M. J., & Francis, J. (2014). *Low density development: Impacts on physical activity and associated health outcomes*. Melbourne: Heart Foundation (Victorian Division).
Giles-Corti, B., Knuiman, M., Timperio, A., Van Niel, K., Pikora, T. J., Bull, F. C., Shilton, T., & Bulsara, M. (2008). Evaluation of the implementation of a state government community design policy aimed at increasing local walking: Design issues and baseline results from RESIDE, Perth Western Australia. *Preventive Medicine, 46*, 46–54.
Giles-Corti, B., Ryan, K., & Foster, S. (2012). *Increasing density in Australia: Maximising the benefits and minimising the harm*. Melbourne: National Heart Foundation of Australia.
Giles-Corti, B., Sallis, J. F., Sugiyama, T., Frank, L. D., Lowe, M., & Owen, N. (2015). Translating active living research into policy and practice: One important pathway to chronic disease prevention. *Journal of Public Health Policy, 36*, 231–243.
Giles-Corti, B., Timperio, A., Bull, F., & Pikora, T. (2005b). Understanding physical activity environmental correlates: increased specificity for ecological models. *Exercise and Sport Sciences Reviews, 33*, 175–181.
Giles-Corti, B., Vernez-Moudon, A., Reis, R., Turrell, G., Dannenberg, A. L., Badland, H., Foster, S., Lowe, M., Sallis, J. F., Stevenson, M., & Owen, N. (2016). City planning and population health: A global challenge. *Lancet, 388*, 2912–2924.

Goldstein, H. (2009). Translating research into public policy. *Journal of Public Health Policy, 30*, S16–S20.
Goodman, A., Sahlqvist, S., Ogilvie, D., & iConnect, c. (2013). Who uses new walking and cycling infrastructure and how? Longitudinal results from the UK iConnect study. *Preventive Medicine, 57*, 518–524.
Goodman, A., Sahlqvist, S., Ogilvie, D., & iConnect, C. (2014). New walking and cycling routes and increased physical activity: One- and 2-year findings from the UK iConnect Study. *American Journal of Public Health, 104*, e38–e46.
Goodspeed, R., Pelzer, P., & Pettit, C. (in press). Planning our future cities: The role computer technologies can play. In T. W. Sanchez (Ed.), *Urban planning knowledge and research*. London: Informa UK Limited.
Hajna, S., Ross, N. A., Brazeau, A. S., Belisle, P., Joseph, L., & Dasgupta, K. (2015). Associations between neighbourhood walkability and daily steps in adults: A systematic review and meta-analysis. *BMC Public Health, 15*, 768.
Heinen, E., Panter, J., Dalton, A., Jones, A., & Ogilvie, D. (2015a). Sociospatial patterning of the use of new transport infrastructure: Walking, cycling and bus travel on the Cambridgeshire guided busway. *Journal of Transport and Health, 2*, 199–211.
Heinen, E., Panter, J., Mackett, R., & Ogilvie, D. (2015b). Changes in mode of travel to work: A natural experimental study of new transport infrastructure. *International Journal of Behavioral Nutrition and Physical Activity, 12*, 81.
Hirsch, J. A., Diez Roux, A. V., Moore, K. A., Evenson, K. R., & Rodriguez, D. A. (2014a). Change in walking and body mass index following residential relocation: The multi-ethnic study of atherosclerosis. *American Journal of Public Health, 104*, e49–e56.
Hirsch, J. A., Moore, K. A., Barrientos-Gutierrez, T., Brines, S. J., Zagorski, M. A., Rodriguez, D. A., & Diez Roux, A. V. (2014b). Built environment change and change in BMI and waist circumference: Multi-ethnic study of atherosclerosis. *Obesity (Silver Spring), 22*, 2450–2457.
Hirsch, J. A., Moore, K. A., Clarke, P. J., Rodriguez, D. A., Evenson, K. R., Brines, S. J., Zagorski, M. A., & Diez Roux, A. V. (2014c). Changes in the built environment and changes in the amount of walking over time: Longitudinal results from the multi-ethnic study of atherosclerosis. *American Journal of Epidemiology, 180*, 799–809.
Hooper, P., Giles-Corti, B., & Knuiman, M. (2014). Evaluating the implementation and active living impacts of a state government planning policy designed to create walkable neighborhoods in Perth, Western Australia. *American Journal of Health Promotion, 28*, S5–S18.
Hooper, P., Knuiman, M., Foster, S., & Giles-Corti, B. (2015). The building blocks of a 'liveable neighbourhood': Identifying the key performance indicators for walking of an operational planning policy in Perth, Western Australia. *Health & Place, 36*, 173–183.
Hooper, P., Middleton, N., Knuiman, M., & Giles-Corti, B. (2013). Measurement error in studies of the built environment: Validating commercial data as objective measures of neighborhood destinations. *Journal of Physical Activity and Health, 10*, 792–804.
Hooper, P., et al. (n.d.) Living Liveable? RESIDE's evaluation of the liveable neighbourhoods planning policy on the health and wellbeing of residents (under review).
Humpel, N., Owen, N., & Leslie, E. (2002). Environmental factors associated with adults' participation in physical activity. *American Journal of Preventive Medicine, 22*, 188–199.
Jenks, R. H., & Malecki, J. M. (2004). GIS--a proven tool for public health analysis. *Journal of Environmental Health, 67*, 32–34.
Kaczynski, A., & Henderson, K. (2007). Environmental correlates of physical activity: A review of evidence about parks and recreation. *Leisure Sciences, 29*, 315–354.
Kaczynski, A. T., Besenyi, G. M., Stanis, S. A., Koohsari, M. J., Oestman, K. B., Bergstrom, R., Potwarka, L. R., & Reis, R. S. (2014). Are park proximity and park features related to park use and park-based physical activity among adults? Variations by multiple socio-demographic characteristics. *International Journal of Behavioral Nutrition and Physical Activity, 11*, 146.
Kahlmeier, S., Kelly, P., Foster, C., Gotschi, T., Cavill, N., Dinsdale, H., Woodcock, J., Schweizer, C., Rutter, H., Lieb, C., Oja, P., & Racioppi, F. (2014). *Health economic assessment tools*

(HEAT) for walking and for cycling. Methodology and user guide. Economic assessment of transport infrastructure and policies. 2014 update [Online]. Copenhagen: WHO Regional office for Europe. Retrieved Jul 7 2014, from, http://www.euro.who.int/__data/assets/pdf_file/0010/256168/ECONOMIC-ASSESSMENT-OF-TRANSPORT-INFRASTRUCTURE-AND-POLICIES.pdf.

Kamruzzaman, M., Baker, D., Washington, S., & Turrell, G. (2016). Determinants of residential dissonance: Implications for transit-oriented development in Brisbane. *International Journal of Sustainable Transportation, 10*, 960–974.

Karvonen, A., & Van Heur, B. (2014). Urban laboratories: Experiments in reworking cities. *International Journal of Urban and Regional Research, 38*, 379–392.

Kerr, J., Emond, J. A., Badland, H., Reis, R., Sarmiento, O., Carlson, J., Sallis, J. F., Cerin, E., Cain, K., Conway, T., Schofield, G., Macfarlane, D. J., Christiansen, L. B., Van Dyck, D., Davey, R., Aguinaga-Ontoso, I., Salvo, D., Sugiyama, T., Owen, N., Mitas, J., & Natarajan, L. (2016). Perceived neighborhood environmental attributes associated with walking and cycling for transport among adult residents of 17 cities in 12 countries: The IPEN study. *Environmental Health Perspectives, 124*, 290–298.

Kerr, J., Rosenberg, D., & Frank, L. (2012). The role of the built environment in healthy aging: Community design, physical activity, and health among older adults. *Journal of Planning Literature, 27*, 43–60.

Kerr, J., Sallis, J. F., Owen, N., De Bourdeaudhuij, I., Cerin, E., Sugiyama, T., Reis, R., Sarmiento, O., Fromel, K., Mitas, J., Troelsen, J., Christiansen, L. B., Macfarlane, D., Salvo, D., Schofield, G., Badland, H., Guillen-Grima, F., Aguinaga-Ontoso, I., Davey, R., Bauman, A., Saelens, B., Riddoch, C., Ainsworth, B., Pratt, M., Schmidt, T., Frank, L., Adams, M., Conway, T., Cain, K., Van Dyck, D., & Bracy, N. (2013). Advancing science and policy through a coordinated international study of physical activity and built environments: IPEN adult methods. *Journal of Physical Activity & Health, 10*, 581–601.

Kim, J. (2000). Creating communities: Does the Kentlands live up to its goals? (The promise of new urbanism). *Places, 13*, 48–55.

Knuiman, M. W., Christian, H. E., Divitini, M. L., Foster, S. A., Bull, F. C., Badland, H. M., & Giles-Corti, B. (2014). A longitudinal analysis of the influence of the neighborhood built environment on walking for transportation: The RESIDE study. *American Journal of Epidemiology, 180*, 453–461.

Koohsari, M. J., Badland, H., & Giles-Corti, B. (2013a). (Re)designing the built environment to support physical activity: Bringing public health back into urban design and planning. *Cities, 35*, 294–298.

Koohsari, M. J., Badland, H., Sugiyama, T., Mavoa, S., Christian, H., & Giles-Corti, B. (2015a). Mismatch between perceived and objectively measured land use mix and street connectivity: associations with neighborhood walking. *Journal of Urban Health, 92*, 242–252.

Koohsari, M. J., Karakiewicz, J. A., & Kaczynski, A. T. (2013b). Public open space and walking: The role of proximity, perceptual qualities of the surrounding built environment, and street configuration. *Environment and Behavior, 45*, 706–736.

Koohsari, M. J., Mavoa, S., Villanueva, K., Sugiyama, T., Badland, H., Kaczynski, A. T., Owen, N., & Giles-Corti, B. (2015b). Public open space, physical activity, urban design and public health: Concepts, methods and research agenda. *Health & Place, 33*, 75–82.

Krizek, K. J. (2003). Residential relocation and changes in urban travel – Does neighborhood-scale urban form matter? *Journal of the American Planning Association, 69*, 265–281.

Kyu, H. H., Bachman, V. F., Alexander, L. T., Mumford, J. E., Afshin, A., Estep, K., Veerman, J. L., Delwiche, K., Iannarone, M. L., Moyer, M. L., Cercy, K., Vos, T., Murray, C. J. L., & Forouzanfar, M. H. (2016). Physical activity and risk of breast cancer, colon cancer, diabetes, ischemic heart disease, and ischemic stroke events: systematic review and dose-response meta-analysis for the Global Burden of Disease Study 2013. *BMJ, 354*, i3857.

Learnihan, V., Van Niel, K. P., Giles-Corti, B., & Knuiman, M. (2011). Effect of scale on the links between walking and urban design. *Geographical Research, 49*, 183–191.

Lee, C., & Moudon, A. (2004). Physical activity and environment research in the health field: Implications for urban and transportation planning practice and research. *Journal of Planning Literature, 19*, 147–181.

Lich, K. H., Ginexi, E. M., Osgood, N. D., & Mabry, P. L. (2013). A call to address complexity in prevention science research. *Prevention Science, 14*, 279–289.

Lowe, M., Whitzman, C., Badland, H., Davern, M., Aye, L., Hes, D., Butterworth, I., & Giles-Corti, B. (2015). Planning healthy, liveable and sustainable cities: How can indicators inform policy? *Urban Policy and Research, 33*, 131–144.

Unit, M. C. (2010). *State of Australian cities, 2010*. Canberra: Major Cities Unit, Infrastructure Australia.

Martin, A., Panter, J., Suhrcke, M., & Ogilvie, D. (2015). Impact of changes in mode of travel to work on changes in body mass index: Evidence from the British Household Panel Survey. *Journal of Epidemiology and Community Health, 69*, 753–761.

McCormack, G., Giles-Corti, B., & Bulsara, M. (2008). The relationship between destination proximity, destination mix and physical activity behaviors. *Preventive Medicine, 46*, 33–40.

McCormack, G., & Shiell, A. (2011). In search of causality: A systematic review of the relationship between the built environment and physical activity among adults. *International Journal of Behavioral Nutrition and Physical Activity, 8*, 125.

McLeroy, K. R., Bibeau, D., Steckler, A., & Glanz, K. (1988). An ecological perspective on health promotion programs. *Health Education Quarterly, 15*, 351–377.

Merom, D., Bauman, A., Vita, P., & Close, G. (2003). An environmental intervention to promote walking and cycling – The impact of a newly constructed Rail Trail in Western Sydney. *Preventive Medicine, 36*, 235–242.

Mokhtarian, P., & Cao, X. (2008). Examining the impacts of residential self-selection on travel behavior: A focus on methodologies. *Transportation Research Part B, 42*, 204–228.

Mueller, N., Rojas-Rueda, D., Cole-Hunter, T., de Nazelle, A., Dons, E., Gerike, R., Götschi, T., Int Panis, L., Kahlmeier, S., & Nieuwenhuijsen, M. (2015). Health impact assessment of active transportation: A systematic review. *Preventive Medicine, 76*, 103–114.

Mulley, C., Tyson, R., McCue, P., Rissel, C., & Munro, C. (2013). Valuing active travel: Including the health benefits of sustainable transport in transportation appraisal frameworks. *Research in Transportation Business & Management, 7*, 27–34.

Murphy, M., Badland, H., Koohsari, M. J., Astell-Burt, T., Trapp, G., Villanueva, K., Mavoa, S., Davern, M., & Giles-Corti, B. (2017). Indicators of a health-promoting local food environment: a conceptual framework to inform urban planning policy and practice. *Health promotion journal of Australia, 28*, 82.

Newell, B., & Proust, K. (2009). *I see how you think: using influence diagrams to support dialogue*. Canberra: Australian National University (ANU) Centre for Dialogue.

Newell, B., Proust, K., Dyball, R., & McManus, P. (2008). Seeing obesity as a systems problem. *New South Wales Public Health Bulletin, 18*, 214–218.

Ogilvie, D., Bull, F., Cooper, A., Rutter, H., Adams, E., Brand, C., Ghali, K., Jones, T., Mutrie, N., Powell, J., Preston, J., Sahlqvist, S., Song, Y., & iConnect, C. (2012). Evaluating the travel, physical activity and carbon impacts of a 'natural experiment' in the provision of new walking and cycling infrastructure: Methods for the core module of the iConnect study. *BMJ Open, 2*, e000694.

Ogilvie, D., Bull, F., Powell, J., Cooper, A. R., Brand, C., Mutrie, N., Preston, J., Rutter, H., & iConnect, C. (2011). An applied ecological framework for evaluating infrastructure to promote walking and cycling: the iConnect study. *American Journal of Public Health, 101*, 473–481.

Ogilvie, D., Foster, C., Rothnie, H., Cavill, N., Hamilton, V., Fitzsimons, C., & Mutrie, N. (2007). Interventions to promote walking: Systematic review. *BMJ, 334*, 1204–1213.

Oliver, K., Innvar, S., Lorenc, T., Woodman, J., & Thomas, J. (2014). A systematic review of barriers to and facilitators of the use of evidence by policymakers. *BMC Health Services Research, 14*, 2.

Orton, L., Lloyd-Williams, F., Taylor-Robinson, D., O'Flaherty, M., & Capewell, S. (2011). The use of research evidence in public health decision making processes: systematic review. *PLoS One, 6*, e21704.
Panter, J. R., Heinen, E., Mackett, R., & Ogilvie, D. (2016). Impact of new transport infrastructure on walking, cycling, and physical activity. *American Journal of Preventive Medicine, 50*, e45–e53.
Panter, J. R., & Jones, A. (2010). Attitudes and the environment as determinants of active travel in adults: What do and don't we know? *Journal of Physical Activities and Health, 7*, 551–561.
Panter, J. R., Jones, A. P., & van Sluijs, E. M. (2008). Environmental determinants of active travel in youth: A review and framework for future research. *International Journal of Behavioral Nutrition and Physical Activity, 5*, 34.
Pearce, J. R., & Maddison, R. (2011). Do enhancements to the urban built environment improve physical activity levels among socially disadvantaged populations? *International Journal for Equity in Health, 10*, 28.
Pelzer, P., Arciniegas, G., Geertman, S., & de Kroes, J. (2013). *Using MapTable® to learn about sustainable urban development. Planning support systems for sustainable urban development.* New York: Springer.
Petticrew, M., Cummins, S., Ferrell, C., Findlay, A., Higgins, C., Hoyd, C., Kearns, A., & Spark, L. (2005). Natural experiments: An underused tool for public health? *Public Health, 119*, 751–757.
Pettit, C. J., Barton, J., Goldie, X., Sinnott, R., Stimson, R., & Kvan, T. (2015). The Australian urban intelligence network supporting smart cities. In G. SSJ, J. Ferreira, & J. Goodspeed (Eds.), *Planning support systems and smart cities, lecture notes in geoinformation and cartography*. Cham: Springer.
Pettit, C. J., Tice, A., & Randolph, B. (2017). Using an online spatial analytics workbench for understanding housing affordability in Sydney. Seeing cities through big data: Research, methods and applications. In P. Thakuriah, N. Tilahun, & M. Zellner (Eds.), *Urban informatics*. Cham: Springer International Publishing.
Reis, R. S., Salvo, D., Ogilvie, D., Lambert, E. V., Goenka, S., Brownson, R. C., & Lancet Physical Activity Series 2 Executive, C. (2016). Scaling up physical activity interventions worldwide: stepping up to larger and smarter approaches to get people moving. *Lancet, 388*, 1337–1348.
Renalds, A., Smith, T., & Hale, P. (2010). A systematic review of built environment and health. *Fam Comm Health, 33*, 68–78.
Ross, S. M. (2014). *Introduction to probability models*. New York: Academic.
Saelens, B., & Handy, S. (2008). Built environment correlates of walking: A review. *Medicine & Science in Sports & Exercise, 40*, S550–S566.
Sahlqvist, S., Goodman, A., Cooper, A. R., Ogilvie, D., & iConnect, c. (2013). Change in active travel and changes in recreational and total physical activity in adults: Longitudinal findings from the iConnect study. *International Journal of Behavioral Nutrition and Physical Activity, 10*, 28.
Sahlqvist, S., Goodman, A., Jones, T., Powell, J., Song, Y., Ogilvie, D., & iConnect, c. (2015). Mechanisms underpinning use of new walking and cycling infrastructure in different contexts: mixed-method analysis. *International Journal of Behavioral Nutrition and Physical Activity, 12*, 24.
Sallis, J. F. (2008). Angels in the details: comment on "The relationship between destination proximity, destination mix and physical activity behaviors". *Preventive Medicine, 46*, 6–7.
Sallis, J. F. (2009). Measuring physical activity environments: A brief history. *American Journal of Preventive Medicine, 36*, S86–S92.
Sallis, J. F., Bull, F., Burdett, R., Frank, L., Griffiths, P., Giles-Corti, B., & Stevenson, M. (2016a). Using science to guide city planning policy and practice to promote health. *Lancet, 388*(10062), 2936–2947.

Sallis, J. F., Cerin, E., Conway, T. L., Adams, M. A., Frank, L. D., Pratt, M., Salvo, D., Schipperijn, J., Smith, G., Cain, K. L., Davey, R., Kerr, J., Lai, P. C., Mitas, J., Reis, R., Sarmiento, O. L., Schofield, G., Troelsen, J., Van Dyck, D., De Bourdeaudhuij, I., & Owen, N. (2016b). Physical activity in relation to urban environments in 14 cities worldwide: a cross-sectional study. *Lancet, 387*(10034), 2207–2217.

Sallis, J. F., Cervero, R. B., Ascher, W., Henderson, K. A., Kraft, M. K., & Kerr, J. (2006a). An ecological approach to creating active living communities. *Annual Review of Public Health, 27*, 297–322.

Sallis, J. F., Cervero, R. B., Ascher, W., Henderson, K. A., Kraft, M. K., & Kerr, J. (2006b). An ecological approach to creating active living communities. *Annual Review of Public Health, 27*, 297–322.

Sallis, J. F., Floyd, M. F., Rodriguez, D. A., & Saelens, B. E. (2012). Role of built environments in physical activity, obesity, and cardiovascular disease. *Circulation, 125*, 729–737.

Stephenson, J., & Bauman, A. (2000). *The cost of illness attributable to physical inactivity in Australia*. Canberra: CDHAC and Australian Sports Commission.

Story, M., Giles-Corti, B., Yaroch, A., Cummins, S., Frank, L., Huang, T. T.-K., & Lewis, L. (2009). Work group IV: Future directions for measures of the food and physical activity environments. *American Journal of Preventive Medicine, 36*, S182–S188.

Sugiyama, T., Francis, J., Middleton, N. J., Owen, N., & Giles-Corti, B. (2010). Associations between recreational walking and attractiveness, size, and proximity of neighborhood open spaces. *American Journal of Public Health, 100*, 1752–1757.

Sugiyama, T., Gunn, L. D., Christian, H., Francis, J., Foster, S., Hooper, P., Owen, N., & Giles-Corti, B. (2015). Quality of public open spaces and recreational walking. *American Journal of Public Health, 105*, 2490–2495.

Sugiyama, T., Neuhaus, M., Cole, R., Giles-Corti, B., & Owen, N. (2012). Destination and route attributes associated with adults' walking: A review. *Medicine and Science in Sports and Exercise, 44*, 1275–1286.

Talen, E. (1996). Do plans get implemented? A review of evaluation in planning. *Journal of Planning Literature, 10*, 248–259.

Taylor, B. T., Fernando, P., Bauman, A. E., Williamson, A., Craig, J. C., & Redman, S. (2011). Measuring the quality of public open space using Google Earth. *American Journal of Preventive Medicine, 40*, 105–112.

Board, T. R. (2005). *Does the built environment influence physical activity? Examining the evidence*. Washington: TRB.

United Nations. (2011). *Sixty seventh session political declaration of the high level meeting of the general assembly on the prevention and control of non-communicable diseases [online]*. New York: United Nations. Retrieved from http://www.un.org/ga/search/view_doc.asp?symbol=A/66/L.1.

United Nations General Assembly. (2015). *Resolution adopted by the general assembly: Transforming our world: the 2030 agenda for sustainable development A/RES/70/1*. New York: United Nations.

Urban, J. B., Osgood, N. D., & Mabry, P. L. (2011). Developmental systems science: Exploring the application of systems science methods to developmental science questions. *Research in Human Development, 8*, 1–25.

Van Cauwenberg, J., De Bourdeaudhuij, I., De Meester, F., Van Dyck, D., Salmon, J., Clarys, P., & Deforche, B. (2011). Relationship between the physical environment and physical activity in older adults: A systematic review. *Health & Place, 17*, 458–469.

Van Dyck, D., Cerin, E., Conway, T., De Bourdeaudhuij, I., Owen, N., Kerr, J., Cardon, G., Frank, L. D., Saelens, B., & Sallis, J. (2012). Associations between perceived neighborhood environmental attributes and adults' sedentary behavior: Findings from the USA, Australia and Belgium (Report). *Social Science and Medicine, 74*, 1375.

Van Dyck, D., Cerin, E., Conway, T., De Bourdeaudhuij, I., Owen, N., Kerr, J., Cardon, G., Frank, L. D., Saelens, B., & Sallis, J. (2013). Perceived neighborhood environmental attributes associated with adults' leisure-time physical activity: Findings from Belgium, Australia and the USA (Report). *Health and Place, 19*, 59.

Van Dyck, D., Cerin, E., Conway, T. L., De Bourdeaudhuij, I., Owen, N., Kerr, J., Cardon, G., & Sallis, J. F. (2014). Interacting psychosocial and environmental correlates of leisure-time physical activity: A three-country study. *Health Psychology, 33*, 699–709.

Van Holle, V., Deforche, B., Van Cauwenberg, J., Goubert, L., Maes, L., Van de Weghe, N., & De Bourdeaudhuij, I. (2012). Relationship between the physical environment and different domains of physical activity in European adults: a systematic review. *BMC Public Health, 12*, 807.

Veitch, J., Salmon, J., Carver, A., Timperio, A., Crawford, D., Fletcher, E., & Giles-Corti, B. (2014). A natural experiment to examine the impact of park renewal on park-use and park-based physical activity in a disadvantaged neighbourhood: The REVAMP study methods. *BMC Public Health, 14*, 600.

Veitch, J., Salmon, J., Giles-Corti, B., Crawford, D., Dullaghan, K., Carver, A., & Timperio, A. (2017). Challenges in conducting natural experiments in parks-lessons from the REVAMP study. *International Journal of Behavioral Nutrition and Physical Activity, 14*, 5.

Von Bertalanffy, L. (1969). *General system theory: Foundations, development, applications (revised edition)*. Westminster: Penguin Books.

Vos, T., Carter, R., Barendregt, J., Mihalopoulos, C., Veerman, J., Magnus, A., Cobiac, L., M Bertram & Wallace A. 2010. Retrieved Feb 1 2017, from, https://public-health.uq.edu.au/files/571/ACE-Prevention_final_report.pdf.

Walker, D., & Daniels, T. L. (2011). *The planners guide to community viz: The essential tool for a new generation of planning*. Chicago: Planners Press, American Planning Association.

Watts, N., Adger, W. N., Agnolucci, P., Blackstock, J., Byass, P., Cai, W., Chaytor, S., Colbourn, T., Collins, M., Cooper, A., Cox, P. M., Depledge, J., Drummond, P., Ekins, P., Galaz, V., Grace, D., Graham, H., Grubb, M., Haines, A., Hamilton, I., Hunter, A., Jiang, X., Li, M., Kelman, I., Liang, L., Lott, M., Lowe, R., Luo, Y., Mace, G., Maslin, M., Nilsson, M., Oreszczyn, T., Pye, S., Quinn, T., Svensdotter, M., Venevsky, S., Warner, K., Xu, B., Yang, J., Yin, Y., Yu, C., Zhang, Q., Gong, P., Montgomery, H., & Costello, A. (2015). Health and climate change: policy responses to protect public health. *Lancet, 386*(10006), 1861–1914.

Wells, N. M., & Yang, Y. (2008). Neighborhood design and walking. A quasi-experimental longitudinal study. *American Journal of Preventive Medicine, 34*, 313–319.

Western Australian Planning Commission. (1997). *Liveable neighbourhoods. Community design code: A Western Australian government sustainable cities initiative* (1st ed.). Perth: Western Australian Planning Commission.

Western Australian Planning Commission. (2000). *Liveable neighbourhoods: A Western Australian government sustainable cities initiative*. Perth: Western Australian Planning Commission.

Western Australian Planning Commission. (2007). *Liveable neighbourhoods : a Western Australian Government Sustainable Cities Initiative*. Perth: Western Australian Planning Commission.

Whitzman, C. (2007). Barriers to planning healthier cities in Victoria. *International Journal of Environmental, Cultural, Economic and Social Sustainability, 3*, 146–153.

WHO & UN Habitat. (2016). *Global report on urban health: equitable healthier cities for sustainable development, Italy*. Geneva: WHO.

Wood, S. (2009). *Generalized additive modles: An introduction using R*. Boca Raton: Chapman & Hall.

World Health Organization. (2016a). *Shanghai consensus on healthy cities. WHO 9th global conference on health promotion, healthy cities mayor forum, 2016a Shanghai*. Geneva: WHO.

World Health Organization. (2016b). *Shanghai Declaration on promoting health in the 2030 Agenda for sustainable development. WHO 9th global conference on health promotion. Shanghai*. Geneva: WHO.

Chapter 19
Urban Form and Road Safety: Public and Active Transport Enable High Levels of Road Safety

Paul Schepers, Gord Lovegrove, and Marco Helbich

19.1 Introduction

Road traffic injuries are estimated to be the ninth cause of death with over 1.2 million deaths each year globally (WHO 2015). Over half of the health burden of traffic crashes can be attributed to deaths; the other part results from disability following injuries (Dhondt et al. 2013). Injuries sustained in road crashes are the main causes of death among those aged 15–29 years (WHO 2015). Moreover, some children involved in a traffic crash experience post-traumatic stress disorder, and the death of a child is associated with increased mortality from both natural and unnatural causes in parents (Li et al. 2003; Salter and Stallard 2004). Action is needed to reduce this massive and largely preventable human toll. Investing in road safety would also lower perceived risks that may deter potential cyclists and pedestrians and thereby prevent them from enjoying the associated health benefits of physical activity (Götschi et al. 2016). And it almost goes without saying that there would be significant economic benefits associated with reduced traffic fatalities and improved public health; the current economic burden of traffic crashes is approximately 3% of gross domestic product (GDP) worldwide (WHO 2015). Clearly there is a social and economic imperative to take action in every community to address this road safety problem.

More than half a century ago, Perry (1939) and Buchanan (1963) already suggested that road safety could be improved via urban form, also known as the built environment of our community development patterns; however, the lure and benefits of private auto use historically obscured any social and economic externalities

P. Schepers (✉) · M. Helbich
Faculty of Geosciences, Department of Human Geography and Spatial Planning, Utrecht University, Utrecht, The Netherlands
e-mail: paul.schepers@rws.nl

G. Lovegrove
Faculty of Applied Science, School of Engineering, Sustainable Transport Safety Research Laboratory, University of British Columbia, Kelowna, BC, Canada

M. Nieuwenhuijsen, H. Khreis (eds.), *Integrating Human Health into Urban and Transport Planning*, https://doi.org/10.1007/978-3-319-74983-9_19

caused by traffic crashes. More recently, amid Brundtland Commission (1987), WHO (2004) and other increasingly vocal global health authorities, the need to reconsider and reduce these negative impacts of auto-oriented communities has again been raised. The problem is of course that in the intervening decades since Buchanan (1963) first warned us, much of our urban form, at least in higher-income countries, has already been built in an auto-oriented development pattern. Once built, these sprawling, low-density and freeway-accessed suburbs are not easily reversed. For example, a single intersection or roadway can be retrofitted within a short time for traffic calming, for instance, converted into a safer roundabout and/or 1-way, narrower, lower-speed couplets; however, it takes much longer and costs prohibitively more (not to mention the social upheaval involved) to overnight retrofit the built form of an entire auto-oriented community. It may take perhaps as long as an entire generation to make any substantive changes! Fortunately, there are some general principles that, if employed successfully, may help to address our current built urban form and road safety problems.

Urban form mostly affects road safety indirectly via travel behaviour such as modal choice and route choice (Van Wee 2009; Schepers et al. 2014). Understanding the indirect relationship between road safety and urban form is not only important for traffic safety but also for the health impact of traffic in general. While the optimization of intersection design may have a largely isolated road safety effect, changing urban design and networks affects other externalities of traffic. For example, suppose we build a network of well-designed bicycle infrastructure to travel to a city centre. This would encourage cycling which may affect road safety, but it will also reduce air and noise pollution by cars and increase physical activity (Nieuwenhuijsen and Khreis 2016). A bicycle path physically separated from the vehicle carriageway would increase separation between cars and cyclists, and reduce cyclists' exposure to noise, buffeting, and exhaust (Schepers et al. 2015b). In this chapter we discuss how urban form affects road safety to inform integrated cross-disciplinary planning efforts and health impact assessments, in the hopes that application of this knowledge and its associated planning principles will help to improve the overall health outcomes of traffic (Khreis et al. 2016; van Wee and Ettema 2016).

19.1.1 The Impact of Urban Form on Travel Behaviour

It is well established that travel behaviour is affected by urban design (Ewing and Cervero 2010). As urban form encompasses a myriad of components, the impact is often described using the '5Ds': land use density and diversity, design (of networks and street patterns), distance to transit, and destination accessibility (Cervero and Kockelman 1997; Ewing and Cervero 2010; Handy 2017). Lower densities (e.g. a sprawling city) and monofunctional land use increase travel distance and thereby travel time to potential destinations and vice versa (depending on the available mode network continuity and infrastructure systems). Unless high-quality public transit is heavily subsidized, the private car is the only economically viable form of transport to travel longer distance criss-cross trips in low-density suburban sprawl communities within travel time budgets typically available (Mokhtarian and Chen 2004). Apart from typical geographic considerations (e.g. terrain, climate), increased travel distance is a daunting factor for cyclists and even more for pedestrians (Heinen et al. 2010). On the other hand, more sustainable forms of development, such as higher densities and land use diversity, decrease trip length and thereby contribute to more active transport.

As walking (and to a lesser extent cycling) is an important access mode for public transport, the modal shares of active transport modes tend to be correlated (Pucher 2004). Mavoa et al. (2012) found that, with proper community planning and design, walking accessibility to destinations via public transit can be significant. So-called transit-oriented development (TOD) aims for high densities and mixed (diverse) land use within 5–10 min walking distance of key public transport stations (Pucher 2004; Bach et al. 2006). Not surprisingly, cities everywhere with high population densities have a substantial model share of public transport, e.g. the share excluding walking amounts to 81% in Hong Kong and 64% in Paris (Sun et al. 2014).

Urban design also must consider network design and street network structure (Handy 2017) and their influence on travel time via resistance along their links. High-end mobility facilities such as freeways, arterial roads, bus rapid transit (BRT), and heavy rail systems allow for high capacity, speed, and performance resulting from low intersection/stop densities over longer distances within the same travel time budget. Network design can also be used to influence the competitiveness of each mode compared to others in terms of travel distance and time, e.g. shortcuts for walking and cycling where drivers have to make detours (Rietveld and Daniel 2004; Frank and Hawkins 2008; Levinson 2012; Schepers et al. 2013).

Individual travel behaviour results in traffic volumes, a modal split, and distribution of traffic over time and space (Van Wee 2009). The 5Ds help describe and predict to which transport mode(s) an urban area is most oriented.

19.1.2 Indicators to Study the Impact of Urban Form on Road Safety

Which indicator is best used for a study of road safety depends on the objective of the comparison (i.e. the policy issue) at hand (Hakkert and Braimaister 2002; Götschi et al. 2016). At an aggregated level, numbers of fatalities per capita, per-total road lane kilometres (TLKM), or per-vehicle kilometre travelled (VKT) are common indicators for road safety (Lovegrove and Sayed 2006). Fatalities per TLKM or per VKT can be misleading as these exposure measures in the denominator (of the risk indicator) can change, without any real safety improvements. For example, if fatal collision counts remained constant while more roads and/or more vehicle kilometres were driven, it would reduce fatalities per VKT and per TLKM. And if fatalities rose at the same rate as exposure, it would appear that the level of safety was unchanged when in reality fatalities were rising. Perhaps more alarming, if fatalities remained constant while vehicle-kilometres decreased, for example, due to transit-oriented developments and reductions in driving, it would actually appear as if safety were worsening, as the ratio of fatalities per VKT would increase! Clearly, fatalities per VKT and per TLKM could mask, or completely discount, the potential safety and sustainability benefits of more dense transit- and bicycle-oriented developments that reduce VKT, serious collisions, pollution emissions, energy consumption, and space needed for roads. Moreover, planning for cycling and walking that relies on off-road paths and/or car-free exclusive networks would offer comparable or even higher levels of human scale and safer accessibility using less vehicle kilometres (Cervero and Day 2008; Karou and Hull 2014).

So what is the fairest way to measure road safety that also considers the benefits of active transport at an aggregated, area-wide, and macro-level? Macro-level, collision prediction models (CPMs) that evaluate road safety in terms of vehicle-related collisions per unit time period are dominated by a non-linear, exponential association with these same exposure-based independent variables VKT and to a lesser extent TLKM (Lovegrove and Sayed 2006; Lovegrove 2007). These widely accepted definitions of traffic crashes focus on (motor) vehicle crashes occurring on public roads but exclude those crashes where vehicles have not been involved, i.e. pedestrian falls (SWOV 2010b). Methorst et al. (2017) have recently criticized this exclusion. They estimated that 86 pedestrians died on public roads in the Netherlands in 2011 due to falls where no vehicles were involved, which is significant when compared to the 520 police-recorded vehicle crash-related deaths (SWOV 2017). However, to date, these CPMs have only been able to be reliably fit to, and be used to predict, data related to collisions involving vehicles, for two reasons. First, data on pedestrian falls and, to a lesser extent, single-bicycle crashes is scarce. Second, exposure measures for these non-vehicular modes differ from traditional VKT and TLKM and have been difficult to predict and assess (Elvik 2009; Wei and Lovegrove 2012). Thus, other measures are needed to evaluate the road safety benefits of increased walking and bicycling.

One possibility is to recall that average daily travel time remains constant, with many researchers observing that, regardless of travel mode, humans often travel a minimum of 10–20 min to reach desired destinations (Mokhtarian and Chen 2004). And while urban sprawl and more road mileage and more driving within this time budget have created additional exposures to risk (e.g. VKT), on the other hand, increased population density and land use diversity and quality transit service can also improve the walking and cycling accessibility of important activities (e.g. parks), facilities (e.g. jobs, schools), and services (shopping, doctors) within this same time budget. As such, a more inclusive measure of road safety that accounts for the human scale walking and bicycling modes, at an aggregated level, would be deaths per 100,000 population and is the main indicator used to study the impact of urban form on road safety in this context.

19.2 Modal Split and Developments Oriented Towards Specific Transport Modes

To illustrate road safety for the aforementioned transport mode orientations, we collected data on a sample of large, high-income cities from developed countries around the world (see Appendix for an explanation of data methodology). Figure 19.1 and Table 19.1 show that the number of fatalities per 100,000 population in large cities is much lower than the overall rate of the country in which they are situated. This result and an association between density and road safety were also found in other studies (Ewing and Dumbaugh 2009; Houwing et al. 2012; Kegler et al. 2012; Ryb et al. 2012), suggesting that as density increases, so does safety. Moreover, we observe that safer cities tend to be located in safer countries and vice versa, that is, countries with fewer overall fatal crashes per 100,000 population tend to see that same trend in their cities. For comparison, we have shown the middle-sized Dutch new towns including Almere, Houten, Lelystad, Nieuwegein, and Zoetermeer; these are discussed in more detail in Sect. 19.3.4. Transit-oriented and bicycle-oriented cities in Table 19.1 seem to have reached comparable levels of road safety, whereas road safety in car-oriented cities seems worst. Drawing conclusions from this small sample of cities should be done with some caution, but larger studies also found lower death rates in cities with higher modal shares of public transport and cycling (Marshall and Garrick 2011; Litman 2012; American Public Transportation Association 2016).

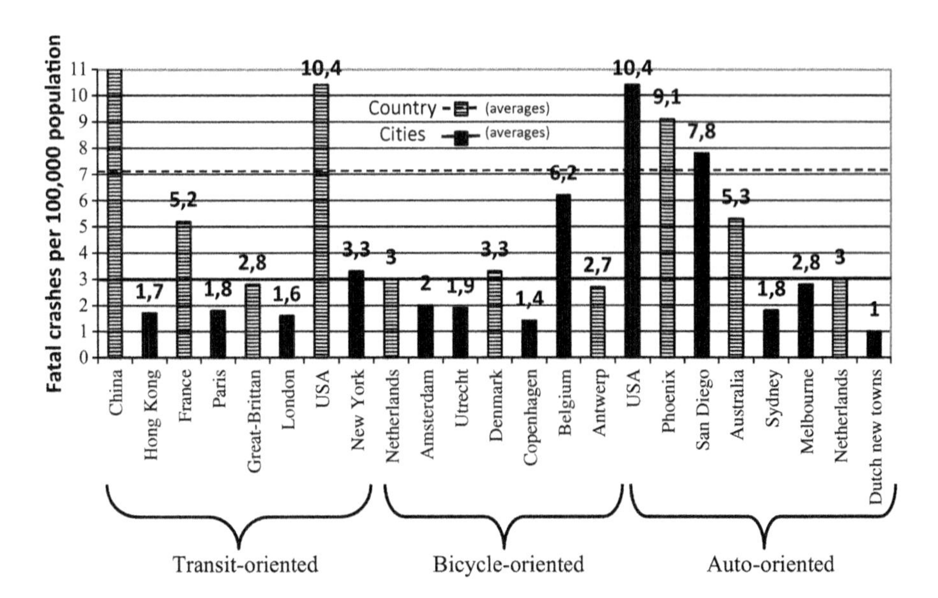

Fig. 19.1 Recorded deaths per 100,000 population in 2011–2015 in large, high-income cities, and, for comparison, in Dutch middle-sized, new towns (far right)

Table 19.1 Modal share, population statistics, and death rate in large, high-income cities and in middle-sized, Dutch new towns

Jurisdiction	Modal shares				Pop. (×1000)	Density per km² (×1000)	Fatalities per 100,000 pop'n
	Walk, %	Bike, %	Transit, %	Auto, %			
Public transport (transit)-oriented cities							
Hong Kong	NA	1	**81**	18	7167	6.5	1.7
Paris	15	5	**59**	20	2241	21.3	1.8
London	21	2	**44**	34	8428	5.5	1.6
New York	10	1	**55**	29	8550	10.9	3.7
Bicycle-oriented cities							
Amsterdam	4	**40**	29	27	838	5.0	2.0
Utrecht	3	**34**	24	39	334	3.6	1.9
Copenhagen	10	**30**	36	26	565	7.0	1.4
Antwerp	20	**24**	17	40	510	2.5	2.7
Auto-oriented cities							
Phoenix	2	1	3	**88**	1512	1.1	8.5
San Diego	3	1	4	**85**	1322	1.5	5.3
Melbourne	4	2	14	**80**	4318	0.4	2.7
Sydney	5	1	21	**74**	4738	0.4	1.8
Dutch new towns					502	1.9	1.0
Almere		**31**			194		
Houten		**37**			48		
Lelystad		**25**			76		
Nieuwegein		**27**			61		
Zoetermeer		**27**			123		
Netherlands	19	**27**	5	49	16,779	0.4	3.0

19.2.1 Cycling Versus Driving

Elvik (2009) was one of the first to study the road safety effects of shifts from driving to cycling and walking using CPMs, in which a non-linear relationship between crashes and volumes is modelled. His results revealed that cycling, walking, and driving tend to become safer as volumes increase, which is known in the literature as Smeed's law on motor vehicle safety, and as the 'safety in numbers' effect in the literature on cyclist and pedestrian safety (Smeed 1949; Jacobsen 2003; Elvik and Bjørnskau 2017). Other benefits of increased cycling volumes may include driver-related factors. That is, as these same individuals who now cycle more drive a car past other cyclists, they likely become more sensitive to, and aware of the needs for, safe passing speed and separation distance, which would reinforce the 'safety in numbers' effect (Schepers et al. 2017b).

Elvik also observed that as the non-linearity of risk ('safety in numbers') applies to volumes of both cyclists and drivers, more cycling does, without changes to circumstances such as infrastructure, not automatically reducing the total number of

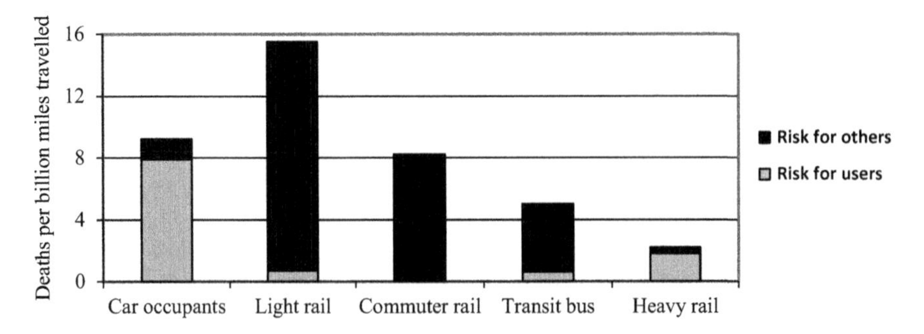

Fig. 19.2 Crash fatality rates per billion passenger miles travelled Litman (2012)

traffic casualties, (Elvik 2013). The CPMs account for the fact that after shifting from car driving to cycling (or walking), road users are less hazardous to other cyclists and pedestrians because of, among other things, the lower speeds and hence lower amounts of kinetic energy released in the event of a crash leading to less severe injuries. Other studies found similar results and also distinguished between age groups (Stipdonk and Reurings 2012). For instance, Schepers and Heinen (2013) expected the number of fatalities above the age of 65 to increase, while the number in the 18–64 age group was expected to decrease.

Therefore, the research suggests that more than just a mode shift is required, since a higher cycling volumes appear to have only a modest impact on the total number of traffic deaths.

19.2.2 Public Transport Versus Driving

While public transport is much safer than driving for transit passengers, it can be relatively risky for other road users without proper system design, as shown in Fig. 19.2 (SWOV 2011b). CPMs can integrate such risks but have not yet been widely applied to estimate the road safety impact of a shift from using private autos to using public transport. Lovegrove et al. (2010) predicted significant 2% safety benefits of transit investment versus 'business-as-usual' (BAU) autocentric approaches using an application of community-level, macro-level CPMs, as part of an evaluation of both urban and rural areas of the Greater Vancouver Regional District's 3-Year Regional Transportation Strategy. Using similar methodology and macro-level CPMs, Alam (2010) predicted an 8% decrease in collision severity due to improving transit in Kelowna, BC, Canada, versus a BAU approach.

It bears repeating that the aforementioned CPMs used autocentric exposure and road safety measures, which by definition are biased and ignore active transport oriented measures, thus discounting overall levels of safety. For example, fatal pedestrian falls are not included in international traffic crash statistics (Methorst et al. 2017). Elvik et al. (2009) studied the modal shift from driving to public

Table 19.2 Road classification and speed limits in the Netherlands in urban areas

Road classes	Speed limits in urban areas, km/h	Location of cyclists
Access roads	30	Mixed with other traffic
Distributor roads	50 or 70	Separated from motorized traffic by bicycle paths or bicycle lanes[a]
Through roads	100 or 120	Cycling not allowed

[a]Bike paths are physically separated from the carriageway; bike lanes are a delineated space on the carriageway

transport on existing infrastructure (i.e. no changes to circumstances), including the potential impact on the number of injuries due to pedestrian falls and concluded 'the unrecorded injuries from falls will, however, increase so much that no overall gain in safety can be expected if car users start using buses or trains.' As the elderly are particularly at risk of severe pedestrian falls (Schepers et al. 2017a), it can be expected that the benefits of a modal shift to transit found by Alam (2010) also apply to the overall level of transport safety for younger age groups. However, more than improved public transit service alone is needed to improve transport safety for all age groups; a system-based approach to transport safety is needed.

19.3 Network Design and Distribution of Traffic Along Infrastructure Networks

19.3.1 Network Design and the Safe System Approach

Organizations such as OECD and the World Road Association endorse the Safe System approach for road safety policies (World Road Association 2014). It recognizes that the network must eventually be forgiving of routine human (road user) errors and must maintain navigation tasks of all components of a system design at a level that will not exceed the user's mental and physical capacity. Examples of Safe System approaches are the Swedish Vision Zero and Dutch Sustainable Safety Vision (Koornstra et al. 2002). Two key principles of the Safe System approach for network design are homogeneity and functionality (World Road Association 2014). Homogeneity implies that differences in speed, direction, and mass should not be too large, for instance, a safe speed to mix cyclists with motorized traffic is no higher than 30 km/h (Tingvall and Haworth 1999). Functionality refers to classification of roads in a hierarchical road network and aims for roads to have only one function, i.e. a flow function or access function. The Netherlands has adopted a hierarchical road classification by which roads are classified as either access, distributor, or through roads.

Table 19.2 lists the recommended maximum speed limits and cyclist location on these three classes of road. Access roads are within 30 km/h zones only meant for direct access to homes and local destinations. Through traffic should avoid 30 km/h zones. Weijermars and Wegman (2011) indicate that categorization of the road

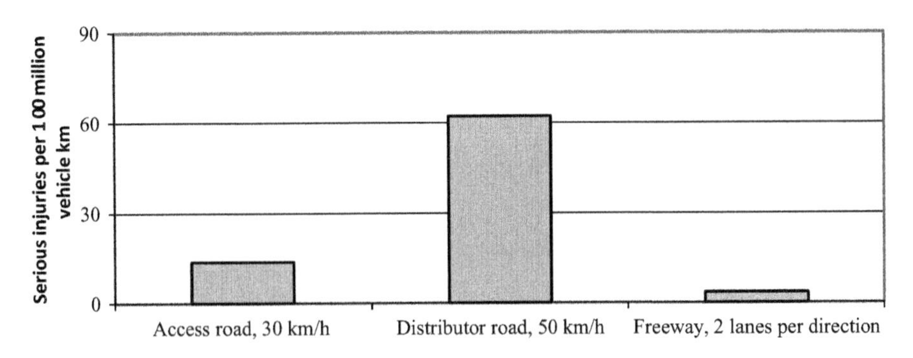

Fig. 19.3 Serious injuries per 100 million motor vehicle kilometres in 2007–2009 in urban areas (Witteveen+Bos 2015)

network and traffic calming measures such as the construction of 30-km/h zones was the most important measure to implement the Dutch sustainable safety approach, accounting for a significant share of the 60% drop in fatalities since the 1990s.

Due to their mixed function character, distributor roads have higher risks than through and access roads; hence, the Dutch safe system approach recommends that the portion of a trip on distributor roads be as low as possible (Dijkstra 2013). Figure 19.3 depicts risk observed on urban roads and freeways in the Netherlands. Other researchers have also found negative road safety effects for an increased length of distributor roads, number of lanes of major roads, and lane-kilometres (Dumbaugh and Rae 2009; Marshall and Garrick 2011). More recently, other researchers have sought to assess the risk of these distributor roads on active transport and found that they are indeed more hazardous for pedestrians and cyclists, and that the likelihood of collisions increases with the number of lanes and road width (Miranda-Moreno et al. 2011; Schepers et al. 2011; Ukkusuri et al. 2012).

19.3.2 Lower Levels of the Hierarchy

Elvik (2001) describes area-wide urban traffic calming as the implementation of a hierarchical road system comprised of physical restrictions to reduce non-local, through-traffic volumes and speeds on residential (access) streets. While this is done in response to resident complaints, a truly effective system solution also requires improvement of distributor and through roads to carry a larger traffic volume without additional delays or more accidents. In a meta-analysis, Elvik (2001) found a collision reduction of 25% for traffic-calmed residential access roads and a 10% reduction for distributor roads, consistent with later research by Lovegrove and Sayed (2006) that found closer to a 30% or 50% safety benefit when done as part of a neighbourhood-wide, systematic road network redesign (Lovegrove and Sayed 2006).

Several studies have analysed the neighbourhood road network patterns depicted in Fig. 19.4a, b using community-based, macro-level CPMs (Lovegrove and Sayed 2006;

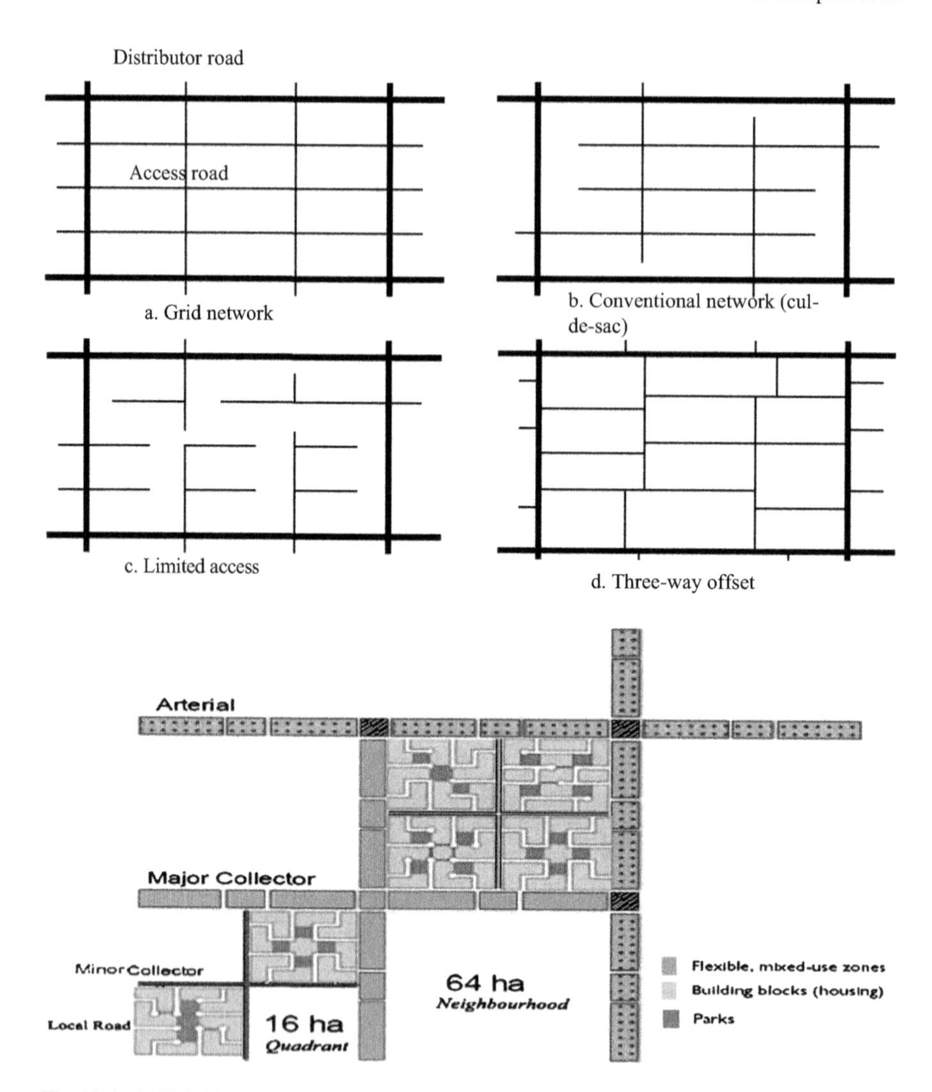

Fig. 19.4 (**a**) Neighbourhood road network patterns (Wei and Lovegrove 2012). (**b**) Neighbourhood road network patterns (Wei and Lovegrove 2012)

Wei and Lovegrove 2012; Sun and Lovegrove 2013; Masoud et al. 2015). The benefits of convenient vehicular connectivity of traditional grid network have been offset by the cost of shortcutting problems. The benefits of the contemporary response, cul-de-sac network designs that prelude shortcutting, have been offset by the cost of severed vehicular connectivity that results in increased VKT (and collisions) and, if not carefully designed for, loss of active transport network connectivity. A three-way offset network combines vehicular connectivity and safety benefits using offset 3-way intersections to calm traffic and thereby prevent shortcutting. When a safely designed vehicle network is 'fused' or integrated with a coordinated, connected, and compacte

Fig. 19.5 Cauliflower network pattern with winding roads and limited access (left); example of a winding road in a cauliflower neighbourhood (right)

land use design, results appear promising, as suggested by recent North American research on a design known as the Fused Grid (FG) sustainable neighbourhood, or SMARTer Growth Grid (SGG) neighbourhood as it has come to be known.

The SGG concept springs from a study of many historic, human-scale communities worldwide, including many in Europe, to identify common patterns that contribute to a sustained higher than average quality of life for its residents and businesses. With its ubiquitous off-road active transport paths conveniently connected to nearby corner parks and a protected core green space, the SGG neighbourhood can be expected to experience up to 60% fewer collisions than conventional grid and cul-de-sac road network patterns, as shown in Fig. 19.4b (Grammenos and Lovegrove 2015; Masoud et al. 2015).

Unlike the recent SGG 'rediscovery' by North American community planners and designers and their struggles to apply it and retrofit urban and suburban sprawl in the face of physical sprawl realities, European community planners have for decades understood and successfully applied safer transport networks and healthier community designs as living models of the SGG design principles. There are many road network patterns in practice, including many so-called cauliflower neighbourhoods built to provide for population growth in the 1970s and 1980s in the Dutch new towns and even earlier, as shown in Fig. 19.5 (Wegman 2014). The 'cauliflower' structure is characterized by winding roads and a maze-like grouping of little courtyards or 'woonerfs' (literally, 'residential-premises' or 'home zones') intended to preclude through traffic, provide a social play zone, and keep speeds at a walking pace. This would be similar to what is only now being recommended as a 'residential shared street' by the North American Association of City Transportation Officials (NACTO) in their recently released Urban Street Design Guide (NACTO 2013). Separated and stand-alone paths built for cyclists and pedestrians conveniently connect between woonerfs and other activity destinations, more directly and shorter than those offered to car drivers, thus making walking and cycling the most convenient mode choices for local trips (Wagenaar et al. 2008).

There are many design measures to achieve the lowered design speed set at the network level for access roads. According to PIARC's Human Factors Guidelines for Safer Road Infrastructure (Birth et al. 2009), drivers tend to accelerate on straight, smooth, wide roads with long sight distances and a monotonous environment. Some of their recommended measures to disrupt drivers and reduce speeding include (Bach et al. 2006; Birth et al. 2009):

1. Surface treatments, such as colour and brick pavers instead of asphalt pavement
2. Physically narrowed roads, such as less road right way, and/or planting flower boxes
3. Visually narrowed roads, such as staggered parking

19.3.3 Higher Levels of the Hierarchy

An example helps to explain the potential safety advantage of the highest ranked mobility or through-road category in the hierarchy, freeways. Similar to Fig. 19.3 for the Netherlands, Bayliss (2009) reports much lower risk of fatal and severe collisions on freeways compared to lower ranked roads (less than 25% of the number of fatalities per VKT observed on distributors). Of course, the external costs of building more lane-kilometres of freeways include induced traffic (and associated energy consumption and pollution emissions) as in the longer term when people relocate or change jobs to travel more kilometres within their travel time budget (Goodwin 1996). If a freeway allows a person to travel twice as far in the same amount of time, simple math suggests that fewer kilometres travelled formerly on the lower category road (e.g. distributor) may be replaced by twice as many kilometres on the new freeway. If this were true, for example, a doubling would halve the expected road safety benefits noted above, suggesting that a shift from urban distributor road to a freeway would reduce the number of deaths by much less than a factor of 4 (e.g. $1/4 \times 2 = 1/2$). However, if a 50% reduction could be realized, this would still be a large safety improvement and illustrates the reasoning behind designing for the highest share of the trip distance to take place on relatively safer mobility-oriented, through-roads (Dijkstra 2011). Schepers et al. (2017b) mention the dense Dutch freeway network, covering about half of the country's motor vehicle kilometres, as one of the success factors for the country's high level of cycling safety. In North American parlance, many Dutch freeways resemble higher speed, restricted access arterial highways, of usually not more than one or two lanes in each direction (see Fig. 19.6).

Cyclists are not allowed on these roads and therefore not exposed to freeway traffic. However, for lower population densities, such as in sprawling communities in North America, it becomes difficult to provide a limited access, safe freeway network dense enough to cover a high share of the distance travelled by motor vehicles. For example, the two hypothetical villages in Fig. 19.7 have similar populations and network design, but the left village has a much higher population density. Obviously, the residents in the village with the highest population density need to travel less kilometres to enter the freeway. A similar line of reasoning would apply to the exposure to risk while accessing public transport, as per Fig. 19.7. If we

Fig. 19.6 Bicycle bride over a freeway (the A50 near Arnhem)

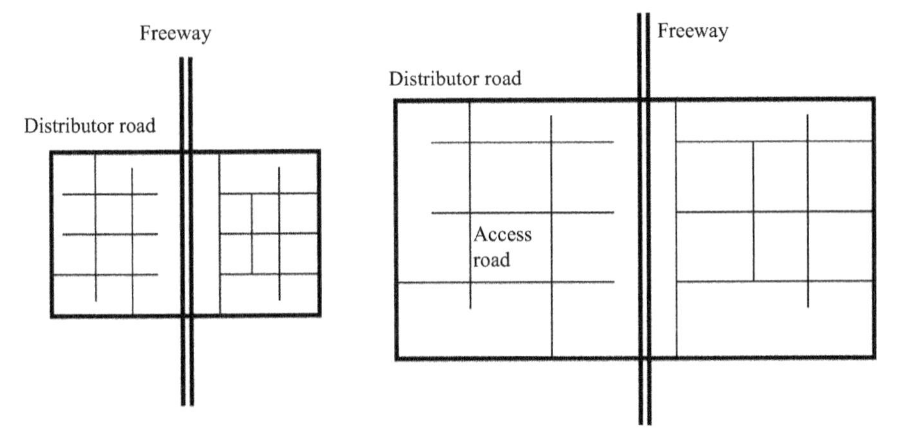

Fig. 19.7 Two hypothetical villages along a freeway with similar populations and road networks but different population densities, i.e. high in the left village and low in the right village

replace the freeway in Fig. 19.7 by heavy rails, we can conclude that less relatively unsafe kilometres using light rail or transit buses would be needed to travel to a railway station to take a relatively safe train.

19.3.4 Alignment of Network, Land Use, and Different Modes

There are different design practices regarding how to position land uses relative to the road network. Slop and Van Minnen (1994) describe conflicts between crossing pedestrians and cyclists and high-speed motor vehicles on distributor roads as a key road safety problem. They suggest the frequency of such conflicts is minimized if traffic-calmed areas enclose neighbourhood-oriented land uses. In the core of residential neighbourhoods, for example, the designers should locate walking and cycling opportunities for primary school, child care, grocery, social meeting spaces, parks, churches, fitness, seniors, library, and so on. Lovegrove and Sayed (2006) were able to quantify this safety benefit of area-wide traffic calming in North American neighbourhoods, including the need to protect the core of each neighbourhood from shortcutting and higher class roads and traffic, a finding consistent with the research on Dutch neighbourhoods that found traffic-calmed areas can be as large as possible, up to some 2 km^2 (SWOV 2010c).

Despite these findings, Dumbaugh and Rae (2009) found that the more common practice in North America is to restrict land uses in residential areas to only parks, schools, and churches. Any other uses placed inside communities would create too much traffic in the auto-oriented North American culture, and thus they get relocated onto perimeter distributors that bound residential areas. This planning approach is intended to divert traffic from invading residential areas while still allowing residents to conveniently access retail stores on their trips to and from work. The problem with this autocentric approach is that it also increases the frequency of conflicts between two distinct streams of traffic: (1) higher-speed, through-vehicles on perimeter distributor roads and (2) lower-speed, access-vehicles, and crossing pedestrians and cyclists going to and from the retail stores (Lovegrove and Sayed 2006; Dumbaugh and Rae 2009). Moreover, it violates the Safe System principle of functionality because the flow (through) and access functions are mixed (SWOV 2010a). The impressive safety results achieved in Dutch new towns (i.e. 1.0 fatal crash per 100,000 residents) seem to confirm that the design approach formulated by Slop and Van Minnen (1994) works, while North American communities does not (i.e. 6.8 fatalities per 100,000 residents in auto-oriented cities).

Finding ways to retrofit and address these safety concerns towards Vision Zero are underway within these auto-oriented North American communities, and pockets of hope do exist as beacons of a vision of what could be; for example, transit-oriented New York experiences only 3.7 fatalities per resident. Others are researching retrofits with safer intersection designs (e.g. roundabouts, three-way offsets) and reinvented, calmed 1-way couplet hybrid land use schemes (Masoud et al. 2015). In these hybrid, calmed 1-way couplet schemes, each carriageway is narrower and controlled by roundabouts to improve visibility and reduce vehicle speeds and task demands (i.e. only one-way vehicle flows) for drivers, cyclists, and pedestrians at crossing conflict points. Moreover, each couplet's carriageway is moved away from the other by an 80-m wide boulevard on which local retail and service land uses occur. While clearly not ideal, this hybrid design recognizes and tries to adapt the pre-existing North American built form of communities that mix access and flow traffic streams. Emerging research findings appear promising that lower vehicle speeds, improved visibility, and reduced driver task demands can reduce risk (and associated consequences) of vulnerable road users that must cross these pre-existing perimeter roads to access commercial areas (Masoud et al. 2017).

A significant barrier to retrofits remains, however, in the fundamental land use planning that encourages 'big box' urban and suburban destination retail developments in North America, which is a foreign concept to planning community-friendly smaller retail establishments in most European communities. For instance, the Netherlands established rules at the national level to curb the growth of (peripheral) hypermarkets and shopping malls (Evers 2002). Box developments are the norm in North America and will remain for at least a generation as artefacts of historic car-centric BAU sprawl. Some evidence suggests (as shopping malls decline in popularity

2.5km

Railway station
City centre
Small shopping centre
Distributor road (50 or 70 km/h)
Residential area (access roads; 30 km/h)
Industrial zone/ office buildings

Fig. 19.8 The city of Houten with a population nearing 50,000; southern railway station (bottom left) and stand-alone bicycle paths near the northern railway station (bottom right)

in favour of more community-oriented shops and activity hubs within local communities) that they will be reabsorbed and transformed as part of community retrofits and SMARTer Growth Grid (SGG) neighbourhood redesigns. However, wherever and whenever SGG retrofits occur, there is significant evidence that employing more sustainability-oriented community design principles in a systematic approach can be successful, as witnessed by Dutch new town successes over the decades even before SMART Growth, FG, and, most recently, SGG labels became 'in vogue'!

For over 30 years since their development, the Dutch new towns included in Table 19.1 have experienced low death rates, even though these cities have a much smaller population and density (greater and denser populated cities tend to be safer) than the larger bicycle and public transport-oriented cities listed in the same table. Interestingly, the modal share of cycling in new towns is similar to that in larger bicycle-oriented cities. To explore why this might be, we point to Fig. 19.8 which shows the perimeter distributor road network of the smallest Dutch new town, Houten, with only 50,000 inhabitants. Houten consists of two traffic-calmed neighbourhoods roughly 5 km^2 in area, in which all facilities including railway stations can be reached by foot or by bicycle in less than 10 min and without having to cross a distributor road. Where access across perimeter roads is needed, for example, when residents in one neighbourhood travel to reach the city centre in the other, they do so using shallow bicycle tunnels and 2% slopes to pass under it. In this way, Houten designers have provided a central, tree-shaped, car-free pedestrian and cycle network, a 'green backbone' that is heavily used by businesses and residents of all ages (Bach et al. 2006).

Designing for people first has been a critical success factor in the Netherlands. In Almere, the largest of the Dutch new towns, the main and direct cycle routes to the city centre were set at an early stage in the design process. Cycle routes run alongside and at the level of waterways, while cars cross via bridges without interrupting cyclist traffic (i.e. grade separated). Almere also has dedicated bus lanes for fast bus travel (Bach et al. 2006). Dutch new towns have large traffic-calmed areas and car-free cores, which are characterized by high ped/bike network connectivity and low auto network connectivity for local access within and high auto connectivity on perimeters. A study by Schepers et al. (2013) shows that cities where cyclists travel more kilometres through traffic-calmed areas (either along access roads or stand-alone bicycle paths), such as in Dutch new towns, are safer for cyclists but also for road users in general.

Bach et al. (2006) describes this people-first, 'reverse', or 'bottom-up' design process as the way to achieve safe and pedestrian- and bicycle-friendly neighbourhoods, such as in Houten, containing all frequent destinations for these road and non-road user groups. His recommended design process is as follows:

1. Locate quality-of-life sustaining daily activities and amenities as priority locations, not forgetting the need to identify plentiful opportunities for 'green' areas of restorative, spiritual, sport, and social space.
2. Layout direct walking and cycling routes between these locations and core residential areas.
3. Identify and protect the cores and main routes for traffic calming priority, crossing safety and pedestrian precincts.
4. Integrate 'people pumps' such as public transport stops, rail and metro stations, and the pedestrian entrances to public buildings and parking garages.
5. Develop lines and networks for public transport.
6. Evaluate whether the objectives have been met.
7. Last, superimpose the network for private cars (on the remaining blank space).

This formalized design process epitomizes the Dutch critical success 'people-first' factor. It replicates how traditional communities everywhere started (and at least in Europe for the most part have organically evolved and been preserved), as a community focused on and scaled by the human, not the private car. First, it starts by identifying the main daily activities and amenities that sustain quality of life, including schools, playgrounds, parks, markets, homes for the elderly, and starting points for walking routes. After this, routes for pedestrians, cyclists, and elderly are to be laid out from home address clusters (neighbourhoods) to important pedestrian and bicycle destinations. Zones such as traffic-calmed areas and pedestrian precincts can be marked off around frequent routes. Public transport stops and networks are planned in the next step. Routes for private car access are designed as the last step with a perimeter distributor road around the residential area and short-looped local access roads serving each residential cluster that minimize crossings and speeds.

Using a similar line of reasoning, several researchers have suggested to map frequent pedestrian and cyclists routes relative to distributor roads to identify potential safety problems (Bach et al. 2006). Researchers should give specific account of the needs and abilities of children and the elderly who may have specific problems with complex traffic situations such as crossing a busy road with high-speed motor traffic (Räsänen and Summala 1998; Van Beek and Schreuders 2002). Multimodal traffic models will support such analyses as pedestrian and cycle traffic is modelled at a microsimulation level of detail (Masoud et al. 2017).

19.4 Discussion

19.4.1 Road Safety

To summarize, we have found that the number of fatalities per 100,000 population is significantly lower in large cities of denser populations, more diverse land uses, greater use of public transport, and more cyclists and pedestrians. However, we cannot confirm that this is a causal relationship, as neither high-density nor more active transport use can directly explain improved road safety (Elvik 2009; Schepers and Heinen 2013). Despite many studies to date, the link between road safety and dense, mixed development oriented towards active transport is complex and difficult to reliably model in a predictive approach (Wei and Lovegrove 2012). We can say that increased volumes of cycling and walking are associated with reduced traffic speeds, increased awareness of cyclists, and increased travel times for all road users, especially at intersections. Intuitively, one would expect public demand for dedicated infrastructure for cyclists and pedestrians to increase with increasing levels of cycling and walking. We see this happening in North America as community governments must respond to climate change, energy conservation, and other sustainability-oriented challenges, whether handed down as legislated mandates or inherited as fiscal realities. Whether or not road authorities concur and take action, community retrofits will take time, during which casualties continue to occur. Unfortunately, what we are seeing to date in North America are very few governments that seem able to find ways out of their traditional patterns of political behaviour and unaffordable BAU cycles that expend precious tax dollars to increase TLKM and maintain travel time budgets between origins and destinations (see Fig. 19.7).

On the other hand, Dutch governments and designers have been able to create new towns, but not without overcoming many of these same traditional patterns. Dutch designers also planned for car traffic after World War II, but the post-war reconstruction took decades to complete, and levels of cycling were still high when the 1971 oil crisis hit (Schepers et al. 2017b), high enough to be recognized by policymakers as a solution for issues associated with motorization such as congestion, air pollution, and noise (De la Bruhèze and Veraart 1999). This may explain why only the first of the five Dutch new towns planned post-war in the 1960s, Lelystad, has a grid network of mostly 70 km/h distributor roads. It was planned first for cars, and separated cyclist path networks were not planned nor built until after the road network was built; hence, Lelystad has a less convenient bicycle network, and often cyclists get lost due to its nondirect networks and longer distances to local destinations. The city has the lowest bicycle modal share of the Dutch new towns in Table 19.1. The lesson to be learned for North American town planners is that despite the fact that dedicated infrastructure did not get priority treatment, and is only now being thought about given existing auto networks, safety levels can still be improved significantly when dedicated cycling and walking infrastructure is retrofitted into existing communities.

Most Dutch cities were built before World War II with higher densities and mixed land use. In cities such as Amsterdam, where most of the built environment survived the war, heavy roads were planned such as a four-lane arterial (Lastageweg) though the Nieuwmarkt neighbourhood. While most of the buildings were demolished in the 1960s, a broad coalition of squatters, residents, and conservation advocates were able to stop the demolition of the monumental Pinto house. On January 5, 1972, the Amsterdam city council decided by a narrow one vote majority to abandon the plan in favour of more dense and diverse human-first, cycling-oriented planning. Summarizing, key factors that may have contributed to this human-first urban planning approach in some European countries are the relatively late, post-war motorization, strong pre-war walking and cycling cultures, awareness of emerging problems in American 'auto-oriented cities', and opposition to demolition of existing neighbourhoods (De la Bruhèze and Veraart 1999; Ebert 2004).

In both Europe and North America, governments and planners are no doubt aware of these studies, yet North American leaders appear afraid to take economic and/or political risks. The research shows conclusive evidence of the economic benefits that with higher densities and mixed land uses (not high rises, just modestly higher densities), more money becomes available to invest in (sustainably) safer transport systems, both between communities (heavy rail, metro systems, and freeways having only grade-separated intersections) and within each community (lower speed limits, roundabouts, separated paths). Moreover, in denser communities, all of these safer transport systems (i.e. safe public transport and freeway systems) are closer to origins and destinations, which contribute to a higher share of kilometres travelled as safely as possible. This is a significant result when one recalls that in low-density, sprawling communities, higher shares of kilometres travelled will need to occur along distributor roads where risks are higher due to the multifunction, multi-purpose nature of the traffic on these distributor roads. To add more ammunition with which designers can help change leaders, we suggest that there is a second avenue beyond economic benefits to be considered, that being the significant health benefits of SGG (aka Dutch new town) community design. For comparison, the annual health benefits of cycling in the Netherlands are estimated at €19 billion (approximately 3% of the Dutch GDP), while annual capital investments by all levels of Dutch government in road and parking infrastructure for cycling amount to only €0.5 billion (Fishman et al. (2015). The estimated health benefits are controlled for negative side-effects due to cyclists' increased exposure to road safety and air pollution risks.

19.4.2 Public Health

In response to the adverse health impacts of urban transport that are the result of accelerated and auto-oriented urbanization and motorization (Khreis et al. 2016), many cities are beginning to shift their mobility strategies. Official community plans and growth management strategies are shifting away from auto-oriented and towards more human-focused, active transport-oriented means to reduce greenhouse gas emissions and improve public health (Nieuwenhuijsen and Khreis 2016). Road safety also benefits from urban designs oriented towards active transport (and vice versa). Moreover, it is no coincidence that the sustainably safer North American SMARTer Growth Grid and the Dutch new town community designs have in common the same traits found in traditionally sustainable communities, which also parallel human-first design principles by Bach et al. (2006).

Due to arguments surrounding these health benefits, the SMARTer Growth Grid design may be starting to catch hold across North America in new developments (e.g. Calgary, Alberta, Canada) and in retrofits of existing areas (e.g. Kelowna, BC, Canada). Robust, science-based public-health research is lending voice to the need to retrofit existing and design new urban form (Handy et al. 2002; Frank et al. 2005; Levine and Frank 2007) These built form and public health linkage experts have concluded that the following urban design principles paralleling Bach et al. (2006) will result in addressing public health problems, for example, related to obesity that in NA is commonly referred to now as 'the new nicotine'.

19.4.3 Study Limitations

In this chapter we mainly focused on the linkages between urban form and road safety in terms of traffic deaths. However, it should be noted that cyclists and pedestrians are frequently injured in falls by themselves, regardless of urban form design (Methorst et al. 2017). For instance, single-bicycle crashes cause almost three-quarters of all serious injuries among hospitalized cyclists, partly due to flaws in associated infrastructure (Schepers et al. 2015a). A shift from driving to cycling may result in increased numbers of serious road injuries until infrastructure design details catch up (Stipdonk and Reurings 2012), but we know that the health benefits of increased physical activity do substantially outweigh morbidity due to these active transport road injuries (Mueller et al. 2015). Moreover, while the net health benefits of cycling in the Netherlands (after adjusting for mortality due to air pollution and traffic fatalities) are estimated at €19 billion per year (and are even higher if the benefits of walking are added), the total road crash costs of all transport modes (including minor injuries and property damage) amount to €13 billion per year (SWOV 2014; Fishman et al. 2015). Increasing the quality and dedicated infrastructure for pedestrians and cyclists would further reduce the likelihood of injurious falls and help to keep elderly people mobile and therefore also need to be a priority in any discussion about urban form design (Fabriek et al. 2012; Methorst et al. 2017; Schepers et al. 2017a).

19.4.4 Recommendations

The Safe System approach (see the Road Safety Manual for more details; World Road Association 2014) is recognized as a solid basis for safe urban road design; however, the results of our exploration suggest that urban form can be better optimized by putting road safety policy in the context of a broader policy focussed also on improved public health. In this chapter, we have acknowledged that integrating urban planning and infrastructure network design for sustainably safer communities is easier said than done; however, we have also cited many examples that demonstrate it can be done, to great economic benefit, especially when linked to public health benefits. Relying on historically successful community designs worldwide, sustainably safer community design starts with a 'bottom-up, humans-first' design process of development patterns that focus on the needs of pedestrians and cyclists first, supported with high-quality public transport, and only consider motor traffic in the last stage. Retrofitting existing, or building new, communities towards a SMARTer Growth Grid and sustainably safer design will be a radical departure for many designers, just as it was for the Dutch new town planners, but it does work when properly done. Therefore, as a critical first step, we urge all urban planners and designers to work in a multidisciplinary team under a strong leadership that demands forward-thinking, evidence-based input from all disciplines (Van Beek and Schreuders 2002). The Safe System approach is based on the ethical premise that preventable loss of human life and long-term disability due to traffic crashes are unacceptable and transport designers ('system designers') are responsible to take preventive measures (Fahlquist 2006). There is strong science-based evidence—economic, safety, and health—that applying this ethical vision to transport and health in general is the proper planning and design approach, both in the immediate and long terms.

19.5 Conclusion

Dense and diverse land use and urban form oriented towards active transport (walking, cycling, and transit) have been demonstrated to be associated with higher levels of road safety (and public health) relative to 'business-as-usual' (BAU) sprawling, lower-density cities. However, we found that this relationship between urban form and road safety is indirect and the aforementioned urban form characteristics are likely to contribute by creating favourable preconditions for road safety (and public health in general). It appears that increased volumes of walking and cycling are associated with reduced traffic speeds, increased awareness of pedestrians and cyclists, and increased public support for large scale traffic calming and dedicated pedestrian and cycling infrastructure. Moreover, with dense and diverse land use, more money becomes available to invest in (sustainably) safer transport systems such as metro and rail, and more origins and destinations can be served by these systems that contribute to a higher share of kilometres travelled as safe as possible. To achieve higher levels of road safety and public health in general, we conclude that the most sustainable community and urban form design will be based on a 'bottom-up, humans-first' design process that focuses on the needs of local residents and pedestrians and cyclists first, supported with high-quality public transport, and that only considers private auto traffic in the last stage.

Appendix: Description of the Data Used for Fig. 19.1 and Table 19.1

Recorded deaths per 100,000 population in 2011–2015 in Fig. 19.1 are either calculated using road death and population statistics from governments such as statistical agencies or taken from OECD (OECD/ITF 2014, 2016) or WHO (WHO 2015). Phoenix or San Diego rates were from Kegler et al. (2012) and adjusted for the reduction in the USA between 2009 and 2011–2015.

Modal share of trips in Table 19.1 data was taken from the Wikipedia (2017) page about modes of transport in large cities (https://en.wikipedia.org/wiki/Modal_share) and supplemented with data for Hong Kong from Sun et al. (2014), Fietsberaad (2010) for cycling in Dutch new towns, and Statistics Netherlands (2017) for the average modal split in the Netherlands. Population densities from the Wikipedia pages of the cities are also added to Table 19.1.

Sources Table 19.1 and Fig. 19.1: Fietsberaad (2010), SWOV (2011a), Kegler et al. (2012), Préfecture de Police (2013), OECD/ITF (2014), Sun et al. (2014), NSW Centre for Road Safety (2015), Office for National Statistics (2015), Transport Accident Commission (2015), Transport Department Hong Kong (2015), WHO (2015), Department for Transport (2016), OECD/ITF (2016), Transport Accident Commission (2016), Statistics Belgium (2017), Statistics Denmark (2017), Statistics Netherlands (2017), Transport Analysis (2017).

References

Alam, A. (2010). *Quantifying the road safety benefits of sustainable transportation: Transit*. Kelowna: University of British Columbia.

American Public Transportation Association. (2016). *The hidden traffic safety solution: Public transportation*. Washington: APTA.

Bach, B., Van Hal, E., Jong, M. I., & De Jong, T. M. (2006). *Urban design and traffic; a selection form bach's toolbox*. Ede: CROW.

Bayliss, D. (2009). *Accident trends by road type*. London: Royal Automobile Club Foundation.

Birth, S., Pflaumbaum, M., Potzel, D., & Sieber, G. (2009). *Human factors guideline for safer road infrastructure*. Paris: World Road Association.

Brundtland Commission. (1987). *Our common future: Report of the world commission on environment and development. UN Documents Gatheringa Body of Global Agreements*. Oxford: Oxford University Press.

Buchanan, C. (1963). *Traffic in towns*. London: Her Majesty's Stationery Office.

Cervero, R., & Day, J. (2008). Suburbanization and transit-oriented development in china. *Transport Policy, 15*(5), 315–323.

Cervero, R., & Kockelman, K. (1997). Travel demand and the 3ds: Density, diversity, and design. *Transportation Research Part D, 2*(3), 199–219.

De La Bruhèze, A. A., & Veraart, F. C. A. (1999). *Het fietsgebruik in negen west-europese steden in de twintigste eeuw (historical comparison of bicycle use in 9 European towns)*. Eindhoven: Stichting Historie der Techniek.

Department for Transport. (2016). *Reported road casualties great britain: 2011–2015*. London: DFT.

Dhondt, S., Macharis, C., Terryn, N., Van Malderen, F., & Putman, K. (2013). Health burden of road traffic accidents, an analysis of clinical data on disability and mortality exposure rates in flanders and brussels. *Accident Analysis and Prevention, 50*, 659–666.

Dijkstra, A. (2011). *En route to safer roads; how road structure and road classification can affect road safety*. Leidschendam: Institute for Road Safety Research.

Dijkstra, A. (2013). Assessing the safety of routes in a regional network. *Transportation Research Part C, 32*, 103–115.

Dumbaugh, E., & Rae, R. (2009). Safe urban form: Revisiting the relationship between community design and traffic safety. *Journal of the American Planning Association, 75*(3), 309–329.

Ebert, A. K. (2004). Cycling towards the nation: The use of the bicycle in germany and the netherlands, 1880–1940. *European Review of History, 11*(3), 347–364.

Elvik, R. (2001). Area-wide urban traffic calming schemes: A meta-analysis of safety effects. *Accident Analysis and Prevention, 33*(3), 327–336.

Elvik, R. (2009). The non-linearity of risk and the promotion of environmentally sustainable transport. *Accident Analysis and Prevention, 41*(4), 849–855.

Elvik, R. (2013). Can a safety-in-numbers effect and a hazard-in-numbers effect co-exist in the same data? *Accident Analysis and Prevention, 60*, 57–63.

Elvik, R., & Bjørnskau, T. (2017). Safety-in-numbers: A systematic review and meta-analysis of evidence. *Safety Science, 92*, 274–282.

Elvik, R., Høye, A., Vaa, T., & Sørensen, M. (2009). *The handbook of road safety measures*. Bingley: Emerald.

Evers, D. (2002). The rise (and fall?) of national retail planning. *Journal of Economic and Social Geography, 91*(1), 107–113.

Ewing, R., & Cervero, R. (2010). Travel and the built environment: A meta-analysis. *Journal of the American Planning Association, 76*(3), 265–294.

Ewing, R., & Dumbaugh, E. (2009). The built environment and traffic safety a review of empirical evidence. *Journal of Planning Literature, 23*(4), 347–367.

Fabriek, E., De Waard, D., & Schepers, J. P. (2012). Improving the visibility of bicycle infrastructure. *International Journal of Human Factors and Ergonomics, 1*, 98–115.

Fahlquist, J. N. (2006). Responsibility ascriptions and vision zero. *Accident Analysis and Prevention, 38*(6), 1113–1118.

Fietsberaad. (2010). *Cijfers over fietsgebruik per gemeente 2004–2008*. Utrecht: CROW/Fietsberaad.

Fishman, E., Schepers, P., & Kamphuis, C. B. M. (2015). Dutch cycling: Quantifying the health and related economic benefits. *American Journal of Public Health, 105*(8), e13–e15.

Frank, L., & Hawkins, C. (2008). *Assessing travel and environmental impacts of contrasting levels of vehicular and pedestrian connectivity: Assessing aspects of the fused grid*. Ottawa: Canada Mortgage and Housing Corporation.

Frank, L., Kavage, S., & Litman, T. (2005). *Promoting public health through smart growth: Building healthier communities through transportation and land use policies and practices.* Vancouver: SmartGrowthBC.

Goodwin, P. B. (1996). Empirical evidence on induced traffic. *Transportation, 23*(1), 35–54.

Götschi, T., Garrard, J., & Giles-Corti, B. (2016). Cycling as a part of daily life: A review of health perspectives. *Transport Reviews, 36*(1), 45–71.

Grammenos, F., & Lovegrove, G. (2015). *Remaking the city street grid: A model for urban and suburban development.* Jefferson: McFarland.

Hakkert, A. S., & Braimaister, L. (2002). *The uses of exposure and risk in road safety studies.* Leidschendam: Institute for Road Safety Research.

Handy, S. (2017). Thoughts on the meaning of mark stevens's meta-analysis. *Journal of the American Planning Association, 83*(1), 26–28.

Handy, S. L., Boarnet, M. G., Ewing, R., & Killingsworth, R. E. (2002). How the built environment affects physical activity: Views from urban planning. *American Journal of Preventive Medicine, 23*(2), 64–73.

Heinen, E., Van Wee, B., & Maat, K. (2010). Commuting by bicycle: An overview of the literature. *Transport Reviews, 30*(1), 59–96.

Houwing, S., Aarts, L. T., Reurings, M. C. B., & Bax, C. A. (2012). *Verkennende studie naar regionale verschillen in relatie tot verkeersveiligheid (exploratory study on regional differences in relation with road safety).* Leidschendam: SWOV Institute for Road Safety Research.

Jacobsen, P. L. (2003). Safety in numbers: More walkers and bicyclists, safer walking and bicycling. *Injury Prevention, 9*, 205–209.

Karou, S., & Hull, A. (2014). Accessibility modelling: Predicting the impact of planned transport infrastructure on accessibility patterns in Edinburgh, UK. *Journal of Transport Geography, 35*, 1–11.

Kegler, S. R., Beck, L. F., & Sauber-Schatz, E. K. (2012). Motor vehicle crash deaths in metropolitan areas—united states, 2009. *Morbidity and Mortality Weekly Report (MMWR), 61*(28), 523–528.

Khreis, H., Warsow, K. M., Verlinghieri, E., Guzman, A., Pellecuer, L., Ferreira, A., Jones, I., Heinen, E., Rojas-Rueda, D., Mueller, N., Schepers, P., Lucas, K., & Nieuwenhuijsen, M. (2016). The health impacts of traffic-related exposures in urban areas: Understanding real effects, underlying driving forces and co-producing future directions. *Journal of Transport and Health, 3*(3), 249–267.

Koornstra, M., Lynam, D., Nilsson, G., Noordzij, P., Pettersson, H. E., Wegman, F., & Wouters, P. (2002). *Sunflower: A comparative study of the development of road safety in sweden, the United Kingdom, and the Netherlands.* Leidschendam: SWOV Institute for Road Safety Research.

Levine, J., & Frank, L. D. (2007). Transportation and land-use preferences and residents' neighborhood choices: The sufficiency of compact development in the atlanta region. *Transportation, 34*(2), 255–274.

Levinson, D. (2012). Network structure and city size. *PLoS One, 7*(1), e29721.

Li, J., Precht, D. H., Mortensen, P. B., & Olsen, J. (2003). Mortality in parents after death of a child in denmark: A nationwide follow-up study. *The Lancet, 361*(9355), 363–367.

Litman, T. (2012). *Evaluating public transit benefits and costs.* Victoria: Victoria Transport Policy Institute.

Lovegrove, G., Lim, C., & Sayed, T. (2010). Community-based, macrolevel collision prediction model use with a regional transportation plan. *Journal of Transportation Engineering, 136*(2), 120–128.

Lovegrove, G., & Sayed, T. (2006). Using macrolevel collision prediction models in road safety planning applications. *Transportation Research Record, 1950*, 73–82.

Lovegrove, G. R. (2007). *Road safety planning: New tools for sustainable road safety and community development.* Berlin: VDM Verlag Dr. Müller.

Marshall, W. E., & Garrick, N. W. (2011). Does street network design affect traffic safety? *Accident Analysis and Prevention, 43*(3), 769–781.

Masoud, A. R., Idris, A. O., & Lovegrove, G. (2017). *Modelling the influence of fused grid neighborhood design principles on active transportation use: Part 1–street connectivity*. Washington: Annual General Meeting of the Transportation Research Board.
Masoud, A. R., Lee, A., Faghihi, F., & Lovegrove, G. (2015). Building sustainably safe and healthy communities with the fused grid development layout. *Canadian Journal of Civil Engineering, 42*(12), 1063–1072.
Mavoa, S., Witten, K., Mccreanor, T., & O'sullivan, D. (2012). Gis based destination accessibility via public transit and walking in auckland, new zealand. *Journal of Transport Geography, 20*(1), 15–22.
Methorst, R., Schepers, P., Christie, N., Dijst, M., Risser, R., Sauter, D., & Van Wee, B. (2017). 'Pedestrian falls' as necessary addition to the current definition of traffic crashes for improved public health policies. *Journal of Transport and Health, 6*, 10–12.
Miranda-Moreno, L. F., Morency, P., & El-Geneidy, A. M. (2011). The link between built environment, pedestrian activity and pedestrian–vehicle collision occurrence at signalized intersections. *Accident Analysis and Prevention, 43*(5), 1624–1634.
Mokhtarian, P. L., & Chen, C. (2004). Ttb or not ttb, that is the question: A review and analysis of the empirical literature on travel time (and money) budgets. *Transportation Research Part A, 38*(9–10), 643–675.
Mueller, N., Rojas-Rueda, D., Cole-Hunter, T., De Nazelle, A., Dons, E., Gerike, R., Götschi, T., Int Panis, L., Kahlmeier, S., & Nieuwenhuijsen, M. (2015). Health impact assessment of active transportation: A systematic review. *Preventive Medicine, 76*, 103–114.
Nacto. (2013). *Urban street design guide*. Washington: National Association of City Transportation Officials.
Nieuwenhuijsen, M. J., & Khreis, H. (2016). Car free cities: Pathway to healthy urban living. *Environment International, 94*, 251–262.
NSW Centre for Road Safety. (2015). *Road traffic casualty crashes in new south wales; statistical statement for the year ended 2011–2015*. Sydney: Centre for Road Safety, Haymarket.
Oecd/Itf. (2014). *Road safety annual report 2014*. Paris: OECD Publishing.
Oecd/Itf. (2016). *Road safety annual report 2016*. Paris: OECD Publishing.
Office for National Statistics. (2015). *Estimated resident population mid-year by single year of age 1999–2015*. London: ONS.
Perry, C. A. (1939). *Housing for the machine age*. New York: Russel Sage Foundation.
Préfecture De Police, 2013. Bilan sécurité routière de la préfecture de police 2013; blessés graves.
Pucher, J. (2004). Public transportation. In S. Hanson & G. Giuliano (Eds.), *The geography of urban transportation* (pp. 199–236). New York: The Guilford Press.
Räsänen, M., & Summala, H. (1998). Attention and expectation problems in bicycle–car collisions: An in-depth study. *Accident Analysis and Prevention, 30*(5), 657–666.
Rietveld, P., & Daniel, V. (2004). Determinants of bicycle use: Do municipal policies matter? *Transportation Research Part A, 38*(7), 531–550.
Ryb, G. E., Dischinger, P. C., Mcgwin, G., Jr., & Griffin, R. L. (2012). Degree of urbanization and mortality from motor vehicular crashes. *Annals of Advances in Automotive Medicine, 56*, 183–190.
Salter, E., & Stallard, P. (2004). Posttraumatic growth in child survivors of a road traffic accident. *Journal of Traumatic Stress, 17*(4), 335–340.
Schepers, J. P., Hagenzieker, M. P., Methorst, R., Van Wee, G. P., & Wegman, F. (2014). A conceptual framework for road safety and mobility applied to cycling safety. *Accident Analysis and Prevention, 62*, 331–340.
Schepers, J. P., & Heinen, E. (2013). How does a modal shift from short car trips to cycling affect road safety? *Accident Analysis and Prevention, 50*(1), 1118–1127.
Schepers, J. P., Heinen, E., Methorst, R., & Wegman, F. C. M. (2013). Road safety and bicycle usage impacts of unbundling vehicular and cycle traffic in dutch urban networks. *European Journal of Transport and Infrastructure Research, 13*(3), 221–238.

Schepers, J. P., Kroeze, P. A., Sweers, W., & Wust, J. C. (2011). Road factors and bicycle-motor vehicle crashes at unsignalized priority intersections. *Accident Analysis and Prevention, 43*(3), 853–861.

Schepers, P., Agerholm, N., Amoros, E., Benington, R., Bjørnskau, T., Dhondt, S., De Geus, B., Hagemeister, C., Loo, B. P. Y., & Niska, A. (2015a). An international review of the frequency of single-bicycle crashes (sbcs) and their relation to bicycle modal share. *Injury Prevention, 21*, e138–e143.

Schepers, P., Den Brinker, B., Methorst, R., & Helbich, M. (2017a). Pedestrian falls: A review of the literature and future research directions. *Journal of Safety Research, 62*, 227–234.

Schepers, P., Fishman, E., Beelen, R., Heinen, E., Wijnen, W., & Parkin, J. (2015b). The mortality impact of bicycle paths and lanes related to physical activity, air pollution exposure and road safety. *Journal of Transport and Health, 2*(4), 460–473.

Schepers, P., Twisk, D., Fishman, E., Fyhri, A., & Jensen, A. (2017b). The dutch road to a high level of cycling safety. *Safety Science, 92*, 264–273.

Slop, M., & Van Minnen, J. (1994). *Duurzaam veilig voetgangers-en fietsverkeer (sustainable safe pedestrian and cycle traffic)*. Leidschendam: Institute for Road Safety Research.

Smeed, R. J. (1949). Some statistical aspects of road safety research. *Journal of the Royal Statistical Society A, 112*(1), 1–34.

Statistics Belgium. (2017). *Loop van de bevolking; verkeersongevallen 2011–2015*. Brussel: FOD Economie; Algemene Directie Statistiek en Economische Informatie.

Statistics Denmark. (2017). *Statbank*. Copenhagen: Statistics Denmark.

Statistics Netherlands, 2017. Statline.

Stipdonk, H., & Reurings, M. (2012). The effect on road safety of a modal shift from car to bicycle. *Traffic Injury Prevention, 13*(4), 412–421.

Sun, G., Gwee, E., Chin, L. S., & Low, A. (2014). Passenger transport mode shares in world cities. *Journeys; Sharing Urban Transport Solutions, 12*, 54–64.

Sun, J., & Lovegrove, G. (2013). Comparing the road safety of neighbourhood development patterns: Traditional versus sustainable communities. *Canadian Journal of Civil Engineering, 40*(1), 35–45.

Swov. (2010a). *Factsheet background of the five sustainable safety principles*. Leidschendam: Institute for Road Safety Research.

Swov. (2010b). *Factsheet international comparability of road safety data*. Leidschendam: Institute for Road Safety Research.

Swov. (2010c). *Factsheet zones 30: Urban residential areas*. Leidschendam: Institute for Road Safety Research.

Swov. (2011a). *Cognos*. Leidschendam: Institute for Road Safety Research.

Swov. (2011b). *Road safety hazards of public transport*. Leidschendam: SWOV Institute for Road Safety Research.

Swov. (2014). *Factsheet road crash costs*. The Hague: SWOV Institute for Road Safety Research.

Swov. (2017). *Cognos*. Leidschendam: Institute for Road Safety Research.

Tingvall, C., Haworth, N. (1999). Vision zero – an ethical approach to safety and mobility. *6th ITE International Conference Road Safety & Traffic Enforcement: Beyond 2000*, Melbourne.

Transport Accident Commission. (2015). *Road safety statistical summary 2013/2015*. Geelong: TAC.

Transport Accident Commission. (2016). *Road safety quarterly statistics – June 2016*. Geelong: TAC.

Transport Analysis, 2017. Road traffic injuries. Stockholm.

Transport Department Hong Kong, 2015. Road traffic accident statistics; year 2015. Hong Kong.

Ukkusuri, S., Miranda-Moreno, L. F., Ramadurai, G., & Isa-Tavarez, J. (2012). The role of built environment on pedestrian crash frequency. *Safety Science, 50*(4), 1141–1151.

Van Beek, P., & Schreuders, M. (2002). *Opstap naar de mobiliteitstoets: Ruimtelijke ordening in relatie tot verkeersveiligheid*. Rotterdam: Rijkswaterstaat.

Van Wee, B. (2009). Verkeer en transport (traffic and transport). In B. Van Wee & J. Anne Annema (Eds.), *Verkeer en vervoer in hoofdlijnen (outlining traffic and transport)*. Bussum: Coutinho.

Van Wee, B., & Ettema, D. (2016). Travel behaviour and health: A conceptual model and research agenda. *Journal of Transport and Health, 3*(3), 240–248.

Wagenaar, C., Mens, N., Singelenberg, J., Visser, A., & Sparenberg, S. (2008). *De toekomst van de bloemkoolwijken (the future of cauliflower neighbourhoods)*. Rotterdam: SEV.

Wegman, F. (2014). Sustainable communities: The dutch example. *Canadian Civil Engineer, Canadian Society of Civil Engineers Winter, 2014*, 17–19.

Wei, V. F., & Lovegrove, G. (2012). An empirical tool to evaluate the safety of cyclists: Community based, macro-level collision prediction models using negative binomial regression. *Accident Analysis and Prevention, 61*, 129–137.

Weijermars, W. A. M., & Wegman, F. C. M. (2011). Ten years of sustainable safety in the netherlands. *Transportation Research Record, 2213*, 1–6.

WHO. (2004). World report on road traffic injury prevention. In M. Peden, R. Scurfield, D. Sleet, D. Mohan, A. A. Hyder, E. Jarawan, & C. Mathers (Eds.), *World report on road traffic injury prevention*. Geneva: World Health Organization.

WHO. (2015). *Global status report on road safety 2015*. Geneva: World Health Organization.

Wikipedia, 2017. Modal share.

Witteveen+Bos. (2015). *Effectstudie verkeersveiligheid blankenburgverbinding*. Rotterdam: Rijkswaterstaat.

World Road Association. (2014). *Road safety manual; guide for practitioners*. Paris: PIARC.

Chapter 20
Green Space and Health

Payam Dadvand and Mark Nieuwenhuijsen

20.1 Overview

During the last century, the world experienced a rapid urbanization which is still ongoing in different parts of the world. Nowadays, more than half of the global population lives in cities and this proportion is projected to rise to two-third by 2050 (UN Department of Economic and Social Affairs 2015). Cities are recognized as the powerhouses of innovation and wealth creation where people usually have better access to healthcare (Bettencourt et al. 2007). However, urban areas are often associated with higher levels of a number of environmental hazards such as air pollution, noise, and heat and limited access to nature, including green spaces. At the same time, urban lifestyle is predominantly associated with lower levels of physical activity and higher exposure to crime and psychological stress (Bettencourt et al. 2007). These environmental and lifestyle factors could contribute to the existing higher prevalence of a wide range of adverse health conditions such as psychological disorders and non-communicable diseases in urban areas (Cyril et al. 2013). Natural environments, including green spaces, have been associated with improved mental and physical health and well-being and are increasingly recognized as a mitigation measure to buffer the aforementioned adverse health effects of urban living. This chapter provides an overview of (1) urban green spaces, (2) the methods that are applied to characterize exposure to these spaces, (3) the potential mechanisms through which green spaces could exert their health effects, (4) the health effects associated with contact to green spaces, and (5) the role of socioeconomic status (SES) in such effects.

P. Dadvand (✉) · M. Nieuwenhuijsen
ISGlobal, Barcelona, Spain
e-mail: payam.davand@isglobal.org; mark.nieuwenhuijsen@isglobal.org

M. Nieuwenhuijsen, H. Khreis (eds.), *Integrating Human Health into Urban and Transport Planning*, https://doi.org/10.1007/978-3-319-74983-9_20

20.2 Urban Green Spaces

The US Environmental Protection Agency (EPA) defines green spaces as the *land that is partly or completely covered with grass, trees, shrubs, or other vegetation* which includes *parks, community gardens, and cemeteries* (US Environmental Protection Agency 2017). The abundance and availability of green spaces in urban areas could be a function of several factors from which the climate and urban planning play key roles. For example, a survey of 386 European cities (2009) revealed that while there was a general north-south decreasing gradient in the percentage of green space coverage within these cities, still there were greener cities in south and less green cities in north (Fuller and Gaston 2009). The amount of green space available to people in cities also varies considerably from, for example, 1.9 m^2 per person in Buenos Aires, Argentina to 52.0 m^2 in Curitiba, Brazil. There are many types of green in cities including parks, street green, and natural green which can be captured by different maps or remote sensing methods (Gascon et al. 2016a).

20.3 Characterization of Contact with Green Spaces

The methods to assess contact to green spaces are currently evolving. Different methods have been developed to characterize the following different aspects of such contact:

1. *Surrounding greenness*: A major part of the available evidence on health effects of green spaces has relied on characterization of greenness surrounding home addresses, and to less extent surrounding school or workplace, as an indicator of general greenness at living environment of the study subjects. These studies have either relied on (a) remote sensing-based indices of greenness (e.g. Normalized Vegetation Difference Index (NDVI)) to quantify the amount of photosynthetically active vegetation in a certain buffer around or within boundaries of home, school, or workplace or (b) available land-cover maps to abstract the percentage of green land covers in a certain buffer around the aforementioned places.
2. *Physical access to green spaces*: Proximity to green spaces has often been used as an indicator of access to these spaces. There are two approaches to characterize proximity to green spaces: (a) objective proximity to green spaces based on quantifying the distance (either Euclidian or network distance through available road network) between the address of interest and the closest green space with whatever size or those larger than a certain size. These studies have used distance as either a continuous variable or have dichotomize them using certain cut-offs. For example, some studies have applied the European Commission's recommendation for access to open spaces (including green spaces) defined as living within 300 m of an open/green space with a minimum area of 5000 m^2 (WHO Regional Office for Europe 2016). To date, a few studies (e.g. Dadvand et al. (2016)) have used (b) subjective proximity to green spaces to characterize access to green

spaces, by asking study subjects whether they have green spaces within a certain distance (e.g. 15-min walk) from their homes.

3. *Visual access to green spaces*: To date, few epidemiological studies have evaluated health effects of visual access to green spaces. These studies have applied either questionnaires asking study subjects about the proportion of green view through their window(s) or have rated the green view in the photos taken by study subjects or fieldworkers from the windows of interest.
4. *Use of green spaces*: Two approaches have been applied by the studies of health effects of green spaces to characterize use of green spaces: (a) questionnaires asking participants to report the time that they have spent and the type of physical activity they have conducted in green spaces over a certain period of time and (b) Global Positioning System (GPS) or smartphones to objectively measure the time that study subjects spend in different microenvironments including green spaces.
5. *Quality of green spaces*: Quality characteristics of green spaces such as aesthetics, biodiversity, walkability, sport/play facilities, safety, and organized social events have been suggested to predict the use of green spaces (McCormack et al. 2010); however, so far, most studies evaluating health effects of green spaces have overlooked these characteristics. Quality of green spaces has been often characterized based on systematic observation (audits) of these spaces by fieldworkers/study participants using tools developed for this aim (e.g. Van Dillen et al. (2012)). Recently, there has been a limited effort to use remote sensing images (e.g. Google Earth Pro (Taylor et al. 2011)) to characterize quality of green spaces which showed a strong correlation with the assessments made by in-person audits.

20.4 Potential Underlying Mechanism

The mechanism underlying health effects of green spaces is yet to be established, but stress reduction/cognitive restoration; mitigation of the exposure to air pollution, noise, and heat; enhancing social cohesion/interactions; increasing physical activity; and enriching micro- and macro-biodiversity and environmental microbial input have been suggested to be involved.

20.4.1 Stress Reduction/Cognitive Restoration

A substantial body of experimental and observational evidence has consistently showed the capability of green spaces in reducing stress and restoring cognition function. The *stress reduction theory* suggests that green spaces, through properties such as spatial openness, curving sightlines, and the presence of water, induce recovery from stress and help to diminish states of arousal and negative thoughts through psychophysiological pathways (Ulrich 1984). *Attention restoration theory* proposes that contact with nature with its inherently delightful stimuli could modestly invoke indirect (i.e. effortless) attention and in time minimize the need for directed attention that together could restore the directed attention mechanisms (Kaplan and Kaplan 1989; Kaplan 1995; Berman et al. 2008). These pathways have been indicated to play important roles in the health benefits of green spaces (de Vries et al. 2013; Dadvand et al. 2016).

20.4.2 Mitigating Environmental Exposures

The impact of green spaces on air pollution is complex and context-specific. On one hand, vegetations have been proposed to reduce air pollution by direct and indirect mechanisms (Givoni 1991). The direct mechanism is via filtering of air pollution by vegetations, principally based on dry deposition of pollutants (both particles and gases) through stomata uptake or non-stomata deposition on plant surfaces (Paoletti et al. 2011; Givoni 1991; Akbari 2002; Nowak et al. 2006). The indirect effect is mediated through cooling effects of plants that in turn reduces smog formation (Givoni 1991). A study on the effects of greenness surrounding residential address on personal exposure to air pollution using personal air pollution monitors reported that higher residential surrounding greenness was associated with reduced personal exposure to particulate air pollution but not nitric oxides (Dadvand et al. 2012c). Another study also showed that higher greenness within and surrounding schools is associated with lower indoor and outdoor levels of traffic-related air pollutants at school (Dadvand et al. 2015b). The ability of vegetations to reduce air pollution is thought to be type-specific with trees being the most efficient and grasses being the least efficient types (Givoni 1991). Studies on the capacity of canopies to remove air pollution in continental USA (Nowak et al. 2006) and Greater London (Tallis et al. 2011) estimated that about 1–2% of air pollution in these areas is removed by canopies. Experimental studies on mitigation effects of roadside vegetation on air pollution have reported inconsistent results with some reports that do not support such an effect (Baldauf et al. 2011; Hagler et al. 2012). Simulation studies of such effects have also indicated that roadside trees are able to generate a canyon effect with higher air pollution levels on the downwind and lower air pollution levels on the upwind side of the street (Buccolieri et al. 2009; Baldauf et al. 2011). Moreover, biodegradation of vegetation residues generates volatile organic compounds (VOCs), a family of air pollutants with potential health effects on humans. VOCs can also engage in complex photo chemical reactions with other air pollutants such as ozone and nitric oxides and participate in generation of biogenic secondary organic aerosols (Kesselmeier and Staudt 1999; Hoyle et al. 2011). Although the interaction between green spaces and air pollution appears to be multifaceted and complex and the available evidence on such interaction is still limited and inconsistent, the available studies evaluating the mediator role of air pollution in the observed health benefits of green spaces are suggestive for such a mediation. For example, a recent study of the effects of green spaces on cognitive development estimated that up to 60% of these effects could be explained by the reduction of traffic-related air pollutants by green spaces (Dadvand et al. 2015a).

The effect of green spaces on reducing temperature is well established. Evapotranspiration (release of water vapour into atmosphere), shading, and micro-regulating air movements and heat exchange are among the mechanisms through which vegetations could ameliorate the temperature (Bowler et al. 2010). A systematic review and meta-analysis of the available literature on such effect concluded that the temperature in urban parks is on average 1 °C less than that of other nongreen areas in the city (Bowler et al. 2010). Given the existence of heat island effect in urban areas, the capability of green spaces to reduce temperate is of importance for promoting resilience in cities, especially in the context of the occurring climate change.

The available evidence on mitigation of noise exposure by green spaces is still limited. However, these studies are suggestive for the buffering of the noise exposure/reduction of noise annoyance by residential surrounding greenness and green facades (De Ridder et al. 2004; Gidlöf-Gunnarsson and Öhrström 2007).

20.4.3 Enhancing Social Cohesion/Interaction

A cohesive society is defined as a society that *works towards the well-being of all its members, fights exclusion and marginalisation, creates a sense of belonging, promotes trust, and offers its members the opportunity of upward mobility* (Organization for Economic Cooperation and Development (OECD) 2011). Social cohesion/interaction have been associated with improved perceived general health (Kawachi et al. 2008), lower morbidity, more longevity, and reduced inequality in health (Marmot et al. 2012). The body of evidence on the association between contact with green spaces and social cohesion/interaction is still limited; however, it is generally supportive for such an association (Sugiyama et al. 2008; Maas et al. 2009a; de Vries et al. 2013; Dadvand et al. 2016), with a few exceptions (Triguero-Mas et al. 2015). Few studies have also shown the mediation of the association between green spaces and perceived general health by the improvement of social cohesion/interaction (Maas et al. 2009a; de Vries et al. 2013; Dadvand et al. 2016).

20.4.4 Increasing Physical Activity

Many of the studies have focused on physical activity as an important mechanism for the health of benefits. However, the available evidence on the impact of green spaces on physical activity is inconsistent with a considerable heterogeneity in the reported direction and strength of associations (Lachowycz and Jones 2011; McGrath et al. 2015) (Bancroft et al. 2015). A part of this inconsistency could be because of not accounting for the quality of green spaces in most of these studies, while these aspects are shown to affect the use of green spaces for physical activity (McCormack et al. 2010). The few studies evaluating the mediation of health benefits of green spaces by physical activity are also suggestive for a modest mediation role of physical activity in these benefits (de Vries et al. 2013; Dadvand et al. 2016).

20.4.5 Enriching Environmental Biodiversity

Plants are able to directly modulate the microbiome present in the rhizosphere (the below-ground microbial habitat provided by plant root systems) and phyllosphere (above-ground microbial habitats provided by plants) (Berendsen et al. 2012; Vorholt 2012) and therefore indirectly modulate the environmental microbiome to which humans are exposed. Studies have shown that bacterial diversity in humans' faeces decreases with the level of urbanization, which is strongly associated with reduced environmental biodiversity (De Filippo et al. 2010; Yatsunenko et al. 2012). Human microbiome including gut microbiome has been shown to interact with the host tissue, regulate systemic immune response, and prevent chronic inflammation (Martinez 2014). Therefore, the ability of urban green spaces to enhance immunoregulation-inducing microbial input from the environment (Rook 2013) could be a potential mechanism underlying the association between green spaces and human health. A study in adolescents, for example, showed that living near a forest increases the diversity of the skin microbiome which in turn was associated with reduced risk of allergic sensitization later in life (Hanski et al. 2012).

20.5 Health Benefits

Exposure to green spaces has been associated with improved physical and mental health and well-being. This exposure, for example, has been associated with improved perceived general health, better pregnancy outcomes (e.g. birth weight), enhanced brain development in children, better cognitive function in adults, improved mental health, lower risk of a number of chronic diseases (e.g. diabetes and cardiovascular conditions), and reduced premature mortality.

20.5.1 Pregnancy Outcomes

Higher greenness surrounding maternal residential address during pregnancy has been associated with increased birth weight in offspring (Dzhambov et al. 2014). The available evidence for such an association for the length of pregnancy is inconsistent. While some studies are suggestive for an increased length of gestation (i.e. reduced risk of preterm birth) associated with higher greenness surrounding maternal residential address (Laurent et al. 2013; Hystad et al. 2014; Grazuleviciene et al. 2015; Nichani et al. 2017), other studies have not supported this association (Dadvand et al. 2012a, b; Agay-Shay et al. 2014).

20.5.2 Brain Development

The "biophilia hypothesis" proposes evolutionary bonds of humans to nature (Wilson 1984; Kellert and Wilson 1993). Accordingly, contact with nature including green spaces is thought to have a crucial role in brain development in children (Kahn and Kellert 2002; Kellert 2005). Experimental studies have shown that playing in green spaces could reduce severity of symptoms and improve attention in children with attention deficit/hyperactivity disorder (ADHD) in short-term (Taylor et al. 2001; Kuo and Taylor 2004; Taylor and Kuo 2009; van den Berg and van den Berg 2011). Observational studies have revealed that higher residential surrounding greenness and more time spent playing in green spaces in the long run could reduce risk of behavioural and emotional problems including ADHD (Amoly et al. 2014; Markevych et al. 2014b) and enhance cognitive development including attention and working memory (Wells 2000; Dadvand et al. 2015a).

20.5.3 Cognitive Function

Exposure to green spaces has been associated with improved cognitive functions including better direct attentional capacity and lower concentration problems in adults (de Keijzer et al. 2016). The available evidence on the potential impact of this exposure in decelerating cognitive decline in elderly is still scarce and inconsistent (de Keijzer et al. 2016).

20.5.4 Perceived General Health

More contact with green spaces has been consistently associated with improved perceived general health (Gascon et al. 2015). Studies have shown that improved mental health and social cohesion and, to less extent, enhanced physical activity are among the main mechanisms underlying this association (de Vries et al. 2013; Dadvand et al. 2016).

20.5.5 Mental Health

The effect of green spaces on mental health is one of the most studied health effects of green spaces. More contact with green space has been associated with lower risk of psychological distress and psychiatric conditions such as depression and anxiety and less likelihood of use of psychiatric medicine (Gascon et al. 2015).

20.5.6 Other Non-communicable Diseases

The available evidence on the impacts of green spaces on non-communicable diseases other than asthma and allergy is still limited but is suggestive for a beneficial impact. More contact with green spaces has been associated with lower risk of cardiovascular conditions, diabetes, and low back pain (Maas et al. 2009b; Dalton et al. 2016). A recent study has also associated this contact with lower blood pressure in children (Markevych et al. 2014a).

20.5.7 Mortality

A recent systemic review and meta-analysis of the available literature on the impact of contact with green spaces on mortality have shown that higher residential surrounding greenness is associated with reduced all-cause premature mortality as well as cardiovascular mortality (Gascon et al. 2016b). Lower exposure to air pollution, higher physical activity, stronger perceived social engagement, and reduced risk of depression have been reported to mediate the association between exposure to green spaces and mortality (James et al. 2016).

20.6 Health Risks

Green spaces could potentially impose a number of health risks including increasing risk of asthma and allergic conditions, enhancing exposure to herbicides and pesticides, hosting reservoirs and vectors of infectious diseases, and increasing risk of accidental injuries.

20.6.1 Asthma and Allergy

The available evidence on the impact of green spaces on asthma and allergic conditions in children is inconsistent. While some studies have associated higher residential surrounding greenness with increased risk of asthma and allergic conditions (DellaValle et al. 2012; Lovasi et al. 2013; Andrusaityte et al. 2016), others have not shown such an association or have even shown protective effects (Lovasi et al. 2008; Maas et al. 2009b; Hanski et al. 2012; Pilat et al. 2012; Hind et al. 2017). The type of green space and the bioclimatic properties of the study region could explain, in part, such an inconsistency. One study, for example, has shown that while urban parks were associated with higher risk of asthma and allergic attacks, natural green spaces (e.g. forests) did not show such an association (Dadvand et al. 2014). Another study conducted in seven birth cohorts in Australia, Canada, Germany, Sweden, and the Netherlands showed a notable between-centre heterogeneity in terms of the direction and strength of associations (Fuertes et al. 2016).

20.6.2 Herbicide and Pesticide Exposure

Application of herbicides and pesticides in green spaces could expose individuals living in proximity of these spaces or those who use these spaces to these chemicals. Such an exposure could in turn lead to a range of health outcomes including cancers as well as adverse conditions in nervous, reproductive, endocrine, and immune systems (Blair et al. 2015).

20.6.3 Vector-Based and Zoonotic Infections

Green spaces could host vectors and reservoirs of infectious diseases, which could increase the risk of vector-borne diseases transferred by mosquitoes (e.g. malaria or dengue fever), ticks (e.g. Lyme disease and tick-borne encephalitis), or sandflies (e.g. leishmaniasis) (WHO Regional Office for Europe 2016). Exposure to animal faeces in green spaces can also result in zoonotic infections such as toxocariasis or toxoplasmosis (WHO Regional Office for Europe 2016).

20.6.4 *Accidental Injuries*

Users of green spaces, especially children, could experience accidental injuries such as falls or drowning while they are in these spaces. However, at population level, the injuries that occur in green spaces account for a very tiny proportion of accidental injuries (WHO Regional Office for Europe 2016).

20.7 Role of Socioeconomic Status

SES can be associated with both contact with greenness (e.g. high SES groups are more likely to live in greener neighbourhoods) and health status (e.g. high SES groups generally have better health status) making SES a potentially strong confounder of the analyses of the health benefits of green spaces. In addition to be a confounder, SES can also act as a modifier of the health effects of green spaces. Available studies are suggestive for greater benefits of green spaces for lower SES groups. This could be partly because groups with lower SES generally have poorer health status and live in areas with more environmental problems, and combination of these could make them more prone to benefit from health promotion interventions such as developing new green spaces (De Vries et al. 2003; Bolte et al. 2010; Su et al. 2011). Furthermore, lower SES groups are more likely to benefit from green spaces in proximity of their homes because they spend more time in the vicinity of their homes and availability of green spaces close to their homes can therefore increase the likelihood of their use of these spaces (Schwanen et al. 2002; Maas 2008). On the other hand, higher SES groups are more likely to use the green spaces farther away because of higher mobility (Greenspace Scotland 2008; Bell et al. 2010) and consequently their use of green spaces is less dependent on having green spaces close to their homes.

Given the greater benefits of green spaces for lower SES groups, these spaces have the potential to reduce inequality in health. A landmark study conducted in the entire England has showed that the income-related inequality in mortality is less evident in greener neighborhoods compared to less green neighborhoods (Mitchell and Popham 2008; Marmot 2010).

20.8 Green Space as a Pathway to Healthy Urban Living

Given the many benefits of green spaces, health of citizens in cities where there is a lack of green space can be improved by increasing the amount of green space (Nieuwenhuijsen et al. 2017; van den Bosch and Nieuwenhuijsen 2017). Cities can be made healthier and more equitable for people, not by painting trees on walls but by having a nearby park where people live, planting trees in the streets, and introducing urban gardens. Urban gardens may have additional benefits in terms of local food production and economy and, if done at a sufficiently large scale, can contribute to the sustainability and self-sufficiency of cities. Many cities need more parks, which can also become part of the identity and attraction of cities. Also, green roofs may transform the city, not only in terms of resilience but also in terms of visual attractiveness. Our current cites are too car dominated, and car infrastructures such as roads and parking lots take up much space that can be used for planting trees and other green. Reducing space for cars and the number of cars may have the additional advantage that people have to switch to public and active transportation and thereby reducing, e.g. air pollution, heat, and noise levels in cities and increasing physical activity in citizens (Nieuwenhuijsen and Khreis 2016). Although greening our cities is not the only solution to improving health of urban residents, it can certainly make an important contribution. Green cities, healthy people.

References

Agay-Shay, K., Peled, A., Crespo, A. V., Peretz, C., Amitai, Y., Linn, S., et al. (2014). Green spaces and adverse pregnancy outcomes. *Occupational and Environmental Medicine, 71*(8), 562–569.

Akbari, H. (2002). Shade trees reduce building energy use and CO_2 emissions from power plants. *Environmental Pollution, 116*, S119–S126.

Amoly, E., Dadvand, P., Forns, J., López-Vicente, M., Basagaña, X., Julvez, J., et al. (2014). Green and blue spaces and behavioral development in Barcelona schoolchildren: The BREATHE project. *Environmental Health Perspectives, 122*(12), 1351–1358.

Andrusaityte, S., Grazuleviciene, R., Kudzyte, J., Bernotiene, A., Dedele, A., & Nieuwenhuijsen, M. J. (2016). Associations between neighbourhood greenness and asthma in preschool children in Kaunas, Lithuania: A case–control study. *BMJ Open, 6*, e010341.

Baldauf, R., Jackson, L., Hagler, G., Vlad, I., McPherson, G., Nowak, D., et al. (2011). The role of vegetation in mitigating air quality impacts from traffic emissions. *EM: Air and Waste Management Associations Magazine for Environmental Managers, 2011*, 30–33.

Bancroft, C., Joshi, S., Rundle, A., Hutson, M., Chong, C., Weiss, C. C., et al. (2015). Association of proximity and density of parks and objectively measured physical activity in the United States: A systematic review. *Social Science & Medicine, 138*, 22–30.

Bell, M. L., Belanger, K., Ebisu, K., Gent, J. F., Lee, H. J., Koutrakis, P., et al. (2010). Prenatal exposure to fine particulate matter and birth weight: Variations by particulate constituents and sources. *Epidemiology, 21*(6), 884–891.

Berendsen, R. L., Pieterse, C. M. J., & Bakker, P. A. H. M. (2012). The rhizosphere microbiome and plant health. *Trends in Plant Science, 17*(8), 478–486.

Berman, M. G., Jonides, J., & Kaplan, S. (2008). The cognitive benefits of interacting with nature. *Psychological Science, 19*(12), 1207–1212.

Bettencourt, L. M. A., Lobo, J., Helbing, D., Kühnert, C., & West, G. B. (2007). Growth, innovation, scaling, and the pace of life in cities. *PNAS, 104*(17), 7301–7306.

Blair, A., Ritz, B., Wesseling, C., & Beane, F. L. (2015). Pesticides and human health. *Occupational and Environmental Medicine, 72*(2), 81–82.

Bolte, G., Tamburlini, G., & Kohlhuber, M. (2010). Environmental inequalities among children in Europe—Evaluation of scientific evidence and policy implications. *European Journal of Public Health, 20*(1), 14–20.

Bowler, D. E., Buyung-Ali, L., Knight, T. M., & Pullin, A. S. (2010). Urban greening to cool towns and cities: A systematic review of the empirical evidence. *Landscape and Urban Planning, 97*(3), 147–155.

Buccolieri, R., Gromke, C., Di Sabatino, S., & Ruck, B. (2009). Aerodynamic effects of trees on pollutant concentration in street canyons. *Science of the Total Environment, 407*(19), 5247–5256.

Cyril, S., Oldroyd, J. C., & Renzaho, A. (2013). Urbanisation, urbanicity, and health: A systematic review of the reliability and validity of urbanicity scales. *BMC Public Health, 13*, 513.

Dadvand, P., Bartoll, X., Basagaña, X., Dalmau-Bueno, A., Martinez, D., Ambros, A., et al. (2016). Green spaces and general health: Roles of mental health status, social support, and physical activity. *Environment International, 91*, 161–167.

Dadvand, P., de Nazelle, A., Figueras, F., Basagaña, X., Sue, J., Amoly, E., et al. (2012a). Green space, health inequality and pregnancy. *Environment International, 40*, 110–115.

Dadvand, P., de Nazelle, A., Triguero-Mas, M., Schembari, A., Cirach, M., Amoly, E., et al. (2012c). Surrounding greenness and exposure to air pollution during pregnancy: An analysis of personal monitoring data. *Environmental Health Perspectives, 120*(9), 1286–1290.

Dadvand, P., Nieuwenhuijsen, M. J., Esnaola, M., Forns, J., Basagaña, X., Alvarez-Pedrerol, M., et al. (2015a). Green spaces and cognitive development in primary schoolchildren. *Proceedings of the National Academy of Sciences of the United States of America, 112*(26), 7937–7942.

Dadvand, P., Rivas, I., Basagaña, X., Alvarez-Pedrerol, M., Su, J., De Castro, P. M., et al. (2015b). The association between greenness and traffic-related air pollution at schools. *Science of the Total Environment, 523*, 59–63.

Dadvand, P., Sunyer, J., Basagaña, X., Ballester, F., Lertxundi, A., Fernández-Somoano, A., et al. (2012b). Surrounding greenness and pregnancy outcomes in four Spanish birth cohorts. *Environmental Health Perspectives, 120*(10), 1481–1487.

Dadvand, P., Villanueva, C. M., Font-Ribera, L., Martinez, D., Basagaña, X., Belmonte, J., et al. (2014). Risks and benefits of green spaces for children: A cross-sectional study of associations with sedentary behavior, obesity, asthma, and allergy. *Environmental Health Perspectives, 122*(12), 1329–1325.

Dalton, A. M., Jones, A. P., Sharp, S. J., Cooper, A. J. M., Griffin, S., & Wareham, N. J. (2016). Residential neighbourhood greenspace is associated with reduced risk of incident diabetes in older people: A prospective cohort study. *BMC Public Health, 16*, 1171.

De Filippo, C., Cavalieri, D., Di Paola, M., Ramazzotti, M., Poullet, J. B., Massart, S., et al. (2010). Impact of diet in shaping gut microbiota revealed by a comparative study in children from Europe and rural Africa. *Proceedings of the National Academy of Sciences of the United States of America, 107*(33), 14691–14696.

de Keijzer, C., Gascon, M., Nieuwenhuijsen, M. J., & Dadvand, P. (2016). Long-term green space exposure and cognition across the life course: A systematic review. *Current Environmental Health Reports, 3*(4), 468–477.

De Ridder, K., Adamec, V., Bañuelos, A., Bruse, M., Bürger, M., Damsgaard, O., et al. (2004). An integrated methodology to assess the benefits of urban green space. *Science of the Total Environment, 334–335*, 489–497.

de Vries, S., van Dillen, S. M. E., Groenewegen, P. P., & Spreeuwenberg, P. (2013). Streetscape greenery and health: Stress, social cohesion and physical activity as mediators. *Social Science & Medicine, 94*, 26–33.

De Vries, S., Verheij, R. A., Groenewegen, P. P., & Spreeuwenberg, P. (2003). Natural environments-healthy environments? An exploratory analysis of the relationship between greenspace and health. *Environment & Planning A, 35*(10), 1717–1732.

DellaValle, C. T., Triche, E. W., Leaderer, B. P., & Bell, M. L. (2012). Effects of ambient pollen concentrations on frequency and severity of asthma symptoms among asthmatic children. *Epidemiology, 23*(1), 55–63.

Dzhambov, A. M., Dimitrova, D. D., & Dimitrakova, E. D. (2014). Association between residential greenness and birth weight: Systematic review and meta-analysis. *Urban Forestry & Urban Greening, 13*(4), 621–629.

Fuertes, E., Markevych, I., Bowatte, G., Gruzieva, O., Gehring, U., Becker, A., et al. (2016). Residential greenness is differentially associated with childhood allergic rhinitis and aeroal-

lergen sensitization in seven birth cohorts. *Allergy, 71*(10), 1461–1471. https://doi.org/10.1111/all.12915.
Fuller, R. A., & Gaston, K. J. (2009). The scaling of green space coverage in European cities. *Biology Letters, 5*(3), 52–355.
Gascon, M., Cirach, M., Martínez, D., Dadvand, P., Valent'in, A., Plasència, A., et al. (2016a). Normalized difference vegetation index (NDVI) as a marker of surrounding greenness in epidemiological studies: The case of Barcelona city. *Urban Forestry & Urban Greening, 19*, 88–94.
Gascon, M., Triguero-Mas, M., Martínez, D., Dadvand, P., Forns, J., Plasància, A., et al. (2015). Mental health benefits of long-term exposure to residential green and blue spaces: A systematic review. *International Journal of Environmental Research and Public Health, 12*(4), 4354–4379.
Gascon, M., Triguero-Mas, M., Martínez, D., Dadvand, P., Rojas-Rueda, D., Plasència, A., et al. (2016b). Residential green spaces and mortality: A systematic review. *Environment International, 86*, 60–67.
Gidlöf-Gunnarsson, A., & Öhrström, E. (2007). Noise and well-being in urban residential environments: The potential role of perceived availability to nearby green areas. *Landscape and Urban Planning, 83*(2), 115–126.
Givoni, B. (1991). Impact of planted areas on urban environmental quality: A review. *Atmospheric Environment Part B Urban Atmosphere, 25*(3), 289–299.
Grazuleviciene, R., Danileviciute, A., Dedele, A., Vencloviene, J., Andrusaityte, S., Uzdanaviciute, I., et al. (2015). Surrounding greenness, proximity to city parks and pregnancy outcomes in Kaunas cohort study. *International Journal of Hygiene and Environmental Health, 218*(3), 358–365.
Greenspace Scotland. (2008). *Health impact assessment of greenspace a guide (Eilidh Johnston)*. Stirling: Greenspace Scotland.
Hagler, G. S. W., Lin, M.-Y., Khlystov, A., Baldauf, R. W., Isakov, V., Faircloth, J., et al. (2012). Field investigation of roadside vegetative and structural barrier impact on near-road ultrafine particle concentrations under a variety of wind conditions. *Science of the Total Environment, 419*, 7–15.
Hanski, I., von Hertzen, L., Fyhrquist, N., Koskinen, K., Torppa, K., Laatikainen, T., et al. (2012). Environmental biodiversity, human microbiota, and allergy are interrelated. *Proceedings of the National Academy of Sciences of the United States of America, 109*(21), 8334–8339.
Hind, S., Mieke, K., Lillian, T., & Michael, B. (2017). Asthma trajectories in a population-based birth cohort. Impacts of air pollution and greenness. *American Journal of Respiratory and Critical Care Medicine, 195*(5), 607–613.
Hoyle, C. R., Boy, M., Donahue, N. M., Fry, J. L., Glasius, M., Guenther, A., et al. (2011). A review of the anthropogenic influence on biogenic secondary organic aerosol. *Atmospheric Chemistry and Physics, 11*(1), 321–343.
Hystad, P., Davies, H. W., Frank, L., Van Loon, J., Gehring, U., Tamburic, L., et al. (2014). Residential greenness and birth outcomes: Evaluating the influence of spatially correlated built-environment factors. *Environmental Health Perspectives, 122*(10), 1095–1102.
James, P., Hart, J. E., Banay, R. F., & Laden, F. (2016). Exposure to greenness and mortality in a nationwide prospective cohort study of women. *Environmental Health Perspectives, 124*(9), 1344–1352.
Kahn, P. H., & Kellert, S. R. (2002). *Children and nature: Psychological, sociocultural, and evolutionary investigations*. Cambridge: Massachusetts Institute of Technology.
Kaplan, R., & Kaplan, S. (1989). *The experience of nature: A psychological perspective*. New York: Cambridge University Press.
Kaplan, S. (1995). The restorative benefits of nature: Toward an integrative framework. *Journal of Environmental Psychology, 15*(3), 169–182.
Kawachi, I., Subramanian, S. V., & Kim, D. (2008). *Social capital and health*. New York: Springer.
Kellert, S. R. (2005). *Building for life: Designing and understanding the human-nature connection*. Washington: Island Press.
Kellert, S. R., & Wilson, E. O. (1993). *The biophilia hypothesis*. Washington: Island Press.

Kesselmeier, J., & Staudt, M. (1999). Biogenic volatile organic compounds (VOC): An overview on emission, physiology and ecology. *Journal of Atmospheric Chemistry, 33*(1), 23–88.
Kuo, F. E., & Taylor, A. F. (2004). A potential natural treatment for attention-deficit/hyperactivity disorder: Evidence from a national study. *American Journal of Public Health, 94*(9), 1580–1586.
Lachowycz, K., & Jones, A. P. (2011). Greenspace and obesity: A systematic review of the evidence. *Obesity Reviews, 12*(5), e183–e189.
Laurent, O., Wu, J., Li, L., & Milesi, C. (2013). Green spaces and pregnancy outcomes in Southern California. *Health & Place, 24*, 190–195.
Lovasi, G. S., O'Neil-Dunne, J. P. M., Lu, J. W. T., Sheehan, D., Perzanowski, M. S., MacFaden, S. W., et al. (2013). Urban tree canopy and asthma, wheeze, rhinitis, and allergic sensitization to tree pollen in a new York City birth cohort. *Environmental Health Perspectives, 121*(4), 494–500. https://doi.org/10.1289/ehp.1205513.
Lovasi, G. S., Quinn, J. W., Neckerman, K. M., Perzanowski, M. S., & Rundle, A. (2008). Children living in areas with more street trees have lower prevalence of asthma. *Journal of Epidemiology and Community Health, 62*(7), 647–649.
Maas, J. (2008). *Vitamin G: Green environments—Healthy environments*. Utrecht: Nivel.
Maas, J., Van Dillen, S. M. E., Verheij, R. A., & Groenewegen, P. P. (2009a). Social contacts as a possible mechanism behind the relation between green space and health. *Health & Place, 15*(2), 586–595.
Maas, J., Verheij, R. A., de Vries, S., Spreeuwenberg, P., Schellevis, F. G., & Groenewegen, P. P. (2009b). Morbidity is related to a green living environment. *Journal of Epidemiology and Community Health, 63*(12), 967–973.
Markevych, I., Thiering, E., Fuertes, E., Sugiri, D., Berdel, D., Koletzko, S., et al. (2014a). A cross-sectional analysis of the effects of residential greenness on blood pressure in 10-year old children: Results from the GINIplus and LISAplus studies. *BMC Public Health, 14*, 477.
Markevych, I., Tiesler, C. M. T., Fuertes, E., Romanos, M., Dadvand, P., Nieuwenhuijsen, M. J., et al. (2014b). Access to urban green spaces and behavioural problems in children: Results from the GINIplus and LISAplus studies. *Environment International, 71*, 29–35.
Marmot, M. (2010). *Sustainable development: The key to tackling health inequalities*. London: Sustainable Development Commission.
Marmot, M., Allen, J., Bell, R., Bloomer, E., & Goldblatt, P. (2012). WHO European review of social determinants of health and the health divide. *Lancet, 380*(9846), 1011–1029.
Martinez, F. D. (2014). The human microbiome. Early life determinant of health outcomes. *Annals of the American Thoracic Society, 11*(Suppl 1), S7–S12.
McCormack, G. R., Rock, M., Toohey, A. M., & Hignell, D. (2010). Characteristics of urban parks associated with park use and physical activity: A review of qualitative research. *Health & Place, 16*(4), 712–726.
McGrath, L. J., Hopkins, W. G., & Hinckson, E. A. (2015). Associations of objectively measured built-environment attributes with youth moderate-vigorous physical activity: A systematic review and meta-analysis. *Sports Medicine, 45*(6), 841–865.
Mitchell, R., & Popham, F. (2008). Effect of exposure to natural environment on health inequalities: An observational population study. *Lancet, 372*(9650), 1655–1660.
Nichani, V., Dirks, K., Burns, B., Bird, A., Morton, S., & Grant, C. (2017). Green space and pregnancy outcomes: Evidence from growing up in New Zealand. *Health & Place, 46*, 21–28.
Nieuwenhuijsen, M. J., & Khreis, H. (2016). Car free cities: Pathway to healthy urban living. *Environment International, 94*, 251–262.
Nieuwenhuijsen, M. J., Khreis, H., Triguero-Mas, M., Gascon, M., & Dadvand, P. (2017). Fifty shades of green: Pathway to healthy urban living. *Epidemiology, 28*(1), 63–71.
Nowak, D. J., Crane, D. E., & Stevens, J. C. (2006). Air pollution removal by urban trees and shrubs in the United States. *Urban Forestry & Urban Greening, 4*(3–4), 115–123.
Organisation for Economic Cooperation and Development (OECD). (2011). *Perspectives on global development 2012: Social cohesion in a shifting world*. Paris: OECD Publishing.

Paoletti, E., Bardelli, T., Giovannini, G., & Pecchioli, L. (2011). Air quality impact of an urban park over time. *Procedia Environmental Sciences, 4*, 10–16.

Pilat, M. A., McFarland, A., Snelgrove, A., Collins, K., Waliczek, T. M., & Zajicek, J. (2012). The effect of tree cover and vegetation on incidence of childhood asthma in metropolitan statistical areas of Texas. *HortTechnology, 22*(5), 631–637.

Rook, G. A. W. (2013). Regulation of the immune system by biodiversity from the natural environment: An ecosystem service essential to health. *Proceedings of the National Academy of Sciences of the United States of America, 110*(46), 18360–18367.

Schwanen, T., Dijst, M., & Dieleman, F. M. (2002). A microlevel analysis of residential context and travel time. *Environment & Planning A, 34*(8), 1487–1508.

Su, J. G., Jerrett, M., de Nazelle, A., & Wolch, J. (2011). Does exposure to air pollution in urban parks have socioeconomic, racial or ethnic gradients? *Environmental Research, 111*(3), 319–328.

Sugiyama, T., Leslie, E., Giles-Corti, B., & Owen, N. (2008). Associations of neighbourhood greenness with physical and mental health: Do walking, social coherence and local social interaction explain the relationships? *Journal of Epidemiology and Community Health, 62*(5), e9.

Tallis, M., Taylor, G., Sinnett, D., & Freer-Smith, P. (2011). Estimating the removal of atmospheric particulate pollution by the urban tree canopy of London, under current and future environments. *Landscape and Urban Planning, 103*(2), 129–138.

Taylor, A. F., & Kuo, F. E. (2009). Children with attention deficits concentrate better after walk in the park. *Journal of Attention Disorders, 12*(5), 402–409.

Taylor, A. F., Kuo, F. E., & Sullivan, W. C. (2001). Coping with ADD the surprising connection to green play settings. *Environment and Behavior, 33*(1), 54–77.

Taylor, B. T., Fernando, P., Bauman, A. E., Williamson, A., Craig, J. C., & Redman, S. (2011). Measuring the quality of public open space using Google earth. *American Journal of Preventive Medicine, 40*(2), 105–112.

Triguero-Mas, M., Dadvand, P., Cirach, M., Martínez, D., Medina, A., Mompart, A., et al. (2015). Natural outdoor environments and mental and physical health: Relationships and mechanisms. *Environment International, 77*, 35–41.

Ulrich, R. (1984). View through a window may influence recovery. *Science, 224*(4647), 224–225.

UN Department of Economic and Social Affairs. (2015). *World urbanization prospects; the 2014 revision*. New York: United Nations.

US Environmental Protection Agency. 2017. Green streets and community open space. Retrieved Jul 30, 2017, from https://www.epa.gov/G3/green-streets-and-community-open-space.

van den Berg, A. E., & van den Berg, C. G. (2011). A comparison of children with ADHD in a natural and built setting. *Child: Care, Health and Development, 37*(3), 430–439.

van den Bosch, M., & Nieuwenhuijsen, M. (2017). No time to lose – Green the cities now. *Environment International, 99*, 343–350.

Van Dillen, S. M. E., de Vries, S., Groenewegen, P. P., & Spreeuwenberg, P. (2012). Greenspace in urban neighbourhoods and residents' health: Adding quality to quantity. *Journal of Epidemiology and Community Health, 66*, e8.

Vorholt, J. A. (2012). Microbial life in the phyllosphere. *Nature Reviews Microbiology, 10*(12), 828–840.

Wells, N. M. (2000). At home with nature effects of greenness on children's cognitive functioning. *Environment and Behavior, 32*(6), 775–795.

WHO Regional Office for Europe. 2016. Urban green spaces and health – A review of evidence. Retrieved Jul 31 2017, from http://www.euro.who.int/en/health-topics/environment-and-health/urban-health/publications/2016/urban-green-spaces-and-health-a-review-of-evidence-2016.

Wilson, E. O. (1984). *Biophilia*. Cambridge: Harvard University Press.

Yatsunenko, T., Rey, F. E., Manary, M. J., Trehan, I., Dominguez-Bello, M. G., Contreras, M., et al. (2012). Human gut microbiome viewed across age and geography. *Nature, 486*(7402), 222–227.

Chapter 21
Air Pollution in Cities: Urban and Transport Planning Determinants and Health in Cities

Barbara Hoffmann

21.1 Introduction

The city is the most frequently inhabited environment for most Europeans and even for most people worldwide; more than 70% of Europeans and more than 50% of the global population live in cities (European Environment Agency 2015). The proportion of city dwellers is expected to increase in the future (Vallance and Perkins 2010). The design of the urban environment plays an important role in shaping human lifestyle and in influencing health and disease of city dwellers.

Cities can influence health and disease in many ways (Mueller et al. 2015). On the one hand, cities can be beneficial for health in terms of supplying resources and education within short distances, health-care services and social interactions. On the other hand, urban living often coincides with risk factors for increased morbidity and mortality; among them are sedentary lifestyles and exposures to environmental pollutants such as air pollution and noise, heat islands and lack of green space.

One of the most important environmental pollutants, which influence morbidity and mortality, is ambient air pollution. In the global burden of disease study, air pollution was ranked fourth as risk factor for premature mortality, only exceeded by hypertension, smoking and dietary risks. Among environmental risk factors, ambient air pollution was the most important cause of disease, leading to more than 4 million premature deaths and more than 100 million DALYs annually worldwide (Global Burden of Disease Study 2016). In comparison, low physical activity due to a sedentary lifestyle is responsible for about 1.6 million premature deaths and 34 million DALYs annually, according to the GBD study (GBD 2016). Important health effects of ambient air pollution include, but are not limited to, cardiovascular

B. Hoffmann (✉)
Environmental Epidemiology Group, Institute of Occupational, Social and Environmental Medicine, Center for Health and Society, Medical Faculty, University Hospital of Düsseldorf, Heinrich-Heine-University of Düsseldorf, Düsseldorf, Germany
e-mail: b.hoffmann@uni-duesseldorf.de

M. Nieuwenhuijsen, H. Khreis (eds.), *Integrating Human Health into Urban and Transport Planning*, https://doi.org/10.1007/978-3-319-74983-9_21

and cerebrovascular disease, chronic respiratory disease and infections and lung cancer, with ischaemic heart disease being responsible for most of the estimated annual premature deaths (GBD 2016).

Important sources of air pollution are anthropogenic as well as natural, with road traffic being one of the most important within-city sources (Nieuwenhuijsen and Khreis 2016). Due to the proximity of emissions, which usually occur where people are living, working and commuting, air pollution hot spots are highly prevalent, and personal behaviour might contribute to high-exposure situations. However, road traffic also determines other aspects of the urban environment. It requires road construction with fixed surfaces, leading to urban heat islands, uptake of urban space and potential green space for roads and parking, promotes sedentary behaviour and social isolation, emits traffic noise and leads to motor vehicle crashes (Nieuwenhuijsen and Khreis 2016).

In this chapter, we will present current knowledge about sources, exposure levels and important drivers of personal air pollution exposure within cities. We will look into health consequences of air pollution and its impact on the population. In doing so, we will specifically focus on potential urban-planning decisions about transportation and road traffic and how they might influence health. Furthermore, using examples, we will present approaches to tackle the health consequences of transport-related air pollution within cities to help decision-makers in shaping the city's future environment in a healthy way.

21.2 Components and Sources of Urban Air Pollution

Ambient air pollution is a complex mixture of particulate and gaseous components. These include airborne particulate matter (PM), which can be divided by size and includes soot, and gaseous pollutants such as ozone (O_3), nitrogen oxides (NO_2, NO), sulphur dioxide (SO_2), volatile organic compounds (VOCs) and carbon monoxide (CO). Pollutants can be primarily emitted from their source, such as diesel soot and NO_2 from diesel-powered combustion engines (referred to as primary pollutants), or they can be formed in the atmosphere from precursor substances (referred to as secondary pollutants). An example for a gaseous secondary pollutant is ozone, which is formed through complex photochemical reactions of nitrogen oxides and volatile organic components. A secondary particle is ammonium sulphate, which is formed from ammonia, emitted by fertilizing agents in agriculture, and SO_2, which is emitted, for example, by coal burning.

Since ambient particulate matter in a specific city comes from a variety of sources, it can differ widely in composition and extent to other cities or even neighbourhoods within one city. A detailed characterization of air pollution is therefore useful when examining the consequences and potential abatement strategies for exposure reduction. By convention, airborne particles are often classified into three major groups by their size, irrespective of their sources or chemical composition, and measured as mass concentration. Particulate matter 10 (PM_{10}) is the mass of all particles with an aerodynamic diameter of <10 μm; $PM_{2.5}$ includes particles with an aerodynamic diameter of <2.5 μm. Ultrafine particles are defined as 100 nm or less and measured as particle number concentration, since they contribute only little to particle mass (Health Effects Institute 2013). In most countries, particle mass concentrations for either PM_{10} or $PM_{2.5}$ are regulated and therefore included in national monitoring networks. Ultrafine particles are not yet measured on a routine basis but only included in dedicated measurement campaigns, often in connection with scientific studies. Next to fine particulate matter, NO_2, SO_2, O_3 and CO are often included in national or regional air quality guidelines and therefore routinely measured at state-operated monitoring stations. In addition, elemental components like metals or specific polycyclic aromatic hydrocarbons (PAH) are measured according to local or state-wide regulations or in the framework of scientific studies to enable an identification of sources.

Important sources of air pollution are anthropogenic as well as natural. Naturally occurring particulate and gaseous air pollutants include sea salt, wind-blown sand dust, combustion products from wild fires, resuspended earth crustal material from volcanic eruptions and biological material such as pollen or secondary particles formed from naturally occurring evaporated precursor gases (EEA 2015). Anthropogenic sources of air pollution, which dominate urban areas due to the high population density, include combustion products from energy production; motorized traffic; household heating with wood, coal or oil; waste incineration, tyre and brake wear; industrial emissions; and emissions from surrounding agricultural areas (EEA 2015). Particles, which have already deposited on surfaces, can be resuspended by wind or moving vehicles. Air pollutants can also be transported over long distances up to hundreds of kilometres, influencing urban environments even if local sources are less important. Combustion processes usually result in smaller particles ($PM_{2.5}$ down to ultrafine particles), whereas particles originating from mechanical processes such as brake and tyre wear will usually be found in the larger particle size fraction (PM_{10}) (WHO Regional Office for Europe 2006).

21.3 Levels of Air Pollution in Cities Worldwide

Ambient air pollution varies in time and space. Recent ground-based air pollution measurements in cities are available for about 3000 human settlements in 103 countries for the years 2008–2015 (WHO 2016). Especially in low- and middle-income countries, air pollution measurements are not readily available, even though the number of monitoring stations has increased substantially within the last years.

Worldwide, air pollution varies strongly by region (Fig. 21.1). While most of North America and Australia shows low to moderate $PM_{2.5}$ concentrations, central Europe has substantially higher air pollution. The highest concentrations can be found in megacities in India and Asia. Very few information is available for South America and Africa; nevertheless, the few monitoring stations imply high air pollution levels in large cities.

$PM_{2.5}$: Fine particulate matter of 2.5 microns or less.

Fig. 21.1 Measured $PM_{2.5}$ in nearly 3000 settlements worldwide, 2008–2015 (WHO 2016)

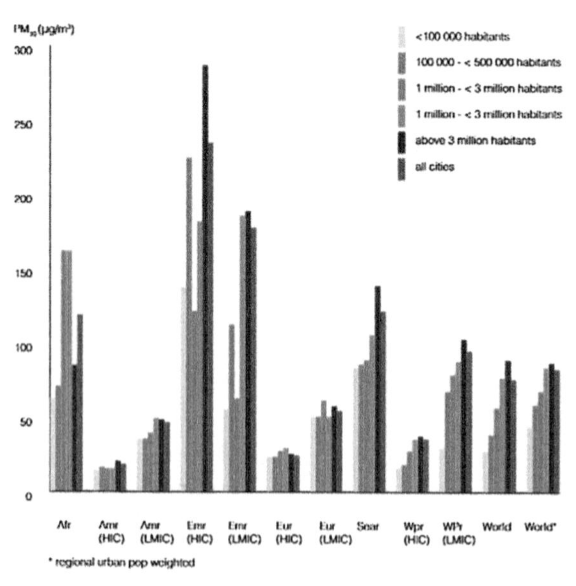

PM_{10}: Particulate matter of 10 microns or less; Afr: Africa; Amr: Americas; Emr: Eastern Mediterranean; Eur: Europe; Sear: South-East Asia; Wpr: Western Pacific; LMIC: low- and middle-income countries; HIC: high-income countries. PM_{10} values for the world are regional urban population-weighted.

Fig. 21.2 PM_{10} measurements by world region and size of city, 2008–2015 (WHO 2016)

PM_{10} measurements by city size show a clear gradient for increasing air pollution by increasing size of the city (Fig. 21.2). Once again, the highest exposure can be found in the WHO Eastern Mediterranean world region, including cities such as Abu Dhabi, Doha and Riyadh.

For Europe, the European Environment Agency presents regular assessments of spatial and temporal variation of ambient air pollution based on routine monitoring networks across Europe (http://www.eea.europa.eu/themes/air), showing a north-south gradient of air pollution with higher pollutant concentrations in the Mediterranean cities and conurbations. This was also observed in the European Study of Cohorts on Air Pollution Effects (ESCAPE), showing substantial differences in $PM_{2.5}$ exposure between Scandinavian and Mediterranean study areas (Eeftens et al. 2012).

While spatial variation of air pollution is mostly related to the presence of local and regional sources, temporal variation of daily exposures is mostly related to meteorological conditions. Barometric pressure, wind speed and direction, temperature and sunlight, rainfall and the height of the mixing layer of air affect the formation, dispersion and dilution of emissions in the atmosphere. To a smaller degree, cyclic changes in measured pollutant concentration also depend on changes of emission patterns during the course of a year, a week or even a day. For example, rush-hour traffic in cities has a cyclic pattern with emission peaks of traffic-related pollutants in the early morning and late afternoon. Working days have different emission patterns than weekend days, and household heating as well as energy production might change in the course of a year due to seasonal requirements.

21.4 Guideline Values and Regulation

Since air pollution is detrimental to health, the WHO has issued air quality guidelines including guideline values and interim targets to promote healthy environments (Table 21.1). Using modelled ambient $PM_{2.5}$ values to estimate the annual

Table 21.1 WHO air quality guidelines for particulate matter, ozone, nitrogen dioxide and sulphur dioxide

Pollutant	Concentration, μg/m^3	Averaging period
PM_{10}	50	24 h
	20	1 year
$PM_{2.5}$	25	24 h
	10	1 year
NO_2	200	1 h
	40	1 year
SO_2	500	10 min
	20	24 h
Ozone	100	Maximal daily 8-h mean

Global update 2005 (WHO Regional Office 2006)

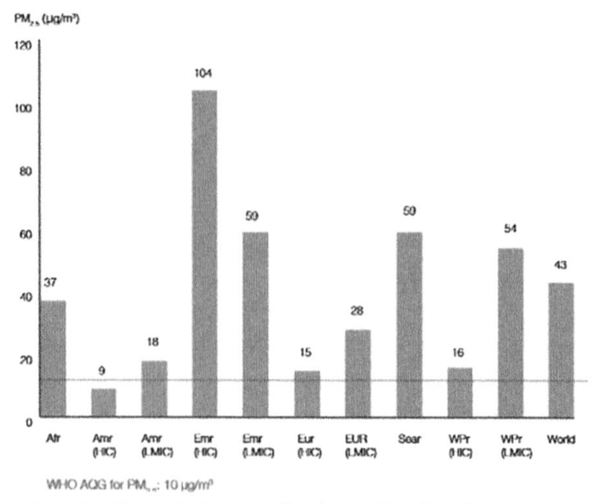

Afr: Africa; Amr: Americas; Emr: Eastern Mediterranean; Eur: Europe; Sear: South-East Asia, Wpr: Western Pacific; LMIC: low- and middle-income countries; HIC: high-income countries. $PM_{2.5}$: particulate matter with an aerodynamic diameter of 2.5 μ$PM_{2.5}$: pa. WHO AQG: WHO Air Quality Guidelines.

Fig. 21.3 Modelled annual median exposure to $PM_{2.5}$ by region. Urban population only, year 2014 (WHO 2016)

median exposure by region, only few world regions can actually reach these guideline values for their urban population (Fig. 21.3). Even within Europe, the EU-funded study ESCAPE revealed non-attainment of WHO guideline values for most of the study areas included in this international project (Eeftens et al. 2012). Air pollution has increased in more than half of the cities specifically in the Western Pacific, South-East Asia and Eastern Mediterranean world regions during the last 5 years, while in Europe and North America, a general trend towards decreased exposures is seen (WHO 2016).

21.5 Important Determinants Driving Population Exposure in Cities

Air pollutant concentrations are typically higher in cities than in the surrounding semiurban or rural areas, a fact that is due to the higher density of emission sources in cities. One specific aspect of urban air pollution is the proximity of emissions to the residential population, resulting in high population exposures. This holds true in an exceptional way for traffic-related pollutants, since people live, work and commute in close proximity to traffic. In the European Study of Cohorts on Air Pollution Effects (ESCAPE), $PM_{2.5}$ concentrations at traffic sites were on average 14% higher than at urban background sites (Eeftens et al. 2012). Large urban-rural differences were also found for soot (average 38% higher) and PM_{coarse} (42% higher). Ultrafine particles peak at the road side and decrease exponentially with increasing distance perpendicular to the road (Karner et al. 2010), leading to exposure hot spots along highly trafficked roads. Air pollution from diesel exhaust, rich in NO_x and cancerogenic diesel particles, is high in countries with a large proportion of diesel engine-powered cars such as Germany. Household heating with wood, oil or coal also leads to high emissions directly next to where people live and work. Small wood stoves and open fireplaces placed within the main living quarters in the house, which are increasingly popular even in cities, additionally increase indoor exposures substantially.

21.6 Health Effects of Air Pollution

Air pollution as a complex mixture of particulate and gaseous constituents has multiple health effects. Due to the concurrent exposure to several pollutants and the moderate to high correlation of many pollutants across time and space, it is often not possible to tease apart the health effects of a single pollutant or component of the mixture.

Air pollution affects the cardiovascular system, respiratory organs, metabolism, immune responses, the unborn child and the brain (Table 21.2) (Thurston et al. 2017). Depending on duration and intensity of exposure, the severity of health effects ranges from subclinical findings, such as slightly elevated inflammatory markers that are not noticed by the individual, to clinical symptoms and findings, such as respiratory complaints and decrease in lung function, to overt clinical disease. The worst case is the elicitation of myocardial infarction, stroke or lung cancer, eventually culminating in increased mortality.

Table 21.2 Summary of health effects of urban air pollution

Time patter	Health effect
Acute effects	Increased mortality
	Increase of cardiovascular and cerebrovascular events (myocardial infarction, stroke, congestive heart failure)
	Exacerbation of lung disease, asthma attacks, lung function decline
	Increased perinatal mortality
Chronic effects	Decrease of life expectancy
	Increased incidence of cardiovascular and cerebrovascular disease, atherogenesis
	Worsening of lung disease, lung function decline
	Lung cancer
	Increased incidence of metabolic disease (diabetes mellitus)
	Accelerated cognitive decline in adults
	Decreased birth weight

21.6.1 Thresholds for Health Effects and Susceptible Populations

Since there are no known lower thresholds for most air pollutants on human health, the whole population is exposed to the adverse effects of air pollution. However, not every person is equally susceptible to adverse health effects. Specific population subgroups who have been shown to be especially susceptible are children, the elderly and chronically ill people. In the industrialized high-income western societies, the proportion of elderly and chronically ill individuals is increasing. Genetic factors, lifestyle and additional environmental or other external risk factors might superimpose their effects and may further increase susceptibility to air pollution effects.

21.6.2 Biological Mechanisms

The biological mechanisms, by which air pollutants cause health effects, comprise three major pathways (Franklin et al. 2015). (1) After inhalation of particulate or gaseous air pollutants, the substances will be deposited in the respiratory tract depending on their size and/or solubility. For example, smaller particles will penetrate deeper into the lungs, with ultrafine and submicron particles being able to reach the alveoli. At the place of deposition, pollutants will elicit a local inflammatory reaction, which can already have consequences for lung health. Furthermore, this localized inflammation leads to the secretion of inflammatory mediators into the lung tissue, from where it can access the systemic circulation. Eventually, these mediators can promote a subclinical systemic inflammation, a process for which the term "spillover hypothesis" has been coined. This subclinical systemic inflammation can have a multitude of health relevant consequences: the endothelial function of the blood vessels decreases, leading to a state of increased vasoconstriction and a rise in blood pressure. The propensity of blood to coagulate is increased, while fibrinolysis is decreased—increasing the risk for thromboembolic complications such as myocardial infarction and stroke. Systemic inflammation also promotes inflammation of the adipose tissue, which can result in decreased insulin sensitivity, making the development of type 2 diabetes mellitus (T2DM) more likely. (2) A second biological pathway involves the activation of pulmonary receptors of the autonomic nervous system. The delicate balance between the sympathetic and the vagal autonomic tone will be disturbed by up-regulating the sympathetic response. This up-regulation induces an increase in heart rate, blood pressure and risk of arrhythmias. (3) A third biological mechanism involves the direct intrusion of particles into the blood stream and eventually into specific organs. Ultrafine particles and specific pollutant components that were carried into the lungs attached to particle surfaces (e.g. cancerogenic PAHs attached on the carbon core of diesel soot) can cross biological membranes, enter the blood stream and reach organ tissues. Ultrafine particles have also been shown to migrate from the nasal respiratory epithelium alongside the olfactory nerve into the frontal brain regions (Block et al. 2012).

21.6.3 Latency of Health Effects

The latency period and required time for the above-mentioned biological mechanisms varies from minutes and hours to months and years (Langrish et al. 2012a). For example, the activation of pulmonary reflexes with resulting consequences for autonomous nervous system control will be accomplished within minutes. Many studies have shown acute increases in blood pressure after a few hours of increased exposure (Brook 2005). Other effects may be achieved within days, such as the elicitation of a local or systemic inflammatory response. Yet again, other mechanisms, such as the development of cancer after exposure to cancerogenic PAHs, are a matter of years.

21.6.4 Mortality

The by far most important and alarming effect of air pollution exposure is increased mortality. Short-term changes in exposure from day to day, which is a natural consequence of changes in meteorology and in part also due to daily emission patterns resulting from commuting and industrial activity, lead to an increase in daily total mortality of about 0.4–1.0% per increase in PM_{10} of 10 μg/m^3 (WHO REVIHAAP 2013). Per 10 μg/m^3 increase in 8-h average ozone, daily mortality increases by 0.3% (WHO fact sheet 2016). Per 20 ppb NO_2 (24-h average), daily mortality increases by 0.5–8.1% (Integrated science assessment, U.S. EPA 2016). The short-term effects of ultrafine particles on mortality have been less well studied so far with inconsistent results. First studies point to increased mortality within a few days after exposure increases (Kettunen et al. 2007). One multicentre study reported an exceptionally high but imprecise effect with an increase of 9.9% (95% CI 6.3–28.8%) in respiratory mortality per increase of 2750 particles/ml in the 6-day average (Lanzinger et al. 2016).

Living in a place with higher air pollution will also increase mortality compared to others who live in cleaner areas of the city. The assessment of health effects of chronic exposure is difficult and very cost-intensive but is very important because it reveals cumulative health effects over a long time period. Early studies conducted in the USA yielded a 4% increase in total mortality per 10 μg/m^3 $PM_{2.5}$ increase (Pope III et al. 2002). The recent ESCAPE study was specifically designed to measure and estimate long-term air pollution exposures in more than 20 European mostly urban study regions and to assess long-term health effects on mortality and a multitude of other health outcomes. This study showed that chronic exposure is associated with an increase in natural cause mortality by 7% (95% CI 2–13%) per 5 μg/m^3 $PM_{2.5}$

(Beelen et al. 2014). A large cohort study by Crouse et al. (2012) showed that even below current air quality guidelines, the association is linear. This finding underscores that the population will benefit from lowering air pollution, regardless of the current absolute level of air pollution. Pope et al. (2009) quantified the benefit by showing an increase in average life expectancy of about 9 months per decrease in average $PM_{2.5}$ exposure by 10 μg/m^3 over a period of 20 years.

21.6.5 Cardiovascular and Cerebrovascular Effects

Short-term changes in air pollution concentrations are associated with incidence of fatal and non-fatal myocardial infarctions, stroke, acute deterioration of congestive heart failure and hospital admissions for cardiovascular disease (Cesaroni et al. 2013; WHO 2013). In Copenhagen, the risk for incidence of stroke increased by 21% (95% CI 4–14%) per increase in mean daily ultrafine particle number by 3918 particles/ml (Andersen et al. 2010). On a subclinical level, air pollution causes systemic inflammation, endothelial dysfunction, increased blood clotting and reduced fibrinolysis (Franklin et al. 2015). Blood pressure and the likelihood for arrhythmias increase, and heart rate variability, a measure of a well-functioning autonomous nervous system, decreases. While these subclinical effects can lead to manifest detrimental sequelae within hours to days, like triggering an acute myocardial infarction or stroke, they also illustrate the biologic mechanisms by which long-term consequences of air pollution exposure lead to the development of cardiovascular disease in the long run.

The long-term consequences of air pollution exposure on cardiovascular health are detrimental and are the major contributor to overall air pollution effects on premature mortality and morbidity. According to the GBD study, ischaemic heart disease and stroke are responsible for 2.4 million premature deaths annually worldwide, which represent about 57% of all deaths and 50.1% of DALYs caused by ambient air pollution in 2015 (GBD 2015). In the multicentre European ESCAPE study, the probability to suffer from a cardiovascular event rises by 12% (HR: 1.12 95% CI 1.01–1.25) per 10 μg/m^3 PM_{10} long-term exposure at the residence (Cesaroni et al. 2013). In the same study, risk for stroke was also increased, specifically in the elderly (Stafoggia et al. 2014). In the Heinz Nixdorf Recall Study, which is located in the German metropolitan Ruhr Area, the risks for stroke, given an increase from the 5th to the 95th percentile of exposure, were 2.61 [95% CI 1.13–6.00] for PM_{10} and 3.20 [95% CI 1.26–8.09] for $PM_{2.5}$.

Several cohort studies suggest that long-term exposure to air pollution promotes the development and progression of atherosclerosis, the underlying pathology for most cardiovascular events and one of the major contributors to many chronic diseases of the elderly (Hoffmann et al. 2007). Animal experiments have supported this observational finding. The above-mentioned biological mechanisms point to different pathways, by which air pollution can exert its influence on vascular pathology. One of the most important risk factors for the development and progression of atherosclerosis is arterial hypertension. Recently, it has been shown in the ESCAPE study that individuals exposed to higher levels of air pollution are more likely to develop hypertension during a mean follow-up of 9 years (Fuks et al. 2016). This is a particular important finding, as the potential effect of road traffic noise was accounted for in the analysis and because hypertension is the number one global risk factor for premature morbidity and mortality (Lim et al. 2012).

21.6.6 Respiratory Effects

Air pollution can cause short-term effects after transient high-exposure situation as well as long-term consequences for respiratory health after years of living in an area with high exposure. Within hours of high exposure, specifically vulnerable people (e.g. with pre-existing asthma or chronic obstructive pulmonary disease (COPD)) can experience shortness of breath, cough and a decrease in lung function, possibly leading to a hospital admission (McCreanor et al. 2007). In the long run, living in dirty areas of a city can cause a decrease in lung growth in children and adolescents, an accelerated decrease in lung function in adults and possibly even the development of asthma and COPD, a disease predominantly affecting year-long smokers (Gauderman et al. 2004; Khreis et al. 2017a). A reduction in traffic-related air pollution on the other hand can lead to an improvement of respiratory health in children as well as in adults as was shown in two studies that observed an increased development of lung function in children and an attenuated decline of lung function in adults when air pollution decreased (Gauderman et al. 2015; Imboden et al. 2009).

Ambient air pollution also leads to the development of lung cancer. In the European ESCAPE study, we observed an increased risk for lung cancer of 22% per 10 $\mu g/m^3$ PM_{10} (Raaschou-Nielsen et al. 2013). The International Agency for Research on Cancer (IARC) has classified diesel soot as carcinogenic to humans in 2012—this is one reason among others, why high exposure to traffic-related air pollution can lead to lung cancer over many years.

21.6.7 Emerging Health Effects of Air Pollution

Further well-founded evidence is available for air pollution effects on metabolic health, such as type 2 diabetes mellitus and insulin resistance, impaired cognitive development in children and accelerated cognitive decline and dementia in adults, as well as influences on the unborn child such as decreased birth weight and preterm birth (Clifford and Zaman 2016). New studies are under way to investigate these emerging health effects and quantify their impact on the population. However, already it is quite clear that the current impact of air pollution on human health is underestimated, because important health effects such as these are not included in the calculations. We can expect that the long-term consequences of living in bad air will be larger than currently estimated.

21.6.8 Co-exposures

Air pollution is but one of several contextual urban exposures with the ability to influence human health. To adequately capture all health effects related to urban living and transport, a broader view must be taken. However, so far only few studies have taken a more comprehensive approach to investigate health effects of urban living and transport. This is a growing area of research, and we are likely to see more studies including air pollution, noise, green space, temperature and the social environment in the near future. First evidence shows that co-exposure of air pollution and road traffic noise might actually synergize in their deleterious effects on cognitive function in adults in a manner that people with double exposure experience more cognitive impairment than if they had received the two exposures separately (Tzivian et al. 2017). This hypothesized synergism will put even more pressure on regulators and city planners to provide a safe and healthy environment.

21.7 Evidence for Beneficial, Neutral or Adverse Effects on Health by Specific Interventions

Many potential interventions are available (Khreis et al. 2017b), and urban-planning decisions that will have a positive effect on air pollution-related health effects in cities can be made. Most of these interventions do not target only one specific air pollutant or one specific health outcome but are broad and intersect with positive effects on climate change, heat stress and noise exposure abatement, leading to broad health and environmental benefits (Münzel et al. 2017; Nieuwenhuijsen et al. 2017).

Interventions to reduce health effects of urban air pollution can be accomplished on different regulatory levels. Many emission reduction interventions will be implemented by state or regional authorities, for example, emission reductions through improved combustion engine technology or new and improved fuels. Strict air quality limit values will support the development of electric transport combined with a sustainable power generation strategy. This will synergize with positive effects on greenhouse gas emissions (Tobollik et al. 2016). The already implemented clean air policies are expected to reduce air pollution by approximately half between 2010 and 2020 (Tobollik et al. 2016).

Another approach that has been taken in some exemplary situations is active emission removal through vegetation. It has been suggested that specific mosses, ornamental plants or trees may act as a sink to improve air quality in highly polluted areas (Saxena and Ghosh 2013). The positive effect of cooling and reduction of greenhouse gas emissions can be another positive effect; however, many questions are still open, and the success of this strategy remains to be seen. Specifically, trees or other plants may inhibit effective ventilation of dense inner-city roads, which could be disadvantageous to air quality.

Comprehensive city-specific policy mixes have shown the greatest benefit for air quality and population exposure in the most dense and most polluted areas of the city (Silveira et al. 2016). In London, next to the EU-wide emission standards and air quality regulations, buses were equipped with diesel particulate filters; a Low Emission Zone, limiting access for highly polluting vehicles, was implemented; new low-emission hybrid buses acquired; and further local-scale schemes were implemented. This comprehensive intervention lead to relevant reductions in air pollution exposure (Font and Fuller 2016). However, achieved reductions were heterogeneous across the city through changes in traffic flow patterns; even increases in air pollution exposure are possible. It is therefore necessary to take a comprehensive multilevel approach to regulation and traffic limitations.

Restricted access zones are currently a widely used approach in Europe to reduce air pollution exposure of the highly exposed urban population. Different models exist, from small incoherent areas or streets closed to specific vehicles (e.g. heavy-duty vehicles) during specific times of the day or week to larger coherent areas with multiple emission reduction measures (i.e. London). The resulting improvements are variable: In Venice, for example, single car-free days did not have a positive effect on air pollution, while the complex intervention of the London Low Emission Zone and comprehensive actions during the Olympic Summer Games 2008 in Beijing lead to substantial reductions of several air quality indicators (Font and Fuller 2016). Another form of traffic-related intervention is the reorganization of traffic into arterial roads and quiet quarters. This can lead to a decrease in road-side pollution through better traffic flows and less polluted and quiet islands within the city through reduction of traffic (Thaker and Gokhale 2016).

The most sustainable, challenging and promising intervention is a modal shift for transportation, which means a decreased reliance on individual combustion engine-powered cars and an increased use of public transport, bikes and other forms of active transport. This modal shift will lead to many positive effects in addition to its direct improvement of air quality. Through less dependence on combustion engine-powered vehicles, emissions, including greenhouse gas emissions, will be reduced, with a resulting positive health effect; noise exposure decreases; less space is taken up by more and wider roads, leaving more room for green spaces and leisure time activities in an improved environment; and physical activity of residents increases. Cities have implemented a variety of measures to promote this beneficial and necessary modal shift, ranging from parking restrictions in inner cities to improvement of public transport (Kwan et al. 2016), construction of bike paths and walking trails, implementation of public bike rental stations and promoting car sharing. An early intervention, which may lead to long-lasting beneficial behavioural effects, is the premium for high school students who ride their bikes to school instead of using publicly financed bus system in the city of Münster in North Rhine-Westphalia,

Germany, and the free public transport and railway tickets provided to all university students in North Rhine-Westphalia, Germany.

Active transport by itself is a highly beneficial effect of the surge towards a sustainable modal shift, because it will increase physical activity, with benefits outweighing by far any possible increase in air pollutant uptake by increased ventilation. Mueller et al. (2015) emphasize that physical inactivity is one of the most important risk factors for morbidity and mortality. When compared to international recommendations, the greatest impact on preventable mortality could be achieved by increasing population physical activity. Further factors were air pollution, noise and heat stress. Through the provision of a functional green infrastructure within cities, several positive effects can be achieved simultaneously—increasing physical activity through active transport and leisure time physical activity as well as a reduction in traffic (Mueller et al. 2015).

Personal protective measures are a final last resort, which can be helpful for those with pre-existing medical conditions that predispose to adverse acute effects of air pollution. The American Heart Association, for example, recommends limiting outdoor activities on high-pollution days, specifically in those patients with coronary artery disease (Brook et al. 2010). Filtering household air or wearing face masks is now increasingly prevalent in Asian cities. However, the surgical masks, which are mostly sold and used in drugstores, do not fit tight and will allow polluted air to enter through the sides. High-efficiency filters which are effectively used in the occupational realm are uncomfortable and will seldom be worn long time (Langrish et al. 2012b). Moreover, the costs of personal protection material will increase health disparities between the affluent and the less privileged populations, increasing existing environmental inequity.

21.8 Conclusions

The city is the most frequently inhabited environment for most Europeans and even for most people worldwide. Cities can influence health and disease in many ways, including ambient air pollution as one of the most important environmental pollutants. Sources of urban air pollution include traffic, industry and household heating. Due to the proximity of emissions to the population, road traffic emissions are particularly important drivers of exposure. Depending on the dominant sources in a specific city or area of the city, the amount, chemical components and physical characteristics of the complex air pollution mixture can vary substantially. Air pollution concentrations vary from day to day depending on meteorology and diurnal variations in source emissions. Short-term increases in air pollution can have acute effects such as triggering myocardial infarctions and stroke or the elicitation of asthma attacks. Long-term exposures can lead to the development and progression of many chronic diseases, including coronary artery disease, lung cancer, diabetes mellitus and dementia. Through combined interventions, it is possible to substantially decrease urban air pollution. Many potential interventions will also have synergistic effects in terms of climate change, heat stress and noise exposure mitigation and increasing physical activity. The promotion of a modal shift to more public and active transportation seems to be the most promising avenue for healthy city planning.

References

Andersen, Z. J., Olsen, T. S., Andersen, K. K., Loft, S., Ketzel, M., & Raaschou-Nielsen, O. (2010). Association between short-term exposure to ultrafine particles and hospital admissions for stroke in Copenhagen, Denmark. *European Heart Journal, 31*, 2034–2040.

Beelen, R., Stafoggia, M., Raaschou-Nielsen, O., Andersen, Z. J., Xun, W. W., Katsouyanni, K., et al. (2014). Long-term exposure to air pollution and cardiovascular mortality: An analysis of 22 European cohorts. *Epidemiology, 25*, 368–378.

Block, M. L., Elder, A., Auten, R. L., Bilbo, S. D., Chen, H., Chen, J. C., et al. (2012). The outdoor air pollution and brain health workshop. *Neurotoxicology, 33*, 972–984.

Brook, R. D. (2005). You are what you breathe: Evidence linking air pollution and blood pressure. *Current Hypertension Reports, 7*, 427–434.

Brook, R. D., Rajagopalan, S., Pope, C. A., III, Brook, J. R., Bhatnagar, A., Diez-Roux, A. V., et al. (2010). Particulate matter air pollution and cardiovascular disease: An update to the scientific statement from the American Heart Association. *Circulation, 121*, 2331–2378.

Cesaroni, G., Badaloni, C., Gariazzo, C., Stafoggia, M., Sozzi, R., Davoli, M., et al. (2013). Long-term exposure to urban air pollution and mortality in a cohort of more than a million adults in Rome. *Environmental Health Perspectives, 121*, 324–331.

Clifford, K. L., & Zaman, M. H. (2016). Engineering, global health, and inclusive innovation: Focus on partnership, system strengthening, and local impact for SDGS. *Global Health Action, 9*, 30175.

Crouse, D. L., Peters, P. A., van Donkelaar, A., Goldberg, M. S., Villeneuve, P. J., Brion, O., et al. (2012). Risk of nonaccidental and cardiovascular mortality in relation to long-term exposure to low concentrations of fine particulate matter: A Canadian national-level cohort study. *Environmental Health Perspectives, 120*, 708–714.

Eeftens, M., et al. (2012). Spatial variation of PM2.5, PM10, PM2.5 absorbance and PMcoarse concentrations between and within 20 European study areas and the relationship with NO_2 – Results of the ESCAPE project. *Atmospheric Environment, 62*, 303–317.

European Environment Agency. (2015). About urban environment. https://www.eea.europa.eu/themes/sustainability-transitions/urban-environment. Retrieved Aug 2, 2017, from https://www.eea.europa.eu/themes/sustainability-transitions/urban-environment.

Font, A., & Fuller, G. W. (2016). Did policies to abate atmospheric emissions from traffic have a positive effect in London? *Environmental Pollution, 218*, 463–474.

Franklin, B. A., Brook, R., & Arden Pope, C., 3rd. (2015). Air pollution and cardiovascular disease. *Current Problems in Cardiology, 40*, 207–238.

Fuks, K. B., Weinmayr, G., Hennig, F., Tzivian, L., Moebus, S., Jakobs, H., et al. (2016). Association of long-term exposure to local industry- and traffic-specific particulate matter with arterial blood pressure and incident hypertension. *International Journal of Hygiene and Environmental Health, 219*, 527–535.

Gauderman, W. J., Avol, E., Gilliland, F., Vora, H., Thomas, D., Berhane, K., et al. (2004). The effect of air pollution on lung development from 10 to 18 years of age. *The New England Journal of Medicine, 351*, 1057–1067.

Gauderman, W. J., Urman, R., Avol, E., Berhane, K., McConnell, R., Rappaport, E., et al. (2015). Association of improved air quality with lung development in children. *The New England Journal of Medicine, 372*, 905–913.

GBD, & Risk Factors Collaborators. (2015). Global, regional, and national comparative risk assessment of 79 behavioural, environmental and occupational, and metabolic risks or clusters of risks in 188 countries, 1990–2013: A systematic analysis for the Global Burden of Disease Study 2013. *Lancet, 386*, 2287–2323.

GBD, & Mortality and Causes of Death Collaborators. (2016). Global, regional, and national life expectancy, all-cause mortality, and cause-specific mortality for 249 causes of death, 1980–2015: a systematic analysis for the Global Burden of Disease Study 2015. *Lancet, 388*, 1459–1544.

HEI Review Panel on Ultrafine Particles. (2013). *Understanding the health effects of ambient ultrafine particles. HEI perspectives 3*. Boston: Health Effects Institute.

Hoffmann, B., Moebus, S., Mohlenkamp, S., Stang, A., Lehmann, N., Dragano, N., et al. (2007). Residential exposure to traffic is associated with coronary atherosclerosis. *Circulation, 116*, 489–496.

Environmental Protection Agency, United States, U.S. E.P.A. (2016). Integrated science assessment. Retrieved Aug 2, 2017, from https://www.epa.gov/isa.

Imboden, M., Schwartz, J., Schindler, C., Curjuric, I., Berger, W., Liu, S. L., et al. (2009). Decreased pm10 exposure attenuates age-related lung function decline: Genetic variants in p53, p21, and ccnd1 modify this effect. *Environmental Health Perspectives, 117*, 1420–1427.

Karner, A. A., Eisinger, D. S., & Niemeier, D. A. (2010). Near-roadway air quality: Synthesizing the findings from real-world data. *Environmental Science & Technology, 44*, 5334–5344.

Kettunen, J., Lanki, T., Tiittanen, P., Aalto, P. P., Koskentalo, T., Kulmala, M., et al. (2007). Associations of fine and ultrafine particulate air pollution with stroke mortality in an area of low air pollution levels. *Stroke, 38*, 918–992.

Khreis, H., Kelly, C., Tate, J., Parslow, R., Lucas, K., & Nieuwenhuijsen, M. J. (2017a). Exposure to traffic-related air pollution and risk of development of childhood asthma: A systematic review and meta-analysis. *Environment International, 100*, 1–31.

Khreis, H., May, A. D., & Nieuwenhuijsen, M. J. (2017b). Health impacts of urban transport policy measures: A guidance note for practice. *Journal of Transport & Health, 6*, 209–227.

Kwan, S. C., Tainio, M., Woodcock, J., & Hashim, J. H. (2016). Health co-benefits in mortality avoidance from implementation of the mass rapid transit (mrt) system in Kuala Lumpur, Malaysia. *Reviews on Environmental Health, 31*, 179–183.

Langrish, J. P., Bosson, J., Unosson, J., Muala, A., Newby, D. E., Mills, N. L., et al. (2012a). Cardiovascular effects of particulate air pollution exposure: Time course and underlying mechanisms. *Journal of Internal Medicine, 272*, 224–239.

Langrish, J. P., Li, X., Wang, S., Lee, M. M., Barnes, G. D., Miller, M. R., Cassee, F. R., Boon, N. A., Donaldson, K., Li, J., Li, L., Mills, N. L., Newby, D. E., & Jiang, L. (2012b). Reducing personal exposure to particulate air pollution improves cardiovascular health in patients with coronary heart disease. *Environmental Health Perspectives, 120*(3), 367–372.

Lanzinger, S., Schneider, A., Breitner, S., Stafoggia, M., Erzen, I., Dostal, M., et al. (2016). Ultrafine and fine particles and hospital admissions in Central Europe. Results from the fire study. *American Journal of Respiratory and Critical Care Medicine, 194*, 1233–1241.

Lim, S. S., Vos, T., Flaxman, A. D., Danaei, G., Shibuya, K., Adair-Rohani, H., et al. (2012). A comparative risk assessment of burden of disease and injury attributable to 67 risk factors and risk factor clusters in 21 regions, 1990–2010: A systematic analysis for the global burden of disease study 2010. *Lancet, 380*, 2224–2260.

McCreanor, J., Cullinan, P., Nieuwenhuijsen, M. J., Stewart-Evans, J., Malliarou, E., Jarup, L., et al. (2007). Respiratory effects of exposure to diesel traffic in persons with asthma. *The New England Journal of Medicine, 357*, 2348–2358.

Mueller, N., Rojas-Rueda, D., Cole-Hunter, T., de Nazelle, A., Dons, E., Gerike, R., et al. (2015). Health impact assessment of active transportation: A systematic review. *Preventive Medicine, 76*, 103–114.

Münzel, T., Sorensen, M., Gori, T., Schmidt, F. P., Rao, X., Brook, J., et al. (2017). Environmental stressors and cardio-metabolic disease: Part i-epidemiologic evidence supporting a role for noise and air pollution and effects of mitigation strategies. *European Heart Journal, 38*, 550–556.

Nieuwenhuijsen, M. J., & Khreis, H. (2016). Car free cities: Pathway to healthy urban living. *Environment International, 94*, 251–262.

Nieuwenhuijsen, M. J., Khreis, H., Verlinghieri, E., Mueller, N., & Rojas-Rueda, D. (2017). Participatory quantitative health impact assessment of urban and transport planning in cities: A review and research needs. *Environment International, 103*, 61–72.

Pope, C. A., III, Burnett, R. T., Thun, M. J., Calle, E. E., Krewski, D., Ito, K., et al. (2002). Lung cancer, cardiopulmonary mortality, and long-term exposure to fine particulate air pollution. *JAMA, 287*, 1132–1141.

Pope, C. A., Ezzati, M., & Dockery, D. W. (2009). Fine-particulate air pollution and life expectancy in the United States. *The New England Journal of Medicine, 360*(4), 376–386.

Raaschou-Nielsen, O., Andersen, Z. J., Beelen, R., Samoli, E., Stafoggia, M., Weinmayr, G., et al. (2013). Air pollution and lung cancer incidence in 17 European cohorts: prospective analyses from the European Study of Cohorts for Air Pollution Effects (ESCAPE). *The Lancet Oncology, 14*(9), 813–822.

Saxena, P., & Ghosh, C. (2013). Ornamental plants as sinks and bioindicators. *Environmental Technology, 34*, 3059–3067.

Silveira, C., Roebeling, P., Lopes, M., Ferreira, J., Costa, S., Teixeira, J. P., et al. (2016). Assessment of health benefits related to air quality improvement strategies in urban areas: An impact pathway approach. *Journal of Environmental Management, 183*, 694–702.

Stafoggia, M., Cesaroni, G., Peters, A., Andersen, Z. J., Badaloni, C., Beelen, R., et al. (2014). Long-term exposure to ambient air pollution and incidence of cerebrovascular events: Results from 11 European cohorts within the escape project. *Environmental Health Perspectives, 122*, 919–925.

Thaker, P., & Gokhale, S. (2016). The impact of traffic-flow patterns on air quality in urban street canyons. *Environmental Pollution, 208*, 161–169.

Thurston, G. D., Kipen, H., Annesi-Maesano, I., Balmes, J., Brook, R. D., Cromar, K., et al. (2017). A joint ERS/ATS policy statement: What constitutes an adverse health effect of air pollution? An analytical framework. *The European Respiratory Journal, 49*(1), pii: 1600419. https://doi.org/10.1183/13993003.00419-2016.

Tobollik, M., Keuken, M., Sabel, C., Cowie, H., Tuomisto, J., Sarigiannis, D., et al. (2016). Health impact assessment of transport policies in Rotterdam: Decrease of total traffic and increase of electric car use. *Environmental Research, 146*, 350–358.

Tzivian, L., Jokisch, M., Winkler, A., Weimar, C., Hennig, F., Sugiri, D., et al. (2017). Associations of long-term exposure to air pollution and road traffic noise with cognitive function – An analysis of effect measure modification. *Environment International, 103*, 30–38.

Vallance, S., & Perkins, H. (2010). Is another city possible? Towards an urbanised sustainability. *City, 14*, 448–456.

WHO. (2016). World health Organization. Ambient (outdoor) air quality and health. Fact sheet. Retrieved Aug 2, 2017, from http://www.who.int/mediacentre/factsheets/fs313/en/.

WHO Regional Office for Europe. (2006). Air quality guidelines global update 2005: particulate matter, ozone, nitrogen dioxide and sulfur dioxide. Copenhagen. Retrieved Aug 2, 2017, from http://www.euro.who.int/__data/assets/pdf_file/0005/78638/E90038.pdf?ua=1. Accessed 02.08.2017.

WHO Regional Office for Europe. (2013). Review of evidence on health aspects of air pollution – Revihaap project: Technical report. Copenhagen. Retrieved 2 Aug, 2017, from http://www.euro.who.int/__data/assets/pdf_file/0004/193108/REVIHAAP-Final-technical-report-final-version.pdf?ua=1.

WHO Regional Office for Europe. (2016). Ambient air pollution: A global assessment of exposure and burden of disease. In: *Ambient air pollution: A global assessment of exposure and burden of disease*. Geneva. http://apps.who.int/iris/bitstream/10665/250141/1/9789241511353-eng.pdf. Accessed 02.08.2017.

Chapter 22
Noise in Cities: Urban and Transport Planning Determinants and Health in Cities

Peter Lercher

22.1 The Reasons and the Scope of the Noise Problem

Historically, urbanization is clearly one of the drivers of increasing noise problems in the society. However, there are other more important drivers, like the growing general mobility demands, paired with an enormous increase in urban freight transport (De Vos and Van Beek 2011). Although a trend for suburbanization of freight center activity is observed—the goods need to be brought to the shops and the customer in the city. Therefore, around megacities both centrifugal and centripetal processes are at work (Cidell 2010; Malecki 2014). The size of each force depends on monocentric or polycentric development—but all lead to increased person and freight traffic. In contrast to other environmental stressors (air pollution), noise exposure is often still increasing in European cities, in spite of large efforts with the advent of the Environmental Noise Directive.

Supranational institutions and governments approach the noise problem mainly by two means. Either by opinion surveys about satisfaction with noise climate and noise annoyance in cities or by physical noise level mapping. Most complete information is available from Europe, where the Environmental Noise Directive (END) required member states to carry out systematic noise mapping. Non-systematic information is available in Asia (from megacities), North America (Metropolitan areas), South America (Brasilia), and less from Africa—although data sources and awareness increase.

P. Lercher (✉)
Medical University Innsbruck, Innsbruck, Austria
e-mail: peter.lercher@i-med.ac.at

M. Nieuwenhuijsen, H. Khreis (eds.), *Integrating Human Health into Urban and Transport Planning*, https://doi.org/10.1007/978-3-319-74983-9_22

22.1.1 Satisfaction with Noise Climate

The latest Eurobarometer survey on the "Perception of Quality of life in European Cities" indicates a wide variation in satisfaction with the noise climate (European Commission 2016). The levels of total satisfaction (highly + rather satisfied) in cities range from 88% (Oulu, FI) down to 28% (Istanbul, TR). On the dissatisfaction side, the highest percentage comes from Istanbul (72%) and the lowest from Oulu and Newcastle (11%). Less is known about how people with certain health conditions rate their city environment. People with learning, speech, and mental disorders and dyslexia were more dissatisfied with their current neighborhood environments (Shiue 2016).

22.1.2 The Contribution of the Major Noise Sources to the Dissatisfaction with the Noise Climate

Considering transportation noise across Europe, by far the major reason for dissatisfaction is road traffic noise (80%) both inside and outside agglomerations, followed by rail traffic and air traffic. In urban agglomerations of the EU, about 94 million people are exposed to more than 55 dBA, Lday from road traffic noise, nearly 6 million by railway, and 2.25 million by airport noise. Note: neighbor noise ranks second in the list of reasons for dissatisfaction with the noise climate, and obviously complaints about this source are higher in cities with higher population density and higher shares of apartment housing when compared to suburban or rural areas.

22.1.3 The Measured Noise Levels in Cities

The EU noise-mapping data for European cities (Source: First round of noise mapping, Brussels 2008) show—like the satisfaction data (Satisfaction survey 2009/2012, Brussels 2015)—substantial variation between cities. However, the correlation between the two sets of data is not very high.

In Table 22.1 the noise exposure at night in European capitals is shown in the left column. The two other columns indicate the percentage of people satisfied with the noise climate or the neighborhood. The observable differences are not unexpected: while the noise indicator measures the physical intensity of the sound reasonably correct, the perceived affectedness by noise depends rather on the soundscape characteristic of the city and other contextual factors (more in Sects. 22.2 and 22.3). For example, Paris and Madrid show below 20% with Lnight > 55 dBA, but satisfaction with noise is in the lower range.

Table 22.1 Nighttime sound levels in European capitals and satisfaction: % of exposed population above 55 dBA, Lnight compared to satisfaction with noise climate, and the beauty of the neighborhood

City	Noise level Lnight>55 dBA[a] (%)	Satisfied with noise[a] (%)	Satisfied with neighborhood[b] (%)
Stockholm SE	4	69	45
Berlin DE	9	60	46
Amsterdam NL	11	69	46
Paris FR	11	43	48
London UK	12	73	n.a.
Vilnius LT	12	68	36
Helsinki FI	14	74	52
Wien AT	14	69	44
Madrid SP	15	31	48
Bratislava SK	16	53	45
Warsaw PL	18	46	47
Copenhagen DK	20	73	n.a.
Prague CZ	21	49	46
Dublin IE	22	83	48
Tallinn EE	26	60	47
Bucharest BG	31	27	34
Budapest HU	36	52	46
Sofia BL	39	40	18
Athens GR	n.a.	33	21
Lisbon PT	n.a.	51	37
Roma IT	n.a.	37	36

Green colors indicate better noise or satisfaction values; orange means less favorite conditions
n.a. not available
[a]2012
[b]2009

A study in Hong Kong (Lam and Chung 2012) determined that 36.2% of its population suffers from road traffic noise exceeding governmental guidelines. A recent monitoring campaign found mean daytime $L_{eq,1\text{-}h}$ values ranged between 54.4 and 70.8 dBA, while the mean nighttime $L_{eq,1\text{-}h}$ values ranged from 52.6 to 67.9 dBA (To et al. 2015).

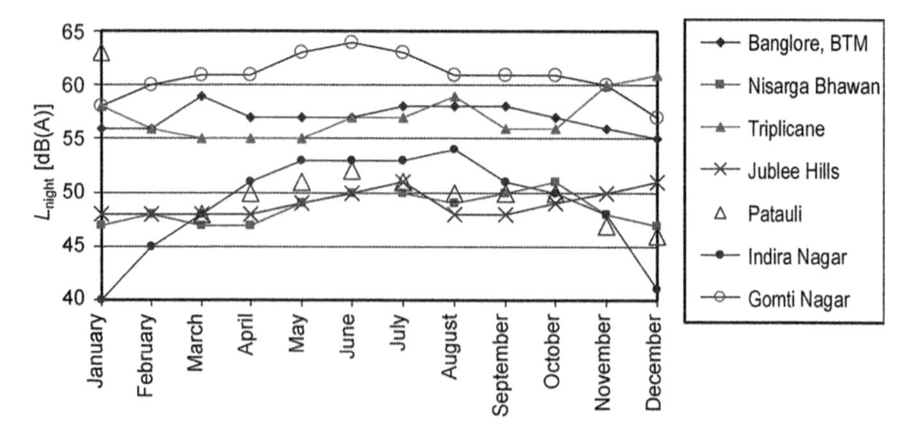

Fig. 22.1 Monthly variation in equivalent sound pressure level, Lnight for residential areas in seven Indian cities in year 2014. Source: Garg N et al. (2017)

A representative monitoring program in India also found large variations in noise levels across seven major cities. The night noise levels of three cities (Fig. 22.1) continuously exceed the 55 dBA level. This level should not be exceeded to avoid sleep disturbance and other health effects (Kim and van den Berg 2010). Figure 22.1 indicates also that most Indian cities show higher noise levels in summer.

Due to the large data and missing legal requirements, noise-mapping initiatives outside the EU are usually limited to a restricted (most exposed) area of a city or areas close to traffic arteries.

Large difference in noise exposure are not only observed between cities but also within cities. A huge noise-mapping effort in Seoul, Korea, makes this important difference visible (Lee et al. 2014b). The Lnight exceedance of 55 dBA varies from 15% to 65% between districts of Seoul (mean = 35%).

Note: although noise levels are generally highest around roadways and near transportation terminals, the exposure of the large majority of city inhabitants is determined by a city's overall background. Within the framework of the EU-URGENCHE project (http://www.urgenche.eu/project/), the study in Basle indicated (Keuken et al. 2014) that about 90% of the city's population is affected by urban background levels, the rest by the somewhat higher levels from direct dwellings' to road, rail, or airport noise. A study in New York estimated the additional street noise exposure to contribute approximately 4% to an average individual's annual noise dose (McAlexander et al. 2015).

An additional issue in this context is the distribution of environmental justice related to exposure to noise in the light of segregation tendencies. On average—across cities or within city areas—often no difference or ambiguous empirical results regarding differential noise exposure were found (Havard et al. 2011; Bocquier et al. 2013; Carrier et al. 2016a, b; Dale et al. 2015; Guo et al. 2017). In Hong Kong, for instance, differential results were obtained: the socially deprived groups tend to live in building groups exposed to higher noise levels. This is particularly pronounced in private housing estates where older, less educated, those engaged in craftsman and elementary jobs, and nonowners of their dwellings are exposing to higher levels of traffic noise—while in public housing estates, this relationship is not found to be a significant factor (Lam and Chung 2012). The same differential relationship was found for air pollution in Hong Kong (Fan et al. 2012).

Even though, the correlation between air pollution and noise may vary strongly (0.25–0.85) between cities (Allen et al. 2009) and across city areas (Foraster et al. 2011), the relation is strongest at the road side (Kim et al. 2012), differs across the time of the day (Kim et al. 2012; Ross et al. 2011), and depends on the emission indicator used (Can et al. 2011; Gan et al. 2012). Note: there is a substantial difference between exposure to noise and air pollution depending on traffic volume. While already a small traffic volume increases noise up to 55 dBA (Seto et al. 2007), higher traffic volumes are required to increase ultrafine particulate exposure (Shu et al. 2014). The possible reasons for the observed large variation in levels of noise between and within cities will be discussed in the next section. The reasons for the discrepancy between objective exposure levels and subjective citizen assessment are related to accompanying co-exposures (air pollution, vibration), the differing potential to cope (sleeping room to the quiet side), and restoration options (quiet and green areas).

22.2 The Main Known Determinants of Noise Levels in Cities

In the literature, a number of factors which can determine noise levels were investigated at different scales: the underlying city form/morphology, aspects of the built environment, population and household density, noise regulations, traffic network density and composition, amount of traffic jams, the number and type of sound sources (e.g., scooters, tramways), the quality of the tire-road interaction, the average insulation of the homes, and the motorized driving behavior.

The following subsections will shortly inform about relevant indicators and determinants and its relationship with noise exposure in cities.

22.2.1 Macro-Scale Structures (Space, Infrastructure, Population, Geographic Context) and Noise

The recent trend to increase urban densities and avoid sprawl certainly was paralleled by changes in urban form and structure. All changes may have possible implications for noise levels experienced by inhabitants at different scales of the city. Some intended planning changes may work well on average at the macro scale but have possible adverse effects at the neighborhood or dwelling scale, where people actually spend most of their time.

22.2.1.1 Population Density

Urban density is expressed in various ways, such as building density, road network density, and population density. The variation of density between cities is very large. A comparison between Greater Manchester in the UK (19.5 inhabitants/ha) and Wuhan in China (48.8 inhabitants/ha) showed lower average ground noise levels in Wuhan: Lavg and L50 is by 5.6–6.5 dBA lower than that of the Greater Manchester samples (Wang and Kang 2011). Note: at the building façade, levels of Lavg and L50 are even higher in Greater Manchester (7.7 and 9.1 dBA, respectively). This is in line with data from the International Association of Public Transport (UITP) which showed that population densities of 50–60 inhabitants/ha are required to increase the share of journeys on foot, by bicycle, and by public transport (UITP 2002).

22.2.1.2 General Urban Form and Morphology

The comparison between Greater Manchester in the UK and Wuhan in China revealed also that the density of buildings in Wuhan is much higher than that in Greater Manchester, creating a huge area that is lowvehicle accessible and with high-rise street canyons (Wang and Kang 2011). Conversely, the road coverage in the Wuhan cases is much lower than that in the Greater Manchester area (1.7% vs 11.3%). It should, however, be noted that some Wuhan areas have equally high noise levels (wide-open space and high-density residential areas). Urban morphology can also determine the amount of green space and the consecutive noisescape. The analysis of (Margaritis and Kang 2017) found that the effect of green spaces on traffic noise pollution varies according to the scale of analysis. Only at the kernel level (small areas of 500 × 500 m) a positive relation between green and noise was found, supported by a further cluster analysis.

22.2.1.3 Traffic Network Density and Traffic Elasticity: Traffic Composition

Traffic volume is typically represented by vehicle kilometers (VKM) driven on the road network. Traffic volume and (population) density is somehow related. However, the relation is not as uniform as some positions in urban planning let us believe.

In the US substantial variations in traffic and noise within and between cities were observed in a comparison between larger cities (Lee et al. 2014a). The total number of vehicle counts explained a substantial amount of variation in measured ambient noise in Atlanta (78%), Los Angeles (58%), and New York City (62%).

For the cities Amsterdam and Rotterdam, it was found that the average sound level increased with road network density and vehicle kilometers per square kilometer per 24 h and decreased with ground space index (building coverage), floor space index (building intensity), and population density (Salomons and Berghauser Pont 2012). This is only true in these two cases (with traffic elasticity of −1). In the author's simulations with other values of traffic elasticity (e = 0 and −0.5 = high car use in densely populated areas), the average sound level increases with increasing FSI and GSI. This means that restrictive measures (e.g., environmental zones) need to be implemented in such cases to protect inhabitants of dense inner city areas from increases in traffic noise. An investigation into the temporal and spatial variability of noise in the city of Toronto made clear that noise variability was predominantly spatial in nature. Spatial variability accounted for 60% of the total observed variations in traffic noise (Zuo et al. 2014), and traffic volume, length of arterial road, and industrial area were the three most important variables. Another study into the spatial determinants of traffic noise in the Taiwanese city of Taichung found road width, traffic flow rates, and land-use types (commercial and industrial sites) significantly associated with annual average LAeq, 24 h. Furthermore, noise levels at 125 Hz revealed the strongest prediction in the multiple regression model—indicating that low-frequency components from vehicles constitute a large proportion of road traffic noise (Wang et al. 2016). This result is supported by specific investigations into the contribution of traffic composition to observed noise levels. A study in San Francisco points to the special importance of heavy truck vehicles and buses (Seto et al. 2007). Trucks and buses emit also strong low-frequency noise and vibrations (see Sect. 22.4.1.2). Eventually, pass-by noises of motorbikes and other two-wheelers are especially annoying due to its spectral and temporal signal properties (Paviotti and Vogiatzis 2012; Morel et al. 2012; Gille et al. 2016).

22.2.1.4 Traffic and Noise Level Assessment: Modeling and Mapping at Different Scales

Traffic modelers found the variability of sound levels in cities largely explained by the following variables: traffic flow, type of vehicles, and average vehicle speed (Kumar et al. 2014). In the city of Toronto, spatial variability accounted for 60% of the total observed variations in traffic noise, and traffic volume, length of arterial road, and industrial area explained the majority of the spatial variability (70%). Only a fine-tuned hourly model found that regions with roads of high traffic volumes, high share of heavy goods vehicles, and being close to activity centers have larger impact on the prevailing ambient noise (Alam et al. 2017). This is only the picture at the macro scale.

The spatial heterogeneity of the built-up environment in cities is substantial, and the local traffic flow, the temporal distribution, the street design and road maintenance, and the building facades induce certain errors and require the inclusion of mesoscale and microscale factors to carry out a proper assessment. In the SONORUS project, the measured noise levels of an inner yard in the city of Gothenburg were compared with typical sound-mapping software: an underestimation of 13 dBA was observed. With an improved software (QSIDE), the differences were reduced to 3 dBA difference (Kropp et al. 2016). In addition, perception-related aspects need to be considered. Sound sources which produce prevalent noticed-sound events (NSE) will produce higher annoyance. Cities with a high share of two-wheelers but also cities with frequent horn use in traffic differ (Phan et al. 2009, 2010; Nassiri et al. 2013) and need other modeling approaches (Torija and Ruiz 2012, 2016). Significant differences in sound frequency components were observed between land-use types, with noise levels at low frequencies (125 Hz), had the highest correlation with total traffic, and revealed highest prediction in the multiple linear regression (up to 80%).

Moreover, a cautious note: noise estimations from land-use regression models, widely used in air pollution assessment, are not yet reliable enough—even when supplemented with short-term noise measurements. While road traffic noise estimates correlate reasonably enough with annoyance reports (Aguilera et al. 2015)—this is not the case with railway and aircraft noise assessments (Ragettli et al. 2015). An evaluation against improved noise-mapping software (e.g., QSIDE) is necessary before it's widely use in health effect studies—especially, when combined effects of transportation noise are an issue. Another important issue—relevant for both planning and epidemiologic studies: noise and air pollution levels correlate moderately on average across cities but poorly at smaller scales (Seto et al. 2007; Fecht et al. 2016).

22.2.1.5 Land Use

Studies studying city design effects on the macro scale reveal often controversial results regarding noise levels or impact on health. Most land-use effect studies have methodological limitations due to its ecological nature, have not investigated noise levels, and cover mainly cities of North America. Only few studies include information on land use and noise level. One study found statistically significantly higher levels of environmental noise in mixed-use neighborhoods (Lday = 8 dBA, Lnight = 6 dBA) when compared with classical residential neighborhoods in Halifax, Canada (King et al. 2012).

A further Canadian study found elevated risk of shortened sleep and sleep problems associated with living in dense commercial, residential areas, and living near industrial sites with subjective noise measures (Chum et al. 2015). A related study also found that the level of urbanicity was negatively associated with infant sleep duration (Bottino et al. 2012). Another study found that living in green spaces reduces the risk of short sleep duration (Astell-Burt et al. 2013)—but noise levels were not investigated. Likewise, an ecological study of 49 large US cities found no association between greenness and mortality from heart disease, diabetes, lung cancer, or automobile accidents—but did not control for noise (Richardson et al. 2012). It is likely that the macro-scale level of analysis did average out the spatial variation necessary to detect effects. Also greener cities cannot always be considered as quieter, as the effect of green spaces on traffic noise pollution varies according to the scale of analysis (Margaritis and Kang 2017). Eventually, how does land use and the built environment affect health equity and environmental justice and avoid segregation? See the notes to this subject in Sect. 22.3.5.2. Ecological and social effects.

22.2.2 *Meso- and Microscale Structure (Interface Between Building Structure, Neighborhood, and Human Perception) and Noise*

The large spatial heterogeneity of urban environments makes it difficult to calculate reliably noise levels with one program (Hornikx 2016). Apartments that are exposed to the same level of road traffic noise at the most exposed façade often have very different neighborhood soundscapes. A Norwegian study showed substantial variation, depending whether you included also the immediate neighborhood (75 m radius) where people live. A later study showed: when, e.g., 50-m radius buffers are compared with 400-m buffers, the assigned $L_{\mathrm{Aeq,24\,h}}$ level can vary across buildings from −9.4 to +22.3 dBA (Tenailleau et al. 2015). Lam and Ma expanded this area even further (600 m) and called it "community" activity space. Therefore, in noise health effects studies, it should be noted that noise-mapping results can be substantially distorted by the local neighborhood shape and size which is included in noise exposure assessment (Kropp et al. 2016).

22.2.2.1 Building Block, Street Design, and Neighborhood Quality

Several investigations have found certain building block and façade designs to provide better shielding and avoid scattering than others (Onaga and Rindel 2007; Lam et al. 2013; Salomons and Berghauser Pont 2012). From these investigations it can be concluded:

- Closed building blocks lead to lower noise levels at quiet facades than open building blocks.
- The percentage of quiet area is negatively related to the street canyon index, both for the whole development site and at the periphery.
- In high-façade streets, the absorption of the façade is determining reverberation between buildings—while in low-facade streets, scattering by the buildings is equally important.

Two extreme examples about the amount of shielding of building block shapes in the high-rise city of Hong Kong are shown in Fig. 22.2. It has been found that the so-called sky-view factor can assist in predicting the shadowing effect (Silva et al. 2017).

An analysis of residential blocks development in Tianjin (China) found low-rise small blocks having the largest spread (~5 dBA) in observed sound level—although on average they show slightly lower levels than other block designs. In the same block design, a larger lot area (size of land where the block is built) was associated with lower noise levels, a higher street coverage ratio with higher noise levels. Further building layout indicators such as ground space index (GSI), street interface density (SID), and floor space index (FSI) influenced noise levels at the buildings (Zhou et al. 2017).

Fig. 22.2 Building block shapes in Hong Kong with different shading of noise. Shading from green color (lower sound levels) to red (high sound levels) (courtesy of KC Lam)

In Hong Kong, Lam and Ma (2012) found that residential complexes built in the 1990s showed lower noise levels at all scales of their analysis (community, neighborhood, dwelling).

22.2.2.2 Building Form and Location of Sleeping and Living Rooms

What is often underestimated by planners, but the shape, design, and position of the façade relative to the roads, the slope of the roof, the greening of the roof (Van Renterghem and Botteldooren 2009), eaves/louvers (Sakamoto and Aoki 2015), and the form of the balconies (Tang 2005; Lee et al. 2007; Naish et al. 2012), can make a noticeable difference to noise levels (see Fig. 22.2). The possible cumulative effect of such factors should be considered in the planning and the design of city buildings. Of overarching importance is the form, the position, and the shape of the building which determines the amount of shielding the building provides to guarantee a quiet façade toward the backyard, where sleeping and living rooms should be located.

22.3 The Main Effects on Health and Health-Related Quality of Life (HRQoL) in Cities

Typical transportation noise in modern cities acts as an ambient stressor via the nervous system (Evans 2001). Although, we can live and adapt with noise to some extent, the efforts of coping with noise come with some costs. Even in the short term, after effects will occur (Glass and Singer 1972).

In the long term, the possible health costs arrive via our evolutionary heritage. It prevents our body to fully habituate during sleep (although we adapt on the surface by waking up less often). Notably, the cardiovascular system is still responding without observed parallel changes in the brain (Muzet 2007) as measured by the electroencephalography (EEG). The way we react depends on our genetic makeup, the coping abilities, and the restoration options in and around our home. The physical intensity of the noise—as measured by a conventional noise meter in dB and with A-weighting—is a reasonable indicator of noise effects at high levels (>65 dBA). Already at intermediate sound levels (50–65 dBA) and particularly at levels below 50 dBA, the signal to noise ratio (the "noticed events") is the decisive element in how humans respond to the sound environment (Schomer and Wagner 1996; Sneddon et al. 2003; De Coensel et al. 2009). The human sensory system reacts relative to the quiet (the L95 or L90 percentile level) as experienced in the surroundings: the so-called adaptation level (Helson 1964). Only a difference of 10 dBA between background and peak level is required to trigger autonomous reactions (Chang et al. 2015). Equivalent research shows that the "startle response" can be reduced when the signal to noise ratio is below 15 dBA (Franklin et al. 2007). For instance, already 100 cars/h can result in equivalent noise levels around 55 dBA with approximate peak levels of up to 70 dBA (e.g., Fig. 7 in Seto et al. 2007). Considering a background level of 40–45 dBA, a high signal to background noise ratio results and creates already well noticeable events for cardiovascular reactions and can interfere with sleep. Therefore, planning needs to protect the city dweller not only against high peak levels but also against an increase in background levels and noticeable events.

22.3.1 Severe Noise Annoyance

In health risk assessment, the estimated effects of noise on annoyance in studies to arrive at disability adjusted life years (DALYs) are determined by assigning a disability weight (DW) to the health outcome in question (high annoyance). In the WHO DALY assessment (Fritschi et al. 2011), a low value (DW = 0.02) was used. Another study revealed a DW of 0.033 (median, 0.03; range, 0.01–0.12) for communication disturbance. Thus, any choice implies a high uncertainty for annoyance estimates in risk assessment. In a European national scale study, the authors even left out severe annoyance as health outcome (Hänninen et al. 2014), which leads to a substantial underestimation of the problem and is scientifically not well justified.

Annoyance is measured in field studies by means of two ICBEN or ISO 15666 defined questionnaire items (Fields et al. 2001). Either using a 5-point verbal or an 11-point numerical scale (0–10), with end points "not annoyed" up to "extremely annoyed". With the 11-point scale, the highly annoyed indicator (cut-off point = 72% of scale length = 8 + 9 + 10) is derived. With the verbal 5-point scale (equal steps), the upper two steps (60% scale length) are—in accordance with the ICBEN recommendation—used predominantly to indicate severe annoyance. The hitherto available exposure-response relations (Miedema and Oudshoorn 2001; EEA 2010) are becoming outdated with the advent of the new WHO evidence reviews on this subject (Guski et al. 2018). Figure 22.3 shows the summary of the updated exposure-response relations observed from available studies. Based on the GRADE system, 112 studies were rated with respect to the risk of bias, and overall 57 studies were included in the quantitative meta-analysis. We limit the discussion to the

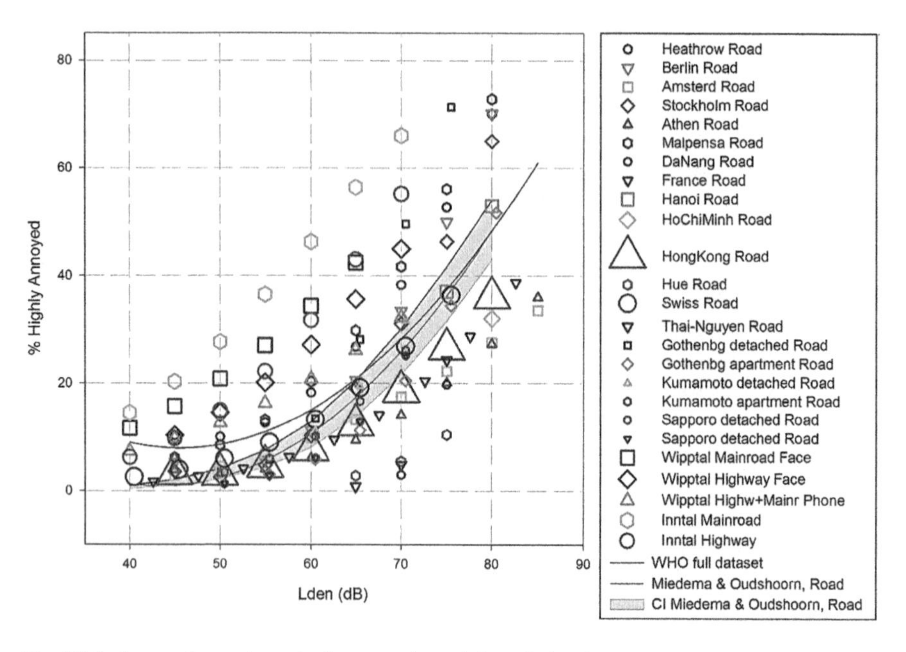

Fig. 22.3 Scatterplot and quadratic regression of the relation between Lden and the calculated percentage highly annoyed for 25 road traffic noise studies, together with the old exposure-response function by Miedema and Oudshoorn (2001) in red with confidence interval

results of the road traffic findings (34.211 respondents). The aggregated regression line is strongly influenced by the very large Hong Kong study ($N = 10.077$) at the lower end of the annoyance response. The observed heterogeneity in the noise annoyance range from 50 to 75 dBA; Lden is large. Although among the highest scorer are three studies from predominantly rural-suburban areas (transalpine freight traffic routes), there is also a wide spread among road traffic annoyance in larger cities. The lowest annoyance response being from Athens and Hong Kong—the highest originate from London-Heathrow, Milan-Malpensa, Stockholm, Berlin, and Gothenburg (with detached housing). All those higher responses level off against the WHO full set from 60 dBA onward.

A further meta-analysis of 21 available studies using Pearson correlations between L_{den} or L_{dn} and road traffic noise annoyance raw scores yields a summary correlation of 0.325 (95% CI 0.273–0.375). This means that the explained variance of the relationship between noise level (Lden) and high annoyance is only 10.6%. Thus, the contribution of the noise level alone is low. (Fidell et al. 2011) coined the term "community tolerance level" (CTL) for the observed large between community differences in annoyance prevalence rates. These authors hypothesize that the large spread is caused by a combination of acoustic and non-acoustic factors. Schomer et al. (2013) added that non-acoustic factors and non A-weighted factors (such as fluctuation, emergent sound events) reflect the context in which the sonic environment is perceived. This is not captured by the physical noise level indicator alone, and the "community tolerance level" is therefore an indicator for the local distortion of the community soundscape through the road traffic load. Apart from individual factors (noise sensitivity or attitude to the source), reviews of non-acoustic factors (Fields 1993; Lercher 1996; Guski 1999; Flindell and Stallen 1999; Miedema and Vos 1999; Schomer 2005) underline the significant contribution of policy relevant situational (e.g., bedroom position, house type, quiet facade), environmental (e.g., air pollution, vibration, quiet and green space), and traffic variables (traffic composition, road type, road surface, speed) as effect-modifying factors of the annoyance response. Furthermore, studies having included such non-acoustic factors with extended analyses (multiple regression or SEM) revealed much higher variance explanation (up to 60%) of annoyance effects (YANO et al. 2002; Hong and Jeon 2015; Oiamo et al. 2015; von Lindern et al. 2016; Lercher et al. 2017). There is also indication from specialized research that taking account of certain acoustic features (non-A-weighted factors), not captured by the standard physical noise indicators, such as strong low frequency, tonal, impulsive components, modulations, noticed events, can increase further the variance explanation of annoyance. The advancement and lower cost of recent development in sensor technique (Sevillano et al. 2016; Zappatore et al. 2017) will make the use of such advanced sound monitoring also feasible for larger studies in the near future.

22.3.2 Severe Sleep Disturbance

Noise-induced sleep disturbances are mainly caused by transportation noise (Finegold et al. 2008). Sleep disturbances show the largest share of disability-adjusted life-years (DALYs) lost from environmental noise in Europe, based on a WHO expert report (Fritschi et al. 2011). The disability weight applied (DW = 0.07) was also more conservative. The upper uncertainty of the DALY assessment would be estimated with a DW of 0.10. Research on the effects of environmental noise on sleep distinguishes three major types of possible effects. Primary effects are related to the actual noise exposure (difficulty to fall asleep, waking up, difficulty to regain sleep, too early awakening, feeling tired). Secondary effects (after effects) are predominantly related to performance, mood, and proneness to accidents on the upcoming day. Tertiary effects (long-term effects) are related to the development of chronic diseases. The last line of evidence is the most difficult to observe and detect. It would require longitudinal studies of 20 years of duration. Nevertheless, uninterrupted sleep of sufficient length is also vital for next day functioning, cognitive performance, physical fitness, and mood. Therefore, the public health importance of primary and secondary short-term effects should not be underestimated.

The difficulties to obtain reliable estimates are mainly related to the fact that sleep disturbance is highly prevalent among adults (Ford et al. 2015; Cao et al. 2017; Pallesen et al. 2014) (up to 20%) and associated with a high comorbidity (Morin and Jarrin 2013; Levenson et al. 2015). As both noise and sleep loss are independently associated with the development of hypertension, noise (but also air pollution) is thought to act as a possible moderator (Akinseye et al. 2015). But there

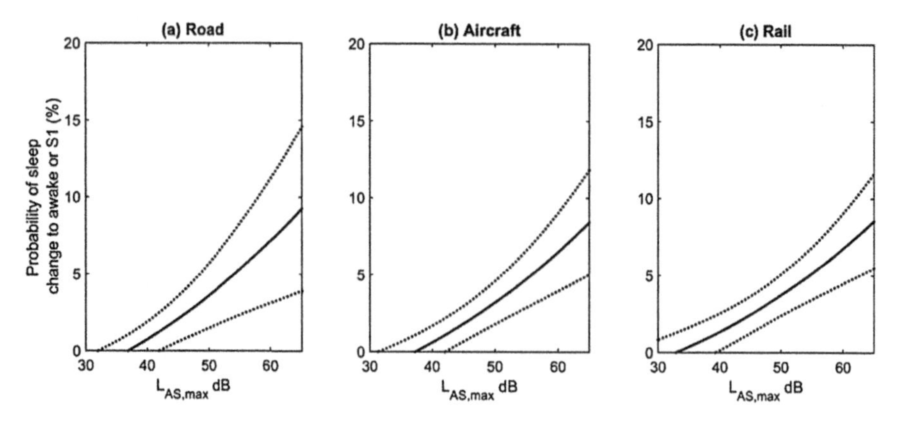

Fig. 22.4 Probability of additional sleep stage changes to wake or S1 in a 90 s time window following noise event (spontaneous awakenings removed) related to maximum indoor sound pressure level (L AS, max). Ninety-five percent confidence intervals (dashed lines), unadjusted models

are additional health-related stressful life conditions (family, and work stress, financial stress, etc.) which can considerably affect sleep (Vedaa et al. 2016). Thus, to reliably extract the single contribution of the noise exposure, sleep laboratory or small field studies are the only option where you can apply the "gold standard" of polysomnography (which is invasive to some degree) to separate out the "true" noise effect and account also for spontaneous awakenings (~5–8%). This means to impose restrictions with regard to health (mostly younger and healthy people) and selection of the sample investigated and to lose representability with respect to the actual population exposed, where illnesses and noise exposure occur simultaneously. The current WHO evidence update on sleep (Basner and McGuire 2018) provides moderate quality evidence (with dose response, $N = 2$) for an increase in cortical awakenings with measured indoor LAS, max sound levels (OR 1.36 (CI 1.19–1.55)) per 10 dB increase of road traffic noise (starting at 30 dB). Similar results were obtained also for aircraft and train noise (Fig. 22.4).

Studies using less invasive motility measures proved not to be an alternative, revealed inconsistent results and were of lower quality. Moderate evidence was also found for self-reported measures of sleep disturbance (OR 2.12 (CI 1.82–2.48)) with a 10 dBA increase in road traffic sound (Lnight, outdoors); however, only in those studies ($N = 12$) where the sleep question was asked with reference to noise sources. The estimates for train noise were higher (OR 3.06 (CI 2.38–3.93)). When the sleep question was asked without reference to noise as potential source, the odds ratio for sleep disturbance was increased but not significant (all traffic sources). This supports two possible interpretations: either people are not aware of the traffic noise associations of their sleep problems (underestimation) until they are reminded or it is a systematic bias (overestimation) by introducing noise in the question. Probably, both factors are at work in practice. People habituate to some degree at the behavioral level (less conscious awakenings), but at the EEG and the level of the autonomous nervous system, the effect is still detectable. No results were reported for cardiovascular effects of sleep disturbance. This may contribute to another possible underestimation of noise-related sleep impairments on health. It is well known that repeated autonomic arousals habituate much less to noise than cortical arousals (Muzet 2007). Thus, the true estimate for road traffic in the general population may be between the estimates for cortical arousals and self-reported sleep disturbance, when mentioning noise sources in the interview in the range of an OR of 1.5–1.7 per 10 dB increase.

Neither secondary nor tertiary effects were subject of this evidence review. Also studies on sleep medication use associated with noise exposure were not included in this review.

The WHO intervention evidence review (Brown and van Kamp 2017) reports several types of interventions. High-quality evidence (GRADE) for a decrease in sleep disturbance came from six studies, where various interventions were implemented (façade insulation, tunnel, quiet-side concept) which resulted in noise level reductions of various degrees (7–12 dBA).

Low-quality evidence (GRADE, only cross-sectional studies) when noise was lower in the bedroom. Specifically, location of the bedroom at the quiet site resulted in decreased annoyance at night Beta = 0.81 (CI 0.65–0.97) in the large sample of the HYENA study (Babisch et al. 2012).

22.3.3 *Effects on Cardiovascular and Metabolic Health*

In the WHO burden of disease (BoD) from environmental noise assessment (Fritschi et al. 2011), the specific attributable fraction for ischemic heart disease (IHD) in people exposed to road traffic noise in agglomerations with more than 250,000 inhabitants was estimated to be 1.8%. For people with higher exposure, it was: 65–69 dBA (6.3%), 70–74 dBA (13.9%), and >75 dBA (23.2%). This was responsible for 2.7% of the DALY's, while 58% were attributable to sleep disturbance and 38% to severe annoyance. No estimates could be derived for hypertension and associated illnesses at that time. In the new WHO evidence assessment, the evidence base increased substantially, and further related health endpoints could be considered (van Kempen et al. 2018). The most robust evidence (GRADE, high quality) was extracted from studies of road traffic noise and ischemic heart disease. It was based on three cohort and four cross-sectional studies (7033 cases). The risk increase per 10 dBA (Lden) for contracting IHD was 1.08 (1.01–1.15). This was supported by an estimate from eight cross-sectional studies of 1.24 (1.08–1.42). The evidence for a relation between noise and hypertension was based on 26 cross-sectional studies (18,957 cases). The risk increase per 10 dBA (Lden) was 1.05 (1.02–1.08). No significant association was observed in one larger cohort study: 0.97 (0.90–1.05).

Additional evidence was found for an association with stroke incidence (based on one larger cohort study with 1881 cases). The per 10 dBA (Lden) risk increase for stroke being 1.14 (1.03–1.25). No increase was found with stroke mortality in three cohort studies: 0.87 (0.71–1.06).

New evidence was found for diabetes incidence with road traffic noise (based on one larger cohort study with 2752 cases). It revealed a 1.08 (1.02–1.14) risk increase per 10 dBA (Lden).

Note: this new assessment will substantially increase the DALY assessment for road traffic noise for two major reasons: more health outcomes are included, and the threshold values at which the increase in risk are taking place are lowered to noise levels, which larger city populations are exposed to (50–55 dBA, Lden).

22.3.4 *Other Effects on Health and Health-Related Quality of Life (HRQoL)*

22.3.4.1 Cognition and Performance in Children (Mostly Aged 8–12 Years)

The last WHO BoD of environmental noise assessment concluded that each year, 45,000 DALY's are lost due to cognitive impairment in children in agglomerations with more than 250,000 inhabitants across Europe. The attributable proportion calculated for the area (WHO EUR-A) was 2%.

The new WHO evidence review on cognitive effects on children (Clark and Paunovic 2018) shows that the evidence has increased, but most studies being cross-sectional and therefore downgraded by the GRADE approach adopted by WHO for this assessment. A full meta-analysis was not possible. The effects observed were of moderate quality for reading and oral comprehension, standardized school test scores (SATs), and long- and short-term memory for aircraft exposure—but only of low or very low quality for road traffic noise. This was mainly due to the number of studies (aircraft 32, road 11) and the higher methodological quality (6 longitudinal studies with aircraft). Other reviewed outcomes were children's attention and executive functioning with low to very low evidence for both aircraft and road traffic noise effects. Cognitive effects of transportation noise on elderly people (Tzivian et al. 2015) were not included. Too few studies were available for railway exposure.

22.3.4.2 Health-Related Quality of Life (HRQoL) and Mental Health

Hitherto, these outcomes were not assessed with a systematic review. The new WHO evidence review on environmental noise and quality of life, well-being, and mental health considered 29 studies (Clark and Paunovic 2018 in press). Most studies were cross-sectional (83%). Most of the evidence was rated as very low quality. The downgrading of the obtained evidence is due to the lack of longitudinal designs (only two) and the inconsistency of the observed results. A limitation of this review is the restriction of the time period 2005–2015 searched for studies. Several papers appeared at the end of this time period or later and were not assessed (Héritier et al. 2014; Riedel et al. 2015; Dreger et al. 2015; Oiamo et al. 2015, b; Tzivian et al.

2015; von Lindern et al. 2016). A few of those later studies have also made further adjustments for air pollution and other confounding facets of the environment.

22.3.4.3 Adverse Birth Outcomes

Three recent systematic reviews (Hohmann et al. 2013; Ristovska et al. 2014; Dzhambov et al. 2014) found conflicting evidence for the effect of occupational and environmental noise exposure on pregnancy outcomes. The new WHO evidence review (Nieuwenhuijsen et al. 2017) focused exclusively on environmental noise and adverse birth outcomes. The review period is extended until December 2016 and included the quality selected studies from the three previous reviews. The evidence for road traffic noise and low birth weight, preterm delivery, and small for gestational age was of low quality (GRADE). Too few studies ($N = 8$) were available for a meta-analysis, and three of those were of ecological design. Air pollution was controlled for in most of the selected studies.

22.3.5 *Less Studied, Subtle, and Other Indirect Effects on City Dwellers*

We need to keep in mind that it is extremely difficult to separate such type of—predominantly—indirect effects from other determinants embedded in various neighborhood socioeconomic compositions (Diez Roux and Mair 2010; Diez Roux 2016). The evidence-base from studies, where attempts were made to isolate noise effects, is small and mixed. Positive associations were found only in older experimental studies, where control over the studied environment is better, but generalization is limited.

22.3.5.1 Behavioral Effects on Streets

In the tradition of Milgram's "the experience of living in cities" research, short-term noise was used as a potent stressor in earlier psychological research (Cohen and Spacapan 1984; Evans 2001). In naturalistic settings, evidence was found for reduced helping behavior, under high noise conditions equivalent to exposure in city roads within 7.5 m. The contribution of noise to stimulus overload and masking, which reduces speech intelligibility and auditory feedback, attention, and empathy and increases in escape and avoidance behavior, is discussed as main explanatory factors. Other experimental studies found higher levels of aggression, decreases in frustration tolerance and attention after high noise exposure. A recent field study found various noise indicators (low-frequency sounds, high peaks) associated with displaced aggression (Dzhambov and Dimitrova 2014). Another experimental study observed decreases in cooperative behavior in a standard linear public good game under acute noise exposure (Diederich 2014). A prospective study on children did find an association of green space exposure (0.5–3 years) with reduced aggression (Child Behavior Checklist), considering surrogate measures of noise exposure in a sensitivity analysis. However, the aggressions scores of the highest quartile were higher for children being exposed to high traffic density within a 150–300 m area (Younan et al. 2016).

These possible effects need to be interpreted cautiously in a conceptual multi-stressor framework (Evans and Lepore 1997; Evans and Kim 2010; Riedel et al. 2015) which includes a restoration perspective (Hartig 2008; von Lindern et al. 2017)—as numerous factors influence and moderate the complex relationships between noise, aggression, violence and crime, and neighborhood quality including green space (Cozens 2007; Bogar and Beyer 2016; Younan et al. 2016). Discussed contributing factors to these possible effects are stimulus overload, reduced attention/frustration tolerance, and empathy induced by high noise exposure or traffic density.

22.3.5.2 Ecological and Social Effects: Segregation, Reduction of Quiet Areas, and Change in Social Relations

Only few of these studies on environmental justice included measured noise levels. Inequality can be introduced through segregation of people with lower financial resources into areas exposed to high noise levels—often coupled with high air pollution exposure and other unfavorable social conditions (Riedel et al. 2015; Carrier et al. 2016b). Two London studies found evidence for inequality at several levels: NHS hospitals in Inner London are considerably noisier than those in Outer London (by 5.4 dBA on average). The household size is smaller, and there are more one-person households. Such flats are more likely to be found in noisier areas, associated with unemployment. Also the pupil/teacher ratio has a significant negative correlation with noise levels (Xie and Kang 2009). In the second study, unemployment rates and academic achievement of school children did not significantly correlate with noise at the borough level. At the neighborhood level, areas with a higher noise level tend to be more deprived, especially in terms of health, barriers to housing and services, and living environment.

In Hong Kong, where nearly 50% of the population is accommodated in public housing, also a differential picture was observed (Lam and Chung 2012). On average, at all scales (community, neighborhood, dwelling level) private housing experienced higher noise levels—however, median income was negatively associated with average noise exposure and exceedances: the reason being that luxurious condominiums of higher-income groups are placed in quieter areas, while low-income people lives in tiny flats in crowded inner city conditions served by private housing. This points to the need to consider further aspects (the multi-stressor dimension) of segregated communities with high noise exposure (Evans and Kim 2010). In such segregated neighborhoods, a reduction of quiet areas takes place and imposes an additional constraint of restorative qualities of the living environment (Lercher et al.

2015; von Lindern et al. 2017, 2016). Apart from the direct effects of noise on communication and social interactions in the streets themselves, the nonavailability of nearby quiet and green areas has an additional negative effect on the likelihood to engage in social relations.

22.3.6 Summary of Health Impact Assessment of Noise Exposure in Cities

One limitation of this "single-effect approach" is the disregard of possible cumulative effects (multiple sound sources) or combined effects (noise and vibration/air pollution). Secondly, these methods assess only the direct pathways. The meta-analyses do not provide insight about indirect pathways through vulnerability vs adaptability in multi-exposure stressor designs—as data are not available for most studies (Lercher 2016).

However, with this limitation in mind, and based on a conceptual model of transport policies and health, several multifactor integrating health impact assessments (HIA) for cities have been conducted (Mueller et al. 2015; Barceló et al. 2016; Tainio 2015; Tobías et al. 2015) and gave slightly different answers. The last study from Madrid found the magnitude of the health impact of noise among people aged ≥65 years similar to reducing average $PM_{2.5}$ levels by 10 μg/m^3 (Tobías et al. 2015). The Warsaw study found about equal contributions (air pollution 44%, 46% due to noise) to DALYs (Tainio 2015). Overall, the negative effect of transport was counteracted to a certain degree (30%) by health benefits of active travel (walking, cycling). The external costs, calculated for Switzerland (Vienneau et al. 2015), amounted to equal parts for air pollution (higher mortality) and noise (higher on HRQoL).

22.4 Classical Preventive and Mitigation Measures: Noise Reduction

The classical approach in noise control aiming exclusively at noise reduction based on noise mapping has experienced several limitations (Kang 2013; Alves et al. 2015; Hornikx 2016). The main reasons are outdated technical textbooks, where often still a 3 dB reduction criterion (LAeq) is mentioned as requirement for a successful implementation of a noise abatement measure (Ortscheid and Wende 2004). This view underestimates the role subjective perception plays in integrated assessments of the human response to noise from transportation (Riedel et al. 2014; Nilsson et al. 2015; Lercher et al. 2015). First, the perception (and judgement of annoyance) of a noise abatement depends not only on the physical intensity but also on frequency and temporal aspects of the changed sound experience (Schorer 1989;

Zwicker and Fastl 1999). For example, a reduction of emergent sound events or specific disturbing sound features (modulation, LFN, sharpness, roughness) is already well perceived—although the average sound level may be reduced by only 1 dBA or less. Secondly, with the implementation of a traffic reduction (e.g., traffic calming, speed reduction), not only the noise is reduced but also other perceptible aspects (vibration, air pollution, safety, residential satisfaction) are simultaneously affected and influence the integrated judgement of the exposed human. Therefore, enough examples exist, where the sound reduction was only small (around 1 or 2 dBA, Leq), but the observed effect on annoyance or residential satisfaction was significant. Eventually, the use of generalized exposure-response relationships may also underestimate the achievable effect due to the different structure and contextual factors of cities. At least for annoyance, the reduction in annoyance in most studies is larger than could be expected based on noise level change (Brown and van Kamp 2017). Thus, a positive excess response is observed compared to responses predicted from steady-state exposure-response relationships, and this excess response does not show a relevant decrease over time (Brown and van Kamp 2009).

22.4.1 Source Oriented: "Technoscape" Measures

There are many single options. Obviously, only integrated legislation efforts, which combine emission reduction of vehicle noise limits with effective traffic management, speed limits, improved road surfaces, better tires, and adapted driver behavior, will ultimately lead to the theoretically possible improvements in the city environment.

22.4.1.1 Classical Means of Noise Control

Noise Control at Source: Emission-Reduction of Transportation Means

There are several options. Propulsion noise, speed, and tire/road noise are the main contributors of the noise emission of a single vehicle. Harmonizing emission regulations worldwide is a long-term goal. The type approval procedures need to be more realistic of daily traffic routines.

Whether all sorts of traffic calming schemes and congestion pricing reduce noise is still an open question. Other and more efficient options are available. For example, setting speed limits and elimination of heavy vehicles from certain roads (at low speeds in urban streets, a heavy-duty trucks have a noise impact levels which is equivalent to 8–16 cars (Seto et al. 2007; Heinz Steven 2012). Reducing speed from 50 to 30 km/h in urban areas achieves up to 4 dBA reductions, provided tight speed monitoring is enforced. Also the annoyance triggering acoustic features (acceleration, gear changes) are reduced. If a permanent setting of speed limits or vehicle restriction is not feasible, a temporary setting during evening and nighttime is an alternative option to prevent sleep disturbances.

Tire-road surface interactions (Sandberg and Ejsmont 2007): smaller and softer tires reduce noise. This effect can be additionally enhanced through a combination of quieter tires with sound absorbing road surfaces. Recent experience reveals that low-noise pavements can constitute an effective noise abatement measure at the source (Heutschi et al. 2016; Tonin 2016). Various types of surface layers can reduce sound levels from 4 up to 12 dBA (depending on type and costs). The maintenance for an ongoing good performance over time is still a problem of any pavement surface.

The "natural" spread of sound emissions from passenger car tires (6–7 dBA) and traction tires from heavy vehicles (4–5 dBA) could be reduced. Low-noise tires exhibit a significant noise abatement potential for passenger cars already at urban speed regimes (Heutschi et al. 2016). The EU-project FOREVER found that in the 125 Hz octave bands, the rolling noise is underestimated (Czuka et al. 2016)—even under ideal test conditions. A further problem is that the "low-noise" tires—as labeled by the EU tire label—do not fit: the tire with the lowest label and those with the highest labels yields similar noise levels, whereas the largest noise levels are due to tires with an intermediate label (Czuka et al. 2016). Eventually, rolling noise from light electric vehicles does not differ from conventional vehicles. The overall pass-by noise level came evidently from the tire-road noise, even at speeds of 20 km/h (Czuka et al. 2016).

Noise Control in the Transmission Path: Greening Instead of Classical Walls

An overview of options and effect sizes is provided by the EU-project HOSANNA (Nilsson et al. 2014). Often little efforts show an effect. For example, the presence of thick, sufficiently dense hedges (about 2 m) in combination with soil is expected to give about a 2 dBA reduction in realistic road traffic conditions at low vehicle speeds. The presence of a strip of grass increases the effect slightly. The interaction with soil is important for the reduction of noise frequencies below 2 kHz. Above 2 kHz, the foliage effect of the hedges becomes relevant. Higher reduction (9–11 dBA) can be achieved with a 5 m vegetation barrier of moderate to high density (Ow and Ghosh 2017). Vertical green wall systems are another option, where horizontal space is limited (Ismail 2013; Davis et al. 2017). In addition, landscape plants provide excess noise attenuating effects through subjects' emotional processing, as indicated by EEG-recordings (Yang et al. 2011).

These measures can be combined (Fig. 22.5) with other measures of the road design (sinking the road, quiet asphalt, low-level sound barriers) and a speed limit to achieve the equivalent effect of a typical (unattractive) 4 m-high noise barrier (European Soundscape Award 2011).

Noise Control at the Receiver Point

In difficult situations corrective measures can be applied through "canyonisation" (bordering roads with continuous apartment buildings) to reclaim high noise impact zones for residential living (de Ruiter 2005). This can be enhanced by several

Fig. 22.5 Combining several abatement measures. Source: European Soundscape Award 2011

architectural means (atrium façade, balcony design), vegetation belts, or vertical green, when horizontal space is limited (Ow and Ghosh 2017; Ismail 2013; Davis et al. 2017).

Moreover, such a strategy needs to be combined with a reduction of vibration entering the buildings and an improvement of the perceptual quality of the backyard side of "canyonised" building rows (as outlined in Sect. 22.5.2).

Other architectural options to improve the "canyonisation" strategy are oriented toward direct façade (Neuhaus and Neuhaus 2016; Sakamoto and Aoki 2015) and balcony design modifications (Hossam El Dien and Woloszyn 2004; Lee et al. 2007; Naish et al. 2012, 2014) which improves both the visual appearance of the façade and reduces sound intensity and reflections.

Another option is to implement suitable buffer zones to eliminate negative impacts of "infill development" on environment and public health (Qiao et al. 2016)

22.4.1.2 The Forgotten Dimension: Vibration and Low-Frequency Noise (LFN) from Traffic

Buses, heavy trucks, tramways, and underground rail systems emit structure-borne sound, low-frequency noise, and perceptible vibrations into nearby buildings (Crocker 2007). The weight of tramways increased substantially over the past two decades, and due to increasing demand for public transportation in a 24/7 hour society, night and morning hours of operation were extended. Sleep research has found these times especially critical regarding interference with the sleep and restoration process (Hume et al. 2012; Basner et al. 2015). Noise and perceptible vibrations from railways and roads can exert combined effects on annoyance (Ohrstrom and Öhrström 1997; Zapfe et al. 2012; Lee and Griffin 2013; Trollé et al. 2015; Lercher et al. 2017). The importance of LFN from transportation is widely underestimated (Berglund et al. 1996; Schreurs et al. 2008; Can et al. 2010; Ascari et al. 2015). Due to the nonlinearity for facade attenuation, higher frequencies are more effectively attenuated than lower frequencies which enter buildings easier. This effect is underestimated by the A-weighting scheme (Kjellberg et al. 1997). Notably, speech perception is strongly impaired by LFN. The relevance of LFN for health assessments can be estimated quickly by the dBC-dBA difference (as recommended by DIN

45680). The presence of strong low-frequency components can enhance vibration perceptions through cross-modality interactions (Takahashi 2013). Few studies have analyzed the relation between LFN components of road traffic noise and health (Chang et al. 2014; Dzhambov and Dimitrova 2014; Baliatsas et al. 2016).

Perceptible vibrations can be reduced at the track, the propagation path to the building, and hindered to reach upper floors of the building by the so-called blocking floor (Sanayei et al. 2014).

22.4.1.3 The Forgotten Dimension: Effects of Combined Exposures to Traffic Sources

Urban densification increases the likelihood of exposure to multiple environmental stressors. It has been questioned whether it is appropriate to use standard exposure-response curves also for mixed exposures in health impact assessment (Gille et al. 2016). Although a not well investigated area, some evidence from experimental and field studies is available—but most include two sources only. The studies show that annoyance (Lam et al. 2009; Lercher 2011; Di et al. 2012; Ragettli et al. 2015; Bodin et al. 2015; Lercher et al. 2017) and sleep disturbance (Basner et al. 2011; Perron et al. 2016) are increased in a mixed sound exposure situation, depending on the source combination. Fewer studies investigated multisensory effects: the presence of air pollution or odors affect annoyance and HRQoL in a cumulative way (Lercher et al. 1995, 2017; Klæboe et al. 2000; Oiamo et al. 2015; Pedersen 2015; Shepherd et al. 2016; von Lindern et al. 2016)

22.4.2 Planning, Built Environment, and Land Use: "Enviroscape" Measures

Sustainability and resilience are the buzz words of innovative conceptual frameworks in city planning. In a review outlining the multidisciplinary and complex nature of urban sustainability and resilience, noise is not mentioned at all (Jabareen 2013). Many conceptual ideas (the compact city, new urbanism, the transit village) are oriented toward mixing housing development for greater density, less segregation by taming the ubiquitous automobile. This should lead to more compact, dense, and more walkable cities with less noise. In reality, some experts say, such developments often lack the high-density development needed (~50–60 inhabitants/ha) which enables sustainable mobility (UITP 2002; Harata 2017) and mixed-use options. The approaches are often weak in reducing ecological impacts or promoting more sustainable lifestyles (Beatley 2012), livability, and neighborhood satisfaction (Howley et al. 2009). Only, when compact cities provide "vital" neighborhoods which are self-supporting in terms of facilities, having good travel accessibility, being safe, and having high interactions between neighbors, then can car travel be significantly reduced (UITP 2002) and noise levels as well.

However, a review summarized in 2004: "the spatial resolution of present (city) models is still too coarse to model neighborhood scale policies and effects," and the integration of environmental sub-models for air quality, traffic noise, land take, and biotopes is likely to play a prominent role in the future (Wegener 2004). This lack of city models with appropriate spatial resolution still hampers planning. Specifically, the proper classification of quiet areas suffers from this low spatial resolution of noise-mapping software (Kropp et al. 2016). Classical noise prediction through static traffic flow analysis (as used in noise mapping) tends to underestimate noise exposure (see Sect. 22.2.1.4), especially at the meso- and much more at the microscale level where the transport dynamics (fluctuation-emergence of events) is of outmost importance. Higher spatial resolution is increasingly important in planning for the reduction of possible health effects—especially during nighttime—as there are trends for an increase of noise during nighttime hours.

22.5 The Promotive Approach: Improve the Soundscape in General

A soundscape is defined as "acoustic environment as perceived or experienced and/or understood by a person or people, in context" (ISO/FDIS 12913–1 2014). "What sounds are noticed, become meaningful or disturbing is a multisensory process which involves visual, olfactory (odors), and somatosensory (vibrations) input from the overall city environment which codetermines the real effect on annoyance and perceived HRQoL. In practice, often, only small acoustic improvements are feasible. By improving in addition the environmental context, the expected effects are often enhanced. Absolute quietness ("silence") is not the ultimate goal (Job 1999)—the improvement of the acoustic quality of cities for the main human activities (work, communication, sleep, recreation) is the central aim of this approach: so-called hi fidelity soundscapes by the founder of this notion (Schafer 1977). The focus is directed at a balance between unwanted (sounds of discomfort) and wanted (sounds of preference) sounds in the environment (Berglund 2006; Brown 2010; Andringa and Lanser 2013; Kang and Schulte-Fortkamp 2015; Kang et al. 2016a, 2016b). For this purpose, both quantitative and qualitative aspects of sounds need to be assessed (Genuit and Fiebig 2016; Botteldooren et al. 2016; Kang et al. 2016a, 2016b). The use of psychoacoustic parameters (Genuit and Fiebig 2006) and the detection of temporal and spectral patterns through specialized sound indicators (Lercher et al. 2012; De Coensel et al. 2009; Morel et al. 2012; Trollé et al. 2013; Klein et al. 2015; Can et al. 2015; Wunderli et al. 2015) support soundscape evaluations and can lead to improved noise assignments in epidemiologic studies.

Fig. 22.6 A schematic illustration of the two types of quiet places in cities. Source: QSIDE-project

Eventually, soundscape management is a holistic task. Therefore, the knowledge of positive *and* adverse effects of exposure combinations (see Sects. 22.4.1.2 and 22.4.1.3) in a multisensory environment on human reactions (Viollon et al. 2002; Botteldooren and Lercher 2004; Hong and Jeon 2015; Jahncke et al. 2015) are prerequisites for implementing effective improvements in an urban soundscape.

On the health side, the orientation is toward outcomes at the lower end of the health spectrum (subjective well-being, HRQoL to annoyance, and self-reported or perceived health). Physiologic indicators of health are rarely studied (Hume and Ahtamad 2013; Lercher et al. 2013).

22.5.1 Providing Relative Quietness for Coping with Noise and Restoration

22.5.1.1 Relative Quietness and Restoration

Even in urban residential settings with high road traffic noise exposure (60–68dBA, Leq24), the provision of a quiet site (Fig. 22.6) with access to a green area reduces annoyance and psychosocial symptoms and increases the use of outdoor spaces (Veisten et al. 2012; Van Kamp et al. 2015). Among other studies, a study in Amsterdam revealed that *relative quietness* provided by a less exposed façade is an independent predictor of less annoyance (de Kluizenaar et al. 2013). If the difference between the most exposed and the least exposed façade is larger than 10 dBA, the effect on reduction of annoyance is higher (equivalent to a 5 dBA level). When associated with a nearby quiet and green area, noise-induced stress can be reduced (Ward Thompson et al. 2012), and additional restorative (Alvarsson et al. 2010; Jahncke et al. 2015) or stress-buffering effects (Wells and Evans 2003; Dzhambov and Dimitrova 2015) can be achieved both in children and adults. These improvements will further foster physical activity and social ties, resulting in improved HRQoL. Moreover, it will support children to do homework and develop cognitive skills in quieter places and the elderly to walk around in the neighborhood without fear. Eventually, children and persons with hearing aids will get better oral comprehension in such places, as the spatial separation of noise and speech is improved ("cocktail party effect" reduced) and the annoying recruitment phenomenon is less likely to occur. Also the blind will feel more comfortable (Rychtáriková et al. 2012). A soundscape assessment tool for the evaluation of urban residential areas can support urban planning or reshaping processes (Berglund and Nilsson 2006) for such places of relative quietness.

22.5.1.2 Innovative Architectural Implementations

Architectural concepts and options for new developments have been described for several scales of planning. Double or deaf façades, shielding galleries, terrace wall structures, curtain walls, glazed conservatories (green house), and loggias are a few of those options in the planning process. More principles are outlined in specific EU-projects (QSIDE, SONORUS, CityHush), described in the literature (de Ruiter 2005; Pheasant et al. 2010; Hellström et al. 2014; Kang and Schulte-Fortkamp 2015; Kang et al. 2016a, 2016b; Schulte-Fortkamp and Jordan 2016), or proposed by sound-oriented architectural groups (Hellström 2003; Neuhaus and Neuhaus 2016). The additional consideration and use of biophilic design patterns (Gillis and Gatersleben 2015; Ryan et al. 2014) may be promising in such planning processes. Similar sound attenuation options are also available at older buildings or on the propagation path to these buildings or walkways (Van Renterghem et al. 2015). More money for evaluations of architectural implementations is needed.

22.5.1.3 Providing Social Cohesion and Safety for the Neighborhood

Planning oriented toward a higher acoustic quality in cities is simultaneously associated with more green places and parks for repose and social communication. Such developments foster a sense of place, increase social networks and the responsibility for this place, and work against segregation. These factors as a group have been found to reduce opportunities for crime, less litter, and lead to a better integration of the elderly (Kweon et al. 1998). "Sounding Brighton" is an example, where adding positive sound into the streets of a large event resulted in greater short-term safety (Lavia et al. 2016). The reshaping of the Nauener Platz in Berlin is an example where a rundown neighborhood park (litter, graffiti, needles from drug users, loitering) was reclaimed with a few soundscape measures (Kang et al. 2016a, 2016b) and brought back neighbors to use it again (https://www.eea.europa.eu/highlights/berlin-park-wins-award-for).

22.5.2 Improving the Soundscape by Masking or Adding Positive Sound

The reduction of road traffic noise near urban roads is often not feasible. An alternative concept is to improve the soundscape of confined places (places, parks) in the neighborhood of these roads (Hong and Jeon 2013; De Coensel et al. 2011).

Soundscape research shows that introducing sounds of preference to directly mask (energetic masking) or distract attention away (informational masking) from road noise is a feasible strategy (Brown and Muhar 2004; Di et al. 2011; Lavia et al. 2016). Sounds of moving water or nature were found to be preferred sounds across cultures, and most examples have analyzed the suitability of such sounds to mask road traffic noise (You et al. 2010; Yang and Kang 2013; Jeon et al. 2010; De Coensel et al. 2011). A wide diversity of waterscapes were developed along the Gold Route in Sheffield in course of the regeneration of the Sheffield City Centre and studied by psychoacoustic means (Kang 2012). Masking water which had 0 to 3 dB lower sound pressure level was evaluated as preferable when the levels of road traffic noise were 55 or 75 dBA (Galbrun and Ali 2013; You et al. 2010). Even the sound level remains unchanged; also an improvement of the spectral balance as well additional nature features working via visual pathways can reduce annoyance. For example, a vegetation belt against a road led to an overrating of noise reduction by the belt through the participants—but this overrating correlated with the results of the (portable) EEG measurements, indicating a buffering effect ("psychological noise reduction") in this group behind the vegetation belt compared with the group without such a green belt (Yang et al. 2011).

22.5.3 *Policy Integration*

A recent analysis of the EEA shows that noise action plan measures around major roads are still dominated (nearly 80%) by classical means such as measures on the propagation path, at the receiver site, or traffic management (EEA 2014). Only 10% of measures are oriented toward land use and planning. Examples of good practice are collected by the EEA through the European Soundscape Award and outlined in large EU projects (QSIDE, SONORUS, HOSANNA, CityHush). Methodological tools for the assessment and improvement of city soundscapes were developed during the past decade (Kang 2010, 2011; Rychtáriková and Vermeir 2013; Davies 2013, 2014; van Kempen et al. 2014; Kang et al. 2016a, b; Aletta et al. 2016; Genuit and Fiebig 2016; Bento Coelho 2016), although further practice research needs to be done. The few examples that arrived at the city planning level in Europe are still dominated by the classical abatement approach.

In Dublin, Ireland, the city council combined noise modeling and measurement to identify long-term average noise levels below noise exposure that harms health. This led to the designation and protection of eight areas in the city. In Gothenburg, Sweden, a policy regarding new dwellings in central areas was established. When the expected noise exposure is above 65 dBA, then no dwellings should be built. Instead, other less sensible buildings should be used as barriers. In cases between 55 and 65 dBA, a quiet side of 45 dBA (which we call a quiet façade) or at least one below 50 dBA, (called a "silenced" side) will be projected. In Zoetermeer, the Netherlands, a curved-shaped building was implemented, which provides a 45–50 dBA quiet backyard side—even when the road exposed façade has noise levels above 65 dBA. Examples of typical urban soundscape implementations are published in two books (Kang et al. 2013; Kang and Schulte-Fortkamp 2015). Specific example for historic city centers or city waterfronts are included. At the planning level, acousticians are frequently called at a late stage in the design process. A sustainable environmental design requires that consideration is given to all the human senses, including sound. It is timely to integrate innovative classical and soundscape planning in transportation and land use planning policies in all stages and at all scales to improve public health in cities worldwide (Nieuwenhuijsen and Khreis 2016).

References

Aguilera, I., et al. (2015). Application of land use regression modelling to assess the spatial distribution of road traffic noise in three European cities. *Journal of Exposure Science and Environmental Epidemiology, 25*(1), 97–105.

Akinseye, O. A., et al. (2015). Sleep as a mediator in the pathway linking environmental factors to hypertension: A review of the literature. *International Journal of Hypertension, 2015*, 1–15.

Alam, M. S., et al. (2017). Modelling of intra-urban variability of prevailing ambient noise at different temporal resolution. *Noise Mapping, 4*(1), 20–44.

Aletta, F., Kang, J., & Axelsson, Ö. (2016). Soundscape descriptors and a conceptual framework for developing predictive soundscape models. *Landscape and Urban Planning, 149*, 65–74.

Allen, R. W., et al. (2009). The spatial relationship between traffic-generated air pollution and noise in 2 US cities. *Environmental Research, 109*(3), 334–342.

Alvarsson, J. J., Wiens, S., & Nilsson, M. E. (2010). Stress recovery during exposure to nature sound and environmental noise. *International Journal of Environmental Research and Public Health, 7*(3), 1036–1046.

Alves, S., et al. (2015). Towards the integration of urban sound planning in urban development processes: The study of four test sites within the SONORUS Project. *Noise Mapping, 2*, 57–85.

Andringa, T. C., & Lanser, J. J. L. (2013). How pleasant sounds promote and annoying sounds impede health: A cognitive approach. *International Journal of Environmental Research and Public Health, 10*(4), 1439–1461.

Ascari, E., et al. (2015). Low frequency noise impact from road traffic according to different noise prediction methods. *Science of the Total Environment, 505*, 658–669.

Astell-Burt, T., Feng, X., & Kolt, G. S. (2013). Does access to neighbourhood green space promote a healthy duration of sleep? Novel findings from a cross-sectional study of 259 319 Australians. *BMJ Open, 3*(8), e003094–e003094.

Babisch, W., et al. (2012). Exposure modifiers of the relationships of transportation noise with high blood pressure and noise annoyance. *The Journal of the Acoustical Society of America, 132*(6), 3788–3808.

Baliatsas, C., et al. (2016). Health effects from low-frequency noise and infrasound in the general population: Is it time to listen? A systematic review of observational studies. *Science of the Total Environment, 557–558*, 163–169.

Barceló, M. A., et al. (2016). Long term effects of traffic noise on mortality in the city of Barcelona, 2004–2007. *Environmental Research, 147*, 193–206.

Basner, M., & McGuire, S. (2018). WHO environmental noise guidelines for the European region: A systematic review on environmental noise and effects on sleep. *International Journal of Environmental Research Public Health*, 15, 519.

Basner, M., Müller, U., & Elmenhorst, E.-M. (2011). Single and combined effects of air, road, and rail traffic noise on sleep and recuperation. *Sleep, 34*(1), 11–23.

Basner, M., et al. (2015). ICBEN review of research on the biological effects of noise 2011–2014. *Noise and Health, 17*(75), 57–82.
Beatley, T. (2012). Green cities of Europe. In T. Beatley (Ed.), *Global lessons on green urbanism*. Washington: Island Press/Center for Resource Economics.
Bento Coelho, J. L. (2016). Approaches to urban soundscape management, planning, and design. In J. Kang & B. Schulte-Fortkamp (Eds.), *Soundscape and the built environment* (pp. 197–214). Boca Raton: CRC Press.
Berglund, B. (2006). *From the WHO guidelines for community noise to healthy soundscapes. Proceedings of the Institute of Acoustics, 2006* (pp. 1–8). St Albans: Institute of Acoustics.
Berglund, B., Hassmén, P., & Job, R. F. S. (1996). Sources and effects of low-frequency noise. *The Journal of the Acoustical Society of America, 99*(5), 2985–3002.
Berglund, B., & Nilsson, M. E. (2006). On a tool for measuring soundscape quality in urban residential areas. *Acta Acustica United with Acustica, 92*(6), 938–944.
Bocquier, A., et al. (2013). Small-area analysis of social inequalities in residential exposure to road traffic noise in Marseilles, France. *The European Journal of Public Health, 23*(4), 540–546.
Bodin, T., et al. (2015). Annoyance, sleep and concentration problems due to combined traffic noise and the benefit of quiet side. *International Journal of Environmental Research and Public Health, 12*(2), 1612–1628.
Bogar, S., & Beyer, K. M. (2016). Green space, violence, and crime. *Trauma, Violence, & Abuse, 17*(2), 160–171.
Botteldooren, D., & Lercher, P. (2004). Soft-computing base analyses of the relationship between annoyance and coping with noise and odor. *The Journal of the Acoustical Society of America, 115*(6), 2974.
Botteldooren, D., et al. (2016). From sonic environment to soundscape. In J. Kang & B. Schulte-Fortkamp (Eds.), *Soundscape and the built environment* (pp. 17–41). Boca Raton: CRC Press.
Bottino, C. J., et al. (2012). The association of urbanicity with infant sleep duration. *Health & Place, 18*(5), 1000–1005.
Brown, A. L. (2010). Soundscapes and environmental noise management. *Noise Control Engineering Journal, 58*(5), 493–500.
Brown, A. L., & Muhar, A. (2004). An approach to the acoustic design of outdoor space. *Journal of Environmental Planning and Management, 47*(6), 827–842.
Brown, A. L., & van Kamp, I. (2009). Response to a change in transport noise exposure: A review of evidence of a change effect. *The Journal of the Acoustical Society of America, 125*(5), 3018.
Brown, A., & van Kamp, I. (2017). WHO environmental noise guidelines for the European region: A systematic review of transport noise interventions and their impacts on health. *International Journal of Environmental Research and Public Health, 14*(8), 873.
Can, A., Guillaume, G., & Gauvreau, B. (2015). Noise indicators to diagnose urban sound environments at multiple spatial scales. *Acta Acustica United with Acustica, 101*(5), 964–974.
Can, A., et al. (2010). Traffic noise spectrum analysis: Dynamic modeling vs. experimental observations. *Applied Acoustics, 71*(8), 764–770.
Can, A., et al. (2011). Correlation analysis of noise and ultrafine particle counts in a street canyon. *Science of the Total Environment, 409*(3), 564–572.
Cao, X.-L., et al. (2017). The prevalence of insomnia in the general population in China: A meta-analysis. *PLoS ONE, 12*(2), e0170772.
Carrier, M., Apparicio, P., & Séguin, A.-M. (2016a). Road traffic noise geography during the night in Montreal: An environmental equity assessment. *The Canadian Geographer, 60*(3), 394–405.
Carrier, M., Apparicio, P., & Séguin, A.-M. (2016b). Road traffic noise in Montreal and environmental equity: What is the situation for the most vulnerable population groups? *Journal of Transport Geography, 51*, 1–8.
Chang, T.-Y., et al. (2014). Road traffic noise frequency and prevalent hypertension in Taichung, Taiwan: A cross-sectional study. *Environmental Health, 13*(1), 37.
Chang, S. S., et al. (2015). The effects of different noise types on heart rate variability in men. *Yonsei Medical Journal, 56*(1), 235–243.

Chum, A., O'Campo, P., & Matheson, F. (2015). The impact of urban land uses on sleep duration and sleep problems. *The Canadian Geographer, 59*(4), 404–418.
Cidell, J. (2010). Concentration and decentralization: The new geography of freight distribution in US metropolitan areas. *Journal of Transport Geography, 18*(3), 363–371.
Clark, C., & Paunovic, K. (2018). WHO environmental noise guidelines for the European region: A systematic review on environmental noise and cognition. *International Journal of Environmental Research and Public Health, 15*(2), 285.
Clark, C., & Paunovic, K. (2018b). WHO environmental noise guidelines for the European region: A systematic review on environmental noise and environmental noise and quality of life, well-being and mental health. *International Journal of Environmental Research and Public Health.* in press on http://www.mdpi.com/journal/ijerph/special_issues/WHO_reviews.
Cohen, S., & Spacapan, S. (1984). The social psychology of noise. In D. Jones & A. J. Chapman (Eds.), *Noise and society* (pp. 221–245). Chichester: Wiley.
Cozens, P. (2007). Planning, crime and urban sustainability. In H. H. Al-Kayiem (Ed.), *Sustainable development and planning III. WIT transactions on ecology and the environment* (Vol. 102, pp. 187–196). Southampton: WIT Press.
Crocker, M. J. (2007). *Handbook of noise and vibration control*. Hoboken: John Wiley & Sons Ltd.
Czuka, M., et al. (2016). Impact of potential and dedicated tyres of electric vehicles on the tyre-road noise and connection to the EU noise label. *Transportation Research Procedia, 14*, 2678–2687.
Dale, L. M., et al. (2015). Socioeconomic status and environmental noise exposure in Montreal, Canada. *BMC Public Health, 15*(1), 205.
Davies, W. J. (2013). Special issue: Applied soundscapes. *Applied Acoustics, 74*(2), 223.
Davies, W. J., Bruce, N. S., & Murphy, J. E. (2014). Soundscape reproduction and synthesis. *Acta Acustica United with Acustica, 100*(2), 285–292.
Davis, M. J. M., et al. (2017). More than just a green facade: The sound absorption properties of a vertical garden with and without plants. *Building and Environment, 116*, 64–72.
De Coensel, B., Vanwetswinkel, S., & Botteldooren, D. (2011). Effects of natural sounds on the perception of road traffic noise. *The Journal of the Acoustical Society of America, 129*(4), EL148–EL153.
De Coensel, B., et al. (2009). A model for the perception of environmental sound based on notice-events. *The Journal of the Acoustical Society of America, 126*(2), 656–665.
de Kluizenaar, Y., et al. (2013). Road traffic noise and annoyance: A quantification of the effect of quiet side exposure at dwellings. *International Journal of Environmental Research and Public Health, 10*(6), 2258–2270.
de Ruiter, E. P. (2005). *Reclaiming land from urban traffic noise impact zones, "The great canyon"*. Delft: TU Delft.
De Vos, P., & Van Beek, A. (2011). Environmental noise. In J. O. Nriyagu (Ed.), *Encyclopedia of environmental health* (pp. 476–488). Burlington: Elsevier.
Di, G., et al. (2011). Adjustment on subjective annoyance of low frequency noise by adding additional sound. *Journal of Sound and Vibration, 330*(23), 5707–5715.
Di, G., et al. (2012). The relationship between urban combined traffic noise and annoyance: An investigation in Dalian, north of China. *Science of the Total Environment, 432*, 189–194.
Diederich, J. (2014). *The effect of ambient noise on cooperation in public good games*. Heidelberg: University of Heidelberg.
Diez Roux, A. V. (2016). Neighborhoods and health: What do we know? What should we do? *American Journal of Public Health, 106*(3), 430–431.
Diez Roux, A. V., & Mair, C. (2010). Neighborhoods and health. *Annals of the New York Academy of Sciences, 1186*(1), 125–145.
Dreger, S., et al. (2015). Environmental noise and incident mental health problems: A prospective cohort study among school children in Germany. *Environmental Research, 143*, 49–54.
Dzhambov, A., & Dimitrova, D. (2014). Neighborhood noise pollution as a determinant of displaced aggression: A pilot study. *Noise and Health, 16*(69), 95–101.

Dzhambov, A. M., & Dimitrova, D. D. (2015). Green spaces and environmental noise perception. *Urban Forestry & Urban Greening, 14*(4), 1000–1008.
Dzhambov, A. M., Dimitrova, D. D., & Dimitrakova, E. D. (2014). Noise exposure during pregnancy, birth outcomes and fetal development: Meta-analyses using quality effects model. *Folia Medica, 56*(3), 204–214.
EEA. (2010). *Good practice guide on noise exposure and potential health effects*. Copenhagen: EEA.
EEA. (2014). *Noise in Europe 2014*. Luxembourg: Publications Office of the European Union.
European Commission, D.-G. for R. and U.P. (2016). *Quality of life in European cities. FLASH EURO*. Brussels: European Union.
Evans, G. W. (2001). Environmental stress and health. In A. Baum, T. Revenson, & J. E. Singer (Eds.), *Handbook of health psychology* (pp. 365–385). Hove: Psychology Press.
Evans, G. W., & Kim, P. (2010). Multiple risk exposure as a potential explanatory mechanism for the socioeconomic status-health gradient. *Annals of the New York Academy of Sciences, 1186*(1), 174–189.
Evans, G., & Lepore, S. J. (1997). Moderating and mediating processes in environment-behavior research R. In G. T. Moore & R. W. Marans (Eds.), *Advances in environment, behavior and design* (pp. 255–285). New York: Springer US.
Fan, X., Lam, K., & Yu, Q. (2012). Differential exposure of the urban population to vehicular air pollution in Hong Kong. *Science of the Total Environment, 426*, 211–219.
Fecht, D., et al. (2016). Spatial and temporal associations of road traffic noise and air pollution in London: Implications for epidemiological studies. *Environment International, 88*, 235–242.
Fidell, S., et al. (2011). A first-principles model for estimating the prevalence of annoyance with aircraft noise exposure. *The Journal of the Acoustical Society of America, 130*(2), 791–806.
Fields, J. M. (1993). Effect of personal and situational variables on noise annoyance in residential areas. *The Journal of the Acoustical Society of America, 93*(5), 2753–2763.
Fields, J. M., et al. (2001). Standardized general-purpose noise reaction questions for community noise surveys: Research and a recommendation. *Journal of Sound and Vibration, 242*(4), 641–679.
Finegold, F., Muzet, A. & Berry, B., 2008. Sleep disturbance due to transportation noise exposure. In: Crocker MJ Handbook of noise and vibration control Wiley Hoboken 308–315.
Flindell, I. H., & Stallen, P. J. M. (1999). Non-acoustical factors in environmental noise. *Noise & Health, 1*(3), 11–16.
Foraster, M., et al. (2011). Local determinants of road traffic noise levels versus determinants of air pollution levels in a Mediterranean city. *Environmental Research, 111*(1), 177–183.
Ford, E. S., et al. (2015). Trends in insomnia and excessive daytime sleepiness among US adults from 2002 to 2012. *Sleep Medicine, 16*(3), 372–378.
Franklin, J. C., Moretti, N. A., & Blumenthal, T. D. (2007). Impact of stimulus signal-to-noise ratio on prepulse inhibition of acoustic startle. *Psychophysiology, 44*, 339–342.
Fritschi, L., et al. (2011). *Burden of disease from environmental noise—quantification of healthy life years lost in Europe*. Bonn: WHO European Centre for Environment and Health.
Galbrun, L., & Ali, T. T. (2013). Acoustical and perceptual assessment of water sounds and their use over road traffic noise. *The Journal of the Acoustical Society of America, 133*(1), 227–237.
Gan, W. Q., et al. (2012). Modeling population exposure to community noise and air pollution in a large metropolitan area. *Environmental Research, 116*, 11–16.
Genuit, K., & Fiebig, A. (2006). Psychoacoustics and its benefit for the soundscape approach. *Acta Acustica United with Acoustica, 92*(6), 952–958.
Genuit, K., & Fiebig, A. (2016). Human hearing–related measurement and analysis of acoustic environments. In J. Kang & B. Schulte-Fortkamp (Eds.), *Soundscape and the built environment* (pp. 133–160). Boca Raton: CRC Press.
Gille, L.-A., Marquis-Favre, C., & Klein, A. (2016). Noise annoyance due to urban road traffic with powered-two-wheelers: Quiet periods, order and number of vehicles. *Acta Acustica United with Acustica, 102*(3), 474–487.

Gille, L.-A., Marquis-Favre, C., & Morel, J. (2016). Testing of the European Union exposure-response relationships and annoyance equivalents model for annoyance due to transportation noises: The need of revised exposure-response relationships and annoyance equivalents model. *Environment International, 94*, 83–94.
Gillis, K., & Gatersleben, B. (2015). A review of psychological literature on the health and wellbeing benefits of biophilic design. *Buildings, 5*(3), 948–963.
Glass, D., & Singer, J. (1972). *Urban stress. Experiments on noise and social stressors*. New York: Academic.
Guo, C., Schwarz, N., & Buchmann, C. M. (2017). Exploring the added value of population distribution indicators for studies of European urban form. *Applied Spatial Analysis and Policy*. https://doi.org/10.1007/s12061-017-9225-7.
Guski, R. (1999). Personal and social variables as co-determinants of noise annoyance. *Noise & Health, 1*(3), 45–56.
Guski, R., Schreckenberg, D., & Schuemer, R. (2017). WHO environmental noise guidelines for the European region: A systematic review on environmental noise and annoyance. *International Journal of Environmental Research and Public Health, 14*(12), 1539.
Hänninen, O., et al. (2014). Environmental burden of disease in Europe: Assessing nine risk factors in six countries. *Environmental Health Perspectives, 122*(5), 439–446.
Harata, N. (2017). Sustainable urban structure and transport policy in the metropolitan region. In M. Yokohari et al. (Eds.), *Science for Sustainable Societies* (pp. 39–47). Tokyo: Springer.
Hartig, T. (2008). Green space, psychological restoration, and health inequality. *The Lancet, 372*(9650), 1614–1615.
Havard, S., et al. (2011). Social inequalities in residential exposure to road traffic noise: An environmental justice analysis based on the RECORD Cohort Study. *Occupational and Environmental Medicine, 68*(5), 366–374.
Hellström, B. (2003). *Noise design. Architectural modeling and the aesthetics of urban acoustic space*. Bo Ejeby Förlag: Gothenburg.
Hellström, B., et al. (2014). Acoustic design artifacts and methods for urban soundscapes: A case study on the qualitative dimensions of sounds. *Journal of Architectural and Planning Research, 31*, 57–71.
Helson, H. (1964). *Adaptation-level theory*. New York: Harper & Row.
Héritier, H., et al. (2014). The association between road traffic noise exposure, annoyance and health-related quality of life (HRQOL). *International Journal of Environmental Research and Public Health, 11*(12), 12652–12667.
Heutschi, K., Bühlmann, E., & Oertli, J. (2016). Options for reducing noise from roads and railway lines. *Transportation Research Part A: Policy and Practice, 94*, 308–322.
Hohmann, C., et al. (2013). Health effects of chronic noise exposure in pregnancy and childhood: A systematic review initiated by ENRIECO. *International Journal of Hygiene and Environmental Health, 216*(3), 217–229.
Hong, J. Y., & Jeon, J. Y. (2013). Designing sound and visual components for enhancement of urban soundscapes. *The Journal of the Acoustical Society of America, 134*(3), 2026–2036.
Hong, J. Y., & Jeon, J. Y. (2015). Influence of urban contexts on soundscape perceptions: A structural equation modeling approach. *Landscape and Urban Planning, 141*, 78–87.
Hornikx, M. (2016). Ten questions concerning computational urban acoustics. *Building and Environment, 106*, 409–421.
Hossam El Dien, H., & Woloszyn, P. (2004). Prediction of the sound field into high-rise building facades due to its balcony ceiling form. *Applied Acoustics, 65*(4), 431–440.
Howley, P., Scott, M., & Redmond, D. (2009). Sustainability versus liveability: An investigation of neighbourhood satisfaction. *Journal of Environmental Planning and Management, 52*(6), 847–864.
Hume, K., & Ahtamad, M. (2013). Physiological responses to and subjective estimates of soundscape elements. *Applied Acoustics, 74*(2), 275–281.

Hume, K. I., Brink, M., & Basner, M. (2012). Effects of environmental noise on sleep. *Noise & Health, 14*(61), 297–302.
Ismail, M. R. (2013). Quiet environment: Acoustics of vertical green wall systems of the Islamic urban form. *Frontiers of Architectural Research, 2*(2), 162–177.
Jabareen, Y. (2013). Planning the resilient city: Concepts and strategies for coping with climate change and environmental risk. *Cities, 31*, 220–229.
Jahncke, H., Naula, S., & Eriksson, K. (2015). The effects of auditive and visual settings on perceived restoration likelihood. *Noise and Health, 17*(74), 1–10.
Jeon, J. Y., et al. (2010). Perceptual assessment of quality of urban soundscapes with combined noise sources and water sounds. *The Journal of the Acoustical Society of America, 127*(3), 1357.
Job, R. F. S. (1999). Internoise '98 – "Sound and Silence": Setting the balance. *Noise & Health, 2*, 78–79.
Kang, J. (2010). From understanding to designing soundscapes. *Frontiers of Architecture and Civil Engineering in China, 4*(4), 403–417.
Kang, J. (2011). Noise management: Soundscape approach. In J. O. Nriagu (Ed.), *Encyclopedia of environmental health* (pp. 174–184). Burlington: Elsevier.
Kang, J. (2012). On the diversity of urban waterscape. In *Acoustics 2012* (pp. 1–7). Nantes: Societe Francaise d'Acoustique.
Kang, J. (2013). Urban acoustic environment. In R. Yao (Ed.), *Design and management of sustainable built environments* (pp. 99–118). London: Springer.
Kang, J., & Schulte-Fortkamp, B. (2015). *Soundscape and the built environment*. London: CRC Press.
Kang, J., et al. (Eds.). (2013). *Soundscape-COST. Soundscape of European cities and landscapes soundscape*. Oxford: COST Office Through Soundscape-COST.
Kang, J., et al. (2016a). *Mapping of soundscape. Soundscape and the built environment* (pp. 161–195). Boca Raton: CRC Press.
Kang, J., et al. (2016b). Ten questions on the soundscapes of the built environment. *Building and Environment, 108*, 284–294.
Keuken, M., et al. (2014). *URGENCHE. Deliverable: 5.1*. Brussels: Seventh Framework Programme.
Kim, R., & van den Berg, M. (2010). Summary of night noise guidelines for Europe. *Noise & Health, 12*(47), 61–63.
Kim, K.-H., et al. (2012). Some insights into the relationship between urban air pollution and noise levels. *Science of the Total Environment, 424*, 271–279.
King, G., et al. (2012). Noise levels associated with urban land use. *Journal of Urban Health, 89*(6), 1017–1030.
Kjellberg, A., et al. (1997). Evaluation of frequency-weighted sound level measurements for prediction of low-frequency noise annoyance. *Environment International, 23*(4), 519–527.
Klæboe, R., et al. (2000). Oslo traffic study – part 1: An integrated approach to assess the combined effects of noise and air pollution on annoyance. *Atmospheric Environment, 34*(27), 4727–4736.
Klein, A., et al. (2015). Spectral and modulation indices for annoyance-relevant features of urban road single-vehicle pass-by noises. *The Journal of the Acoustical Society of America, 137*(3), 1238–1250.
Kropp, W., Forssén, J., & Mauriz, L. E. (2016). In C. U. of Technology (Ed.), *Urban sound planning – the SONORUS project*. Gothenburg: Chalmers University of Technology.
Kumar, P., Nigam, S. P., & Kumar, N. (2014). Vehicular traffic noise modeling using artificial neural network approach. *Transportation Research Part C: Emerging Technologies, 40*, 111–122.
Kweon, B.-S., Sullivan, W. C., & Wiley, A. R. (1998). Green common spaces and the social integration of inner-city older adults. *Environment and Behavior, 30*(6), 832–858.
Lam, K., & Chung, Y. T. (2012). Exposure of urban populations to road traffic noise in Hong Kong. *Transportation Research Part D: Transport and Environment, 17*(6), 466–472.

Lam, K.-C., & Ma, W.-C. (2012). Road traffic noise exposure in residential complexes built at different times between 1950 and 2000 in Hong Kong. *Applied Acoustics, 73*(11), 1112–1120.
Lam, K.-C., et al. (2009). Annoyance response to mixed transportation noise in Hong Kong. *Applied Acoustics, 70*(1), 1–10.
Lam, K.-C., et al. (2013). Relationship between road traffic noisescape and urban form in Hong Kong. *Environmental Monitoring and Assessment, 185*(12), 9683–9695.
Lavia, L., et al. (2016). Applied soundscape practices. In J. Kang & B. Schulte-Fortkamp (Eds.), *Soundscape and the built environment*. Boca Raton: CRC Press.
Lee, P. J., & Griffin, M. J. (2013). Combined effect of noise and vibration produced by high-speed trains on annoyance in buildings. *The Journal of the Acoustical Society of America, 133*(4), 2126–2135.
Lee, P. J., et al. (2007). Effects of apartment building facade and balcony design on the reduction of exterior noise. *Building and Environment, 42*(10), 3517–3528.
Lee, E. Y., et al. (2014a). Assessment of traffic-related noise in three cities in the United States. *Environmental Research, 132*, 182–189.
Lee, J., et al. (2014b). Estimation of populations exposed to road traffic noise in districts of seoul metropolitan area of Korea. *International Journal of Environmental Research and Public Health, 11*(3), 2729–2740.
Lercher, P. (1996). Environmental noise and health: An integrated research perspective. *Environment International, 22*(1), 117–129.
Lercher, P. (2011). Combined noise exposure at home. In J. O. Nriagu (Ed.), *Encyclopedia of environmental health* (pp. 764–777). Burlington: Elsevier.
Lercher, P. (2016). Systematic reviews in noise epidemiology. Limitations and chances from a public health view. In *Inter-noise and noise-con congress and conference proceedings* (pp. 6208–6219). The Hague: Institute of Noise Control Engineering.
Lercher, P., Evans, G. W., & Widmann, U. (2013). The ecological context of soundscapes for children's blood pressure. *The Journal of the Acoustical Society of America, 134*(1), 773–781.
Lercher, P., Schmitzberger, R., & Kofler, W. (1995). Perceived traffic air pollution, associated behavior and health in an alpine area. *Science of the Total Environment, 169*(1–3), 71–74.
Lercher, P., et al. (2012). The application of a notice-event model to improve classical exposure-annoyance estimation. *The Journal of the Acoustical Society of America, 131*(4), 3223–3223.
Lercher, P., et al. (2015). Perceived soundscapes and health-related quality of life, context, restoration, and personal characteristics. In J. Kang & B. Schulte-Fortkamp (Eds.), *Soundscape and the built environment* (pp. 89–131). London: CRC Press.
Lercher, P., et al. (2017). Community response to multiple sound sources: Integrating acoustic and contextual approaches in the analysis. *International Journal of Environmental Research and Public Health, 14*(6), 663.
Levenson, J. C., Kay, D. B., & Buysse, D. J. (2015). The pathophysiology of insomnia. *Chest, 147*(4), 1179–1192.
Malecki, E. J. (2014). Connecting the fragments: Looking at the connected city in 2050. *Applied Geography, 49*, 12–17.
Margaritis, E., & Kang, J. (2017). Relationship between green space-related morphology and noise pollution. *Ecological Indicators, 72*, 921–933.
McAlexander, T. P., Gershon, R. R. M., & Neitzel, R. L. (2015). Street-level noise in an urban setting: Assessment and contribution to personal exposure. *Environmental Health, 14*(1), 18.
Miedema, H. M., & Oudshoorn, C. G. (2001). Annoyance from transportation noise: Relationships with exposure metrics DNL and DENL and their confidence intervals. *Environmental Health Perspectives, 109*(4), 409–416.
Miedema, H. M., & Vos, H. (1999). Demographic and attitudinal factors that modify annoyance from transportation noise. *The Journal of the Acoustical Society of America, 105*(6), 3336–3344.
Morel, J., et al. (2012). Road traffic in urban areas: A perceptual and cognitive typology of pass-by noises. *Acta Acustica United with Acustica, 98*(1), 166–178.

Morin, C. M., & Jarrin, D. C. (2013). Epidemiology of insomnia. *Sleep Medicine Clinics, 8*(3), 281–297.
Mueller, N., et al. (2015). Health impact assessment of active transportation: A systematic review. *Preventive Medicine, 76*, 103–114.
Muzet, A. (2007). Environmental noise, sleep and health. *Sleep Medicine Reviews, 11*(2), 135–142.
Naish, D. A., Tan, A. C. C., & Demirbilek, F. N. (2014). Simulating the effect of acoustic treatment types for residential balconies with road traffic noise. *Applied Acoustics, 79*, 131–140.
Naish, D. A., Tan, A. C. C., & Nur Demirbilek, F. (2012). Estimating health related costs and savings from balcony acoustic design for road traffic noise. *Applied Acoustics, 73*(5), 497–507.
Nassiri, P., et al. (2013). Traffic noise prediction and the influence of vehicle horn noise. *Journal of Low Frequency Noise, Vibration and Active Control, 32*(4), 285–292.
Neuhaus, I., & Neuhaus, F. (2016). *Akustisch gute Architektur für Strassenräume*. Basel.
Nieuwenhuijsen, M. J., & Khreis, H. (2016). Car free cities: Pathway to healthy urban living. *Environment International, 94*, 251–262.
Nieuwenhuijsen, M. J., Ristovska, G., & Dadvand, P. (2017). WHO environmental noise guidelines for the European region: A systematic review on environmental noise and adverse birth outcomes. *International Journal of Environmental Research and Public Health, 14*(10), 1252.
Nilsson, M., Bengtsson, J., & Klaeboe, R. (2014). *Environmental methods for transport noise reduction*. London: CRC Press.
Nilsson, M. E., et al. (2015). Perceptual effects of noise mitigation. In M. Nilsson, J. Bengtsson, & R. Klaeboe (Eds.), *Environmental methods for transport noise reduction* (pp. 195–220). London: CRC Press.
Ohrstrom, E., & Öhrström, E. (1997). Effects of exposure to railway noise—a comparison between areas with and without vibration. *Journal of Sound and Vibration, 205*(4), 555–560.
Oiamo, T. H., Baxter, J., et al. (2015). Place effects on noise annoyance: Cumulative exposures, odour annoyance and noise sensitivity as mediators of environmental context. *Atmospheric Environment, 116*, 183–193.
Oiamo, T. H., Luginaah, I. N., & Baxter, J. (2015). Cumulative effects of noise and odour annoyances on environmental and health related quality of life. *Social Science & Medicine, 146*, 191–203.
Onaga, H., & Rindel, J. H. (2007). Acoustic characteristics of urban streets in relation to scattering caused by building facades. *Applied Acoustics, 68*(3), 310–325.
Ortscheid, J., & Wende, H. (2004). Sind 3 dB wahrnehmbar? Eine Richtigstellung. *Zeitschrift für Lärmbekämpfung, 51*(3), 80–85.
Ow, L. F., & Ghosh, S. (2017). Urban cities and road traffic noise: Reduction through vegetation. *Applied Acoustics, 120*, 15–20.
Pallesen, S., et al. (2014). A 10-year trend of insomnia prevalence in the adult Norwegian population. *Sleep Medicine, 15*(2), 173–179.
Paviotti, M., & Vogiatzis, K. (2012). On the outdoor annoyance from scooter and motorbike noise in the urban environment. *Science of the Total Environment, 430*, 223–230.
Pedersen, E. (2015). City dweller responses to multiple stressors intruding into their homes: Noise, light, odour, and vibration. *International Journal of Environmental Research and Public Health, 12*(3), 3246–3263.
Perron, S., et al. (2016). Sleep disturbance from road traffic, railways, airplanes and from total environmental noise levels in Montreal. *International Journal of Environmental Research and Public Health, 13*(8), 809.
Phan, H. A. T., et al. (2009). Annoyance caused by road traffic noise with and without horn sounds. *Acoustical Science and Technology, 30*(5), 327–337.
Phan, H. Y. T., et al. (2010). Community responses to road traffic noise in Hanoi and Ho Chi Minh City. *Applied Acoustics, 71*(2), 107–114.
Pheasant, R. J., et al. (2010). The importance of auditory-visual interaction in the construction of "tranquil space". *Journal of Environmental Psychology, 30*(4), 501–509.

Qiao, F., et al. (2016). Effects and frameworks to estimate the environmental and public health impacts of infill development for urban planning. In W. KCP (Ed.), *International conference on transportation and development 2016*. Houston: American Society of Civil Engineers.
Ragettli, M., et al. (2015). Annoyance from road traffic, trains, airplanes and from total environmental noise levels. *International Journal of Environmental Research and Public Health, 13*(1), 90.
Richardson, E. A., et al. (2012). Green cities and health: A question of scale? *Journal of Epidemiology and Community Health, 66*(2), 160–165.
Riedel, N., et al. (2014). Assessing the relationship between objective and subjective indicators of residential exposure to road traffic noise in the context of environmental justice. *Journal of Environmental Planning and Management, 57*(9), 1398–1421.
Riedel, N., et al. (2015). Objective exposure to road traffic noise, noise annoyance and self-rated poor health – framing the relationship between noise and health as a matter of multiple stressors and resources in urban neighbourhoods. *Journal of Environmental Planning and Management, 58*(2), 336–356.
Ristovska, G., Laszlo, H., & Hansell, A. (2014). Reproductive outcomes associated with noise exposure — A systematic review of the literature. *International Journal of Environmental Research and Public Health, 11*(8), 7931–7952.
Ross, Z., et al. (2011). Noise, air pollutants and traffic: Continuous measurement and correlation at a high-traffic location in New York City. *Environmental Research, 111*(8), 1054–1063.
Ryan, C. O., et al. (2014). BIOPHILIC DESIGN PATTERNS: Emerging nature-based parameters for health and well-being in the built environment. *International Journal of Architectural Research: ArchNet-IJAR, 8*(2), 62–76.
Rychtáriková, M., Herssens, J., & Heylighen, A. (2012). Towards more inclusive approaches in soundscape research: The soundscape of blind people. In *Inter-noise and noise-con congress and conference proceedings*. New York: Institute of Noise Control Engineering.
Rychtáriková, M., & Vermeir, G. (2013). Soundscape categorization on the basis of objective acoustical parameters. *Applied Acoustics, 74*(2), 240–247.
Sakamoto, S., & Aoki, A. (2015). Numerical and experimental study on noise shielding effect of eaves/louvers attached on building façade. *Building and Environment, 94*, 773–784.
Salomons, E. M., & Berghauser Pont, M. (2012). Urban traffic noise and the relation to urban density, form, and traffic elasticity. *Landscape and Urban Planning, 108*(1), 2–16.
Sanayei, M., et al. (2014). Measurement and prediction of train-induced vibrations in a full-scale building. *Engineering Structures, 77*, 119–128.
Sandberg, U., & Ejsmont, J. A. (2007). Tire/road noise—Generation, measurement, and abatement. In M. J. Crocker (Ed.), *Handbook of noise and vibration control* (pp. 1054–1071). Hoboken: Wiley.
Schafer, R. (1977). *The tuning of the world*. New York: Alfred A. Knopf.
Schomer, P. (2005). Criteria for assessment of noise annoyance. *Noise Control Engineering Journal, 53*(4), 132–144.
Schomer, P. D., & Wagner, L. R. (1996). On the contribution of noticeability of environmental sounds to noise annoyance. *Noise Control Engineering Journal, 44*(6), 294–305.
Schomer, P., et al. (2013). Respondents' answers to community attitudinal surveys represent impressions of soundscapes and not merely reactions to the physical noise. *The Journal of the Acoustical Society of America, 134*(1), 767–772.
Schorer, E. (1989). Vergleich eben erkennbarer Unterschiede und Variationen der Frequenz und Amplitude von Schallen. *Acta Acustica United with Acoustica, 68*(3), 183–199.
Schreurs, E., Koeman, T., & Jabben, J. (2008). *Low frequency noise impact of road traffic in the Netherlands* (pp. 1943–1948). Paris: European Acoustics Association.
Schulte-Fortkamp, B., & Jordan, P. (2016). When soundscape meets architecture. *Noise Mapping, 3*(1). https://doi.org/10.1515/noise-2016-0015.

Seto, E. Y. W., et al. (2007). Spatial distribution of traffic induced noise exposures in a US city: An analytic tool for assessing the health impacts of urban planning decisions. *International Journal of Health Geographics, 6*(1), 24.
Sevillano, X., et al. (2016). DYNAMAP – Development of low cost sensors networks for real time noise mapping. *Noise Mapping, 3*(1). https://doi.org/10.1515/noise-2016-0013.
Shepherd, D., et al. (2016). The covariance between air pollution annoyance and noise annoyance, and its relationship with health-related quality of life. *International Journal of Environmental Research and Public Health, 13*(8), 792.
Shiue, I. (2016). People with dyslexia and heart, chest, skin, digestive, musculoskeletal, vision, learning, speech and mental disorders were more dissatisfied with neighbourhoods: Scottish Household Survey, 2007–2008. *Environmental Science and Pollution Research, 23*(23), 23840–23853.
Shu, S., Yang, P., & Zhu, Y. (2014). Correlation of noise levels and particulate matter concentrations near two major freeways in Los Angeles, California. *Environmental Pollution, 193*, 130–137.
Silva, L. T., et al. (2017). Assessing the influence of urban geometry on noise propagation by using the sky view factor. *Journal of Environmental Planning and Management*, 1, 535–18, 552.
Sneddon, M., Pearsons, K., & Fidell, S. (2003). Laboratory study of the noticeability and annoyance of low signal-to-noise ratio sounds. *Noise Control Engineering Journal, 51*, 300–3005.
Steven, H. (2012). The role of vehicles in road traffic noise, effects of type approval limit value reductions. In *Sound level of motor vehicles* (pp. 39–48). Brussels: European Union.
Tainio, M. (2015). Burden of disease caused by local transport in Warsaw, Poland. *Journal of Transport & Health, 2*(3), 423–433.
Takahashi, Y. (2013). Vibratory sensation induced by low-frequency noise: The threshold for "vibration perceived in the head" in normal-hearing subjects. *Journal of Low Frequency Noise, Vibration and Active Control, 32*(1), 1–10.
Tang, S. K. (2005). Noise screening effects of balconies on a building facade. *The Journal of the Acoustical Society of America, 118*(1), 213–221.
Tenailleau, Q. M., et al. (2015). Assessing residential exposure to urban noise using environmental models: Does the size of the local living neighborhood matter. *Journal of Exposure Science and Environmental Epidemiology, 25*(1), 89–96.
To, W., Mak, C., & Chung, W. (2015). Are the noise levels acceptable in a built environment like Hong Kong? *Noise and Health, 17*(79), 429–439.
Tobías, A., et al. (2015). Health impact assessment of traffic noise in Madrid (Spain). *Environmental Research, 137*, 136–140.
Tonin, R. (2016). Quiet road pavements: Design and measurement—State of the art. *Acoustics Australia, 44*(2), 235–247.
Torija, A. J., & Ruiz, D. P. (2012). Using recorded sound spectra profile as input data for real-time short-term urban road-traffic-flow estimation. *Science of the Total Environment, 435–436*, 270–279.
Torija, A. J., & Ruiz, D. P. (2016). Automated classification of urban locations for environmental noise impact assessment on the basis of road-traffic content. *Expert Systems with Applications, 53*, 1–13.
Trollé, A., Marquis-Favre, C., & Klein, A. (2013). Acoustical indicator of noise annoyance due to tramway in in-curve operating configurations. In *Proceedings of meetings on acoustics* (pp. 040023–040023). Boston: Acoustical Society of America.
Trollé, A., Marquis-Favre, C., & Parizet, É. (2015). Perception and annoyance due to vibrations in dwellings generated from ground transportation: A review. *Low Frequency Noise, Vibration and Active Control, 34*(4), 413–458.
Tzivian, L., et al. (2015). Effect of long-term outdoor air pollution and noise on cognitive and psychological functions in adults. *International Journal of Hygiene and Environmental Health, 218*(1), 1–11.

UITP. (2002). *Public transport for sustainable mobility*. Brussels: International Association of Public Transport.
Van Kamp, I., Klæboe, R., Brown, A. L., & Lercher, P. (2015). Soundscapes, human restoration, and quality of life. In J. Kang & B. Schulte-Fortkamp (Eds.), *Soundscape and the built environment* (pp. 43–68). London: CRC Press.
van Kempen, E., et al. (2014). Characterizing urban areas with good sound quality: Development of a research protocol. *Noise and Health, 16*(73), 380–387.
van Kempen, E., et al. (2018). WHO environmental noise guidelines for the European region: A systematic review on environmental noise and cardiovascular and metabolic effects: A summary. *International Journal of Environmental Research and Public Health, 15*(2), 379. Full report available on https://doi.org/10.21945/RIVM-2017-0078.
Van Renterghem, T., & Botteldooren, D. (2009). Reducing the acoustical façade load from road traffic with green roofs. *Building and Environment, 44*(5), 1081–1087.
Van Renterghem, T., et al. (2015). Using natural means to reduce surface transport noise during propagation outdoors. *Applied Acoustics, 92*, 86–101.
Vedaa, Ø., et al. (2016). Prospective study of predictors and consequences of insomnia: Personality, lifestyle, mental health, and work-related stressors. *Sleep Medicine, 20*, 51–58.
Veisten, K., et al. (2012). Valuation of green walls and green roofs as soundscape measures: Including monetised amenity values together with noise-attenuation values in a cost-benefit analysis of a green wall affecting courtyards. *International Journal of Environmental Research and Public Health, 9*(11), 3770–3788.
Vienneau, D., et al. (2015). Years of life lost and morbidity cases attributable to transportation noise and air pollution: A comparative health risk assessment for Switzerland in 2010. *International Journal of Hygiene and Environmental Health, 218*(6), 514–521.
Viollon, S., Lavandier, C., & Drake, C. (2002). Influence of visual setting on sound ratings in an urban environment. *Applied Acoustics, 63*(5), 493–511.
von Lindern, E., Hartig, T., & Lercher, P. (2016). Traffic-related exposures, constrained restoration, and health in the residential context. *Health & Place, 39*, 92–100.
von Lindern, E., Lymeus, F., & Hartig, T. (2017). The restorative environment: A complementary concept for salutogenesis studies. In *The handbook of salutogenesis* (pp. 181–195). Cham: Springer International Publishing.
Wang, B., & Kang, J. (2011). Effects of urban morphology on the traffic noise distribution through noise mapping: A comparative study between UK and China. *Applied Acoustics, 72*(8), 556–568.
Wang, V.-S., et al. (2016). Temporal and spatial variations in road traffic noise for different frequency components in metropolitan Taichung, Taiwan. *Environmental Pollution, 219*, 174–181.
Ward Thompson, C., et al. (2012). More green space is linked to less stress in deprived communities: Evidence from salivary cortisol patterns. *Landscape and Urban Planning, 105*(3), 221–229.
Wegener, M. (2004). Overview of land-use transport models. In K. Button, D. A. Hensher, K. E. Haynes, & P. R. Stopher (Eds.), *Transport geography and spatial systems* (pp. 127–146). Kidlington: Pergamon.
Wells, N. M., & Evans, G. W. (2003). Nearby nature: A buffer of life stress among rural children. *Environment & Behavior, 35*(3), 311–330.
Wunderli, J. M., et al. (2015). Intermittency ratio: A metric reflecting short-term temporal variations of transportation noise exposure. *Journal of Exposure Science and Environmental Epidemiology, 26*(6), 575–585.
Xie, H., & Kang, J. (2009). Relationships between environmental noise and social–economic factors: Case studies based on NHS hospitals in Greater London. *Renewable Energy, 34*(9), 2044–2053.
Yang, F., Bao, Z. Y., & Zhu, Z. J. (2011). An assessment of psychological noise reduction by landscape plants. *International Journal of Environmental Research and Public Health, 8*(4), 1032–1048.

Yang, M., & Kang, J. (2013). Psychoacoustical evaluation of natural and urban sounds in soundscapes. *The Journal of the Acoustical Society of America, 134*(1), 840–851.

YANO, T., et al. (2002). Comparison of community response to road traffic noise in Japan and Sweden—Part II: Path analysis. *Journal of Sound and Vibration, 250*(1), 169–174.

You, J., Lee, P. J., & Jeon, J. Y. (2010). Evaluating water sounds to improve the soundscape of urban areas affected by traffic noise. *Noise Control Engineering Journal, 58*(5), 477–483.

Younan, D., et al. (2016). Environmental determinants of aggression in adolescents: Role of urban neighborhood greenspace. *Journal of the American Academy of Child & Adolescent Psychiatry, 55*(7), 591–601.

Zapfe, J. A., Saurenman, H., & Fidell, S. (2012). Human response to groundborne noise and vibration in buildings caused by rail transit: Summary of the TCRP D-12 study. In B. Schulte-Werning et al. (Eds.), *Noise and vibration mitigation for rail transportation systems* (pp. 25–32). New York: Springer.

Zappatore, M., Longo, A., & Bochicchio, M. A. (2017). Crowd-sensing our smart cities: A platform for noise monitoring and acoustic urban planning. *Journal of Communications Software and Systems, 13*(2), 53–67.

Zhou, Z., et al. (2017). Analysis of traffic noise distribution and influence factors in Chinese urban residential blocks. *Environment and Planning B: Urban Analytics and City Science, 44*(3), 570–587.

Zuo, F., et al. (2014). Temporal and spatial variability of traffic-related noise in the City of Toronto, Canada. *Science of the Total Environment, 472*, 1100–1107.

Zwicker, E., & Fastl, H. (1999). Just-noticeable sound changes. In E. Zwicker & H. Fastl (Eds.), *Psychoacoustics* (pp. 175–201). Berlin: Springer International Publishing.

Chapter 23
Heat Islands/Temperature in Cities: Urban and Transport Planning Determinants and Health in Cities

Xavier Basagaña

23.1 Introduction

The city structure, its land uses, and the materials used in buildings or streets can all alter the ambient temperature experienced in urban areas compared to the temperature registered in surrounding rural areas. Heat islands are a well-studied phenomenon by which urban areas can register temperatures that are several degrees hotter than surrounding rural areas. In this chapter, we describe the heat island phenomenon, its causes, its consequences, and proposed solutions to alleviate it. Appropriate urban planning and architecture are two key elements that can contribute to alleviate this phenomenon. The chapter will also devote a section to health, as exposure to extremely hot or cold temperatures is a well-recognized health threat. We will review what is known on the effects of temperatures on health, how the heat island effect can intensify the health effects, and how several studies have shown that measures to reduce the heat island phenomenon can be translated into health benefits.

23.2 Heat Island Definition

It has been known for some time that large urban areas can register hotter air temperatures than surrounding rural areas, a phenomenon that is known as the urban heat island, which is driven by the replacement of green or open natural areas with buildings, roads, and parking areas built with materials that absorb and store more heat. Such differences in temperature can reach values of more than 10 °C at nighttime, although in most cities, the daytime difference is between 2 and 4 °C

X. Basagaña (✉)
Barcelona Institute for Global Health (ISGlobal), Barcelona, Spain
e-mail: xavier.basagana@isglobal.org

M. Nieuwenhuijsen, H. Khreis (eds.), *Integrating Human Health into Urban and Transport Planning*, https://doi.org/10.1007/978-3-319-74983-9_23

Table 23.1 Magnitude of urban heat island in different cities around the world. Data obtained from Tzavali et al. (2015) and Santamouris (2015)

City	Country	Magnitude of heat island
Melbourne	Australia	2–4 °C, with daily peaks of 7 °C
Tokyo	Japan	8 °C
Delhi	India	8.6–10.7 °C in dense and commercial built-up areas, 3.1–6.9 °C in open and green areas
Tel Aviv	Israel	3–5 °C, 5.2 °C
Cairo	Egypt	0.5–2 °C
Moscow	Russia	0.8–0.9 °C, to 14 °C under strong anticyclone conditions
Athens	Greece	6–12 °C
London	UK	0.7–0.8 °C during the day, 2.5–3.0 °C at night, with peaks of 8 °C
Johannesburg	South Africa	2.8 °C, reaching 11 °C during strong-inversion winter nights
Chicago	USA	2.34–2.68 °C

(Table 23.1) (CalEPA 2017). The urban heat island represents one of the most significant human-induced changes to Earth's surface climate (Zhao et al. 2014).

There are several reasons that contribute to the increased temperatures in cities, including a large area covered by heat-absorptive surfaces, the lack of vegetation and moisture, the generation of heat by human activities, and the city geometry, all of which can influence the city energy balance. However, there are many other factors that play a role, such as the size of the city, topography, elevation, latitude, location relative to the sea or vicinity to bodies of water, local climate, land use, presence of urban canyons, presence of industry, building density, and urban air pollution, which make the intensity of the heat island effect a very site-specific issue (Burkart et al. 2016; Heaviside et al. 2017; Koppe et al. 2004; Santamouris 2015; Tzavali et al. 2015). Even within a city, there might be variations in the heat island intensity, with the presence of micro-urban heat islands that are usually located in parts of the city covered by buildings and roadways and with a small presence of green space.

The heat island effect can be present throughout the year and at any time of the day, although it is usually more noticeable at night, and in days with clear skies and light winds, which prevent the dispersal of the warm air accumulated in the cities. The nighttime pattern is explained by the fact that several urban materials, such as concrete or pavement, absorb the sun energy they receive during the day and gradually release it during the night. The surrounding rural areas, without those materials with high heat capacity, cool down faster during the night. The heat island effect is particularly evident during heat waves, i.e., during episodes with several consecutive days of extreme heat. Thus, urban areas are usually hit harder by heat waves than rural areas. The heat island effect is also influenced by humidity, with lower intensity of the phenomenon in regions or days with higher relative humidity (Santamouris 2015).

23.3 Causes of Heat Islands

As mentioned, heat islands are created by a combination of reasons. In this section, we review some of the most important ones. The built environment of cities usually leads to a reduction of evaporative cooling (e.g., by the reduction of vegetation), which is one of the main factors involved in the heat island phenomenon. The artificial materials used (e.g., dark pavement or roofing) can also reduce the albedo (sun reflectance), thus storing more radiation energy during daytime that can then be released during the night. Street canyon geometry also has influences on the radiation balance and on the redistribution of energy through convection with the atmospheric boundary layer. Other human activities that generate heat, such as those involving engines or generators, including emissions from vehicles, can also contribute the warming of cities.

23.3.1 Urban Area Materials

The materials used to build urban areas absorb and reflect solar radiation, store heat, emit infrared radiation, and release sensible and latent heat through convection and evaporation processes, and they do that at different rates than rural areas. On hot and sunny days, urban surfaces can become up to 50 °C hotter than the air, while shaded or moist surfaces remain close to air temperatures. For example, dark and dry surfaces can reach temperatures of up to 88 °C when exposed to the sun (Tzavali et al. 2015).

23.3.2 Vegetation

Differences in vegetation can influence the difference in temperature between rural and urban areas (Burkart et al. 2016; Heaviside et al. 2017; Koppe et al. 2004). Vegetation absorbs solar radiation, so reflected radiation is small. Green spaces also provide cooling through evapotranspiration. Instead of increasing air temperatures, solar energy is used to evaporate water from plants, which cools the plant and reduces the amount of energy that is left to warm the air. In addition, vegetation has lower heat capacity and thermal conductivity than the materials used in buildings. Another important way in which vegetation can contribute to cooling cities is through shading, which reduces the incident shortwave radiation, leads to lower ground and wall surface temperatures, and reduces radiant temperature. Different trees have different shading capacity, which is one of the criteria that urban planners need to take into account when deciding to plant trees in an urban area. Finally, green spaces can also reduce wind speed within a city. This may be undesirable in terms of reducing temperature in the summer, as it leads to reduced cooling through

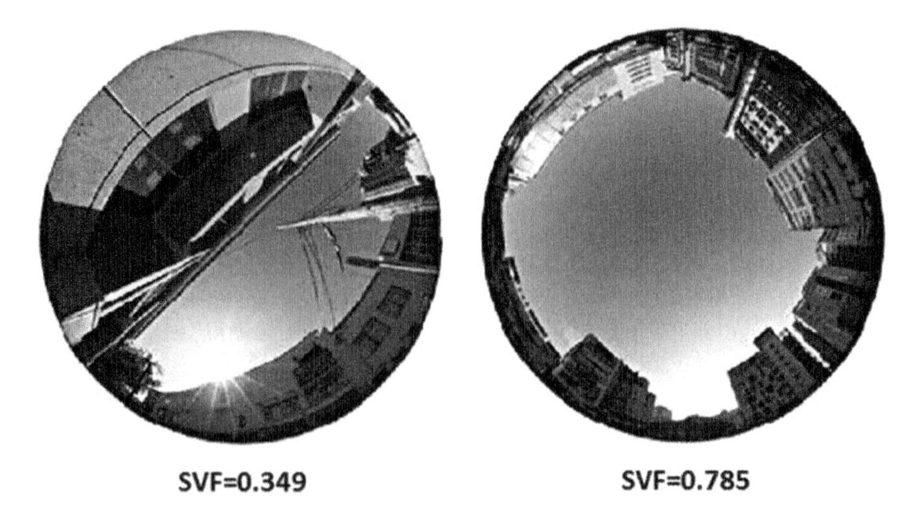

Fig. 23.1 Examples of Sky View Factors (SVF) obtained from fisheye photographs in two different locations. Figure adapted from Fig. 1 in: Landscape and Urban Planning; Vol 125; Sookuk Park, Stanton E. Tuller, Myunghee Jo; Application of Universal Thermal Climate Index (UTCI) for microclimatic analysis in urban thermal environments; pages 146–155; Copyright 2014; with permission from Elsevier

wind. For example, in hot and humid areas, it is recommended to use trees with high canopies that allow for ventilation at ground level (Koppe et al. 2004).

23.3.3 City Form and Topography

City morphology influences the air temperature of cities in complex ways, by modifying, for example, the area of exposed external surfaces and surface reflectance (Koppe et al. 2004; Zhao et al. 2014). As an example, a cubical form can collect three times more radiation than unbuilt ground. Thus, things like the orientation, layout and width of streets, the height, shape, and location of buildings, and the resulting shading patterns can all influence the heat transfer of the urban area by modifying the magnitude of shortwave radiation reaching the street level and the longwave radiation escaping from the canyon (Santamouris 2015). Although it is complex to capture all these aspects in a single index, the Sky View Factor (SVF), defined as the fraction of sky visible from the ground up, is a good marker of the heat island effect, as it is related to the long-wave heat radiation losses. Figure 23.1 shows two examples of SVF. Indeed, several studies have documented a negative correlation between SVF and the increase in temperature comparing built-up and open areas (Svensson 2004). It is important to note, though, that wind patterns can modify those associations.

Topography is also important in modulating the heat island effect. Having mountains around can prevent air circulation and heat dispersion and can prevent the circulation of the sea-land breeze. The San Francisco Bay Area is an example of ocean and mountains modulating the heat island effect (CalEPA 2017). The breeze from the ocean cools coastal cities, but the inland mountains trap warm air and displace it to other areas, thus warming areas that are upwind of the heat island. Dark-colored mountains can also absorb shortwave radiation and contribute to the warming of cities, as shown for Muscat (Oman) (Santamouris 2015). It is well-known that areas that are close to large bodies of water show less extreme temperatures, in part because of the land-sea breeze generated by temperature and pressure differences between the land and the sea. The heat island can prevent the inland movement of the sea breeze. Then, the cooling impact of the breeze can only be noticeable at the coastal front. For example, the waterside districts in Lisbon (Portugal) register temperatures that are up to 3–4 °C lower than in the city center (Burkart et al. 2016).

23.3.4 *Size of the Urban Area*

Several studies have reported that the larger and the more populated an urban area is, the more pronounced is the urban heat island intensity. In 1973, Oke reported that the maximum urban-rural difference of 2.5 °C for towns of 1000 inhabitants increases to 12 °C for cities of one million inhabitants (Oke 1973).

23.3.5 *Wind*

As mentioned, wind can influence heat islands, as in the case of sea breezes or wind systems in mountain valleys (Koppe et al. 2004; Santamouris 2015). Usually, faster winds imply a higher rate of heat dissipation by convective cooling. When the rural areas surrounding the cities are open country, wind speed is lower in the city than in the rural areas and so is the heat dissipation. However, street canyons can create complex airflow patterns that may alter this pattern, increasing wind speeds in the city. Winds can also contribute to warming areas downwind the city. A study in Birmingham (UK) found a warming of 1.2 °C up to 12 km downwind of the city, and this phenomenon can reach up to 40 km for a large city such as London (UK) (Bassett et al. 2016).

23.3.6 *Heat-Generating Activities*

Several human activities that generate heat can contribute to the heat island phenomenon, although the anthropogenic heat is much smaller than that received from the solar radiation in summer (Santamouris 2015; Tzavali et al. 2015). Those activities include transportation, industry, or energy consumption at homes. The average anthropogenic heat flux in urban areas has been suggested to be 100 W/m^2, although they can reach 250 W/m^2 in European and North American cities. In Tokyo (Japan), the anthropogenic heat flux exceeds 400 W/m^2, with a maximum reaching 1500 W/m^2 in winter (Santamouris 2015). The translation of those fluxes to air temperatures depends on many features of the city, including prevailing climatic conditions and geographical location. In Taipei, the anthropogenic sources were estimated to contribute 0.2 °C to air temperature, while in an industrial zone in Malaysia, they contributed 1.1 °C (Santamouris 2015).

23.3.7 Building Types

Different housing types have been found to have different risks of overheating, also in part due to their location within a city. For example, in a study in the West Midlands (UK), Macintyre et al. found that the type of housing that is more likely to overheat (flats in buildings) tends to be located in the warmest parts of the city (Macintyre et al. 2018). They registered ambient temperatures around buildings and terraced houses were 0.1 °C higher than the average for all housing types, while detached houses, which tended to be in the suburbs, had 0.2 °C lower temperatures. Interestingly, care homes and hospitals tended to be exposed to higher ambient temperatures than average.

23.3.8 Daytime Cool Islands

Apart from heat islands, cool islands, i.e., urban areas with cooler temperatures than surrounding rural areas, have also been documented (Santamouris 2015). This is mostly a daytime phenomenon that may occur in deep and dense urban canyons due to the extensive shading provided by the buildings, along with their high heat storage capacity. Thus, deep canyons can present a heat island effect during nighttime and a cool island effect during daytime. In Beijing, urban cool islands were observed during the morning period and were attributed to reduced solar radiation because of high concentrations of aerosols and to the possible advection of warmer air to rural areas.

23.4 Consequences of Heat Islands

Heat islands have several adverse effects (CalEPA 2017; Heaviside et al. 2017; Koppe et al. 2004; Kyriakodis and Santamouris 2017; Santamouris 2015; Tzavali et al. 2015). The increase in temperature can exacerbate the health consequences of heat waves, as detailed below, and reduce human comfort. In addition, the warmer temperatures favor the creation of ground-level ozone, which is associated with the exacerbation of several health conditions such as respiratory and cardiovascular problems. Air pollution and high temperatures may act synergistically during heat waves to produce stronger health effects. The warmth of urban sites can also favor the spread of vector-borne diseases. Apart from the health-related problems, high temperatures also accelerate the degradation of materials, e.g., road surfaces. In addition, one of the most studied consequences of heat islands is the increase in electricity consumption because of the increasing use of air conditioning.

The widespread use of air conditioning contributes to peak electricity loads. The increase in energy consumption leads to increases in greenhouse gas emissions. It has been calculated that the peak electricity demand increases from 0.45% to 4.6% for each 1 °C increase in maximum temperature during the summer (Kyriakodis and Santamouris 2017; Tzavali et al. 2015). A study based mainly on cities in the USA, Europe, and Australia estimated that the increase of the demand for cooling in buildings has increased by 23% in the last four decades (Tzavali et al. 2015). A study in Athens estimated that the urban heat island may double the cooling load and triple the peak electricity load of buildings designed for having low cooling and heating needs (Koppe et al. 2004). Projections estimate that future energy consumption for cooling in residences will increase up to 750% (Kyriakodis and Santamouris 2017). This is due not only to global warming but also to the increase of population in cities and the increased penetration of air conditioning systems in many parts of the world. An economic study in Los Angeles, USA, estimated that the costs of the heat island were about $150,000 per hour, resulting in $100 million for cooling (Koppe et al. 2004). The use of air conditioning also contributes to the increase of anthropogenic heat in cities. Although the use of air conditioning may be beneficial for health, relying on air conditioning to combat heat can widen inequalities, and it makes the population vulnerable in the event of power outages, which occur in some cities during hot summer days with large energy demands.

On the contrary, the heat island effect can have some benefits in winter. In particular, the study in Athens estimated that the heat island reduced the heating load of the buildings by 30% (Koppe et al. 2004). The beneficial effects of heat islands have received less attention in the literature.

23.5 Proposed Mitigation Actions

The heat island phenomenon can be moderated by appropriate urban and transport planning actions or by using certain technologies. The most popular measures include replacing urban materials by others that reflect more solar radiation ("cool" materials) and increasing the vegetation in the city. However, there are more potential actions, such as reducing anthropogenic heat. This section reviews some of the suggested actions. More detailed information can be found, for example, in the US Environmental Protection Agency report *Reducing Urban Heat Islands: Compendium of Strategies* (2008).

23.5.1 Increasing the Albedo of Cities

Increasing the albedo or reflectance of the city can be achieved in several ways. Simple solutions such as painting roofs or building walls or pavement with white colors can already achieve increases in reflectance that can be translated into lower air temperatures. This is an old solution applied for many years in many South European and North African towns characterized by white houses. The effectiveness of this measure is due to the fact that white surfaces directly reflect much of the sun radiation back to space, unlike dark materials, which absorb the sunlight and then reradiate it at much longer wavelenghts, which is then trapped by the greenhouse gases in the atmosphere.

There are several materials available for "cool roofs," the term used for the use of highly reflective materials in roofs of buildings (see US Environmental Protection Agency (2008)). While traditional roofing materials in the USA have a reflectance of 5–15% (i.e., they absorb 85–95% of solar radiation), some cool roof materials can reach more than 65% reflectance. In the summer sun, black asphalt roofs can reach temperatures of 85 °C, metallic roofs can reach 165 °C, and cool roofs can stay below 46 °C. This can be translated into reductions in air temperature. A simulation conducted for New York City found that replacing 50% of roofs with cool roofs would achieve reductions of 0.2 °C in the average temperature of the city, with reductions of up to 0.8 °C in some parts of the city (US Environmental Protection Agency 2008). Increasing the albedo of a city would not directly affect the nighttime urban heat island, but it can have indirect effects by reducing the amount of heat that is absorbed during the day and, therefore, the amount of heat that is released at night (Zhao et al. 2014).

Cool roofs are not necessarily more expensive than traditional ones. The degree of success of implementing cool roofs will depend on several conditions, including local climate and the building design, but several studies have reported annual cooling savings ranging from 10% to 69% (US Environmental Protection Agency 2008). A detailed cost-benefit analysis is needed before implementing cool roofs, as, for example, having a cool roof may require consuming more energy for heating in winter. One study found that, in California, cool roofs had a net benefit of up to $0.66 per square foot. Implementing cool roofs may have other unintended consequences for a city. For example, the increased reflectance may favor ozone formation (Heaviside et al. 2017). Thus, appropriate planning of such policies is recommended.

Acting on pavements is another way to improve the reflectance of a city and reduce the heat island effect. For example, a study in Athens evaluated a large-scale intervention to introduce cool asphalt in a major traffic street (Kyriakodis and Santamouris 2017). They concluded that the intervention could reduce the surface temperature by 11.5 °C and the ambient temperature by up to 1.5 °C, although the efficacy of the asphalt will be reduced by aging. Many other studies have evaluated similar interventions and found reductions in temperatures of around 0.6 °C.

23.5.2 *Increasing Vegetation*

As explained above, vegetation can reduce the heat island effect through multiple pathways, especially through evapotranspiration, increased shading, and decreasing radiant temperature, which is a key parameter for thermal stress in humans (US Environmental Protection Agency 2008; Koppe et al. 2004). Increasing vegetation in an urban area needs to be carefully planned taking into account issues such as water demand and availability, maintenance costs, the different shading provided by different species, or their allergenic potential. Vegetation can have other benefits apart from reducing the heat island effect, such as reducing air pollution levels through dry deposition; reducing noise levels; carbon sequestration; reducing exposure to UV radiation; increasing physical activity levels of residents around green areas; and reducing the volume of storm water received and thus preventing sewer overflow.

Several cost-benefit analyses have shown a net benefit of planting trees. For example, studies in several cities in the USA have calculated benefits of $1.50–$3.00 per dollar invested, or of around between $15.50 and $85 per tree (US Environmental Protection Agency 2008; Koppe et al. 2004). Other studies in the USA have also documented that increasing the urban tree coverage by 25% can lead to savings of up to 40% in the cooling load of a city, although results depend, among other factors, on the city current spending on cooling (Koppe et al. 2004). Another study estimated that the peak temperature in summer in New York City could be reduced by 0.2 °C by adding trees to increase shade in 6.7% of the city area and estimated a reduction of 0.5 °C if 31% of the city was changed to have trees and green roofs (Mills and Kalkstein 2012).

Green roofs, i.e., having a layer of vegetation on top of roofs, and green walls, the same concept applied to vertical surfaces, are other proposed solutions to counteract the heat island phenomenon. The cooling potential of green roofs depends on several factors, including the local climate, the type of vegetation and soil, irrigation, and maintenance. Green roofs also provide insulating effects in winter (Heaviside et al. 2017). A study in Toronto, Canada, estimated that adding green roofs to 50% of the downtown surfaces would reduce ambient temperatures by between 0.1 and 0.8 °C, and those reductions could be increased to 2 °C by irrigating those green roofs (US Environmental Protection Agency 2008).

Installation of green roofs may have large initial costs, but full life-cycle analyses have shown that green roofs can result in net benefits in densely populated areas, especially through reducing the energy needs of the building. When incorporating the public benefits, those investments are even more favorable.

23.5.3 *Other Strategies*

Another strategy that could be considered is trying to reduce the anthropogenic heat of a city. However, a study estimated that such a strategy would result in almost no changes in the heat island effect (Zhao et al. 2014). Other strategies that focus on managing the convection efficiency of the city by changing its morphology could bring important benefits, but they are unfeasible as they would require, for example, citywide changes in building height.

23.6 Effects of Heat on Health

It is well-known that temperatures influence human health. In particular, it is well documented that episodes of extreme heat and cold result in increases in mortality and hospital admissions (Basu 2009; Basu and Samet 2002; Gasparrini et al. 2015). To cite a few, heat waves produced 5000–10,000 extra deaths in the USA in 1988, 70,000 additional deaths in Western Europe in 2003, 11,000 extra deaths in Russia in 2010, and 3500 additional deaths in India and Pakistan in 2015. However, even non-extreme episodes of heat and cold produce such increases in mortality. Actually, results from an international study using mortality data from 384 locations in 13 countries estimated that 0.42% of all deaths can be attributed to heat every year, even if no heat waves occur (Gasparrini et al. 2015). The relationship between mortality and temperature is U-shaped as displayed in Fig. 23.2 for the city of Barcelona (Spain). The same relationship, with different slopes for the effects of heat and cold, has been documented in many other parts of the world. The temperature of minimum mortality also changes by location, showing the adaptation of the population to its local climate (e.g., the effect of heat on mortality starts to be detected at hotter temperatures in hot cities in comparison to colder cities).

Humans need to maintain a constant body temperature of around 37 °C regardless of the outdoor temperature. To do so under hot ambient temperatures, the body activates several thermoregulatory mechanisms, such as increased sweating leading to loss of water and minerals, diverting blood from internal organs toward the skin to dissipate body heat, changing blood viscosity, or increasing the heart and respiratory rate (Koppe et al. 2004). These changes are well tolerated in healthy people, but they may put an additional stress to the body of vulnerable populations such as the elderly or those suffering from chronic diseases. Ultimately, a heat episode may trigger acute events such as heart attacks or renal failure in a percentage of that frail population, leading to the observed excess deaths during hot episodes. Other vulnerable populations include children, pregnant women, the obese, those with limited mobility and little social contact, those with psychiatric conditions, those taking medications that alter thermoregulation, and the socially disadvantaged (Basu and Samet 2002; Klein Rosenthal et al. 2014).

Apart from the increases in mortality and hospital admissions, other studies have shown that heat can trigger delivery and thus increase the number of preterm births

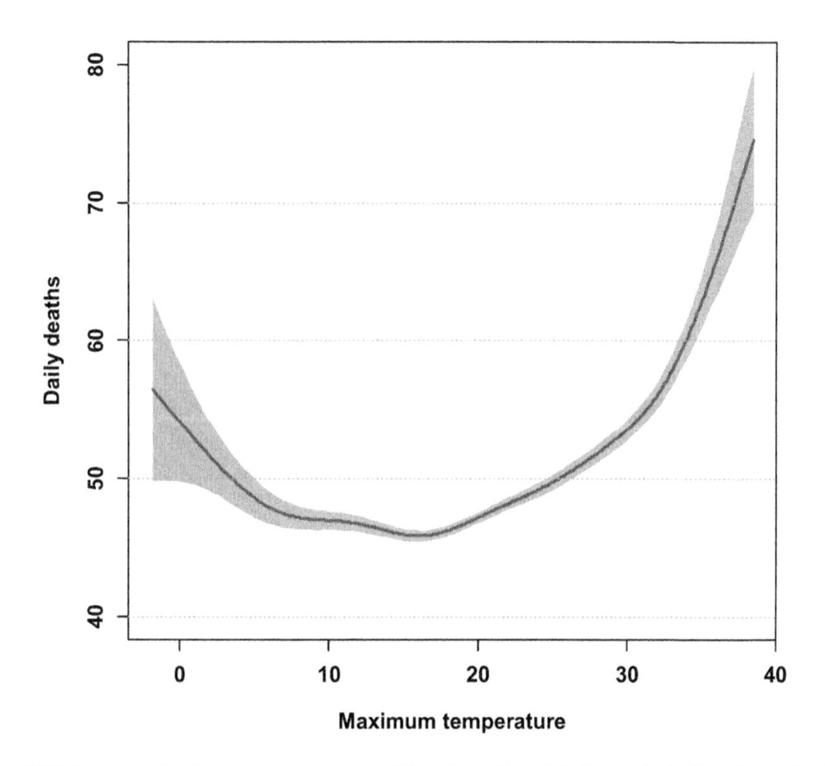

Fig. 23.2 Relationship between temperature (°C) and number of daily deaths in Barcelona, Spain. The shaded region indicates 95% confidence intervals

(Zhang et al. 2017) and that heat increases the likelihood of accidents (including traffic crashes) and injuries to occur (Otte im Kampe et al. 2016).

23.7 Influence of the Heat Island on Health Effects

Heat islands can aggravate the health consequences of hot temperatures and heat waves. In situations of extreme heat, the increased temperatures of urban areas put an extra burden to the human body. In particular, it has been reported that high nighttime temperatures are especially sensitive for health, as they limit the ability of individuals to cool down. As explained above, the heat island effect is especially noticeable at night. Some studies also have shown that those living on the top floors of buildings are at increased risk of health events during heat waves (Klein Rosenthal et al. 2014; Laaidi et al. 2011). A number of papers have also documented that the hotter parts of the city show increased mortality rates during heat waves, indicating an extra effect of heat islands (Goggins et al. 2012; Heaviside et al. 2016, 2017; Smargiassi et al. 2009; Vandentorren et al. 2006).

A study in Montreal, Canada, found that heat increased mortality by 28% in the areas with highest surface temperatures, while this increase was only 13% in the areas with lowest surface temperatures (Smargiassi et al. 2009). Similarly, a study in Hong Kong found that the increase in mortality for every 1 °C increase in temperature over 29 °C in the areas with high urban heat island index was 4.1%, while it was only 0.7% in areas with low urban heat island index (Goggins et al. 2012). A study in the West Midlands (UK) estimated that 50% of the total heat-related mortality during the 2003 heat wave could be attributed to the heat island effect (Heaviside et al. 2016). In London (UK), however, a study suggested almost complete acclimatization of the heat island effect (Milojevic et al. 2016). A study in the USA estimated that the heat island phenomenon was associated with an increase of 1.1 deaths per million population (Lowe 2016). In terms of projections and economic costs, a study in Melbourne, Australia, estimated that the heat island is expected to add 2.2 days with temperatures over 35 °C per year and an additional heat wave every 10 years compared to the areas that are outside of the heat island. This was associated with an extra cost of $300 million (AECOM Australia Pty Ltd 2012).

As mentioned above, the heat island effect can bring some benefits in winter, although these have been less studied. However, a study in the USA estimated that the urban heat island could reduce cold-related mortality by four deaths per million, in comparison with an increase of 1.1 deaths per million due to heat-related mortality (Lowe 2016).

23.8 Influence of Green and Blue Spaces on Health Effects

Given the potential of green and blue spaces to reduce ambient temperatures, several studies have investigated the variation of the health effects of heat within a city by these characteristics. A study in Lisbon, Portugal, found that a 1 °C increase in temperature over the 99th percentile was associated with a 14.7% increase in mortality in the area with lowest greenness, while it was associated with a 3% increase in the greener areas. Likewise, in areas that were more than 4 km away from a water body, the increase in mortality was 7.1%, while it was only 2.1% in the areas close to blue spaces (Burkart et al. 2016). Thus, the study supported that both green and blue spaces are able to mitigate the effects of extreme heat on mortality. A study in Seoul, Korea, also found decreases in heat-related mortality from 4.1% to 2.2% as greenness of the area increased (Son et al. 2016). In Michigan (USA), the increase in cardiovascular mortality during extreme heat was 17% in the areas with low green space, while there was no increase in greener areas (Gronlund et al. 2015). A study that included the Medicare beneficiaries in the USA found that the effect of extreme temperature on mortality was 3% higher in areas with less green space (Zanobetti et al. 2013). On the contrary, a study in Australia, one in Philadelphia (USA), and one in Massachusetts (USA) did not find differences on heat-related mortality by vegetation (Madrigano et al. 2013; Uejio et al. 2011; Vaneckova et al. 2010).

Other studies have estimated what the health benefits of different vegetation interventions could be. Although these studies are extremely useful, it is often difficult to validate their estimated impacts. A study in Melbourne, Australia, estimated that reductions in heat-related mortality could go from 5% to 28% if the vegetation cover of the business district increased from 15% to 33% (Chen et al. 2014). Transforming the entire business district into a forested park could reduce heat-related mortality between 37% and 99%. A study in Arizona, USA, estimated a 48% reduction on heat-related emergency service calls associated with simultaneous improvement in emissivity, vegetation, thermal conductivity, and albedo (Silva et al. 2010). Finally, a study based on metropolitan areas in the USA estimated that changes in vegetation cover and surface albedo could reduce the projected increases in heat-related deaths due to global warming by 40–99% (Stone et al. 2014).

23.9 Conclusions

Heat islands have been documented in many large urban areas around the world. They can contribute 2–4 °C to the ambient temperature of cities during daytime and to over 10 °C in certain places and times, especially at night. Certain actions related to urban planning, such as using cool materials or increasing the vegetation of cities, can have an important influence in ameliorating the intensity of heat islands. These actions have been shown to produce benefits, especially in terms of energy savings and also in the reduction of the health effects of heat. Given the complexities associated with the multiple benefits and costs of such interventions, implementations to correct the effects of urban heat islands should be based on full life-cycle studies involving public planners, climatologists, experts in air pollution, botanists, and public health experts.

References

AECOM Australia Pty Ltd. (2012). Economic assessment of the urban heat island effect. Retrieved Sept 6, 2017, from https://www.melbourne.vic.gov.au/SiteCollectionDocuments/eco-assessment-of-urban-heat-island-effect.pdf.

Bassett, R., Cai, X., Chapman, L., Heaviside, C., Thornes, J. E., Muller, C. L., Young, D. T., & Warren, E. L. (2016). Observations of urban heat island advection from a high-density monitoring network: Urban heat advection. *Quarterly Journal of the Royal Meteorological Society, 142*, 2434–2441. https://doi.org/10.1002/qj.2836.

Basu, R. (2009). High ambient temperature and mortality: A review of epidemiologic studies from 2001 to 2008. *Environmental Health: A Global Access Science Source, 8*, 40. https://doi.org/10.1186/1476-069X-8-40.

Basu, R., & Samet, J. M. (2002). Relation between elevated ambient temperature and mortality: A review of the epidemiologic evidence. *Epidemiologic Reviews, 24*, 190–202.

Burkart, K., Meier, F., Schneider, A., Breitner, S., Canário, P., Alcoforado, M. J., Scherer, D., & Endlicher, W. (2016). Modification of heat-related mortality in an elderly urban population by vegetation (urban green) and proximity to water (urban blue): Evidence from Lisbon, Portugal. *Environmental Health Perspectives, 124*, 927–934. https://doi.org/10.1289/ehp.1409529.

CalEPA, California Environmental Protection Agency. (2017). Understanding the Urban Heat Island Index. Retrieved Sept 6, 2017, from https://calepa.ca.gov/climate/urban-heat-island-index-for-california/understanding-the-urban-heat-island-index/.

Chen, D., Wang, X., Thatcher, M., Barnett, G., Kachenko, A., & Prince, R. (2014). Urban vegetation for reducing heat related mortality. *Environmental Pollution, 192*, 275–284. https://doi.org/10.1016/j.envpol.2014.05.002.

Gasparrini, A., Guo, Y., Hashizume, M., Lavigne, E., Zanobetti, A., Schwartz, J., Tobias, A., Tong, S., Rocklöv, J., Forsberg, B., Leone, M., De Sario, M., Bell, M. L., Guo, Y.-L. L., Wu, C., Kan, H., Yi, S.-M., de Sousa Zanotti Stagliorio Coelho, M., Saldiva, P. H. N., Honda, Y., Kim, H., & Armstrong, B. (2015). Mortality risk attributable to high and low ambient temperature: A multicountry observational study. *Lancet (London, England), 386*, 369–375. https://doi.org/10.1016/S0140-6736(14)62114-0.

Goggins, W. B., Chan, E. Y. Y., Ng, E., Ren, C., & Chen, L. (2012). Effect modification of the association between short-term meteorological factors and mortality by urban heat islands in Hong Kong. *PLoS One, 7*, e38551. https://doi.org/10.1371/journal.pone.0038551.

Gronlund, C. J., Berrocal, V. J., White-Newsome, J. L., Conlon, K. C., & O'Neill, M. S. (2015). Vulnerability to extreme heat by socio-demographic characteristics and area green space among the elderly in Michigan, 1990–2007. *Environmental Research, 136*, 449–461. https://doi.org/10.1016/j.envres.2014.08.042.

Heaviside, C., Macintyre, H., & Vardoulakis, S. (2017). The urban heat island: Implications for health in a changing environment. *Current Environmental Health Reports, 4*(3), 296–305. https://doi.org/10.1007/s40572-017-0150-3.

Heaviside, C., Vardoulakis, S., & Cai, X.-M. (2016). Attribution of mortality to the urban heat island during heatwaves in the West Midlands, UK. *Environmental Health, 15*, 27. https://doi.org/10.1186/s12940-016-0100-9.

Klein Rosenthal, J., Kinney, P. L., & Metzger, K. B. (2014). Intra-urban vulnerability to heat-related mortality in New York City, 1997–2006. *Health Place, 30*, 45–60. https://doi.org/10.1016/j.healthplace.2014.07.014.

Koppe, C., Kovats, S., Menne, B., Jendritzky, G., Wetterdienst, D., & World Health Organization. (2004). *Heat-waves: Risks and responses*. Geneva: World Health Organization.

Kyriakodis, G.-E., & Santamouris, M. (2017). Using reflective pavements to mitigate urban heat island in warm climates – Results from a large scale urban mitigation project. *Urban Climate*. https://doi.org/10.1016/j.uclim.2017.02.002.

Laaidi, K., Zeghnoun, A., Dousset, B., Bretin, P., Vandentorren, S., Giraudet, E., & Beaudeau, P. (2011). The impact of heat islands on mortality in Paris during the August 2003 heat wave. *Environmental Health Perspectives, 120*, 254–259. https://doi.org/10.1289/ehp.1103532.

Lowe, S. A. (2016). An energy and mortality impact assessment of the urban heat island in the US. *Environmental Impact Assessment Review, 56*, 139–144. https://doi.org/10.1016/j.eiar.2015.10.004.

Macintyre, H. L., Heaviside, C., Taylor, J., Picetti, R., Symonds, P., Cai, X.-M., & Vardoulakis, S. (2018). Assessing urban population vulnerability and environmental risks across an urban area during heatwaves – Implications for health protection. *Science of the Total Environment, 610–611*, 678–690. https://doi.org/10.1016/j.scitotenv.2017.08.062.

Madrigano, J., Mittleman, M. A., Baccarelli, A., Goldberg, R., Melly, S., von Klot, S., & Schwartz, J. (2013). Temperature, myocardial infarction, and mortality: Effect modification by individual- and area-level characteristics. *Epidemiology (Cambridge, Mass), 24*, 439–446. https://doi.org/10.1097/EDE.0b013e3182878397.

Mills, D. M., & Kalkstein, L. (2012). Estimating reduced heat-attributable mortality for an urban revegetation project. *Journal of Heat Island Institute International, 7*(2), 18–24.

Milojevic, A., Armstrong, B. G., Gasparrini, A., Bohnenstengel, S. I., Barratt, B., & Wilkinson, P. (2016). Methods to estimate acclimatization to urban heat island effects on heat- and cold-related mortality. *Environmental Health Perspectives, 124*, 1016–1022. https://doi.org/10.1289/ehp.1510109.

Oke, T. R. (1973). City size and the urban heat island. *Atmospheric Environment, 1967*(7), 769–779. https://doi.org/10.1016/0004-6981(73)90140-6.

Otte im Kampe, E., Kovats, S., & Hajat, S. (2016). Impact of high ambient temperature on unintentional injuries in high-income countries: A narrative systematic literature review. *BMJ Open, 6*, e010399. https://doi.org/10.1136/bmjopen-2015-010399.
Santamouris, M. (2015). Analyzing the heat island magnitude and characteristics in one hundred Asian and Australian cities and regions. *Science of the Total Environment, 512–513*, 582–598. https://doi.org/10.1016/j.scitotenv.2015.01.060.
Silva, H. R., Phelan, P. E., & Golden, J. S. (2010). Modeling effects of urban heat island mitigation strategies on heat-related morbidity: A case study for Phoenix, Arizona, USA. *International Journal of Biometeorology, 54*, 13–22. https://doi.org/10.1007/s00484-009-0247-y.
Smargiassi, A., Goldberg, M. S., Plante, C., Fournier, M., Baudouin, Y., & Kosatsky, T. (2009). Variation of daily warm season mortality as a function of micro-urban heat islands. *Journal of Epidemiology and Community Health, 63*, 659–664. https://doi.org/10.1136/jech.2008.078147.
Son, J.-Y., Lane, K. J., Lee, J.-T., & Bell, M. L. (2016). Urban vegetation and heat-related mortality in Seoul, Korea. *Environmental Research, 151*, 728–733. https://doi.org/10.1016/j.envres.2016.09.001.
Stone, B., Vargo, J., Liu, P., Habeeb, D., DeLucia, A., Trail, M., Hu, Y., & Russell, A. (2014). Avoided heat-related mortality through climate adaptation strategies in three US cities. *PLoS One, 9*, e100852. https://doi.org/10.1371/journal.pone.0100852.
Svensson, M. K. (2004). Sky view factor analysis – Implications for urban air temperature differences. *Meteorological Applications, 11*, 201–211. https://doi.org/10.1017/S1350482704001288.
Tzavali, A., Paravantis, J. P., Mihalakakou, J., Fotiadi, A., & Stigka, E. (2015). Urban heat island intensity: A literature review. *Fresenius Environmental Bulletin, 24*, 4535–4554.
Uejio, C. K., Wilhelmi, O. V., Golden, J. S., Mills, D. M., Gulino, S. P., & Samenow, J. P. (2011). Intra-urban societal vulnerability to extreme heat: The role of heat exposure and the built environment, socioeconomics, and neighborhood stability. *Health & Place, 17*, 498–507. https://doi.org/10.1016/j.healthplace.2010.12.005.
U.S. Environmental Protection Agency. (2008). Reducing urban heat islands: Compendium of strategies. Retrieved Sept 6, 2017, from https://www.epa.gov/heat-islands/heat-island-compendium.
Vandentorren, S., Bretin, P., Zeghnoun, A., Mandereau-Bruno, L., Croisier, A., Cochet, C., Riberon, J., Siberan, I., Declercq, B., & Ledrans, M. (2006). August 2003 heat wave in France: Risk factors for death of elderly people living at home. *European Journal of Public Health, 16*, 583–591. https://doi.org/10.1093/eurpub/ckl063.
Vaneckova, P., Beggs, P. J., & Jacobson, C. R. (2010). Spatial analysis of heat-related mortality among the elderly between 1993 and 2004 in Sydney, Australia. *Social Science and Medicine, 1982*(70), 293–304. https://doi.org/10.1016/j.socscimed.2009.09.058.
Zanobetti, A., O'Neill, M. S., Gronlund, C. J., & Schwartz, J. D. (2013). Susceptibility to mortality in weather extremes: Effect modification by personal and small-area characteristics. *Epidemiology (Cambridge, Mass), 24*, 809–819. https://doi.org/10.1097/01.ede.0000434432.06765.91.
Zhang, Y., Yu, C., & Wang, L. (2017). Temperature exposure during pregnancy and birth outcomes: An updated systematic review of epidemiological evidence. *Environmental Pollution, 225*, 700–712. https://doi.org/10.1016/j.envpol.2017.02.066.
Zhao, L., Lee, X., Smith, R. B., & Oleson, K. (2014). Strong contributions of local background climate to urban heat islands. *Nature, 511*, 216–219. https://doi.org/10.1038/nature13462.

Part V
Society and Participation

Chapter 24
Aiming at Social Cohesion in Cities to Transform Society

Mikael Stigendal

24.1 Introduction

Does social cohesion exist? Obviously, it does as a term. Otherwise I would not have been able to write this chapter. But does it also exist as a concept? That is not the same as a term. To exist as a concept, it has to be defined, or in other words, the term has to be provided with a content: an idea. The difference between term and concept means that social cohesion can be defined in different ways. But yet, to exist as a concept, it has to be defined in one way or the other. Is it normally used as a concept and not just as a term? Far too seldom, in my view. In the common use of it, I would describe its content as floating. People use the term but not with any precise meaning or with a meaning which is taken for granted (Eizaguirre et al. 2012: 2007).

The choice of words and terms matter. They put our minds on a certain track and direct our attention. They can, for example, put us on track to take things for granted. It also matters how we define them, what we mean by them, more or less intentionally. In addition to term and concept, social cohesion also exists in a third way, namely as something which we can refer to by using the term and concept of social cohesion (Fairclough et al. 2002). But how can we refer to something which we have no idea about? That is unfortunately the situation with the normal use of the term social cohesion. People use it but do not really know what they are talking about. Then, it gets filled with a content which is taken for granted.

In this chapter, I will start to analyse the use of the term in the EU context. It is part of the EU cohesion policy which directs the structural funds, and through measures funded by them, the definition of the term has had a profound impact on cities. Thereafter, I will draw on recent research to present a preliminary definition of the concept. Finally, I will suggest a further elaboration of that definition, based on an explanation of the problems that social cohesion could become a solution to.

M. Stigendal (✉)
Malmö University, Malmo, Sweden
e-mail: mikael.stigendal@mau.se

M. Nieuwenhuijsen, H. Khreis (eds.), *Integrating Human Health into Urban and Transport Planning*, https://doi.org/10.1007/978-3-319-74983-9_24

To highlight the required changes, I will use the concept of social innovation but in the sense of something transformative, which replaces existing societal forms and not only reforms them. The chapter will finish with the presentation of five criteria which social innovations should comply with.

24.2 The EU Cohesion Policy

The EU's position on social cohesion is part of its cohesion policy with roots in the reassessments of the 1980s. The experience of the 1970s had convinced many politicians that the Keynesian welfare policies of the nation-states did not work anymore. In the second half of that decade, influential organisations such as the OECD had also begun to advocate changes in a neoliberal direction (see, e.g. Jessop 2002). The British Government after the election of the Tories in 1979 with Margaret Thatcher as the new Prime Minister came to spearhead such a change.

The neoliberals at the time argued, and the contemporary ones still do, that high unemployment and low growth occur because the market is not free enough. Deregulation has thus been demanded as remedy. The markets must be freed and people must be encouraged or forced to apply for a job, regardless of the job's quality. Activation, it has been called. Safety is less important. The state should focus on supporting the supply side and not to strengthening demand, whereof the latter was characteristic of the Keynesian policy (Morel et al. 2012). In *Encyclopedia of Globalization*, Bob Jessop (2012) defined neoliberalism as a "political project that is justified on philosophical grounds and seeks to extend competitive market forces, consolidate a market-friendly constitution and promote individual freedom".

For a start, many politicians on the continent opposed the neoliberal policies represented by the Thatcher government. The President of the European Commission Jacques Delors became a rallying figure for those who wanted to defend a regulated capitalism, and he was instrumental in developing a response to the crisis of Keynesianism and the emerging neoliberalism (Novy et al. 2012: 1876). Cohesion policy was central to the EU's reorientation. When established in 1988, it was to replace the nation-state Keynesianism with something else at the European level, but not with neoliberalism. As Liesbeth Hooghe (1998: 459) writes, the reform of 1988 constituted "the bedrock of the anti-neoliberal programme. Though the immediate goal is to reduce territorial inequalities in the European Union, its larger objective is to institutionalise key principles of regulated capitalism in Europe".

However, it was not a question of replacing a national Keynesianism with a European one. In Keynesianism, social policy was concerned with the redistribution between people. That orientation was shifted in cohesion policy to a redistribution between regions, i.e. from a people-based to a place-based redistribution. That could be seen against the background of the new member states from southern Europe joining the EU during the first half of the 1980s, which highlighted the large differences between the countries. That is the orientation that cohesion policy in Europe has maintained ever since. The view of differences has remained place based and as a regional problem. Redistribution should thus take place between regions and places, while redistribution between people at the national level is increasingly restricted. As current cohesion policy does not primarily concern the relationship between rich and poor people (see also Novy 2011), it is therefore not the same as social policy before the 1980s.

The structural funds have been the main tools for implementing cohesion policy. The largest ones are ERDF (European Regional Development Fund) and ESF (European Social Fund). Cities and regions across Europe have attracted huge amounts of money from these funds. The rules for applying, how to spend the money and how to think of progress have been set by the EU Cohesion policy. In this way, what the EU means by cohesion has had a profound impact on urban and transport planning as well as on policies dealing with health inequalities. Therefore, we need to know what the EU means by cohesion.

24.3 What Does the EU Mean by Cohesion?

The European Commission published its first report on cohesion policy in 1996 (European Commission 1996). It begins commendably with a chapter called "What do we mean by cohesion?" In line with the regulated capitalism that the initiators of cohesion policy advocated, the report refers to a European social model based on the values of the social market economy:

This seeks to combine a system of economic organisation based on market forces, freedom of opportunity and enterprise with a commitment to the values of internal solidarity and mutual support which ensures open access for all members of society to services of general benefit and protection. (European Commission 1996*: 13*)

This is the model of society that the European Union should seek to preserve; it is said in the first report on cohesion. Deepening European integration means that the European Union must take greater responsibility and share it with member states, something which should be done with the help of the cohesion policy. Cohesion must therefore be not only economic but also social, and as stated in the quotation, these two forms of cohesion should be combined. With economic cohesion, the report means "convergence of basic incomes through higher GDP growth, of competitiveness and of employment" (European Commission 1996: 13).

Social cohesion is said in the report to be a little harder to define. Based on the notion of solidarity, it is associated with "universal systems of social protection, regulation to correct market failure and systems of social dialogue" (European Commission 1996: 14). That creates favourable conditions for economic development, the report stresses. Economic and social cohesion appear in the report as interdependent.

The key term to social cohesion, however, is solidarity, the word of the day at the time of the World Exhibition in Paris in 1900. In fact, as Sven-Eric Liedman (1999) claims in his book on solidarity, its importance at that time corresponds to today's "sustainable development", which was also the theme of the World Expo in Hanover in 2000. Solidarity had also its leading theorist, namely, Emile Durkheim, one of the classics in sociology. Durkheim saw solidarity as a fact and not primarily as an ideal. It was built into the functioning of society. Earlier forms of society were characterized by a solidarity which Durkheim calls mechanical, based on similarity. In such a form of society, individuals are similar because they experience the same feelings, cherish the same values and hold the same things sacred. Eventually, an opposite form of solidarity emerged, called organic by Durkheim, based on diversity and differentiation (Novy et al. 2012: 1875).

The use of the word solidarity in contemporary European cohesion policy has its roots in those decades around the turn of the previous century. It is easy to trace the word solidarity back there, also because it stems from French. It was written into the Code Civil already in 1804. In the 1840s, it spread to German and English as a consequence of the revolutionary sentiments of that decade. Marx used it at a later stage but then only for the class, not for the whole of society; partial or collectivist solidarity as Liedman calls it. In contrast, the approach advocated by Durkheim referred to the whole of society.

For both Marx and Durkheim, however, solidarity meant a reciprocity. That is a different definition than the one that simply relate solidarity to those who suffer, normative solidarity as Liedman (1999: 86) calls it, which means that the fortunate should care about the poor and oppressed, while they in their turn should accept the privileged position of the fortunate. It is a definition of solidarity that has come to dominate, to the extent it is whatsoever spoken about solidarity nowadays. The strength of the normative notion of solidarity also depends on its foundation in Catholicism since the 1930s. Hence, the boundary between solidarity and charity has been relaxed. At the same time, the word has become associated with history and not with the future. It is not a new society that is foretold in the Catholic use of the word, but rather a return to the principles of an old one, as described in the Bible.

24.4 The Vanishing Substance of Social Cohesion as a Concept

The EU cohesion policy expresses such a normative notion of solidarity, although not between groups of people but between states as well as regions. The fortunate regions and states should show solidarity with "the poorest Member States and regions and with the most disadvantaged regions and people in the more prosperous Member States" (European Commission 1996: 11). In the most recent cohesion reports, the word solidarity does not only lack a clear meaning, but it disappears almost entirely. In the EU's fifth Cohesion Report (European Commission 2010), a normative concept of solidarity is purely reflexively expressed in the preface, written by the two Commissioners for Regional and Urban Policy and Employment, Social Affairs and Inclusion: "Our efforts will in particular support development in the poorest regions in line with our commitment to solidarity".

The current cohesion policy is clearly a cohesion on market conditions. It is primarily through our roles in the market such as consumers and wage workers that cohesion should be achieved. And it does happen. Even neoliberals have an idea about cohesion, Liedman (1999: 44) reminds, as "the market links producer and consumer". But some of these social relations cause inequalities. Neoliberals have not much to say about such social relations within what they see as the economy, grounded as they are in neoclassical economics (Fine 2007).

The influence from neoclassical economics seems obvious in the distinction between economic and social cohesion. The neoclassical revolution of the 1870s consolidated an emerging division between economics and the other social sciences, as Ben Fine (2007: 48) puts it, "with the former tending to offer an individualistic approach (economic rationality) to market relations, and the latter focusing on holistic approaches to non-market relations (and/or "irrationality" in individual behaviour)". In this way, the economists made economics their own subject while other sciences were expected to deal with the rest, understood by the economists as the social.

Well in accordance with this scientific background, the cohesion reports do not say anything at all about the content of economic cohesion, but only about its preconditions. What the earlier reports do say a little about is the content of social cohesion, but that rather reflects a conservative thinking and in the later reports the term has almost disappeared. That, I would say, reflects the success of neoliberalism. As Gamble (2016: 50) says: "Many of the challenges to the survival of the welfare state, whether in its conservative or socialist form, are now expressed in market libertarian language".

This leads me to the quite paradoxical conclusion that the European cohesion policy lacks an idea about cohesion. It really deals with the preconditions for cohesion, but a cohesion whose meaning one talks quietly about. This is the outlook that has made its mark on the operation of the Structural Funds across Europe and metropolitan measures which have become the transmission belts for it. Fully in line with neoliberalism, it is taken for granted that cohesion policy creates the preconditions for a cohesion that is in everyone's interest.

In the absence of an idea about social cohesion, the referents (see Sect. 24.1) have become increasingly important. Social cohesion has become a signifier for the situation in certain neighbourhoods (referents), where it does not look socially cohesive, at least not in comparison with the life of the included middle-class people in other neighbourhoods. The latter have, thus, become the norm and yardstick. "Be like us", the included call out to the excluded, "and the city will become socially cohesive". This expresses an empiricist approach, which means that knowledge becomes a matter of putting names (signifier) on observations (referents), without spending much time on developing ideas (signified) of what hidden patterns they may express (see also Jessop 2015: 252).

24.5 Social Cohesion as a Problématique

A profoundly different approach to social cohesion ran through "Social Polis - Social Platform on Cities and Social Cohesion", the largest research project up to date on these issues, that lasted from late 2007 until the end of 2010. The project had borrowed the term polis from ancient Greece where it meant a city. In Social Polis, it rather meant urban democracy. Social Polis involved 11 research institutions from across Europe (even one from Canada), a network consisting of experienced researchers and more than a hundred stakeholders from governments, the civil society and private organisations at different levels.

The overall objective with Social Polis was to develop a research agenda on the role of cities in social cohesion and key-related policy questions. This was approached on two fronts: by critical analysis of research to date and by constructing a social platform of networks for information gathering, dialogue and agenda setting. In contrast to the approach implicit in the EU cohesion policy, Social Polis did not locate the issue of social cohesion to some kind of existence beside economic cohesion but treated it as overarching, as here explained in an information leaflet:

> Since social cohesion concerns society as a whole, at multiple spatial scales, beyond issues of inequality, exclusion and inclusion, and across public, market and voluntary sectors, 'Social Polis' assembles multiple dimensions of relevant debates (on economy, polity, society, culture, ethics) across the city and a variety of life spheres. To facilitate analysis of the highly interlinked dynamics involved in social cohesion, a range of specific fields affecting people's existence are focused upon.

Thirteen such so-called existential fields were identified. In relation to each one of these fields, research to date were analysed critically in survey papers. These papers were then further elaborated, and in the end, some of them were published in a Special Issue of the journal *Urban Studies* in 2012, with the objectives of "conceptualising, exploring and operationalising different meanings of social cohesion to make them useful for studying the dynamics of 'cities and social cohesion' in urban Europe" (Miciukiewicz et al. 2012: 1855).

I was myself part of Social Polis and I had the privilege to popularise the results in a report with the theme "Cohesion of the city as a whole" (Stigendal 2010). Thinking back on Social Polis, now some years later, its most important achievement, as I see it, can be summarised as the suggestion to treat social cohesion as a problématique. To understand that, we need to recall the distinction that I presented above between signifier, signified and referents (Jessop 2015: 251). Social cohesion is first of all a name (signifier). In the EU, social cohesion has become the name of a referent, i.e. something which can be pointed out, often the life of middle-class people in certain parts of cities. Social Polis, in contrast, made it the name of an idea (signified) by treating it as a problématique. This middle position of signifieds has been short-cut in the EU cohesion policy. Instead, social cohesion becomes the name of an existing condition, which is then taken for granted. In contrast, Social Polis addressed and problematized the ideas and understandings of cities.

In the EU, social cohesion is the name of a condition that parts of cities are in, while in Social Polis it signifies a mode of understanding. The only thing we have got to know from the EU on social cohesion is what the previous cohesion reports say on solidarity. And when that disappeared, social cohesion becomes the state of events and conditions as they seems to be. Its self-evidence emerges from referring to situations that seem not to be social cohesion.

When Miciukiewicz et al. (2012: 1857) said that "no clear progress in understanding 'urban social cohesion' has been made ever since the publication of the EC DG Employment and Social Policy's analysis of Urban Social Development (1992) …", I don't find that strange, given the restrictions of the EU perspective to the referents. The EU expresses an empiricist perspective on social cohesion and the lack thereof, inspired by the life of the included. It takes for granted that something called cohesion more or less exists in the middle-class way of life. And it urges us to mainly talk about the lack of cohesion in certain situations and places. When, for example, young people set cars on fire, they break the norm and do not seem to comply with a condition of social cohesion. Thus, social cohesion becomes part of the justification for stopping these young people from setting cars on fire. This so called referent-definition of social cohesion is incompatible with a critical perspective. It does not go behind the obvious and ask what it may give expression to. This is an almost forbidden question, which I now still want to deal with.

24.6 If Social Cohesion Is the Solution, What Is the Problem?

Whatever we think of contemporary cities, we cannot call them particularly cohesive. On the contrary, cities are characterized by big cleavages between, for example, rich and poor or different ethnic groups. Often these categories of people live in different parts of cities which makes the lack of social cohesion highly visible. It is a reality which we can experience by just walking around, comparing, for example, Lozells and East Handsworth in Birmingham, Trinitat Nova in Barcelona, Feyenoord in Rotterdam, Elefsina in Athens or Herrgården in Malmö with other parts of these cities. This reality may also intrude on us when we, for example, hear about riots or unemployment figures. It makes us appalled to read that the gap in life expectancy between the Calton and Lenzie areas in Glasgow amounts to 28 years (Therborn 2013: 112). In terms of the philosophy of science that I represent, critical realism, such experiences and figures belong to the level of reality called the empirical.

We respond to these empirical experiences by calling them something, i.e. labelling them with a signifier. Cleavages like the one between rich and poor are usually called inequality. Many of us also want to make sense of our experiences by providing the name of them with a content: a signified. In science, the signified usually means a definition. By defining a term, it turns into a concept. A common and useful definition of inequality is the one suggested by Sen (2013). On this basis, inequalities should be seen as differences which violate the human rights of the disadvantaged. Recent research has highlighted the existence of different types of inequality, for example, with regard to income (Piketty 2014).

Others have focused on inequality in terms of health, defined as systematic differences in health between socially determined groups of individuals on the basis of class, gender, ethnicity, age and residency (Marmot 2015; Wilkinson and Pickett 2009). Health inequality is particularly detrimental to social cohesion. As Sen (2013) has emphasised, health constitutes a human right that comes before other human rights since those rights can only be fully enjoyed if the individual first enjoys good health.

Göran Therborn (2013: 49) makes a distinction between three general types of inequality: vital inequality, existential inequality and inequality in terms of resources. These three types are described as applying to us in our capacities as biological beings, persons and actors, respectively. According to Therborn (2013: 35), "very little theoretical reflection on the meanings and implications of inequality, and of equality, has come to the fore". He calls for a multidimensional and global approach to inequality (2013: 4).

In my view, such an approach should be underpinned by critical realism (Bhaskar 1989; Sayer 1992, 2000; Danermark 2002). This means a treatment of what we see and experience as not the whole story, but in the first place as belonging to a certain level of reality, called the empirical. We make sense of these impressions and experiences by understanding them as expressing something else, i.e. by treating them as symptoms. As such, the empirical express a specific content associated, in turn, with another level of reality, called the actual. The forms of inequality mentioned above belong to that level. They certainly exist but express themselves at the level of the empirical.

To explain the causes of inequality, however, we need to understand the existence of the most profound level, that is, the level of the real, which embraces it all, i.e. not only what appears to be and what has been actualised but also the potentials. Before we can establish the actual causes of inequality, we need to know what has the potential to cause inequality. These potential causes of inequality should be seen as the main problems that efforts to achieve social cohesion should address.

This understanding of inequality draws on the broad societal perspective called Cultural Political Economy (CPE) (Sum and Jessop 2013), underpinned philosophically by critical realism. In line with this perspective, a further distinction can be made of existential inequality into cultural and structural inequality (Stigendal 2016, 2018). This can be motivated by the ontological assumption of our social existence as being both cultural and structural. We become part of the social world by on the one hand making sense and meaning of it and on the other hand by structuring social relations. According to CPE, this is how we reduce the complexity of the world.

Both cultures and structures have to be produced and reproduced by actors. Otherwise they would not exist (Jessop 2005). Thereby, actors may also reproduce causes of inequality, even though it is not their intention. In line with critical realism (Sayer 2000), causes should be understood first of all as potentials, inherent in systems, institutions, actors, cultures, etc. The actualisation of potentials always takes place in specific, contextual and concrete situations. There, different potentials combine to produce effects which should be regarded as emergent as they cannot be derived from only one of the potentials. An explanation of a certain event needs to take the context into consideration. We need to examine the potentials and their properties at higher levels of abstraction, apart from the concrete situations and contexts where they get actualised. By doing that, we can understand that causes are inherent in individuals and other single objects, as mentioned above, but also in how both social structures and contexts of meaning work, i.e. their so to speak structural and cultural mechanisms. But then we need to carry out empirical investigations of concrete situations in order to explain what happens and why.

So what has the potential to cause the inequality actualised in cities and appearing at the empirical level as, for example, differences in health? Such potential can be inherent in transport practices, according to the extensive review by Khreis et al. (2017) of 64 different transport policy measures. Similarly, the WHO Commission on Social Determinants of Health (WHO 2008: 35) "stated that a toxic combination of poor social policies, unfair economics, and bad politics is responsible for much of health inequity". These toxic combinations constitute, what the WHO Commission calls, the social determinants of health.

As I see it, it is not only about certain policies, procedures and practices. In a more fundamental sense, it is about how systems work. Accordingly, I claim that the most important potential causes of inequality are inherent in capitalism (Stigendal 2018). One of the main effects of neoliberalism is that it has brought about new structural forms for the actualisation of these potential causes, associated with the dominance of finance capital, and with a global coverage, not national. This has led to a major redistribution of wealth, confirmed by, for example, OECD (2011) and, of course, Piketty (2014).

As part of this new global economic order since the 1980s (Arrighi 2007; Becker et al. 2010), the responsibilities of cities have been much enhanced. On the one hand, globalisation has put cities at the forefront, strengthening them with regard to their regional and national territories and generating new prospects. Every employment hub, city or region, has to compete with the others to attract and keep investments in its area. As Doreen Massey (2007: 12) describes it, neoliberalism is made in the city. On the other hand, cities have become concentrations of inequality, where different forms coincide, often also with segregation.

The EU cohesion policy has been instrumental in this development, supporting the market forces in the hope that it would be beneficial for all. Purporting to improve social cohesion, the current neoliberal EU cohesion policy has instead aggravated health inequalities, which Stuckler and Basu (2013) show when they highlight the effects of austerity measures. In terms of economic cohesion, conveyed by the structural funds, the EU policies have favoured a transport design, planning and policy-making "that are operating separately from public health delivery at some level, resulting in adverse impacts on the population's health, many of which only manifest on the long term" (Khreis et al. 2016: 255). It has paved the way, literally speaking, for the car-centred urbanisation and mass motorisation, exacerbating the negative health impacts of the transport systems.

On the one hand, society has turned into a condition of social inclusion with surrounding borders consisting of demands on what to have, to do and to mean in order to be included. On the other hand, spatial concentrations have emerged of people who do not have, do or mean what is demanded to participate in society. The socially included and the socially excluded thus tend to live in different parts of cities. In between, borders have arisen, which I prefer to see as societal (Graham 2011). Transport and urban planning measures have contributed to such community severance by favouring the more profitable transport modes, regarding, for example, road constructions and rail services (Khreis et al. 2017). The calm neighbourhoods in the cities where most people go to work in the morning and thus appear as examples of social cohesion represent in fact the included side in the development of society to a condition of social inclusion. What the EU in its cohesion policy has come to see as the solution of social cohesion should from the perspective that I represent, thus, be seen as part of the problem.

24.7 Prospects for Social Cohesion in Cities

The most comprehensive project to date on the issue of cities and social cohesion, Social Polis, ended up by treating social cohesion as a problématique. As explained above, I see this as the most important legacy of Social Polis. That is because it restored social cohesion as an idea, in contrast to using the term as a signifier to a natural referent, i.e. what seems to be. The use of the term problématique can also be justified in order to treat social cohesion in the first place as a problem, not something which necessarily exists. Cities are not socially cohesive and that is the problem. They can perhaps become socially cohesive, but in that case how and what does it mean?

I have tried to develop this perspective further in the chapter, by in the first place highlighting potentials. To explain inequality and thus the lack of social cohesion, we need to understand the involved potentials, their ability to cause inequality. Let us call these potentials negative, because they cause problems of inequality. In such a perspective, we can also identify positive potential. These are, for example, the ones of the young people who get to know each other across ethnicities and cross the borders of social inclusion everyday they go to school or work. By that, many of them develop an intercultural competence; really a scarce resource among policy-makers in contemporary societies.

To understand this, a perspective is needed which can be called potential-oriented. I want to use this term (signifier) because it can bring hope. It urges us to look after the possible and is in itself an example of what I wrote in the introduction about how the use of signifiers can put us on a certain track. Potential-oriented is the name (signifier) of the perspective (signified) that I have pursued in this chapter. Adopting this perspective is the first of five criteria that I have suggested on the basis of the project CITISPYCE, spelled out as "Combating Inequalities through Innovative Social Practices of and for Young People in Cities across Europe", where

I also participated, which lasted for 3 years, 2013–2015, and just like Social Polis was funded by the seventh framework programme. In total, CITISPYCE involved 13 partner organisations from 10 cities in just as many countries across the EU: Athens, Barcelona, Birmingham, Brno, Hamburg, Krakow, Malmö, Rotterdam, Sofia and Venice. In my own conclusions from this project, I have suggested five criteria, which should be met for initiatives to combat the causes of inequality. In accordance with the objective of the CITISPYCE research project, the criteria are designed to concern young people and I will stick to that here, also because young people have been highlighted as one of the categories worst affected by inequality.

In contrast, the notion of social cohesion, as part of the EU cohesion policy, deals with the inclusion of people in existing society, without addressing the causes excluding them. It presupposes that there is nothing inherently wrong with contemporary society. It simply has to improve its measures to include young people in it. This perspective is also characterized by a depoliticisation of collective action, instrumentalisation of local initiatives, localism and individualism. It is limited to what critical realism calls the empirical and thus associated with empiricism. What appears to be a problem is also seen as the problem that has to be solved. That means that the problems are taken for granted. No particular effort is made to define the problems. Instead, they are perceived as self-defining, as if people appear as excluded, that is, seen as the problem, not the societal causes of their exclusion. This problem-oriented perspective should not be seen as guidance towards solutions but on the contrary as one of the causes of inequality and, thus, part of the problem.

Various efforts are needed to combat the causes of inequality and at different scale levels. It requires not only reforms but a transformation of society. This should not, however, be seen as a call for revolution. The old dualism of reform versus revolution has to be overcome, in line with the idea of a double transformation, which, according to Novy (2017), "describes the twofold challenge of civilizing capitalism by overcoming the neoliberal mode of regulation while at the same time taking the first steps towards transcending capitalism and its unsustainable social forms …". That makes the work of local actors particularly important and enables me to put the five criteria in the context of not only combatting the causes of inequality but at the same time also a transformation of society.

Firstly, projects, initiatives, practices and indeed services in general should adopt a potential-oriented approach to young people which means that young people should be seen and approached in the first place as potentials and not only for solving their own problems but for tackling the causes of inequality and developing society. The potential-oriented perspective means to understand causes first of all as potentials. To have an effect, they must be actualised. That always happens somewhere, in a specific context, at different levels, and it is made by actors who always have a discretion to make a difference.

Secondly, on the basis of a potential-oriented perspective, it becomes imperative to produce knowledge on the causes of inequality by taking advantage of the experiences, views and cultural expressions of young people. Producing such knowledge in collaboration with young people and establishing contexts which favour creativity can be called knowledge alliances. In principle, they should include all the ones with an interest in combatting inequality, e.g. practitioners in general, public sector workers, citizens, policy-makers, young people, politicians and volunteers. The term knowledge alliance is also used by the European Commission in its strategy Europe 2020. The definition, however, tends to be limited to alliances between "education and business". A broader definition was suggested in Social Polis (Stigendal 2010), but also by the influential report Cities of Tomorrow: Challenges, Visions, Ways Forward (Hermant-de-Callataÿ and Svanfeldt 2011).

Furthermore, the term was picked up and redefined by the commission for a socially sustainable Malmö (the "Malmö Commission"), which was set up in 2011 against the backdrop of increasing health inequalities. Its main source of inspiration was the WHO Commission on Social Determinants of Health, led by Michael Marmot, which had published its final report, "Closing the Gap in a Generation", in 2008. To reduce health inequalities, the Malmö Commission (Stigendal and Östergren 2013), in the first of two overarching recommendations, proposes that the City of Malmö pursues a social investment policy that can reduce inequities in living conditions and make societal systems more equitable. In the second overarching recommendation, the Commission proposes changes to the processes embedded in these systems through the creation of knowledge alliances (see also Novy et al. 2013).

Thirdly, this recognition of young people's knowledge does not of course imply that they know enough. Young people have a lot to learn, but education systems should draw much more on their potential and experience of inequalities. The cultures of young people should be used as tools for education. Young people should be encouraged to express injustices in processes that educate them to become critical citizens with a wish to be politically engaged. To be critical means to learn how to not take things for granted, but instead question the obvious and reveal what lies behind.

Fourthly, wherever young people are involved attention should be paid to how the work is organised. As work makes us who we are, the demand on the quality of jobs should be raised. Just to get a job should not be seen as sufficient. The EU should fund projects which show how the potential of young people could be taken advantage of, enabling young people to learn and grow on the job. Such projects should then serve and be highlighted as sources of inspiration for the development of work organisations in general.

Fifthly, practices that aspire to become innovative should open up opportunities for young people from different parts of a neighbourhood, city and even across Europe to get to know each other and about each other's situations as well as working together. Young people should be empowered to deal collectively with the problems of inequality, in line with the idea of collective empowerment, based on the notion of collective solidarity. That should be regarded as a concern for all of us and the future of societies. This is the social cohesion to aim at, which also may turn cities into strongholds for the needed societal transformations. The extent to which this social cohesion succeeds should be possible to assess regarding, for example, participation in political organisations/party or community-/environmentally oriented organisations as well as in political elections.

References

Arrighi, G. (2007). *Adam Smith in Beijing: Lineages of the twenty-first century*. London: Verso.

Becker, J., Jäger, J., Leubolt, B., & Weissenbacher, R. (2010). Peripheral financialization and vulnerability to crisis: A regulationist perspective. *Competition and Change, 14*(3–4), 225–247.

Bhaskar, R. (1989). *Reclaiming reality. A critical introduction to contemporary philosophy*. Verso: London.

Danermark, B. (2002). *Explaining society: Critical realism in the social sciences*. London: Routledge.

Eizaguirre, S., Pradel, M., Terrones, A., Martinez-Celorrio, X., & García, M. (2012). Multilevel governance and social cohesion: Bringing back conflict in citizenship practices. *Urban Studies, 49*, 9.

European Commission. (1996). *First cohesion report*. Brussels: European Commission (Directorate-General Regio).

European Commission. (2010). *Fifth report on economic, social and territorial cohesion*. Brussels: European Commission (Directorate-General Regio).

Fairclough, N., Jessop, B., & Sayer, A. (2002). Critical realism and semiosis. *Alethia, 5*(1), 2–10. Retrieved Jun 20, 2017, from http://www.lancaster.ac.uk/fass/resources/sociology-online-papers/papers/fairclough-jessop-sayer-critical-realism-and-semiosis.pdf.

Fine, B. (2007). Eleven hypotheses on the conceptual history of social capital: A response to James Farr. *Political Theory, 35*(1), 47–53.

Gamble, A. (2016). *Can the welfare state survive?* Cambridge: Polity Press.

Graham, S. (2011). *Cities under siege: The new military urbanism* (Pbk ed.). London: Verso.

Hermant-de-Callataÿ, C., & Svanfeldt, C. (2011). *Cities of tomorrow. Challenges, visions, ways forward. Luxembourg:* European Commission - Directorate General for Regional Policy.

Hooghe, L. (1998). EU cohesion policy and competing models of European capitalism. *Journal of Common Market Studies, 36*(4), 457–477.

Jessop, B. (2002). *The future of the capitalist state*. Cambridge: Polity Press.

Jessop, B. (2005). Critical realism and the strategic-relational approach. *New Formations, 56*, 40–53.

Jessop, B. (2012). Neoliberalism. In G. Ritzer (Ed.), *The Wiley-Blackwell encyclopedia of globalization*. Chichester: Wiley Blackwell.

Jessop, B. (2015). The symptomatology of crises, reading crises and learning from them: Some critical realist reflections. *Journal of Critical Realism, 14*(3), 238–271.

Khreis, H., Warsow, K. M., Verlinghieri, E., Guzman, A., Pellecuer, L., Ferreira, A., Jones, I., Heinen, E., Rojas-Rueda, D., & Mueller, N. (2016). Health impacts of urban transport policy measures: A guidance note for practice. *Journal of Transport & Health, 3*, 249–267.

Khreis, H., May, A. D., & Nieuwenhuijsen, M. J. (2017). Health impacts of urban transport policy measures: A guidance note for practice. *Journal of Transport & Health, 2006*, 209–227.

Liedman, S.-E. (1999). *Att se sig själv i andra. Om solidaritet*. Bonnier: Stockholm.

Marmot, M. (2015). *The health gap: The challenge of an unequal world*. London: Bloomsbury.

Massey, D. (2007). *World city*. Cornwall: Polity Press.

Miciukiewicz, K., Moulaert, F., Novy, A., Musterd, S., & Hillier, J. (2012). Introduction: Problematising urban social cohesion: A transdisciplinary endeavour. *Urban Studies, 49*(9), 1855–1872.

Morel, N., Palier, B., & Palme, J. (Eds.). (2012). *Towards a social investment welfare state?: Ideas, policies and challenges*. Bristol: Policy.

Novy, A., Swiatek, D. C., & Moulaert, F. (2012). Social cohesion: A conceptual and political elucidation. *Urban Studies, 49*(9), 1873–1889.

Novy, A., Habersack, S., & Schaller, B. (2013). Innovative forms of knowledge production: Transdisciplinarity and knowledge alliances. In F. Moulaert et al. (Eds.), *The international handbook on social innovation. collective action, social learning and transdisciplinary research*. Cheltenham: Edward Elgar.
Novy, A. (2011). Unequal diversity – On the political economy of social cohesion in Vienna. *European Urban and Regional Studies, 18*, 239–253.
Novy, A. (2017). Emancipatory economic deglobalisation: A polanyian perspective. *Brazilian Journal of Urban and Regional Studies, 19*, 558–579.
OECD. (2011). *Divided we stand – Why inequality keeps rising*. Paris: OECD Publishing.
Piketty, T. (2014). *Capital in the twenty-first century*. Cambridge: Belknap Press of Harvard University Press.
Sayer, A. (1992). *Method in social science. A realist approach*. Worcester: Routledge.
Sayer, A. (2000). *Realism and social science*. London: SAGE.
Sen, A. (2013[1999]). *Development as freedom*. Oxford: Oxford University Press.
Stigendal, M., & Östergren, P.-O. (2013). *Malmö's path towards a sustainable future*. Malmö: Malmö Stad.
Stigendal, M. (2010). *Cities and social cohesion. Popularizing the results of Social Polis*. Malmö: MAPIUS 6/Malmö högskolas publikationer i urbana studier.
Stigendal, M. (2016). *Samhällsgränser. Ojämlikhetens orsaker och framtidsmöjligheterna i en storstad som Malmö*. Kina: Liber.
Stigendal, M. (2018). *Combatting the causes of inequality affecting young people across Europe*. London: Routledge.
Stuckler, D., & Basu, S. (2013). *The body economic: Why austerity kills : Recessions, budget battles, and the politics of life and death*. New York: Basic Books.
Sum, N.-L., & Jessop, B. (2013). *Towards a cultural political economy: Putting culture in its place in political economy*. Cheltenham: Edward Elgar.
Therborn, G. (2013). *The killing fields of inequality*. Cambridge: Polity Press.
World Health Organization. (2008). *Closing the gap in a generation: Health equity through action on the social determinants of health. Final report of the Commission on Social Determinants of Health*. Geneva: World Health Organization.
Wilkinson, R., & Pickett, K. (2009). *The spirit level : Why more equal societies almost always do better*. London: Allen Lane.

Chapter 25
Establishing Social Equity in Cities: A Health Perspective

Carme Borrell, Mercè Gotsens, and Ana M. Novoa

25.1 Introduction

The percentage of the world's population living in urban areas is projected to increase from 54% in 2015 to 60% in 2030 and to 66% by 2050. In absolute terms, more than 1 billion people were added to urban areas between 2000 and 2014. It is important to recognize that cities are not just economic drivers for countries but are hubs of ideas, culture, science, productivity, etc. Cities are centres of innovation that have to manage and respond to dramatic demographic and epidemiological transitions. However, cities can also produce great health inequalities, for instance, 828 million people live in slums today, and this number will increase (WHO/UN-HABITAT 2016).

C. Borrell (✉)
Agència de Salut Pública de Barcelona, Barcelona, Spain

Universitat Pompeu Fabra, Barcelona, Spain

CIBER Epidemiología y Salud Pública (CIBERESP), Madrid, Spain

Institut d'Investigació Biomèdica (IIB Sant Pau), Barcelona, Spain
e-mail: cborrell@aspb.cat

M. Gotsens · A. M. Novoa
Agència de Salut Pública de Barcelona, Barcelona, Spain

Institut d'Investigació Biomèdica (IIB Sant Pau), Barcelona, Spain
e-mail: mgotsens@aspb.cat; anovoa@aspb.cat

M. Nieuwenhuijsen, H. Khreis (eds.), *Integrating Human Health into Urban and Transport Planning*, https://doi.org/10.1007/978-3-319-74983-9_25

Social health inequalities are differences in health which are 'unnecessary and avoidable but, in addition are also considered unfair and unjust' (Whitehead 1992). People of working class, immigrants of poor countries have worse health and higher mortality. Women have worse self-perceived health although their life expectancy is larger than men's (World Health Organization 2016). The WHO Commission on Social Determinants of Health (CSDH) concluded that social inequalities in health arise from inequalities in the conditions of daily life and the fundamental drivers that give rise to them: inequities in power, money and resources. These social and economic inequalities underpin the determinants of health—the range of interacting factors that shape health and wellbeing (CSDH 2008).

Following the Millenium Development Goals, in 2016 the Sustainable Development Goals (SDFs) of the 2030 Agenda for Sustainable Development were established. They are 17 goals adopted by world leaders in September 2015 at a UN Summit. These new goals aim to fight inequalities and tackle climate change while ensuring that no one is left behind. The new goals call for action by all countries to promote prosperity while protecting the planet and fighting poverty. Goal number 11 specifically refers to 'Make cities inclusive, safe, resilient and sustainable' (Table 25.1). These SDGs are not legally binding, but governments should establish national frameworks to achieve them and monitor their implementation.

This chapter starts with reviewing the social determinants of health in urban areas, followed by a section on surveillance of health inequalities in cities including some examples. Finally, policies to tackle health inequalities in cities are presented with the example of Urban HEART as a useful tool.

25.2 Social Determinants of Health in Urban Health

The factors or determinants of health are important to understand what causes health inequalities. These factors can vary in urban areas because cities do not have the same capacity as countries to implement policies. Figure 25.1 shows a conceptual framework with the factors or determinants related to health and health inequalities in urban areas (Borrell et al. 2013). These factors are:

Urban governance: This refers not only to the political power of the government (local, regional, national) but also to other actors who play important roles, such as the private sector and the civil society through community groups. To tackle the root causes of health inequalities implies building good governance for action on social determinants and to implement collaborative action between sectors ('intersectoral action'), and therefore governance is related with physical and socioeconomic environment policies.

Physical environment: This refers to the natural context (climate or geography). Also, the built environment is based on urban planning and housing policies and usually depends on local authorities. Urban planning determines the public infrastructure (communications, sewerage system), the general regulations (concerning buildings and the use of public space) and the equipments in a neighbourhood (sport, health and education facilities, markets, libraries, etc.). A good housing

Table 25.1 Targets of Goal 11 of Sustainable Development Goals of United Nations: 'Make cities inclusive, safe, resilient and sustainable'

• By 2030, ensure access for all to adequate, safe and affordable housing and basic services and upgrade slums
• By 2030, provide access to safe, affordable, accessible and sustainable transport systems for all, improving road safety, notably by expanding public transport, with special attention to the needs of those in vulnerable situations, women, children, persons with disabilities and older persons
• By 2030, enhance inclusive and sustainable urbanization and capacity for participatory, integrated and sustainable human settlement planning and management in all countries
• Strengthen efforts to protect and safeguard the world's cultural and natural heritage
• By 2030, significantly reduce the number of deaths and the number of people affected and substantially decrease the direct economic losses relative to global gross domestic product caused by disasters, including water-related disasters, with a focus on protecting the poor and people in vulnerable situations
• By 2030, reduce the adverse per capita environmental impact of cities, including by paying special attention to air quality and municipal and other waste management
• By 2030, provide universal access to safe, inclusive and accessible, green and public spaces, in particular for women and children, older persons and persons with disabilities
• Support positive economic, social and environmental links between urban, peri-urban and rural areas by strengthening national and regional development planning
• By 2020, substantially increase the number of cities and human settlements adopting and implementing integrated policies and plans towards inclusion, resource efficiency, mitigation and adaptation to climate change, resilience to disasters, and develop and implement, in line with the Sendai Framework for Disaster Risk Reduction 2015–2030, holistic disaster risk management at all levels
• Support least developed countries, including through financial and technical assistance, in building sustainable and resilient buildings utilizing local materials

Source: Sustainable Development Goals. Goal 11. http://www.un.org/sustainabledevelopment/cities/ [Accessed 2 June 2017]

policy (taxation, social housing for rent or for sale) can make high-quality housing affordable for everybody and particularly for low-income households, while the absence of this policy can convert housing into a means of speculation. This policy can heavily influence income distributions by offering housing for a relatively cheap or moderate price. On the other hand, housing quality, informal settlements or overcrowding are also factors to be taken into account. Another aspect of the physical context is transport planning and transport mobility, affecting the capacity to walk and to use public or private transport and impacting land use, the public infrastructure and also housing development. The factors mentioned can influence environmental characteristics, as, for example, water and air quality and noise pollution, which are important determinants of health in urban areas. Food security and access to healthy food (availability and price) are also important. Finally, emergency management, for example, to terrorism, bioterrorism, climate change and natural disasters, is included in this section.

Socioeconomic environment: This refers to different aspects such as economic factors (taxes), employment and working conditions, domestic and family environment, public services (education, health care, care services for families, etc.) and

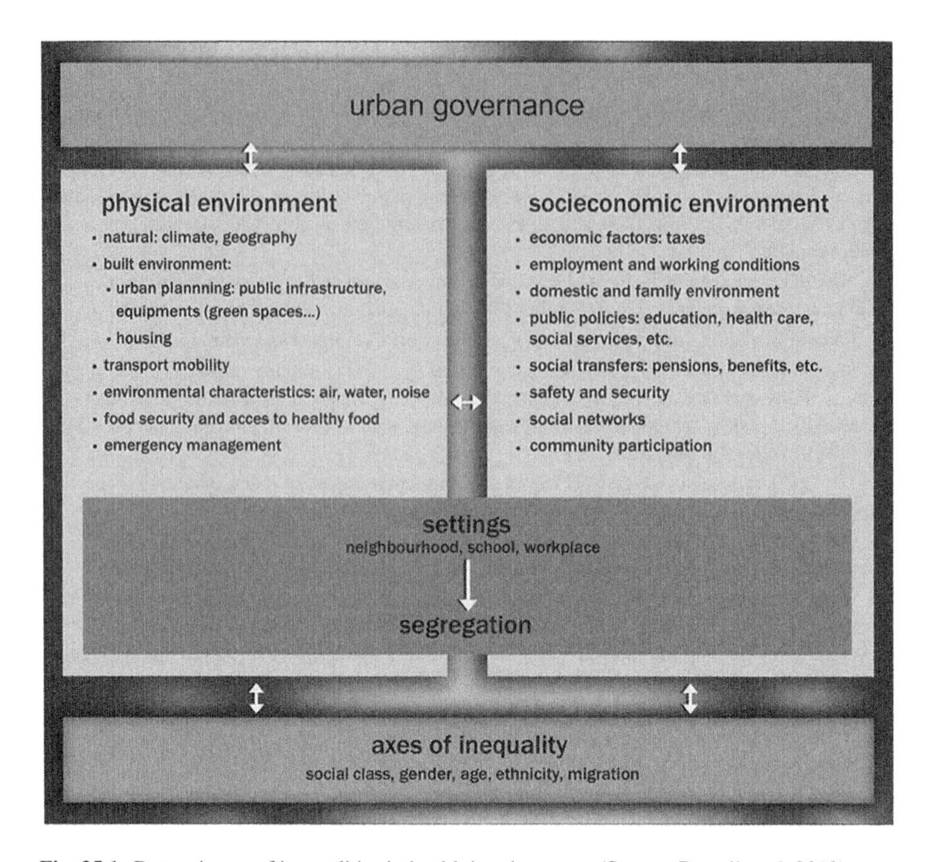

Fig. 25.1 Determinants of inequalities in health in urban areas (Source: Borrell et al. 2013)

social transfers (pensions, unemployment benefits, etc.). Although most of these factors are the responsibility of the country's government, cities have the capacity to modify them. For example, city governments can try to improve living conditions of the poor population by redistributing the income collected through city taxes or distributing income pensions to people under the poverty threshold. Other aspects included are safety and security, social networks and community participation, aspects that have also been associated with health inequalities.

Settings: A setting is where people actively use and shape the environment; thus it is also where people create or solve problems relating to health. Settings included in the framework are neighbourhoods, schools and workplaces, although there are many other settings as, for example, places of worship, clubs and community activities and places for recreation and leisure. Settings are related with the physical and socioeconomic contexts.

Segregation may be important in settings, and it refers to the separation of humans into groups determined by axes of inequality as, for example, social class or race. Segregation can be referred to where people live (residential segregation) but also can refer to other settings (school segregation).

Axes of inequalities: The different determinants commented above may change according to the different social axes of inequality such as social class, gender, age or ethnicity/migration which are social constructs that determine the social structure. For example, working conditions are not the same for men and women or for people of different social classes, and it can be expected that there will be differences in the degree of physical and social working exposure experienced by different genders and social classes as well.

25.3 Surveillance of Health Inequalities in Cities

Public health surveillance is the continuous and systematic collection, analysis and interpretation of information related to the health of the population and their determinants, information that is essential to the planning, implementation and evaluation of public health interventions.

Therefore, surveillance of inequalities in health involves monitoring inequalities over time and across the different axes of inequality with the objective to follow them over time or to compare different groups (Espelt et al. 2016). To achieve this objective, it is necessary to have adequate urban health metrics that allow to measure both inequalities in health and in its determinants because it is essential for surveillance but also for promoting more healthy and equitable cities (Corburn and Cohen 2012). Therefore surveillance is a useful tool to implement and evaluate policies. The 2010 WHO and UN-HABITAT report, *Hidden Cities* (WHO/UN-HABITAT 2010), notes: 'Understanding urban health begins with knowing which city dwellers are affected by which health issues, and why. To achieve this understanding, available information must be disaggregated according to defining characteristics of city dwellers, such as their socioeconomic status or place of residence…'.

Metrics usually are based on indicators. An indicator is a variable with characteristics of quality, quantity and time used to measure, directly or indirectly, changes in health and health-related situations. It is important that urban indicators are based on a conceptual framework of social determinants of health as the model shown in Fig. 25.1. To obtain these indicators, several sources of information have to be used as, for example, the census, demographic and health surveys, information based on the different registries of the municipality, environmental measurement stations, health registries, vital statistics, etc. In order to measure socioeconomic inequalities, it will be necessary that the data allows to calculate indicators at the small area level (ideally, data should be geo-referenced) and/or in different social groups (Prasad et al. 2016). However, many health information systems do not allow to obtain indicators at this level of disaggregation, since they are not geo-referenced and/or do not include the necessary information (such as the neighbourhood or social class). A very important source are health surveys, since they collect a wealth of information on health, use of health services and health determinants, and they usually include information on social variables.

Metrics for the surveillance of health inequalities should be analyzed and disseminated. The analysis is usually based on analysis of trends over time or analysis comparing different areas or different socioeconomic groups. This can be done by simply comparing the numbers obtained across the different groups as, for example, comparing the rates of disease of social class V (more disadvantaged) with social class I (more advantaged). Moreover, it is possible to estimate an inequality indicator such as the Relative and Slope Indexes of Inequality which can be interpreted as the ratio or absolute differences of rates between the two extremes of the socioeconomic spectrum, among others (World Health Organization 2013a). The dissemination is very important in surveillance. Sometimes it has been neglected. It is important to use the tools to disseminate taking into account the target population. Some tools are reports (for public health workers), leaflets and videos (for general population), etc.

25.3.1 Examples of Surveillance of Health Inequalities in Cities

Different initiatives have been launched to monitor health inequalities and their determinants in urban areas around the world. Some of them are explained in this section.

At worldwide level, the Global Health Observatory (GHO) is an initiative of the World Health Organization (WHO) with the objective to share data on global health and, in particular, on urban health (World Health Organisation 2017). It includes statistics by region and by country on specific health determinants, service coverage, risk factors and outcomes and disaggregated by socio-demographic variables such as wealth and sex. Data are presented in the form of global situation and trends, graphs and maps and country profiles. These profiles have been developed for the purpose of visualizing urban health issues, especially focused on urban health inequalities. Table 25.2 presents some examples of health indicators and health determinants by urban wealth quintile. The infant mortality rate for children in the poorest 20% urban households is approximately twice as high as that among children in the richest 20% urban households, globally. The highest infant mortality rate occurs in African region where it is 70.0 per 1000 live births in the poorest 20% urban households, while in the richest 20% is 46.0 per 1000 live births. The same pattern is observed in the under-five mortality rate. The under-five mortality rate is 114.5 per 1000 live births in the African poorest 20% urban households while is 57.5 in the African richest 20% urban households. In contrast, this rate is 27.3 in the Eastern Europe poorest 20% urban households and 13.5 in the Eastern Europe richest 20% urban households. The prevalence of smoking among men is higher in the poorest 20% than in the richest 20% urban population in all regions of the world except in the Asia Pacific. On the other hand, the prevalence of stunting among children in the poorest urban 20% households in Africa and Asia-Pacific are more than twice as high (around 35% in the poorest urban household and 15% in the richest urban household), and nearly four times as high in Latin America and the Caribbean, as their richest 20% counterparts, respectively (19.3% in the poorest urban household and 5.8% in the richest urban household). In addition, the antenatal care coverage among urban women in the richest 20% households in Africa (82.7%) and Asia-Pacific (92.3%) are about 1.5 times as high as that among women in the poorest

Table 25.2 Health indicators and determinants of health in urban areas by urban health quintile, 2005–2013

	African		Asia-Pacific		Eastern Europe		Latin American and Caribbean	
	Urban poorest 20%	Urban richest 20%	Urban poorest 20%	Urban richest 20%	Urban poorest 20%	Urban richest 20%	Urban poorest 20%	Urban richest 20%
Infant mortality rate per 1,000 live births	70.0	46.0	40.6	20.7	23.0	13.5	35.0	15.6
Under-five mortality rate per 1,000 live births	114.5	57.5	56.5	23.0	27.3	13.5	42.0	19.0
Prevalence of cigarette smoking among men (%)	20.5	15.8	43.3	49.4	53.8	50.0	24.8	16.4
Prevalence of stunting in children (%)	35.3	14.0	36.1	16.2	19.4	11.5	19.3	5.8
Antenatal care coverage (%)	51.1	82.7	56.3	92.3	90.3	97.6	88.9	97.5
Households using a piped onto premises water source (%)	3.8	80.0	27.9	70.2	79.6	93.2	61.3	81.8

Source: Global Health Observatory, urban health (World Health Organisation 2016b)

20% urban households (51.1% and 56.3%, respectively). These inequalities are smaller in the rest of regions. Finally, in Africa only, 3.8% of the poorest urban 20% households use a drinking water source that is piped into their premises compared to 80.0% in the richest 20% urban households (World Health Organisation 2016a).

Another initiative of WHO Kobe Centre is the development of the Urban Health Index, published in 2014, for assessments of intra-urban inequalities in health determinants and outcomes, as well as for intercity comparisons (World Health Organisation 2016b). This index is constructed using several indicators that belong to different domains or categories such as income, education, access to water, living in slum conditions or crowding. However, the selection of exact indicators forming the Urban Health Index varies with each region depending on the availability and purpose of the index (World Health Organization 2014). The Urban Health Index provides information about the level of health and health inequalities in small areas within cities. In addition, the Urban Health Index guides users through a flexible approach in the selection of indicators, tailored to local needs. This is particularly useful since priorities can vary substantially across cities and contexts. This index adopts a method used to construct the Human Development Index, using standardized indicators for small area units on a (0, 1) interval and combining them using their geometric mean. Inequalities are assessed using the ratio of the highest to lowest decile and measuring the slope of the eight middle deciles (middle 80%) of the data. The Urban Health Index has been successfully computed for Atlanta, USA; Rio de Janeiro, Brazil; Shanghai, China; and Tokyo, Japan, among other places (Rothenberg et al. 2014; Bortz et al. 2015; Stauber et al. 2003).

In the UK and Ireland exists the Association of Public Health Observatories (APHO) which is made up of a network of 12 public health observatories (PHOs). The APHO produces information, data and intelligence on people's health and health determinants for practitioners, policy makers and the wider community (Wilkinson 2015). One of the most popular observatory is the London Health Observatory which provides up-to-date information on health determinants and health of the people who live in London (World Health Organization 2013b). One of the products that the London Health Observatory produces is the Local Health Profile (Public Health England 2016). It provides an annual snapshot of the overall health and the differential patterns of health of every local population in England through detailed data maps and spine charts, all accessible via an interactive tool. For example, Southwark is one of the 20% most deprived areas in England, and about 28% (15,000) of children live in low-income families. Its profile reports that life expectancy is 8.3 years lower for men and 6.2 years lower for women in the most deprived areas of Southwark than in the least deprived areas.

In 2015, the Equality Indicators project was launched in New York City. The purpose of this project was to investigate whether New York City (NYC) and, in the future, also other cities are making progress in reducing inequalities on an annual basis (Institute for State and Local Governance 2017). The Equality Indicators framework consists of 96 indicators to measure progress towards equality across six thematic areas: economy, education, health, housing, justice and services. Static Scores capture findings for a given year; and Dynamic Scores capture change from 1 year to the next. These scores will provide more meaningful measures since the primary purpose of the Equality Indicators is to capture change over time. Each of the 96 indicators is scored on a scale from 1 (highest possible inequality) to 100 (highest possible equality) using a conversion table. Two groups, generally the most- and least-likely to be disadvantaged for each issue, were compared to calculate the ratios. The new report indicates that the 2016 NYC Equality Score was 46.01 out of a possible 100, an increase of +0.56 from the 2015 score of 45.45. These scores suggest that NYC continues to be characterized by vast inequalities and that, when looking at the city as a whole, little has changed. Generally speaking this score means that overall, disadvantaged groups such as children, immigrants, poor people or with low education, racial minorities, elderly, single parents, women, etc. are almost twice as likely as those not disadvantaged to experience negative outcomes in fundamental areas of life, as measured by the Equality Indicators (Lawson et al. 2016).

Another initiative is The San Francisco Indicator Project (San Francisco Department of Health 2014). This project was created in 2007 by the San Francisco Department of Health, and it is a neighbourhood-level data system that measures how San Francisco performs as an equitable community in eight dimensions of health (environment, transportation, community cohesion, public realm, education, housing, economy and health systems). Each dimension contains multiple indicators which are presented in the form of maps and tables. The goal of this project is to support collaboration, planning, decision-making, and advocacy for social and physical environments that meet the needs of all citizens (San Francisco Department of Health 2014). Furthermore, there are Neighbourhood Profiles which provide a summary of the indicators by neighbourhood. These profiles can be used to assess data for a particular neighbourhood in comparison to other neighbourhoods and for San Francisco as a whole. In addition, the metrics and methods used in the San Francisco Indicator Project have been used and adapted by a number of other cities including Richmond, California; Denver, Colorado; Galveston, Texas; Oakland, California; Philadelphia, Pennsylvania; and Geneva, Switzerland (San Francisco Department of Health 2014).

In 2011 the Canadian Institute for Health Information published the Urban Physical Environments and Health Inequalities report which explored two aspects of the urban physical environment known to negatively affect health: outdoor air pollution and heat extremes (Canadian Institute for Health Information 2011). This report showed that those who are already more vulnerable to poor health may be at increased risk of being exposed to the effects of air pollution and heat extremes because of the areas in which they live. Moreover, the analyses of green space data for five Canadian cities showed that there were large differences in the percentage of green space between the lowest and highest socioeconomic areas within those cities. In general, a clear positive gradient existed, i.e. the higher the socioeconomic level, the more green space there was in the area (Canadian Institute for Health Information 2011).

25.3.2 *Examples of Research of Inequalities in Mortality in Cities*

Moreover, in recent years there has been an increase in the number of research studies that analyse socioeconomic health inequalities in small geographical areas, since area of residence has been recognized as a factor affecting health, independently of individual determinants. These studies are useful to advance research but also in the monitoring of health inequalities at small area level. In this sense, the Ineq-Cities project (https://www.ucl.ac.uk/ineqcities/), led by the Barcelona Public Health Agency in the period 2009–2013, aimed to study socioeconomic inequalities in mortality in the census tracts of 15 European cities at the beginning of the twenty-first century. This project revealed the existence of socioeconomic inequalities in total mortality and in some specific cause of death. In general, socioeconomic inequalities in mortality were more pronounced for men than for women, and relative inequalities were greater in Eastern and Northern European cities (Borrell et al. 2014; Marí-Dell'Olmo et al. 2015).

Another example is the Euro-Healthy project (http://www.euro-healthy.eu/about-euro-healthy), a 3-year Horizon 2020 research which started in 2015 that, among its objectives, aimed to analyse the evolution of socioeconomic inequalities in mortality in nine European metropolitan areas. In the majority of the metropolitan areas, socioeconomic inequalities in mortality tended to be stable during the three periods studied (2000–2003, 2004–2008 and 2009–2014) in men and women. In addition, these inequalities are higher for men than for women. The results for Barcelona are

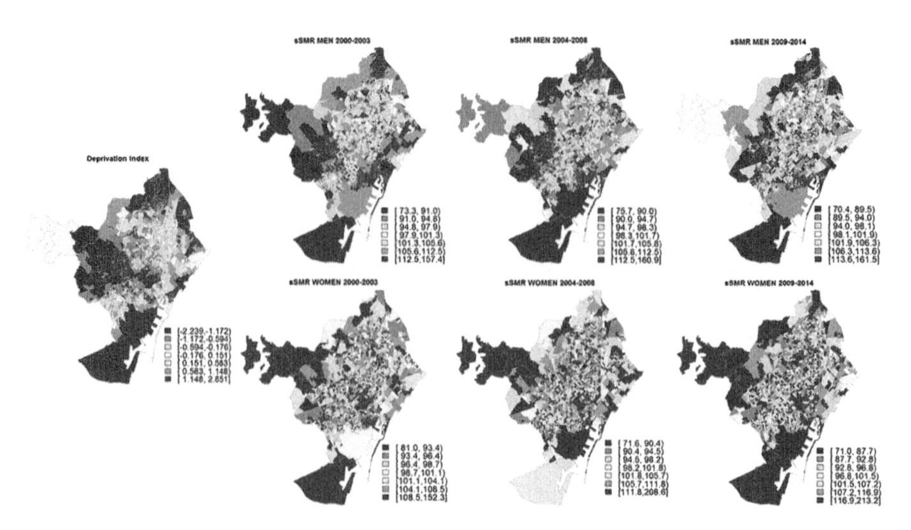

Fig. 25.2 Maps of socioeconomic deprivation and smoothed age standardised mortality ratio (sSMR) for total mortality in the three periods in men and woman. Barcelona, 2000–2014 (Source: Created by the authors on the basis of mortality register of Barcelona and inhabitants register of Barcelona)

shown in Fig. 25.2. In men, areas with relatively higher smoothed age standardized mortality ratios correspond to those areas that are more deprived, indicated by the clusters of brown areas. Similarly, areas with relatively lower smoothed age standardized mortality ratios correspond to those areas that are less deprived, indicated by the clusters of green areas. This pattern remains constant over time. In women, the pattern is not clear in any period.

25.4 Policies to Tackle Health Inequalities in Cities

In October 2016, the United Nations Conference on Housing and Sustainable Urban Development established the New Urban agenda by the world leaders, in order to rethink the way cities are built, managed as places to live. The agenda also provides guidance for achieving the Sustainable Development Goals. Table 25.3 shows this New Urban Agenda where an emphasis is the importance to ensure that all citizens have access to equal opportunities and face no discrimination, which mean equity. Other aspects mentioned are related with social determinants of health.

Table 25.3 Key commitments of the New Urban Agenda

Provide basic services for all citizens: These services include access to housing, safe drinking water and sanitation, nutritious food, healthcare and family planning, education, culture and access to communication technologies
Ensure that all citizens have access to equal opportunities and face no discrimination: Everyone has the right to benefit from what their cities offer. The New Urban Agenda calls on city authorities to take into account the needs of women, youth and children, people with disabilities, marginalized groups, older persons, indigenous people, among other groups
Promote measures that support cleaner cities: Tackling air pollution in cities is good both for people's health and for the planet. In the agenda, leaders have committed to increase their use of renewable energy, provide better and greener public transport and sustainably manage their natural resources
Strengthen resilience in cities to reduce the risk and the impact of disasters: Many cities have felt the impact of natural disasters, and leaders have now committed to implement mitigation and adaptation measures to minimize these impacts. Some of these measures include better urban planning, quality infrastructure and improving local responses
Take action to address climate change by reducing their greenhouse gas emissions: Leaders have committed to involve not just the local government but all actors of society to take climate action taking into account the Paris Agreement on climate change which seeks to limit the increase in global temperature to well below 2 °C. Sustainable cities that reduce emissions from energy and build resilience can play a lead role
Fully respect the rights of refugees, migrants and internally displaced persons regardless of their migration status: Leaders have recognized that migration poses challenges, but it also brings significant contributions to urban life. Because of this, they have committed to establish measures that help migrants, refugees and IDPs make positive contributions to societies
Improve connectivity and support innovative and green initiatives: This includes establishing partnerships with businesses and civil society to find sustainable solutions to urban challenges
Promote safe, accessible and green public spaces: Human interaction should be facilitated by urban planning, which is why the agenda calls for an increase in public spaces such as sidewalks, cycling lanes, gardens, squares and parks. Sustainable urban design plays a key role in ensuring the liveability and prosperity of a city

United Nations Conference on Housing and Sustainable Urban Development, 2016
Source: Sustained Development Goals. http://www.un.org/sustainabledevelopment/blog/2016/10/newurbanagenda/ [Accessed 2 June 2017]

25.4.1 Principles for Action for Tackling Health Inequalities

When implementing policies to reduce health inequalities, it is necessary to take into account the following principles for action (modified from Whitehead and Dahlgren 1991, 2006; Dahlgren and Whitehead 2006):

1. Polices should strive to level up, not level down: The levels of poor health of disadvantaged populations have to be brought up to the levels of the groups who are better off. Levelling down is not an option.
2. In order to reduce social inequalities in health, it is necessary to use the three main approaches, which are (a) focusing on people in poverty only, which implies to target the most disadvantaged groups; this is a selective approach

(One example can be to focus a programme in poor neighbourhoods), (b) narrowing the health divide between the better off and the worst off and (c) reducing social inequalities throughout the whole population: as inequalities in health affect the whole population (tend to increase with declining socio-economic status), universal policies targeting everybody are also necessary.

3. Actions should be concerned with tackling the social determinants of health inequalities. For example, social conditions that change across different social classes (as, e.g. working conditions) have to be taken into account. Tackling such social determinants, therefore, requires a greater understanding of the processes that generate and maintain social inequalities and then intervening in these processes at the most effective points.
4. It is necessary to have appropriate tools and metrics to measure health inequalities in order to monitor them. Moreover, it is important to monitor actions doing harm as, for example, the budget cuts on the National Health System that have occurred in the financial crisis that started in 2008.
5. Participation: It is necessary to make efforts to give a voice to the voiceless. Participation has to be encouraged in all the steps of program implementation. Usually the better off have more chances to participate in society and have more power to influence decisions. It is important to reverse this situation.
6. Wherever possible, social inequalities in health should be described and analysed separately for men and women and also by ethnic background: these analyses are important in order to understand the health differences in these social groups.

25.4.2 *Tackling Health Inequalities in Cities*

Policies to tackle health inequalities in urban areas have to be related with the conceptual framework of the determinants of health inequalities in these areas. Taking into account the conceptual framework presented in Fig. 25.1, policies should address the following aspects:

(a) Governance: Policies to improve good governance and democracy have to be implemented. For example, the participatory budgeting, a process of democratic deliberation and decision-making, in which residents of a city decide how to allocate part of a municipal budget. Participatory budgeting allows citizens to identify, discuss and prioritize public spending projects and gives them the power to make real decisions about how money is spent.

(b) Physical environment: As has been explained in Sect. 25.2, policies related to the physical environment are very important in urban areas. For example, municipalities have the possibility to decide the urban planning of the city, some housing policies and transport mobility. These policies should promote equity in results, for example, making all the neighbourhoods a healthy place to live with green areas, places to walk, to go by bicycle, etc. Also, it is important that policies promote the accessibility to housing for everybody, for example, promoting public social houses.

(c) Socioeconomic environment: Policies related with the socioeconomic environment are not always a responsibility of the municipal government. However, there are some examples in which the municipality can be important. For example, jobs created by the city government (direct or indirect) should have good working conditions. Some cities have the responsibility for education or health care; it is important that these services are universal and accessible to everybody to promote equity. Also, the municipality can promote safety and community participation.

(d) Settings: Selective policies implemented in settings can tackle socioeconomic inequalities in health, as programmes focussed in poor neighbourhoods, as the example shown for Barcelona in the next section.

(e) Axes of inequality: It is important that policies take into account the different axes of inequality.

The next section presents, as an example, a tool designed by the World Health Organization that is useful to monitor and to intervene in urban areas to tackle socioeconomic inequalities in health.

25.4.2.1 Urban HEART, a Tool for Action to Tackle Health Inequalities

The Urban Health Equity Assessment and Response Tool (Urban HEART) is a tool to measure urban health inequalities, and, most importantly, it is a tool for action to help tackle these inequalities (World Health Organization 2010; Kumaresan et al. 2010; WHO Kobe Centre 2010). Urban HEART was developed by the World Health Organization and launched in 2010. Although initially designed for low- and middle-income countries (LMIC), it has also been used in high-income countries. Implementation of Urban HEART implies several steps (Fig. 25.3). First, a local set of indicators must be chosen which will be used to identify and monitor inequalities in health and in its social determinants using the Urban HEART matrix and monitor, which depict inequalities in a simple and user-friendly manner (assessment stage). This set of indicators must be actionable through interventions and policies aimed at reducing inequalities. Afterwards, based on the identified inequalities, a list of interventions should be proposed to reduce them, which will be later prioritized and budgeted. So as to ensure that health equity issues are included in the political debate and that interventions are adequately budgeted, linkage with other sectors throughout the whole process is essential.

Since 2010, Urban HEART has been used in several cities. Among those from LMIC, most have used Urban HEART to prioritize those neighbourhoods in most need and implement interventions to reduce inequalities (Prasad et al. 2017). The response component can vary considerably across settings. For instance, while some cities decided to create new programmes to tackle inequalities, others prioritized and strengthened existing initiatives. In Guarulhos, Brazil, it was decided that

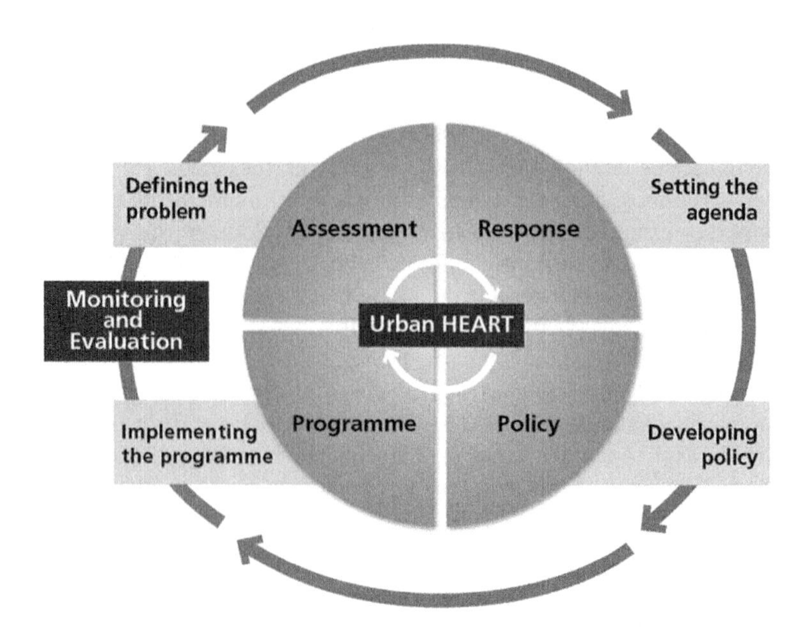

Fig. 25.3 Urban HEART planning and implementation cycle (World Health Organisation 2016a)

Urban HEART would guide annual action plans and would be integrated with the planning of other city departments (Prasad et al. 2013). In Tehran, Urban HEART was used as a knowledge transfer tool to inform neighbourhood stakeholders about the presence of inequalities in health and in its social determinants at their localities to foster the implementation of interventions aimed at reducing them. A Neighbourhood Development Committee, made up of politicians, technicians and local residents, was created in all neighbourhoods. The committee was responsible for developing an evidence-based local action plan for reducing inequalities on a set of health and health determinants previously prioritized (Asadi-Lari et al. 2013).

Toronto (Canada) and Barcelona (Spain) are two cities of a high-income country which have implemented Urban HEART. In Toronto, Urban HEART was used to develop a Neighbourhood Equity Index which allowed to identify 31 Neighbourhood Improvement Areas (Social Policy Analysis and Research City of Toronto 2014). The City Council included these neighbourhoods in the Toronto Strong Neighbourhoods Strategy 2020 (City of Toronto 2014). Residents and stakeholders decided what actions would be implemented in those neighbourhoods based on brainstorming exercises, discussion groups, video diaries and art projects.

Implementing Urban HEART in Barcelona

In Barcelona, socioeconomic health inequalities have been measured and studied since long ago. Different studies have analysed inequalities in mortality as well as in other health outcomes (Borrell et al. 1991; Brugal et al. 1999; Nebot et al. 1997), and some interventions have been implemented to tackle them, from tuberculosis or drug control programmes to urban renewal programmes (Diez et al. 1996; Manzanera et al. 2000; Mehdipanah et al. 2014). When Urban HEART was launched, several public health technicians advocated for its implementation in Barcelona. However, the local government at that moment (a coalition of three left-wing parties) only allowed to use it as a diagnosis tool, without assigning it any budget. Technicians decided to postpone its implementation until the response phase was adequately budgeted. It was not until 2015, when a new left-wing party based on social movements won the elections that implementing Urban HEART became possible. The new City Council, which had reducing health inequalities as a main priority, required for the immediate publication of Urban HEART, which would be used to prioritize those city neighbourhoods in most need for interventions aimed at reducing health inequalities.

Due to a tight time schedule, a provisional Urban HEART matrix was developed and showed how some neighbourhoods in the city systematically fared worse than others in most health and social determinants indicators. The provisional matrix was used to calculate an index which identified 18 neighbourhoods as a priority for intervention. Both the matrix and the prioritization were included in the 2014 Barcelona Health Report and had an important impact in the media, which helped to include health inequalities in the public debate. The prioritized neighbourhoods were included in the Health in the Barcelona Neighbourhoods programme and in the Neighbourhoods Plan, for which the City Council assigned a specific budget (300.000€ and 150 million €, respectively) (Novoa et al. 2017).

Later on, this first version of the matrix was reviewed and improved by the Urban HEART Barcelona Working Group, composed of public health technicians as well as technicians of non-health areas (social services, statistics, urbanism, environment) of the City Council and other institutions. The final matrix is shown in Fig. 25.4.

The Health in the Barcelona Neighbourhoods programme, ongoing since 2007, attempts to reduce health inequality through a community action strategy. It is a selective approach implemented in some settings (neighbourhoods) of the city. It includes five stages, the first one consisting of creating a Steering Committee made up of technicians from the public health, health services and social services areas, medical doctors as well as neighbours from local organisations such as the Neighbour's Association, among others. In the second stage, the neighbourhood's health and health determinant needs and assets are identified using both quantitative and qualitative data. The third stage consists of the prioritisation of health needs and the implementation of evidence-based interventions, which are assessed for effectiveness in the fourth stage. Finally, in the fifth stage, maintenance of the interventions is ensured by professionals working in the district, preferably based on local resources. Neighbour empowerment and participation must be guaranteed throughout the whole process (Díez et al. 2012).

With regard to the Neighbourhoods Plan, among its objectives, it aims to recover and promote local economy and reduce unemployment, to tackle urban infrastructure needs (green areas, public space, housing and public equipment, among others) and to improve life conditions (such as education, sports or health). It also considers the empowerment of neighbours as a main aspect of the plan.

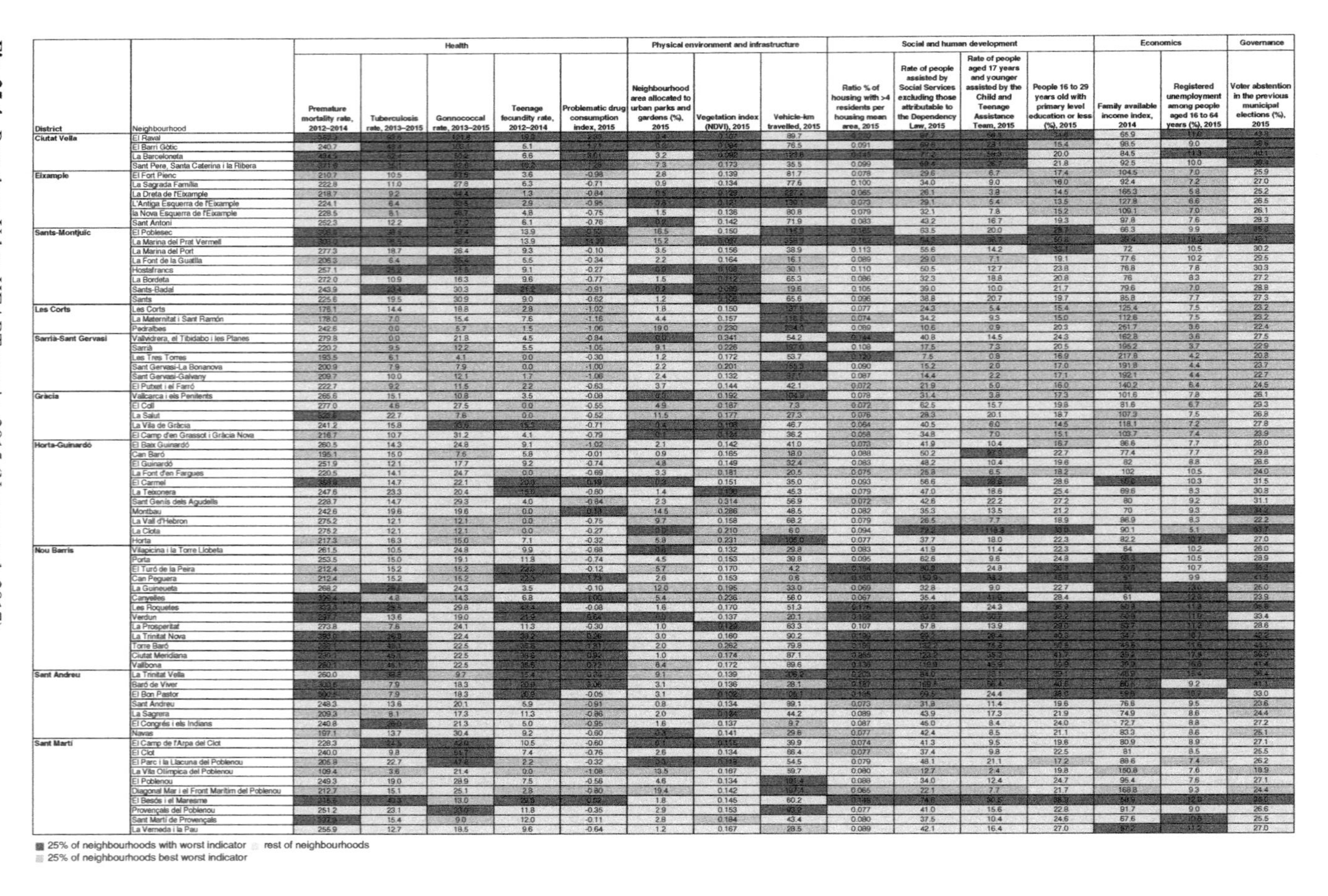

District	Neighbourhood	Health					Physical environment and infrastructure			Social and human development				Economics		Governance
		Premature mortality rate, 2012–2014	Tuberculosis rate, 2013–2015	Gonnococcal rate, 2013–2015	Teenage fecundity rate, 2012–2014	Problematic drug consumption index, 2015	Neighbourhood area allocated to urban parks and gardens (%), 2015	Vegetation index (NDVI), 2015	Vehicle-km travelled, 2015	Ratio % of housing with >4 residents per housing mean area, 2015	Rate of people assisted by Social Services excluding those attributable to the Dependency Law, 2015	Rate of people aged 17 years and younger assisted by the Child and Teenage Assistance Team, 2015	People 16 to 29 years old with primary level education or less (%), 2015	Family available income index, 2014	Registered unemployment among people aged 16 to 64 years (%), 2015	Voter abstention in the previous municipal elections (%), 2015
Ciutat Vella	El Raval	[illegible]	[illegible]	121.8	19.3	2.33	0.4	0.107	89.7	[illegible]	97.7	56.3	34.8	65.9	11.6	43.8
	El Barri Gòtic	240.7	[illegible]	[illegible]	5.1	[illegible]	[illegible]	0.194	76.5	0.091	69.6	23.1	15.4	98.5	9.0	38.9
	La Barceloneta	434.5	[illegible]	50.2	6.6	3.61	3.2	0.096	128.8	0.145	77.2	34.3	20.0	84.5	11.3	40.1
	Sant Pere, Santa Caterina i la Ribera	371.6	[illegible]	62.8	19.3	1.28	7.3	0.173	35.5	0.099	69.4	26.7	21.8	92.5	10.0	39.4
Eixample	El Fort Pienc	210.7	10.5	51.5	3.6	-0.98	2.8	0.139	81.7	0.078	29.6	6.7	17.4	104.5	7.0	25.9
	La Sagrada Família	222.8	11.0	27.8	6.3	-0.71	0.9	0.134	77.6	0.100	34.0	9.0	16.0	92.4	7.2	27.0
	La Dreta de l'Eixample	218.7	9.2	44.4	1.3	-0.84	0.5	0.129	227.2	0.065	26.1	3.8	14.5	165.3	5.8	25.2
	L'Antiga Esquerra de l'Eixample	224.1	6.4	52.5	2.9	-0.95	0.8	0.121	130.1	0.073	29.1	5.4	13.5	127.8	6.6	26.5
	la Nova Esquerra de l'Eixample	228.5	8.1	46.7	4.8	-0.75	1.5	0.136	80.8	0.079	32.1	7.8	15.2	109.1	7.0	26.1
	Sant Antoni	252.3	12.2	57.2	6.1	-0.76	0.2	0.142	71.9	0.083	43.2	16.7	19.3	97.8	7.6	28.3
Sants-Montjuïc	El Poblesec	[illegible]	[illegible]	47.4	13.9	0.52	16.5	0.150	[illegible]	[illegible]	63.5	20.0	28.7	66.3	9.9	[illegible]
	La Marina del Prat Vermell	[illegible]	[illegible]	[illegible]	13.9	[illegible]	15.2	0.067	[illegible]	[illegible]	[illegible]	[illegible]	[illegible]	[illegible]	[illegible]	[illegible]
	La Marina del Port	277.3	18.7	26.4	9.3	-0.10	3.5	0.156	38.9	0.113	55.6	14.2	33.1	72	10.5	30.2
	La Font de la Guatlla	206.3	6.4	35.4	5.5	-0.34	2.2	0.164	16.1	0.089	29.0	7.1	19.1	77.6	10.2	29.5
	Hostafrancs	257.1	25.2	31.5	9.1	-0.27	0.9	0.108	30.1	0.110	50.5	12.7	23.8	76.8	7.8	30.3
	La Bordeta	272.0	10.9	16.3	9.6	-0.77	1.5	0.112	65.3	0.086	32.3	18.8	20.8	76	8.3	27.2
	Sants-Badal	243.9	23.4	30.3	21.2	-0.91	0.2	0.089	19.6	0.105	39.0	10.0	21.7	79.6	7.0	28.8
	Sants	225.6	19.5	30.9	9.0	-0.62	1.2	0.105	65.6	0.096	38.8	20.7	19.7	85.8	7.7	27.3
Les Corts	Les Corts	176.1	14.4	18.8	2.8	-1.02	1.8	0.150	137.5	0.077	24.3	5.4	15.4	125.4	7.5	23.2
	La Maternitat i Sant Ramón	178.0	7.0	15.4	7.6	-1.16	4.4	0.157	118.5	0.074	34.2	9.3	15.0	112.6	7.5	23.2
	Pedralbes	242.6	0.0	5.7	1.5	-1.06	19.0	0.230	234.1	0.069	10.6	0.9	20.3	261.7	3.6	22.4
Sarrià-Sant Gervasi	Vallvidrera, el Tibidabo i les Planes	279.8	0.0	21.8	4.5	-0.84	0.0	0.341	54.2	0.144	40.8	14.5	24.3	162.8	3.6	27.5
	Sarrià	220.2	9.5	12.2	5.5	-1.05	9.1	0.226	157.0	0.108	17.5	7.3	20.5	195.2	3.7	22.9
	Les Tres Torres	193.5	6.1	4.1	0.0	-0.30	1.2	0.172	53.7	0.120	7.5	0.8	16.9	217.8	4.2	20.8
	Sant Gervasi-La Bonanova	200.9	7.9	7.9	0.0	-1.00	2.2	0.201	153.3	0.090	15.2	2.0	17.0	191.8	4.4	23.7
	Sant Gervasi-Galvany	209.7	10.0	12.1	1.7	-1.06	2.4	0.132	37.1	0.087	14.4	2.2	17.1	192.1	4.4	22.7
	El Putxet i el Farró	222.7	9.2	11.5	2.2	-0.63	3.7	0.144	42.1	0.072	21.9	5.0	16.0	140.2	6.4	24.5
Gràcia	Vallcarca i els Penitents	265.6	15.1	10.8	3.5	-0.08	6.5	0.192	104.9	0.078	31.4	3.8	17.3	101.6	7.8	26.1
	El Coll	277.0	4.6	27.5	0.0	-0.55	4.9	0.187	7.3	0.072	62.5	15.7	19.8	81.6	6.7	29.3
	La Salut	322.5	22.7	7.6	0.0	-0.52	11.5	0.177	27.3	0.076	28.3	20.1	18.7	107.3	7.5	26.8
	La Vila de Gràcia	241.2	15.8	30.5	15.3	-0.71	0.4	0.108	46.7	0.064	40.5	6.0	14.5	118.1	7.2	27.8
	El Camp d'en Grassot i Gràcia Nova	216.7	10.7	31.2	4.1	-0.79	0.1	0.124	36.2	0.058	34.8	7.0	15.1	103.7	7.4	23.9
Horta-Guinardó	El Baix Guinardó	260.5	14.3	24.8	9.1	-1.02	2.1	0.142	41.0	0.073	41.9	10.4	16.7	86.6	7.7	28.0
	Can Baró	195.1	15.0	7.6	5.8	-0.01	0.9	0.165	18.0	0.088	50.2	27.9	22.7	77.4	7.7	29.8
	El Guinardó	251.9	12.1	17.7	9.2	-0.74	4.8	0.149	32.4	0.083	48.2	10.4	19.6	82	8.8	28.6
	La Font d'en Fargues	220.5	14.1	24.7	0.0	-0.69	3.3	0.181	20.5	0.075	25.8	6.5	18.2	102	10.5	24.0
	El Carmel	359.8	14.7	22.1	20.8	0.19	0.3	0.151	35.0	0.093	56.6	29.6	28.6	55.0	10.3	31.5
	La Teixonera	247.6	23.3	20.4	15.0	-0.80	1.4	0.136	45.3	0.079	47.0	18.6	25.4	69.6	8.3	30.8
	Sant Genís dels Agudells	228.7	14.7	29.3	4.0	-0.84	2.3	0.314	56.9	0.072	42.6	22.2	27.2	80	9.2	31.1
	Montbau	242.6	19.6	19.6	0.0	0.11	14.5	0.286	48.5	0.082	35.3	13.5	21.2	70	9.3	34.2
	La Vall d'Hebron	275.2	12.1	12.1	0.0	-0.75	9.7	0.158	68.2	0.079	26.5	7.7	18.9	86.9	8.3	22.2
	La Clota	275.2	12.1	12.1	0.0	-0.27	0.2	0.210	6.0	0.094	72.2	119.3	30.0	90.1	5.1	33.7
	Horta	217.3	16.3	15.0	7.1	-0.32	5.8	0.231	105.0	0.077	37.7	18.0	22.3	82.2	10.7	27.0
Nou Barris	Vilapicina i la Torre Llobeta	261.5	10.5	24.8	9.9	-0.68	0.6	0.132	29.8	0.083	41.9	11.4	22.3	64	10.2	26.0
	Porta	253.5	15.0	19.1	11.8	-0.74	4.5	0.153	39.8	0.095	62.6	9.6	24.8	58.3	10.5	28.9
	El Turó de la Peira	212.4	15.2	15.2	22.8	-0.12	5.7	0.170	4.2	0.154	80.9	24.8	30.1	50.6	10.7	35.2
	Can Peguera	212.4	15.2	15.2	22.5	1.73	2.6	0.153	0.6	0.153	150.9	33.2	45.0	51	9.9	41.5
	La Guineueta	268.2	28.6	24.3	3.5	-0.10	12.0	0.195	33.0	0.069	32.8	9.0	22.7	56	13.0	25.0
	Canyelles	306.4	4.8	14.3	6.8	1.00	5.4	0.236	56.0	0.067	35.4	41.9	28.4	61	12.8	23.9
	Les Roquetes	333.3	29.5	29.8	43.4	-0.08	1.6	0.170	51.3	0.176	27.9	24.3	36.9	50.8	11.8	35.8
	Verdun	297.7	13.6	19.0	21.9	5.64	0.0	0.137	20.1	0.122	30.0	30.7	32.2	50.9	11.9	33.4
	La Prosperitat	273.8	7.6	24.1	11.3	-0.30	1.0	0.122	63.3	0.107	57.8	13.9	29.0	53.7	11.2	28.6
	La Trinitat Nova	393.0	26.0	22.4	38.2	0.26	3.0	0.160	90.2	0.190	99.2	28.4	40.3	34.7	16.7	42.2
	Torre Baró	[illegible]	46.1	22.5	35.6	1.81	2.0	0.262	79.8	0.186	132.2	75.8	56.5	45.6	11.6	46.1
	Ciutat Meridiana	[illegible]	45.1	22.5	36.6	0.92	1.0	0.174	87.1	0.265	123.2	35.2	41.7	39.2	17.8	36.9
	Vallbona	280.1	46.1	22.5	35.5	0.72	6.4	0.172	89.6	0.136	119.9	45.9	55.9	39.9	16.6	41.4
Sant Andreu	La Trinitat Vella	260.0	38.3	9.7	15.4	0.24	9.1	0.139	206.2	0.203	84.0	33.1	29.1	40.9	15.4	36.4
	Baró de Viver	303.6	7.9	18.3	20.9	3.06	3.1	0.136	28.1	0.187	169.8	56.4	40.8	80.6	9.2	41.1
	El Bon Pastor	302.5	7.9	18.3	26.5	-0.05	3.1	0.102	106.1	0.185	69.5	24.4	38.0	59.8	10.7	33.0
	Sant Andreu	248.3	13.6	20.1	5.9	-0.91	0.8	0.134	89.1	0.073	31.8	11.4	19.6	76.6	9.5	23.6
	La Sagrera	209.3	8.1	17.3	11.3	-0.86	2.0	0.124	44.2	0.089	43.9	17.3	21.9	74.9	8.6	24.4
	El Congrés i els Indians	240.8	26.0	21.3	5.0	-0.95	1.6	0.137	9.7	0.087	45.0	8.4	24.0	72.7	8.8	27.2
	Navas	197.1	13.7	30.4	9.2	-0.60	0.3	0.141	29.8	0.077	42.4	8.5	21.1	83.3	8.6	25.1
Sant Martí	El Camp de l'Arpa del Clot	228.3	24.5	42.0	10.5	-0.60	0.1	0.115	39.9	0.074	41.3	9.5	19.6	80.9	8.9	27.1
	El Clot	240.0	9.8	51.7	7.4	-0.76	2.6	0.134	66.4	0.077	37.4	9.8	22.5	81	8.5	25.5
	El Parc i la Llacuna del Poblenou	205.8	22.7	47.8	2.2	-0.32	0.3	0.118	54.5	0.079	48.1	21.1	17.2	88.6	7.4	26.2
	La Vila Olímpica del Poblenou	109.4	3.6	21.4	0.0	-1.08	13.5	0.187	59.7	0.080	12.7	2.4	19.8	150.8	7.6	18.9
	El Poblenou	249.3	19.0	28.9	7.5	-0.56	4.6	0.134	101.4	0.088	34.0	12.4	24.7	95.4	7.6	27.1
	Diagonal Mar i el Front Marítim del Poblenou	212.7	15.1	25.1	2.8	-0.80	19.4	0.142	107.1	0.065	22.1	7.7	21.7	168.8	9.3	24.4
	El Besòs i el Maresme	315.6	40.3	13.0	29.9	0.52	1.8	0.145	60.2	0.148	74.6	30.5	38.9	58.9	12.9	28.0
	Provençals del Poblenou	251.2	23.1	33.0	11.8	-0.35	2.9	0.153	93.2	0.077	41.0	15.6	22.8	91.7	9.0	26.6
	Sant Martí de Provençals	337.9	15.4	9.0	12.0	-0.11	2.8	0.184	43.4	0.080	37.5	10.4	24.6	67.6	10.6	25.5
	La Verneda i la Pau	255.9	12.7	18.5	9.6	-0.64	1.2	0.167	28.5	0.089	42.1	16.4	27.0	67.2	11.2	27.0

25% of neighbourhoods with worst indicator; rest of neighbourhoods; 25% of neighbourhoods best worst indicator

Fig. 25.4 Barcelona Urban HEART matrix, 2015 (Novoa et al. 2017)

Limitations and Strengths of Urban HEART

Unavailability of disaggregated and quality data, mostly in low-income settings, can undermine the implementation of Urban HEART (Elsey et al. 2016; Prasad et al. 2017; WHO Kobe Centre 2011). In these settings, to overcome this problem, household surveys are used, although they can be under-representing the urban poorest, as well as census-based data although they can exclude illegal settlements and the homeless. To overcome this limitation, a simplified Urban HEART methodology has been proposed and put into practice in some settings, such as in several Indian cities (Elsey et al. 2016). In addition, engagement of non-health sectors has been identified as a barrier by some of the countries that have implemented Urban HEART (Prasad et al. 2013). As previously mentioned, this is an essential part of Urban HEART. Not only is this important with regards to non-health departments but also to local communities; the implication of these groups is fundamental to ensure sustainability of actions (Prasad et al. 2017).

One of the strengths of Urban HEART is that it is a comprehensive tool, including indicators and domains absent in other area-based deprivation indices. This facilitates generating debate, about the structural factors related to urban health, both at the political level and among the public. This debate can help in implementing evidence-based policies and interventions (Pakeman 2017). In addition, the Urban HEART matrix and monitor facilitate effective communication of results to non-experts, since they use very simple visual techniques (i.e. colour coding to classify neighbourhoods according to levels of achievement with respect to specific targets) (WHO Kobe Centre 2011).

Experience from high-, low- and middle-income countries shows that Urban HEART can be useful to identify and monitor health inequalities and, most importantly, to help implementing interventions aimed at reducing these inequalities. This is achieved as the tool allows to introduce health inequalities in the public debate and in the political agenda. However, experience also shows that political will and institutional and financial support from local authorities are essential for implementing both the assessment and the response components of Urban HEART.

25.5 Conclusions

The factors or determinants of health are important to understand what causes health inequalities. In urban areas, they belong to the following aspects: urban governance, physical and socioeconomic environment, settings and axes of inequalities.

Surveillance of inequalities in health involves monitoring inequalities over time and across the different axes of inequality. To achieve this objective, it is necessary to have adequate urban health metrics that allow to measure both inequalities in health and in its determinants.

Policies to tackle health inequalities in urban areas have to be related with the conceptual framework of the determinants of health inequalities in these areas. The Urban Health Equity Assessment and Response Tool (Urban HEART) is a tool to measure urban health inequalities and, most importantly, it is a tool for action to help tackle these inequalities. This tool has been used in many cities to advance on including health inequalities in the political agenda.

References

Asadi-Lari, M., et al. (2013). Response-oriented measuring inequalities in Tehran: Second round of Urban Health Equity Assessment and Response Tool (Urban HEART-2), concepts and framework. *Medical journal of the Islamic Republic of Iran, 27*(4), 236–248.

Borrell, C., et al. (2013). Factors and processes influencing health inequalities in urban areas. *Journal of Epidemiology and Community Health, 67*(5), 389–391.

Borrell, C., et al. (2014). Socioeconomic inequalities in mortality in 16 European cities. *Scandinavian Journal of Public Health, 42*(3), 245–254.

Borrell, C., Plasència i Taradach, A., & Pañella i Noguera, H. (1991). Excess mortality in an inner-city area: the case of Ciutat Vella in Barcelona. *Gaceta Sanitaria, 5*(27), 243–253.

Bortz, M., et al. (2015). Disaggregating health inequalities within Rio de Janeiro, Brazil, 2002–2010, by applying an urban health inequality index. *Cadernos de Saúde Pública, 31*(Suppl 1), 107–119.

Brugal, M. T., et al. (1999). A small area analysis estimating the prevalence of addiction to opioids in Barcelona, 1993. *Journal of Epidemiology and Community Health, 53*(8), 488–494.

Canadian Institute for Health Information. (2011). *Urban physical environments and health inequalities—Data and analysis methodology*. Ottawa: Canadian Institute for Health Information.

City of Toronto. (2014). *Toronto strong neighbourhoods strategy 2020, Toronto*. Toronto: City of Toronto.

Corburn, J., & Cohen, A. K. (2012). Why we need urban health equity indicators: Integrating science, policy, and community. *PLoS Medicine, 9*(8), e1001285.

CSDH. (2008). *Closing the gap in a generation: health equity through action on the social determinants of health. Final report of the Commission on Social Determinants of Health*. Geneva: World Health Organization.

Dahlgren, G., & Whitehead, M. (2006). *Levelling up (part 2): A discussion paper on European strategies for tackling social inequities in health, Copenhagen*. Geneva: World Health Organization.

Diez, E., et al. (1996). Evaluation of a social health intervention among homeless tuberculosis patients. *Tubercle and Lung Disease, 77*(5), 420–424.

Díez, E., et al. (2012). "Salut als barris" en Barcelona, una intervención comunitaria para reducir las desigualdades sociales en salud. *Comunidad, 14*(2), 121–126.

Elsey, H., et al. (2016). Addressing inequities in urban health: Do decision-makers have the data they need? Report from the urban health data special session at international conference on urban health Dhaka 2015. *Journal of Urban Health, 93*(3), 526–537.

Espelt, A., et al. (2016). La vigilancia de los determinantes sociales de la salud. *Gaceta Sanitaria, 30*, 38–44.

Institute for State and Local Governance. (2017). Equality indicators – Institute for State and Local Governance. Retrieved May, 31, 2017, from http://equalityindicators.org/.

Kumaresan, J., et al. (2010). Promoting health equity in cities through evidence-based action. *Journal of Urban Health: Bulletin of the New York Academy of Medicine, 87*(5), 727–732.

Lawson, V., et al.. (2016). Equality indicators. Measuring change toward greater equality in New York City, New York.
Manzanera, R., et al. (2000). Coping with the toll of heroin: 10 years of the Barcelona Action Plan on Drugs, Spain. *Gaceta Sanitaria, 14*(1), 58–66.
Marí-Dell'Olmo, M., et al. (2015). Socioeconomic inequalities in cause-specific mortality in 15 European cities. *Journal of Epidemiology and Community Health, 69*(5), 432–441.
Mehdipanah, R., et al. (2014). The effects of an urban renewal project on health and health inequalities: A quasi-experimental study in Barcelona. *Journal of Epidemiology and Community Health, 68*(9), 811–817.
Nebot, M., Borrell, C., & Villalbí, J. R. (1997). Adolescent motherhood and socioeconomic factors: An ecologic approach. *European Journal of Public Health, 7*(2), 144–148.
Novoa, A. M., et al. (2017). The experience of implementing Urban HEART Barcelona, a tool for action. *Journal of Urban Health*. https://doi.org/10.1007/s11524-017-0194-6.
Pakeman, K. (2017). *Checking Kingston's equity pulse: An application and critical evaluation of the Urban HEART @Toronto methodology to investigate intra-city social and health inequities of Kingston, ON*. Kingston: QSPACE.
Prasad, A., et al. (2013). Linking evidence to action on social determinants of health using Urban HEART in the Americas. *Pan American Journal of Public Health, 34*(6), 407–415.
Prasad, A., et al. (2016). Metrics in urban health: Current developments and future prospects. *Annual Review of Public Health, 37*(1), 113–133.
Prasad, A., et al. (2017). Prioritizing action on health inequities in cities: An evaluation of Urban Health Equity Assessment and Response Tool (Urban HEART) in 15 cities from Asia and Africa. *Social Science & Medicine (1982), 145*, 237–242.
Public Health England, (2016). Public health profiles. Retrieved May 31, 2017, from https://fingertips.phe.org.uk/profile/health-profiles.
Rothenberg, R., et al. (2014). A flexible urban health index for small area disparities. *Journal of Urban Health: Bulletin of the New York Academy of Medicine, 91*(5), 823–835.
San Francisco Department of Health. (2014). The San Francisco Indicator Project. Retrieved May 31, 2017, from http://www.sfindicatorproject.org/.
Social Policy Analysis and & Research City of Toronto. (2014). *TSNS 2020 Neighbourhood Equity Index methodological documentation* (p. 2020). Toronto: Social Policy Analysis and & Research City of Toronto.
Stauber, C., et al. (2003). *Comparison of cities using the Urban Health Index: An analysis of demographic and health survey data from prepared for the WHO Centre for Health Development*. Geneva: WHO.
Whitehead, M. (1992). The concepts and principles of equity and health. *International Journal of Health Services: Planning, Administration, Evaluation, 22*(0020-7314; 3), 429–445.
Whitehead, M., & Dahlgren, G. (2006). *Levelling up (part 1): A discussion paper on concepts and principles for tackling social inequities in health, Copenhagen*. Geneva: World Health Organization.
Whitehead, M., & Dahlgren, G. (1991). What can be done about inequalities in health? *Lancet (London), 338*(8774), 1059–1063.
WHO/UN-HABITAT. (2016). Global report on urban health: Equitable, healthier cities for sustainable development. In *Geneva*.
WHO/UN-HABITAT. (2010). Hidden cities: Unmasking and overcoming health inequities in urban settings. In *Geneva*.
WHO Kobe Centre. (2011). *Report of consultation meeting on urban health metrics research*. Kobe: WHO Kobe Centre.
WHO Kobe Centre. (2010). *Urban HEART urban health equity assessment and response tool*. Kobe: WHO Kobe Centre.
Wilkinson, J. (2015). Public health observatories in England: Recent transformations and continuing the legacy. *Cadernos de Saude Publica, 31*(Suppl 1), 269–276.

World Health Organization. (2013a). *Handbook on health inequality monitoring: With a special focus on low- and middle-income countries*. Geneva: WHO.
World Health Organization. (2013b). *Urban health observatories: A possible solution to filling a gap in public health intelligence*. Kobe: WHO.
World Health Organization. (2014). *The Urban Health Index: A handbook for its calculation and use*. Kobe: WHO.
World Health Organization. (2016). *World health statistics 2016: Monitoring health for the SDGs, sustainable development goals*. Geneva: WHO.
World Health Organisation. (2016a). *Global Health Observatory, urban health*. Geneva: WHO. Retrieved May 30, 2017, from http://www.who.int/gho/urban_health/en/.
World Health Organisation. (2016b). *Urban Health Index*. Geneva: WHO. Retrieved May 30, 2017, from http://www.who.int/kobe_centre/measuring/innovations/urban_health_index/en/.
World Health Organisation. (2017). *Global Health Observatory (GHO) data*. Geneva: WHO. Retrieved May 30, 2017, from http://www.who.int/gho/en/.
World Health Organization. (2010). *Urban HEAR T: urban health equity assessment and response tool: User manual*. Kobe: WHO.

Chapter 26
Participating in Health: The Healthy Outcomes of Citizen Participation in Urban and Transport Planning

Ersilia Verlinghieri

26.1 The Ringland Project: A Live Example of Citizen Participation, Urban and Transport Planning, and Health

In 1998 the Flemish government proposed to complete the ring road around the city of Antwerp. This strategic transport infrastructure investment aimed at solving the lasting congestion issues and improving traffic flow. Following this proposal, an intense debate that involved different planning authorities; local, national and international bodies; academics; and citizens has arose. Almost 20 years later, the project is still in the drawer. This is not the consequence of lack of funding or institutional capacity but rather the strong effect of what I would call in this paper *citizens' participation in urban and transport planning for health.*

Since the 1998 proposal, a variety of citizen-led initiatives have opposed the new road, resulting in 2013 with the birth of the Ringland project (Ringland n.d.) (https://ringland.be/about/the-project/). Ringland is "crowd-brained and crowd-funded" infrastructure project for a six billion euro investment "that has been completely initiated and developed bottom-up by local citizens" (Ringland n.d.). It consists of a road tunnelling project that would highly mitigate the health impacts associated with the originally proposed new ring road (Brusselen et al. 2016; Van Brussel et al. 2016). The people involved in Ringland used participation in an unconventional manner, demanding referenda on the decisions regarding the ring road and producing community plans. They contested not only the construction of a new road but also demanded for the existing road to be tunnelled in order to reduce health impacts (Wolf and Van Dooren 2017). As such, they successfully brought health at the centre

E. Verlinghieri (✉)
Transport Studies Unit, University of Oxford, Oxford, UK
e-mail: ersilia.verlinghieri@ouce.ox.ac.uk

M. Nieuwenhuijsen, H. Khreis (eds.), *Integrating Human Health into Urban and Transport Planning*, https://doi.org/10.1007/978-3-319-74983-9_26

of the Flemish transport planning agenda and reframed the public debate about transport and health.

Based on a number of feasibility studies and on a full Health Impact Assessment (HIA) that were crowdfunded (Van Brussel et al. 2016), the Ringland project has completely reshaped the government agenda and is now scheduled to be implemented as proposed by citizens (De Oosterweelverbinding 2015). Accordingly to its HIA, the Ringland project will result in a significant reduction in premature mortality and increase in life expectancy (Brusselen et al. 2016). Specifically, improved health is predicted for children whose schools are located near to the road. Further benefits are the increased availability of recreational and green spaces and reduction of noise and pollution (Wolf and Van Dooren 2017).

Although not unique, as a variety of anti-road building initiatives have took place worldwide in the last decades (Rawcliffe 1995; Paterson 2007), the Ringland project stands out because it opposes road infrastructures in name of public health. This is in line with an increasing public concern with the polluted state of cities and natural environment. It shows that health has to become an urgent and central objective of both urban and transport planning. It also reminds us that we still need to better link planning practices and health-related decisions (Khreis et al. 2016; Nieuwenhuijsen et al. 2017).

The Ringland project is also a peculiar example of the variety of forms, meanings, benefits and limits of participation in urban and transport planning in relation to health. It shows how participatory planning can take place both inside and outside the institution, being initiated by and involving a variety of actors, from citizens to planners and academics, and often promoting new forms of governance. It also shows how the benefits of participation are much broader than just making quicker decisions.

What can we, as academics, practitioners and decision-makers, learn from this example? How can we use participation to implement innovative decision-making practices that contribute to the construction of healthy cities, also in contexts in which there are no previous citizens initiatives calling for health in urban and transport planning?

In this chapter I explore possible answers to these questions. Considering in more details the various aspect of the Ringland project and building on the literature on participation in urban and transport planning, I explore the connections between citizens' participation and health, showing their potentials and limits in an increasingly complex world. After giving some definitions, I consider the wide benefits and limitations of participation recognised by the literature. Subsequently, I provide a summary of the main planning traditions and consider how they approach participation in different ways. I then consider the specific benefits that participation can offer to health and reflect on which would be the most appropriate planning settings and practices to allow these to take place. I propose that we build a culture of participation across society in order to do so. I conclude with a reflection on the role of academics and of participatory research to support the construction of such culture of participation.

26.1.1 Definitions

Following the WHO (2003, p. 10) report on *Community participation in local health and sustainable development*, I define participation as "a process by which people are enabled to become actively and genuinely involved in defining the issues of concern to them, in making decisions about factors that affect their lives, in formulating and implementing policies, in planning, developing and delivering services and in taking action to achieve change". Specifically, in considering "participation in health planning" and "participation in urban and transport planning for health", I broadly refer to the involvement of citizens in the design of health-related policy and planning intervention, as well as to the broad participation of citizens in urban planning and its effects on health. This definition does not specifically focus on the narrower notion of involvement of individual patients in the design of their health care (Conklin et al. 2015).

I consider participation to be an institutional process organised by local authorities, councils, researchers, international bodies to involve local communities, stakeholders and a variety of actors in planning processes. At the same time I consider participation in broader terms. If, following Friedmann (1987), we consider planning to be an activity that takes place in the public domain *in the course of social transformation*, then participation can be defined as an arena of transition from the private sphere to the public sphere that can be initiated and opened by different actors (Padilla et al. 2007). Therefore, participation including all those processes that commenced from the bottom up by actors such as grassroots initiatives, social movements or citizens groups impacts, to a small or great extent, on the way we perceive and plan our cities. As such, the proposed definition of participation includes grassroots initiatives like the Ringland project.

Finally, I concentrate on the urban realm. This is not only because it is the most common condition for human settlement but also the location where the effects of current environmental and social crises appear clearest (Khreis et al. 2016). Furthermore, cities, both from a governance setting perspective and from a cultural view, have the potential and are generally, even if often only formally, widely committed to open up to citizens participation, enabling spaces in which different actors can meet and exchange knowledge (Dooris and Heritage 2013).

26.2 The Benefits of Participation

Before examining in which way participation can support healthy cities, it is worth considering in more general terms what aspects of planning and decision-making can be improved by citizens' participation. Beyond the discourse on participation being part of a "proper conduct" of democratic processes (Renn et al. 1995; Tickner 2001; Crawford et al. 2002), its benefits can be listed as follows:

- Firstly, substantively involving citizens into a decision-making process allows greater understanding of phenomena, making space for other viewpoints to emerge and for different inputs to contribute (Irwin 1995; Tickner 2001). This is specifically meaningful when approaching complex phenomena such as urban and transport planning, whose understanding requires data from multiple sources. Through participation, planners and decision-makers can have a clearer picture of the citizens' desires and preferences: participation is an invaluable tool to build effective evaluation and assessment procedures and allow better informed decisions (Lowndes et al. 2001; Innes and Booher 2004; Rowe and Frewer 2016). Moreover, including citizens and other actors can back up the work of research teams or experts that might not have the appropriate sensibilities or time resources to gather deep knowledge of specific issues and contexts (Becker et al. 2004).

  For example, it is thanks to citizens' participation that health has become one of the core goal of the planning process for the Antwerp ring road (Van Brussel et al. 2016). The Ringland project revisited completely the concept of ring road (Van Brussel et al. 2016) showing how an infrastructural project could result into increased health and wellbeing (Verbeek 2014).
- Secondly, according to a more instrumental-oriented assumption (Renn et al. 1995; Mumpower 2001; Crawford et al. 2002), participation, when appropriately set up, allows to increase public acceptability of decisions and, in some cases, to build stronger consensus, reduce conflicts and produce shared projects and visions (Innes and Booher 2004; van de Kerkhof 2006). As such, it can be used, in a positive or negative sense, to legitimate and reinforce support for public decisions and overcome governance barriers (Reed 2008). Reversely, not doing so greatly slows the process. If the city of Antwerp had involved the residents from the start of the planning process, it would have possibly avoided waiting 10 years or more to implement its ring road project.
- Thirdly, in a more transformative perspective, participation can be used as a process to inform, educate and empower citizens, ensuring conditions for fair and just decision-making (Renn et al. 1995; Innes and Booher 2004; Bailey and Grossardt 2010; Bailey et al. 2012). Participation can be an important space for change in planning practice (Hilbrandt 2017) and one of the most viable pathways towards sustainable transportation (Banister 2008). Similarly it is a way of mobilising new resources and energies and possibly ensuring ownership and sustainability of decisions (Dooris and Heritage 2013). It can be a process of information diffusion, knowledge exchange and creation in which communities and marginalised group increase their voice and power (Campbell and Jovchelovitch 2000; Dooris and Heritage 2013). In this way participation can also catalyse behavioural and social change, becoming a space in which practices can be transformed and social learning built (Kesby 2005; Stringer et al. 2006).

  The Ringland project is a great example of this more transformative aspect of participation. Demanding to be able to participate in the planning process, citizens pushed institutions to change their practices, introduced a greater attention on health at all levels of planning and in the public domain (Flandersnews.be 2016; Flanderstoday 2017) and mobilised a variety of new actors in the discussion and planning, such as the academics that prepared the HIA (Brusselen et al. 2016).

Clearly, further to these positive aspects, participation also offers numerous challenges that I explore in the next section.

26.3 The Challenges of Participation

A vast body of interdisciplinary literature has highlighted the limitations of participation. Generally, criticisms concern the actual settings of participation or more broadly its validity in political, ethical and scientific terms (Reed et al. 2017). As Cooke and Kothari (2001) summarise in their famous book *Participation: the new Tyranny?*, within the current political and institutional settings, the main issues are linked to:

- Potential conflict arising between the outcomes of participatory processes and decisions taken within formal democratic settings
- The risk for group dynamics to reinforce within participation the existing patterns of inequality and exclusion
- The idea that a certain form of participation is good per se, often preventing more appropriate approaches to be used

When not carefully set up, participatory spaces might fail to effectively guarantee communication among different actors, due to power unbalances and risk to reproduce or reinforce social exclusion (Elvy 2014). Participation can also miss accounting for diversity, inclusivity and differentiation, proposing often a too homogeneous definition of what a community is and what its views and needs are (Cooke and Kothari 2001). As such participatory arenas can become spaces in which decisions already made are brought to be legitimised instead of being discussed in depth (Tickner 2001). Participation can also become an instrument for external actors to exploit local resources, knowledges and willingness to improve conditions to their own benefits, off-setting costs of running specific programmes (Guareschi and Jovchelovitch 2004). Careful set-up of the participatory processes and assessment of the power balances within them could prevent these issues to emerge (Willson 2001; Bailey et al. 2012; Elvy 2014).

More fundamental challenges appear when attempting to establish participatory arenas as truly transformational and creative processes of co-production. As Irwin (1995) suggested, participatory spaces need to open up discussions and not be only seen as mandatory ticking boxes exercises or events in which a specific policy direction is publicised. This might be an important challenge especially in times of economic crisis, when participation arenas can be limited by time restraints, lack of resources, political divisions and inexperience in public consultation processes

(Verlinghieri 2016). Whilst considering participation in broader terms might allow for more resources to join and more actors to initiate it as a process, awareness is needed on the risk of participation to become a tyranny and progressively exhaust community resources (Cooke and Kothari 2001). I will explore later in the chapter methods and resources to overcome these barriers.

Finally, participation should not be considered as a way to necessarily build consensus but a tool to embrace diversity, complexity and conflicts, grounding a better understanding of inequalities and the type of intervention needed from an economic but also social perspective.

Having considered these points, it is clear that setting up appropriate participatory spaces for health planning or to allow health to become the centre of urban and transport planning might pose major challenges. Which modes of planning are able to set up and accommodate appropriate participatory arenas? Which of these modes are more appropriate when aiming at planning for healthier cities? In order to answer these questions, it is worth considering how participation has been conceived in the planning literature.

26.4 Climbing a Ladder of Citizen's Empowerment: Participation in the Planning Literature

One of the foundation stones in the literature on participation and planning is the famous Arnstein's (1969) ladder. The ladder (Fig. 26.1) classifies the different degrees of participation in planning as a ladder of citizen empowerment.[1] Starting

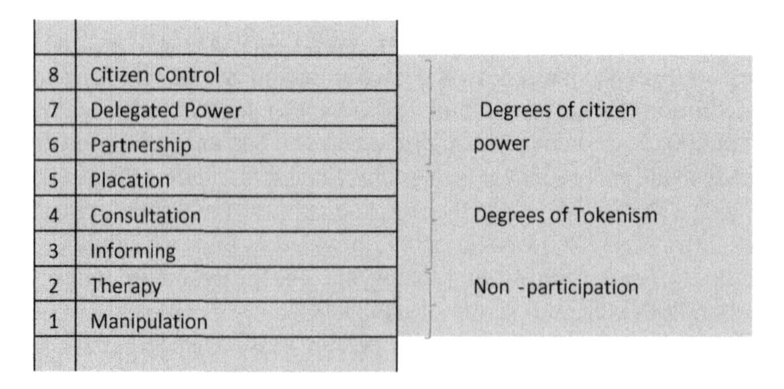

Fig. 26.1 Arnstein's ladder of participation (readapted from Arnstein (1969))

[1]Empowerment is a complex term whose definition is linked to the specific vision on power adopted. Dooris and Heritage (2013) report various points on the debate. For the sake of this chapter, without entering in too many details on the nature of power, we adopt the blurred definition of empowerment being both an individual and collective process of obtaining control over own destiny, lives, resources and capabilities (Freire 1970, 2013; Campbell and Jovchelovitch 2000).

from non-participation, different levels of power delegation permit to reach higher levels of participation, in which citizens assume total control of the planning process.

Different steps of the ladder correspond to different "participatory arenas" and settings that span from information provision activities, websites, surveys, interviews and online interaction to more dialogic spaces such as focus groups, citizen juries and community planning initiatives (Lowndes et al. 2001).

Several authors have used the ladder metaphor to analyse different types of participation considering the flux of power from the planning authority to the hands of citizens (e.g. Bailey et al. 2012; Tippett et al. 2007; Lane 2005; Souza 2001; Bickerstaff et al. 2002). Whilst on the bottom five steps the planning institutions retain all their power, higher up the ladder citizens are progressively empowered, in a movement from top-down to bottom-up participation. As such, the ladder represents a useful tool to understand the different types of participation in place or as desired. A typical top-down organised meeting on health planning might fall in any of the steps between 3 and 6 and depends on the effects of the outcomes and the role given to citizens. The Ringland project, with its exceptional settings and outcomes, is higher up the ladder and could be classified as a partnership or delegated power depending on the final outcome of the project.

However, critiques have been posed to the ladder and its classification of participation. Specifically, authors have stressed how its linear approach to power misses the complexity and dynamicity of the interpersonal relations in the decision-making process, the shades between the potential and actual power held by different actors and the effects of the specific context on the participatory process (Lawrence 2006). Power is not a unique "commodity" that can be transferred or exchanged (Buchy and Race 2001; Collins and Ison 2009) but a property of different relations that depends on the type of actions performed (Gallagher 2008). The discourse of empowerment needs then to be framed in terms of challenging existing power structures and not in terms of "transferring" power (Kaufmann 1997; Buchy and Race 2001). In certain participatory arenas, citizens can have potential, but not actual power. This can be judged by looking at the outcomes of the planning process rather than the formal settings of the decision-making. The fact that, on the higher steps of the ladder, citizens are formally guaranteed "more power" doesn't necessarily ensure their impact on the decisions; similarly, citizens can have actual power on influencing decisions also on steps of the ladder where they are theoretically just passively consulted (Hilbrandt 2017).

For example, the citizens that initiated Ringland might have well met each other in an information meeting and have used this purely information-gathering exercise to connect to each other and develop more nuanced forms of engagement. At the same time a project like the Ringland project, if it does not result in any outcome in terms of its ability to actually implement the process or to change the local culture on health and planning, it might show a high level of disempowerment, despite being guided by citizens power.

To account for these elements when designing participation arenas, we need to look not only at the arrangements but also at the outcomes and effects of the participatory processes and consider the different power relations in play between the different actors (Gallagher 2008). This would reveal that real empowerment is reached only when a suited worldview is in place and adopted by all the actors involved. The effects of different worldviews on the settings participation are clear when looking at how participation is conceptualised and practised in different planning traditions. Traditions differ for their understanding of what is planning, what it is the role of the planner in society and how it should use its knowledge to propose actions to shape the city (Friedmann 1987). As such, as also stressed by Ferreira in this book (Chap. 17), traditions deeply shape planners' practices. Several other authors have provided insightful analysis of planning traditions, such as the capital work "*Planning in the Public Domain*" by Friedmann (1987). Different urban and transport planning traditions have evolved and cohabit at the same time, as I briefly explore in the following paragraphs.

The more classical planning tradition, normally named *rational or instrumental planning*, sees planning as a rational, value-free activity directed by the State authority. This tradition believes in an "unitary public interest" (Lane 2005, p. 290), where the planner has the technical skills to decide the good for society and decision-making is a matter of the policymaking arena. The planner is a professionally trained actor in charge of producing objective scientific knowledge to inform reforms (Sager 1992), maximising welfare and solving problems (Innes 1995). The decision-making process is centred on the evaluation of goals and objectives, and planning activity is guided often by the use of econometric and forecasting models such as cost-benefit analysis. As a broad literature has stressed, this is the planning tradition mostly adopted also in the transport planning realm (Willson 2001; Stangl 2008; Lindelöw et al. 2016; Verlinghieri 2016). Here, despite the coexistence of different worldviews (Willson 2001; Ferreira et al. 2009), decisions are predominantly taken following the principles of economic efficiency of transport as an activity that can be predicted and provided (Schiefelbusch 2010; Levine 2013; Hickman and Banister 2014). Accordingly with these ideas, in this tradition participation is "only required to validate and legitimise the goals of planning" (Lane 2005, p. 290) and normally performed as information gathering activities such as surveys or information provision (e.g. websites or information campaigns).

To find planning views that support participation at higher steps of the ladder beyond the non-participation or tokenism stages, we ought to refer to planning traditions that Friedmann (1987) names *planning as social learning* and *planning as social mobilization*. These traditions stress the political face of planning and its responsibility with regard to social justice and environmental issues (Healey 1992). Specifically, the first tradition conceives planning as a process of transformation and cyclical learning in which a variety of actors beyond professional planners participates. Planning evolves to adapting to the different contexts (Friedmann 1987). Participatory and collaborative processes become then integral part of the knowledge production and decision-making processes that form planning. The second one includes more radical approaches to planning, which fully challenge the idea of professional planning. They conceive planning as an activity aiming at transforming society that is performed in the "public domain" and "from below". Planning is not only an institutional process but a political act, in which science loses its central role as a unique force to drive change and becomes a tool for emancipation. Urban social movements and other grassroots actors are fully recognised as planning actors.

These two traditions, having questioned the traditional role of planners, consider participation not as a technique but as the main decision-making mode. Planning is based on a continuous exchange, in which choices are determined through communication and participation between different actors. The professional planner cannot purely assume the role of expert but needs to develop new skills and planning techniques (Hall 2014). For example, he can become a supporter of disadvantaged groups (Davidoff 1965) or the facilitator of a participatory debate around the future of the city in which citizens are the protagonist (Forester 1989; Healey 1992, 1997). In this conceptualization of participatory planning, however, a debate has emerged between authors that believe that consensual decisions can emerge from democratic participatory processes and others that see this position as unrealistic and not accounting for power dynamics, plurality, difference and complexity of society (Flyvbjerg 1998; Yiftachel 1998; Yiftachel and Huxley 2000; Albrechts 2003). As seen in the general debate about participation, also in the debate on participation in the planning literature, there is an emphasis on the importance of accounting for power dynamics and giving space to different voices, especially supporting the least strong ones.

It is evident that different planning traditions, proposing a different view on who is the planner and what she does, would support different forms of participation. Specifically, only the ones that see planning as a multi-actor and dynamic process seem to support worldviews that can accommodate participation at the higher steps of the ladders. In the next sections, I will consider which of these types of participation, and thus planning traditions, are the most appropriate when looking at urban and transport planning and health.

26.5 The Potential Health Benefits of Participation

Health planning and urban and transport planning for health could potentially benefit from all the aspects of participation previously highlighted (Rifkin 2014; George et al 2015; Farmer et al. 2017). Yet, participation in health planning and urban and transport planning for health are exposed to similar or even more complex challenges and very context-specific effects (Rifkin 2009, 2014; George et al. 2015; Pagatpatan and Ward 2017).

As stressed by Gallent and Ciaffi (2014), approaches to participation belonging to the lower steps of the Arnestein's ladder can allow citizens to have more nuanced information about existing services and better awareness of health risks and health prevention strategies. This can directly impact on lifestyles and behavioural changes. Participation is a powerful mean to increase citizens' awareness on the health impacts of motorised transport. Such awareness is increasingly needed to implement meaningful changes in everyday mobility choices (Khreis et al. 2016).

In addition to these benefits, looking at higher steps of the ladders, there are three main areas in which participation can contribute not only to improving health planning but also to grounding more comprehensive and holistic approaches and behaviours for health: (1) design tailored interventions and value alternative resources for health; (2) generate empowerment-related community wellbeing; (3) redefine authoritative knowledge on the base of co-production. I explore these three aspects in the following.

26.5.1 Design Tailored Interventions and Value Alternative and Complementary Resources for Health

Research has shown that consulting citizens before and whilst implementing health policies can increase their success in responding to local issues (Laverack and Labonte 2000; den Broeder et al. 2017; Farmer et al. 2017). At the same time, participatory arenas in which health risks and health prevention strategies are discussed can substantially facilitate the uptake of new behaviours, at the individual and collective level (Woodall et al. 2010; George et al. 2015; den Broeder et al. 2017). As shown by the initiatives in Antwerp, citizens can also have an important role in challenging policy directions taken and interventions, bringing to the agenda values and themes that they feel are relevant to their livelihoods. They can "supervise" the implementation of health-centred and environmentally friendly solutions as well as uncover business-as-usual practices that have little efficacy in tackling crises (Irwin 1995; Kickbusch et al. 2014). With these goals in mind, new tools for health can be built, such as participatory HIA (Nieuwenhuijsen et al. 2017).

Moreover, participatory practices initiated by citizens can help respond to the lack of services and provide complementary/alternative community-based health services. For example, Gallent and Ciaffi (2014) have reported the benefit of citizen-led mental health first aid and community support initiatives. Furthermore, as seen with the role of the Ringland project in demanding for referenda and setting up public discussions about the project, citizens and their organisations have a crucial role in guaranteeing the conditions for participation to be widely adopted in urban and transport planning for health. As Campbell and Jovchelovitch (2000, p. 256) highlight: "it is only through the participation and representation of grassroots communities in planning and implementing health programs that such programs are likely to have an impact". This is evident when looking at the role that social and grassroots movements have historically played in reshaping the discourse around health promotion. Kickbusch et al. (2014, p. 6) remark, "health activism [...] has been pivotal in bringing about changes in how societies govern health and disease. Citizens changed the ways they approached health and governance as individuals, civil society communities and organizations during the 20th century". Citizens have contributed to build an idea of participation that goes beyond individualised life-management exercises but embedded into processes of empowerment and advocacy (Laverack and Labonte 2000). It is very likely that their contribution needs to still be promoted for urban health. Citizens' initiatives are valuable community resources that can be nurtured and supported. However, they also risk to become ground for utilitarian uses of participation (Guareschi and Jovchelovitch 2004). For this reason they need to be carefully designed and monitored, and researchers, practitioners and policymakers can have an important role in guaranteeing and maintaining effective participatory arenas and processes (Verlinghieri 2016; Khreis et al. 2016).

Having acknowledged these benefits of participatory arenas, as shown in more general terms before, the literature highlights the importance of focussing not only on their outcomes but also on the procedural effects and abilities to mobilise different stakeholders, creating interaction and exchanges, allowing citizens to experience empowerment and deliberation, creating identities and gaining control over their capabilities (Laverack and Labonte 2000; Woodall et al. 2010; Rifkin 2014). In this context Conklin et al. (2015), reviewing more than 2000 studies on public involvement in health-care policy, have shown how there were recurring procedural effects such as increased citizen knowledge, nuanced decision-makers' and service providers' awareness and/or changed perceptions by the stakeholders involved, although the direct outcomes of involving citizens in health planning varied in geographical scale and magnitude and were difficult to determine (similar points are made by Dooris and Heritage (2013)). Similar results are reported by Crawford et al. (2002) who analysed 40 patients involvement studies and showed how this generally resulted in improved quality of health services and people's health. Participation has then the intrinsic value of establishing new relations and networks among stakeholders. These can reshape the priorities of the health agenda making it more suitable for specific needs and more attentive to the local resources available. At the same time, participation can spread healthy practices both in terms of prevention and treatment of a variety of diseases. These phenomena directly connect to the next point.

26.5.2 Generate Empowerment-Related Community Wellbeing

There is a strict connection between health and wellbeing, and research has shown the healing properties of perceived subjective wellbeing (Lewis et al. 2014; Steptoe et al. 2015). In broader terms, research has asserted that wellbeing is constituted by an overlap of *hedonic* or *eudaimonic* factors (Díaz et al. 2015), the former being the perceived happiness and the latter the active realisation of oneself potentials and aspirations (Schwanen and Ziegler 2011). As Lewis et al. (2014) reported, people that score low in eudaimonic wellbeing have higher risk of depression and poorer physical health. Eudaimonic wellbeing is composed by dimensions such as personal growth, autonomy, self-acceptance, purpose in life, etc. (Schwanen and Ziegler

2011). The dimension of autonomy strictly connects to the ability to be actively involved and participate in decision-making processes.

As such, participation has been included in the list of human needs fundamental to ensure overall wellbeing (Doyal and Gough 1991). There is sufficient evidence on the detrimental mental and physical health effects of not being able to impact decision-making regarding people's own environments and communities (Albrecht 2005) and lack of control over own destinies (Campbell and Jovchelovitch 2000). These points are specifically important when considering the health problems for marginalised groups, which are also the groups with less input in decision-making, poorest health and lacking access to health facilities, health-related knowledge and support (Baum et al. 2000). Research has shown the intimate connection between empowerment and positive health effects at individual and community levels (Campbell and Jovchelovitch 2000).

In line with the literature on eudaimonic wellbeing, participation provides also positive outcomes at the individual level that are connected to increased self-esteem, self-efficacy, sense of coherence, social identities and consciousness (Campbell and Jovchelovitch 2000; Ryan and Deci 2001; Guareschi and Jovchelovitch 2004; Attree et al. 2011). This applies also at a broader level, when individual and community empowerment and participation are able to instigate transformative actions with positive health outcomes and reduced health inequalities (Abel and Frohlich 2012).

Specifically, the presence of social networks and the availability of social capital, as well as the involvement in community activities and the existence of institutional spaces that allow interaction, are important factors in promoting health and wellbeing, especially in the case of isolated communities (Campbell and Jovchelovitch 2000; Gallent and Ciaffi 2014). Here, the possibility to participate and interact within the community and with other external actors "can influence health decision-making and create opportunities for access to health information, programmes and services to achieve good health and wellbeing" (Gallent and Ciaffi 2014, p. 81). Similarly, there are evidences on how health programmes designed with the collaboration of specific groups help improving social support, self-esteem and perceived power (Laverack and Labonte 2000; Attree et al. 2011). Overall, health-related decisions impact citizens lifestyles as well as the ecosystems in which they live; involving citizens in those decisions is involving them on deciding about their own lives (Tickner 2001).

26.5.3 Redefine Authoritative Knowledge on the Base of Co-production

Participatory arenas that go beyond consultation can allow knowledge co-production. This is specifically important in the context of health and urban and transport planning for health (Khreis et al. 2016; Nieuwenhuijsen et al. 2017). Health is a highly hard-science-driven topic regarding which there are growing debates on morality, reliability and validity[2] (De Vries and Lemmens 2006; Light et al. 2013). Thanks to co-production, health-related facts, assumptions and interpretations can be re-discussed, possibly leading to the construction of an authoritative common knowledge ground on the base of which deliberation might be easier (Heiman 1997; Edelenbos et al. 2011; Boswell et al. 2015). Co-production with different actors can also overcome the limits of interdisciplinarity highlighted by Ferreira in this book (Chap. 17). Participation can allow a debate on the value and importance of the established hierarchy of knowledges in places that prioritises expert knowledge over other forms of knowledge (Irwin 1995; Campbell and Jovchelovitch 2000). Research has indeed shown how local knowledges, appropriately balanced and embedded in a critical dialogue with science, are fundamental sources for effective health interventions, especially in more deprived communities (Campbell and Jovchelovitch 2000; Guareschi and Jovchelovitch 2004; den Broeder et al. 2017). The collaboration with other type of knowledges in shaping health interventions can also improve the overall approach to health. As, for example, shown by research with indigenous communities in various urban settings, traditional approaches to health brought by participants can complement traditional medicine and pursue a more holistic outlook to human wellbeing in which environmental, psychological and physical components are accounted for together (Auger et al. 2016). With this setting, participation can help reduce health inequalities and propose solutions to debates on the validity of health interventions (Dally and Barr 2008; Woodall et al. 2010).

In light of this, it emerges that health could truly benefit from participation when it is conceived not only as a consultation exercise but as a lasting-in-time practice that involves a variety of actors and stakeholders, towards the higher steps of the Arnstein's ladder (Rifkin 2014). At those levels one can positively attempt to tackle existing inequalities and power unbalances and their effects on urban health. Moreover, as I demonstrated with my research, in order for cities to fully benefit from the potentials of participation for health, participatory arenas need to be knowledge-based and ethically grounded (Verlinghieri 2016). They need to allow for a continuous co-production of knowledges taking into account issues of inclusivity, representativeness, fairness and impact on decision-making. In line with what was considered in the previous section, planning modes in between planning as societal learning and planning as social mobilisation with participatory settings towards the higher steps of the ladder might effectively be the preferred ground for planning healthy cities.

However, as various studies have demonstrated, despite the 1986 WHO Health Promotion Charter inviting for participatory approaches to health (World Health Organization 1986), current practices are normally only performed at the information stage of the Arnestein's ladder, in the forms of exhibitions, municipal newsletters, internet information, campaigns, conferences and similar (Dooris and Heritage 2013; Conklin et al. 2015). As Ocloo and Matthews (2016) found in a more recent

[2] See, for example, debates on vaccinations or alternative medicine (Clark 2000; Loe Fisher 2017).

review, participatory settings for health can be often characterized by tokenism and exclusivity. Similar observations can be made with regard to health planning practices, especially within transport planning (Verlinghieri 2016). Similarly, there is often a lack in political will to accept the value of participation and gather resources for it to function (Nieuwenhuijsen et al. 2017; Pagatpatan and Ward 2017). Despite new studies and trends in which participation is acknowledged as fundamental also in the assessment of health impacts – as, for example, in the development of participatory HIA (Nieuwenhuijsen et al. 2017) – new knowledge production and exchange practices as well as decision-making settings and tools still need developing to accommodate more nuanced forms of co-production (Khreis et al. 2016; Nieuwenhuijsen et al. 2017).

In the next sections I give a few insights on some principles on which more participatory approaches to urban and transport planning for health could be based on, looking at the importance of building a culture of participation across society and at the potentials of Participatory Action Research (PAR) as an approach that could guide academics, practitioners, policymakers and planners to facilitate its emergence.

26.6 Building a Culture of Participation

Differently from what the current practice of participation for health seems to be based on, I showed the importance of practising participation at the higher levels of the Arnstein's ladder to effectively be included in pathways towards healthy cities. Participation should not be limited to one-off exercises, but continuously practised on a number of fronts (World Health Organization 2003; Rifkin 2014; Verlinghieri 2016). This demands a considerable number of resources in place as well as appropriate institutional, political and societal conditions. I argue that, in order to allow those to be in place, it is useful to think about creating a *culture of participation*.

A culture of participation can be defined as "an attitude of openness towards and a capability of supporting the existence of participatory decision-making processes at the 'citizen power' levels of the ladder" (Verlinghieri 2016, p. 218). As I found out with my PhD research, this culture is a fundamental requirement for participation to have true impacts and meanings. When a culture of participation is in place, participatory arenas can be embedded into democratic decision-making processes as participation becomes a usual practice of decision-making. It is the increasingly high culture of participation in place in Antwerp that allows the Ringland road project to start and achieve its results. A culture of participation can support establishing new governance models in which decisions are made on the base of a continuous dialogue among different viewpoints and knowledges, having made explicit the underlying value assumptions and grounding the search for solutions into accountable and transparent deliberation (Stringer et al. 2006; Kickbusch et al. 2014).

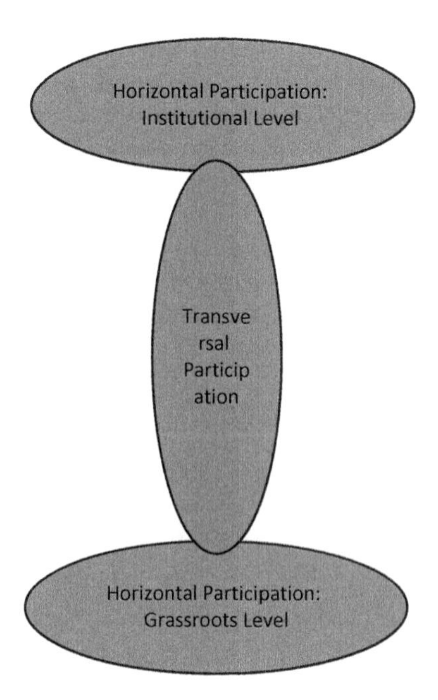

Fig. 26.2 A model of horizontal and transversal participation (Verlinghieri 2016, p. 204)

Adopting the extended conception of participation and planning proposed in this chapter can allow to identify a variety of actors that can contribute to build this culture at the urban level, both at an institutional and grassroots level (Verlinghieri 2016). Specifically, the diagram proposed in Fig. 26.2 is useful to understand the different types of actions required and the way in which participation at the higher steps of the ladder could work. In the diagram, I summarised a model of participation originally developed by a transport grassroots/advocacy group I worked with during my PhD, the Move Your City group from L'Aquila, Italy (Castellani 2014; Verlinghieri 2016). The model highlights how independent functioning both at the institutional and grassroots level is essential to allow a connection and co-production between the two levels. When a culture of participation is in place, participation can mobilise resources at different levels of society, working as horizontal participation, at the grassroots and at the institutional level, as well as a process of transversal participation in between institutions and citizens. This means that both institutions and grassroots actors need to have the ability to work participatorily within themselves and in collaboration among each other.

For example, in the Ringland project, grassroots actors built a highly functioning culture of participation at the grassroots level. The Ringland project is developed first of all as a collaboration and convergence among different grassroots initiatives. At the same time, it has an open and continuous dialogue with different institutional actors, from the local government to universities.

As also reported in a report by the World Health Organization (2003), in the specific context of health, different activities can be used to support and develop this culture, such as institutional exercises, community-level activities and support to community organisation. To do so, crucial processes are community-level work, networking and organisational development. Similarly, it is important to nurture a political committment to participation (Pagatpatan and Ward 2017). Initiating these processes, academics, practitioners, policymakers and planners can support the construction of a culture of participation at both the institutional and grassroots level, enabling coordination and co-production (Khreis et al. 2016). Moreover, they can make sure that adequate settings are in place for citizens to be fully engaged and recognised as legitimate actors in shaping health policies. Only then they can propose alternative solutions from the bottom up and independently implement health-centred behaviours and practices. To do so, academics need to critically approach the epistemological discussion around co-production, as, for example, proposed by Khreis et al. (2016).

26.7 A Research Approach to Build a Culture of Participation: Participatory Action Research for Health

If as academics, practitioners, policymakers and planners we aim to support the development of a culture of participation, we need to challenge current knowledge production, planning and research practices and propose new frameworks (Khreis et al. 2016). We need to support the emergence of a culture of participation at each level of the model proposed. In order to do so, especially when approaching the grassroots level, I believe that useful insights can come from the literature on participatory research approaches such as PAR, emancipatory research and feminist standpoint theory.

These approaches assume that co-produced knowledge has validity comparable or superior to traditional scientific knowledge and that knowledge can be co-produced with citizens and other actors in society, towards emancipation and well-being. Participatory research approaches aim and allow exploring issues and phenomena from the "standpoint of the impacted".[3] "Impacted" people are considered to have, as theorised in feminist literature (e.g. Harding 1993), an epistemically privileged position to understand the issues[4] and how to solve them. Including their views allows accounting for specific social, cultural and historical factors that shape the impacts. With a careful methodological setting, participatory research approaches

[3] In the feminist literature, a standpoint is considered to be an "achieved collective identity or consciousness" (Bowell n.d.). For extension the term can be used to include groups that, more or less in forms organised before the impact assessment exercises, are mobilising as "impacted" or "potentially impacted".

[4] According with feminist standpoint theories, the point of view of the marginalised/oppressed/impacted gives both a deeper account of their problems and also is a privileged position to look at everybody's necessities and broad societal and political processes that might be otherwise neglected utilising a traditional conceptual framework (Bowell n.d.; Harding 1993).

can ensure the production of objective and valid knowledge and explore solutions otherwise not available (Bowell n.d.).

Among the participatory research approaches, PAR specifically focuses on praxis and concrete solutions to real-life problems through collaboration and co-production, aiming to produce knowledge with a moral humanistic goal towards social justice (Fals-Borda and Rahman 1991; Morgan 2012) and amelioration of living conditions, with a clear reference to environment protection and social wellbeing (Smith et al. 1997; Reason and Bradbury 2001).

Three principles are the basis of PAR: (1) belief on popular knowledge and on the possibility of a community-based approach to crises, (2) shared ownership of knowledge and resources and (3) authentic commitment of participant and researcher and focus on community action (Smith and al. 1997; Kemmis and McTaggart 2005). All of this is performed under a careful consideration of power issues among participants and participants and researchers (Kemmis and McTaggart 2005).

The PAR approach uses participation to actively involve all the people directly and indirectly affected by an issue, allowing them to perform actions to challenge it (Kemmis and McTaggart 2005). The research participants (as patients, communities, citizens or health and planning practitioners) become co-researchers (Gallacher and Gallagher 2008). Different from traditional methods that use consultation to research *on* a certain group or community, PAR is an approach that researches *with*, *by* or *for* them (Nind 2014). To the relation of subject/object, action research substitutes a new relation of subject/subject of research (Fals-Borda and Rahman 1991), surpassing the hierarchy between researchers and researched and making research a "cooperative experiential enquiry" (Kiernan 1999, p. 44). PAR is based on deep epistemological reflection both on how knowledge is produced (participatorily) and on what the purpose of the knowledge produced is (help the people included).

Under a PAR approach, research and knowledge production are structured as a process of planning, acting and observing and reflecting, as visible in Fig. 26.3. This structure is dynamic and based on positive feedback loops: the PAR cycle is repeated several times until the outcome is commonly agreed as the best. In this way research has precise practical outcomes that are designed and agreed between the participants, building a sense of legitimacy and involvement (Kindon et al. 2007). Research questions and contexts are designed and agreed by the research participants, allowing for producing knowledge on issues highly relevant to them. The flexibility of the approach also implies the possibility of adopting a wide range of strategies depending on the issues considered and the specific contextual needs. As McIntyre (2008, p. 6) stresses: "participant-generated actions can range from changing public policy, to making recommendations to government agencies, to making informal changes in the community that benefit the people living there, to organizing a local event, to simply increasing awareness about an issue native to a particular locale". In this way, findings generated via PAR are more likely to have higher dissemination beyond the research communities and therefore affect behaviours (Bach et al. 2017). At the same time, grounded on the "experiences and values of those concerned" (Nind 2014, p. 24), PAR poses "greater emphasis on process and on seeing people as agents of change" (Nind 2014, p. 9), aiming at empowerment, not seeing power

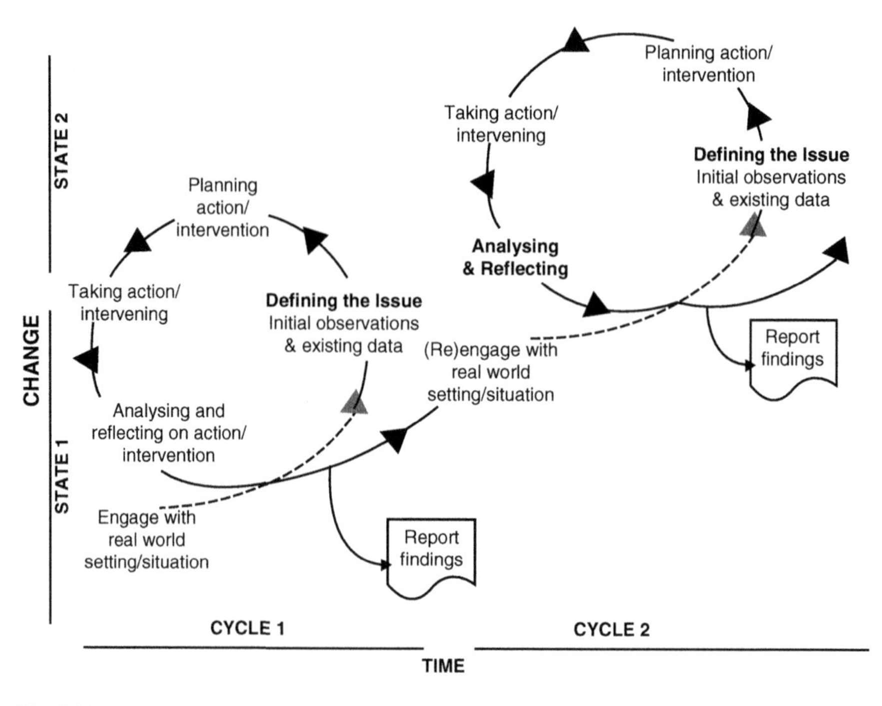

Fig. 26.3 The PAR cycle (Velasco 2013)

as incremental but built via interactions (Baum et al. 2006). It gives to communities control over their environments, resources and health. PAR gives ownership of knowledge produced to the impacted, directly enhancing those processes of building self-esteem, self-efficacy and sense of coherence, social identities and consciousness that form part of wellbeing. At the same time, in bringing together different actors to challenge issues and research possible solutions, it helps reinforcing those social networks and social capital necessary for wellbeing (Campbell and Jovchelovitch 2000; Ryan and Deci 2001; Guareschi and Jovchelovitch 2004).

PAR has been used as research approach in health practices and research since the 1990s especially in the development context and in health research with marginalised or indigenous communities. It has been proved to be a research approach able to challenge health inequalities (Cacari-Stone et al. 2014) and has gained increasing support in nursing, medicine, psychology and health research (Baum et al. 2006; Minkler and Wallerstein 2011; Munn-Giddings and Winter 2013; Wright et al. 2013a, b; Baum 2016), also with the creation of the International Collaboration for Participatory Health research (ICPHR) (Wright et al. 2013a, b). Among these, Baum et al. (2006, p. 855) have considered PAR as a "mechanism through which to put the rhetoric of participation [as suggested by the WHO] into action" and Ekirapa-Kiracho et al. (2016) as a way to increase capacity of community groups to improve their health conditions. As the literature has shown, grounded in the idea of research as a critical and reflective practice, PAR improves the professional work of health and planning practitioners (Munn-Giddings and Winter 2013; Frahsa et al. 2014) and allows a process of building awareness of the impacts of political, social and economic conditions on their lives and health (Guareschi and Jovchelovitch 2004). Specifically, applied to epidemiology, PAR has been shown to overcome the limits of the discipline classical methodological individualism and bridging between individual-level and social- and environmental-level risk factors (Bach et al. 2017).

Involving communities and practitioners, PAR generates a learning process that builds a culture of participation at all level required. It can result in increased capacities, in policy and governance structural changes and in shifting roles of professionals and decision-makers (Frahsa et al. 2014; Verlinghieri 2016). Specifically using a PAR approach, academics, practitioners, policymakers and planners can facilitate the development of processes that, being highly participatory and towards the higher steps of the Arnestein's ladder, can fully support the emergence and establishment of participatory arenas and processes in urban and transport planning for health.

Researchers adopting a PAR approach can both support the development of citizen-led initiatives, such as the Ringland project, or set up spaces in which participatory practices can emerge (Verlinghieri 2016). I have myself, using a PAR approach, supported public transport advocacy groups, working with them towards the research for more sustainable transport solutions and at the same time putting in practice innovative decision-making processes that would make participation a mean to better public health.

26.8 Conclusion

Shaping a comprehensive and unilateral definition of what a fully "healthy city" would entail is extremely challenging. However, it is clear that its establishment is linked to transport and urban planning decisions and grounded into the search for universal rights, social and environmental justice and equity, as well as better ecological balances and access to public goods. This will be a step towards guaranteeing that the groups in society that most suffer of the effect of the current crisi are protected and able to reach dignity and wellbeing (Kickbusch and Gleicher 2013) whilst putting also in place actions towards the construction of a healthy city in which human health can be nurtured (Albrecht 2005).

Learning from the Ringland project, in this chapter I highlighted the importance of new decision-making processes in improving the health outcomes of urban and transport planning. I have followed the recommendation that authors from a variety of disciplines have made for more participatory settings and multilateral discussions to tackle increasingly complex and dynamic challenges in the urban environments that are progressively overcoming national borders and becoming global issues (Kickbusch et al. 2014).

Increasingly participatory settings and more open forms of governance are today an emerging response to the changing landscape of knowledge production. With increased access to knowledge sources for citizens and the use of new media, new meanings are produced around questions of epistemology, validities, objectivity and scientific value. At the same time, with increasing awareness of the complexity of the current social and environmental crises and of their extensive impacts, emerges an urge for involving multiple knowledges and actors in describing the issues and seeking solutions. As the World Health Organization (WHO) recognises, we are moving towards a "knowledge society" in which new actors, not traditionally charged with the authority to do so, have a growing role in shaping the governance of health (Kickbusch and Gleicher 2013; Kickbusch et al. 2014). As emerged clearly in the Ringland project, citizens themselves are increasingly challenging the authority of expert knowledge and demanding for shared evidences of the benefits of projects (Petts and Brooks 2006). At the same time there is an increasing awareness on how the behaviours of the whole society are accountable for health outcomes: individual mobility choices as well as consumer behaviours highly impact on pollution and global health.

However, whilst we witness a growing number of open channels for participation in knowledge production as well as a constant decrease in public faith in information coming from "single-trustfully-sources", decisions are still made on the basis of traditional evaluation and assessments methods that might leave little space for multiple points of view and voices to be effectively heard or to collaborate (Irwin 1995; Nowotny et al. 2001). There are contrasting calls for opening knowledge to more voices and ensure plurality and at the same time maintaining a grounded hierarchy in the decision-making processes and in the validation of knowledge (Petts and Brooks 2006). New knowledge production and exchange practices as well as decision-making settings and tools still need developing to accommodate more nuanced forms of co-production that also account for the ecological and social impacts of the choices made.

This applies also to the realm of health in which new governance structures, grounded into more collaborative and adaptive structures in which policy makers, experts and citizens are in constant dialogue, are being called for, but only limitedly established and with mixed outcomes (Kickbusch and Gleicher 2013; Kickbusch et al. 2014). Already in 1986 the WHO Health Promotion Charter (World Health Organization 1986, p. 3) recommended participation and empowerment as crucial tools for healthier communities and cities:

> "Health promotion works through concrete and effective community action in setting priorities, making decisions, planning strategies and implementing them to achieve better health. At the heart of this process is the empowerment of communities – their ownership and control of their own endeavours and destinies".

More recently participation has been a core principle of the WHO European Healthy Cities initiative which required cities to increase capacity for participation and empowerment (Dooris and Heritage 2013; de Leeuw et al. 2014), building "support for community-level action and capacity-building; strengthening of infrastructures and networks; and meaningful organisational development and change" (Dooris and Heritage 2013, p. S76).

Building health requires a socio-ecological approach that considers together people and their environment, and in which all parties, from academics, practitioners, policymakers, planners and citizens interact in the attempt to redesign lifestyles and societal patterns (World Health Organization 1986). This would entail a holistic approach to health that considers human beings as needing a complementary interaction of environmental, psychological, social and physical factors to ensure their wellbeing.

In this piece, I proposed some reflections on how participatory planning practices and participatory research – when carefully set up; monitoring power unbalances and ensuring that it is a continuous process of culture building – could contribute to moving towards this more holistic approach. I showed the potential for participation to support the design of tailored interventions and value alternative resources for health; generate empowerment-related community-wellbeing; and redefine authoritative knowledge on the base of co-production.

Participation, as a process of negotiating new worldviews, can create new approaches to knowledge and constitute a space in which health-related knowledges are brought in dialogue and advanced. It can expand the health research's perspectives, including popular and lay interpretations and views, responding to the increasing calls for democratisation of knowledge, in a process of "contextualization of science" in which scientific problems ought to be recognised intertwined and complementary to social problems (Mowat 2011).

At the same time, participation can open new spaces and build networks of social interaction through which planners, health practitioners, decision-makers, academics and citizens can openly commit to sustainability, justice and wellbeing. Bridging health and participation goes hand in hand with a social justice approach that can ensure that the needs and wills of marginalised groups that are often the ones mostly suffering of poor health can be included and considered in planning.

In order to fully take advantage of these potentials of participations, I showed the need for fundamental cultural changes and the importance of new research methods and processes. In this I acknowledged how, as already stressed by Kickbusch et al. (2014), planners, health practitioners, decision-makers, academics and citizens need to engage together in two fundamental processes. We need to both design new decision-making structures for a "shared governance for health", and work towards implementing a "shared health and care" in which all parties contribute, with different roles, to caring for citizens, their health and environments.

References

Abel, T., & Frohlich, K. L. (2012). Capitals and capabilities: Linking structure and agency to reduce health inequalities. *Social Science & Medicine, 74*(2), 236–244. https://doi.org/10.1016/j.socscimed.2011.10.028.

Albrecht, G. (2005). "Solastalgia": A new concept in health and identity. *PAN: Philosophy, Activism, Nature, 3*, 41–55.
Albrechts, L. (2003). Planning and power: Towards an emancipatory planning approach. *Environment and Planning C: Government and Policy, 21*(6), 905–924. https://doi.org/10.1068/c29m.
Arnstein, S. R. (1969). A ladder of citizen participation. *Journal of the American Institute of Planners, 35*(4), 216–224. https://doi.org/10.1080/01944366908977225.
Attree, P., French, B., Milton, B., Povall, S., Whitehead, M., & Popay, J. (2011). The experience of community engagement for individuals: a rapid review of evidence. *Health & Social Care in the Community, 19*(3), 250–260.
Auger, M., Howell, T. and Gomes, T. (2016) 'Moving toward holistic wellness, empowerment and self-determination for indigenous peoples in Canada: Can traditional indigenous health care practices increase ownership over health and health care decisions?', Canadian Journal of Public Health 107(4–5) e393–e398. Retrieved May 12, 2017, from https://www.ncbi.nlm.nih.gov/labs/articles/28026704/.
Bach, M., et al. (2017). Participatory epidemiology: The contribution of participatory research to epidemiology. *Emerging Themes in Epidemiology, 14*(1), 2. https://doi.org/10.1186/s12982-017-0056-4.
Bailey, K., & Grossardt, T. (2010). Toward structured public involvement: Justice, geography and collaborative geospatial/geovisual decision support systems. *Annals of the Association of American Geographers, 100*(1), 57–86. https://doi.org/10.1080/00045600903364259.
Bailey, K., Grossardt, T., & Ripy, J. (2012). Toward environmental justice in transportation decision making with structured public involvement. *Transportation Research Record: Journal of the Transportation Research Board, 2320*, 102–110
Banister, D. (2008). The sustainable mobility paradigm. *Transport Policy, 15*(2), 73–80. https://doi.org/10.1016/j.tranpol.2007.10.005.
Baum, F. E., et al. (2000). Epidemiology of participation: An Australian community study. *Journal of Epidemiology & Community Health, 54*(6), 414–423. https://doi.org/10.1136/jech.54.6.414.
Baum, F. E. (2016). Power and glory: Applying participatory action research in public health. *Gaceta Sanitaria, 30*(6), 405–407. https://doi.org/10.1016/j.gaceta.2016.05.014.
Baum, F., MacDougall, C., & Smith, D. (2006). Glossary: Participatory action research. *Journal of Epidemiology and Community Health (1979), 60*(10), 854–857.
Becker, D. R., et al. (2004). A comparison of a technical and a participatory application of social impact assessment. *Impact Assessment and Project Appraisal, 22*(3), 177–189. https://doi.org/10.3152/147154604781765932.
Bickerstaff, K., Tolley, R., & Walker, G. (2002). Transport planning and participation: The rhetoric and realities of public involvement. *Journal of Transport Geography, 10*(1), 61–73.
Boswell, J., Settle, C., & Dugdale, A. (2015). Who speaks, and in what voice? The challenge of engaging 'The Public' in health policy decision-making. *Public Management Review., 17*(9), 1358–1374.
Bowell, T. (n.d.) Feminist standpoint theory, Internet Encyclopedia of Philosophy Retrieved May 9, 2017, form http://www.iep.utm.edu/fem-stan/.
Brusselen, D. V., et al. (2016). Health impact assessment of a predicted air quality change by moving traffic from an urban ring road into a tunnel. The case of Antwerp, Belgium. *PLoS One, 11*(5), e0154052. https://doi.org/10.1371/journal.pone.0154052.
Buchy, M., & Race, D. (2001). The twists and turns of community participation in natural resource management in Australia: What is missing? *Journal of Environmental Planning and Management, 44*(3), 293–308. https://doi.org/10.1080/09640560120046070.
Cacari-Stone, L., et al. (2014). The promise of community-based participatory research for health equity: A conceptual model for bridging evidence with policy. *American Journal of Public Health, 104*(9), 1615–1623.
Campbell, C., & Jovchelovitch, S. (2000). Health, community and development: Towards a social psychology of participation. *Journal of Community & Applied Social Psychology, 10*, 255–270.

Castellani, S. (2014). Participation as a possible strategy of post-disaster resilience: Young people and mobility in L'Aquila (Italy). In L. M. Calandra, G. Forino, & A. Porru (Eds.), *Multiple geographical perspectives on hazards and disasters* (pp. 105–117). Rome: Valmar.
Clark, P. A. (2000). The Ethics of alternative medicine therapies. *Journal of Public Health Policy, 21*(4), 447.
Collins, K., & Ison, R. (2009). Jumping off Arnstein's ladder: Social learning as a new policy paradigm for climate change adaptation. *Environmental Policy and Governance, 19*(6), 358–373. https://doi.org/10.1002/eet.523.
Conklin, A., Morris, Z., & Nolte, E. (2015). What is the evidence base for public involvement in health-care policy?: Results of a systematic scoping review. *Health Expectations, 18*(2), 153–165. https://doi.org/10.1111/hex.12038.
Cooke, B., & Kothari, U. (2001). *Participation: The new tyranny?* Zed Books.
Crawford, M. J., et al. (2002). Systematic review of involving patients in the planning and development of health care. *BMJ, 325*(7375), 1263.
Dally, J., & Barr, A. (2008). *Understanding a community-led approach to health improvement.* Glasgow: Scottish Community Development Centre. Retrieved from http://www.scdc.org.uk/media/resources/what-we-do/mtsc/Understanding%20a%20community-led%20approach%20to%20health%20improvement.pdf.
Davidoff, P. (1965). Advocacy and pluralism in planning. *Journal of the American Institute of Planners, 31*(4), 331–338. https://doi.org/10.1080/01944366508978187.
De Oosterweelverbinding. (2015). *The Oosterweel link, oosterweelverbinding.be*. Retrieved Sept 3, 2017, from https://www.oosterweelverbinding.be/oosterweel-link.
De Vries, R., & Lemmens, T. (2006). The social and cultural shaping of medical evidence: Case studies from pharmaceutical research and obstetric science. *Social Science & Medicine. (Part Special Issue: Gift Horse or Trojan Horse? Social Science Perspectives on Evidence-based Health Care), 62*(11), 2694–2706. https://doi.org/10.1016/j.socscimed.2005.11.026.
den Broeder, L., Uiters, E., ten Have, W., Wagemakers, A., & Schuit, A. J. (2017). Community participation in Health Impact Assessment. A scoping review of the literature. *Environmental Impact Assessment Review, 66*, 33–42.
Díaz, D., et al. (2015). The eudaimonic component of satisfaction with life and psychological well-being in Spanish cultures. *Psicothema, 27*(3), 247–253.
Dooris, M., & Heritage, Z. (2013). Healthy cities: Facilitating the active participation and empowerment of local people. *Journal of Urban Health, 90*(S1), 74–91. https://doi.org/10.1007/s11524-011-9623-0.
Doyal, L., & Gough, I. (1991). *A theory of human need.* New York: Palgrave Macmillan. Retrieved May 24, 2017, from http://www.palgrave.com/home/.
Edelenbos, J., van Buuren, A., & van Schie, N. (2011). Co-producing knowledge: Joint knowledge production between experts, bureaucrats and stakeholders in Dutch water management projects. *Environmental Science & Policy, 14*(6), 675–684. https://doi.org/10.1016/j.envsci.2011.04.004.
Ekirapa-Kiracho, E., et al. (2016). Unlocking community capabilities for improving maternal and newborn health: Participatory action research to improve birth preparedness, health facility access, and newborn care in rural Uganda. *BMC Health Services Research, 16*(S7), 638. https://doi.org/10.1186/s12913-016-1864-x.
Elvy, J. (2014). Public participation in transport planning amongst the socially excluded: An analysis of 3rd generation local transport plans. *Case Studies on Transport Policy. (Social Exclusion in the Countries with Advanced Transport Systems), 2*(2), 41–49. https://doi.org/10.1016/j.cstp.2014.06.004.
Fals-Borda, O., & Rahman, M. A. (1991). *Action and knowledge: Breaking the monopoly with participatory action research.* New York: Apex Press.
Farmer, J., Taylor, J., Stewart, E., & Kenny, A. (2017). Citizen participation in health services co-production: A roadmap for navigating participation types and outcomes. *Australian Journal of Primary Health, 23*(6), 509.

Ferreira, A., Sykes, O., & Batey, P. (2009). Planning theory or planning theories? The hydra model and its implications for planning education. *Journal for Education in the Built Environment, 4*(2), 29–54. https://doi.org/10.11120/jebe.2009.04020029.

Flandersnews.be. (2016). *Concerns about air quality in Antwerp, flandersnews.be*. Retrieved Sept 3, 2017, from http://deredactie.be/cm/vrtnieuws.english/Antwerp/1.2696243.

Flanderstoday. (2017). Largest-ever European air quality tests carried out in Antwerp. Retrieved Sept 3, 2017, www.flanderstoday/innovation/largest-ever-european-air-quality-tests-carried-out-antwerp.

Flyvbjerg, B. (1998, 210). Habermas and Foucault: Thinkers for civil society? *The British Journal of Sociology, 49*(2). https://doi.org/10.2307/591310.

Forester, J. (1989). *Planning in the face of power*. University of California Press.

Frahsa, A., et al. (2014). Enabling the powerful? Participatory action research with local policy-makers and professionals for physical activity promotion with women in difficult life situations. *Health Promotion International, 29*(1), 171–184. https://doi.org/10.1093/heapro/das050.

Freire, P. (1970). *Pedagogy of the oppressed*. New York: Herder and Herder.

Freire, P. (2013). *Education for critical consciousness*. London: Bloomsbury Publishing.

Friedmann, J. (1987). *Planning in the public domain: From knowledge to action*. Princeton, NJ: Princeton University Press.

Gallagher, M. (2008). Foucault, power and participation. *The International Journal of Children's Rights, 16*(3), 395–406. https://doi.org/10.1163/157181808X311222.

Gallacher, L. A., & Gallagher, M. (2008). Methodological immaturity in childhood research? *Thinking Through 'Participatory Methods Childhood, 15*(4), 499–516.

Gallent, N., & Ciaffi, D. (Eds.). (2014). *Community action and planning: Contexts, drivers and outcomes*. Bristol: Policy Press. https://doi.org/10.1332/policypress/9781447315162.001.0001.

George, A. S., Mehra, V., Scott, K., Sriram, V., & Li, X. (2015). Community participation in Health Systems Research: A systematic review assessing the state of research, the nature of interventions involved and the features of engagement with communities. *PLOS ONE, 10*(10), e0141091.

Guareschi, P. A., & Jovchelovitch, S. (2004). Participation, health and the development of community resources in southern Brazil. *Journal of Health Psychology, 9*(2), 311–322. https://doi.org/10.1177/1359105304040896.

Hall, P. (2014). *Cities of tomorrow: An intellectual history of urban planning and design since 1880*. Hoboken: John Wiley & Sons.

Harding, S. (1993). Rethinking standpoint epistemology: What is strong objectivity. In L. Alcoff & E. Potter (Eds.), *Feminist epistemologies* (pp. 49–82). New York: Routledge.

Healey, P. (1992). Planning through debate: The communicative turn in planning theory. *The Town Planning Review, 63*(2), 143–162.

Healey, P. (1997). *Collaborative planning: Shaping places in fragmented societies*. Vancouver: UBC Press.

Heiman, M. K. (1997). Science by the people: Grassroots environmental monitoring and the debate over scientific expertise. *Journal of Planning Education and Research, 16*(4), 291–299. https://doi.org/10.1177/0739456X9701600405.

Hickman, R., & Banister, D. (2014). *Transport, climate change and the City*. London: Routledge.

Hilbrandt, H. (2017). Insurgent participation: Consensus and contestation in planning the redevelopment of berlin-Tempelhof airport. *Urban Geography, 38*(4), 537–556. https://doi.org/10.1080/02723638.2016.1168569.

Innes, J. E. (1995). Planning theory's emerging paradigm: Communicative action and interactive practice. *Journal of Planning Education and Research, 14*(3), 183–189.

Innes, J. E., & Booher, D. E. (2004). Reframing public participation: Strategies for the 21st century. *Planning Theory & Practice, 5*(4), 419–436. https://doi.org/10.1080/1464935042000293170.

Irwin, A. (1995). *Citizen science: A study of people, expertise and sustainable development*. London: Psychology Press.

Kaufmann, M. (1997). Differential participation: Men, women and popular power. In M. Kaufmann & H. D. Alfonso (Eds.), *Community power and grassroots democracy: The transformation of social life* (pp. 151–171). London: Zed Books. Retrieved Sept 4, 2017, from http://www.michaelkaufman.com/wp-content/uploads/2009/01/differentialparticipation.pdf.

Kemmis, S., & McTaggart, R. (2005). Participatory action research : Communicative action and the public sphere. In N. K. Denzin & Y. S. Lincoln (Eds.), *The Sage handbook of qualitative research* (pp. 559–603). Thousand Oaks: Sage.

van de Kerkhof, M. (2006). Making a difference: On the constraints of consensus building and the relevance of deliberation in stakeholder dialogues. *Policy Sciences, 39*(3), 279–299. https://doi.org/10.1007/s11077-006-9024-5.

Kesby, M. (2005). Retheorizing empowerment-through-participation as a performance in space: Beyond tyranny to transformation. *Signs: Journal of Women in Culture and Society, 30*(4), 2037–2065. https://doi.org/10.1086/428422.

Khreis, H., et al. (2016). The health impacts of traffic-related exposures in urban areas: Understanding real effects, underlying driving forces and co-producing future directions. *Journal of Transport & Health, 3*(3), 249–267. https://doi.org/10.1016/j.jth.2016.07.002.

Kickbusch, I., et al. (2014) *Smart governance for health and well-being: The evidence*. Retrieved May 18, 2017, from http://apps.who.int/iris/bitstream/10665/131952/1/Smart%20governance%20for%20health%20and%20well-being%20the%20evidence.pdf.

Kickbusch, I., & Gleicher, D. (2013). *Governance for health in the 21st century*. Copenhagen: World Health Organization, Regional Office for Europe.

Kiernan, C. (1999). Participation in research by people with learning disability : Origins and issues. *British Journal of Learning Disabilities., 27*(2), 43–47.

Kindon, S., Pain, R., & Kesby, M. (Eds.). (2007). *Participatory action research approaches and methods: Connecting people, participation and place*. New York/London: Routledge.

Lane, M. B. (2005). Public participation in planning: an intellectual history. *Australian Geographer, 36*(3), 283–299.

Laverack, G., & Labonte, R. (2000). A planning framework for community empowerment goals within health promotion. *Health Policy and Planning, 15*(3), 255–262.

Lawrence, A. (2006). 'No personal motive?' Volunteers, biodiversity, and the false dichotomies of participation. *Ethics Place and Environment, 9*(3), 279–298.

Leeuw, E. de, et al. (2014). *Healthy cities, promoting health and equity – Evidence for local policy and practice: Summary evaluation of Phase V of the WHO European Healthy Cities Network*. Retrieved May 18, 2017, from http://apps.who.int/iris/bitstream/10665/137512/1/Healthy%20Cities%20-%20promoting%20health%20and%20equity.pdf.

Levine, J. (2013). Urban transportation and social equity: Transportation-panning paradigms that impede policy reform. In N. Carmon & S. S. Fainstein (Eds.), *Policy, planning, and people: Promoting justice in urban development* (pp. 141–160). Philadelphia, PA: University of Pennsylvania Press.

Lewis, G. J., et al. (2014). Neural correlates of the "good life": Eudaimonic well-being is associated with insular cortex volume. *Social Cognitive and Affective Neuroscience, 9*(5), 615–618. https://doi.org/10.1093/scan/nst032.

Light, D. W., Lexchin, J., & Darrow, J. (2013). Institutional corruption of pharmaceuticals and the myth of safe and effective drugs. *The Journal of Law, Medicine and Ethics, 41*(3), 590–600.

Lindelöw, D., Koglin, T., & Svensson, Å. (2016). Pedestrian planning and the challenges of instrumental rationality in transport planning: Emerging strategies in three Swedish municipalities. *Planning Theory & Practice, 17*(3), 405–420.

Loe Fisher, B. (2017). *Mass protests in italy highlight global vaccination agenda and the resistance movement*. Vaccine Impact. [Online]. Accessed 3 Sep 2017. Available from: https://vaccineimpact.com/2017/mass-protests-in-italy-highlight-global-vaccination-agenda-and-the-resistance-movement/

Lowndes, V., Pratchett, L., & Stoker, G. (2001). Trends in public participation: Part 1 – Local government perspectives. *Public Administration, 79*(1), 205–222. https://doi.org/10.1111/1467-9299.00253.

McIntyre, A. (2008). *Participatory action research*. Thousand Oaks: SAGE.

Minkler, M., & Wallerstein, N. (2011). *Community-based participatory research for health: From process to outcomes*. Hoboken: John Wiley & Sons.

Morgan, S. T. (2012). *Justifying a methodological approach : Action research*. [Online]. Accessed 10 Dec 2015. Available from: http://www.stmorgan.co.uk/justifying-a-methodology-action-research.html

Mowat, H.. (2011). Alan Irwin, citizen science, *Opticon1826*, (10). Retrieved Jun 10, 2017, from http://ojs.lib.ucl.ac.uk/index.php/up/article/view/1368.

Mumpower, J. L. (2001). Selecting and evaluating tools and methods for public participation. *International Journal of Technology, Policy and Management, 1*(1), 66. https://doi.org/10.1504/IJTPM.2001.001745.

Munn-Giddings, C., & Winter, R. (2013). *A handbook for action research in health and social care*. New York: Routledge.

Nieuwenhuijsen, M. J., et al. (2017). Participatory quantitative health impact assessment of urban and transport planning in cities: A review and research needs. *Environment International, 103*, 61–72. https://doi.org/10.1016/j.envint.2017.03.022.

Nind, M. (2014). *What is inclusive research?* Edinburgh: A&C Black.

Nowotny, H., Scott, P., & Gibbons, M. T. (2001). *Re-thinking science: Knowledge and the public in an age of uncertainty*. Hoboken: Wiley.

Ocloo, J., & Matthews, R. (2016). From tokenism to empowerment: Progressing patient and public involvement in healthcare improvement. *BMJ Quality & Safety, 25*(8), 626–632. https://doi.org/10.1136/bmjqs-2015-004839.

Padilla, M. A. S., et al. (2007). Approaches to participation: Some neglected issues. In *5ª Conferência critical management studies*. Retrieved May 25, 2017, https://www.researchgate.net/profile/Miguel_Sahagun/publication/264839037_Approaches_to_participation_some_neglected_issues/links/54e773830cf277664ffa9b5e.pdf.

Pagatpatan, C. P., & Ward, P. R. (2017). Understanding the factors that make public participation effective in health policy and planning: A realist synthesis. *Australian Journal of Primary Health, 23*(6), 516.

Paterson, M. (2007). *Automobile politics: Ecology and cultural political economy*. Cambridge: Cambridge University Press.

Petts, J., & Brooks, C. (2006). Expert conceptualisations of the role of lay knowledge in environmental decisionmaking: Challenges for deliberative democracy. *Environment and Planning A, 38*(6), 1045–1059. https://doi.org/10.1068/a37373.

Rawcliffe, P. (1995). Making inroads: Transport policy and the british environmental movement. *Environment: Science and Policy for Sustainable Development, 37*(3), 16–36. https://doi.org/10.1080/00139157.1995.9929227.

Reason, P., & Bradbury, H. (Eds.). (2001). *Handbook of action research : Participative inquiry and practice*. Thousand Oaks: Sage.

Reed, M. S. (2008). Stakeholder participation for environmental management: A literature review. *Biological Conservation, 141*(10), 2417–2431. https://doi.org/10.1016/j.biocon.2008.07.014.

Reed, M. S., et al. (2017). A theory of participation: What makes stakeholder and public engagement in environmental management work? *Restoration Ecology*. https://doi.org/10.1111/rec.12541.

Renn, O., Webler, T., & Wiedemann, P. M. (1995). *Fairness and competence in citizen participation: Evaluating models for environmental discourse*. New York: Springer Science & Business Media.

Rifkin, S. B. (2009). Lessons from community participation in health programmes: A review of the post Alma-Ata experience. *International Health, 1*(1), 31–36.

Rifkin, S. B. (2014). Examining the links between community participation and health outcomes: A review of the literature. *Health Policy and Planning, 29*(suppl 2), ii98–ii106.
Ringland. (n.d.). Grassroots design of a large infrastructure project. *Ringland*. Retrieved Aug 10, 2017, https://ringland.be/grassroots-design-a-large-infrastructure-project/.
Rowe, G., & Frewer, L. J. (2016). Evaluating public-participation exercises: A research Agenda. *Science, Technology, & Human Values, 29*(4), 512–556.
Ryan, R. M., & Deci, E. L. (2001). On happiness and human potentials: A review of research on hedonic and eudaimonic well-being. *Annual Review of Psychology, 52*(1), 141–166.
Sager, T. (1992). Why plan? A multi-rationality foundation for planning. *Scandinavian Housing and Planning Research, 9*(3), 129–147. https://doi.org/10.1080/02815739208730300.
Schiefelbusch, M. (2010). Rational planning for emotional mobility? The case of public transport development. *Planning Theory, 9*(3), 200–222. https://doi.org/10.1177/1473095209358375.
Schwanen, T., & Ziegler, F. (2011). Wellbeing, independence and mobility: An introduction. *Ageing and Society, 31*(05), 719–733. https://doi.org/10.1017/S0144686X10001467.
Smith, S. E., Willms, D. G., & Johnson, N. A. (1997). *Nurtured by knowledge : Learning to do participatory action-research*. New York: Apex Press.
Souza, M. J. L. (2001). *Mudar a cidade: uma introdução crítica ao planejamento e à gestão urbanos*. Rio de Janeiro: Bertrand Brasil.
Stangl, P. (2008). Evaluating the pedestrian realm: Instrumental rationality, communicative rationality and phenomenology. *Transportation, 35*(6), 759–775.
Steptoe, A., Deaton, A., & Stone, A. A. (2015). Subjective wellbeing, health, and ageing. *The Lancet, 385*(9968), 640–648. https://doi.org/10.1016/S0140-6736(13)61489-0.
Stringer, L. et al. (2006) Unpacking "participation" in the adaptive management of social–ecological systems: A critical review', Ecology and Society, 11(2). Retrieved Jun 10, 2017, from http://www.ecologyandsociety.org/vol11/iss2/art39/main.html
Tickner, J. A. (2001). Democratic participation: A critical element of precautionary public health decision-making. *New Solutions: A Journal of Environmental and Occupational Health Policy, 11*(2), 93–111.
Tippett, J., Handley, J. F., & Ravetz, J. (2007). Meeting the challenges of sustainable development—A conceptual appraisal of a new methodology for participatory ecological planning. *Progress in Planning., 67*(1), 9–98.
Van Brussel, S., Boelens, L., & Lauwers, D. (2016). Unravelling the Flemish mobility Orgware: The transition towards a sustainable mobility from an actor-network perspective. *European Planning Studies, 24*(7), 1336–1356. https://doi.org/10.1080/09654313.2016.1169248.
Velasco, X. C. (2013) Participatory action research (PAR) for sustainable community development. Post Growth Institute Ashland. Retrieved Jun 10, 2017, from http://postgrowth.org/participatory-action-research-par-for-sustainable-community-development/.
Verbeek, T. (2014). Reconnecting urban planning and public health: An exploration of a more adaptive approach. In *Annual congress: From control to co-evolution (AESOP-2014)*. Retrieved Aug 10, 2017, from https://www.researchgate.net/profile/Thomas_Verbeek/publication/288668035_Reconnecting_urban_planning_and_public_health_an_exploration_of_a_more_adaptive_approach/links/5721ff5208ae586b21d3b816.pdf.
Verlinghieri, E.. (2016). *Planning for resourcefulness: Exploring new frontiers for participatory transport planning theory and practice in Rio de Janeiro and L'Aquila*. Ph.D. University of Leeds. Retrieved May 19, 2017, from http://etheses.whiterose.ac.uk/16709/.
Willson, R. (2001). Assessing communicative rationality as a transportation planning paradigm. *Transportation, 28*(1), 1–31. https://doi.org/10.1023/A:1005247430522.
Wolf, E. E. A., & Van Dooren, W. (2017). How policies become contested: A spiral of imagination and evidence in a large infrastructure project. *Policy Sciences, 50*, 449. https://doi.org/10.1007/s11077-017-9275-3.
Woodall, J. et al. (2010) Empowerment & health and well-being: Evidence review. Retrieved May 25, 2017, from http://eprints.leedsbeckett.ac.uk/2172/.

World Health Organization. (1986). *Ottawa charter for health promotion*. Retrieved May 12, 2017, from http://apps.who.int/iris/bitstream/10665/70578/5/9788499841335_cat.pdf.

World Health Organization. (2003). *Community participation in local health and sustainable development: Approaches and techniques/text editing: David Breuer*. Copenhagen: World Health Organization, Regional Office for Europe.

Wright, M. T., Cook, T., et al. (2013a) 'The International Collaboration for Participatory Health Research', in Action research, innovation and change, International perspectives across disciplines: Routledge, pp. 56–72.

Wright, M. T., Brito, I., et al. (2013b). What is participatory health research. *Position paper* no. 1. Retrieved Jun 10, 2017, http://www.icphr.org/uploads/2/0/3/9/20399575/ichpr_position_paper_1_defintion_-_version_may_2013.pdf.

Yiftachel, O. (1998). Planning and social control: Exploring the dark side. *CPL Bibliography, 12*(4), 395–406. https://doi.org/10.1177/088541229801200401.

Yiftachel, O., & Huxley, M. (2000). Debating dominence and relevance: Notes on the "communicative turn" in planning theory 'debating dominence and relevance: Notes on the "communicative turn" in planning theory. *International Journal of Urban and Regional Research, 24*(4), 907–913. https://doi.org/10.1111/1468-2427.00286.

Chapter 27
On the Front Line of Community-Led Air Quality Monitoring

Muki Haklay and Irene Eleta

27.1 Introduction

In a way, the challenge of ensuring that city air is breathable is a consistent feature of city governance and management. Early regulations to prohibit the burning of coal because it was "prejudicial to health" appeared in London in 1273 (Laundon 1967), though were later relaxed. The systematic governance of air pollution through regulations and enforcement started to appear in the second part of the nineteenth century, as a result of the increased industrialisation and use of coal (Lowenthal 1990; Heidorn 1978). Air quality is also intimately linked to the modern awareness of environmental issues and to government action to address it. A prime example of it is the December 1952 smog event in London, which killed about 4000 people, leading to a regulatory response and the Clean Air Act of 1956 (Fenger 2009). Beyond the UK, air quality regulations are some of the early examples of the European Union (EU) environmental regulations, dating back to 1979 with regulations to address transboundary pollution (Kuklinska et al. 2015).

Yet, from the mid-1950s, when public awareness of air pollution and its harmful health impacts was raised, to the 1990s, air quality issues were addressed through top-down regulations, usually mandating a centralised network of monitoring stations, run under the control of public authorities, to check that progress was being

M. Haklay (✉)
Extreme Citizen Science group, Department of Geography,
University College London (UCL), London, UK
e-mail: m.haklay@ucl.ac.uk

I. Eleta
ISGlobal, Centre for Research in Environmental Epidemiology (CREAL), Barcelona, Spain

Universitat Pompeu Fabra (UPF), Barcelona, Spain

CIBER Epidemiología y Salud Pública (CIBERESP), Madrid, Spain
e-mail: irene.eleta@isglobal.org

M. Nieuwenhuijsen, H. Khreis (eds.), *Integrating Human Health into Urban and Transport Planning*, https://doi.org/10.1007/978-3-319-74983-9_27

made. Addressing such environmental concerns through top-down scientific management was somewhat expected, since environmental problems require expertise from multiple areas such as atmospheric sciences, public health, toxicology, chemistry and so on, and every measure to address environmental problems requires a discussion with many stakeholders—from community leaders to various industries that need to be involved in addressing pollution issues to health workers.[1] Thus, environmental decision-making was seen as an area that required scientific expertise and, therefore, involved experts and decision-makers without much engagement with the public (Haklay 2017).

Since the late 1980s, there has been a growing awareness of the importance of public engagement in the process of environmental decision-making (see Haklay 2017). The demand for public participation in environmental decision-making in a meaningful way emerged in the 1980s, receiving a notable mention in the 1987 Brundtland report (WCED and Brundtland 1987) and the acceptance of the sustainable development principles at the Rio conference in 1992. Significantly, the principle of public participation in decision-making received due attention in the outcomes of the Earth Summit, with Principle 10 which highlights access to information, participation in decision-making and access to justice. As the 1990s progressed, the principle was enshrined in conventions and laws (e.g. the Aarhus Convention in 1998), turning Principle 10 into practical commitments by government and the subsequent EU directives that implement it—2003/4/EC and 2003/35/EC (see Haklay 2003). Efforts to release environmental information became common during this period, and since air monitoring systems were already in place, they became one of the early sources of environmental information that was published, publicly, on the Internet some 20 years ago, with repositories such as the UK Air Quality Archive (which maintains data going back to the 1970s), providing its information over the web and many other systems following these early examples.

It is noteworthy that, although the Aarhus Convention and Principle 10 clearly call for engaging the public in the environmental decision-making process, they are not challenging the underlying assumption of who will produce the information. The opening declaration of Chap. 40 in Agenda 21 (the official outcome of the Rio conference), which is dedicated to information for decision-making, stated:

> In sustainable development, everyone is a user and provider of information considered in the broad sense. That includes data, information, appropriately packaged experience and knowledge. The need for information arises at all levels, from that of senior decision makers at the national and international levels to the grass-roots and individual levels. (United Nations 1992)

However, the rest of Chap. 40 and Principle 10 assumed that environmental information was produced by experts, and the public was granted access to it in order to facilitate participation in decision-making. It took more than a decade until the suggestion that the public could actively participate in the generation of

[1] Notably, the Clean Air Council that was created following the 1956 legislation included representatives from local government and industry (e.g. Guinness, coal producers, Unilever), a public health specialist and national trade union representative, engineers and scientists.

environmental information which was declared by the leading environmental information experts (see McGlade 2008).

Yet, the 1990s also saw another form of environmental awareness, with the emergence of environmental justice as an important facet of environmental policy. The evidence for different exposure to environmental harms as a result of racial discrimination appeared in the early 1990s (Bullard and Wright 1993). By the end of the decade, methodologies to support community-led measurement of the harms that they were exposed to started to appear. Community-led environmental monitoring is now recognised as a specific part of the wider landscape of public participation in collecting and sharing environmental information—which is now termed citizen science (Bonney et al. 2014; Haklay 2013).

Most citizen science is contributory (Shirk et al. 2012), which means that scientists set the project and define the data collection protocols, and participants, who are not professional scientists, join the project and assist the scientists in data collection or in basic analysis tasks. For example, one of the earliest examples of an activity that used the term "citizen science" was carried out in 1989 by the Audubon society in the USA, which provided 225 volunteers from across the country with equipment and guidance so they could collect rain samples to monitor acid rain. The project was led by scientists who guided participants in how to contribute their observations and time.

In contrast, community-led environmental monitoring activities have emerged from concerns of a community member or members (Table 27.1). Here, again, the scientific nature of environmental decision-making plays a major part. Since the environmental problem can be discovered through scientific measurement, and addressed through the use of the resulting information in discussions with authorities, the use of scientific methods that engage community members in information collection and analysis is becoming a clear route to address the concerns. Air and water monitoring are good examples of this—although community members may suspect that a water source is contaminated or that the level of chemicals in the air is leading to ill health, only through laboratory analysis of water samples can they provide evidence for this. The nature of community-led projects is that they address localised problems, rely on local knowledge and use community resources to carry out the investigation. By resources, we mean the availability of people from the community to carry out measurements at different locations throughout the day, using residents' balconies or windowsills as a basis for equipment, or possibly raising money from a large group of residents to pay for laboratory costs.

Examples of community-led air pollution measurement efforts to collect the scientific data needed to make the regulatory case started to emerge in the 1990s (Corburn 2005). This, in part, was made possible by the proliferation of equipment for sensing the environment as it became part of routine, large-scale monitoring programmes at local and national levels. This proliferation and reduction in costs meant that instruments and science measurements came within the reach of non-governmental organisations and community groups. An example of this is the Global Community Monitor—an organisation that, since 1998, has developed a method to allow communities to monitor air quality near polluting factories (Scott

Table 27.1 Main characteristics of community-led monitoring

Aspect	Detail
Organisation	Ad hoc or an organisation with other remit (e.g. residents' association, friends of a local park)
Project design	The problem is set by the community, and the information that emerges from it is being used by them, although the analysis might be carried out by external expert. The project will be action oriented
Geographical scope	Local or hyperlocal—a neighbourhood and sometimes a specific street Territoriality can play a role in the location of sampling and decisions on where the measurements will take place
Length of time	Limited—the participants use their free time for participation and usually expect results within a framework of 1 or 2 months
Resources—personnel	Community-led studies can benefit from the availability of local volunteers who are motivated in addressing the problem. There is a potential for a wide range of skills within the community (e.g. a resident with engineering or science background) although that could be a challenge. Usually a small group of highly committed residents will lead the project and carry it out
Resources—equipment	Usually a limiting factor in the ability to access and operate environmental monitoring equipment, beyond widely available technologies (e.g. smartphones), and therefore need to work with experts and universities to access them and operate them
Resources—funding	Usually very limited, with potential for very limited sums to be raised locally in the case of marginalised groups. Effective use of small grants to carry out monitoring
Problem solving, learning, creativity	Integral part of community-led projects is an accelerated learning of the environmental issue that is the focus of the investigation, identifying locations for sampling and solving problems in the different stages of the project
Impact	The impact of the project will depend on the community's ability to mobilise wider political and social resources to address the local problem. Can be especially problematic when the environmental issue is addressed at national level (e.g. regulations to encourage recycling or incineration) while the problems are hyperlocal

and Barnett 2009). The sampling is done by members of the affected community using widely available plastic buckets and bags, followed by analysis in an air quality laboratory with the appropriate knowledge and ability to analyse the sample. Finally, the community is provided with guidance on how to understand the results. This activity is termed "bucket brigade" and is used across the world in environmental justice campaigns, for example, in the struggle of local African-American residents in Diamond, Louisiana, against a polluting Shell Chemical plant (Ottinger 2010).

Another community tool for air monitoring is a passive diffusion tube. This is a simple yet widespread tool for local level air quality monitoring. Passive diffusion tubes or, simply, diffusion tubes sense through a simple mechanism (see Fig. 27.1a). A plastic tube is sealed with a rubber cup, within which there is a metal mesh on which there is a chemical absorbent of a known quantity (Fig. 27.1b). Figure 27.1 shows a nitrogen dioxide—NO_2—diffusion tube, and the white cup, which is used

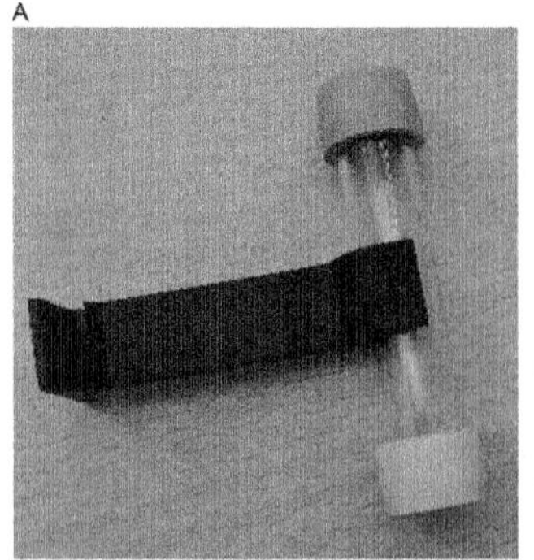

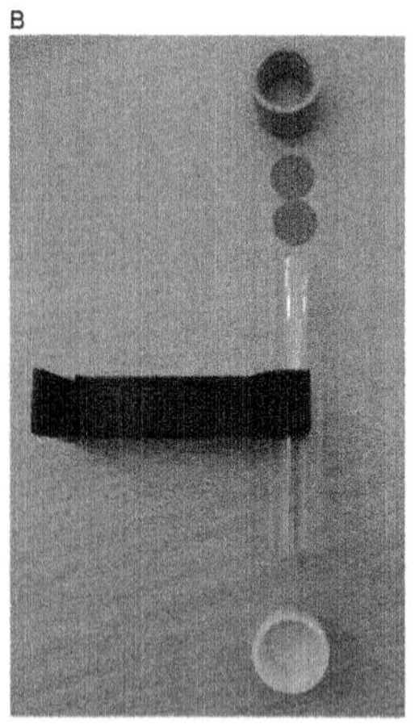

Fig. 27.1 (**a**) An assembled passive diffusion tube for measuring nitrogen dioxide; (**b**) The parts of the diffusion tube—the grey cup contains the absorbent, and the white cup is taken off during the measurement period

for transportation to and from the laboratory, is removed to allow the absorbent to react with the specific pollutant that is being measured (in other types of diffusion tubes for other pollutants the white cup acts as a filter and is not removed). Once the measurement period ends, the tube is sent to the laboratory, and the concentration of the pollutant can be calculated. Diffusion tubes are easy to use—they require simple installation that does not need much more than a stepladder and brief training, do not require any maintenance while installed, gradually became cheaper to build and analyse, are evaluated in an accredited laboratory that can ensure data quality and are also well established as a tool that has been used by local authorities to monitor pollution since their invention in the 1970s by Edward Palmes (Sella 2016). These characteristics make them highly suitable for community-led air monitoring, and they have been in use in community-led campaigns in Sheffield, UK, since 1998, in collaboration with the city's authorities (Parry and Rimmington 2013). The Sheffield East End Quality of Life Initiative explicitly links health with social and environmental justice. Its air quality monitoring involves using tubes in five locations and has influenced a range of local decision-making, including halting a planning process for an increase in the size of a major supermarket in the area due to concerns about increased traffic and the evidence from the monitoring effort that the area is already exposed to high levels of traffic-related pollution. Moreover, it continually monitors the literature and publications in the area of linkage between health, air pollution and noise and, since March 2013, has published regular summaries of the latest research on a monthly basis, thus providing its community with an accessible summary of the latest findings (see http://www.sheffieldeastend.org.uk/Reports.htm).

This type of community-led monitoring is the topic of this chapter. In the next section, we briefly review several cases of the application of high-density diffusion tube campaigns in London. We then turn to the core of the chapter and a detailed account of the experience of carrying out a community-led diffusion tube campaign.

Finally, we summarise with some of the lessons learnt from our engagement with the process.

27.2 Mapping for Change Diffusion Tube Campaigns

By 2010, the cost of diffusion tubes had continued to drop, and it became possible to provide communities with a larger number of passive diffusion tubes that they could install in a dense network of tubes over a period of a month or several months and therefore reveal the localised pollution picture. Mapping for Change, a social enterprise that was created by UCL and London 21 Sustainability Network, seized the opportunity to use diffusion tubes that became available within the Open Air Laboratories air pollution work (Tregidgo et al. 2013) and on the Pepys Estate in London.

The Pepys Estate in Deptford, South London, is a predominately 1960s housing estate on the banks of the Thames, characterised by tower blocks and social housing. Situated near a busy thoroughfare and surrounding an industrial site, the estate suffers a variety of urban environmental issues. Following a noise pollution study of a local scrapyard in 2008, residents, in 2010, expressed concerns over the local air quality and wanted to assess how good or poor the air quality was. They had particular concerns over the mechanical break-up of vehicles and large goods vehicles servicing the scrapyard, as well as local traffic, which were seen as potential sources of pollution. The possible impact of a planned housing development further heightened their desire to assess local air quality. The survey was initiated with the Pepys Community Forum and commissioned by the London Sustainability Exchange (LSx), a charity geared towards creating a more sustainable London.

After a meeting in the local community centre, Mapping for Change provided residents with instructions and survey equipment to carry out their investigations. The area of the estate was divided up into 100-m grids to obtain a good distribution of samples taken. Activities commenced with a series of diffusion tubes being set out on lamp posts around the area. Wipe samples were taken to assess the quantity and type of metals being deposited on vertical surfaces. Ozone levels were measured using Eco-badge™ ozone detection kits (see Fig. 27.2), and leaf samples were collected and analysed by Lancaster University. All the data collected was analysed and compiled into a series of maps. A public meeting was held to provide feedback on the findings. As a result, the local authority (Lewisham) installed diffusion tube monitoring devices at the main junctions identified by the project as having higher levels of NO_2. Levels of NO_2 in London are largely from vehicle exhausts and are also a strong indicator of the presence of other air pollutants derived from vehicle emissions. The council also installed a PM_{10} particulate monitoring station in the neighbourhood to monitor the local situation. Previously, the closest fixed monitoring station was just over a kilometre away.

Fig. 27.2 Testing Eco-Badge ozone detection kits on the Pepys Estate

Of the methodologies that were used at the Pepys Estate, the diffusion tubes proved the most effective in terms of communicating with local decision-makers as well as installation and confidence in the quality of the data.

A year later, as part of the project EveryAware (everyaware.eu), which was dedicated to participatory sensing and collective awareness, the use of diffusion tubes was developed further. The EveryAware project was set to develop real-time, portable air quality sensors, but in the early stages of the project, there was a need to evaluate the level of interest among the public in the area of participatory monitoring. The aim was to provide communities with a way to measure air quality using low-technical methods that could be replicated easily and would engage all sectors of the community. A number of interested groups and individuals concerned about harmful levels of air pollution came forward via Twitter and through local community events and talks held at seminars. This led to a pilot that comprised seven locations across London and involved residents from Putney and Highbury and volunteers from Sustrans, a charity promoting sustainable transport.

Each community had slightly different motivations for wanting to carry out the investigation, which were explored in the initial discussion. The Putney Society had long been concerned about poor air quality in the area, which had consistently breached EU limits for NO_2 and PM_{10}. Monitoring activities in Highbury were led by members of the local Green Party who had been lobbying for 20 mph zones in the area. In addition, communities wanted reliable localised data, which they could use to lobby local government, raise awareness, generate a better understanding of the issues and with which they could compare other relevant datasets. Sustrans, a sustainable transport charity, was keen to collect data to demonstrate the difference in pollution levels between routes not accessible to motorised vehicles and adjacent roads.

NO_2 was selected as the focus, primarily because of the affordability of the monitoring equipment. Levels of NO_2 largely arise from vehicle exhausts in London and are also a strong indicator of the presence of other air pollutants derived from vehicle emissions. In each location, a series of diffusion tubes, used to measure NO_2 levels, were set out across the area. Based on lessons from previous studies, the areas were divided up into grid squares to ensure there was sufficient coverage; Putney, with a total of 38 sites, had the most monitoring locations. After a period of 4 weeks, the diffusion tubes were collected and analysed, and the results mapped for each location. The results from both Putney and Highbury indicated that levels along the main road network were up to 75% higher than EU limits for the period. They also highlighted several residential back roads used as "rat runs" and, therefore, showing high levels of pollution. The remaining five monitoring locations across London each comprised one of London's Greenways (safe, quiet routes through parks, green spaces and streets with light traffic) compared with an adjacent busy road. The results showed significantly higher NO_2 levels on the roads compared with the Greenways, despite their close proximity.

In addition to these studies, a year-long community-led monitoring was carried out on the Barbican Estate in Central London, and a 6-month, multisite study was carried out with communities in London. Among the outcomes of the Barbican study, participating residents made suggestions to the City of London Corporation, which is the local authority of the area. These suggestions included aspects of urban planning and transportation, like extending the Ultra Low Emissions Zone, only allowing green buses; creating green areas and living walls; adjusting the phasing of traffic lights; introducing penalties for idling taxis, delivery vehicles and buses; and closing or improving the ventilation of the Beech Street tunnel (Francis and Stockwell 2015).

In December 2015, Mapping for Change ran a crowdfunding campaign to support community-led air quality projects, raising enough funding to support six communities. One of these was in the vicinity of University College London's main campus, in Somers Town, near Euston and King's Cross railway stations in Central London. We now turn to the experience of community-led monitoring of one of the authors (Irene).

27.3 On the Ground: Somers Town Study

A cold wind was bringing dark clouds over London on that Wednesday morning of March. I hastened to enter University College London. In one of the offices—an open space full of desks—researchers and doctoral students were already busy at their computers; this is the main office that is shared by Mapping for Change and the Extreme Citizen Science group.

I crossed the main room, heading towards the kitchen. As Louise Francis, the co-founder and managing director of Mapping for Change had told me, I looked under the counter on the right, where I found a small fridge. Inside, Louise had left a zipped plastic bag with two diffusion tubes. I put them in my backpack and walked quickly to Euston train station.

I passed Euston train station, and I took out a map of the Somers Town neighbourhood. It was a printed version of the Mapping for Change visualisation—the Community Maps system—of the neighbourhood, with numbered spots on it.

During the previous weeks, I had witnessed an email exchange between Louise and Tracey,[2] one of the Somers Town residents who had taken the lead of monitoring air pollution in the neighbourhood. Tracey and other residents had chosen ten locations where they wanted to measure NO_2 levels. They explained the characteristics and reasons for choosing these particular spots: crossroads with a lot of traffic in a residential area, in front of schools, in a park, etc. They were using their knowledge of the area and their understanding of the way they use it—either as the route to school or the playground in the afternoon—to consider where the information will help them most. Louise suggested adding another two locations close to main roads so as to understand the scale of pollution in main thoroughfares nearby. Initially, Tracey and her team only had ten diffusion tubes, and they chose ten spots carefully. Louise decided to give them two additional diffusion tubes to include her suggested spots, which I was carrying that morning to Somers Town. My map had 12 numbered marks, where the diffusion tubes would be placed.

The wind was getting colder and the sky was darkening. After passing the bustling surroundings of the train station, Somers Town looked relatively quiet. I wandered for a few minutes up and down a street until I asked an elderly resident who was walking his dog for the exact address. He pointed to a white apartment building.

It was almost 10 am and I rang the doorbell. Tracey and I were introduced to each other for the first time. She looked young and energetic. She was getting ready: she had a bag full of diffusion tubes, tube holders, a laptop, a mobile phone, a notebook, a pen, scissors, white stickers about the activity and a ladder.

Tracey, her friend and I went out to the street and joined two other neighbours for a team photo. By then, an intermittent drizzle announced that the weather was not going to be cooperative. We started by a lamp post next to a school, at a crossing of streets. Using the ladder, Tracey and her friend tightened the plastic belt to the lamp post.

Another neighbour came to say hello and told Tracey how great her initiative to measure air pollution in the neighbourhood was. The neighbour, an elderly lady, chatted with us about her concerns with the air pollution entering her house and explained that there were plans to make the street where we were a lorry passage. Meanwhile, we ran under the shelter of a building nearby because the drizzle was becoming more intense.

---

[2] Name changed for ethical reasons

Fig. 27.3 Installing a diffusion tube in Somers Town and leaving a note about the activity

Tracey connected to the Mapping for Change application to enter the data. She selected the location on the map and entered the date, distance from road (less than 1 m), height (2.5–3 m), time, etc. Afterwards, she went under the rain to attach the diffusion tube to the plastic belt on the lamp post with a tube holder. The grey top of the diffusion tube (see Fig. 27.3) was facing up, and she took out the white cup, facing down. Then, she left a sticker about the activity, with a contact number, on the lamp post by the diffusion tube. She explained to me that the sticker would inform city council workers, like cleaners, and prevent their taking it down. The rain discouraged the neighbours and they left.

At 10:25 we moved to another location. It was an awkward crossing of streets. Tracey realised that the ideal lamp post at the central area was too large for the plastic belts, so they chose a thinner lamp post on a side of the crossing instead. Using the ladder again, they tightened one plastic belt and the tube holder to the lamp post. For a few minutes, the rain stopped. Then, Tracey recorded the data and attached the diffusion tube. She also made notes in her notebook and put the sticker on the plastic belt.

While we headed to the next location, it started to rain again. It was a crossing of streets next to a school. They chose a lamp post next to a bike parking rack, by the road. The rain became heavier, and there was a gusty wind that made us feel very cold. Tracey's hands became red and numb from the cold while tightening the plastic belt. She proceeded to enter the data in the application form. I tried to cover both of us with my umbrella, but the wind was blowing too strongly, and we were getting wet. We sought shelter in the school. Tracey asked for permission to stay inside for a few minutes, and she briefly explained the project. However, we were not allowed to stay in the school building because of the rules. We went outside in the cold and rain. I was disappointed to see that not everybody in the neighbourhood was so enthusiastic or cooperative.

By 10:45, the rain became lighter and the wind stopped. Tracey's friend attached the tube following her instructions: grey cup up, white cup out facing down. They put the sticker on and we moved to the next location. The location was a green area between two schools. Tracey explained that there were plans to renovate that space. They identified a lamp post by the school's fence. Tracey's friend went up the ladder to tighten the plastic belt, but it broke. Tracey gave him another one and cheered him up saying "good work". The rain became heavier again, while Tracey was trying to enter the data in the application and making notes in her notebook. Hail and wind burst into a sudden storm. They rushed to put the diffusion tube and sticker on, under the hail, and we all ran to find shelter in a nearby community centre.

It was 11 am. We ordered tea in the community centre bar. Our hands were very cold; our clothes were wet. We chatted while drinking hot tea near the radiator. Tracey and her friend told me why it is worth braving the weather and setting up these little plastic tubes: "The air quality is very bad around here". She had been a researcher for 5 years and worked in a hospital for asthma and respiratory diseases. Now she was doing health communication. While we were enjoying this break, Tracey finished entering the data about the last location. Tracey's phone rang: another person would join the team soon.

The rain had stopped and we went out to put up another diffusion tube. The new location was in a street crossing around the corner from the community centre. Tracey had become efficient: she stepped up the ladder, tightened the plastic belt with the tube holder, put the sticker near the plastic belt, recorded the details and held the diffusion tube up, taking the white cup out.

Right after that, we met the new team member at the community centre: a young lady from the Science Museum, who would be documenting the experience and taking pictures. It was time for my goodbye. We had put five diffusion tubes up of a total of 12, and they had planned to stay until 2 pm. They would be using the community centre as a base between locations, given the bad weather. At 11:34, I left the group and walked to St. Pancras train station with many new questions in my head. There was a cold breeze, but the rain had stopped.

A few days later, I met Louise at her office in University College London. She told me about the preparatory session she organised for this particular study, where she taught participants how to use the Mapping for Change application and how to install the diffusion tubes. Tracey had attended that session, representing the Somers Town neighbourhood in the Camden district, and three other community leaders attended from other neighbourhoods.

The next step after having installed the diffusion tubes was to collect them after 4 weeks, by 30th March. Louise explained that the dates chosen for installing and collecting the diffusion tubes were not random—they coincided with the official cycle of measurement. Once the diffusion tubes were collected, the participating residents could give them to Louise for her to send them to the lab for analysis, or they could send the diffusion tubes directly to the lab with the stickers that Louise had provided. The lab would send the results of the analysis in about 2 weeks upon receiving them. The results would show the average NO_2 concentration during the 4 weeks of exposure on the spot where each diffusion tube was placed, and they could be compared among the different locations. The participating community leaders, like Tracey, would have to input the data from the lab into the Mapping for Change system. Then, everybody would be able to see the results on an online map; they would be able to zoom in to their street and see if there was air pollution data nearby. The Mapping for Change system already had public data from several boroughs in London that had been participating in these community-led initiatives.

"Is it possible to compare this data to the official sources? What are residents going to do with it?" I asked her. Louise replied to me with an engaging story. One of the groups that had done the air quality monitoring during the previous month had contacted the local authorities of their district to tell them about the results and their plan to make the data public on the Mapping for Change system. Louise clarified to me that they do not need permission from anyone to make their data public. However, the district authorities were reluctant: they asked the group not to publish the results because they feared that the data was not accurate. After this disappointing reaction from the district authorities, the community leader explained to them that they were advised and taught by researchers from University College London and detailed the procedures they followed. She shared these details to demonstrate that their data was accurate. The district authorities replied that they would have to annualise the results to make them comparable to their official measurements. Louise confirmed to me that there is a problem of comparison because the residents only have 1 month of data.

It is important to note that a precedent has been set by the Barbican neighbourhood, where participants collected data every month during an entire year (Francis and Stockwell 2015), which could be compared with the official measurements. As noted above, the Barbican study was carried out in collaboration with the local authority, and it is a story of success (see the report—Francis and Stockwell 2015 or view in an online documentary https://www.youtube.com/watch?v=YvSuPCxl188). The project "Science in the City" had the financing and support of the City of London, a motivated community leader, and the sustained logistic support of Mapping for Change with follow-up workshops that engaged around 50 local residents. After this successful experience, in summer 2015, Mapping for Change had funding to do a 1-month data collection with several groups in the city. But, later, the lack of funding to continue these local initiatives forced the researchers at UCL and Mapping for Change to organise a crowdfunding campaign. Only those communities that paid for the materials and analysis could conduct their air quality monitoring activities. Louise told me that one of the groups that had participated in the summer 2015 initiative repeated the experience and was very organised through a Facebook page. Other groups were new, or there was only one motivated person conducting the bulk of the work, like the case of Tracey in Somers Town.

27.4 Discussion

So far, in this chapter, we have only mentioned in passing the use of real-time low-cost sensors. This is a deliberate decision in the context of a community-led, action research framework. Low-cost sensors are an important area of research, and millions of euros are invested in their development, both by public and private bodies. Yet, despite the ongoing investment over the past decade, they are still not suitable for regular, everyday use that will yield robust, accurate and consistent results—see Kumar et al. (2015) and Lewis and Edwards (2016) for a discussion of their issues. Moreover, their installation is more complex—they require electricity, connectivity and, at this stage of development, the device to be shielded from adverse weather such as the rain and the wind that frequented Somers Town on that wintry day. Moreover, because of the need for calibration, and the amount of data that the device produces, there are major challenges in making the outputs suitable for non-specialists to interpret in a way that will make the information useful for them. For that reason, the focus of community-led projects, ever since they emerged in the mid-1990s, has been to use reliable methodologies that are easy to use but, at the same time, comparable with official information. It is this aspect that makes the Palmes diffusion tubes so powerful. This is a stable technology that gains acceptance by regulators, local environmental officers and decision-makers. By aligning the dates of the monitoring with the official period being used in government-mandated programmes, the results of the community effort are readily comparable with those that are obtained by the local authority. Indeed, as our description above shows, there can be a discussion around the length of the observation (the need for annual monitoring or careful seasonal measurements as offered by Cyrys et al. (2012)), but the methodology itself and its results are not questioned or challenged, whereas community-led low-cost sensing can be easily dismissed on the basis of the sensors that have been used (Lewis and Edwards 2016).

Moreover, within the environmental monitoring and decision-making processes, these seemingly simple plastic tubes act as more than just a sensor—they are a policy instrument that is used by central and local government in their local air quality management plans. This aspect provides the tubes with political power in the discussion of the result, which other inexpensive methods cannot offer.

Our detailed story of the experience of Tracey in Somers Town is indicative of many issues that are common in community-led air quality campaigns and which we experienced elsewhere.

Firstly, it illustrates some of the difficulties that local residents have in influencing local authorities about their environmental, public health, urban planning and transportation decisions, starting with the doubts about data quality on the part of local authorities and the lack of sustained funding to address environmental issues by local government bodies that are operating under severe budgetary restrictions. This issue of trust and feeling that the measurements are taken in the wrong locations compared to the experience of the people who are living in the area and to the local knowledge about the routes and places of activities is a motivating factor in community-led efforts. For example, notice how Tracey and her neighbours commented on their concerns about potential plans to make a lorry passage in front of a school and the renovation of a space that was a green area between two schools. By taking control over the positioning of the tubes and the installation process itself, the participants gain an understanding of the process of monitoring and the tools that are being used. Arguably, this can increase the understanding of where the data from the local authorities is coming from and the dialogue between communities and local authorities, as happened in the City of London and the London Borough of Lambeth.

Secondly, it provides an illustration of the local focus of the monitoring—noteworthy is the need to guide the participants to position diffusion tubes in a busy road. The reason for this is that, in many communities, the residents perceive the busy road as the boundary of their area and at the edge of the local neighbourhood area. The need for comparison and benchmarking can be achieved by using local authority monitoring data, but, as an intermediary organisation, Mapping for Change guides participants to carry out such data collection so they can compare and contrast the results that they receive from within their area of interest with a nearby pollution source.

Thirdly, we should highlight the nature of the process—it is very common for very few participants to be highly engaged and to be active in the process, in the way that Tracey braved the weather and carried out the installation. However, since it is her locality, other people within the area got involved in either direct help for a short while or by noticing Tracey and asking about her effort. In multiple cases, we have seen that, in feedback sessions, many residents, well beyond those that were active in the installation, join in to see the results and discuss them (see Fig. 27.4). The effort of installing the diffusion tubes is not trivial and can mean dedication of a significant part of a day in unpleasant weather conditions. It therefore requires well-motivated individuals to lead such efforts, which is the case in many locations. Yet, it is this wider community interest and support that makes their effort worthwhile and sustainable, in the case of donating to a common fund to maintain the activity.

Finally, the Somers Town installation demonstrates the way in which local problem solving and technology are being used in these community-led campaigns. Notice how Tracey confronted and solved the issue of installing tubes in wide and narrow lamp posts or the use of the Mapping for Change community maps system

Fig. 27.4 Feedback session at UCL with community members

to record the location of the installation—using her smartphone as a Wi-Fi hotspot and her computer to enter the data while sheltering from the rain (the resulting map is shown in Fig. 27.4a). Over the years, we have seen many examples of local problem solving and innovations—such as decisions to install diffusion tubes on the balconies in the Barbican. This is an important aspect of community-led citizen science efforts, which provides background to both learning and creativity (see Jennett et al. 2016, 2017).

27.5 Conclusions

In this chapter, we provided a detailed account of the experience of community-led air quality monitoring that is done within the context of community concern over local air pollution from traffic and its impact on their health. We have discussed how changes in the context of environmental decision-making, awareness of environmental justice and in technology enabled a new wave of community-led monitoring efforts.

As we have seen, the use of accessible and reliable sensing tools, such as the air capture buckets or diffusion tubes, provides the ability for community members to use their local knowledge to collect the information at the location that matters to them and to understand local conditions. A recent campaign by Friends of the Earth demonstrated that diffusion tubes can also be scaled to a national campaign in the UK (see FoE 2017), while the London Sustainability Exchange provided guidance for their use to primary schools (LSx 2013).

We also pointed to the deceiving surface simplicity that the use of these tools provides—they allow for deep local engagement—an understanding of the process of measuring and understanding air pollution issues and a potential foundation for a discussion between communities and local decision-makers on the basis of scientific evidence that both sides accept. In what might seem contradictory, at first sight, the use of agreed-upon devices through citizen science and community-led monitoring can reduce the conflict between residents and local environmental management authorities, improve the process of air quality monitoring and offer new solutions to local issues.

Community-led air quality monitoring is not being offered as a panacea or as a replacement for government-mandated monitoring but as an enhancement and as a way to address specific emerging issues, such as an increase in traffic due to a temporary construction project, or in addressing ongoing concerns such as the impact of queuing traffic. This form of citizen and stakeholder participation can also integrate into health impact assessments of urban and transport planning projects to address power unbalance, select scenarios for monitoring and evaluation and identify health effects and vulnerable populations (Nieuwenhuijsen et al. 2017).

Acknowledgements Thanks to Louise Francis, managing director of Mapping for Change, for sharing her valuable insights and practical knowledge acquired during many years of working with communities in London and to Tracey for allowing us to join her on the day. This chapter benefited from the European Union's Seventh Framework Programme (FP7/2007–2013) under grant agreement EveryAware (award 265432), the EU Horizon 2020 research and innovation programme under the Marie Sklodowska-Curie grant (award 656439) and Doing it Together Science (award 709433) as well as the UK's Engineering and Physical Sciences Research Council (award EP/I025278/1).

References

Bonney, R., Shirk, J. L., Phillips, T. B., Wiggins, A., Ballard, H. L., Miller-Rushing, A. J., & Parrish, J. K. (2014). Next steps for citizen science. *Science, 343*(6178), 1436–1437.

Bullard, R. D., & Wright, B. H. (1993). Environmental justice for all: Community perspectives on health and research. *Toxicology and Industrial Health, 9*(5), 821–841.

Corburn, J. (2005). *Street science: Community knowledge and environmental health justice (urban and industrial environments)*. Cambridge, MA: MIT Press.

Cyrys, J., Eeftens, M., Heinrich, J., Ampe, C., Armengaud, A., Beelen, R., Bellander, T., Beregszaszi, T., Birk, M., Cesaroni, G., & Cirach, M. (2012). Variation of NO 2 and NO x concentrations between and within 36 European study areas: Results from the ESCAPE study. *Atmospheric Environment, 62*, 374–390.

Fenger, J. (2009). Air pollution in the last 50 years—From local to global. *Atmospheric Environment, 43*(1), 13–22.

Francis, L., & Stockwell, H. (2015). *Science in the city: Monitoring air quality in the Barbican. [online]* (p. 33, 34). London: Mapping for Change. Retrieved August 2017, from https://www.cityoflondon.gov.uk/business/environmental-health/environmental-protection/air-quality/Documents/barbican-final-report-13012015.pdf.

Friends of the Earth. (2017). *Unmasked: The true story of the air you're breathing* (p. 23). London: Friends of the Earth. Retrieved August 2017, from https://cdn.foe.co.uk/sites/default/files/downloads/FoE_Unmasked_Report_2017.pdf.

Haklay, M. E. (2003). Public access to environmental information: Past, present and future. *Computers, Environment and Urban Systems, 27*(2), 163–180.

Haklay, M. (2013). Citizen science and volunteered geographic information—Overview and typology of participation. In D. Z. Sui, S. Elwood, & M. F. Goodchild (Eds.), *Crowdsourcing geographic knowledge: Volunteered geographic information (VGI) in theory and practice* (pp. 105–122). Berlin: Springer.

Haklay, M. (2017). The three eras of environmental information: The roles of experts and the public. In V. Loreto, M. Haklay, A. Hotho, V. D. P. Servedio, G. Stumme, J. Theunis, & F. Tria (Eds.), *Participatory sensing, opinions and collective awareness* (pp. 163–179). Cham: Springer International Publishing.

Heidorn, K. C. (1978). A chronology of important events in the history of air pollution meteorology to 1970. *Bulletin of the American Meteorological Society, 59*(12), 1589–1597.

Jennett, C., Kloetzer, L., Schneider, D., Iacovides, I., Cox, A., Gold, M., Fuchs, B., Eveleigh, A., Methieu, K., Ajani, Z., & Talsi, Y. (2016). Motivations, learning and creativity in online citizen science. *Journal of Science Communication, 15*(3).

Jennett, C., Kloetzer, L., Cox, A. L., Schneider, D., Collins, E., Fritz, M., Bland, M. J., Regalado, C., Marcus, I., Stockwell, H., & Francis, L. (2017). Creativity in citizen cyberscience. *Human Computation, 3*(1), 181–204.

Kuklinska, K., Wolska, L., & Namiesnik, J. (2015). Air quality policy in the US and the EU—A review. *Atmospheric Pollution Research, 6*(1), 129–137.

Kumar, P., Morawska, L., Martani, C., Biskos, G., Neophytou, M., Di Sabatino, S., Bell, M., Norford, L., & Britter, R. (2015). The rise of low-cost sensing for managing air pollution in cities. *Environment International, 75*, 199–205.

Laundon, J. R. (1967). A study of the lichen flora of London. *The Lichenologist, 3*(3), 277–327.

Lewis, A., & Edwards, P. (2016). Validate personal air-pollution sensors. *Nature, 535*(7610), 29–32.

London Sustainability Exchange (LSx). (2013). *Cleaner Air 4 Primary Schools toolkit*. Retrieved August 2017, from https://www.london.gov.uk/sites/default/files/ca4s_toolkit.pdf

Lowenthal, D. (1990). Awareness of human impacts: Changing attitudes and emphases. In B. L. Turner (Ed.), *The earth as transformed by human action: Global and regional changes in the biosphere over the past 300 years* (pp. 121–135). Cambridge: Cambridge University Press with Clark University.

McGlade, J. (2008). *Environmental information and public participation*. Retrieved February 2014, from http://www.eea.europa.eu/media/speeches/environmental-information-and-public-participation

Nieuwenhuijsen, M. J., Khreis, H., Verlinghieri, E., Mueller, N., & Rojas-Rueda, D. (2017). Participatory quantitative health impact assessment of urban and transport planning in cities: A review and research needs. *Environment International, 103*, 61–72.

Ottinger, G. (2010). Buckets of resistance: Standards and the effectiveness of citizen science. *Science, Technology, & Human Values, 35*(2), 244–270.

Parry, N., & Rimmington, B. (2013). Citizen science—Local air quality; local action. *Chemical Hazards and Poisons Report*, p. 45.

Scott, D., & Barnett, C. (2009). Something in the air: Civic science and contentious environmental politics in post-apartheid South Africa. *Geoforum, 40*(3), 373–382.

Sella, A. (2016). Palmes' tube, Chemistry World, 1 November 2016.

Shirk, J. L., Ballard, H. L., Wilderman, C. C., Phillips, T., Wiggins, A., Jordan, R., McCallie, E., Minarcheck, M., Lewenstein, B. V., Krasny, M. E., & Bonney, R. (2012). Public participation in scientific research: A framework for deliberate design. *Ecology and Society, 17*(2), 29.

Tregidgo, D. J., West, S. E., & Ashmore, M. R. (2013). Can citizen science produce good science? Testing the OPAL Air Survey methodology, using lichens as indicators of nitrogenous pollution. *Environmental Pollution, 182*, 448–451.

United Nations. (1992). *Agenda 21*. Rio de Janeiro: United Nations.

WCED, & Brundtland, G. H. (1987). *Our common future*. Oxford, UK: Oxford University Press.

Part VI
Health Impact Assessment and Policies

Chapter 28
Transport Policy Measures for Climate Change as Drivers for Health in Cities

Haneen Khreis, Andrew Sudmant, Andy Gouldson, and Mark Nieuwenhuijsen

28.1 Introduction

Climate action in the urban transport sector presents a unique and daunting challenge. The transport sector accounted for 23% of global greenhouse gas (GHG) emissions in 2010 and remains one of the fastest growing sources of global emissions, despite advances in vehicle efficiency (Sims et al. 2014). In contrast with most other major sources of global emissions, fossil fuels remain the dominant final energy source in transport, with oil accounting for over 90% of the final energy demand (IEA 2016). The lack of progress with electricity decarbonization limits the prospects of potential emission, climate and air quality benefits of electric vehicles (UK Energy Research Centre 2016), and one of the key components of the decarbonization of vehicle travel to date—the shift to diesel powertrain—is now under significant scrutiny (UK Energy Research Centre 2016; Cames and Helmers 2013).

H. Khreis (✉)
Texas A&M Transportation Institute (TTI), College Station, TX, USA

ISGlobal, Centre for Research in Environmental Epidemiology (CREAL), Barcelona, Spain

Universitat Pompeu Fabra (UPF), Barcelona, Spain

CIBER Epidemiologia y Salud Publica (CIBERESP), Madrid, Spain

Institute for Transport Studies, University of Leeds, Leeds, UK
e-mail: haneen.khreis@isglobal.org; H-Khreis@tti.tamu.edu

A. Sudmant · A. Gouldson
ESRC Centre for Climate Change Economics and Policy, Priestley International Centre for Climate, University of Leeds, Leeds, UK

M. Nieuwenhuijsen
ISGlobal, Centre for Research in Environmental Epidemiology (CREAL), Barcelona, Spain

Universitat Pompeu Fabra (UPF), Barcelona, Spain

CIBER Epidemiologia y Salud Publica (CIBERESP), Madrid, Spain

M. Nieuwenhuijsen, H. Khreis (eds.), *Integrating Human Health into Urban and Transport Planning*, https://doi.org/10.1007/978-3-319-74983-9_28

Finally, and potentially of greatest importance, established transport networks and systems are costly, technically challenging and time-consuming to change once they are built, leaving current design and planning trends difficult to alter (Khreis et al. 2016). Transport investments over the coming 5 years will therefore substantially dictate the pathway of transport-related emissions for decades to come, adding more pressure for low-carbon transport options to be developed and implemented in the immediate future (Leather 2009; IEA 2015).

At the same time, the benefits of action are also substantial. The impact of climate change on cities is being felt. Rising seas, heat waves, air pollution and weather disasters made more violent by changing weather patterns are already costing millions of lives and billions in livelihoods each year (Harlan and Ruddell 2011). At the same time, research establishes that an economic case exists for many urban transport options that can mitigate climate change. Investments in a range of currently available public and private measures in the world's urban areas could save 2.8 gigatonnes of GHG emissions annually by 2050—just less than the entire GHG emissions of India in 2011—or 7% of the 2011 global GHG emissions (WRI 2012), whilst providing direct net economic benefits between 2015 and 2050 of $10.5 trillion USD (Gouldson et al. 2015; Sudmant et al. 2016).

Perhaps even more compellingly, urban transport investments to mitigate climate change can have a substantial positive impact on public health in cities. In 2015, the Lancet Commission on Health and Climate Change emphasized that the response to climate change could be "the greatest global health opportunity of the 21st century" (Watts et al. 2015, p. 1). The health co-benefits of climate action in the transport sector can be realized through pathways of improved indoor and outdoor air quality, reduced exposures to noise, mitigation of the urban heat island effect, increased active travel and physical activity, reduced motor vehicle crashes, increased green space exposure and reduced social exclusion and inequalities. Up to 74% of air pollution in cities worldwide, and as much as 90% of urban air pollution in some cities of the developing world, may be attributed to motor vehicle emissions (Arup/C40 2014; UNEP 2014). Traffic-related air pollution is responsible for a large burden of global disease and at least 184,000 deaths in 2010 (Bhalla et al. 2014; Khreis et al. 2016). Transport is also responsible for much of the noise in urban areas (Foraster et al. 2011; Bell and Galatioto 2013; Zuo et al. 2014) with traffic density, distance from the location to the sidewalk and building density explaining over 73% of the variability of noise levels (Foraster et al. 2011). Lee et al. (2014) found that the total number of vehicle counts explained a substantial amount of variation in measured ambient noise in Atlanta (78%), Los Angeles (58%) and New York City (62%). The burden of disease attributable to traffic-related noise is comparable to that of air pollution (Hänninen et al. 2014; Tainio 2015; Mueller et al. 2017a, b). Over the last decades, declining active travel and increased vehicle use have been important contributors to overall declines in physical activity and increases in sedentary behaviour (Brownson et al. 2005; Ewing et al. 2003). According to the World Health Organization (2017), one in four adults and over 80% of the world's adolescents are insufficiently physically active. Motor vehicle crashes cause over 1.5 million global deaths annually and 79.6 million healthy years of life lost (Bhalla et al. 2014). Indeed, after the first year of life through age 59, motor vehicle crashes rank amongst the top ten causes of global deaths (Bhalla et al. 2014) and are the number one global cause of death amongst those aged 15–29 years (World Health Organization 2015). Further deleterious impacts from the transport sector that have been linked with public health outcomes include exposure to local temperature increases (the so-called heat island effect), green space reduction and biodiversity loss (Khreis et al. 2016). The cost of these transport externalities is very high, especially in rapidly developing urban areas. For example, the costs of motorized transport's congestion, air pollution, motor vehicle crashes, noise and climate change in Beijing are between 7.5% and 15.0% of its GDP (Creutzig and He 2009). Rapid technological changes and volatility in energy prices, which radically alter the economic case for investments, make predictions and scenario comparisons challenging. Nonetheless, and as we will show next, there is great opportunity for policymakers to develop transport roadmaps that jointly achieve climate change, health and economic objectives.

In this chapter, we provide a state-of-the-art review of the co-benefits of climate action on health at the urban level focusing on five urban transport actions: (1) compact land use planning to reduce motorized passenger travel demand, (2) passenger modal shift and improving transit efficiency, (3) electrification and passenger vehicle efficiency, (4) freight logistics and (5) freight vehicle efficiency and electrification. Health impacts from these actions occur via pathways of reduced air pollution, noise and temperature and increased green space and physical activity. We argue that presenting a more robust health case of low carbon action by assessing the co-benefits of these policy measures may unlock policy support and accelerate action.

28.2 Co-benefits of Climate Action on Health

28.2.1 Urban Planning and Reduced Passenger Travel Demand

28.2.1.1 Air Pollution

Urban planning to reduce passenger travel demand is a key measure that cities can adopt to improve air quality and public health and achieve significant economic savings (Ling-Yun and Lu-Yi 2016; Bartholomew 2007; Stone et al. 2007; Reisi et al. 2016; Guttikunda and Mohan 2014; Frumkin 2002; Grabow et al. 2012; Giles-Corti et al. 2016; Conlan et al. 2016; Stevenson et al. 2016). Designing cities to be compact and mixed-use can lead to shorter distances and easier access to work, school, and other activities and therefore reduce the need for passenger car travel (Guttikunda and Mohan 2014). Reductions in passenger car travel demand are often also accompanied by modal shifts towards more sustainable transport means, such as walking, cycling and the use of public transport. Conversely, the rapid expansion of metropolitan areas, or urban sprawl, and the resulting un-mixed land use and low-density development patterns reinforce the need and convenience for extensive road networks and private car travel (Frumkin 2002). The literature is supportive of this narrative; in the following key articles are described.

Ewing et al. (2008a, b) found that high density can reduce vehicle kilometres by 40% per capita and comparisons of urban centres showed that dense, highly connected urban centres like Hong Kong produce only 1/3 of the carbon emissions per capita of European cities, whilst European cities produce only 1/5 the carbon emissions of sprawling poorly connected cities like Houston (Rode et al. 2013). Recent reviews and large-scale health impact assessments concluded that urban planning measures, unlike other transport policy instruments aimed at reducing traffic-related air pollution (e.g. freight management), have the potential to realise air quality improvements over the longer term and provide additional benefits related to relieving congestion, improving the quality of places and increasing population physical activity levels, all of which associated with many health and wellbeing benefits (Conlan et al. 2016; Giles-Corti et al. 2016; Stevenson et al. 2016).

Many studies have provided quantification of the expected air quality and health benefits from such measures, although these were almost exclusively based on health impacts assessment modelling exercises. For example, reducing the kilometres travelled by Chinese residents by 5% and 10% via increasing cycling would lead to around 1.56% and 3.11% decrease in annual average concentrations of SO_2, respectively, and up to 2.80%, 6.18% and 5.86% decrease in NO_2, $PM_{2.5}$ and PM_{10}, respectively. If these reductions in demand were accompanied with a shift towards public transit, the estimated benefits were higher. The number of associated preventable deaths from air pollution-related disease per year was estimated to range from 569,000 to 4,516,000, depending on the scenario being tested. The estimated health improvements including the reduction in total mortality, cardiovascular and respiratory hospital admissions and asthma attacks would save 3433.25 to 27,337.1027 billion Yuan (Ling-Yun and Lu-Yi 2016).

Grabow et al. (2012) suggested that the elimination of automobile round-trips $\leq$8 km in 11 metropolitan areas in the upper Midwestern United States would reduce $PM_{2.5}$ by 0.1 $\mu g/m^3$ and although summer ozone (O_3) would slightly increase in cities, it would decline regionally, resulting in net health benefits of $4.94 billion/year (95% confidence interval (CI), $0.2 billion, $13.5 billion). If 50% of the eliminated trips were made by bicycle, the health benefits would increase significantly due to increased levels of physical activity reducing mortality by 1295 deaths/year (95% CI: 912–1636). The combined health benefits of improved air quality and increased physical activity were estimated to exceed $8 billion/year (Grabow et al. 2012).

Reisi et al. (2016) selected multiple sustainability indicators including vehicle emissions and mortality effects of air pollution and evaluated these under three urban planning scenarios: a base case scenario based on governmental plans in Melbourne for 2030, activity centres scenario based on compact urban development patterns and a fringe focus scenario based on expansive urban development patterns. The activity centres scenario resulted in the least GHG and other emissions, as well as a reduction of mortality when compared to the other two scenarios.

In a health impact assessment of six cities, land use changes were modelled to reflect a compact city in which land use density and diversity were increased and distances to public transport were reduced to drive a modal shift from private vehicles to walking, cycling and public transport. The modelled compact city scenario resulted in health benefits for all cities (for diabetes, cardiovascular disease and respiratory disease) with overall health gains of 420–826 disability-adjusted life years (DALY) per 100,000 people. However, for moderate to highly motorized cities, such as Melbourne, London and Boston, the compact city scenario predicted a small increase in road trauma for cyclists and pedestrians (health loss of between 34 and 41 DALYs per 100,000 people) (Stevenson et al. 2016).

Overall, these studies demonstrate that compact and mixed-use urban planning which can reduce passenger travel demand will improve air quality and public health in urban areas.

28.2.1.2 Noise

Whilst relatively less research has focused on the effect of urban planning and reduced passenger travel demand on noise and human health, impacts are increasingly recognized. Creutzig et al. (2012) provided scenarios of increasingly ambitious policy packages, reducing GHG emissions from urban transport by up to 80% from 2010 to 2040. Based on stakeholder interviews and data analysis, the main target was a modal shift from motorized individual transport to public transit and nonmotorized individual transport (walking and cycling) in the four European cities of Barcelona, Malmo, Sofia and Freiburg. The authors reported significant concurrent co-benefits of better air quality, reduced noise, less traffic-related injuries and deaths, increased physical activity, alongside less congestion and monetary fuel savings. For the most ambitious scenario explored which included multiple measures and a large shift towards nonmotorized individual transport (referred to as push scenario), a reduction between 10% and 29% in noise levels in cities was reported. The push scenario included congestion charging and aggressive land use policies restricting new development to car-free areas with bicycle infrastructure and public transport access, amongst others. The scenario included policies that were only considered by some stakeholders, and although these were mostly not under consideration by local authorities, they were judged to be plausible.

There is a clear need to fill large gaps in the literature on the effects of urban planning and reduced passenger travel demand on noise and associated health effects.

28.2.1.3 Physical Activity

Multiple climate policies targeting the transport sector have the potential to increase physical activity levels and provide significant health benefits, even after consideration of increased air pollution exposure and motor vehicle crashes (Mueller et al. 2015). Urban planning to reduce passenger travel demand and increase transit mode share has been identified as a key measure which cities can adopt to increase population physical activity and improve public health[1] (D'Haese et al. 2015; Ewing et al. 2014; Frank et al. 2005; Cohen et al. 2006; Sugiyama et al. 2012; Buehler and Pucher 2012; Wong et al. 2011; Committee on Environmental Health 2009; Saelens et al. 2003; Lee and Moudon 2004; Rodriguez et al. 2006; Salon 2016; Stevenson et al. 2016). The following describes the key articles we identified in the literature.

Sugiyama et al. (2012) systematically reviewed 46 quantitative studies examining the associations between walking, as an active travel mean (utilitarian walking), and multiple built environment factors including the presence and proximity of destinations and sidewalks, connectivity, aesthetics and traffic on and safety of the walking routes. Half of the studies came from North America, 11 from Australia, 8 from Europe, 3 from South America and 1 from Japan. The literature synthesized in this review consistently showed (80% of studies) that the presence and proximity of retail and service destinations are conducive to adults walking. Other factors including well-connected streets and the availability of sidewalks also facilitated active travel.

Additional reviews support the findings of Sugiyama et al. (2012). Day et al. (2006) focused their review on 42 empirical studies conducted in China and similarly showed that active travel was most strongly associated with the proximity of non-residential locations. The literature reviewed supported an association between land use mix (proximal non-residential locations) and active travel in Chinese cities. In a review of 65 studies of children from North America, Europe, Australia and Asia, D'Haese et al. (2015) found that walkability, density and accessibility were associated with active travel to school. Buehler and Pucher (2012) reviewed the literature on the advantages and disadvantages of higher urban densities and showed that numerous studies pinpointed the public health benefits stemming from more walkable and cycling-friendly environments alongside decreasing the number and distance of vehicle trips. Similarly, in a systematic review synthesizing 14 studies by Wong et al. (2011), the authors concluded that increasing distance is negatively associated with children's active travel to schools. Lee and Moudon (2004) showed that long distances between destinations and poor accessibility to recreational facilities are the most cited barriers to physical activity, whilst Cohen et al. (2006) showed that every mile a girl lived further to her school translated into 13 fewer weekly minutes of metabolic activity and that the time spent commuting was responsible for this reduction.

Increased urban sprawl and further distances between origins and destinations were also associated with decreased physical activity and increased obesity rates in adults (Ewing et al. 2014; Frank et al. 2005). In a review synthesizing studies conducting neighbourhood comparisons of walkability and physical activity, Saelens et al. (2003) showed that residents of high-walkable neighbourhoods reported two times more walking trips/week as compared to residents of low-walkable neighbourhoods (3.1 vs. 1.4 trips). The difference magnitude between high-walkable and low-walkable neighbourhoods ranged from −0.1 to 5.7 walk

[1] Other policy measures include promoting electric bikes (Ji et al. 2012; MacArthur et al. 2014), which are relevant but were outside the scope of this review.

trips and was partially dependent on the purpose of the trip, with walking to work being consistently more likely in high-walkable compared to low-walkable neighbourhoods. This difference translated into 1–2 km of active travel or about 15–30 min more walking per week. Similarly, Rodriguez et al. (2006) showed that residents of walkable urban neighbourhoods are making twice as many trips walking and cycling compared to residents of suburban neighbourhoods, logging 40–55 extra minutes of weekly physical activity. Like previous observations, these differences were mainly driven by utilitarian travel, rather than leisure travel. Residents of the walkable urban neighbourhoods also travelled fewer vehicle miles. Salon (2016) showed that pedestrian and cyclist road use in urban census tracts is double that in suburban census tracts, which in turn is an order of magnitude greater than that in rural census tracts.

Besides reducing passenger travel demand, urban planning that promotes compact and mixed land use patterns can also make transit more viable and efficient (Boyko and Cooper 2011) and contributes to increasing public transit demand (Buehler and Pucher 2012). When coordinated in packages of other complementary policies (e.g. attractive transit fares, high taxes, restrictions on car usage, etc.), compact cities can increase the likelihood to use public transport by five times (Buehler and Pucher 2012). American people who use public transit spend a median of 19 min a day walking to and from transit, and almost 30% of transit users achieve the daily recommended physical activity levels of ≥30 min (Besser and Dannenberg 2005). There is also evidence that people from ethnic minorities and lower socio-economic classes are more likely to harvest more health benefits of physical activity as these groups were more likely to spend ≥30 min daily commuting to and from transit (Besser and Dannenberg 2005).

Overall, there are plenty of studies and evidence reviews which demonstrate that compact and mixed-use urban planning can reduce passenger travel demand and promote active travel modes, increasing population physical activity and therefore improving public health in urban areas.

28.2.1.4 Motor Vehicle Crashes

Increased traffic flows are associated with increased motor vehicle crashes (Nakahara et al. 2011; Yiannakoulias and Scott 2013); therefore, urban planning measures to reduce traffic and vehicle kilometres travelled are likely to reduce road deaths and injuries (Ewing and Hamidi 2015). Further, more sustainable urban planning patterns accommodating walking, bicycling and transit-friendly neighbourhoods are expected to lead to reduced passenger car demand and use, which in turn has been associated with reduced motor vehicle crashes (Wei and Lovegrove 2012).

Research has shown that traffic volumes and road densities are the main determinants of motor vehicle crash frequency. Dumbaugh and Li (2010) estimated that each additional mile of thoroughfare is associated with a 9.3% increase in motor vehicle-pedestrian crashes. A reduction of 30% in traffic volumes is associated with a 35% reduction in the number of pedestrians injured in motor vehicle crashes and a 50% reduction in the average risk of a pedestrian collision (Miranda-Moreno et al. 2011). Lowering traffic flows was linked to a decline in child pedestrian deaths in New Zealand (Roberts et al. 1992), and when flow reductions were achieved as part of the London congestion charging, they were linked to a substantial reduction in both the number of crashes and crash rates (Green et al. 2016). The London congestion charging zone was associated with 44 less motor vehicle crashes per month which represented a 35% decline.

Density and 'compactness' are shown to be negatively correlated with motor vehicle crashes and fatalities. Ewing et al. (2003) found that all traffic fatalities and pedestrians' fatalities decrease by 1.49% and 1.47%, respectively, with each 1% increase in a compactness index. Godwin and Price (2016) argued that the distinct pattern of low-density urban areas in Southeast USA is likely to be leading to rare and more dangerous walking and cycling resulting in decreased road safety. Yeo et al. (2015) conducted a path analysis to examine the causal linkages between urban sprawl, vehicle miles travelled and traffic fatalities, drawing on data from 147 urbanized areas in the USA. The authors, in line with previous research (Ewing et al. 2014), found that there was an indirect positive effect of urban sprawl on fatalities which occurred through increases in vehicle miles travelled, alongside a more influential direct effect possibly occurring through increased traffic speeds and emergency medical service delays.

However, literature also shows that urban planning interventions which are established to reduce passenger car demand may have different effects on road safety, perhaps mediated by the increase in active travel levels. In a systematic review of 85 quantitative studies investigating the effect of built environment on childhood walking and pedestrian injuries, Rothman et al. (2014) found that higher road density and higher traffic speeds increase injury incidence and severity. Land use mix and proximity to services and facilities were also found to increase injuries incidence and severity but were associated with increased walking. The authors discussed that these urban environment characteristics may not be inherently dangerous but rather be markers for increased exposure to traffic in general. Two urban measures were consistently found associated with increased walking and a reduction in child pedestrian injuries, namely, traffic calming and the diversity of land use including the presence of playgrounds, recreation, parks and open spaces. Further studies similarly showed that land use mix and proximity to retail, although favourably promoting walking and cycling, are associated with increased crash risks in children and adults (Cho et al. 2009; Elias and Shiftan 2014; Yu 2014; Lee et al. 2013), whilst the results from other studies were mixed (Blazquez et al. 2016). Policymakers therefore need to be cautious when applying urban planning measures and take into account contextual factors and the need for complementary measures to protect those shifting or choosing active travel modes.

28.2.2 *Passenger Mode Shifts*

28.2.2.1 Air Pollution

Passenger mode shifts towards more sustainable travel means and improving transit efficiency can also lead to significant public health improvements, not only through improvements in air quality but also importantly through the increase in physical activity as people walk, cycle and move to catch public transport (Nieuwenhuijsen and Khreis 2016; Pathak and Shukla 2016; Rojas-Rueda et al. 2012, 2013; Woodcock et al. 2013, 2009; Sabel et al. 2016; Xia et al. 2015; McKinley et al. 2005; Yang et al. 2016; Rabl and De Nazelle 2012; Mueller et al. 2017a, b). Some of the studies identified in the previous section also fall under this category but were kept in the section whose searches located them. Similarly, some of the studies we identify next on modal shifts and transit efficiency fall under the next category of vehicle efficiency measures and indicate health impacts through pathways other than air pollution (e.g. changes in physical activity).

Nieuwenhuijsen and Khreis (2016) evaluated the radical concept of car-free cities, a model primarily driven by the need to reduce GHG emissions but has impacts on public health. The authors suggested great benefits in terms of reduction in not only air pollution (up to a 40% reduction in NO_2 levels on car-free days) but also noise and heat island effects and potential increases in green space and physical activity. Three health impact assessments cited in the review of Nieuwenhuijsen and Khreis estimated small air quality improvements and health benefits from the replacement of private car journeys by active or public transport (Rojas-Rueda et al. 2012, 2013; Woodcock et al. 2013). For example, 76 annual deaths and 127 cases of diabetes, 44 of cardiovascular diseases, 30 of dementia, 16 minor injuries, 0.14 major injuries, 11 of breast cancer, 3 of colon cancer, 7 of low birth weight and 6 of preterm birth can be prevented each year, if 40% of long-duration car trips were substituted by public transport and cycling. The largest health benefits estimated were in association with increased physical activity, and then with the reduced air pollution. The reductions in air pollution were not as large as was expected because the contribution of private cars was minor compared to trucks, buses and motorbikes, which were not included in the assessment.

Sabel et al. (2016) estimated that the introduction of a new metro in Thessaloniki, Greece, would reduce local deaths from air pollution by about 20%. Xia et al. (2015) estimated that shifting of 40% of vehicle kilometres travelled to alternative transport in Adelaide, South Australia, would reduce annual average $PM_{2.5}$ by a small margin of 0.4 $\mu g/m^3$, preventing 13 deaths a year and 118 DALYS. Woodcock et al. (2009) conducted a health impact assessment of alternative transport scenarios in London, UK, and Delhi, India. The use of lower-carbon-emission motor vehicles, increased active travel and a combination of the two scenarios were tested. The increase in active travel and less use of motor vehicles (i.e. modal shifts) had larger health benefits per million people (7332 DALYs in London and 12,516 in Delhi in 1 year) than the benefits from the increased use of lower-emission motor vehicles (160 DALYs in London and 1696 in Delhi). The combination of active travel and lower-emission motor vehicles resulted in the largest health benefits (7439 DALYs in London and 12,995 in Delhi). Most of the preventable premature deaths were estimated to result from increased physical activity, followed by the reduction of air pollution exposures.

McKinley et al. (2005) quantified cost and health benefits from a subset of air pollution control measures in Mexico City. The control measures tested were taxi fleet renovation, metro expansion and use of new hybrid buses replacing diesel buses. Avoided cases of 11 health outcomes, including premature mortality, chronic bronchitis, hospitalizations and emergency room visits for cardiovascular and respiratory disease, and minor restricted activity days (MRAD) were quantified in association to reductions in O_3 and PM_{10}. The measures were found to have air pollution reductions of approximately 1% for PM_{10} and 3% for O_3. The associated health benefits were substantial, and their sum over the three measures was greater than the measures' investment costs. For the individual scenarios, the benefit to cost ratio was 3.3 for the taxi renovation measure, 0.7 for the metro expansion measure and 1.3 for the new hybrid buses measure. The taxi fleet renovation was the most appealing option, but this was contextual because of the size and age of the taxi fleet in Mexico City.

Overall, these studies demonstrate that passenger modal shifts away from the private car and towards active travel modes and public transit will improve air quality and yield public health benefits in urban areas.

28.2.2.2 Noise

Private vehicles are one of the largest single sources of noise pollution in urban areas. Transport measures that shift transport demand away from motorized vehicles and towards public transport or nonmotorized options therefore present a major opportunity for impacts on noise pollution.

In their health impact assessment exercise, James et al. (2014) projected that a proposed fare increases and service cuts to public transport which would shift 48,600 people from public transport to driving in the Boston region would result in lost time due to congestion, increased air pollution, lower levels of physical activity, additional motor vehicle crashes, increased exposure to high noise levels and increased greenhouse gas emissions. In their scenario representing fare increases by up to 35% and service reductions affecting 53–64 million trips each year, an additional 2000 people on average were estimated to become exposed to over 60 dB of noise per day, which would be detrimental to health.

Sabel et al. (2016) modelled the impact of urban climate change mitigation transport measures in five European and two Chinese cities. The changes in exposure to noise due to the investigated transport measures were modelled in three of the seven cities, and the results suggested that promoting electric cars and reducing the use of personal cars had a very limited effect on reducing noise levels. In part, this was due to the high proportions of traffic on non-urban roads such as motorways which were not subjected to the investigated local traffic reduction policy measures (e.g. up to 35% of motorway traffic in Rotterdam). On the other hand, the introduction of a metro in Thessaloniki was predicted to more significantly reduce noise levels.

Rabl and De Nazelle (2012) presented an estimate of the potential health impacts due to a shift from car to cycling or walking, by evaluating the effects relating to changes in air pollution exposures and accident risk and citing costs for other impacts, namely, from noise to congestion. A driver who switches to cycling for a commute of 5 km (one way) 5 days/week for 46 weeks/year would experience health benefits from the reduction in noise which are worth about €1700 (at a cost of 0.76 h/km). The results for walking (2.5 km) were similar. Applying these estimates to a specific example of a policy measure in Paris, the Velib bike sharing system, the population benefits of the reduction in noise were estimated at €69.9 million/year.

Research of the effects of modal shift on noise pollution on cities, though limited in scope, presents a strong case that climate actions can yield substantial co-benefits.

28.2.2.3 Physical Activity

Promoting passenger modal shift from the private car towards active travel means, including walking and cycling, also has many well-documented health benefits through increased physical activity. These benefits were explored in multiple epidemiological studies, health impact assessments and systematic reviews (Lubans et al. 2011; Saunders et al. 2013; Ogilvie et al. 2004; Mueller et al. 2015; Flint et al. 2016; Matthews et al. 2007; Hu et al. 2007, 2003; Hamer and Chida 2009; Woodcock et al. 2009, 2013; Xia et al. 2015; Rojas-Rueda et al. 2012, 2013; Rabl and De Nazelle 2012; Stevenson et al. 2016).

One of most well regarded such studies followed 47,840 Finnish people, aged 25–64 years old, for an average of 19 years. The results showed that the risk of coronary heart disease was significantly decreased with increasing occupational, leisure time or active commuting-related physical activity (Hu et al. 2007). Daily commuting on foot or by bike was associated with a 20% reduction in risk of new coronary heart disease amongst women, after adjusting for a comprehensive set of confounders. Similarly, walking or cycling to work for ≥30 min/day was associated with a 36% reduction in risk of type 2 diabetes in both women and men, even after adjusting for a comprehensive set of confounders (Hu et al. 2003).

Flint et al. (2016) used longitudinal data for 5861 individuals from the UK Biobank to test the effect of transition from car to active (walking and cycling) or public transport on objectively measured body mass index. After adjusting for a comprehensive set of confounders including baseline body mass index, age, sex, ethnicity, income and education, a mode shift from the car to active or public transport was associated with a 0.30 kg/m^3 decrease in body mass index. Conversely, a mode shift from active or public transport to car commuting was associated with a 0.32 kg/m^3 increase in body mass index. These results were almost identical to a similar study using self-reported body mass index (Martin et al. 2015).

Finally, a recent health impact assessment evaluated the potential effects of a compact city model on physical activity levels in six cities: Melbourne, Australia; Boston, MA, USA; London, UK; Copenhagen, Denmark; São Paulo, Brazil; and Delhi, India (Stevenson et al. 2016). The compact city model tested resulted in increased active travel and public transport use which translated into increases in travel-related physical activity ranging from +72.1% metabolic equivalent per week (in the most motorized city; Melbourne) to +18.5% in the rapidly motorizing Delhi.

28.2.2.4 Motor Vehicle Crashes

Modal shifts can also reduce the number of vehicles on the road, thereby reducing the opportunities for motor vehicle crashes. At the same time, cycling and walking are more vulnerable transport modes. Research on increases in public and nonmotorized transit generally finds reductions in total crashes, but as we will show next, interventions need to be carefully planned to avoid unintended consequences and protect individuals switching to more vulnerable transport modes.

In a systematic review including 21 health impact assessment studies of the effects of passenger modal shift on motor vehicle crashes, Mueller et al. (2015) reported that 14 studies estimated an increase in motor vehicle crashes, 6 studies estimated a decrease in motor vehicle crashes and 1 study estimated no change in motor vehicle crashes with the increase in active travel. However, only eight studies accounted for the non-linear traffic incident risk attributed to the increased safety that comes with the increased number of active travellers, the so-called safety in numbers effect (Jacobsen 2003).

Jacobsen (2003) showed a consistent inverse relation between the amount of walking and cycling and the likelihood of a pedestrian and cyclist being involved in a motor vehicle crash. The author argued that a community doubling its walking can expect a 32% reduction in cycling injuries. Contrasting with this result, another review by Götschi et al. (2016), which considered studies modelling the safety impacts of cycling, showed that a modal shift from car travel to cycling is estimated to increase the number of cyclists' fatalities and seriously injured. The authors, however, also showed that the public health benefits stemming from increases in physical activity well outweigh the estimated risks, for example, shifting a 10-km commute from car to bike will result in an average cost of €50 per person/year from fatal crashes but an average saving of €1300 per person/year from physical activity (Rabl and De Nazelle 2012). Götschi et al. (2016), in line with other investigators (Rabl and De Nazelle 2012), also concluded that impact modelling of cycling crash risks is a highly case- and context-specific matter. Furthermore, the increases in cyclists' and pedestrians' fatalities and injuries following a mode shift from the car might be a representation of a period of increased vulnerable road users' collisions and injuries which will cease when the cycling mode split approaches the 20% level (Wei and Lovegrove 2012).

From a different angle, Tainio et al. (2014) estimated the years lived disabled or injured for pedestrians, cyclists and car occupants and found that injured pedestrians and cyclists sustain 9.4 and 12.8 years lived disabled or injured whilst car occupants sustain a higher period of 18.4 years due to the severities of injuries. The authors estimated that a person who would switch from car use to cycling in an urban area would, on average, have 40% less severe injuries.

Public transit offers similar benefits. Public transport is substantially safer than car travel, both for passengers and for the public (Litman 2011; Kenworthy and Laube 2002; Beck et al. 2007). Research shows that cities with the highest share of public transport users have the lowest share of traffic fatalities (Litman 2012). Cities built on a combination of transit and walking could also mitigate motor vehicle crashes. For comparison, between 2010 and 2014, the four largest Dutch cities, Amsterdam, The Hague, Rotterdam and Utrecht (all with a very high bicycle modal share by international standards) recorded 2.0 road deaths per 100,000 populations (SWOV 2016), whilst Hong Kong and Paris, both centred on mass transit (Sun et al. 2014), recorded 1.5 and 1.6 road deaths per 100,000 populations, respectively (Transport Department Hong Kong 2016; Préfecture de Police 2013). In a more recent analysis, public transit was shown to have 1/10 the per mile traffic casualty rate when compared to car travel (American Public Transportation Association 2016).

Overall, these studies demonstrate that passenger modal shifts away from the private car and towards active travel modes and public transit will improve traffic safety and yield public health benefits in urban areas, although contextual factors are specifically important.

28.2.3 Passenger Car Efficiency and Electrification[2]

28.2.3.1 Air Pollution

A small number of studies investigated the air quality and health impacts of improving passenger car efficiency and electrification (Xue et al. 2015; Pathak and Shukla 2016; Timmers and Achten 2016; Ji et al. 2012; Soret et al. 2014).

Xue et al. 2015 showed that although increasing new energy vehicle proportions including hybrids, CNG and electric vehicles will improve air quality in Xiamen, China, the greatest contribution to air pollution reduction comes from other policy scenarios including alternative fuels and diesel truck decreases contributing to nearly 30% of air pollution reductions, followed by an option to control the intensity of private vehicles.

Incentivizing the electrification of passenger cars is often proposed as a sustainable approach to urban mobility and economic development (Ji et al. 2012). The literature, however, suggests that this may be a simplistic view and that electric vehicles may not result in presumed air quality and health benefits, unless their uptake is optimistically large, and they are complemented by non-exhaust PM

[2] No studies on physical activity or noise relating to passenger car efficiency/electrification and health were identified by this review. This area therefore remains for further research.

reduction measures. A state-of-the-art review by Timmers and Achten (2016) investigated the effects of fleet electrification on non-exhaust PM emissions and found that total PM_{10} emissions from electric vehicles are likely to be higher than their non-electric counterparts, due to non-exhaust PM being increased by the higher weight of electric vehicles. Electric vehicles are, on average, 280 kg or 24% heavier than their non-electric counterparts. Vehicle weight is directly proportional to non-exhaust PM emissions attributable to traffic, as road abrasion and tyre wear, brake wear and resuspension are all influenced by the vehicle's weight. Tyre, brake and road wear increase by 50% when comparing a medium (1600 kg) and small (1200 kg) car. Large cars (2000 kg) emitted more than double the amount of PM_{10}. Further, the reduction in $PM_{2.5}$ emissions from electric vehicles was estimated to be small (1–3%). Research on the health effects of non-exhaust PM is new and perhaps not sufficiently explored, but there are numerous studies, both epidemiological and toxicological, indicating distinct adverse health effects of non-exhaust PM that warrant consideration (Gasser et al. 2009; Gehring et al. 2015).

In a large-scale health impact assessment conducted by Ji et al. (2012), the health impacts of the use of conventional vehicles and electric vehicles in 34 major Chinese cities were modelled. $PM_{2.5}$ emissions from electric cars were estimated to be higher than conventional gasoline vehicles, resulting in more preventable deaths, even when (un)proximity to the emission sources was accounted for. The total annual excess deaths in Shanghai were 9 as attributable to gasoline cars and 26 as attributable to electric cars. When compared to diesel cars, electric cars were shown to provide health benefits, yet the main reason behind these beneficial impacts was not the reduction of total $PM_{2.5}$ emissions from electric cars, but the fact that most of these emissions occurred at power plants instead, away from human receptors. The total annual excess deaths in Shanghai were 90 as attributable to diesel cars and 26 as attributable to electric cars. There was evidence, however, that the urban use of electric vehicles will move adverse exposures and health impacts to rural, non-electric-car users and potentially lower socioeconomic populations.

Overall, the available evidence demonstrated that electric vehicles may only have a small impact on improving air quality and population public health and that conclusions will vary depending on the reference/comparison scenario (e.g. comparing to gasoline versus diesel cars) and whether rural populations are considered in the analysis. The remaining studies which investigated the air quality impacts of improving passenger car efficiency or electrification provided no estimates for the potential health impacts associated with these policies (Pathak and Shukla 2016; Soret et al. 2014).

28.2.3.2 Motor Vehicle Crashes

In relation to the potential effects of fleet electrification on road safety, the only theoretical link currently explored in the literature relates to the effects of increasing vehicle weight due to electrification and decreasing vehicle noise increasing non-detection (Timmers and Achten 2016). Although increased vehicle weight and stiffness may increase the safety of car occupants (Torrao et al. 2016), the opposite is expected for vulnerable and other road users involved in motor vehicle crashes with higher weighted vehicles. For example, Anderson and Auffhammer (2013) estimated that the baseline risk of fatalities increases by 47% when being hit by a vehicle that is 1000 pounds heavier. Studies also suggested that the reduction in noise from electric vehicles may lead to increased motor vehicle crashes because of non-detection, especially to the blind and visually impaired (Mendonça et al. 2013; Verheijen and Jabben 2010; Jabben et al. 2012).

28.2.4 *Freight Logistics Improvements and Freight Vehicle Efficiency and Electrification*[3]

28.2.4.1 Air Pollution

Lee et al. (2012) assessed the air quality and health impacts attributable to a clean truck program in the Alameda corridor, USA, which progressively banned the older and most polluting trucks. The truck replacement was estimated to have reduced NO_x and PM emissions by 48% and 55%, respectively, within a 7-year period. The health benefits from the associated reduction of $PM_{2.5}$ only were equivalent to $428.2 million, but these estimates only incorporated two age groups (age 30–65, >65), and two health endpoints (mortality and chronic bronchitis), and are therefore likely underestimated. To contextualize the results, the authors demonstrated that the payback period for replacing all drayage trucks in the study area (11,000 trucks with an assumed new truck costs $150,000) is less than 4 years.

The published research for this policy measure is scarce. There is clearly a need for more research regarding the effects of freight policies on air quality and public health, but there also is sufficient current evidence showing that freight is a major contributor to local and regional air quality problems. These problems and associated adverse health impacts can be mitigated by upgrading freight fleets and increasing their efficiency (Guttikunda and Mohan 2014; Conlan et al. 2016).

28.3 Discussion

Co-benefits of climate mitigation actions in the transport sector include improved outdoor air quality, increased physical activity, reduced ambient noise and reduced motor vehicle crashes, all of which are associated with reduced morbidity and premature mortality. Generally, there was a lack of costing estimates in the

[3]No studies on noise, physical activity or motor vehicles crashes relating to freight logistics improvements and freight vehicle efficiency and electrification and health were identified. This area therefore remains for further research.

identified literature, and only a few studies monetized the value of the estimated co-benefits. This creates challenges for comparative assessments and for the transfer or good practice in planning and policy.

The most important co-benefits we identified from the literature seem to be the increases in physical activity at the population level, which predominantly result from passenger modal shifts to active travel and from urban planning measures that reduce car travel and promote walking, cycling and public transit use. The least obvious co-benefits, which can also be risks depending on technical improvements, contextual factors and the investigated reference scenarios, seem to be the co-benefits of electrification which may occur through improved outdoor air quality.

Our findings suggest that policymakers should actively pursue urban planning and land use elements to reduce passenger car travel and at the same time support a modal shift away from the private car towards walking, cycling and public transit. Such measures should be complemented with safety increasing measures to avoid unintended consequences by reducing motor vehicle crashes and protecting existing and new pedestrians and cyclists. Further, new interventions should be carefully examined as they can feasibly introduce an additional way by which wealthy neighbourhoods become more fragmented from poorer areas. Compact cities and dense and diverse land use planning can also facilitate transit-oriented development, further adding to the net benefits from such interventions and providing other economic benefits. Despite being expensive and technically challenging to implement in already developed cities, urban planning and land use policies have the potential to realize longer-term and sustainable health improvements and provide additional benefits related to congestion, retail and quality of public space (Timilsina and Dulal 2010; Conlan et al. 2016). Unfortunately, as it stands, transport planning is generally not well integrated with land use planning. This integration is essential—if system transformations are to be made towards sustainable and healthy developments (Khreis et al. 2017).

Although the increases in physical activity from more compact and more connected urban planning patterns may seem small (e.g. 15–55 min of physical activity per week based on the evidence overviewed above), these differences should be considered important and relevant. Unlike interventions which target individuals or specific population segments to increase their physical activity levels, changes in the urban environment are a universal intervention that reaches the whole population of the targeted area and therefore increases population level physical activity rather than select individuals who are motivated and participate in tailored interventions. Further, changes in the urban environment are long lasting, whilst behaviour change campaigns and programs are rarely maintained (Saelens et al. 2003). The benefits of higher urban densities go beyond the climate and public health benefits we discussed and include increasing the potential for urban agriculture which in turn reduces food miles and improves local food security, improving the housing choices for all residents, reducing social exclusion, increasing the opportunities for creative social interaction, reducing crime rates and improving economic efficiency and employment opportunities, amongst others. There are, however, documented disadvantages of increasing density which highlight the need for an integrated policy approach to overcome these potential negative consequences. These include exacerbating traffic congestion and parking problems, reducing an area's capacity to absorb rainfall, reducing green space and negatively impacting the economic development of surrounding rural areas.

There was considerable evidence for an improvement in air quality with the mitigation measures we reviewed, and thereby improvements in health. Most of the benefits come from reducing car use and increasing public and active transport. Technological advances such as electrification have been suggested to improve air quality. However, the literature we identified for electrification is scarce, but some key points emerged. First, the air quality (and associated health) benefits of electric cars need more rigorous quantification, but current evidence suggests that these might be negligible (in the case of $PM_{2.5}$) or might even have the opposite direction (in the case of PM_{10}). Second, the reduction of $PM_{2.5}$ emissions from electric vehicles is also likely to disappear as exhaust emission standards become stricter over time (Timmers and Achten 2016). Particularly, the non-exhaust PM component of air pollution is expected to increase with the adaptation of electric vehicles, primarily driven by the increase in vehicle weight due to electrification. Whilst both emissions control regulation and electrification plans could lead to reductions in exhaust emissions, non-exhaust emissions from road vehicles remain unabated and need more attention (Thorpe and Harrison 2008). Some views expect battery technology to quickly improve making electric vehicles lighter, yet the overall trend over the last decade was a steady increase in vehicle weight in almost all segments (Timmers and Achten 2016). Third, there was evidence that there may be air quality benefits of electric vehicles when compared to diesel vehicles, but this was not true when compared to gasoline vehicles. Improvements in gasoline technology could also lead to important air quality and health benefits, yet the advancement of gasoline engines has received considerably less attention during the past 20 years, when, for example, Europe shifted its research capacity towards the diesel powertrain (Cames and Helmers 2013).

Despite diesel-fuelled freight vehicles being the largest motor vehicle contributors to air pollution within and between cities (Guttikunda and Mohan 2014; Conlan et al. 2016), there is significantly less research on freight interventions to improve air quality and health outcomes. In this research, the main uncertainty is related to uncertainties in the emission factors used which are likely to have underestimated the air quality benefits as these are generally optimistic and not reflective of real-world driving and conditions. The current evidence base suggests that the replacement of old trucks by newer, more efficient models can lead to significant air quality and health benefits, with savings that exceed the cost of investments.

The studies on noise are scarce. These studies used a health impact assessment methodology and mainly focused on the effects of a mode shift. The estimates for noise were crude and were not based on detailed modelling or measurement exercises. In all the studies we found, noise was not the main exposure that was being explored, but was rather a secondary outcome. The literature on the potential health impacts of noise is very limited and did not incorporate any scenario testing to quantify the health effects of specific policy measures or interventions (Hänninen et al. 2014; Mueller et al. 2017a, b; Fritschi et al. 2011). There is a clear need to fill large gaps in the literature on the effects of urban planning; reduced passenger travel demand, vehicles efficiency and electrification; and freight logistic improvements on noise and associated health effects. There is also a need to provide health estimates under different scenarios and for different policy measures.

Although the research is currently lagging behind, policymakers need to be aware that noise can have large adverse health impacts, which are comparable to those of air pollution (in the case of premature mortality) (Mueller et al. 2017a), or exceeding those of air pollution (in the case of morbidity) (Mueller et al. 2017b). Such impacts are currently not seriously considered in policy or in academic circles, as shown by the results of our searches and review. Further, there was evidence in the literature we identified that the scale of the positive impacts of traffic noise reduction measures may be limited to the scale of interventions, where local actions can have little effect due to national sources such as motorway traffic not being subjected to local traffic reduction measures. This highlights the advantage of complementing local policies with national policies. Unintended consequences of traffic noise reduction measures include the potential for increases in motor vehicle crashes, because of non-detection. Various studies have suggested that electric cars or hybrid cars are favourably quieter than conventional cars but that this may lead to increased motor vehicle crashes because of non-detection, especially in the blind and visually impaired (Mendonça et al. 2013; Verheijen and Jabben 2010; Jabben et al. 2012). For this reason, some countries like the USA and Japan are considering minimum noise requirements for such vehicles; however, this can restrict the noise reduction advantages (Verheijen and Jabben 2010).

The increases in motor vehicle crashes with increasing mixed land use may be explained by the increasing walking or cycling population which becomes exposed to a well-established higher crash risk. Policies that encourage densification, mixed land use and increase transit supply are likely to increase pedestrian and potentially cyclists' activity. With no supplementary safety strategies in place, these policies may indirectly increase motor vehicle crashes (Miranda-Moreno et al. 2011). Similarly, compact areas with low-speed, but high-conflict traffic environments, can increase crash exposures (Ewing and Hamidi 2015). Policies that encourage modal shifts from the private car, although clearly demonstrating net public health benefits, can increase the risk of pedestrians and cyclists being involved in motor vehicle crashes. These results, however, were highly context specific and differ substantially depending on the study area, including its motorization levels, baseline active travel commuters and safety measures in place for pedestrians and cyclists. Furthermore, there are numerous methodological limitations which should be considered in this literature including the fact that most studies do not account for the 'safety in numbers' effect and the weaknesses in the indicators used to measure crashes (numerators) and corresponding exposures (denominators) (Mueller et al. 2015; Götschi et al. 2016).

Overall, we show that climate change mitigation measures in the transport sector have a great potential to improve public health in urban areas whilst mitigating climate change and its distal impacts. The scale and magnitude of the health benefits differed across the policy measures and were wider and highest in the case of the two policy measures of (1) compact land use planning to reduce motorized passenger travel demand and (2) encouraging passenger modal shift away from the private car and improving transit efficiency. We conclude that climate change action represents a great opportunity for policymakers to develop transport roadmaps that jointly achieve climate change objectives and improve public health in cities.

Acknowledgement Research for this chapter was originally conducted with support from the New Climate Economy, an international initiative that examines the risks and opportunities of addressing climate change. Special thanks are extended to Sarah Colenbrander for her help in reviewing this analysis.

References

American Public Transportation Association. (2016). *The hidden traffic safety solution: Public transportation.* Retrieved from https://www.apta.com/resources/reportsandpublications/Documents/APTA-Hidden-Traffic-Safety-Solution-Public-Transportation.pdf.

Anderson, M. L., & Auffhammer, M. (2013). Pounds that kill: The external costs of vehicle weight. *Review of Economic Studies, 81*(2), 535–571. Retrieved from https://academic.oup.com/restud/article/81/2/535/1517632.

Arup/C40. (2014). Climate Action in Megacities Version 2.0. Retrieved from http://issuu.com/c40cities/docs/c40_climate_action_in_megacities/11?e=10643095/6541335.

Bartholomew, K. (2007). Land use-transportation scenario planning: Promise and reality. *Transportation, 34*(4), 397–412. Retrieved from https://link.springer.com/article/10.1007/s11116-006-9108-2.

Beck, L. F., Dellinger, A. M., & O'Neil, M. E. (2007). Motor vehicle crash injury rates by mode of travel, United States: Using exposure-based methods to quantify differences. *American Journal of Epidemiology, 166*(2), 212–218. Retrieved from https://academic.oup.com/aje/article/166/2/212/98784.

Bell, M. C., & Galatioto, F. (2013). Novel wireless pervasive sensor network to improve the understanding of noise in street canyons. *Applied Acoustics, 74*(1), 169–180.

Besser, L. M., & Dannenberg, A. L. (2005). Walking to public transit: Steps to help meet physical activity recommendations. *American Journal of Preventive Medicine, 29*(4), 273–280.

Bhalla, K., Shotten, M., Cohen, A., Brauer, M., Shahraz, S., Burnett, R., et al. (2014). *Transport for health: The global burden of disease from motorized road transport.* Retrieved from http://documents.worldbank.org/curated/en/984261468327002120/pdf/863040IHME0T4H0ORLD0BANK0compressed.pdf.

Blazquez, C., Lee, J. S., & Zegras, C. (2016). Children at risk: A comparison of child pedestrian traffic collisions in Santiago, Chile, and Seoul, South Korea. *Traffic Injury Prevention, 17*(3), 304–312. Retrieved from http://www.tandfonline.com/doi/abs/10.1080/15389588.2015.1060555.

Boyko, C. T., & Cooper, R. (2011). Clarifying and re-conceptualising density. *Progress in Planning, 76*(1), 1–61. Retrieved from http://www.sciencedirect.com/science/article/pii/S0305900611000274.

Brownson, R. C., Boehmer, T. K., & Luke, D. A. (2005). Declining rates of physical activity in the United States: What are the contributors? *Annual Review of Public Health, 26*, 421–443. Retrieved from http://www.annualreviews.org/doi/abs/10.1146/annurev.publhealth.26.021304.144437.

Buehler, R., & Pucher, J. (2012). Demand for public transport in Germany and the USA: An analysis of rider characteristics. *Transport Reviews, 32*(5), 541–567. Retrieved from http://www.tandfonline.com/doi/abs/10.1080/01441647.2012.707695.

Cames, M., & Helmers, E. (2013). Critical evaluation of the European diesel car boom-global comparison, environmental effects and various national strategies. *Environmental Sciences Europe, 25*(1), 15. Retrieved from https://enveurope.springeropen.com/articles/10.1186/2190-4715-25-15.

Cho, G., Rodríguez, D. A., & Khattak, A. J. (2009). The role of the built environment in explaining relationships between perceived and actual pedestrian and bicyclist safety. *Accident Analysis & Prevention, 41*(4), 692–702. Retrieved from http://www.sciencedirect.com/science/article/pii/S0001457509000554.

Cohen, D. A., Ashwood, S., Scott, M., Overton, A., Evenson, K. R., Voorhees, C. C., et al. (2006). Proximity to school and physical activity among middle school girls: The trial of activity for adolescent girls study. *Journal of Physical Activity & Health, 3*(Suppl 1), S129–S138. Retrieved from https://activelivingresearch.org/sites/default/files/JPAH_9_Cohen_0.pdf.

Committee on Environmental Health. (2009). The built environment: Designing communities to promote physical activity in children. *Pediatrics, 123*(6), 1591–1598. Retrieved from http://pediatrics.aappublications.org/content/123/6/1591.short.

Conlan, B., Fraser, A., Vedrenne, M., Tate, J., & Whittles, A. (2016). *Evidence review on effectiveness of transport measures in reducing nitrogen dioxide*. London: Department for Environment Food and Rural Affairs (DEFRA). Retrieved from https://uk-air.defra.gov.uk/assets/documents/reports.

Creutzig, F., & He, D. (2009). Climate change mitigation and co-benefits of feasible transport demand policies in Beijing. *Transportation Research Part D: Transport and Environment, 14*(2), 120–131. Retrieved from http://www.sciencedirect.com/science/article/pii/S1361920908001478.

Creutzig, F., Mühlhoff, R., & Römer, J. (2012). Decarbonizing urban transport in European cities: Four cases show possibly high co-benefits. *Environmental Research Letters, 7*(4), 044042. Retrieved from http://iopscience.iop.org/article/10.1088/1748-9326/7/4/044042/meta.

D'Haese, S., Vanwolleghem, G., Hinckson, E., De Bourdeaudhuij, I., Deforche, B., Van Dyck, D., & Cardon, G. (2015). Cross-continental comparison of the association between the physical environment and active transportation in children: A systematic review. *International Journal of Behavioral Nutrition and Physical Activity, 12*(1), 145. Retrieved from https://ijbnpa.biomedcentral.com/articles/10.1186/s12966-015-0308-z.

Day, K., Boarnet, M., Alfonzo, M., & Forsyth, A. (2006). The Irvine–Minnesota inventory to measure built environments: Development. *American Journal of Preventive Medicine, 30*(2), 144–152. Retrieved from http://www.sciencedirect.com/science/article/pii/S0749379705004289.

Dumbaugh, E., & Li, W. (2010). Designing for the safety of pedestrians, cyclists, and motorists in urban environments. *Journal of the American Planning Association, 77*(1), 69–88. Retrieved from http://www.tandfonline.com/doi/abs/10.1080/01944363.2011.536101.

Elias, W., & Shiftan, Y. (2014). Analyzing and modeling risk exposure of pedestrian children to involvement in car crashes. *Accident Analysis & Prevention, 62*, 397–405.

Ewing, R., & Hamidi, S. (2015). Compactness versus sprawl: A review of recent evidence from the United States. *CPL Bibliography, 30*(4), 413–432. Retrieved from http://journals.sagepub.com/doi/abs/10.1177/0885412215595439.

Ewing, R., Bartholomew, K., Winkelman, S., Walters, J., & Anderson, G. (2008a). Urban development and climate change. *Journal of Urbanism, 1*(3), 201–216. Retrieved from http://rsa.tandfonline.com/doi/abs/10.1080/17549170802529316#.Wfoi2WhSyUk.

Ewing, R., Schmid, T., Killingsworth, R., Zlot, A., & Raudenbush, S. (2008b). Relationship between urban sprawl and physical activity, obesity, and morbidity. In *Urban ecology* (pp. 567–582). New York: Springer. Retrieved from https://link.springer.com/chapter/10.1007/978-0-387-73412-5_37.

Ewing, R., Meakins, G., Hamidi, S., & Nelson, A. C. (2014). Relationship between urban sprawl and physical activity, obesity, and morbidity–Update and refinement. *Health & Place, 26*, 118–126. Retrieved from http://www.sciencedirect.com/science/article/pii/S135382921300172X.

Flint, E., Webb, E., & Cummins, S. (2016). Change in commute mode and body-mass index: Prospective, longitudinal evidence from UK Biobank. *The Lancet. Public Health, 1*(2), e46–e55. Retrieved from http://www.sciencedirect.com/science/article/pii/S2468266716300068.

Foraster, M., Deltell, A., Basagaña, X., Medina-Ramón, M., Aguilera, I., Bouso, L., et al. (2011). Local determinants of road traffic noise levels versus determinants of air pollution levels in a Mediterranean city. *Environmental Research, 111*(1), 177–183. Retrieved from http://www.sciencedirect.com/science/article/pii/S0013935110001878.

Frank, L. D., Schmid, T. L., Sallis, J. F., Chapman, J., & Saelens, B. E. (2005). Linking objectively measured physical activity with objectively measured urban form: Findings from SMARTRAQ. *American Journal of Preventive Medicine, 28*(2), 117–125. Retrieved from http://www.sciencedirect.com/science/article/pii/S0749379704003253.

Fritschi, L., Brown, L., Kim, R., Schwela, D., & Kephalopolous, S. (2011). *Burden of disease from environmental noise - quantification of healthy life years lost in Europe*, World Health Organisation.

Frumkin, H. (2002). Urban sprawl and public health. *Public Health Reports, 117*(3), 201. Retrieved from https://www.cdc.gov/healthyplaces/articles/urban_sprawl_and_public_health_phr.pdf.

Gasser, M., Riediker, M., Mueller, L., Perrenoud, A., Blank, F., Gehr, P., & Rothen-Rutishauser, B. (2009). Toxic effects of brake wear particles on epithelial lung cells in vitro. *Particle and Fibre Toxicology, 6*(1), 30. Retrieved from https://particleandfibretoxicology.biomedcentral.com/articles/10.1186/1743-8977-6-30.

Gehring, U., Beelen, R., Eeftens, M., Hoek, G., De Hoogh, K., De Jongste, J. C., Keuken, M., et al. (2015). Particulate matter composition and respiratory health: The PIAMA Birth Cohort study. *Epidemiology, 26*(3), 300–309. Retrieved from http://journals.lww.com/epidem/Abstract/2015/05000/Particulate_Matter_Composition_and_Respiratory.3.aspx.

Giles-Corti, B., Vernez-Moudon, A., Reis, R., Turrell, G., Dannenberg, A. L., Badland, H., et al. (2016). City planning and population health: A global challenge. *The Lancet, 388*(10062), 2912–2924. Retrieved from http://www.sciencedirect.com/science/article/pii/S0140673616300666.

Godwin, A., & Price, A. M. (2016). Bicycling and walking in the Southeast USA: Why is it rare and risky? *Journal of Transport & Health, 3*(1), 26–37. Retrieved from http://www.sciencedirect.com/science/article/pii/S2214140516000074.

Götschi, T., Garrard, J., & Giles-Corti, B. (2016). Cycling as a part of daily life: A review of health perspectives. *Transport Reviews, 36*(1), 45–71. Retrieved from http://www.tandfonline.com/doi/abs/10.1080/01441647.2015.1057877.

Gouldson, A., Colenbrander, S., Sudmant, A., McAnulla, F., Kerr, N., Sakai, P., et al. (2015). Exploring the economic case for climate action in cities. *Global Environmental Change, 35*, 93–105. https://doi.org/10.1016/j.gloenvcha.2015.07.009.

Grabow, M. L., Spak, S. N., Holloway, T., Stone, B., Jr., Mednick, A. C., & Patz, J. A. (2012). Air quality and exercise-related health benefits from reduced car travel in the midwestern United States. *Environmental Health Perspectives, 120*(1), 68. Retrieved from https://www.ncbi.nlm.nih.gov/pmc/articles/PMC3261937/.

Green, C. P., Heywood, J. S., & Navarro, M. (2016). Traffic accidents and the London congestion charge. *Journal of Public Economics, 133*, 11–22. Retrieved from http://www.sciencedirect.com/science/article/pii/S0047272715001929.

Guttikunda, S. K., & Mohan, D. (2014). Re-fueling road transport for better air quality in India. *Energy Policy, 68*, 556–561. Retrieved from http://www.sciencedirect.com/science/article/pii/S0301421514000020.

Hamer, M., & Chida, Y. (2009). Physical activity and risk of neurodegenerative disease: A systematic review of prospective evidence. *Psychological Medicine, 39*(1), 3–11. Retrieved from https://www.cambridge.org/core/journals/psychological-medicine/article/physical-activity-and-risk-of-neurodegenerative-disease-a-systematic-review-of-prospective-evidence/5FB109E05E85CF701F11FB6DBA9AE9B3.

Hänninen, O., Knol, A. B., Jantunen, M., Lim, T. A., Conrad, A., Rappolder, M., et al. (2014). Environmental burden of disease in Europe: Assessing nine risk factors in six countries. *Environmental Health Perspectives, 122*(5), 439. Retrieved from https://www.ncbi.nlm.nih.gov/pmc/articles/PMC4014759/.

Harlan, S. L., & Ruddell, D. M. (2011). Climate change and health in cities: Impacts of heat and air pollution and potential co-benefits from mitigation and adaptation. *Current Opinion in Environmental Sustainability, 3*, 126–134. Retrieved from http://www.sciencedirect.com/science/article/pii/S1877343511000029.

Hu, G., Qiao, Q., Silventoinen, K., Eriksson, J. G., Jousilahti, P., Lindström, J., et al. (2003). Occupational, commuting, and leisure-time physical activity in relation to risk for type 2 diabetes in middle-aged Finnish men and women. *Diabetologia, 46*(3), 322–329. Retrieved from https://link.springer.com/article/10.1007/s00125-003-1031-x.

Hu, G., Jousilahti, P., Borodulin, K., Barengo, N. C., Lakka, T. A., Nissinen, A., & Tuomilehto, J. (2007). Occupational, commuting and leisure-time physical activity in relation to coronary heart disease among middle-aged Finnish men and women. *Atherosclerosis, 194*(2), 490–497. Retrieved from http://www.sciencedirect.com/science/article/pii/S0021915006005363.

IEA. (2015). *World energy outlook special report 2015: Energy and climate change*. Retrieved from https://www.iea.org/publications/freepublications/publication/weo-2015-special-report-2015-energy-and-climate-change.html.

IEA. (2016). *World energy outlook special report 2016: Energy and air pollution*. Retrieved from https://www.iea.org/publications/freepublications/publication/weo-2016-special-report-energy-and-air-pollution.html.

Jabben, J., Verheijen, E., & Potma, C. (2012, August). Noise reduction by electric vehicles in the Netherlands. In *Proceedings of Internoise*.

Jacobsen, P. L. (2003). Safety in numbers: More walkers and bicyclists, safer walking and bicycling. *Injury Prevention, 9*(3), 205–209. Retrieved from http://injuryprevention.bmj.com/content/9/3/205.short.

James, P., Ito, K., Buonocore, J. J., Levy, J. I., & Arcaya, M. C. (2014). A Health Impact Assessment of proposed public transportation service cuts and fare increases in Boston, Massachusetts (USA). *International Journal of Environmental Research and Public Health, 11*(8), 8010–8024. Retrieved from http://www.mdpi.com/1660-4601/11/8/8010/htm.

Ji, S., Cherry, C. R., Bechle M, J., Wu, Y., & Marshall, J. D. (2012). Electric vehicles in China: Emissions and health impacts. *Environmental Science & Technology, 46*(4), 2018–2024. Retrieved from http://pubs.acs.org/doi/abs/10.1021/es202347q.

Kenworthy, J., & Laube, F. (2002). Travel demand management: The potential for enhancing urban rail opportunities & reducing automobile dependence in cities. *World Transport Policy & Practice, 8*(3.) Retrieved from https://trid.trb.org/view.aspx?id=768994.

Khreis, H., Warsow, K. M., Verlinghieri, E., Guzman, A., Pellecuer, L., Ferreira, A., et al. (2016). The health impacts of traffic-related exposures in urban areas: Understanding real effects, underlying driving forces and co-producing future directions. *Journal of Transport & Health, 3*(3), 249–267. Retrieved from http://www.sciencedirect.com/science/article/pii/S2214140516301992.

Khreis, H., May, A. D., & Nieuwenhuijsen, M. J. (2017). Health impacts of urban transport policy measures: A guidance note for practice. *Journal of Transport & Health, 6*, 209–227. Retrieved from http://www.sciencedirect.com/science/article/pii/S2214140516304145.

Leather, J. (2009). *Rethinking transport and climate change*. Asian Development Bank and Clean Air Initiative. Retrieved from http://hdl.handle.net/11540/1403. License: CC BY 3.0 IGO.

Lee, C., & Moudon, A. V. (2004). Physical activity and environment research in the health field: Implications for urban and transportation planning practice and research. *CPL Bibliography, 19*(2), 147–181. Retrieved from http://journals.sagepub.com/doi/abs/10.1177/0885412204267680.

Lee, T., & Van de Meene, S. (2013). Comparative studies of urban climate co-benefits in Asian cities: An analysis of relationships between CO_2 emissions and environmental indicators. *Journal of Cleaner Production, 58*, 15–24. Retrieved from http://www.sciencedirect.com/science/article/pii/S0959652613003211.

Lee, G., Ritchie, S. G., Saphores, J. D., Jayakrishnan, R., & Ogunseitan, O. (2012). Assessing air quality and health benefits of the Clean Truck Program in the Alameda corridor, CA. *Transportation Research Part A: Policy and Practice, 46*(8), 1177–1193. Retrieved from http://www.sciencedirect.com/science/article/pii/S0965856412000808.

Lee, J. S., Christopher Zegras, P., & Ben-Joseph, E. (2013). Safely active mobility for urban baby boomers: The role of neighborhood design. *Accident Analysis & Prevention, 61*, 153–166.

Lee, E. Y., Jerrett, M., Ross, Z., Coogan, P. F., & Seto, E. Y. (2014). Assessment of traffic-related noise in three cities in the United States. *Environmental Research, 132*, 182–189. Retrieved from https://www.ncbi.nlm.nih.gov/pubmed/24792415.

Ling-Yun, H. E., & Lu-Yi, Q. I. U. (2016). Transport demand, harmful emissions, environment and health co-benefits in China. *Energy Policy, 97*, 267–275. Retrieved from http://www.sciencedirect.com/science/article/pii/S0301421516304001.

Litman, T. (2011). *Pricing for traffic safety: How efficient transport pricing can reduce roadway crash risk*. Victoria: Victoria Transport Policy Institute.

Litman, T. (2012). Pricing for traffic safety: How efficient transport pricing can reduce roadway crash risks. *Transportation Research Record, 2318*, 16–22. Retrieved from http://trrjournalonline.trb.org/doi/pdf/10.3141/2318-03.

Lubans, D. R., Boreham, C. A., Kelly, P., & Foster, C. E. (2011). The relationship between active travel to school and health-related fitness in children and adolescents: A systematic review. *International Journal of Behavioral Nutrition and Physical Activity, 8*(1), 5. Retrieved from https://ijbnpa.biomedcentral.com/articles/10.1186/1479-5868-8-5.

MacArthur, J., Dill, J., & Person, M. (2014). Electric bikes in North America: Results of an online survey. *Transportation Research Record, 2468*, 123–130. Retrieved from http://docs.trb.org/prp/14-4885.pdf.

Martin, A., Panter, J., Suhrcke, M., & Ogilvie, D. (2015). Impact of changes in mode of travel to work on changes in body mass index: Evidence from the British Household Panel Survey. *Journal of Epidemiology Community Health*, jech-2014.

Matthews, C. E., Jurj, A. L., Shu, X. O., Li, H. L., Yang, G., Li, Q., et al. (2007). Influence of exercise, walking, cycling, and overall nonexercise physical activity on mortality in Chinese women. *American Journal of Epidemiology, 165*(12), 1343–1350. Retrieved from https://academic.oup.com/aje/article/165/12/1343/125702.

McKinley, G., Zuk, M., Höjer, M., Avalos, M., González, I., Iniestra, R., et al. (2005). *Quantification of local and global benefits from air pollution control in Mexico City*. Retrieved from http://pubs.acs.org/doi/abs/10.1021/es035183e.

Mendonça, C., Freitas, E., Ferreira, J. P., Raimundo, I. D., & Santos, J. A. (2013). Noise abatement and traffic safety: The trade-off of quieter engines and pavements on vehicle detection. *Accident Analysis & Prevention, 51*, 11–17. Retrieved from http://www.sciencedirect.com/science/article/pii/S0001457512003740.

Miranda-Moreno, L. F., Morency, P., & El-Geneidy, A. M. (2011). The link between built environment, pedestrian activity and pedestrian–vehicle collision occurrence at signalized intersections. *Accident Analysis & Prevention, 43*(5), 1624–1634.

Mueller, N., Rojas-Rueda, D., Cole-Hunter, T., de Nazelle, A., Dons, E., Gerike, R., et al. (2015). Health impact assessment of active transportation: A systematic review. *Preventive Medicine, 76*, 103–114. Retrieved from http://www.sciencedirect.com/science/article/pii/S0091743515001164.

Mueller, N., Rojas-Rueda, D., Basagaña, X., Cirach, M., Cole-Hunter, T., Dadvand, P., et al. (2017a). Urban and transport planning related exposures and mortality: A health impact assessment for cities. *Environmental Health Perspectives, 125*(1), 89. Retrieved from https://www.ncbi.nlm.nih.gov/pmc/articles/PMC5226698/.

Mueller, N., Rojas-Rueda, D., Basagaña, X., Cirach, M., Cole-Hunter, T., Dadvand, P., et al. (2017b). Health impacts related to urban and transport planning: A burden of disease assessment. *Environment International, 107*, 243–257. Retrieved from http://www.sciencedirect.com/science/article/pii/S0160412017303665.

Nakahara, S., Ichikawa, M., & Kimura, A. (2011). Population strategies and high-risk-individual strategies for road safety in Japan. *Health Policy, 100*(2), 247–255. Retrieved from http://www.sciencedirect.com/science/article/pii/S0168851010003350.

Nieuwenhuijsen, M. J., & Khreis, H. (2016). Car free cities: Pathway to healthy urban living. *Environment International, 94*, 251–262. Retrieved from http://www.sciencedirect.com/science/article/pii/S0160412016302161.

Ogilvie, D., Egan, M., Hamilton, V., & Petticrew, M. (2004). Promoting walking and cycling as an alternative to using cars: Systematic review. *BMJ, 329*(7469), 763. Retrieved from http://www.bmj.com/content/329/7469/763.short.

Pathak, M., & Shukla, P. R. (2016). Co-benefits of low carbon passenger transport actions in Indian cities: Case study of Ahmedabad. *Transportation Research Part D: Transport and Environment, 44*, 303–316. Retrieved from http://www.sciencedirect.com/science/article/pii/S1361920915001078.

Préfecture de Police. (2013). 'Bilan Sécurité Routière de la Préfecture de Police; Blesses Graves' 2013.

Rabl, A., & De Nazelle, A. (2012). Benefits of shift from car to active transport. *Transport Policy, 19*(1), 121–131.

Reisi, M., Aye, L., Rajabifard, A., & Ngo, T. (2016). Land-use planning: Implications for transport sustainability. *Land Use Policy, 50*, 252–261. Retrieved from http://www.sciencedirect.com/science/article/pii/S0264837715002896.

Roberts, I., Marshall, R., & Norton, R. (1992). Child pedestrian mortality and traffic volume in New Zealand. *BMJ, 305*(6848), 283. Retrieved from https://www.ncbi.nlm.nih.gov/pmc/articles/PMC1882717/.

Rode, P., Floater, G., et al. (2013). *Going green: How cities are leading the next economy*. London: LSE Cities, London School of Economics and Political Science.

Rodriguez, D. A., Khattak, A. J., & Evenson, K. R. (2006). Can new urbanism encourage physical activity?: Comparing a new Urbanist neighborhood with conventional suburbs. *Journal of the American Planning Association, 72*(1), 43–54. Retrieved from http://www.tandfonline.com/doi/abs/10.1080/01944360608976723.

Rojas-Rueda, D., De Nazelle, A., Teixidó, O., & Nieuwenhuijsen, M. J. (2012). Replacing car trips by increasing bike and public transport in the greater Barcelona metropolitan area: A health impact assessment study. *Environment International, 49*, 100–109. Retrieved from http://www.sciencedirect.com/science/article/pii/S0160412012001833.

Rojas-Rueda, D., De Nazelle, A., Teixidó, O., & Nieuwenhuijsen, M. J. (2013). Health impact assessment of increasing public transport and cycling use in Barcelona: A morbidity and burden of disease approach. *Preventive Medicine, 57*(5), 573–579. Retrieved from http://www.sciencedirect.com/science/article/pii/S0091743513002739.

Rothman, L., Buliung, R., Macarthur, C., Teresa, T., & Howard, A. (2014). Walking and child pedestrian injury: A systematic review of built environment correlates of safe walking. *Injury Prevention, 20*(1), 41–49.

Sabel, C. E., Hiscock, R., Asikainen, A., Bi, J., Depledge, M., van den Elshout, S., Friedrich, R., et al. (2016). Public health impacts of city policies to reduce climate change: Findings from the URGENCHE EU-China project. *Environmental Health, 15*(1), S25.

Saelens, B. E., Sallis, J. F., & Frank, L. D. (2003). Environmental correlates of walking and cycling: Findings from the transportation, urban design, and planning literatures. *Annals of Behavioral Medicine, 25*(2), 80–91. Retrieved from https://link.springer.com/article/10.1207%2FS15324796ABM2502_03?LI=true.

Salon, D. (2016). Estimating pedestrian and cyclist activity at the neighborhood scale. *Journal of Transport Geography, 55*, 11–21. Retrieved from http://www.sciencedirect.com/science/article/pii/S0966692316303593.

Saunders, L. E., Green, J. M., Petticrew, M. P., Steinbach, R., & Roberts, H. (2013). What are the health benefits of active travel? A systematic review of trials and cohort studies. *PLoS One, 8*(8), e69912. Retrieved from http://journals.plos.org/plosone/article?id=10.1371/journal.pone.0069912.

Sims, R., Schaeffer, R., Creutzig, F., Cruz-Núñez, X., D'Agosto, M., Dimitriu, D., Figueroa Meza, M. J., Fulton, L., Kobayashi, S., Lah, O., McKinnon, A., Newman, P., Ouyang, M., Schauer, J. J., Sperling, D., & Tiwari, G. (2014). Transport. In O. Edenhofer, R. Pichs-Madruga, Y. Sokona, E. Farahani, S. Kadner, K. Seyboth, A. Adler, I. Baum, S. Brunner, P. Eickemeier, B. Kriemann, J. Savolainen, S. Schlömer, C. von Stechow, T. Zwickel, & J. C. Minx (Eds.),

Climate change 2014: Mitigation of climate change. Contribution of Working Group III to the Fifth Assessment Report of the Intergovernmental Panel on Climate Change. Cambridge, UK: Cambridge University Press.

Soret, A., Guevara, M., & Baldasano, J. M. (2014). The potential impacts of electric vehicles on air quality in the urban areas of Barcelona and Madrid (Spain). *Atmospheric Environment, 99*, 51–63. Retrieved from http://www.sciencedirect.com/science/article/pii/S1352231014007419.

Stevenson, M., Thompson, J., de Sá, T. H., Ewing, R., Mohan, D., McClure, R., et al. (2016). Land use, transport, and population health: Estimating the health benefits of compact cities. *The Lancet, 388*(10062), 2925–2935. Retrieved from http://www.sciencedirect.com/science/article/pii/S0140673616300678.

Stone, B., Jr., Mednick, A. C., Holloway, T., & Spak, S. N. (2007). Is compact growth good for air quality? *Journal of the American Planning Association, 73*(4), 404–418. Retrieved from http://www.tandfonline.com/doi/abs/10.1080/01944360708978521.

Sudmant, A., Millward-Hopkins, J., Colenbrander, S., & Gouldson, A. (2016). Low carbon cities: Is ambitious action affordable? *Climatic Change, 138*(3–4), 681–688. Retrieved from https://link.springer.com/article/10.1007/s10584-016-1751-9.

Sugiyama, T., Neuhaus, M., Cole, R., Giles-Corti, B., & Owen, N. (2012). Destination and route attributes associated with adults' walking: A review. *Medicine and Science in Sports and Exercise, 44*(7), 1275–1286. Retrieved from http://europepmc.org/abstract/med/22217568.

Sun, G., Gwee, E., Chin, L. S., & Low, A. (2014). Passenger transport mode shares in world cities. *Journeys, 12*, 54–64. Retrieved from http://www.lta.gov.sg/ltaacademy/doc/Journeys_Issue_12_Nov_2014.pdf.

SWOV, Institute for Road Safety Research. (2016). *Road deaths and population data in the Netherlands*. Retrieved from http://www.swov.nl/NL/Research/cijfers/Cijfers.htm.

Tainio, M. (2015). Burden of disease caused by local transport in Warsaw, Poland. *Journal of Transport & Health, 2*(3), 423–433. Retrieved from http://www.sciencedirect.com/science/article/pii/S2214140515005125.

Tainio, M., Olkowicz, D., Teresiński, G., De Nazelle, A., & Nieuwenhuijsen, M. J. (2014). Severity of injuries in different modes of transport, expressed with disability-adjusted life years (DALYs). *BMC Public Health, 14*(1), 765. Retrieved from https://bmcpublichealth.biomedcentral.com/articles/10.1186/1471-2458-14-765.

Thorpe, A., & Harrison, R. M. (2008). Sources and properties of non-exhaust particulate matter from road traffic: A review. *Science of the Total Environment, 400*(1), 270–282. Retrieved from http://www.sciencedirect.com/science/article/pii/S004896970800658X.

Timilsina, G. R., & Dulal, H. B. (2010). Urban road transportation externalities: Costs and choice of policy instruments. *The World Bank Research Observer, 26*(1), 162–191. Retrieved from https://academic.oup.com/wbro/article/26/1/162/1728191.

Timmers, V. R., & Achten, P. A. (2016). Non-exhaust PM emissions from electric vehicles. *Atmospheric Environment, 134*, 10–17. Retrieved from http://www.sciencedirect.com/science/article/pii/S135223101630187X.

Torrao, G., Fontes, T., Coelho, M., & Rouphail, N. (2016). Integrated indicator to evaluate vehicle performance across: Safety, fuel efficiency and green domains. *Accident Analysis & Prevention, 92*, 153–167. Retrieved from http://www.sciencedirect.com/science/article/pii/S0001457516300793.

Transport Department Hong Kong. (2016). *Road safety; Summary of key statistics*. Retrieved from http://www.td.gov.hk/en/road_safety/.

UK Energy Research Centre. (2016). *Review of UK energy policy, A UKERC Policy Briefing*. Retrieved November 2016, from http://www.ukerc.ac.uk/news/ukerc-calls-for-urgent-action-on-uk-energy-during-this-parliament.html.

UNEP. 2014. *Urban air pollution*. United Nations Urban Environment Unit. Retrieved from http://www.unep.org/urban_environment/Issues/urban_air.asp.

Verheijen, E., & Jabben, J. (2010). *Effect of electric cars on traffic noise and safety*. Bilthoven: RIVM.

Watts, N., Adger, W. N., Agnolucci, P., Blackstock, J., Byass, P., Cai, W., et al. (2015). Health and climate change: Policy responses to protect public health. *The Lancet, 386*(10006), 1861–1914. Retrieved from http://www.thelancet.com/journals/lancet/article/PIIS0140-6736(15)60854-6/abstract.

Wei, V. F., & Lovegrove, G. (2012). Sustainable road safety: A new (?) neighbourhood road pattern that saves VRU lives. *Accident Analysis & Prevention, 44*(1), 140–148. Retrieved from http://www.sciencedirect.com/science/article/pii/S0001457510003829.

Wong, B. Y. M., Faulkner, G., & Buliung, R. (2011). GIS measured environmental correlates of active school transport: A systematic review of 14 studies. *International Journal of Behavioral Nutrition and Physical Activity, 8*(1), 1. Retrieved from https://ijbnpa.biomedcentral.com/articles/10.1186/1479-5868-8-39.

Woodcock, J., Edwards, P., Tonne, C., Armstrong, B. G., Ashiru, O., Banister, D., et al. (2009). Public health benefits of strategies to reduce greenhouse-gas emissions: Urban land transport. *The Lancet, 374*(9705), 1930–1943. Retrieved from http://www.sciencedirect.com/science/article/pii/S0140673609617141.

Woodcock, J., Givoni, M., & Morgan, A. S. (2013). Health impact modelling of active travel visions for England and Wales using an Integrated Transport and Health Impact Modelling Tool (ITHIM). *PLoS One, 8*(1), e51462. Retrieved from http://journals.plos.org/plosone/article?id=10.1371/journal.pone.0051462.

World Health Organization. (2015). *Global status report on road safety 2015*. Retrieved from http://www.who.int/violence_injury_prevention/road_safety_status/2015/GSRRS2015_Summary_EN_final2.pdf?ua=1.

World Health Organization. (2017). *Physical activity, Fact Sheet, Updated February 2017*. Retrieved from http://www.who.int/mediacentre/factsheets/fs385/en/.

WRI. (2012). *Top emitters in 2012*. Retrieved from http://www.wri.org/resources/charts-graphs/top-10-emitters-2012.

Xia, T., Nitschke, M., Zhang, Y., Shah, P., Crabb, S., & Hansen, A. (2015). Traffic-related air pollution and health co-benefits of alternative transport in Adelaide, South Australia. *Environment International, 74*, 281–290. Retrieved from http://www.sciencedirect.com/science/article/pii/S0160412014002980.

Xue, X., Ren, Y., Cui, S., Lin, J., Huang, W., & Zhou, J. (2015). Integrated analysis of GHGs and public health damage mitigation for developing urban road transportation strategies. *Transportation Research Part D: Transport and Environment, 35*, 84–103. Retrieved from http://www.sciencedirect.com/science/article/pii/S1361920914001746.

Yang, X., Liu, H., & He, K. (2016). The significant impacts on traffic and emissions of ferrying children to school in Beijing. *Transportation Research Part D: Transport and Environment, 47*, 265–275. Retrieved from http://www.sciencedirect.com/science/article/pii/S136192091630339X.

Yeo, J., Park, S., & Jang, K. (2015). Effects of urban sprawl and vehicle miles traveled on traffic fatalities. *Traffic Injury Prevention, 16*(4), 397–403. Retrieved from http://www.tandfonline.com/doi/abs/10.1080/15389588.2014.948616.

Yiannakoulias, N., & Scott, D. M. (2013). The effects of local and non-local traffic on child pedestrian safety: A spatial displacement of risk. *Social Science & Medicine, 80*, 96–104. Retrieved from http://www.sciencedirect.com/science/article/pii/S0277953612008076.

Yu, C. Y. (2014). Environmental supports for walking/biking and traffic safety: Income and ethnicity disparities. *Preventive Medicine, 67*, 12–16. Retrieved from http://www.sciencedirect.com/science/article/pii/S0091743514002291.

Zuo, F., Li, Y., Johnson, S., Johnson, J., Varughese, S., Copes, R., Liu, F., Wu, H. J., Hou, R., & Chen, H. (2014). Temporal and spatial variability of traffic-related noise in the City of Toronto, Canada. *Science of the Total Environment, 472*, 1100–1107. Retrieved from http://www.sciencedirect.com/science/article/pii/S0048969713014423.

Chapter 29
The Role of Health Impact Assessment for Shaping Policies and Making Cities Healthier

Mark Nieuwenhuijsen, Haneen Khreis, Ersilia Verlinghieri, Natalie Mueller, and David Rojas-Rueda

29.1 Introduction

Health impact assessment (HIA) is an important tool to integrate evidence in the decision-making process and introduce health in all policies (WHO 1999; Ståhl et al. 2006; NAS 2011). Multiple international and national organizations proposed HIA as a tool to promote and protect public health in multiple sectors (WHO 1999; Ståhl et al. 2006; NHS 2002; IFC 2009). In urban and transport planning, HIAs have been used generally to assess qualitatively urban interventions rather than offering more useful/powerful estimations to stakeholders through quantitative approaches (Shafiea et al. 2013). Also many HIAs did not entail the stakeholders and citizens' visions and necessities, losing the opportunity of successful implementation or policy utility.

M. Nieuwenhuijsen (✉) · N. Mueller · D. Rojas-Rueda
ISGlobal, Centre for Research in Environmental Epidemiology (CREAL), Barcelona, Spain

Universitat Pompeu Fabra (UPF), Barcelona, Spain

CIBER Epidemiologia y Salud Publica (CIBERESP), Madrid, Spain
e-mail: mark.nieuwenhuijsen@isglobal.org; natalie.mueller@isglobal.org; david.rojas@isglobal.org

H. Khreis
ISGlobal, Centre for Research in Environmental Epidemiology (CREAL), Barcelona, Spain

Universitat Pompeu Fabra (UPF), Barcelona, Spain

CIBER Epidemiologia y Salud Publica (CIBERESP), Madrid, Spain

Institute for Transport Studies, University of Leeds, Leeds, UK
e-mail: haneen.khreis@isglobal.org

E. Verlinghieri
Transport Studies Unit, University of Oxford, Oxford, UK
e-mail: ersilia.verlinghieri@ouce.ox.ac.uk

M. Nieuwenhuijsen, H. Khreis (eds.), *Integrating Human Health into Urban and Transport Planning*, https://doi.org/10.1007/978-3-319-74983-9_29

Comprehensive HIAs can be an important tool as it brings together all the available evidence on the impacts of urban and transport planning on health. As where qualitative approaches provide a more general impact of the potential important exposures and health outcomes, quantitative HIAs provide actual numbers of the people that may end up with disease or die prematurely. HIAs can shape policies and make cities healthier because they examine the impacts of health from different potential scenarios, based on the best available evidence, and provide an overview of what are scenarios with the least health impacts and which have the largest health impacts.

HIAs could answer various pressing questions such as: What are the best and most feasible urban and transport planning policy measures to improve public health in cities? Also the process on how to get there is often as important as the actual output of the HIA, as the process may provide answers to important questions as to how different disciplines/sectors can effectively work together and develop a common language, how to best incorporate citizen and stakeholder, how different modelling and measurement methods can be effectively integrated and whether a public health approach could make changes in urban and transport planning.

Currently there is no overarching HIA framework or model for cities that can deal with multiple exposures and complexities, data limitations, location-specific effects, errors, etc., and we work with separate quantitative and qualitative models/ modules, which from one perspective may be considered as an advantage to reduce the complexity and burden of this work. In our experience, the quantitative estimates that have been produced so far have been useful for policymakers by placing actual numbers of health impacts on different policy scenarios. This policy-specific quantification has been an advantage when compared to a qualitative approach, but both approaches have their merit and can coexist. In particular, we believe that qualitative studies should be performed in preparation of quantitative ones, preparing the terrain—in terms of accessing the data, but also influencing the political and policy discourses—and informing on the necessity and utility of a full quantitative HIA. A qualitative approach may be preferred in cases where there is no good quality quantitative data available.

Quantitative HIAs are unlikely to be conducted for small projects, where generally little funding is available, although if models are previously set-up and exposures are readily available for the area, then undertaking the assessment would be feasible. However, we are also aware that a qualitative approach in some of these cases may suffice or at least be used successfully to prepare the favourable terrain to gather more resources for more comprehensive studies. On the other side, we stress that large infrastructure projects or comprehensive urban planning projects which are long-lasting, highly impactful on the population's exposures and highly consuming of public money could highly benefit from participatory quantitative HIAs. HIAs can be particularly effective especially when different options/scenarios are available for the future, and a political decision has not been made yet or is difficult to reach. Knowing and disseminating the possible scenarios and the associated health impacts can then play an important role, especially bringing the public's attention to what new policies and projects mean for their communities' health. The concept of 'health' is usually a strong argument for the public (or the population affected) to be in favour for a certain policy to be implemented. Being able to demonstrate the magnitude of expected health benefits—as possible with participatory, quantitative HIAs—can influence the acceptance/decision process. Health has not been high on the agenda for such projects, but, as the current national and international debates across Europe and beyond demonstrate, more and more citizens rate health highly. They also would object to large changes when they have not been involved in decision-making (e.g. Barcelona SuperIlles). Although more expensive, a participatory quantitative HIA may pay off in the long term by providing positive and sustainable changes with the least adverse health impacts and greatest acceptability. Relevant stakeholders can contribute in this process with their expertise. A participatory, quantitative HIA ensures that no aspect is forgotten, that the process is inclusive and comprehensive and that a consensus can be reached at the end by weighing the estimated risks against the benefits.

As it stands, the use of quantitative HIA is generally limited to research and academic purposes, and the scenarios used in these models are usually judged to be plausible but are optimistic and often not under consideration by local authorities or policymakers. This, in part, is perhaps a reflection of a communication gap between the sectors where nonacademic stakeholders lack the tools, knowledge and interest to carry out a quantitative HIA, while academics/researchers lack the expertise and understandings as to what extent scenarios are plausible, realizable and acceptable to local authorities or policymakers. In this chapter we focus on quantitative HIAs. The aim of the chapter is to provide an overview of what is currently being done in quantitative HIA of urban and transport planning.

29.2 Quantitative Health Impact Assessment

Quantitative HIAs follow a comparative risk assessment approach estimating first the burden of disease (e.g. cases of disease, injuries, deaths or disability-adjusted life years [DALY]) and then comparing this burden of disease (BOD) with the health impacts of a future change associated with a proposed intervention or policy (Briggs 2008; WHO 2015; Nieuwenhuijsen et al. 2017) (Fig. 29.1). The aim is to provide a quantitative estimate of the expected health impact and the distribution thereof for the exposed population that is attributable to an environmental exposure and/or policy. Quantitative assessments include a number of steps (Table 29.1), and a range of potential scenarios could be evaluated (Table 29.2). Furthermore, the input is obtained from stakeholders including, for example, citizens and businesses we refer to the HIAs as being participatory HIAs. Participatory HIAs have the advantage that make use of much wider views on urban and transport besides the "expert" and, if integrated well, may count on much wider support for any proposed measure.

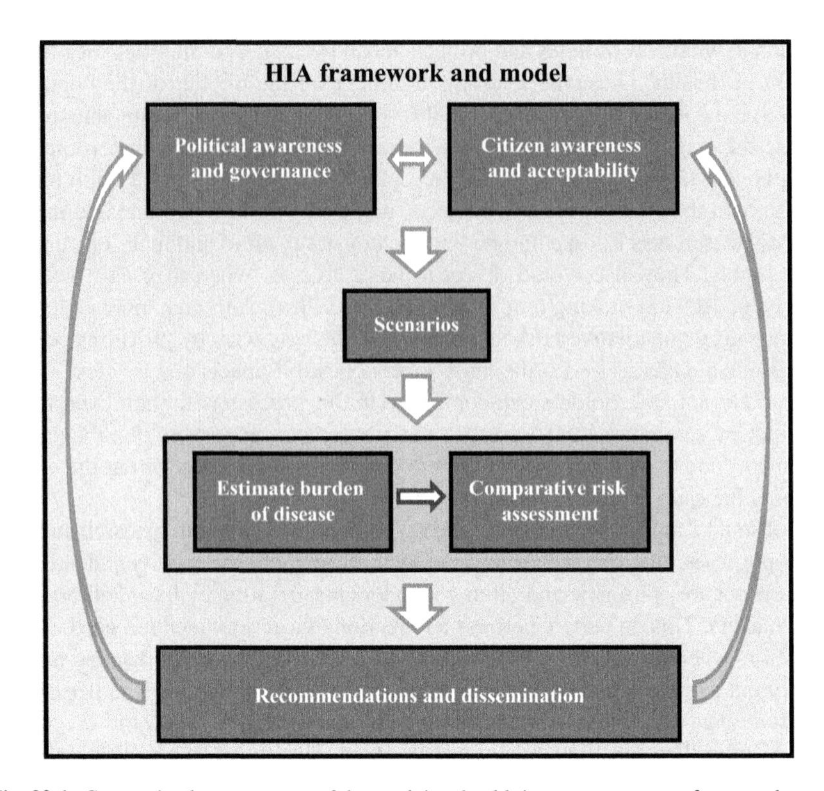

Fig. 29.1 Connection between parts of the work in a health impact assessment framework

29.2.1 Examples of Urban HIA Studies

BOD studies for cities and these are sparse. For example, Tobías et al. (2014) estimated that 470 deaths were attributable to a theoretic traffic noise exposure decrease by solely 1 dB(A) in Madrid. A recent study for Warsaw estimated more than 40,000 DALYs attributable to air pollution, noise and traffic injuries, with traffic noise contributing the largest (Tainio 2015). In Barcelona, Mueller et al. (2017) estimated that almost 3000 deaths, making up almost 20% of annual mortality in Barcelona, could be prevented if international recommendations for performance of physical activity, exposure to air pollution, noise, heat and access to green space were complied with.

A few studies have started evaluating the health impacts of urban planning policy measures in cities. Reisi et al. (2016) evaluated three urban planning scenarios in Melbourne for 2030, base case scenario based on governmental plans, fringe focus scenario based on expansive urban development patterns and activity centres scenario based on compact urban development patterns, and estimated that the latter resulted in the least greenhouse and other emissions, as well as a reduction of mortality when compared to the other scenarios. Stevenson et al. (2016) estimated the population health effects arising from alternative land use and transport policy initiatives in six cities. Land use changes were modelled to reflect a compact city in

Table 29.1 Steps in a HIA and added value in the participatory integrated fullchain (PIF) HIA approach

HIA steps	HIA task	Added value in the PIF-HIA approach
1. Screening	• Selecting proposals	• Proposal selection using stakeholders and citizen participation approaches
2. Scoping	• Identifying health effects and the population groups affected • Describes research questions, data sources, the analytic plan, data gaps and how gaps will be addressed	• Identification of relevant health effects and populations affected including stakeholder and citizen perspectives. • Introduce integrated, full-chain and complex system approach in the analytic plan
3. Appraisal	• Collecting and analyzing quantitative and qualitative data on health effects in various population groups	• Introduce integrated, full-chain and complex system approach • Stakeholders and citizens are informed of the preliminary results, and their feedback is integrated at this stage, too
4. Recommendations	• Identifies alternatives to proposal or actions that could be taken to avoid, minimize or mitigate adverse effects and to optimize beneficial ones • Proposes a health management plan to identify stakeholders who could implement recommendations, indicators for monitoring and systems for verification	• Introduce integrated, full-chain and complex system vision in the alternatives to proposal or actions • Include citizens and stakeholder perspectives to prioritize and implement recommendations, to increase their social acceptation and impact • Propose citizen science approaches to complement monitoring and evaluation processes
5. Reporting and dissemination	• Writing an HIA report based on the results • Disseminating the report	• Introduce novel channels of communications through citizens and stakeholder participation
6. Monitoring and evaluation	• Evaluating the process • Evaluating the outcome (or results) • Evaluating the impact (effectiveness)	• Introduce the stakeholder and citizen perspective in the evaluation of the process, based on an iterative process to strength the citizens and stakeholder capacities • Perform outcome evaluation at the citizen's level, through citizen's science approaches, with an integrated, full-chain and complex system vision • Include an evaluation with an integrated, full-chain and complex system vision

Table 29.2 General description of examples of urban and transport policies/interventions/scenarios that could be modelled using health impact assessment

Policy area	Description	Types	Examples
Land use planning	Land use systems that increase density, diversity of uses and connectivity	Density	Compact cities
		Diversity	Increase horizontal land use
			Increase vertical land use
Built environment	All of the physical parts of where we live and work (e.g. homes, buildings, streets, open spaces and infrastructure)	Design	Improve connectivity
		Maintenance	Control temperature and humidity
		Availability	Increase accessibility
Green spaces	Vegetation in the streets (trees, grass, etc.), squares and parks	Infrastructure	Increase and improve green spaces
		Management	Improve access and quality of green spaces
		Promotion	Promote use of green spaces
Blues paces	Surface water in urban public spaces (fountains, lakes, rivers, sea front, etc.)	Infrastructure	Increase and improve blue spaces
		Management	Improve access and quality of blue spaces
		Promotion	Promote use of blue spaces
Public transport	Investment in and provision of transport network space for rapid transit/public transport infrastructure	Infrastructure	Improvement and increase of public transport infrastructure
		Management	Improve public transport service
			Reduce public transport costs
		Promotion	Public transport promotion
Active transport (walking and cycling)	Investment in and provision of transport network space for pedestrian and cycle infrastructure	Infrastructure	Improvement and increase of active transport infrastructure
		Management	Improve active transport service
		Promotion	Active transport promotion

(continued)

Table 29.2 (continued)

Policy area	Description	Types	Examples
Traffic regulation	Reduce car use	Infrastructure	Reduce public space for cars (car lanes and parking)
		Management	Road use an parking pricing
		Promotion	Promote alternative modes of transport
Vehicles and fuels	Invest in new technologies and fuels	Infrastructure	Create city grid for electric vehicles
		Management	Technology and fuels pricing
		Promotion	Promote technological transitions (e.g. electric car or autonomous cars)
Traffic safety	Engineering and speed reduction measures to moderate the leading hazards of road transport	Infrastructure	Built environment changes to reduce speed
		Management	Speed regulations
		Promotion	Traffic safety campaigns

which land use density and diversity were increased and distances to public transport were reduced with the objective to reduce private motorized transport and promote a modal shift to walking, cycling and public transport use. The modelled compact city scenario resulted in health gains for all cities.

There are a larger number of studies that have evaluated specific transport policy measures in cities. Woodcock et al. (2009) estimated the health effects of alternative urban land transport scenarios for two settings – London, UK and Delhi, India – and found that a combination of active travel and lower-emission motor vehicles would give the largest benefits (7439 DALYs in London, 12,995 in Delhi). Creutzig et al. (2012) provided scenarios of increasingly ambitious policy packages, reducing greenhouse gas emissions from urban transport by up to 80% from 2010 to 2040. Based on stakeholder interviews and data analysis, the main target was a modal shift from motorized individual transport to public transit and nonmotorized individual transport (walking and cycling) in four European cities (Barcelona, Malmö, Sofia and Freiburg). The authors reported significant concurrent co-benefits of better air quality, reduced noise, less traffic-related injuries and deaths, increased physical activity, alongside less congestion and monetary fuel savings. Perez et al. (2015) modelled various scenarios in Basel including particle emission standards for diesel cars, increase in active travel and electric vehicle introduction and estimated that the first measure would result in a reduction of premature mortality by 3%, the second one would have little effect and the third one would have the largest effect, as the electricity would come from renewable resources. Ji et al. (2012) compared emissions (CO_2, $PM_{2.5}$, NO_X, HC) and environmental health impacts (primary $PM_{2.5}$) from the use of conventional vehicles (CVs), electric vehicles (EVs) and electric bikes (e-bikes) in 34 major cities in China. E-bikes yielded lower environmental health impacts per passenger-km than gasoline cars (2×), diesel cars (10×) and diesel buses (5×). McKinley et al. (2005) quantified cost and health benefits from a subset of air pollution control measures (taxi fleet renovation, metro expansion and the use of new hybrid buses replacing diesel buses) in Mexico City and found that the measures reduced air pollution by approximately 1% for PM_{10} and 3% for O_3. The associated health benefits were substantial, and their sum over the three measures was greater than the measures' investment costs (benefit to cost ratio was 3.3 for the taxi renovation measure, 0.7 for the metro expansion measure and 1.3 for the new hybrid buses measure). Xia et al. (2015) estimated that the shifting of 40% of vehicle kilometres travelled to alternative transport in Adelaide, South Australia, would reduce annual average $PM_{2.5}$ by a small margin of 0.4 $\mu g/m^3$, preventing 13 deaths a year and 118 DALYS. A range of HIA studies evaluating mortality and other health effects of increases in active transport were recently reviewed and estimated considerable reductions in premature deaths and other negative health outcomes with most benefits attributable to increases in physical activity and low risks of motor vehicles crashes and air pollution for those who switched to public and active transport (Mueller et al. 2015; Tainio et al. 2016).

Most models have been static (e.g. no feedback loops) and thus insensitive to feedback loops and time delays. Few exceptions emerge in the literature, such as the work of Macmillan et al. (2014). These authors used participatory system dynamics modelling (SDM) to compare the impacts of realistic policies, incorporating feedback effects, nonlinear relationships and time delays between variables in a study on cycling and societal costs including those related to health impacts. Participatory SDM involves citizen, academic and policy stakeholders in a process that explores the dynamic effects of realistic policies (Richardson 2011).

Finally, there is a general lack of integrated full-chain HIA models. In integrated full-chain HIA modelling for urban and transport planning, the work considers the full chain from exposure source, through pathways to health endpoints, considering multiple exposures and complexities, interdependencies and uncertainties of the real world and examining multiple scenarios generally.

Full-chain models are important for policymaking as the contribution of different sources to the overall environmental exposures is often unclear and is highly variable depending on source and context. For example, car traffic contributes to a significant proportion of ambient air pollution in cities, but the extent varies depending on factors such as car density, car fleet make-up, traffic conditions, street design, city design and dispersion factors (e.g. wind speed and direction and cloud cover). Traffic contribution to urban PM_{10} and $PM_{2.5}$ levels in Europe is on average 39% (range 9–53%) and 43% (range 9–66%), respectively, and is up to over 80% for NO_2 (Sundvor et al. 2012).

To improve understanding and target the right sources with the right mitigation policies, it is important to understand the full chain of events from sources, through pathways to health, but very few, if any studies have done so. For example, in the

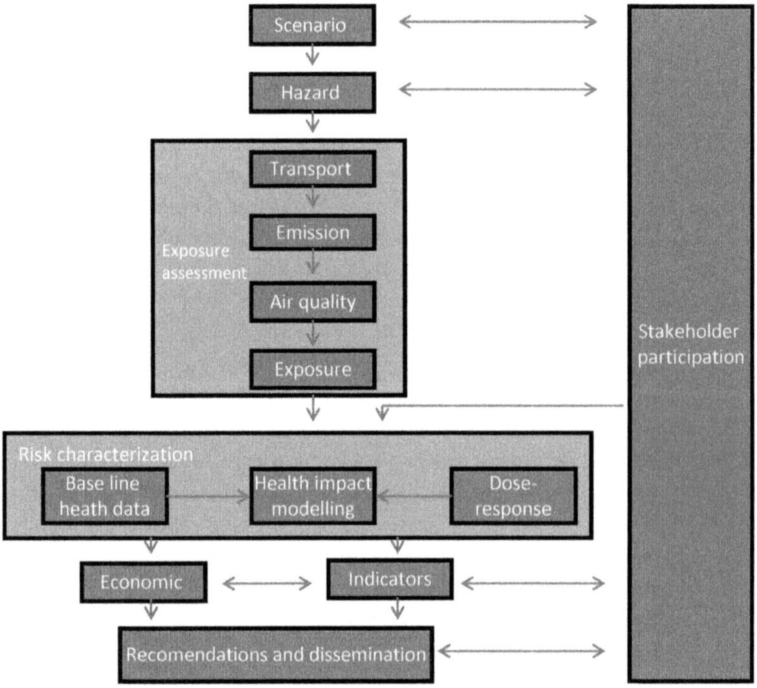

Fig. 29.2 Participatory full-chain health impact assessment with an example for air quality

case of air pollution exposures, the common lack of full-chain assessment limits disentangling the health impacts of traffic-related air pollution (TRAP) from the health impacts of other emission sources, and vice versa (Khreis et al. 2016b). It also limits the comprehensibility of and confidence in recommending fleet-specific and traffic planning or management-specific interventions which would be valuable and desirable for policymakers.

Full-chain assessments could be obtained by coupling existing models of source activity, source emissions and pathways of exposure, to predict final human exposures and associated health impacts. Again, in the case of TRAP, as an example, this can be done by integrating existing models of traffic activity, traffic emissions and air dispersion to predict air quality and people's exposure and subsequently estimate associated health impacts (Fig. 29.2). Data on traffic counts, origin and destination zones and fleet composition can be used to construct traffic activity models for cities (van Vliet 1982; SATURN manual 2015). The outputs from traffic activity models, most importantly the traffic flows and average traffic speeds, are then linked with vehicle emissions models such as the European leading emission model known as COPERT (Gkatzoflias et al. 2007). Vehicle emission inventories are calculated from this data and are then entered into air dispersion models such as ADMS-Urban (McHugh et al. 1997), which uses this data alongside terrain, meteorological and boundary layer data, to estimate seasonal and/or annual air pollution concentrations in cities.

Studies which undertake this full-chain assessment, however, are very few (e.g. Namdeo et al. 2002; Hatzopoulou and Miller 2010; Wang et al. 2016). Often, such assessments can be problematic as the referred to models are data and labour intensive and require expertise from different scientific disciplines. These models are also not easy to obtain or run due to their high commercial prices, their complex set-ups and the occasional need for specific arrangements, e.g. dedicated UNIX workstation. Furthermore, there are challenges regarding the performance and accuracy of the multiple models used in the exposure assessment chains. Traffic activity models such as SATURN which are used to provide input data for emission models tend to underestimate congestion and over-predict the average traffic speeds over the road network lending inaccuracy to the final emission estimates which are generally higher at the lower average speeds. Vehicle emission models, especially at lower average speeds which incorporate significant proportions of high emitting, fuel consuming, stop-start driving, are uncertain (Health Effects Institute 2010; Khreis 2016) and tend to underestimate TRAP emissions. Air dispersion models can over or under predict air pollution levels, in part due to inaccurate traffic and emission inputs, but also due to inherent limitations in these models (Williams et al. 2011) and the incompletion and/or inaccuracy of input data. This causes a propagation of uncertainties and inaccuracies through a complex chain of models involved and is a problem whose implications are not fully understood and is not yet addressed in practice and policy.

29.2.2 Existing Models

There are already a few quantitative HIA models and tools that have been used in specific case studies, for example, the Health Economic Assessment Tool (HEAT) for walking and cycling, the Integrated Transport and Health Impact Modelling Tool (ITHIM), the Transportation, Air Pollution and Physical Activities (TAPAS) model or the Urban and TranspOrt Planning Health Impact Assessment (UTOPHIA) model. Except for HEAT, these tools and models still tend to be research tools that are being further developed and improved but have the potential to be used in practice. The way these studies have been performed also shows how multidisciplinary teams of academics, both from qualitative and quantitative backgrounds, would take the process through all its phases.

29.2.3 Citizen and Other Stakeholder Involvement

HIAs could be conducted just by expert, but changes in city urban and transport planning are difficult to achieve and sustain without direct support of politicians, decision-makers and citizens. Therefore, it is desirable, if not necessary, to involve stakeholders in the process, i.e. conduct participatory HIAs. There is a considerable body of literature that stresses the importance of citizens' participation in improving planning and decision-making in a number of aspects, and today participation is recognized to be a fundamental requirement for sustainable development and environmental decision-making (Banister 2008; Linzalone et al. 2016).

Firstly, involving stakeholders into a decision-making process allows more viewpoint to contribute to the interpretation of complex issues (Mumpower 2001). Participation allows planners and decision-makers to gain a deeper and a more detailed knowledge on stakeholder behaviours, desires, necessities and preferences, becoming an invaluable tool to backup evaluation and assessment procedures such as HIA and allowing better-informed decisions (Lowndes et al. 2001; Innes and Booher 2004; Palerm 2000). Secondly, participation allows to increase public acceptability of decisions and to build stronger consensus, reduce conflicts and produce shared projects and visions (van de Kerkhof 2006; Innes and Booher 2004). As such it can be used to support public decisions. Thirdly, participation can be used as a process to inform and empower citizens developing healthier democratic practices and more fair and just solutions (Palerm 2000; Bailey and Grossardt 2010; Bailey et al. 2012; Innes and Booher 2004; Renn and Webler 1995). Finally, participation can generate in itself spaces of information diffusion, knowledge exchange and creation, becoming a space in which practices and behaviours can be transformed and social learning built (Sagaris 2014; Kesby 2005; Palerm 2000; Reed 2006).

However, the overall implementation of participatory approaches within HIA is not widespread, particularly not in quantitative HIA. The literature reports only a few studies in which stakeholders have been consulted during HIA (e.g. Kearney 2004; Creutzig et al. 2012; Macmillan et al. 2014; Linzalone et al. 2016); among which only very few used participatory methodologies in combination with quantitative assessments. For example, Linzalone et al. (2016) integrated a quantitative HIA based on epidemiological study of plausible causes of mortality and morbidity with the Agenda21 methodology for participation, based on focus groups and meetings with community stakeholders all along the process. With this methodology, the authors prepared the terrain for new forms of HIA to overcome the barriers between various forms of technical knowledge and these local knowledges. Following this, it is clear that there is a need for citizens and stakeholder participation in HIA, especially those parties with vested interest that may be affected by the proposed or investigated scenarios. We advocate for it, being however aware that participation can have its shortcoming and can be not as effective as expected especially when lacking adequate time resources or when not specifically addressing power unbalances and communication issues (Elvy 2014).

In HIA, citizens and stakeholder participation should occur in the selection of the scenarios, identification of health effects and vulnerable populations and selection and periodization of recommendations, identifying the best channels of dissemination and monitoring and evaluation (Table 29.1). Particularly important is to include more vulnerable groups such as those with low socioeconomic status, children, pregnant women and the elderly who all have their specific needs. More and more often, new kinds of citizen participation, as citizen's science, have begun to offer new tools to assess and include the citizens' perspective in the public health arena.

New models of HIA need to take advantage of these innovative citizens' participation approaches to improve the utility, social acceptance and impact of their results. As such they can build a dialogue among different sectors and actors, avoiding to reproduce a pattern in which the use of quantitative HIA is limited to academia. Incorporating other views can indeed enhance the understanding of how plausible, realizable and acceptable scenarios developed by HIA practitioners are and whether authorities would ever consider them for implementation.

29.3 Challenges

BOD assessments and HIAs encounter many challenges in terms of data availability and assumptions that need to be made and are sensitive to the contextual setting and underlying population parameters. Some of the main problems conducting quantitative HIAs on a city level are the lack of availability of baseline data for some of the exposures and health outcomes, the implied need to make assumptions of these parameters and how to deal with uncertainty. Furthermore, there may be a lack of good exposure-response relationship data. Previous models and tools have been solving this challenge accessing input data through multiple official databases from public entities, identifying information on health, exposures and population. What stands out is also the importance of developing a comprehensive search strategy for input data, from different data sources, languages and time periods, and also the importance of data quality assessment and final comprehensive models validation. The input data identification is without doubt one of the most important steps in the quantitative HIA process, and cannot be achieved without a close collaboration with the stakeholders. The participation of different stakeholders, approached with a variety of methods and participatory tools, is crucial for the identification of specific and high-quality data.

With regard to epidemiological input data on the exposure-response functions, performing a systematic review of the literature will be the key point for identifying the most robust evidence to quantify the health impacts. Quantitative HIA has limitations in assessing the complexity of real policies or scenarios, mainly because of the unavailability of the needed amount of quantitative evidence, limiting the results of a quantitative HIA to those exposures and outcomes that can be quantified. Quantitative HIA can highlight these limitations and also be combined with qualitative HIA, so to generate recommendations able to involve and inform the stakeholders in a broader dimension.

A further challenge is how to make models accessible, so that they can be used outside the research community by practitioners and policymakers. Only in this way can we ensure that HIA has the needed wide uptake in cities across countries. Simplification without losing the essence may be the answer, and this is, for example, the approach in the PASTA project (Gerike et al. 2016). Similarly, models that are coupled with qualitative evidences and built with the collaboration of stakeholders might have wider impact on the policy realm and as such be more easily disseminated across different actors and cities, becoming best practice in the process of policy transfer and policy learning.

Finally, most of the work so far has been done in high-income countries. There is a need for this type of work outside high-income countries, where urbanization rates are the highest, where there is the greatest burden of disease related to non-communicable diseases and where many cities are in the process of being shaped leaving room for timely interventions. This also brings forward specific challenges because often there is a real lack of data to conduct the work (Gascon et al. 2016b) in combination with a lack of vision on the future health necessities and the lack of governance and institutional strength. Yet, at the same time low- and medium-income regions and countries have a real opportunity ahead, to improve and consider public health in the urban and transport development, avoiding the mistakes made by developed countries.

29.4 Uncertainty

An important issue is how to deal with uncertainty. Uncertainty may occur when conceptualizing the problem, during analysis and/or while communicating the results (Briggs et al. 2009). Much focus has been placed on characterizing and quantifying the analysis, and various statistical methods have been developed to estimate analytical uncertainties and model their propagation through the analysis (Mesa-Frias et al. 2013). As described before, transport, emission and air quality modelling each have their uncertainties, and these are propagated through the chain. Validation and uncertainty assessment is needed at every stage, but is rarely conducted. On the other hand, larger uncertainties may be associated with the conceptualization of the problem, i.e. the scenarios building and communication of the analytical results, both of which depend on the perspective and viewpoint of the observer (Briggs et al. 2009). Therefore, more participatory approaches to investigation, and more qualitative measures of uncertainty, are needed, not only to define uncertainty more inclusively and completely but also to help those involved better understand the nature of the uncertainties and their practical implications (Briggs et al. 2009).

29.5 Conclusions

Health impact assessment can help urban and transport planning suggest urban and transport scenarios that are better for health. The field is still growing, but already there are quite a number of studies out there. In practice qualitative HIAs are more prevalent, while in research circles quantitative HIAs are generally conducted. Further improvements are needed, for example, through the development of participatory full-chain quantitative HIA methods, models and tools to improve evidence-based decision-making and to obtain and implement the most feasible and acceptable urban and transport policy measures to improve public health in cities.

References

Bailey, K., & Grossardt, T. (2010). Toward structured public involvement: Justice, geography and collaborative geospatial/geovisual decision support systems. *Annals of the Association of American Geographers, 100*, 57–86.
Bailey, K., et al. (2012). Toward environmental justice in transportation decision making with structured public involvement. *Transportation Research Record, 2320*(1), 102–110.
Banister, D. (2008). The sustainable mobility paradigm. *Transport Policy, 15*(2), 73–80.
Briggs, D. J. (2008). A framework for integrated environmental health impact assessment of systemic risks. *Environmental Health, 7*, 61.
Briggs, D. J., Sabel, C. E., & Lee, K. (2009). Uncertainty in epidemiology and health risk and impact assessment. *Environmental Geochemistry and Health, 31*(2), 189–203.
Creutzig, F., Mühlhoff, R., & Römer, J. (2012). Decarbonizing urban transport in European cities: Four cases show possibly high co-benefits. *Environmental Research Letters, 7*(4), 044042.
Elvy, J. (2014). Public participation in transport planning amongst the socially excluded: An analysis of 3rd generation local transport plans. *Case Studies on Transport Policy, 2*(2), 41–49.
European Centre for Health Policy. (1999). *Health impact assessment: Main concepts and suggested approach (Gothenburg Consensus)*. Brussels: European Centre for Health Policy.
Gascon, M., Triguero-Mas, M., Martínez, D., Dadvand, P., Forns, J., Plasència, A., & Nieuwenhuijsen, M. J. (2016a). Green space and mortality: A systematic review and meta-analysis. *Environment International, 2*(86), 60–67.
Gascon, M., Rojas-Rueda, D., Torrico, S., Torrico, F., Manaca, M. N., Plasència, A., & Nieuwenhuijsen, M. J. (2016b). Urban policies and health in developing countries: The case of Maputo (Mozambique) and Cochabamba (Bolivia). *Public Health Open Journal, 1*(2), 24–31.
Gerike, R., de Nazelle, A., Nieuwenhuijsen, M., Panis, L. I., Anaya, E., Avila-Palencia, I., Boschetti, F., Brand, C., Cole-Hunter, T., Dons, E., Eriksson, U., Gaupp-Berghausen, M., Kahlmeier, S., Laeremans, M., Mueller, N., Orjuela, J. P., Racioppi, F., Raser, E., Rojas-Rueda, D., Schweizer, C., Standaert, A., Uhlmann, T., Wegener, S., Götschi, T., & PASTA consortium. (2016). Physical Activity through Sustainable Transport Approaches (PASTA): A study protocol for a multicentre project. *BMJ Open, 6*(1), e009924.
Gkatzoflias, D., Kouridis, C., Ntziachristos, L., & Samaras, Z. (2007). *COPERT 4: Computer programme to calculate emissions from road transport*. Copenhagen: European Environment Agency.
Hatzopoulou, M., & Miller, E. J. (2010). Linking an activity-based travel demand model with traffic emission and dispersion models: Transport's contribution to air pollution in Toronto. *Transportation Research Part D: Transport and Environment, 15*(6), 315–325.
Health Effects Institute. (2010). *Traffic-related air pollution: A critical review of the literature on emissions, exposure, and health effects*, Special Report 17. HEI Panel on the Health Effects of Traffic-Related Air Pollution. Boston, MA: Health Effects Institute.
IFC (International Finance Corporation). (2009). *Introduction to health impact assessment.* Washington, DC: The World Bank.
Innes, J. E., & Booher, D. E. (2004). Reframing public participation: Strategies for the 21st century. *Planning Theory & Practice, 5*(4), 419–436.
Ji, S., Cherry, C. R., Bechle, M. J., Wu, Y., & Marshall, J. D. (2012). Electric vehicles in China: Emissions and health impacts. *Environmental Science & Technology, 46*(4), 2018–2024.
Kearney, M. (2004). Walking the walk? Community participation in HIA. A qualitative interview study. *Environmental Impact Assessment Review, 24*(2004), 217–229.
Kesby, M. (2005). Retheorizing empowerment-through-participation as a performance in space: Beyond tyranny to transformation. *Journal of Women in Culture and Society, 30*(4), 2037–2065.
Khreis, H. 2016. *Critical issues in estimating human exposure to traffic-related air pollution: Advancing the assessment of road vehicle emissions estimates.* Presented at the World Conference on Transport Research, Transportation Research Procedia, WCTR 2016 Shanghai, 10–15 July 2016.

Khreis, H., Kelly, C., Tate, J., Parslow, R., Lucas, K., & Nieuwenhuijsen, M. (2016b). Exposure to traffic-related air pollution and risk of development of childhood asthma: A systematic review and meta-analysis. *Environment International, 100*, 1–31.

Linzalone, N., et al. (2016). Participatory health impact assessment used to support decision-making in waste management planning: A replicable experience from Italy. *Waste Management, 59*(2017), 557–566.

Lowndes, V., et al. (2001). Trends in public participation: Part 1 - Local government perspectives. *Public Administration, 79*(1), 205–222.

Macmillan, A., Connor, J., Witten, K., Kearns, R., Rees, D., & Woodward, A. (2014). The societal costs and benefits of commuter bicycling: Simulating the effects of specific policies using system dynamics modeling. *Environmental Health Perspectives, 122*, 335–344.

McHugh, C. A., Carruthers, D. J., & Edmunds, H. A. (1997). ADMS–Urban: An air quality management system for traffic, domestic and industrial pollution. *International Journal of Environment and Pollution, 8*(3–6), 666–674.

McKinley, G., Zuk, M., Höjer, M., Avalos, M., González, I., Iniestra, R., et al. (2005). Quantification of local and global benefits from air pollution control in Mexico City. *Environmental Science & Technology, 39*(7), 1954–1961.

Mesa-Frias, M., Chalabi, Z., Vanni, T., & Foss, A. M. (2013). Uncertainty in environmental health impact assessment: Quantitative methods and perspectives. *International Journal of Environmental Health Research, 23*(1), 16–30.

Mueller, N., Rojas-Rueda, D., Cole-Hunter, T., de Nazelle, A., Dons, E., Gerike, R., Götschi, T., Panis, L. I., Kahlmeier, S., & Nieuwenhuijsen, M. (2015). Health impact assessment of active transportation: A systematic review. *Preventive Medicine, 76*, 103–114.

Mueller, N., Rojas-Rueda, D., Basagaña, X., Cirach, M., Cole-Hunter, T., Dadvand, P., Donaire-Gonzalez, D., Foraster, M., Gascon, M., Martinez, D., Tonne, C., Triguero-Mas, M., Valentín, A., & Nieuwenhuijsen, M. (2017). Urban and transport planning related exposures and mortality: A health impact assessment for cities. *Environmental Health Perspectives, 125*(1), 89–96.

Mumpower, J. L. (2001). Selecting and evaluating tools and methods for public participation. *International Journal of Technology, Policy and Management, 1*(1), 66–77.

Namdeo, A., Mitchell, G., & Dixon, R. (2002). TEMMS: An integrated package for modelling and mapping urban traffic emissions and air quality. *Environmental Modelling & Software, 17*(2), 177–188.

NAS (National Academy of Sciences). (2011). *Improving health in the United States: The role of health impact assessment*. Washington: National Academy of Sciences.

NHS. (2002). *Health impact assessment: A review of reviews*. London: Health Development Agency.

Nieuwenhuijsen, M. (2016). *Urban and transport planning, environmental exposures and health-new concepts, methods and tools to improve health in cities*. Retrieved from http://ehjournal.biomedcentral.com/articles/supplements/volume-15-supplement-1.

Nieuwenhuijsen, M. J., Khreis, H., Triguero-Mas, M., Gascon, M., & Dadvand, P. (2017). Fifty shades of green: Pathway to healthy urban living. *Epidemiology, 28*(1), 63–71.

Palerm, J. R. (2000). An empirical-theoretical analysis framework for public participation in environmental impact assessment. *Journal of Environmental Planning and Management, 43*(5), 581–600.

Perez, L., Trüeb, S., Cowie, H., Keuken, M. P., Mudu, P., Ragettli, M. S., Sarigiannis, D. A., Tobollik, M., Tuomisto, J., Vienneau, D., Sabel, C., & Künzli, N. (2015). Transport-related measures to mitigate climate change in Basel, Switzerland: A health-effectiveness comparison study. *Environment International, 85*, 111–119.

Reed, M. S. (2006). Unpacking 'participation' in the adaptive management of social–ecological systems: A critical review. *Ecology and Society, 11*(2), 39.

Reisi, M., Aye, L., Rajabifard, A., & Ngo, T. (2016). Land-use planning: Implications for transport sustainability. *Land-Use Policy, 50*, 252–261.

Renn, O., & Webler, T. (1995). *Fairness and competence in citizen participation: Evaluating models of environmental discourse*. London: Springer.

Richardson, G. P. (2011). Reflections on the foundations of system dynamics. *System Dynamics Review, 27*, 219–243.

Sagaris, L. (2014). Citizen participation for sustainable transport: The case of "Living City". *Journal of Transport Geography, 41*, 74–83.

SATURN Manual. (2015). SATURN Manual, April 2015 Version 11.3.12 [Online]. Retrieved April 10, 2016, from http://www.saturnsoftware.co.uk/saturnmanual/pdfs/SATURN%20v11.3.12%20Manual%20(All).pdf.

SDG. (2015). Retrieved November 8, 2016, from https://sustainabledevelopment.un.org/?menu=1300.

Shafiea, F., Omara, D., & Karuppannanb, S. (2013). Environmental health impact assessment and urban planning. *Procedia - Social and Behavioral Sciences, 85*, 82–91.

Ståhl, T., Wismar, M., Ollila, E., Lahtinen, E., & Leppo, E. (2006). *Health in all policies: Prospects and potentials*. Finland: Ministry of Social Affairs and Health.

Stevenson, M., Thompson, J., de Sá, T. H., Ewing, R., Mohan, D., McClure, R. I., Tiwari, G., Giles-Corti, B., Sun, X., & Wallace, M. (2016). Land-use, transport, and population health: Estimating the health benefits of compact cities. *The Lancet*. https://doi.org/10.1016/S0140-6736(16)30067-8.

Sundvor, I., Castell Balaguer, N., Viana, M., Querol, X., Reche, C., Amato, F., Mellios, G., & Guerreiro, C. (2012). *Road traffic's contribution to air quality in European cities*. ETC/ACM Technical Paper 2012/14 November 2012. The European Topic Centre on Air Pollution and Climate Change Mitigation (ETC/ACM) (a consortium of European institutes under contract of the European Environment Agency).

Tainio, M. (2015). Burden of disease caused by local transport in Warsaw, Poland. *Journal of Transport and Health, 2*, 423–433. https://doi.org/10.1016/j.jth.2015.06.005.

Tainio, M., de Nazelle, A. J., Götschi, T., Kahlmeier, S., Rojas-Rueda, D., Nieuwenhuijsen, M. J., de Sá, T. H., Kelly, P., & Woodcock, J. (2016). Can air pollution negate the health benefits of cycling and walking? *Preventive Medicine, 87*, 233–236.

Tobías, A., Recio, A., Díaz, J., & Linares, C. (2014). Health impact assessment of traffic noise in Madrid (Spain). *Environmental Research, 137C*, 136–140. https://doi.org/10.1016/j.envres.2014.12.011.

van de Kerkhof, M. (2006). Making a difference: On the constraints of consensus building and the relevance of deliberation in stakeholder dialogues. *Policy Sciences, 39*(3), 279–299.

Van Vliet, D. (1982). SATURN—A modern assignment model. *Traffic Engineering and Control, 23*(12), 578–581.

Wang, A., Fallah-Shorshani, M., Xu, J., & Hatzopoulou, M. (2016). Characterizing near-road air pollution using local-scale emission and dispersion models and validation against in-situ measurements. *Atmospheric Environment, 142*, 452–464.

WHO. (1999). *European Centre for Health Policy, WHO Regional Office for Europe*. Gothenburg Consensus Paper.

WHO. (2015). *Global health Observatory*. Retrieved 26 October, 2015, from http://www.who.int/gho/road_safety/mortality/en/.

Williams, M., Barrowcliffe, R., Laxen, D. & Monks, P. (2011). *Review of air quality modelling in DEFRA* [Online]. Defra. Retrieved from http://uk-air.defra.gov.uk/assets/documents/reports/cat20/1106290858_DefraModellingReviewFinalReport.pdf.

Woodcock, J., Edwards, P., Tonne, C., Armstrong, B. G., Ashiru, O., Banister, D., Beevers, S., Chalabi, Z., Chowdhury, Z., Cohen, A., & Franco, O. H. (2009). Public health benefits of strategies to reduce greenhouse-gas emissions: Urban land transport. *The Lancet, 374*(9705), 1930–1943.

Xia, T., Nitschke, M., Zhang, Y., Shah, P., Crabb, S., & Hansen, A. (2015). Traffic-related air pollution and health co-benefits of alternative transport in Adelaide, South Australia. *Environment International, 74*, 281–290.

Chapter 30
Health Impact Assessment of Active Transportation

David Rojas-Rueda

30.1 Health Impact Assessment of Active Transportation

Active transportation policies have many implications for population health. Many but not all of the active transport policies proposed are justified based on their health benefits. Active transportation policies often come inside of the transport or urban master plan, which has multiple possible scenarios or policies to be implemented. The understanding of the health implications of those different policies or scenarios is essential to achieve the goal of implementing the most healthy transportation policies. Methods like health impact assessment (HIA) help to characterize and estimate the health implications of different active transport policies and scenarios. In this chapter we will explain the use of health impact assessment in active transportation, including a description of the health determinants and health outcomes in health impact assessment of active transportation, the qualitative and quantitative HIA on active transportation, the development of different methods to monitor and evaluate the impacts of active transport interventions on health, participatory processes in HIA, best practices for policy-makers and stakeholders, and the future of HIA on active transportation.

Health impact assessment is an approach to assess the health risks and benefits of future interventions (or policies) in non-health sectors and the distribution of such impacts within the population (Harris et al. 2007). HIA has been proposed by multiple governments and international agencies (like the World Health Organization) (Mindell et al. 2010), as a tool to include health in all policies and develop evidence-based policy-making. In the transport sector, HIA has often been used to assess the health risks and benefits of transport infrastructure and policies, such as the implementation of roadways, public transport networks, or walking and cycling policies (de Nazelle et al. 2011; Rojas-Rueda et al. 2011, 2012, 2013).

D. Rojas-Rueda (✉)
ISGlobal, Barcelona, Spain
e-mail: David.rojas@isglobal.org

M. Nieuwenhuijsen, H. Khreis (eds.), *Integrating Human Health into Urban and Transport Planning*, https://doi.org/10.1007/978-3-319-74983-9_30

Walking and cycling policies have been most often assessed using an HIA approach. Multiple HIAs on cycling and walking have been performed in Europe, Australia, New Zealand, Canada, and the USA (Mueller et al. 2015). Fewer HIAs on active transportation have been applied in developing countries, mainly because of the lack of knowledge on HIA, the lack of available data to assess potential health impacts, and the lack of evidence-based policymaking (Gascon et al. 2016).

The HIA has six steps: (1) screening, (2) scoping, (3) appraisal, (4) recommendations, (5) reporting, and (6) monitoring and evaluation (Harris et al. 2007). The screening step examines if an HIA is necessary or appropriate for a specific policy, also including an assessment of the feasibility of the HIA. The scoping step includes a description of how the HIA will be done, the type of HIA, the population, setting, temporality, and health determinants. The appraisal step includes the actual assessment of the health impacts (which could be either qualitative or quantitative). In the appraisal step, the health outcomes will be identified. This step also provides an assessment of the distribution of the health impacts in vulnerable populations. The recommendations step develops a list of recommendations for stakeholders and policy-makers. In this step, the identification of relevant stakeholders and policy-makers who will receive the recommendations will also be listed. The reporting step is aimed at developing a report in a language and context for stakeholders and policy-makers. This step is essential as often the results of the assessment are not presented or translated in a language that policy-makers and stakeholders (including citizens) can always understand. Finally, the last HIA step is the monitoring and evaluation process; in this step a list of indicators (of health determinants and health outcomes) will be provided to monitor the impacts of the active transport policy. These steps in the context of active transport policies are described in Table 30.1.

The overall framework of active transportation policies, health determinants, and health outcomes is presented in Fig. 30.1. This conceptual framework presents an example of active transport policies, plans, or programs and their relation with behavioral and environmental health determinants and includes a list of possible health outcomes. This conceptual framework will help to improve understanding of the relation between active transportation and health, especially in the context of an HIA. In the following sections, a more detailed description of each of these steps is provided.

30.2 Active Transportation: Health Determinants and Health Outcomes

Active transportation has been related to multiple health determinants (de Nazelle et al. 2011). The scoping step of the HIA will identify the health determinants related to active transportation. The main health determinants are generally reported in scientific reviews, including traffic incidents, air pollution, noise, physical activity, social interaction, ultraviolet radiation exposure, diet, and heat and cold exposure, among others (de Nazelle et al. 2011; Mueller et al. 2015).

Table 30.1 Health impact assessment steps in the context of active transportation

HIA steps	Interpretation in active transport context	Policy implications
Screening	– Is the active transport policy relevant for the community or the region? – Are there any previous HIAs on active transportation in similar contexts? – Identify stakeholders on active transportation	– Policy framework development to assess active transport interventions – Development of a database of active transport policies and their assessments – Development of a network of stakeholders in active transportation – Promote participatory policy-making
Scoping	– Consult different stakeholders, from different sectors and backgrounds, to establish the scoping – Define the active transport intervention. Where and when it will be implemented? Which modes of transport will be affected? Will it modify the built environment and/or another health determinant? Who will be affected/benefited? Are vulnerable groups implicated? – Identify available databases to define the scope (i.e., mobility and urban plans, travel and health surveys, etc.)	– Promote intersectoriality and health in all policies – Promote open data and governance transparency
Appraisal	– Identify the health determinants related to active transportation interventions – Give special attention to physical activity, traffic incidents, air pollution, etc. – Collect information on health determinants and their distribution between the populations – Identify the availability and assess the weight of the evidence related to important health determinants and their health outcomes – Consult stakeholders to identify preferences in terms of expected outcomes (qualitative, quantitative, economic assessment) – If quantitative, follow a risk assessment approach, and if required/possible, estimate the economic assessment and compare with the transport intervention costs	– Promote open data and governance transparency, with special attention to collect and report health, environmental, and socioeconomic data – Promote public health research – Promote participatory policy-making
Recommendations	– Develop recommendations for each active transport policy and stakeholder – Consider the recommendations also to the different phases of development and implementation of active transport policies	– Provide an open list of the policy plans and description – Provide a list of the stakeholders, policy-makers, and authorities related to the active transport policy.

(continued)

Table 30.1 (continued)

HIA steps	Interpretation in active transport context	Policy implications
Reporting	– Prepare a stakeholder report and dissemination activities with special attention on policy-makers and stakeholders of active transportation – Design report and dissemination activities with a language that stakeholders can understand and in the context of active transport policies	– Develop mechanisms to integrate HIA recommendations in active transport policy frameworks – Promote participatory policy-making
Monitoring and Evaluation	– Develop and collect indicators on health determinants, health outcomes, and policy processes for monitoring and evaluation	– Promote open data and governance transparency, with special attention to reporting health, environmental, and socioeconomic data

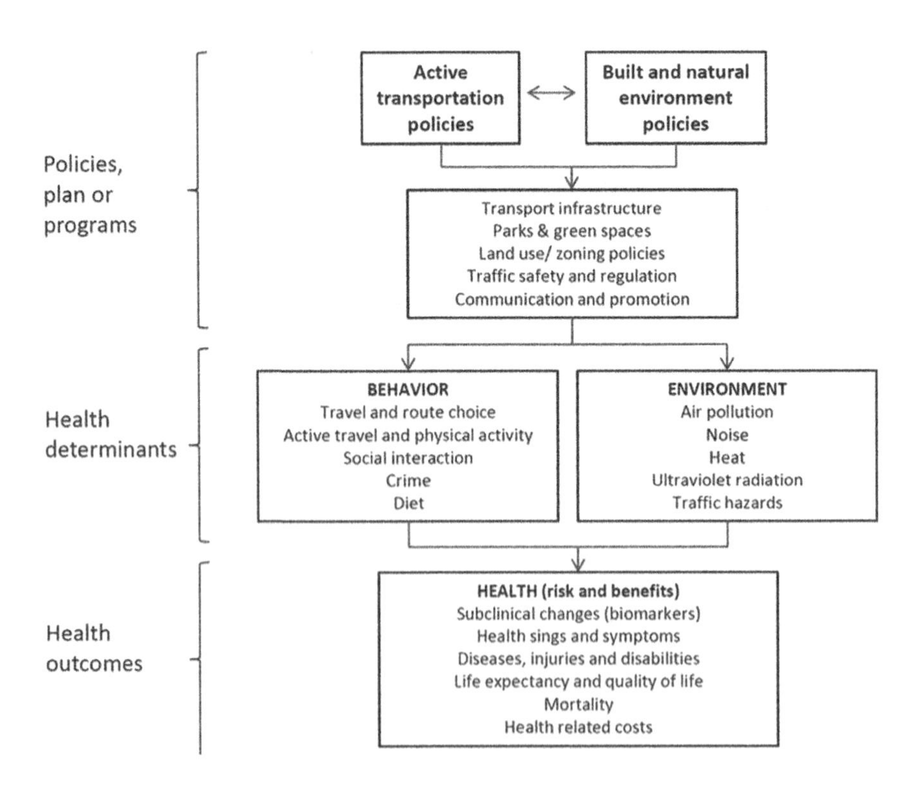

Fig. 30.1 Conceptual framework of active transportation and health

The health determinants of active transportation can be classified based on the population affected. The two main groups are the travelers and the general population (Rojas-Rueda et al. 2012, 2013). The health determinants related to active transport travelers are increasing physical activity, air pollution inhalation during active transportation, traffic incidents (crashes and falls) during active transportation, traffic noise exposure during active transportation, green space exposure in active transportation routes, ultraviolet radiation exposure during active transportation, exposure to heat and cold during active transportation, increasing social interaction during active transportation, exposure to crime during active transport, and finally a change in diet associated with increasing physical activity.

All these health determinants of active travelers have multiple health outcomes that should be included in the HIA (appraisal step) (Mueller et al. 2015). The health outcomes can be classified as subclinical changes; signs and symptoms; diseases, injuries, and disabilities; quality of life and life expectancy; and mortality. Some examples of health outcomes for active travelers related with physical activity are life expectancy, quality of life, breast and colon cancer, cardiovascular and cerebrovascular diseases, diabetes, respiratory diseases, neurodevelopmental disorder, dementia, depression, and bone/joint/muscle diseases, among others; for air pollution, respiratory diseases, cardiovascular and cerebrovascular diseases, diabetes, neurodevelopmental disorder, and birth outcomes, among others; for traffic noise, cardiovascular and cerebrovascular diseases, sleep disturbance, and annoyance; for green spaces, mental health, stress, life expectancy, quality of life, and cardiovascular diseases, among others; for ultraviolet radiation, skin cancer, bone diseases, and immunological disorders, among others; for heat exposure, dehydration and electrolyte disorders, respiratory diseases, cardiovascular and cerebrovascular diseases, and heat stroke, among others; for cold exposure, infectious diseases, respiratory diseases, and cardiovascular and cerebrovascular diseases, among others; for social interaction, mental health, infectious diseases, life expectancy, and quality of life; for crime, injuries, fatalities, mental disorders, and quality of life; and for diet, metabolic and digestive disorders, cardiovascular and cerebrovascular diseases, life expectancy, and quality of life, among others (de Nazelle et al. 2011).

On the other hand, active transportation also modifies health determinants that affect the general population (non-travelers), mainly by the substitution of modes of transport, reduction of air and noise pollution, and improvement of the built environment and public spaces. The most common health determinants related to active transport policies that affect the general population (non-active travelers) are the reduction of air pollution, reduction of traffic noise, increasing of green spaces (and possible reduction of the heat island effect), increasing social capital, reduction of crime, and reduction of traffic incidents between other modes of transport (non-active modes) (de Nazelle et al. 2011; Mueller et al. 2015). The health outcomes related to these general population health determinants are similar to those mentioned for active travelers.

30.3 Qualitative HIA for Active Transportation

As mentioned above there exist two main types of HIA, qualitative and quantitative. Qualitative HIA is the most common approach in active transportation. Qualitative HIAs could provide enough information to create recommendations for policy-makers. Qualitative HIA in active transportation can be applied when sufficient resources (personal and monetary) and/or time are not available. Qualitative HIA is also commonly described as a rapid HIA and should include all the HIA steps (see Table 30.1). Qualitative HIA can also include a participatory approach (see below) where the stakeholders and citizens can be take part in multiple HIA steps.

An example of a qualitative HIA on active transportation is shown in Fig. 30.2 where a cascade of health determinants and health outcomes related to an active transportation intervention are presented. Based on this description and following the HIA steps, the qualitative approach will also generate specific indicators (health determinants or health outcome indicators) to monitor and evaluate the health

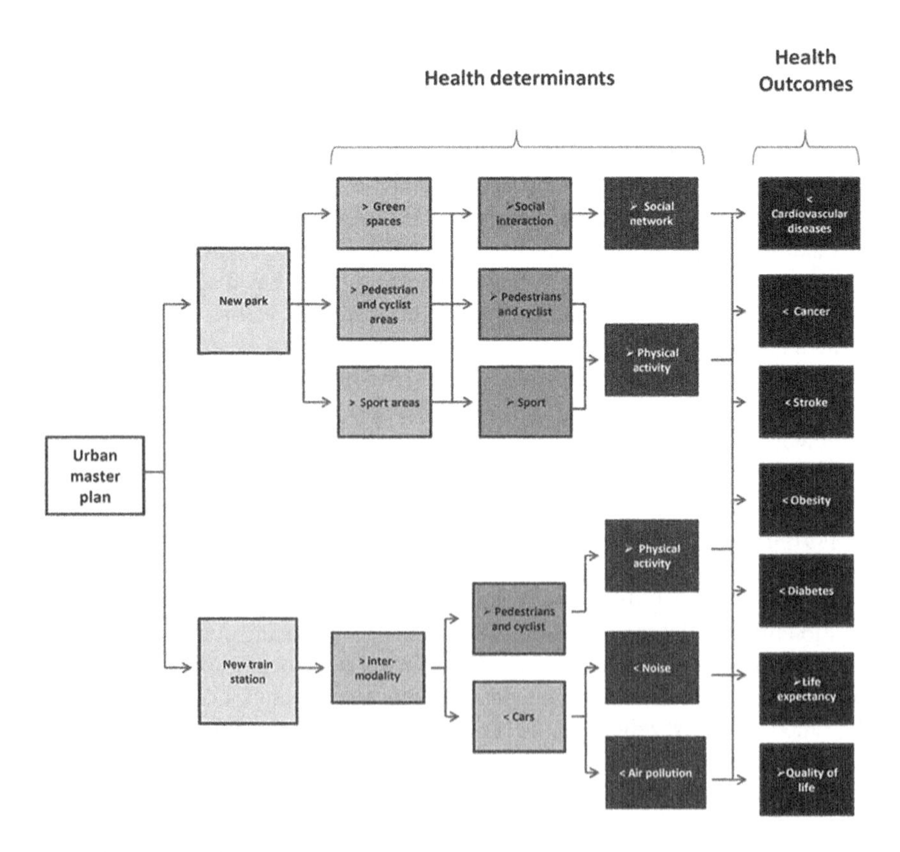

Fig. 30.2 Example of health determinants and health outcomes reported in a qualitative HIA of active transportation

impacts of active transport policies. An example of health determinant indicators could be the annual average concentration of air pollution, daily traffic noise levels, the number of active travel trips per day/year, the duration or distance of active travel trips, the amount of green spaces, the number of traffic crashes or falls, or the percentage of modal shift, among others. Examples of health outcome indicators are incidence or prevalence of diabetes, myocardial infarction, hypertension, colon cancer, asthma, etc.; number of hospital admission for asthma attacks, hypertensive emergencies, etc.; number of people who report annoyance or sleep disturbance; traffic injuries and fatalities; and mental disorders, among others.

HIA recommendations derived from qualitative HIA could be very similar to those derived from a quantitative HIA (Harris et al. 2007). General recommendations for policy-makers will be based on the city context and characteristics (Rojas-Rueda et al. 2016). For example, in a qualitative HIA in Seattle, a list of key recommendations were developed for the transport department focusing on public transit, bike lane, sidewalk, and street improvements creating a complete network that connects residents to community services, green space, and other community assets in line with the Seattle Transit, Bicycle, and Pedestrian Master Plans (Gundersen et al. 2015). Another example is active transport HIA comparing six European cities which found that the risk of traffic fatalities between active transport modes was higher in cities like Warsaw or Prague (Rojas-Rueda et al. 2016). As such, recommendations were focused more on improving traffic safety. On the other hand, cities like Basel or Copenhagen showed a greater benefit to promote more active transportation modes, and the recommendations were focused on promoting active modes especially in those populations that are not currently active transport users.

Recommendation for other audiences including citizens, health practitionersm and researchers can also be generated by qualitative HIAs on active transport (Nieuwenhuijsen et al. 2017). The importance of including recommendations to audiences beyond policy-makers will improve the acceptability and performance of active transport policies. Examples of recommendations for citizens are the integration of active transportation into their daily activities, the reduction of the use of cars, increased intermodality, increased healthy behaviors such as physical activity, and support to sustainable development through active transport activities, among others. For health practitioners, some general recommendations are promoting health in all policies, stimulating intersectoriality, providing health data related to health determinants and outcomes of active transportation, and increasing the use and knowledge of HIA, among others. Finally, HIA recommendations for researchers have increased evidence on active transportation and health, developing studies on different population subgroups (elderly and children) and settings (developed and developing countries), including health equity, sustainability, and governance in research, and promoting intersectoriality and health in all policies, including responsible research and innovative approaches in active transport research, among others.

30.4 Quantitative HIA for Active Transportation

Quantitative HIA refers to those HIAs that include a quantitative assessment of health determinants and health outcomes. Quantitative HIA is a useful complement of an HIA and should be performed when there exists the possibility to measure different health determinants and estimate the health outcomes of different policies. Although the quantitative HIA is dependent on the availability of resources (monetary and personnel) and time to conduct measurements and/or quantitative estimates, a quantitative HIA may be requested by stakeholders and policy-makers. It is important to highlight that a quantitative HIA is not a necessary step, and does not necessarily provide more information (or produce different conclusions) than those already provided by a qualitative HIA. However a quantitative HIA could increase the value, acceptability, support, and precision of HIA conclusions.

In active transportation, multiple quantitative HIAs have been performed, from assessing the impact of promoting cycling or walking to the implementation of active transport infrastructure (Mueller et al. 2015). The main quantitative health determinants and outcomes included in active transportation HIAs are presented in Fig. 30.3. The most common health determinants included in a quantitative HIA for active travelers are physical activity, traffic incidents, and the inhalation of air pollution. Quantitative HIA can also be performed to quantify the health risk and ben-

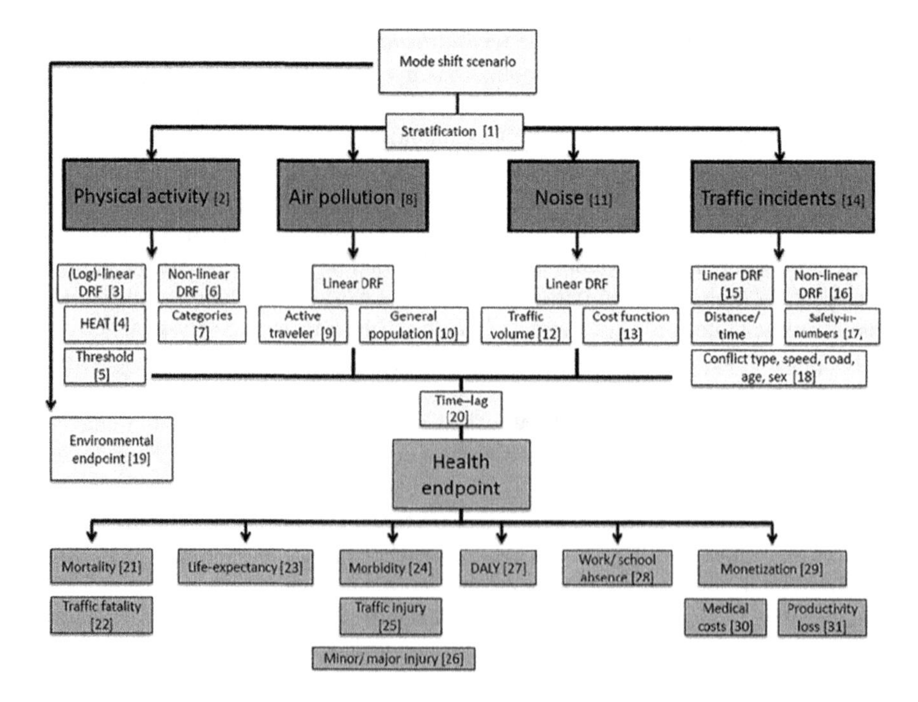

Fig. 30.3 Most common methods and outcomes used in quantitative health impact assessments of active transportation (Mueller et al. 2015)

efits of active transport policies in the general population (non-travelers). In a quantitative HIA, the most common health determinants included for assessing the impact of active transport policies in the general population are air pollution, traffic noise, increase of green spaces, and reduction of the heat island effect.

A quantitative approach to estimating the health risks and benefits is known as comparative risk assessment as proposed by the World Health Organization (Ezzati et al. 2004). Sometimes the comparative risk assessment approach has also been incorrectly labeled "impact modeling," but this can produce confusion between conventional epidemiological analysis and other impact assessments available.

The comparative risk assessment approach proposed by the World Health Organization includes four steps, (a) hazard identification, (b) dose-response assessment, (c) exposure assessment, and (d) risk characterization. Hazard identification includes the identification of health risk/benefits (health determinants) associated with active transportation (physical activity, air pollution, and traffic incidents) and their possible health outcomes. Dose-response assessment refers to the identification of the available risk estimates (e.g., relative risk, odd ratios) from epidemiological studies describing the relationship between the magnitude of a dose and a specific biological response (health outcome). The dose-response functions are specific to each exposure and outcome. There are two main types of dose-response functions, linear or nonlinear. The epidemiological evidence from the main health determinants of active transportation that can be quantified such as air pollution, physical activity, and traffic incidents have been proposed as nonlinear dose-response functions. This means that a constant increase in the dose (or exposure) will not produce a similar constant change in the outcome. As an example, in the case of physical activity, an increase in 10 min of walking or cycling will produce a different mortality risk reduction depending on the existing physical activity level of the population. The third step of a comparative risk assessment is exposure assessment. Exposure assessment focuses on the design and conduct of measurements of health determinants related to active transportation (e.g., measurement of physical activity, air pollution, and traffic incidents). An example is the measurement of walking or cycling using questionnaires asking the number of trips, the trip duration, or distance. Physical activity can also be measured using pedometers, accelerometers, and geographical positioning systems (GPS). This more direct measurement can result in more precise exposure assessment than questionnaires. Another example is the inhalation of air pollution during walking or cycling. Exposure assessment could be done measuring the concentration of air pollution (individual measurement) during walking or cycling combined with physical activity measurements using an accelerometer combined with heart rate or breath rate monitoring. Finally the last step of a comparative risk assessment is risk characterization, when all previous steps are combined in an analysis that results in a relative risk of the specific exposure levels (of health determinants) during walking or cycling in the population studied (using the dose-response functions) and the estimation of the attributable fraction, to estimate the proportion of disease that can be attributed to a particular level of exposure.

Quantitative risk assessment can also be translated to monetary terms, adding an economic assessment (Mueller et al. 2015). The economic assessment translates the health outcomes into health costs associated with each case of disease prevented or increased (morbidity) with the active travel policy, through the estimation of a healthcare cost related with diagnosis or treatment (direct health cost) and the cost related with work absences due to disease incapacity (indirect health costs). The societal costs related with mortality can also be included in this economic assessment, estimating the economic cost related with a premature death based on the value of a statistical life. Many quantitative HIAs have included a combination of economic assessment including morbidity (direct and indirect costs) and/or mortality (societal cost) (Mueller et al. 2015; Rutter et al. 2013). Many stakeholders especially policy-makers are interested to know the economic cost of their policies. So this economic assessment is a valuable step to increase the applicability and usefulness of a quantitative HIA results.

It is important to highlight that not all health determinants related with active transport policies (in travelers and the general population) can be quantified in an HIA. For this reason, quantitative HIA should be considered as a partial part of the assessment. The main reasons that do not allow a quantification of health determinants in a quantitative risk assessment are usually the lack of robust and quantitative epidemiological evidence and the lack of data in the study population. Many of the health determinants related with active transportation such as social capital or ultraviolet radiation and their health outcomes have not been described in quantitative dose-response functions, so this makes it difficult to include them in a quantitative risk assessment. Possibly many other health determinants related with active transport could also not have studied. Another important point to highlight is that results of some cohort studies, for example, of pedestrians or cyclists, have been only adjusted by conventional epidemiological covariates (such as sex, age, socioeconomic status, smoking, etc.), but not other relevant health determinants, such as air pollution, noise, green space, diet, and ultraviolet radiation exposure, among others, so the resulting relative risk estimates (dose-response functions) from these cohort studies include already these multiple health determinants (in one single risk estimate); in this case maybe it will be not necessary to add any other relative risk to assess the overall impact of multiple health determinants. Until now most of the quantitative HIAs have preferred to include different dose-response functions for each health determinant (physical activity, air pollution, and traffic incidents), assuming that double counting will be minimal and that adding relative risk estimates from air pollution or traffic incidents in each city will produce more conservative and realistic estimates. On the other hand, many of the cohort studies have been performed only in high-income countries, where cities have a more homogeneous pattern of exposures. This can also make it difficult to use these studies in other countries with different urban contexts.

30.5 Monitoring and Evaluation for Active Transportation

One main product of any HIA (qualitative or quantitative) is providing a list of indicators for monitoring and evaluation. Monitoring and evaluation are the sixth step of HIA and focus on providing indicators to follow the health determinants and health outcomes included in the HIA (Harris et al. 2007). This step is important for stakeholders and policy-makers, as these indicators will help them to identify if the HIA estimations were correct and if there is any deviation of the health determinants and/or outcomes from the main estimation. If any deviation occurs, the monitoring process will help to prevent future unexpected changes.

There are two different types of indicators, indicators of health determinants and indicators of health outcomes. Health determinant indicators are those factors that summarize the level of exposure. For example, a health determinant is physical activity, and its indicator could be the amount of walking that can be expressed at the population (number of persons walking to work) or individual level (minutes walked per person per day). Another example for health determinants is air pollution, and its indicator could be the concentration of a specific air pollutant in the city/neighborhood/street, expressed as the annual average concentration of particulate matter (or any other pollutant) in the city/neighborhood/street for example.

Health indicators are measurement units of the levels of health outcomes related to health indicators. For example, for physical activity (health determinant), the incidence (new cases) or prevalence (current cases) of stroke in the studied population at a certain time could be considered. Another example is for air pollution (health determinant), specifically for particulate matter $<2.5\ \mu m$ in diameter; a health indicator could be the incidence or prevalence of type 2 diabetes in the study population. As you can note, health determinant indicators (number of pedestrians or the level of air pollution) sometimes could be an easier and more effective indicator to follow, because the change of health determinants (physical activity or air pollution) will happen before any change in the health outcomes (stroke or type 2 diabetes). This mainly applies to those health outcomes where the natural history of the disease is long (years). In other words, this mainly applies to those diseases, such as type 2 diabetes, where it takes many years to see any development of the disease. For other diseases or symptoms with a shorter time development, such as asthmatic crises or myocardial infarction patients with existing cardiovascular diseases, the change in the health indicators (emergency room visits for asthma or infarct) could appear in only a few hours after the change in the health determinant (hourly concentrations of air pollutants).

In summary, health determinant and health outcome indicators for active transport policies should be provided for any HIA. These indicators will help stakeholders and policy-makers to prevent any unexpected changes and detect the benefits of their policy. The indicators should also be presented with a list of possible information sources in the study population where stakeholders could identify these indicators. This will reduce the effort to collect (and invest valuable resources) some indicators that possibly already exist in other data sources. An example of those data sources where health determinants and outcome indicators can be identified includes travel surveys (e.g., number of pedestrians or cyclist trip duration), air quality surveillance systems (e.g., hourly/daily/annual concentration of particulate matter or gases), health surveys (e.g., incidence or prevalence of multiple diseases or levels of physical activity), emergency room visit reports (e.g., specific health conditions, visits per day/month), traffic safety reports (e.g., number of traffic injuries or fatalities in pedestrians or cyclists), and urban noise maps (e.g., level of traffic noise by street), among others.

30.6 Participatory Process in HIA for Active Transportation

Participatory approaches in HIA can increase the utility, applicability, and acceptance for stakeholders (Nieuwenhuijsen et al. 2017). The HIA (qualitative and quantitative) should always include a participatory approach, especially (but not exclusively) in the screening step (see Fig. 30.4). As mentioned above, the screening step provides the justification (relevance and utility) of an HIA for a specific active transport policy or intervention. Not all active transport policies or interventions will require an HIA. Only those that could affect the majority of the population, or a vulnerable population, or have particular societal relevance should be considered for an HIA. This screening process should then be performed by the local stakeholders or policy-maker. Examples of this participatory approach in the screening process could be in-depth interviews with relevant stakeholders or stakeholder workshops. These participatory approaches will provide invaluable information regarding the appropriateness of an HIA for a specific policy or intervention. This participatory approach will also provide information on the different scenarios to be included in the HIA and the necessity to perform a quantitative or economic assessment in the HIA. The participatory approach in the screening step of an HIA will define the necessity and, if so, the type of HIA that should be performed.

The participatory approach can also be implemented in the rest of the HIA as described in Fig. 30.4. In the scoping step of an HIA, the participatory approach could help to identify the main population of interest, if there is a vulnerable population to consider, and the temporality and geographical boundaries of the HIA. In the appraisal step, stakeholder involvement through participatory approaches will help to identify possible health determinants and health outcomes that could have been missed in the first assessment. It is important to highlight that the local stakeholders, citizens, and policy-makers are the experts in the local context. In the scoping step, citizens and stakeholders can also help in the collection of data, for example, they can contribute to recruiting other citizens to answer travel or health surveys and they can also be part of data collection of performance measurements (air quality, noise, traffic counts, etc.). In the recommendation steps, participatory approaches could help to identify priorities for the local population and context that could help to develop more effective recommendations and identify the stakeholders or authorities that should receive the recommendations. In the reporting step, a participatory approach can also improve the language and content of the report; this is especially useful for increasing the efficacy and utility of the HIA resulting in increased understanding between stakeholders and citizens on the issues related to the active transport policies and empowering the population. Finally, participatory approaches could be used in the sixth step of an HIA, monitoring and evaluation. In this last step, participatory approaches can help to design the most valuable and practical indicators. In this last step, stakeholders and citizens can also be active participants collecting the indicators, as in the appraisal step.

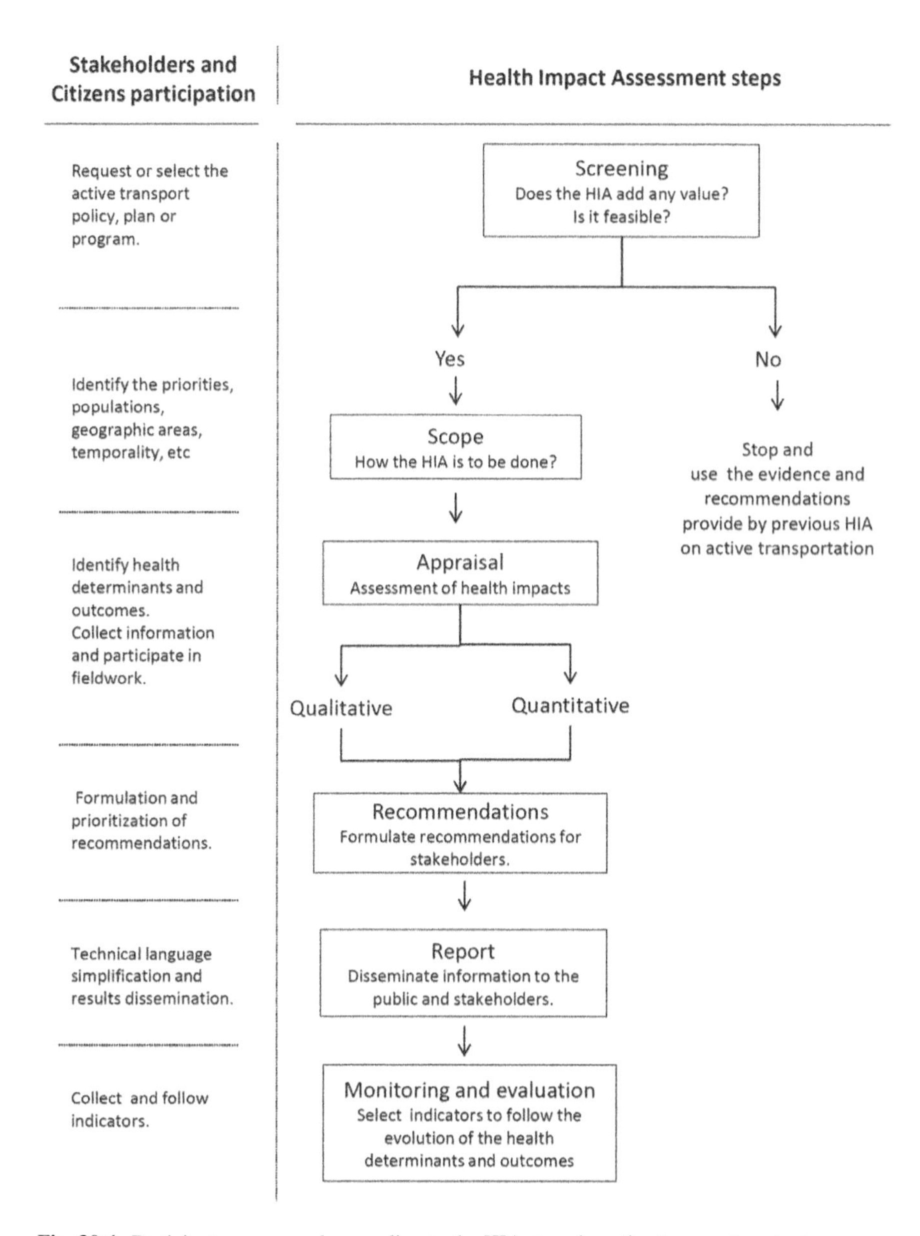

Fig. 30.4 Participatory approach according to the HIA steps in active transport context

30.7 Best Practices for Policy-Makers and Stakeholders

The HIA is a valuable tool for stakeholders and policy-makers to identify and estimate the health implications of their active transport policies or interventions (Mueller et al. 2015; Rojas-Rueda et al. 2016; Nieuwenhuijsen et al. 2017). As described in this chapter, the HIA (qualitative or quantitative) will provide with a list of health determinants and health outcomes related to the interventions and will identify the distribution of the health risks and benefits and also the vulnerable populations that could be affected by the policy or intervention. It will provide a list of recommendations to minimize health risks and maximize health benefits of policies, including a list of indicators to monitor and evaluate the policy or intervention.

The HIA is also a tool to stimulate the collaboration between sectors that previously may have not been working together and between different levels of government or authorities, therefore improving governance in the local context. Also, the HIA is a tool for health in all policies, approach, a mandate proposed by multiple health authorities in different countries.

The HIA has four main values, equity, sustainable development, ethical use of evidence, and a holistic view of health. The HIA will introduce these four values in any recommendation generated, favoring the development of equitable, sustainable, evidence-based, and healthy active transport policies and interventions.

30.8 Tools for Quantitative HIA on Active Transportation

Some researchers have developed some tools to assess the health impacts and costs related to active transport interventions. These tools summarize the evidence on quantitative HIA, but with some differences. They are tools that put more attention to some health determinants than others. Most of them are free and available online or by contacting the tool developers. In the next paragraphs are explained the three main tools to quantify the health impacts of active transport policies.

The most common and known tool available is the "Health Economic Assessment Tool (HEAT) for walking and cycling." The HEAT for walking and cycling was developed by the World Health Organization with the aim to create a tool that can be accessible to any stakeholder, especially without experience in health sciences (mainly urban and transport planners). The HEAT for walking and cycling is an online free tool (http://heatwalkingcycling.org/) focused on assessing the health benefits (in mortality) of walking or cycling interventions, related to physical activity. The HEAT for walking and cycling includes an economic assessment (based on the value of statistical life). The HEAT for walking and cycling has been used by multiple local and national transport authorities around the globe. HEAT for walking and cycling is a simple tool, who asks only a few questions about active transport interventions (like a number of cyclist or pedestrians expected to increase with the active travel intervention) and the characteristics of the trips (trip duration and trip frequency). The tool provides, in a few steps, the annual number of deaths that can be avoided by the active travel intervention and includes an economic assessment related to those estimated deaths. The HEAT for walking and cycling also offers a possibility to assess the benefit-cost ratio of the intervention. Overall the HEAT for walking and cycling is the easiest tool available, offering a robust analysis on physical activity and mortality.

More complex tools to assess the benefits of physical activity, the risks associated with the inhalation of air pollution during walking or cycling, the risk of traffic incidents, and the estimation of carbon emissions associated with active travel interventions are also available. Two of the more known complex tools are the TAPAS (Transportation, Air Pollution, and Physical Activities) tool and the ITHIM (Integrated Transport and Health Impact Modelling) tool, both from European research centers.

The TAPAS tool has recently been updated in the European PASTA project (Physical Activity Through Sustainable Transport Approaches). The TAPAS/PASTA tool includes a quantitative assessment of physical activity, air pollution, and traffic incidents. The TAPAS/PASTA tool provides results in terms of mortality, morbidity (cases of disease), the burden of disease (disability-adjusted life years), and economic costs. The TAPAS/PASTA tool also includes a module to quantify the carbon emission (CO_2), according to the vehicle fleet composition and fuel characteristics. The TAPAS/PASTA tool has been applied in several cities in Europe in walking, cycling, and public transport interventions. These tools have been the first to assess interventions like bike-sharing systems or public transport interventions.

The ITHIM tool has been applied mostly in the UK and the USA. It is a high data-demanding tool, requiring an important amount of work for data collection, especially for traffic injuries. Similar to the TAPAS/PASTA tool, this tool assesses the health impacts of physical activity, traffic injuries, and air pollution. But this tool has been mostly focused on intense characterization of traffic injuries, with a low development of air pollution impact assessment and not carbon assessment.

In general, the TAPAS/PASTA tool and the ITHIM tool are more data-demanding and complex tools than HEAT for walking and cycling. But the results provided by these two complex tools are more specific and comprehensive than results from HEAT for walking and cycling. Each of these three tools has strengths and limitations, and the selection of the toll will depend on the resources (time, money), data availability, and user modeling experience. For a general assessment, the HEAT for walking and cycling will be an excellent tool; if required more specificity and detail in the quantifications, the TAPAS/PASTA tool or ITHIM tool is the most suitable.

30.9 The Future of HIA in Active Transportation

The future of the HIA in active transportation should focus on reducing the lack of general knowledge about HIA by including active stakeholders (including health experts) as a tool for evidence-based policy-making, the lack of active transport HIA in developing countries, and the inclusion of participatory approaches as a common practice and work as a common tool to identify knowledge gaps (in the epidemiological evidence and the local data sources on transport and health) and stimulate the development of studies and quality surveillance systems of health determinants and health outcomes.

References

Ezzati, M., et al. (2004). *Comparative quantification of health risks global and regional burden of disease*. Geneva: WHO.

Gascon, M., Rojas-Rueda, D., & Torrico, S. (2016). Urban policies and health in developing countries: The case of Maputo (Mozambique) and Cocha-bamba (Bolivia). *Public Health Open Journal, 1*(2), 24–31.

Gundersen, G. et al. 2015. Health impact assessment seattle's delridge corridor multimodal improvement project.

Harris, P., et al. (2007). *Health impact assessment: A practical guide*. Sydney, NSW: Centre for Health Equity Training, Research and Evaluation (CHETRE). Part of the UNSW Research Centre for Primary Health.

Mindell, J., et al. (2010). Improving the use of evidence in health impact assessment. *Bulletin of the World Health Organization, 88*(7), 543–550. Retrieved December 8, 2014, from http://www.pubmedcentral.nih.gov/articlerender.fcgi?artid=2897984&tool=pmcentrez&rendertype=abstract.

Mueller, N., et al. (2015). Health impact assessment of active transportation: A systematic review. *Preventive Medicine, 76*, 103–114. https://doi.org/10.1016/j.ypmed.2015.04.010.

de Nazelle, A., et al. (2011). Improving health through policies that promote active travel: A review of evidence to support integrated health impact assessment. *Environment International, 37*(4), 766–777. Retrieved August 22, 2014, from http://www.ncbi.nlm.nih.gov/pubmed/21419493.

Nieuwenhuijsen, M. J., Khreis, H., & Verlinghieri, E. (2017). Participatory quantitative health impact assessment of urban and transport planning in cities: A review and research needs. *Environment International, 103*, 61–72. https://doi.org/10.1016/j.envint.2017.03.022.

Rojas-Rueda, D., de Nazelle, A., et al. (2013). Health impact assessment of increasing public transport and cycling use in Barcelona: A morbidity and burden of disease approach. *Preventive Medicine, 57*(5), 573–579. Retrieved December 17, 2014, from http://www.ncbi.nlm.nih.gov/pubmed/23938465.

Rojas-Rueda, D., et al. (2016). Health impacts of active transportation in Europe. *PLoS One, 11*(3), e0149990.

Rojas-Rueda, D., De Nazelle, A., et al. (2012). Replacing car trips by increasing bike and public transport in the greater Barcelona metropolitan area: A health impact assessment study. *Environment International, 49*, 100–109. Retrieved September 17, 2014, from http://www.ncbi.nlm.nih.gov/pubmed/23938465.

Rojas-Rueda, D., et al. (2011). The health risks and benefits of cycling in urban environments compared with car use: Health impact assessment study. *British Medical Journal, 343*, d4521.

Rutter, H., et al. (2013). Economic impact of reduced mortality due to increased cycling. *American Journal of Preventive Medicine, 44*(1), 89–92. Retrieved December 17, 2014, from http://www.ncbi.nlm.nih.gov/pubmed/23253656.

Chapter 31
Barriers and Enablers of Integrating Health Evidence into Transport and Urban Planning and Decision Making

Rosie Riley and Audrey de Nazelle

31.1 Introduction

By 2050 the global population will reach ten billion, 70% of which is expected to live in urban areas. Whilst cities are centres of education, wealth creation, innovation and progress, they are simultaneously characterised by air and noise pollution, congestion, heat and overcrowding with adverse impacts on human health on a global scale. The relationship between the built environment and human health is a complex adaptive system involving continuous interactions between stakeholders that result in dynamic and unpredictable outcomes with positive and negative feedback loops (Hunter 2015; Plsek and Greenhalgh 2001; Swanson et al. 2012). Pathways of causation between actors in the system and outcomes are not always clear and are rarely linear. Linkages between the actors can be context dependent; policy changes and decisions can result in unintended consequences. Further, health impacts of cities are not equally distributed across the population with certain socio-economic and ethnic groups at greater risk than others to issues such as air pollution (Marmot et al. 2008; Marmot 2011). Despite this complexity, the evidence for how we should be building our cities of the future is well advanced. Comprehensive urban and transport planning and policy that seeks to maximise health outcomes is critical. In response to the growing recognition of the impact of the built environment on health, various pieces of international and national guidance have been published on urban and transport planning and policy (Barton and Tsourou 2000; Chang et al. 2016, 2015; Ross and Chang 2014). Despite these attempts, current urban and transport planning frequently fails to reflect best practice. Examples of effective research translation are rare. With current rates of population growth and urbanisation set to continue, it is critical that the challenges to address the upstream determinants of health through urban and transport planning and policy are identified and overcome.

R. Riley · A. de Nazelle (✉)
Imperial College London, London, UK
e-mail: rosie.riley16@imperial.ac.uk; anazelle@imperial.ac.uk

M. Nieuwenhuijsen, H. Khreis (eds.), *Integrating Human Health into Urban and Transport Planning*, https://doi.org/10.1007/978-3-319-74983-9_31

This chapter highlights some of the barriers to integrating health into urban and transport planning and policy with the aim of suggesting opportunities through which these barriers may be overcome. The chapter starts with a discussion of the barriers, as broken down into four main areas: (1) differing understandings of health between sectors, (2) differing understandings of evidence and evidence translation, (3) governance and politics and (4) the institutional context. The second part of the chapter outlines opportunities to overcome these issues. Enabling factors include improving communication, understanding and collaboration between the different disciplines of urban and transport planning and health. Greater understanding, collaboration and communication are likely to ease challenges originating in the political context and may be enabled through changes to structures in the institutional environment as well as changes to the way in which research is developed and communicated. The chapter starts with an exploration of the challenges as broken down into the above four areas.

31.2 Barriers

31.2.1 Differing Understandings of Health

Health is understood in different ways. For instance, a public health practitioner may conceptualise health differently to an engineer. These differing views can make the integration of health into decision making challenging. Health can be understood in narrow terms as treatment of sickness and disease. Alternatively, more comprehensive understandings of health incorporate well-being, mental health and the importance of social relationships (Davis 2005). The way in which health is understood impacts on how different actors perceive their responsibility in relation to public health promotion. For example, a narrow understanding of health leads to a limited view of the role of actors outside of the traditional health sector. Transport planners tend to display a narrow understanding of health and consequently view their own role within public health promotion as limited. In this sense, their role in health is restricted to the reduction of traffic collisions and the delivery of casualties to hospital, falling short of action to promote physical activity, for example, or the functioning of social support networks (Davis 2005). Similarly, urban planners may display a narrow understanding of health, perceiving their own role in public health promotion as limited to improving access to doctors and dental services and leisure centres. These understandings prevent comprehensive approaches to urban and transport planning that could include improving access to affordable and healthy food and blue or green space or the provision of safer and well-lit streets. The way in which different actors understand health impacts their perceived role in public health promotion and consequently the extent to which health is a factor in their decisions and plans.

In the latter part of the twentieth century, calls to ensure the adoption of a comprehensive understanding of health that incorporated its wider determinants, such as housing, transport and education, grew stronger. As a marker of this change in understanding of health, the Constitution of the World Health Organisation defines health as 'a state of complete physical, mental and social well-being and not merely the absence of disease or infirmity' (World Health Organization 2014). The adoption of this wider understanding was paralleled by the rekindling of interest in public health driven by the work of Lalonde and McKeown in the 1970s (Lalonde 1974; McKeown 1976). They argued that the narrow understanding of health in the years prior and the dominance of the medical perspective had obscured the fact that the origins of much ill health and disease are derived from activities beyond the health sector. By identifying the wider determinants of health as located within sectors considered formally separate, such as education, housing, transport and urban planning, their work necessitated intersectoral collaboration for public health. Nonetheless, despite the now advanced evidence on the relationship between the built environment and health, the narrow understanding of health still prevails. Such an understanding presents a barrier to attempts to address public health through transport and urban planning.

31.2.2 Evidence

Issues around evidence challenge the integration of health into urban and transport planning and policy. These issues relate to (1) the different understandings of evidence, (2) its effective communication, and (3) how it can be applied to policy. Different disciplines understand 'evidence' in different ways (Petticrew and Roberts 2003). Traditionally, public health professionals adopt different standards to lawyers or policymakers, for instance (Autant-bernard et al. 2013; Davis 2014). The public health and biomedical understanding of evidence are known for its robust and thorough approach given the predominance of evidence-based public policy within the field (Parsons 2002). Evidence-based public policy aims to prioritise robust methodologies, such as systematic reviews, randomised controlled trials and cohort studies, in informing policy decision making (Akobeng 2005). Through employing such methodologies, the approach aims to reduce bias and subjectivity in decision making. In contrast to the public health understanding, policymakers often adopt a selective and less critical approach in terms of what counts as evidence (Petticrew and Roberts 2003; Sallis et al. 2016). Policymakers frequently perceive stories and case studies as more compelling evidence than rigorous studies or literature syntheses (Sallis et al. 2016). Historical precedent and ideology often take primacy in terms of influencing the weighing up of policy options (Kent and Thompson 2012). In addition, the range of policy options available is often restricted by an overreliance on preconceived ideas so that option choice frequently does not take into account the variety available (Eddington 2006). Given their accountability to various stakeholders, local government officers tend to favour evidence rooted in localism and empiricism in order to argue, justify and defend policies (Phillips and Green 2015). As public health evidence aims to remove context and case studies of practice, urban and transport planning often fail to employ this evidence and instead rely on that which is less robust, but arguably more compelling, to make decisions.

These differences in understandings of evidence can prevent the integration of health into urban and transport planning as options that appeal to individual decision makers acting in specific contexts are chosen over what the hard evidence stipulates.

Effective evidence translation is difficult. Whilst research on the relationship between the built environment and health is growing, it is rarely picked up by policymakers or practitioners. Frequently, a 'gap' emerges between research on the one hand and those responsible for its implementation on the other. This results in suboptimal decision making in terms of health promotion as the evidence is neglected or misunderstood (Brownson et al. 2006; Giles-corti et al. 2015). This 'gap' emerges in part due to issues with the evidence base. The relationship between the built environment and health is complex, context specific and dynamic. Evidence often fails to effectively communicate this complexity so that decision makers are left with only a partial understanding. This makes translating the evidence into appropriate and effective policies difficult (Giles-corti et al. 2015; Gray et al. 2010; Pilkington et al. 2016). The problem is exacerbated by a lack of research proposing policy options and interventions (Sallis et al. 2016). Whilst research has focused on understanding how features of urban and transport design affect health, little research has been done on appropriate and effective policy interventions. Research is often not policy relevant nor clearly communicated. Finally, research often fails to take into account the complex environments in which decision makers operate. Additional difficulties lie in the different timeframes and pressures of the policy-making process as opposed to the lengthy peer review process of the scientific research sector making further collaboration disjointed (Orton et al. 2011).

31.2.3 Governance and Politics

Issues around governance and the political context exacerbate the disconnect between the health evidence base and urban and transport planning decision making. Whilst evidence-based public policy in principal draws on robust evidence to determine appropriate courses of policy direction, in reality decisions are often made on political grounds (Davis 2014; Juntti et al. 2009; Phillips and Green 2015). As Davis comments 'politics and power take the lead in determining what and how evidence is used' (Davis 2014). Urban and transport planners in local government are accountable to a variety of stakeholder groups in a way that public health practitioners have traditionally not been (Phillips and Green 2015). As part of the decision-making process, urban and transport planners weigh up the interests of various stakeholder groups including politicians, residents, local businesses, internal departments and health services. Often these groups have different competing priorities and interests. Taxi drivers, for instance, may oppose policies to restrict movement of vehicles in city centres, residents may oppose more stringent parking rules and local businesses may challenge attempts to pedestrianise connecting roads for fear of losing business. Therefore, urban and transport planning and policy must be understood as a political process that is not neutral but involves choices regarding what information is deemed relevant, which interests should be prioritised and how resources, such as land, should be allocated and used (Corburn 2004). Decision making around health takes place in this political context. Frequently, the strength and power of certain groups exert greater influence on decision making than the hard evidence (Hill 2017). This means that, unless health professionals engage politically and lobby effectively, decisions are not likely to reflect optimal solutions in terms of health promotion.

The strength of the car lobby is often cited as a barrier to health-optimising decision making. Car owners, be they residents, local business owners or taxi drivers, can be particularly vocal opposition to schemes that prioritise active travel modes, such as walking or cycling (Mead 2015; Carthcart-Keays 2017). Often arguments against these 'anticar' schemes centre on the potentially adverse impacts on footfall into local businesses. At times these proponents are sufficiently vocal to cause schemes to be altered or dropped completely. The power of certain interest groups challenge health promoting urban and transport planning and policy (Davis 2014; Phillips and Green 2015). In addition, this reinforces issues raised in relation to evidence translation and knowledge gaps. For instance, several localised trials in Germany, London and New York, respectively, suggest that pedestrianising retail areas can actually increase footfall into local businesses, restaurants and cafes (Hass-Klau 1993; Tolley 2011; Litman 2014). Other emerging research indicates that walkers spend more money per month on average than people accessing high streets by car. Additional benefits may relate to increased land value of private and commercial rents from changes to the streetscape (Whitehead et al. 2006; Rajé and Saffrey 2016). This emerging evidence is insufficiently developed and even less sufficiently communicated. Further, as public consultation is a critical part of urban and transport planning, in order to reflect and serve the views of stakeholder groups, decisions that receive opposition to or seek to override these groups become politically difficult.

The nature of the political cycle and the lack of political will and leadership present additional challenges. The shortism of the political cycle means that an incoming administration often engenders a shift in policy focus in order to distinguish itself from predecessors. Therefore, whilst one administration may seek to address the upstream determinants of health and encourage the integration of health into urban and transport planning, this policy may stagnate or be reversed with the next administration. Frequently, the extent to which a political leader seeks to address the upstream determinants comes down to personal preferences and understanding of health (PASTA 2016). Political will is often required to push through potentially controversial policies. Due to the potential loss of political capital from pursuing policies that may incite opposition, greater political capital is required in the first place. Politicians are concerned with longevity and survival meaning seemingly unpopular policies are unlikely to be chosen. Therefore, the short-termism of the political cycle as well as the lack of transformational leadership, in the form of visionary and courageous leaders, often results in the deprioritisation of health in decision making (Swanson et al. 2012).

The absence of health as a key consideration in land-use regulation presents a further impediment. Whilst guidance on how to build healthier urban environments is increasing, regulation is weak (Barton et al. 2003; Barton and Tsourou 2000; Chang et al. 2016, 2015; Ross and Chang 2014). Urban and transport planners who challenge proposals for new developments on the grounds that they fail to optimise health benefits have very little to fall back on in terms of regulation. Weak regulation in the field of healthy urban and transport planning and policy makes it difficult for planners to reject new proposals that do not sufficiently optimise health outcomes. In contrast, environmental professionals in Europe, for instance, are able to draw on EU directives to substantiate their arguments, yet health professionals must point to health outcomes, which are often weaker due to the complexity of causality in their relationship to human health.

The lack of regulation mandating the integration of health into urban and transport planning becomes a more acute issue amidst the political drive for more affordable housing and jobs. The push to provide more affordable housing at a greater density and at a greater speed means that provisions to safeguard health are often overlooked. For example, ensuring access to green space and blue space may be ignored alongside adequate provision of safe cycling infrastructure and access to affordable food. In effect, the political drive to address voters' concerns around lack of housing and jobs can systematically deprioritise health-optimising decisions.

31.2.4 Institutional Arrangements

Factors relating to the institutional environment challenge the integration of health into urban and transport planning and policy. Frequently institutional structures embed professional and policy silos, limiting the potential for intersectoral collaboration on health (Carmichael et al. 2013; Kent and Thompson 2012; Sallis et al. 2016; Wooten 2010). For example, transport decisions are made by economists, engineers and signal experts with little or no input from health professionals. Standard practice for larger urban and transport infrastructure policies and plans dictate the use of environmental impact assessment, strategic environmental assessment (SEA) and increasingly health impact assessment (HIA) to ensure health is accounted for. Nonetheless, these appraisals, when done at all, frequently fail to sufficiently account for health (Bond et al. 2013). Policy may dictate that health teams comment on overall strategic direction, through SEA or HIA on bigger schemes, but the general lack of engagement with professionals outside of their own team is likely to lead decision makers to fall back on a business as usual type of approach. The disconnected nature of the institutional environment presents a challenge to the prioritisation of health-optimising solutions. Further challenges relate to the position of public health within the structure of institutions. Often, health professionals do not report to senior executives, which limits their ability to influence decision making or effectively communicate the health agenda.

Other challenges relate to institutional capacity and resources. Typically, programmes that aim to build healthier environments are often overlooked and underfunded. In addition, general lack of budget within departments removes incentives to embark on schemes outside of business-as-usual-type activities. For example, underfunded and under-resourced health departments are often under pressure to provide basic needs such as access to medical and sexual health services or care for the elderly. Within this context of limited resources, schemes that address health determinants within the built environment are often considered a secondary priority and therefore neglected. Institutional capacity exacerbates this problem. Cuts to personnel result in a fewer number of staff responsible for the same amount of workload. Therefore, solutions and programmes that imply additional work in the short term or that seem to detract attention from direct and current health concerns face internal opposition.

In an under-resourced environment, value for money and the business case of different choices become important considerations. Decision makers employ cost-benefit analysis in order to weigh up options. Cost-benefit analysis becomes the proxy upon which decisions are made. Quantifying health outcomes is difficult, and despite advances in metrics such as the Health Economic Assessment Tool, benefits to cost-type evaluations frequently do not sufficiently account for the health risks or benefits (Cavill et al. 2012; Fishman et al. 2015; Rojas-Rueda et al. 2016). Effectively accounting for health benefits of different options is complex leading to suboptimal decision making. Additionally, where the benefit to cost ratio is evident, frequently the positive outcome occurs at a time in the future beyond that stipulated by the funding. Therefore, issues relating to the nature of quantifying the health benefits as well as the timeframes of funding make it difficult for decision makers to choose health-optimising solutions over other policy options.

31.3 Enablers

The chapter so far has presented the multitude of challenges impeding the integration of health into urban and transport planning and decision making. Many of these challenges are not without solutions. However, the viability of solutions depends on the specific context in which they are implemented. Solutions cannot therefore be considered as a one size fits all. For example, the strength of the car lobby will depend on the specific context and make certain solutions more appropriate than others. Similarly, the presence of a strong political leader with the will to push through controversial policies may open up windows of opportunity which are absent in other contexts. Nonetheless, certain solutions are likely to facilitate the integration of health into urban and transport planning decision making regardless of specific context. One such opportunity lies in increasing collaboration between different sectors that affect health such as urban and transport planning, housing and education. Increasing political engagement on the part of health is a further enabler. The rest of the chapter develops these ideas and suggests how some of the challenges may be overcome. Specific attention is drawn to the potential opportunities around changes to the institutional environment, the political context and the evidence base.

31.3.1 Institutional Context

To start, changes can be made to the structure of the institutional environment in which urban and transport planning decisions are made. Intersectoral collaboration could be encouraged through formal and informal structures. Formal structures that encourage collaboration include joint work programmes across departments, joint funding bids, co-location of health professionals within planning departments and intersectoral advisory boards and task forces (Carmichael et al. 2013; Chen and Florax 2010; Wooten 2010). Funding a permanent member of staff with intersectoral responsibilities may encourage the inclusion of health into urban and transport planning (Thomas et al. 2009). Suffusing public health teams across departments can break down silos and assist in building understanding, trust and rapport, therefore, encouraging collaboration. Such institutional changes can encourage the integration of health into decision making.

These positive changes can be compounded by increasing the availability of resources and capacity. Specifically allocating budget for a member of staff with responsibility for the integration of health is likely to be effective (Schwarte et al. 2010). In the UK, the co-location of health professionals within Transport for London, the organisation responsible for the direction and delivery of London's transport system, has driven health up the agenda. In part, as a result of this co-location, the latest Mayor's Transport Plan, a statutory document produced by the Mayor of London, aims to integrate health promotion into its core (GLA 2017). The Mayor's Transport Strategy explicitly adopts the Healthy Streets for London approach that ties funding pots to nine health indicators, with the aim being to encourage local boroughs applying for funding to focus on the health impacts and outcomes of their proposals (GLA 2016). In this example, changes to the institutional context, through co-location and tying funding to health, pushed health up the agenda. In addition, changes made to the funding process and proposal requirements can also increase the likelihood of health-optimising decision making. Extending the timeframe in which funding guidelines expect to see positive outcomes may allow decision makers to adopt a longer-term view facilitating integration of health benefits that often occur further in the future. Funding for training and education to improve understanding of different professions is also needed. Specifically, training for transport and urban planners around the health impacts of their work and the wider determinants of health is likely to increase collaboration and understanding (World Health Organisation 2017). Such training should be included in the urban and transport planning curriculum for further generations.

31.3.2 Political Context

In terms of political context, greater engagement is needed with politics on the part of health (Hunter 2015; Kickbusch 2015; Mackenbach 2013). Public health training should equip professionals with the ability to understand and analyse political context, understand complexity of the decision-making environment and frame arguments effectively to influence policy (Hunter 2015; Kickbusch 2015; Phillips and Green 2015). This may play out through formal public health training or through intersectoral collaboration made possible through changes in the institutional environment. Active engagement with politics in terms of lobbying for the prioritisation of public health, as well as assisting with messaging and communication around more controversial schemes, may enable the integration of health into decision making (Burbidge 2010; Dodson et al. 2009; Salvesen et al. 2008). Increasing communication to effectively display the positive health benefits to wider society can help local government push back on opposition groups (Toy et al. 2014). Increased collaboration between health professionals and planners on the front line of political decision making can improve understanding around which arguments and justifications are likely to be better heard. For example, emerging research suggests that vulnerable and certain socio-economic groups are often at the highest risk of harmful exposures to air pollution, noise and traffic injuries (Bentley 2014; Rundle and Heymsfield 2016; Sehat et al. 2012). Integrating this research into messaging and communications may provide an opportunity to unite opposition groups around health and social justice agenda (Corburn 2004). Better engagement with the political environment could improve understanding and collaboration across sectors as well as offer more effective arguments for health-optimising solutions in decision making.

Stronger regulation to integrate health into land-use considerations may assist planners to push back against suboptimal development proposals in terms of health promotion. Whilst guidance on how planners can integrate health is increasing, regulation that mandates the integration of health into urban and transport planning is sparse. Desirable regulation might include policies banning fast food takeaways within 400 m of schools and building public transport stops and green space within 300 m of schools (World Health Organisation 2017). However, regulation has its limits and will not necessarily overcome challenges that have roots in professional silos and understandings of health (Bond et al. 2013). For example, mandating the use of health impacts assessments will only enable the integration of health if there is sufficient capacity, in terms of qualified personnel and time, to undertake the process. Therefore, stronger regulation on healthy urban and transport planning must be accompanied by sufficient training and capacity building in local government in order to avoid continued challenges that originate in the separation of functions between professions (Bond et al. 2013). Similarly, local regulation should be backed up by national regulation in order to ensure consistency at the country level.

The UK offers a unique example in this regard where regulation in the form of the Health and Social Care Act 2013 (HSCA) relocated public health to local authorities from the National Health Service. A primary driver for the HSCA was recognition of the social determinants of health and an awareness that the levers for improvement of public health were in local government (Carmichael et al. 2012, 2013; Marks et al. 2015; Milne 2012). In this sense, the HSCA provides an opportunity for the integration of health into urban and transport planning through national regulation (Carmichael et al. 2012, 2013; Davis 2015; Marks et al. 2015; Milne 2012). The view is that, through co-location of public health within the local government environment, the HSCA will increase intersectoral collaboration around health, highlight the role of various sectors in public health promotion and improve understanding of health from other professions. So far, little has been done in terms of a review of the effectiveness of this regulation although a 2015 study reached positive conclusions following a series of interviews with urban and transport officials operating in the new environment (Davis 2015). This may be supportive of the suggested use of regulation to facilitate the integration of health into urban and transport planning and policy.

31.3.3 Evidence

Solutions to overcoming challenges of effective evidence translation in part have their roots in collaboration between policymakers, practitioners and researchers. Collaboration between policymakers and researchers to co-identify effective and appropriate research questions that directly assist practitioners faced with challenges in the real world is important (Innaever et al. 2002; Oliver et al. 2014). Researchers could also work to better communicate their findings with policymakers and suggest interventions that are actionable. This may also include a costing of the business case for different policy options. In many circumstances, clearly communicated evidence can help combat more emotive arguments and misperceptions. For instance, local businesses are often concerned that reducing vehicle access will result in a fall in footfall and profits. Despite emerging evidence that refutes this argument, opposition groups frequently display concern in this regard. Clearly communicating the evidence may assist in combating these negative perceptions and reverse opposition to health-promoting schemes (Glasgow et al. 1999; Orton et al. 2011; Spoth et al. 2013). Effective evidence communication may result from improved collaboration between health professionals and planners as each becomes accustomed to different understandings and terminologies. When political windows of opportunity arise, for example, the presence of strong political will and leadership, pre-prepared and well-communicated evidence can then be pushed through.

31.4 Conclusion

This chapter has discussed the barriers and enablers to effectively integrating health into urban and transport planning and policy. The chapter discusses the challenges with reference to four general categories. These categories included (1) differing understandings of health between sectors, (2) differing understandings of evidence and difficulties around evidence translation, (3) governance and politics and (4) institutional context. In addressing these challenges, various solutions were suggested. By adopting particular changes within the institutional environment, collaboration between different disciplines can be encouraged. Engagement with politics on the part of health can assist urban and transport planners to push back against vocal opposition to health-promoting schemes. Stronger regulation to integrate health into land-use considerations is also part of the solution especially when combined with educational and training programmes to improve understanding between sectors and to build institutional capacity. In order to ensure that decisions reflect the evidence and that the evidence in turn reflects the needs and questions of practitioners and policymakers, researchers should actively seek collaboration with policymakers and practitioners to co-identify areas for further research. More effective communication of results will help planners and stakeholders to argue and justify health-promoting schemes against opposing ideological positions. Evidence that provides a costing and an analysis of actionable alternative policies is likely to be picked up. In addition, due to the different timelines for researchers and policymakers (Orton et al. 2011), researchers should keep evidence up to date to ensure they are able to capitalise on the windows of opportunity when they arise.

These enablers around evidence and institutional arrangements provide opportunities to overcome challenges associated with opposition interest groups and lack of political will and leadership. Institutional arrangements and changes to the way evidence is developed and applied can lead to greater collaboration and communication between urban and transport planning and health and result in decision making that optimises health outcomes for those living in cities.

References

Akobeng, A. K. (2005). Principles of evidence based medicine. *Archives of Disease in Childhood, 90*(8), 837–840. https://doi.org/10.1136/adc.2005.071761.

Autant-bernard, C., Fadairo, M., & Massard, N. (2013). Knowledge diffusion and innovation policies within the European regions: Challenges based on recent empirical evidence. *Research Policy, 42*(1), 196–210. https://doi.org/10.1016/j.respol.2012.07.009.

Barton, H., Mitcham, C., & Tsouros, A. (2003). Healthy urban planning. *World Health, 31*(4), 208. Retrieved from http://eprints.uwe.ac.uk/8626/.

Barton, H., & Tsourou, C. (2000). *Healthy urban planning*. London: Spon Press.

Bentley, M. (2014). An ecological public health approach to understanding the relationships between sustainable urban environments, public health and social equity. *Health Promotion International, 29*(3), 528–537. https://doi.org/10.1093/heapro/dat028.

Bond, A., Cave, B., & Ballantyne, R. (2013). Who plans for health improvement?: SEA, HIA and the separation of spatial planning and health planning. *Environmental Impact Assessment Review, 42*, 67–73.

Brownson, R. C., Royer, C., Ewing, R., & Mcbride, T. D. (2006). Researchers and policymakers travelers in parallel universes (Table 1). *American Journal of Preventive Medicine, 30*, 164. https://doi.org/10.1016/j.amepre.2005.10.004.

Burbidge, S. K. (2010). Merging long range transportation planning with public health: A case study from Utah's Wasatch Front. *Preventive Medicine, 50*, S6–S8. https://doi.org/10.1016/j.ypmed.2009.07.024.

Carmichael, L., Barton, H., Gray, S., & Lease, H. (2013). Health-integrated planning at the local level in England: Impediments and opportunities. *Land Use Policy, 31*, 259–266. https://doi.org/10.1016/j.landusepol.2012.07.008.

Carmichael, L., Barton, H., Gray, S., Lease, H., & Pilkington, P. (2012). Integration of health into urban spatial planning through impact assessment: Identifying governance and policy barriers and facilitators. *Environmental Impact Assessment Review, 32*, 187–194. https://doi.org/10.1016/j.eiar.2011.08.003.

Carthcart-Keays, A. (2017, June 13). Oslo's car ban sounded simple enough. Then the backlash began. *The Guardian*. Retrieved from https://www.theguardian.com/cities/2017/jun/13/oslo-ban-cars-backlash-parking.

Cavill, N., Kahlmeier, S., Dinsdale, H., Ĝtschi, T., Oja, P., Racioppi, F., & Rutter, H. (2012). The Health Economic Assessment Tool (HEAT) for walking and cycling: From evidence to advocacy on active transport. *Journal of Science and Medicine in Sport, 15*(2013), S69. https://doi.org/10.1016/j.jsams.2012.11.166.

Chang, M., Green, L., Steinacker, H., & Jonsdottir, S. (2016). *Better Health and Well-being in Wales: A briefing on integrating planning and public health for practitioners working in local planning authorities and health organisations in Wales*. London: Town & Country Planning Association, with Wales Health Impact Assessment Support Unit and Public Health Wales.

Chang, M., Jukes, A., & Egbutah, C. (2015). *Public health in planning: Good practice guide*. London: Town & Country Planning Association. Retrieved from https://www.tcpa.org.uk/healthyplanning.

Chen, S. E., & Florax, R. J. G. M. (2010). Zoning for health: The obesity epidemic and opportunities for local policy intervention. *The Journal of Nutrition, 140*(6), 1181–1184. https://doi.org/10.3945/jn.109.111336.healthy.

Corburn, J. (2004). Confronting the challenges in reconnecting urban planning and public health. *American Journal of Public Health, 94*(4), 541–546. https://doi.org/10.2105/AJPH.94.4.541.

Davis, A. (2005). Transport and health - What is the connection? An exploration of concepts of health held by highways committee Chairs in England. *Transport Policy, 12*(4), 324–333. https://doi.org/10.1016/j.tranpol.2005.05.005.

Davis, A. (2014). Public health evidence to support transport planning. *Town and Country Planning, 83*, 487–489.

Davis, A. (2015). *A healthy relationship: Public health and transport collaboration in local government*. Leeds: Urban Transport Group.

Dodson, E. A., Fleming, C., Boehmer, T., Haire-Joshu, D., & Luke, D. (2009). Preventing childhood obesity through state policy: Qualitative assessment of enablers and barriers. *Journal of Public Health Policy, 30*(1), S161.

Eddington, R. (2006). *The Eddington transport study*. London: Department for Transport.

Fishman, E., Schepers, P., & Kamphuis, C. B. M. (2015). Dutch cycling: Quantifying the health and related economic benefits. *American Journal of Public Health, 105*(8), e13–e15. https://doi.org/10.2105/AJPH.2015.302724.

Giles-corti, B., Sallis, J. F., & Sugiyama, T. (2015). Translating active living research into policy and practice: One important pathway to chronic disease prevention. *Journal of Public Health Policy, 36*(2), 231–243. https://doi.org/10.1057/jphp.2014.53.

Gray, S., Barton, H., Pilkington, P., Lease, H., & Carmichael, L. (2010). *Review 3 Spatial Planning & Health Identifying barriers & facilitators to the integration of health into planning*. Bristol: University of the West of England.

Greater London Authority. (2017). *Mayor's transport strategy: Draft for public consultation*. London: Greater London Authority. Retrieved from https://consultations.tfl.gov.uk/policy/mayors-transport-strategy/user_uploads/pub16_001_mts_online-2.pdf.

Greater London Authority. (2016). *Healthy streets for London*. London: Greater London Authority. Retrieved from http://content.tfl.gov.uk/healthy-streets-for-london.pdf.

Glasgow, R. E., et al. (1999). Evaluating the public health impact of health promotion interventions: The RE-AIM Framework. *American Journal of Public Health, 89*(9), 1322–1327.

Hass-Klau, C. (1993). A review of the evidence from Germany and the UK. *Transport Policy, 1*(1), 21–31. Retrieved March 17, 2017, from http://publiekeruimte.info/Data/Documents/rc5abtiq/39/Pedestrianization---retailing.pdf.

Hunter, D. (2015). Role of politics in understanding complex, messy health systems: An essay by David J Hunter. *BMJ, 350*, h1214. https://doi.org/10.1136/bmj.h1214.

Hill, D. (2017, January 17). The long war of the Mini Holland Scheme. *The Guardian*. Retrieved from https://www.theguardian.com/uk-news/davehillblog/2017/jan/17/the-long-war-of-mini-holland-in-enfield.
Innaever, S., Vist, G., et al. (2002). Health policy-makers' perceptions of their use of evidence: A systematic review. *Journal of Health Services Research & Policy, 7*(4), 239–244.
Juntti, M., Russel, D., & Turnpenny, J. (2009). Evidence, politics and power in public policy for the environment. *Environmental Science & Policy, 12*, 207–215. https://doi.org/10.1016/j.envsci.2008.12.007.
Kent, J., & Thompson, S. (2012). Review article health and the built environment: Exploring foundations for a new interdisciplinary profession. *Journal of Environmental and Public Health, 2012*, 958175. https://doi.org/10.1155/2012/958175.
Kickbusch, I. (2015). The political determinants of health--10 years on. *BMJ, 350*, h81. https://doi.org/10.1136/bmj.h81.
Lalonde, M. (1974). *A new perspective on the health of Canadians*. Ottawa: Minister of Supply and Services Canada.
Litman, T. A. (2014). Economic value of walkability-Victoria Transport Policy Institute. Paper presented at the Transportation Research Board. In 82nd Annual Meeting. Retrieved March 17, 2017, from http://www.vtpi.org/walkability.pdf.
Mackenbach, J. P. (2013). Political determinants of health. *European Journal of Public Health, 24*(1), 2. https://doi.org/10.1093/eurpub/ckt183.
Marks, L., Hunter, D. J., Scalabrini, S., Gray, J., McCafferty, S., Payne, N., et al. (2015). The return of public health to local government in England: Changing the parameters of the public health prioritisation debate? *Public Health, 129*, 1194–1203.
Marmot, M., Friel, S., Bell, R., Houweling, T. A., & Taylor, S. (2008). Closing the gap in a generation: Health equity through action on the social determinants of health. *The Lancet, 372*(9650), 1661–1669. https://doi.org/10.1016/S0140-6736(08)61690-6.
Marmot, M. G. (2011). Closing the gap in a generation. *Bulletin of the World Health Organization, 89*(11), 702.
McKeown, T. (1976). *The role of medicine - Dream, mirage or nemesis*. London: Nuffield Trust.
Mead, N. (2015, October 5). Bike lane blues: Why don't businesses want a £30m cycle-friendly upgrade? *The Guardian*. Retrieved from https://www.theguardian.com/cities/2015/oct/05/bike-lane-blues-london-local-businesses-cycle-enfield-green-lanes.
Milne, E. M. G. (2012). A public health perspective on transport policy priorities. *Journal of Transport Geography, 21*, 62–69. https://doi.org/10.1016/j.jtrangeo.2012.01.013.
Oliver, K., Lorenc, T., & Innvaer, S. (2014). News directions in evidence-based policy-research: A critical analysis of the literature. *Health Research Policy and Systems., 12*, 34.
Orton, L., Lloyd-Williams, F., et al. (2011). The use of research evidence in public health decision making processes: Systematic review. *PLoS One, 6*(7), e21704. https://doi.org/10.1371/journal.pone.0021704.
Parsons, W. (2002). From muddling through to muddling up - Evidence based policy making and the modernisation of British Government. *Public Policy and Administration, 17*(3), 43.
PASTA-Consortium (2016) D2.1 - Baseline analysis of active mobility in case study cities., Final report 03/07.
Petticrew, M., & Roberts, H. (2003). Evidence, hierarchies, and typologies: Horses for courses. *Journal of Epidemiology and Community Health, 57*(7), 527–529. https://doi.org/10.1136/jech.57.7.527.
Phillips, G., & Green, J. (2015). Working for the public health: Politics, localism and epistemologies of practice. *Sociology of Health & Illness, 37*(4), 491–505. https://doi.org/10.1111/1467-9566.12214.
Pilkington, P., Powell, J., & Davis, A. (2016). Evidence-based decision making when designing environments for physical activity: The role of public health. *Sports Medicine, 46*(7), 997–1002. https://doi.org/10.1007/s40279-015-0469-6.
Plsek, P. E., & Greenhalgh, T. (2001). Complexity science: The challenge of complexity in health care. *BMJ, 323*(7313), 625–628.

Rajé, F., & Saffrey, A. (2016) The value of cycling. Retrieved March 17, 2017, from https://www.gov.uk/government/uploads/system/uploads/attachment_data/file/509587/v alue-of-cycling.pdf.

Rojas-Rueda, D., De Nazelle, A., Andersen, Z. J., Braun-Fahrländer, C., Bruha, J., Bruhova-Foltynova, H., et al. (2016). Health impacts of active transportation in Europe. *PLoS One, 11*(3), 1–14. https://doi.org/10.1371/journal.pone.0149990.

Ross, A., & Chang, M. (2014). Planning healthy-weight environments - A TCPA reuniting health with planning project. Retrieved from http://www.tcpa.org.uk/data/files/Health_and_planning/Health_2014/PHWE_Report_Final.pdf.

Rundle, A. G., & Heymsfield, S. B. (2016). Can walkable urban design play a role in reducing the incidence of obesity-related conditions? *JAMA, 315*(20), 2175–2177. https://doi.org/10.1001/jama.2016.5635.

Sallis, J. F., Bull, F., Burdett, R., Frank, L. D., Gri, P., Giles-corti, B., & Stevenson, M. (2016). Urban design, transport, and health. 3. Use of science to guide city planning policy and practice: How to achieve healthy and sustainable future cities. *The Lancet, 388*(10062), 2936–2947. https://doi.org/10.1016/S0140-6736(16)30068-X.

Salvesen, D., Evenson, K. R., Rodriguez, D. A., & Brown, A. (2008). Factors influencing implementation of local policies to promote physical activity: A case study of Montgomery County, Maryland. *Journal of Public Health Management and Practice, 14*(3), 280–288.

Schwarte, L., Samuels, S. E., Boyle, M., Clark, S. E., Flores, G., & Prentice, B. (2010). Local Public Health Departments in California: Changing nutrition and physical activity environments for obesity prevention. *Journal of Public Health Management and Practice, 16*(2), 17–28.

Sehat, M., Naieni, K. H., Asadi-Lari, M., Foroushani, A. R., & Malek Afzali, H. (2012). Socioeconomic status and incidence of traffic accidents in Metropolitan Tehran: A population-based Study. *International Journal of Preventive Medicine, 3*, 181–190.

Spoth, R., Greenberg, M., et al. (2013). Addressing core challenges for the next generation of type 2 translation research, and systems: The translation science to population impact (TSCi Impact) framework. *Prevention Science, 14*(4), 319–351.

Swanson, R. C., Cattaneo, A., Bradley, E., Chunharas, S., Atun, R., Abbas, K. M., et al. (2012). Rethinking health systems strengthening: Key systems thinking tools and strategies for transformational change. *Health Policy and Planning, 27*, iv54–iv61. https://doi.org/10.1093/heapol/czs090.

Thomas, M. M., Hodge, W., & Smith, B. J. (2009). Building capacity in local government for integrated planning to increase physical activity: Evaluation of the VicHealth MetroACTIVE program. *Health Promotion International, 24*(4), 353. https://doi.org/10.1093/heapro/dap035.

Toy, S., Tapp, A., Musselwhite, C., & Davis, A. (2014). Can social marketing make 20 mph the new norm? *Journal of Transport & Health, 1*(3), 165–173. https://doi.org/10.1016/j.jth.2014.05.003.

Tolley, R. (2011). Good for Busine$$: The benefits of making streets more walking and cycling friendly. Discussion Paper. Retrieved March 17, 2017, from https://www.heartfoundation.org.au/images/uploads/publications/Good-for-business.pdf.

Whitehead, T., Simmonds, D., & Preston, J. (2006). The effect of urban quality improvements on economic activity. *Journal of Environmental Management, 80*(1), 1–12. Retrieved March 17, 2017, from http://ac.els-cdn.com/S0301479705001180/1-s2.0-S0301479705001180-main.pdf?_tid=f1f2 6378-0c31-11e7-b3b9-00000aab0f01&acdnat=1489879611_acc06e7cdc808eff9710d5baa56 c7068.

Wooten, H. (2010). Healthy planning in action. *Planning, 76*, 20–23.

World Health Organization. (2014). *Constitution of the world health organization. Basic documents* (48th ed.pp. 1–19). Geneva: WHO. 12571729.

World Health Organisation. (2017). *Draft WHO Global Action plan on physical activity (2018–2030). WHO Discussion Paper*. Geneva: WHO. Retrieved from http://www.who.int/ncds/governance/gappa_version_4August2017.pdf.

Chapter 32
Translating Evidence into Practice

Marcus Grant and Adrian Davis

32.1 Translating Evidence into Practice

The World Health Organization expresses ongoing concerns about the wide gap between the production of scientific evidence and a policy-making response (WHO 2009, 2016). A large gulf remains between what we know and what we practice (WHO 2008). This chapter discusses what the issues generated when attempting to translate public health evidence for urban planning and transport planning into practice. It explores the tensions, the arguments and proposes possible solutions in the world of evidence, policy and practice in the fields of non-communicable disease, health equity and wellbeing in the urban realm. As health policy moves out of its traditional comfort zone of the medical sphere and health-care delivery, so the gap between the models of how the evidence base 'should' work and the complexity in terms of where it needs to work widens, a radical rethink is required (Carmichael et al. 2016; Rutter et al. 2017). In disciplines such as health, insights from policy theory, public administration and organizational studies, which may be of use in explaining the gap, are often overlooked (Cairney et al. 2016). With the increasing focus on city and transport planning providing a response to urban health challenges, something needs to change.

The chapter starts with a major section which unfolds the city as the context and background for population health and its research. We ask, are we actually building illness into the places we live. We rapidly then shift our focus to knowledge, looking at the nature of evidence and practice in cities planning for health. The manifold nature of actors involved means that the issue of different epistemologies, or ways of

M. Grant (✉)
Environmental Stewardship for Health, Bristol, UK
e-mail: marcusxgrant@citieshealth.world

A. Davis
University of the West of England, Bristol, UK
e-mail: Adrian.Davis@uwe.ac.uk

M. Nieuwenhuijsen, H. Khreis (eds.), *Integrating Human Health into Urban and Transport Planning*, https://doi.org/10.1007/978-3-319-74983-9_32

knowing, must be acknowledged. The city as both laboratory and subject throws yet more confusion into the science, and we question whether evidence in itself is enough. The next short section looks at the distance between the way the main protagonist traditions, public health and built environment use evidence. This is followed by a section examining the links made, and broken, between evidence and policy. The elements that might be used to forge a consensus are reviewed. In the penultimate section, we plunge the reader into the disruption caused by wicked problems, complexity and adaptive systems, suggesting that ecological public health might provide a useful guiding framework. In the final section, with evidence translation, cogeneration and co-use in mind, we look at the essential roles for advocacy and leadership. Tactical urbanism is touched on, as is the need for pragmatism in the face of policy and funding agendas. Concluding thoughts may be rather rosy for some, but a form of knowledge collaboration for city wisdom is described and advocated.

32.2 The City as Context

In pursuit of supporting the health of urban populations, two concerns conspire to elevate the importance of examining the issues surrounding translating evidence into practice. The first is an increasing importance of environmental influences on human health in cities and towns. A renewed importance comes from the relentless global trend of urbanization (UN 2015). Most people now live in cities and towns, and forecasts have shown that urbanization will continue. How we build, design, manage, maintain and renew our towns and cities is critical for human health. The second concern a rise in non-communicable disease, with urban environments receiving increased attention as presenting challenges and risks to health and health equity (Braubach and Grant 2010).

The debate has extended beyond the usual suspects of 'environmental bads', e.g. air pollution, noise and poor housing; it is taking on board the concept that some environments hinder or frustrate healthy human behaviours and is now ready to embrace the notion that certain combinations of characteristics of living environments can actually support or even 'create' health.

> Health is created and lived by people within the settings of their everyday life; where they learn, work, play, and love. WHO Ottawa Charter 1986, p. 7

As we will discover in this chapter, it will be difficult to put an exact figure on the degree to which health is, in this holistic manner, influenced by the living environment. But a measure of the influence can be found in the work of the Robert Wood Johnson Foundation (Table 32.1).

Around 12.6 million deaths a year are due to preventable environmental health risks (WHO 2016); this represents 23% of the global burden of disease. These environmental risks are a critical and pressing concern for the public health community. Global action is key to combating this epidemic of environmentally related diseases. Evidence bases need to be refined and translated into local level actions; this must

Table 32.1 The degree to which the environment is a determinant of health

Health behaviours 30%	Socio-economic factors 40%	Clinical care 20%	Built environment 10%
Smoking 10%	Education 10%	Access to care 10%	Environmental quality 5%
Diet/exercise **10%**	Employment **10%**	Quality of care 10%	Built environment 5%
Alcohol use 5%	Income 10%	*Suggested transfer to built environment influences:* *Diet/exercise 5%* *Employment 5%* *Social support 2.5%* *Community safety 2.5%*	
Poor sexual health 5%	Family/social support **5%**		
	Community safety **5%**		
Revised totals:			
Health behaviours 25%	*Socio-economic factors 30%*	*Clinical care 20%*	*Built environment 20%*

Italic text: Suggested transfer to built environment influences
Adapted from Robert Wood Johnson Foundation and University of Wisconsin Population Health Institute in LGA 2016, p. 12

include all levels of actors from municipal authorities to 'bottom up' asset-based community approaches, plus in the private sector with an interest in the design and investment in the built environment.

32.2.1 Pathogenic Urban Form?

In 2007, The Lancet published a diagram showing health problems being investigated for possible links with built environment (Rao et al. 2007). Based on a review of the literature on buildings, public spaces and movement networks (Lavin et al. 2006), this indicated an evidence base for over 60 health problems, covering social, physical and mental health issues, linked to some 15 attributes of the built environment relating to their design, maintenance or availability. Table 32.2 gives a lists of those health problems in alphabetic order.

A steady stream of papers, supported by reviews, has been setting the scene for the argument that in towns and cities, we were creating our own pathogenic environments. Although there is long history of texts, including the often cited "'On air, waters" and places by Hippocrates circa 410 BCE (transl. 1849), we could cite Freeman (1985) and Halpern (1995) and helping focus more recent interest on the role of built environments in pathology, in both these cases mental health. In 2001, Swinburn coined the term 'obesogenic environments', built on later by Townshend and Lake (2009), extending it from a focus on the food environment, to include those

Table 32.2 Health problems investigated for possible links with external built environment

Accidents	Increased mortality in elderly people	Poor child development
Alveolitis	Ischaemic heart disease	Poorer mental wellbeing
Anxiety	Lower physical wellbeing in elderly people	Prolonged cognitive performance in children
Arthritic problems	Lung cancer	Prolonged recovery time from illness
Asbestosis	Meningitis, tuberculosis, slow growth and development	Reduced life expectancy
Asthma	Wheezing in childhood	Reduced physical functional health
Bronchospasm	Mental ill health	Respiratory disease and poor mental wellbeing
Cardiovascular disease	Hypertension	Rhinitis
Colon cancer	Negative self-reported healthy outcomes in short term,	Road traffic injuries
Depression	Respiratory symptoms	Stress
Diabetes	Obesity	Stroke
Eczema	Osteoporosis	
Eye, nose and throat irritations	Overall child development	
Fatigue		
Hypertension		
Hypothermia		
Increased anxiety		
Increased mortality		

Adapted from Rao et al. 2007

aspects of the built environment that also could be seen to suppress everyday physical activity. Francis et al. (2012a) added the term 'depressogenic environments', capturing the impact of urban environments on mental health. Their study highlighted the importance of maintenance, in this case of public open space, not just presence, absence or quantity (Francis et al. 2012b). This theme of 'pathogenic' urban environments continued with the identification of 'toxic high streets' (Townshend 2017); these are high streets, where cheap processed high calorific food is cheap and abundant often combined with poor availability of fresh fruit and vegetable. There are shops selling cut price cigarettes and alcohol, betting shops with highly addictive gambling machines, tanning salons, loan and pawn shops and easy credit. In common with the obesogenic and depressogenic environments, there is a strong social gradient in the distribution of these pathogenic environments. Those with lower socio-economic status are more exposed to these characteristics in their local neighbourhoods. Moreover, as many studies now show, socio-economic status alone does not explain the difference in health outcomes, and there is increasing evidence that the physical environments in which people spend their lives are implicated in the observed health disparities (Ball et al. 2007; Gustat et al. 2012; Townshend 2017).

32.2.2 Evidence and Practice in the City

Those concerned with strengthening the influence of research evidence on policy and practice are faced with a formidable task. Improving the supply of evidence may be necessary but it is not a sufficient means for getting evidence of 'what works' disseminated and implemented. Even in health care, for all the richness and strength of the research base, it remains unclear how best to bring about changes in professional practice that are congruent with the evidence (Davies et al. 1999). Compared to health care and medicine, in the urban health context, we have two added, but parallel, complexities. Firstly, instead of laboratory conditions, the locus of intervention is the open system of our physical living environment, be that city region, city, district, neighbourhood, street or building. Secondly, instead of a relatively tightly knit medically trained fraternity, actors are many different professional and practitioner disciplines each with different relationship to what they call 'evidence'. We can count among these actors are architects and urban developers and investors; landscape architects and urban ecologists; transport planners and town planners; those in the housing, energy, water and waste sectors; public health practitioners; and city leaders and community activists. Many of these to date have also given very little, or no, thought to the health impacts of their actions. Health may be outside their traditional professional roles and their knowledge domains (Giles-Corti et al. 2015).

32.2.3 How Can We Know?

Posing questions about the nature of evidence is vital to its development (Lawrence 2004); it also challenges the notion of 'neutral' or 'pure' evidence. Epistemology is the name we give to the responses and reflections provoked by attempting to answer the question 'how can we know?' This is not just of abstract or philosophical interest. Epistemology guides the science that we then use to determine the methods of research that we hope will provide us with solutions to societal problems. If we get the methods wrong, so will be our solutions. So asking 'how can we know?' becomes the basis of how we will understand it is fundamental. There is a growing consensus that the physical layout and design of our settlements and associated transport infrastructure are contributing factors in public health outcomes (Rydin et al. 2012; Grant et al. 2017). This is supported by research that then attempts to shed light on how better to design and manage urban areas for health. If only it was that simple. This chapter attempts to debunk the all-to-frequent implicit assumption that there is a linear link whereby knowledge flows from the scientific evidence base, through urban policy and into implementation and practice in towns and cities. We need to examine the epistemology when the city itself becomes a laboratory for change.

32.2.4 The Laboratory and the Subject

The city is a complex system, not only composed of many living and nonliving subsystems but also in turn embedded into larger regional, national and global systems. A snapshot of the city reveals its urban form; physical scales, characteristics and locations of all the sub-elements facilitate different functions. Functions include working, learning, shopping, leisure, and being at home. Function also includes the movement network. Every town and city has a unique movement network that accommodates people and goods as they need to move between origins and destinations, between functions. The movement network accommodates a plethora of modes, usually including walking; cycling; using buses, trains, trams and taxis; and driving in private cars, sometimes also including underground metro, cable cars, boats, lifts and escalators. A dynamic survey captures the daily, weekly, seasonal and longer-term cycles and fluxes in urban life, including dereliction and gentrification. The subject itself can seem to shift, is it the city as a place for living, of the population therein, or both? In the midst of spatial and temporal complexity and of flux and change, what do we need to know to support health and reduce health inequalities? And how can we find this out?

32.2.5 Evidence Is Not Enough

Many health outcomes have been associated with characteristics of urban-built environment and natural environments. Table 32.2 lists some of the health outcomes linked to urban environments; the original paper by Rao et al. (2007) cites evidence for each disease; Table 32.3 lists characteristics of urban environments that have been the subject of health research. This is a very limited selection drawn from a very large number of published reviews. Each review will approach the subject from a different angle, having different funding and different audiences and purposes. The intention of the list is to illustrate just some of the wide-range features of the built environment which are under scrutiny. Tables 32.2 and 32.3 are deliberately presented separately. Linking all items across these two tables, especially if attempting to take into account the quality of the evidence and potential causality, is not advised.

As an illustration of the interrelated and systemic nature of the relationship between measureable urban characteristics, lets take an example from research into physical activity. Sallis et al. (2016) found that the design of urban environments has the potential to contribute substantially to physical activity, to the tune of 90 min/week. This implies that daily living can add about half to the recommended weekly guidelines for an individual's physical activity. Four urban design factors, already acknowledged in the literature as important factors in neighbourhood 'walkability', were found in this study to be important in determining levels of everyday physical activity. These four were:

Net residential density—This is the number of residential dwellings (houses and apartments) divided by the residential land area.

Intersection density—Measured as the number of pedestrian-accessible street intersections divided by the area.

Public transport density—Measured as the number of bus, rail or ferry stops and stations divided by the land area.

Number of parks—Number of public parks of any size in or abutting the local area. A public park was designated park of any size that was free of cost and open to the public and maintained by a government agency. Parks included improved or landscaped areas and unimproved or natural areas.

Table 32.3 Some characteristics of urban environments that have been under scrutiny in health research

How the built environment influences health: Broad: Built environment; availability, maintenance, design Specific: Light, noise, safety, humidity, temperature, physical activity, air quality, distance, attractiveness, housing quality, locality, immediate surroundings, accessibility space, social networks Lavin et al. 2006
Association between the built environment, health and well-being and levels of physical activity in neighbourhoods: Walkability and functionality, accessibility, density and land use mix, individual design variables, aesthetics, safety Adapted from Croucher et al. 2007
Urban elements posing risks and challenges to health: Land use pattern, transport, green space, urban design Braubach and Grant 2010
Facilitators of physical activity, strong connected communities and healthy food options: Connected street networks, mixed uses, destinations, network signage, bike parking, walking trails, open space, shade facilities, open street frontages, destinations, reduced travel times, mixed densities, community gardens, public art, street lighting, passive surveillance, prohibition of co-location of unhealthy food vendors, public space for farmers markets, restriction on public advertising for unhealthy foods Kent et al. 2011
Direct impacts of the built environment on health and wellbeing: Air quality (indoor and outdoor), climate, water quantity and quality, noise and traffic-related injuries *Indirect impacts on health and wellbeing*: Housing and buildings, neighbourhoods, social environments, connectivity, density and land use mix, accessibility, amenities and decision-making processes, greenspace Jones and Yates 2013
Five aspects of the local environment that can be influenced by local plan policy for better health: Neighbourhood design, housing, healthier food, natural and sustainable environment, transport PHE 2017

These were all measured within an area of proximity to a person's home, a proxy for their local neighbourhood environment. The study design was cross-sectional and repeated in 14 cities worldwide. It found that the combination of these four factors was more powerful than the factors taken singly, a re-enforcing effect. The effect worked across different socio-economic categories. The study concluded that 'observed effect sizes suggest that designing urban environments to be activity-supportive could have large effects on physical activity and those effects can be expected to generally apply to adults living in the neighbourhoods.' (Sallis et al. 2016, p. 2214)

Of interest now is what happens of we take the conclusions as a starting point for an urban intervention. Housing density, road network design, public transport routes and park provision are all dealt with by different professional actors, and each have different policy drivers and investment streams determining outcomes on the ground. Arguably, in a large new build site or comprehensive regeneration scheme, what amounts to a blank sheet of paper may enable good control over these four factors. But in an existing city, implementation through 'retrofit' will require control and co-ordination of housing policy, highways, parks and public transport, some of will be in the hands of private operators. And just how can control over these four factors be exerted in a rapidly growing informal settlement?

In terms of pathogenic environments, the weight of evidence indicates that a broad effect is proven. However there is much academic interest in the detailed nuance of causal relationships, dosage, vulnerability in individuals or subgroups and other specific research questions to better understand 'how we have built pathology into our urban environments'. There are some tough methodological and conceptual challenges to overcome (Schofield and Das-Munshi 2017) before we understand. But:

Do we need a more complete understanding before we act?

We don't apply the precautionary principle, familiar to the environmental sciences, to the creation of our own human habitat. Are we stretching the application of research appropriate to the medical model to cover research of a kind to which they are poorly suited (Petticrew and Roberts 2003)? We desperately need a more integrated approach between research, policy and place-making.

32.3 Worlds Apart?

Public health and urban planning seem to be naturally complementary as the two professions needed for the creation of healthy communities and healthy places. However, city planners and public health practitioners have developed along different paths in the twentieth century with different epistemologies, governance and policies priorities (Carmichael et al. 2012). In the late twentieth century, the drive for health was articulated through the concept of a public health policy, as reflected in the 1988 Adelaide Charter (WHO 1988). By the early twenty-first, this had evolved into Health in All Policies (HiAP) approach, combined with a concern for health inequalities (Marmot et al. 2008). An underlying logic being:

> to address complex health challenges through an integrated policy response across portfolio boundaries. By incorporating a concern with health impacts into the policy development process of all sectors and agencies, it allows government to address the key determinants of health in a more systematic manner. (Kickbusch 2008, pp. 12–13)

With this in mind, evidence gleaned from the growing area of interdisciplinary research about the role of controlling the built environment for human health is crucial. Nevertheless, despite closely linked origins, the contemporary professions of public health and land use planning have grown apart within a neo-liberal context of academic, political and policy silos (Kent and Thompson 2012). The same disconnect has been identified between public health and transport planning (Litman 2003; Davis 2005). From these disconnects, a lost potential for shared agendas, shared meanings and creation of is born (Giles-Corti et al. 2010). They neither share the same evidence bases nor, through divergent epistemologies, have the ability to share evidence.

It could be argued that given their agency over the management of the built environment and public realm, the emphasis should be on built environment practitioners to learn how health and health equity are influenced by their actions. This may be true but if public health is to achieve its goals of improved health outcomes across populations, an onus is also placed on them to learn to translate and provide their evidence for city and urban practitioners in a more accessible manner. Both sets of sectors need to be able to confidently operate outside their own traditional disciplinary and professional boundaries and develop transdisiplinarity collaborations. In defining transdisiplinarity Lawrence (2010) suggests, transdisiplinarity admits and confronts complexity in science, and it challenges knowledge fragmentation, accepts local contexts and uncertainty in research, implies emergent meanings through valuing communicative action and is often action-oriented.

32.3.1 The Contested Nature of Evidence

It has been posited that scientific knowledge is not different to local knowledge because it has a superior access to 'reality' but its difference lies in being more powerful through being able to act over greater distances (Murdoch and Clarke 1994). Local knowledge, however, often provides the necessary place-based considerations and meanings. Juntti et al. (2009) note that several writers emphasize that the distinction between lay and expert knowledge is a political one and ultimately concerns the allocation of power in policy decision-making. Restricting the scope of evidence, whether in favour of lay or expert knowledge, is of course a powerful way of influencing decision outcomes and sometimes seen as key to achieving any outcome at all (Juntti et al. 2009).

We must bear in mind that evidence, as repeatedly and cogently argued in the literature, is socially constructed (Krieger 1992; Chan and Chan 2000). This is a pragmatic and also useful lynch pin for making progress since it gives rise to a question that needs answering if public health and built environment are to work in concert.

What is accepted as evidence, how much is it valued, and how does this differ across professions and groups?

One of the defining claims for scientific knowledge is its objectivity and freedom from distorting factors that may alter the way that the object under study is detected, measured and reported. Yet, looking at various disciplinary-based literatures, it is evident that the meaning of evidence for some disciplines appears very clear cut, seeming to ignore insights gained in other disciplines. Complicating matters further, when attempting to apply the results of often hard-won research knowledge in cities, what counts as evidence and the rules and criteria for assessing evidence, and indeed whether evidence is valued at all, are all negotiated in 'realpolitik'. A diverse stock of 'evidence' is drawn on by various professions, and by lay people, with a matching diversity of what constitutes 'evidence'. For the sake of universal validity and implementation, scientific evidence cannot be allowed to remain disjointed from contextual factors such as local stakeholder experience and opinions (Murdoch and Clarke 1994). To some of these stakeholders, evidence is understood as a central pillar of public policy decision-making (notably evidence-based medicine and public health), while for others evidence may be considered more of a second-order consideration once the policy direction has been decided. As Rychetnik and Wise (2004) note:

> concepts of evidence vary among professionals, disciplinary and social groups: for example, scientists have traditionally adopted different standards of evidence to lawyers. Since the advent of evidence based medicine in the early 1990s, health professionals, managers and consumers have been debating (and negotiating) what is considered as valuable and credible evidence to support decisions about health services, public health, health promotion and health policy. (Rychetnik and Wise 2004, p. 248)

In areas such as health care, there is a long standing consensus about the need for evidence and the nature of convincing evidence. However, in other areas the very nature of evidence is hotly disputed, and there has been strong resistance to assigning privileged status to one research method over another (Davies et al. 1999).

More emphasis on transdisiplinarity could provide a space for the emergence of a new field connecting disparate actors committed to urban health. In such a journey to transdisciplinarity, we need to develop new conceptual and analytical frameworks allowing for a diversity of methods and approaches. We need to be able to address precise localities, whereby specific cases can demonstrate creative, reflexive and transformative capacity of multi-sector interventions (Lawrence 2015).

32.3.2 Hierarchies of Evidence: The Medical Model

Derived from the medical tradition, where mistakes in application or misunderstandings of evidence can lead to an accountability for death or injury, public health and health care have assumed a strict hierarchy of evidence Table 32.4.

In the medical model the often-quoted 'gold standard' quality of evidence is the double-blind randomized control trial. In its ideal form, this is an experiment

Table 32.4 Public health and health care use a strict hierarchy of evidence (after Hadorn et al. 1996 in Smith et al. 2000)

Health hierarchy of evidence	
I-1	Systematic review and meta-analysis of two or more double-blind randomized control trials
I-2	One or more large double-blind randomized control trials
II-1	One or more well-conducted cohort studies
II-2	One or more well-conducted case-control studies
II-3	A dramatic uncontrolled experiment
III	Expert committee sitting in review; peer leader opinion
VI	Personal experience

conducted to test a hypothesis in which experimental subjects are given the 'intervention' in a trial and other subjects, selected at random, are given an intervention that mirrors that under test as closely as possible but lacks the active component of the treatment under trial. Neither experimenter nor subject knows which subjects are in which group until after the trial has finished. The treatment under trial may be a drug (dose, regime or type), a procedure or other intervention. The double-blind randomized control trial has been designed to prove causality. Proof of causality is vital in many aspects of medical practise, and it is embedded within a professional ethics. However, when we move out from academically focussed practice into policy-making we can find an evidence environment that cannot be more different.

An example of a strong evidence hierarchy in action can be found in a highly proscribed process administered by the government agency responsible for public health guidance in England and Wales (NICE). Here research is converted into national policy for health interventions. In this tightly regulated regime, a focus on the detail of study design and process of decision-making runs the risk of overlooking more fundamental issues such as how health policy governance is framed and represented to the other professionals for implementation. In a study using interviews and observation of key informants, the researchers suggested that in a discourse founded on ideas of clinical and cost-effectiveness data tended to take precedence over more subjective, experientially based perspectives (Milewa and Barry 2005). However, the appropriateness of relying on study design as a marker for credibility of evidence may not be appropriate outside health disciplines (Petticrew and Roberts 2003).

Working in the interdisciplinary space between public health and city planning, it soon becomes evident that professionals and researchers, in different disciplines, all attempt to apply their own disciplinary approach to evidence. When working on multidisciplinary projects, they may take on board evidence from another discipline but use their own disciplinary tradition to ascribe what importance to give it. In many disciplines this is not a chosen or reflexive action but something hardwired in their accepted practice.

32.4 Scientific Evidence Meets Policy and Practice

Different types of research give rise to different types of evidence. It is useful to highlight here difference between a biomedical model of health that often adopts a symptom-treatment interpretation of health, even at population level, and a more holistic or integrated model that combines biological, cultural, economic, political, psychological and social factors in new ways to find the best way forward within an ethical framework of health equity.

When we consider the sectors influencing urban health matters, we find a range of methodological preferences and debates in different policy areas (see Table 32.5).

In town planning and urban design, experimental proof of causality does not sit highly within the epistemology. Validation of an intervention or investment for a

Table 32.5 Methodological preferences and debates in selected policy areas (Adapted from Sorrell 2007 Box 2, p. 1861)

Policy area	Methodological preferences and debates
Health care	Gold standard of randomized controlled trials with additional methodological safeguards. Growing interest in qualitative methods to give a complementary view
Social care	Preference for qualitative methodologies. Quantification and experimentation often viewed with suspicion and hostility. Eclectic use of methods to provide complementary insights
Housing	Predominant use of qualitative and quantitative survey methods. The use of econometric methods for forecasting need
Planning	As for housing, plus quasi-legal method. Some aspects of planning decision-making using evidence of policy adherence and precedent
Urban policy	Recent emergence of more multidisciplinary approaches. Diverse methods employed, but particular reliance on case studies. Little or no experimentation
Landscape and ecology	Survey and layering of geographical and spatial data, the use of design and creativity to obtain the 'least wrong' solution
Architecture	Research into context, function and precedent to advise a design-driven response
Transport	Multidisciplinary. Policy research frequently rooted in economic modelling and forecasting

proposed course of action arises from a different set of parameters. As already stated, the context is not conducive to the 'experimental method' which assumes a controlled and closed system where the scientist is an observer and the method is repeatable. Research activity in this sector needs to follow a 'participative action research method'; an open system in which a reflexive practitioner acts as scientist and is an active participant as both an observer and an agent of change. Yet, due to a poorly articulated interface between different scientific methods, potential evidence for what could be valid interventions is being missed.

There is a world of difference between 'knowing that' (know a concept) and 'knowing how' (understand an operation). Haack (1995) askes whether we become too obsessed with the 'knowledge by description' and ignored 'knowledge by acquaintance'. Fortunately, realist evaluation (Pawson and Tilley 1997) had been developed to answer the twin questions 'what works and why?' It treats an intervention as an instance of theory 'in action'. It addresses what it terms 'programmes', the delivery environment supporting interventions and the contexts into which interventions are placed, both as open systems. And importantly, the methodology can be open to stakeholder expertise at all levels. Realist evaluation provides a platform to synthesize many of the qualitative aspects of an intervention that have been overlooked by quantitatively driven methods.

Public health professionals, strongly influenced by the medical model, are conducting critical systematic reviews of the literature for an evidence base to support healthy urban planning interventions. Time and again they often find there are few studies published that meet their quality criteria. They are now starting to question their review methods. For example, umbrella reviews have recently emerged as a more productive strategy (in comparison to systematic reviews) for assessing existing evidence in an attempt to guide the decisions of policy-makers (Khangura et al. 2012).

Conversely, urban environment professionals, coming from a policy and/or design background, are implementing changes to the city fabric, with little or no pre-evaluation or post-evaluation with regard to their effect on health or health equity.

The medical model is of course not the whole story for public health, as qualitative research makes an important contribution too. It has a particular role in exploring factors affecting the implementation of an intervention, including understanding behaviours and developing theory (Sidhu et al. 2017). As such a mixed-methods approach can prove a good starting point for developing that all-important interface between different ways of creating evidence.

32.4.1 Evidence-Based Policy and Practice

Evidence-based public policy is not so old, and many commentators remind us that the rise of evidence-based policy and practice was first attributed to medicine. 'Evidence-based medicine' became 'fashionable coinage' during the 1990s (Pope 2003; Parsons 2002). Adherence to a mantra of evidence-based policy and practice has now spread across most, if not all, areas of European public policy (Muriel and Autant-Bernard 2013) and is spreading across other areas of the world. This approach indicates a continuing influence of the 'modernist' faith in progress informed by reason (Sanderson 2002). It is not our intention to explore this avenue of thinking, though others have including problematizing the very notion of 'progress' (see for example Gray 2013).

Evidence-based policy and practice has been defined as 'the integration of experience, judgement and expertise with the best available external evidence from systematic research' (Davies 1999). Hence, evidence-based policy and practice implies striking a balance between formal research evidence and professional judgement.

So for evidence to have an impact, there must be agreement as to what counts as evidence and in what circumstances. However, as a distinct area of academic study, there is maybe too much emphasis placed on 'what should happen' and rather less interest in 'what actually does happen'.

32.4.2 Policy-Making, Expertise and Evidence

Policy-makers will often consider expert advice as the most important 'evidence' while relegating research evidence to seventh position in a hierarchy Table 32.6 (contrast this with the strict hierarchy of evidence for health given in Table 32.5). And expertise does still count! In a study of health professionals, Marshall and Godfrey

Table 32.6 Policy-makers' hierarchy of evidence. (Adapted from Davies 2005 with acknowledged developments from by Prof. Hunter 2017, Health in All Policies: Making it Work in Practice—Winter School, Durham University)

Policy-makers' hierarchy of evidence	
1	Expert's advice
2	Evidence from professional associations
3	Opinion-based evidence (including lobbyists and pressure groups)
4	Ideological evidence (including party think tanks and manifestos)
5	Media evidence
6	Internet evidence
7	Research evidence
8	Lay evidence
9	Street evidence (urban myths and conventional wisdom)

(2006) revealed that they rely on a range of different sources of knowledge, including clinical experience; training courses; reading journals, research, policies and guidelines; watching or speaking to colleagues; and personal experience. Accumulation of previous experience also acted as a reference point to test out new information including research findings. At some stage a 'competent practitioner is required' to assess evidence. Whatever the methodological approach, resolution will ultimately require additional considerations such as the moral and ethical issues of universal versus selective action; about informed choices, concerning social inequalities and social justice; or about resource allocation and prioritization (Davies 1999).

This leads to practices that generally concurred with current research evidence (Peavey and Vander Wyst 2017). The much-cited work by Weiss (1979) on 'research utilization' in policy acknowledges this. This short discussion paper presents six models of how policy-makers engage with research. It is important not to overlook these in a drive for 'evidence-based' policy. They may even seem unpalatable to empirically driven researchers. One example is the commonly found pragmatic use of research to add post hoc justification for a route of action driven by political expediency or ideology. We doubt if any one working in a local authority is unable to provide an example of this.

As another source of evidence, evaluation studies are highly sought after both locally and nationally. However, their findings risk being too late to be of value in influencing implementation especially at the cutting edge of innovative urban design practice or in a fast-moving funding or political arena. Even where evidence exists and has been presented to governments by their own agencies, they can choose to ignore it or simply select from it what they want to suit their purposes (Hunter and Shelina 2016). Davis (2016) has portrayed this relationship within national and municipal government as a bounded reality triad where evidence is a much smaller constituent part of a triad (Fig. 32.1). Power and influence lies not with the evidence providers but with politicians tempered by incrementalism, a business as usual approach. Strong advocacy is needed to support even the best evidence.

> … researchers within the different disciplines need to translate their knowledge and understanding into action and actively work together to ensure that the health of the population is

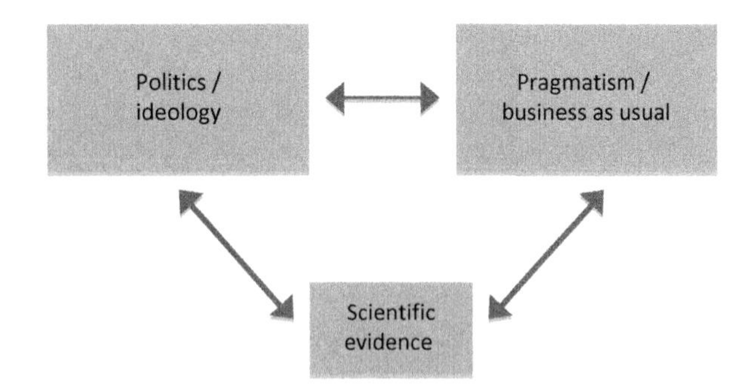

Fig. 32.1 The bounded reality triad of government decision-making. Source: Davis (2016)

at the top of the list of competing priorities for regulatory policy decision-making. (Khreis et al. 2017, p. 60)

As we have discussed, the published public health evidence base is skewed towards studies, whose epistemological origins have led to an attempt to identify simple, often short-term, individual-level health outcomes. Healthy cities action, however, requires an evidence base for complex, multiple, upstream, population-level actions and outcomes. This skew accounts for, and echoes, the prioritization by public health policy-makers of individual-level interventions over system-level responses, in the face of a widening recognition of the need to do the opposite (Rutter et al. 2017).

Policy-makers cannot consider all evidence relevant to policy, so they use two shortcuts: firstly, 'emotions and beliefs' to understand problems and, secondly, 'rational' ways of establishing the best evidence for solutions, to act quickly in complex, multilevel policy-making environments (Cairney et al. 2016). Improving the quality and supply of evidence may help, but it only addresses one side of this equation.

We obviously need to extend the debate about the nature and use of evidence in public health decision-making. Do we really know what is going on? Both political expediency and academic kudos can hamper the evidence to decision-making process. Smith et al. (2000) found that a lack of demand for research findings from policy-makers and practitioners was due to a cultural disbelief in the usefulness of research in the realpolitik of everyday life. At the same time also finding a history where the research agenda was predominantly investigator-led, in which investigators were more concerned with gaining academic kudos than connecting with policy and practice needs.

32.4.3 *Towards a Consensus*

There seems to be an emerging consensus that the 'hierarchy of evidence' is difficult to apply as the intervention arena moves into open settings, such as town and transport planning, and attempts to forge collaborations with practitioners outside the public health fraternity. Moreover, given that trade-offs are always involved, a single hierarchy by which to determine the 'quality' of research design is not helpful (Hammersley 2002). However, Petticrew and Roberts (2003) argue that it may be unhelpful to simply abandon the hierarchy without having a framework or guide to replace it. They emphasize the need to match different secondary research questions to a range of specific research methods. This is the basis of mixed-methods approaches, and arguably there is a need to acknowledge a mixed epistemologies approach too. Thus, emphasis needs to be placed on methodological appropriateness, and on typologies of knowledge, rather than strict hierarchies.

32.5 Applying Complex Interventions to Wicked Problems in Complex Adaptive Systems

So far this chapter has reviewed the context of the city as an open system, looked at the contested nature of evidence in different areas of science and explored 'evidence-based' approaches for policy and practice with its associated shortcomings from an urban health perspective. In this short section, we describe the truly layered complexity of the situation and why looking for a perfect linear alignment of evidence-intervention-solution is itself evident of false understanding.

In the urban nexus, there is a continual dynamic interaction between all living and nonliving elements. In the human sphere alone, there is constant interaction, feedback, accommodation and adjustment between financial, political and social drivers. We are dealing with a complex adaptable system. Every intervention will have unintended consequences and also lead to adaptations of the city system itself. This all occurs in a highly politicized and bias-ridden environment in which 'evidence' or facts need to operate. We even now have the advent of 'post-truth', 'alternative facts' and 'fake news' to contend with (The Guardian 2017).

We need to distinguish between simple interventions akin to following a recipe, complicated interventions which are interventions with lots of parts, and complex interventions where outcomes are uncertain and emergent. For complex interventions, following a formula has limited application (Rodgers et al. 2009). Unfortunately, the medical model (which can be applied to a complicated but not a complex intervention) has strongly influenced our concept of what a 'non-health service' public health intervention might look like (Craig et al. 2011) and how to evaluate it.

32.5.1 Wicked Problems

Faced with the crisis of non-communicable diseases driven by factors such as unhealthy diets, physical inactivity, tobacco use and alcohol consumption, the medical model has strayed into the territory of non-health service public health interventions. Some commentators (see Kelly et al. 2007; Graham and White 2016) suggest tackling these social determinants of disease as a 'wicked problem' (Rittel and Webber 1973). So what can we learn from the literature about 'wicked problems'?

> Rittel and Webber suggested that by the middle of the twentieth century the low-hanging fruit of obvious public health measures had been harvested, leaving urban planners with

> much more intractable problems. These are often symptoms of deeper or more endemic problems, such as social inequality. They allow little room for trial-and-error learning since once communities are demolished, they cannot be restored. Wicked problems are persistent precisely because they lack a clear set of alternative solutions. They cannot be solved definitively, but rather must be managed for better or worse. (RCEP 2007, p. 5)

Although there is no single solution, a better understanding of the mechanisms involved is needed for those working with wicked problems (Parkhurst 2016). Figure 32.2 is a useful graphic. It follows the consequential impacts of increasing car ownership, so demonstrating some of the elements of a 'wicked problem', such as unintended consequences, the unbounded nature of impacts and both negative and positive feedback loops.

Maybe the seemingly simple question 'how to translate scientific evidence into practice' is the wrong question and the wrong starting point.

> The search for scientific bases for confronting problems of social policy is bound to fail, because of the nature of these problems. (Rittel and Webber 1973, p. 155)

Wicked problems share several fundamental characteristics which place multiple challenges in the way of a quest for their solution. In a table listing ten such characteristics (Rittel and Webber 1973), we find:

- There is no definitive formulation of a wicked problem. Each attempt at creating a solution changes your understanding of the problem
- Getting all stakeholders to agree that a resolution is "good enough" can be a challenge

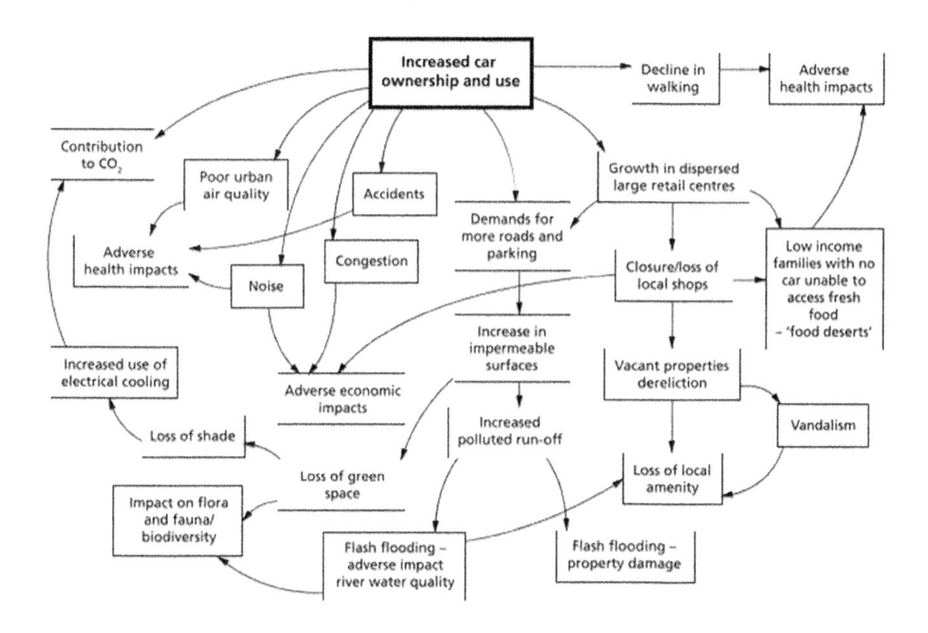

Fig. 32.2 A graphic reflecting many elements of a 'wicked problem'. Part of the web of connections between increased car ownership and use and environmental and social outcomes in urban. Source: RCEP (2007, p. 7)

- Every wicked problem is essentially unique. There are no "classes" of solutions that can be applied to a specific case
- Each wicked problem can be considered a symptom of another problem.

Tools developed to tell us which intervention to choose over another are not appropriate to address complex challenges of wicked problems such as designing healthier cities or creating healthier, more sustainable food systems (Rutter and Glonti 2016). It has been suggested that ongoing and entrenched traditions in the public health approach have resulted in an ongoing tendency to evaluate only discrete projects, in definable settings, thus failing to capture the 'added value' of whole system working (Dooris 2006). And are we even asking the right questions?

- In urban design, let's begin by asking what keeps local people healthy (Burns 2014).
- In neighbourhood development, let's find which physical and social community assets can be optimized for better health (Smith et al. 2001).

32.5.2 Ecological Public Health as an Approach

So, we need to apply complex interventions to wicked problems in complex adaptive systems. How do we use evidence in such an environment? The 'ecological public health' concept presents an integrative framework as a starting point (Hanlon et al. 2012). We now require clarity about the necessary partnerships, themselves having to reflect sophisticated ecological interdependencies, with an understanding of how to identify the relevant knowledge (Lang and Rayner 2012). According to the nature of wicked systems, a single model framework will not fit each circumstance of application. However, the lack of a transparent framework to arrive at solutions may leave advocates for evidence based Policy seeing these situations as representing the misuse of evidence for strategic ends. Conversely, policy critiques may counter that policy decisions are fundamentally about competing values, in which a seemingly blind embrace of technical evidence attempts to depoliticize what are essential political decisions (Parkhurst 2016).

The health map (Barton and Grant 2006) model in Fig. 32.3 has made great headway in bridging silos between and within academia, policy and communities, and assisting a move towards the ecological public health approach.

The health map is a loose fit conceptual framework. It reflects Bentley's (2014) four principles of urban ecological public health: conviviality, equity, sustainability and global responsibility. It also is resonant with Steiner's (2014) four urban ecological design and planning research areas:

1. Application of ecosystem services
2. Adaptation of settlements for natural disasters
3. Ecological renewal of degraded urban places
4. Ability of people to link knowledge to action to affect positive change

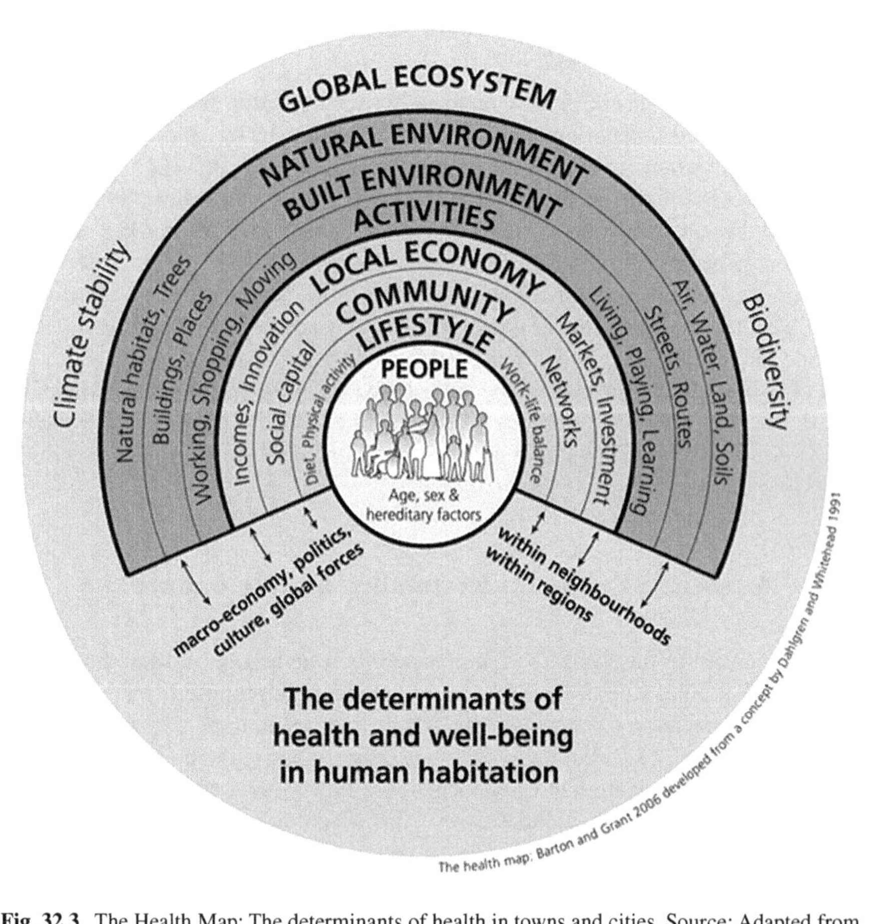

Fig. 32.3 The Health Map: The determinants of health in towns and cities. Source: Adapted from Barton, H. and Grant, M. (2006) developed from the model by Dahlgren and Whitehead (1991)

It is already being used around the world to link evidence to proposals and to garner action across cross-sectoral collaborations.

32.5.3 Knowledge Flow

A key task will be to improve flows of knowledge in all directions. The emergence of knowledge translation (Liyanage et al. 2009; Peirson et al. 2013) has even begun to challenge some forms of evidence-based medicine:

> Prior to the emergence of knowledge translation, approaches to knowledge transfer (such as evidence based medicine and research utilisation) placed limited value on research conducted from certain theoretical and methodological perspectives, such as research utilising qualitative methods. Knowledge translation offers an exciting opportunity to bring knowl-

edge from multiple methodologies and theoretical frameworks into the practice environment. (Baumbusch et al. 2008, p. 137)

Progress to date has been slow and patchy with regard to linking city development to improved population health (Grant 2015). For better evidence translation and more appropriate research infrastructures, including training will be required in order to understand the needs of policy-makers and their timescales (Holmes et al. 2016). At the same time, policy-makers and city leaders must have more access in influencing the research agenda. The research must be seen to add value, have salience and improve health by employing methods and approaches and asking questions that need addressing and not be driven by academics with different agendas and interests (Hunter and Shelina 2016). A difficult balancing act then.

This will require researchers to be both skilled scientists and capable participants in the world of policy and politics—a model of engaged scholarship (Hambleton 2013)—and conversely for policy-makers to assume a reflective stance akin to the reflexive practitioner (Schön 1987).

32.6 Advocacy, Public Understanding and Resonance

Politicians need to find ways to communicate their understanding and their intentions to the public, so that they can better realize what proposed changes, in say lifestyle, will mean to them personally and to their communities in the long term (Khreis et al. 2017). In addition to able to access a ‘good enough’ evidence base, a vital skill for those wanting to use good science for healthier cities is skilled advocacy. Over time, the political landscape within which the initiative is implemented becomes increasingly important in determining the use of evidence (Wright et al. 2007). Another way of putting this is to say that there needs to be a equal emphasis placed on the ‘receptors’ of knowledge (the decision-makers and the public) as on the ‘generators’ of knowledge (the scientific endeavour).

32.6.1 Shared and Collaborative Agendas

The ‘healthy city’ endeavour calls for a process that could be termed evidence-informed, not evidence-based, whereby opportunities are sought for synthesis between policy-makers and the different researcher fields. Prior engagement whereby policy-makers and researchers actively work together in framing the nature of the problem would support this (Campbell 2012). In this type of collaboration, researchers propose and coordinate methodologies, whereas the definition and analysis of research questions and dissemination of results are carried out jointly with non-academic representatives of society (Rosendahl et al. 2015). Knowledge cogeneration has also been used as a term to capture a new settlement between science, lay knowledge and practice. This has some similarity to Smith et al.’s (2000) maxim:

- Ensure the development of an appropriate research evidence base
- Establish mechanisms to push the evidence out to policy-makers and practitioners
- Encourage a situation where there is a pull for evidence among policy-makers and practitioners

32.6.2 Action Research and Imaginative Action

What do we do where evidence is compelling but not complete? We can use the city laboratory to fill in the gaps. The public realm changes in Times Square, New York, are an example of 'tactical urbanism'. Metropolitan authorities, who may role their eyes up at mention of randomized control trials, sometimes respond to case studies and examples as evidence; facts on the ground. This is useful in a context where the longevity of policy attention does not move beyond election cycles at best and the thematic cycles of news agenda at worst. Using the city as a living laboratory, with temporary changes, made permanent if they work, or removed if they don't, is an action research approach. Politicians need to be shown the levers for population level mechanisms for public health within metropolitan regions in a demonstrable manner.

As Polanyi pointed out, science, like many other activities, relies on skilled judgement, the capacity for which is built up through experience (Polanyi 1959, 1966). This will often involve apparently breaking the rules. If one tries to reduce science to the following of explicit procedures, one runs the risk of obstructing its progress (Hammersley 2002). Lovasi et al. (2016) draw our attention to a number of potential study designs and tools which could enhance our understanding of the impact of urban policy initiatives on mental and physical health. Commenting on place-based approaches to mental health, they posit that improving existing research strategies will require partnerships between those specialized in public health, those advancing new methods for place effects on health and those who seek to optimize the design of local environments (Lovasi et al. 2016). Yet methodological and conceptual challenges remain a vexing concern. It may be time to expand the public health research toolbox through the application, with caution, thought and care, of novel methods to urban contexts (Schofield and Das-Munshi 2017). Let us construct teams where epidemiologists, urban tacticians and investors cocreate healthier places and new knowledge concurrently.

Bear in mind that for successful interventions, there can be multiple entry points, in several interconnected realms of action. To jolt the system, a single intervention may have multiple arms and be targeted at urban policy, built environment and transport, public realm management, planning processes, community activism and tactical urbanism and individual behavioural change. Let's remember too that in designing an intervention, even with best evidence, what is most important will depend on which elements of the causal web can be manipulated most successfully; so we need also to be pragmatic (Rutter et al. 2017).

32.6.3 Evolving Knowledge Leadership in the City

Within this melee of a messy, non-linear, multi-stakeholder system, strong city leadership for health has been shown again and again to be essential (Tsouros 2013). When it comes to place-based leadership, this can come in many forms and from many places, not solely from city mayors (Hambleton 2015). In place-based governance for health, we need to recognise, and encourage also, business leadership, community leadership and professional leadership. For a whole population approach to health, city leadership or the lack of it, will trump a robust evidence base; both, however, are necessary but not sufficient.

As a joint endeavour, there is a need for policy-makers, city leaders and activists to all help evolve the broken evidence-intervention model. Jones and Seelig (2005) see the evolution in three stages. In the past we have an 'engineering model', whereby research attempts to solve policy problems by providing empirical evidence. The orientation was technocratic and instrumental; this was shown to have limited efficacy. Currently we now often find the 'enlightenment model'. Research is undertaken for the whole society in a spirit of detachment and scepticism, researchers remain detached from policy processes.

What we now need to co-generate is an 'engagement model'. We need an ongoing interactive relationship through networks and partnerships to address policy issues in a complex political environment (Jones and Seelig 2005). Maybe setting 'knowledge' in context is part of wisdom? (Marshall and Godfrey 2006). So let's brand our aim 'to marry complex science in a complex system for social good', as cogeneration of societal wisdom. And we certainly need more of that in our cities!

References

Ball, K., et al. (2007). Personal, social and environmental determinants of educational inequalities in walking: A multilevel study. *Journal of Epidemiology and Community Health, 61*, 108–114.

Barton, H., & Grant, M. (2006). A health map for the local human habitat. *The Journal of the Royal Society for the Promotion of Health, 126*(6), 252–253. https://doi.org/10.1177/1466424006070466.

Baumbusch, J., et al. (2008). Pursuing common agendas: A collaborative model of knowledge translation between research and practice in clinical settings. *Research in Nursing and Health, 31*, 130–140.

Bentley, M. (2014). An ecological public health approach to understanding the relationships between sustainable urban environments, public health and social equity. *Health Promotion International, 29*(3), 528–537.

Braubach, M., & Grant, M. (2010). Evidence review on the spatial determinants of health in urban settings. In *Annex 2 in urban planning, environment and health: From evidence to policy action. Meeting report* (pp. 22–97). Copenhagen: WHO Regional Office for Europe. Retrieved from http://eprints.uwe.ac.uk/12071.

Burns, H. (2014). What causes health? *The Journal of the Royal College of Physicians of Edinburgh, 44*, 103–105.

Cairney, P., Oliver, K., & Wellstead, A. (2016). To bridge the divide between evi-dence and policy: Reduce ambiguity as much as uncertainty. *Public Administration Review, 76*(3), 399–402.
Campbell, H. (2012). Planning to change the world: Between knowledge and action lies synthesis. *Journal of Planning Education and Research, 32*(2), 135–146.
Carmichael, L., Barton, H., Gray, S., Lease, H., & Pilkington, P. (2012). Integration of health into urban spatial planning through impact assessment: Identifying governance and policy barriers and facilitators. *Environmental Impact Assessment Review, 32*, 187–194.
Carmichael, L., Lock, K., Sweeting, D., Townshend, T., Fischer, T. B., & TCPA. (2016). Evidence base for health and planning lessons from an ESRC seminar series. *Town and Country Planning, 85*(11), 461–464.
Chan, J., & Chan, J. (2000). Medicine in the millennium: The challenge of postmodernism. *Medical Journal of Australia, 17*, 332–334.
Craig, P., Dieppe, P., Macintyre, S., Michie, S., Nazareth, I., & Petticrew, M. (2011). *Developing and evaluating complex interventions*. London: Medical Research Council.
Croucher, K., Myers, L., Jones, R., Ellaway, A., & Beck, S. (2007). *Health and the physical characteristics of urban neighbourhoods: A critical literature review*. Glasgow: Centre for Population Health.
Dahlgren G, Whitehead M (1991). "The main determinants of health" model, version accessible. In: Dahlgren G, and Whitehead M. (2007) European strategies for tackling social inequities in Dahlgren, G. and Whitehead, M., 2006. European strategies for tackling social inequities in health: Levelling up Part 2. Copenhagen: World Health Organization.
Davies, H. T. O. H., Nutley, S. M. S., & Smith, P. C. P. (1999). What works? The role of evidence in public sector policy and practice (Editorial). *Public Money and Management, 19*(1), 3–5. https://doi.org/10.1111/1467-9302.00144.
Davies, P., 2005. Evidence-based policy at the Cabinet Office — ODI: 17 October 2005, Impact and insight series. Retrieved from https://www.odi.org/sites/odi.org.uk/files/odi-assets/events-documents/2866.pdf.
Davies, P. T. (1999). What is evidence-based education? *British Journal of Educational Studies, 47*(2), 108–121.
Davis, A. (2005). Transport and health - What is the connection? An exploration of concepts ofhealth held by highways committee Chairs in England. *Transport Policy, 12*(4), 324–333. https://doi.org/10.1016/j.tranpol.2005.05.005.
Davis, A. 2016 Population strategies and the prevention paradox as applied to road safety in Bristol: A public health approach. Paper presented at Road Safety GB Annual Conference, Bristol, UK
Dooris, M. (2006). Healthy settings: challenges to generating evidence of effectiveness. *Health Promotion International, 21*(1), 55–65.
Francis, J., Giles-Corti, B., Wood, L., & Knuiman, M. (2012a). Creating sense of community: The role of public space. *Journal of Environmental Psychology, 32*, 401–409.
Francis, J., Wood, L. J., Knuiman, M., & Giles-Corti, B. (2012b). Quality or quantity? Exploring the relationship between public open space attributes and mental health in Perth, Western Australia. *Social Science & Medicine, 74*, 1570–1577.
Giles-Corti, B., Foster, S., Shilton, T., & Falconer, R. (2010). The co-benefits for health of investing in active transportation, NSW. *Public Health Bulletin, 21*(5-6), 122–127.
Giles-Corti, B., Sallis, J. F., & Sugiyama, T. (2015). Translating active living research into policy and practice: One important pathway to chronic disease prevention. *Journal of Public Health Policy, 36*, 231–243. https://doi.org/10.1057/jphp.2014.53.
Graham, H., & White, P. C. L. (2016). Social determinants and lifestyles: Integrating environmental and public health perspectives. *Public Health, 141*, 270–278. https://doi.org/10.1016/j.puhe.2016.09.019.
Grant, M. (2015). European Healthy City Network Phase V: Patterns emerging for healthy urban planning. *Health Promotion International, 30*(suppl 1), i54–i70.
Grant, M., Brown, C., Caiaffa, W. T., Capon, A., Corburn, J., Coutts, C., Crespo, C. J., Ellis, G., Ferguson, G., Fudge, C., & Hancock, T. (2017). Cities and health: An evolving global conversation. *Cities and Health, 1*, 1–9.

Gray, J. (2013). *The silence of animals: On progress and other modern myths*. London: Penguin Books.
Gustat, J., et al. (2012). Effect of changes to the neighbourhood built environment on physical activity in a low-income African American neighbourhood. *Preventing Chronic Disease, 9*, 110. https://doi.org/10.5888/pcd9.110165.
Haack, S. (1995). *Evidence and inquiry*. Oxford: Blackwell Publishers.
Hadorn, D. C., Baker, D., Hodges, J. S., & Hicks, N. (1996). Rating the quality of evidence for clinical practice guidelines. *Journal of Clinical Epidemiology, 49*, 749–754.
Halpern, D. (1995). *Mental health and the built environment: More than bricks and mortar*. London: Taylor and Francis Ltd.
Hambleton, R. (2013). Place-based leadership: New possibilities for planning? In *Association of European Schools of Planning/Association of Collegiate Schools of Planning, Dublin, Ireland, 15–19th July 2013*. Retrieved from http://eprints.uwe.ac.uk/21796.
Hambleton, R. (2015). Place-based collaboration: Leadership for a changing world. *Administration, 63*(3), 5–25.
Hammersley, M. (2002). *Systematic or unsystematic: Is that the question? Some reflections on the science, art, and politics of reviewing research evidence*. Public Health Evidence Steering Group of the Health Development Agency.
Hanlon, P., Carlisle, S., Hannah, M., Lyon, A., & Reilly, D. (2012). A perspective on the future public health: An integrative and ecological framework. *Perspectives in Public Health, 132*(6), 313–319. https://doi.org/10.1177/1757913912440781.
Hippocrates. (1849). On air, waters, and places. In *The genuine works of Hippocrates*. New York: W. Wood. Translated with a commentary by Francis Adams. London: The Sydenham Society.
Holmes, B., Best, A., Davies, H., Hunter, D. J., Kelly, M., Marshall, M., & Rycroft-Malone, J. (2016). Mobilising knowledge in complex health systems: A call to action. *Evidence and Policy, 13*, 539–560.
Freeman, H. L. (Ed.). (1985). *Mental health and the environment*. London: Churchill Livingstone.
Hunter, D. J., & Shelina, V. (2016). Better evidence for smarter policy making. *British Medical Journal, 355*, i6399. https://doi.org/10.1063/1.2756072.
Jones, A., & Seelig, T. (2005). *Enhancing research-policy linkages in Australian housing: An options paper* (pp. 1–46). Melbourne: Australian Housing and Urban Research Institute (AHURI).
Jones, R., & Yates, G. (2013). *The built environment and health: An evidence review. Briefing paper 11*. Glasgow: Centre for Population Health.
Juntti, M., Russell, D., & Turnpenny, J. (2009). Evidence, politics and power in public policy for the environment. *Environmental Science and Policy, 12*, 207–215.
Kelly, M. P., Morgan, A., Bonnefoy, J., Butt, J., & Bergman, V. (2007). The social determinants of health: Developing an evidence base for political action.
Kent, J., Thompson, S. M., & Jalaludin, B. (2011). *Healthy built environments: A review of the literature*. Sydney: Healthy Built Environments Program, City Futures Research Centre, UNSW.
Kent, J., & Thompson, S. (2012). Health and the built environment: Exploring foundations for a new interdisciplinary profession. *Journal of Environmental and Public Health, 2012*, 958175. https://doi.org/10.1155/2012/958175.
Khangura, S., Konnyu, K., Cushman, R., Grimshaw, J., & Moher, M. (2012). Evidence summaries: The evolution of a rapid review approach. *Systematic Reviews, 1*, 10.
Khreis, H., van Nunen, E., Mueller, N., Zandieh, R., & Nieuwenhuijsen, M. J. (2017). Commentary: How to create healthy environments in cities. *Epidemiology (Cambridge, Mass.), 28*(1), 60.
Kickbusch, I. (2008). *Policy innovation for health*. New York: Springer.
Krieger, N. (1992). The making of public health data: Paradigms, politics, and policy. *Journal of Public Health Policy, 13*, 412–427.
Lang, T., & Rayner, G. (2012). Ecological public health: The 21st century's big idea? An essay by Tim Lang and Geof Rayner. *British Medical Journal, 345*(1306), e5466. https://doi.org/10.1136/bmj.e5466.

Lavin, T., Higgins, C., Metcalfe, O., & Jordan, A. (2006). *Health impacts of the built environment: A review*. Dublin, Belfast: Institute of Public Health in Ireland.

Lawrence, R. J. (2015). Advances in transdisciplinarity: Epistemologies, methodologies and processes. *Futures, 65*, 1–9. https://doi.org/10.1016/j.futures.2014.11.007.

Lawrence, R. J. (2004). Housing and health: From interdisciplinary principles to transdisciplinary research and practice. *Futures, 36*(4), 487–502.

Lawrence, R. J. (2010). Deciphering interdisciplinary and transdisciplinary contributions. *Transdisciplinary Journal of Engineering & Science, 1*(1), 125–130.

LGA. (2016). *Health in all policies: A manual for local government*. London: Local Government Association.

Litman, T. (2003). Integrating public health objectives in transportation decision-making. *American Journal of Health Promotion, 18*(1), 103–108.

Liyanage, C., Elhag, T., Ballal, T., & Qiuping, L. (2009). Knowledge communication and translation – A knowledge transfer model. *Journal of Knowledge Management, 13*(3), 118–131.

Lovasi, G. S., Mooney, S., DiMaggio, C. J., & Muennig, P. A. (2016). Cause and context: Place-based approaches to investigate how environments affect mental health. *Social Psychiatry and Psychiatric Epidemiology, 51*(12), 1571–1579.

Marmot, M., Friel, S., Bell, R., Houweling, T. A., & Taylor, S. (2008). Closing the gap in a generation: Health equity through action on the social determinants of health. *The Lancet, 372*(9650), 1661–1669.

Marshall, R., & Godfrey. (2006). Using evidence in practice: What do health professionals really do? A study of care and support for breastfeeding women in primary care. *Clinical Effectiveness, 953*, e181–e190.

Milewa, T., & Barry, C. (2005). Health policy and the politics of evidence. *Social Policy and Administration, 39*(5), 498–512. https://doi.org/10.1111/j.1467-9515.2005.00452.x.

Murdoch, J., & Clarke, J. (1994). Sustainable knowledge. *Geoforum, 25*(2), 115–132.

Muriel, F., & Autant-Bernard, C. (2013). Knowledge diffusion and innovation within the European regions: Challenges based on recent empirical evidence. *Research Policy, 41*(1), 196–201.

Parkhurst, J. O. (2016). Appeals to evidence for the resolution of wicked problems: The origins and mechanisms of evidentiary bias. *Policy Sciences, 49*, 373–393. https://doi.org/10.1007/s11077-016-9263-z.

Parsons, W. (2002). From muddling through to muddling up – Evidence based policy making and the modernisiation of British Government. *Public Policy and Administration, 17*, 43–60.

Pawson, R., & Tilley, N. (1997). *Realistic evaluation*. London: Sage.

Peavey, E., & Vander Wyst, K. B. (2017). Evidence-based design and research-informed design: What's the difference? Conceptual definitions and comparative analysis. *Health Environments Research & Design Journal, 10*, 143.

Peirson, L., Catallo, C., & Chera, S. (2013). The registry of knowledge translation methods and tools: A resource to support evidence-informed public health. *International Journal of Public Health, 58*, 493–500.

Petticrew, M., & Roberts, H. (2003). Evidence, hierarchies, and typologies: Horses for courses. *Journal of Epidemiology and Community Health, 57*(7), 527–529.

Polanyi, M. (1959). *Personal knowledge*. Manchester: Manchester University Press.

Polanyi, M. (1966). *The tacit dimension, garden city*. Garden City: Doubleday.

Pope, C. (2003). Resisting evidence: The study of evidence-based medicine as a contemporary social movement. *Health: An Intersectoral Journal for the Social Study of Health, Illness and Medicine, 7*(3), 267–282.

Public Health England. (2017). *Spatial planning for health: An evidence resource for planning and designing healthier places*. London: Public Health England.

Rao, M., Prasad, S., Adshead, F., & Tissera, H. (2007). The built environment and health. *The Lancet, 370*(9593), 1111–1113.

Rittel, H. W. J., & Webber, M. M. (1973). Dilemmas in a general theory of planning. *Policy Sciences, 4*(2), 155–169. https://doi.org/10.1007/BF01405730.

Rodgers, M., et al. (2009). Testing methodological guidance on the conduct of narrative synthesis in systematic reviews effectiveness of interventions to promote smoke alarm ownership and function. *Evaluation, 15*(1), 47–71.
Rosendahl, J., Zanella, M., Rist, S., & Weigelt, J. (2015). Scientists' situated knowledge: Strong objectivity in transdisciplinarity. *Futures, 65*, 17–27.
Royal Commission on Environmental Pollution. (2007). *The Urban environment* (Vol. 26). London: The Stationary Office.
Rutter, H., & Glonti, K. (2016). Towards a new model of evidence for public health. *The Lancet, 388*, S7.
Rutter, H., Savona, N., Glonti, K., Bibby, J., Cummins, S., Finegood, D. T., et al. (2017). Viewpoint: The need for a complex systems model of evidence for public health. *The Lancet, 6736*(17), 9–11. https://doi.org/10.1016/S0140-6736(17)31267-9.
Rychetnik, L., & Wise, M. (2004). Advocating evidence-based health promotion: Reflections and a way forward. *Health Promotion International, 19*(2), 247257.
Rydin, Y., Bleahu, A., Davies, M., Dávila, J. D., Friel, S., De Grandis, G., Groce, N., Hallal, P. C., Hamilton, I., Howden-Chapman, P., & Lai, K. M. (2012). Shaping cities for health: Complexity and the planning of urban environments in the 21st century. *Lancet, 379*(9831), 2079.
Sallis, J. F., Cerin, E., Conway, T. L., Adams, M. A., Frank, L. D., Pratt, M., Salvo, D., Schipperijn, J., Smith, G., Cain, K. L., & Davey, R. (2016). Physical activity in relation to urban environments in 14 cities worldwide: A cross-sectional study. *The Lancet, 387*(10034), 2207–2217.
Sanderson, I. (2002). Evaluation, policy learning and evidence-based policy making. *Public Administration, 80*(1), 1–22.
Schofield, P., & Das-Munshi, J. (2017). New directions in neighbourhood research—A commentary on Lovasi et al. (2016): Cause and context: Place-based approaches to investigate how environments affect mental health. *Social Psychiatry and Psychiatric Epidemiology, 52*(2), 135–137. https://doi.org/10.1007/s00127-016-1332-2.
Schön, D. A. (1987). *Educating the reflective practitioner: Toward a new design for teaching and learning in the professions*. San Francisco: Jossey-Bass.
Sidhu, K., Jones, R., & Stevenson, F. (2017). Publishing qualitative research in medical journals. *The British Journal of General Practice, 67*(658), 229–230.
Smith, N., Baugh Littlejohns, L., & Thompson, D. (2001). Shaking out the cobwebs: Insights into community capacity and its relation to health outcomes. *Community Development Journal, 36*(1), 30–41.
Smith, P. C., Nutley, S. M., & Davies, H. T. O. (Eds.). (2000). *Urban policy: Addressing wicked problems. What works?: Evidence-based policy and practice in public services*. Bristol: The Policy Press.
Sorrell, S. (2007). Improving the evidence base for energy policy: The role of systematic reviews. *Energy Policy, 35*(3), 1858–1871.
Steiner, F. (2014). Frontiers in urban ecological design and planning research. *Landscape and Urban Planning, 125*, 304–311. https://doi.org/10.1016/j.landurbplan.2014.01.023.
Swinburn, B. (2001). Prevention of type 2 diabetes, 2001 Prevention needs to reduce obesogenic environments. *British Medical Journal, 323*(7319), 997.
Jenkins, S. (2017). Post-truth politics will be debunked by online facts. *The Guardian*. Retrieved July 17, 2017, from https://www.theguardian.com/commentisfree/2017/jan/26/post-truth-politics-online-facts-donald-trump-lies.
Townshend, T., & Lake, A. (2009). Obesogenic urban form: Theory, policy and practice. *Health & Place, 15*(4), 909–916.
Townshend, T. G. (2017). Toxic high streets. *Journal of Urban Design, 22*(2), 167–186.
Tsouros, A. (2013). City leadership for health and well-being: Back to the future. *Journal of Urban Health, 90*, 4–13.
UN. (2015). *World urbanization prospects: The 2014 revision (ST/ESA/SER.A/366)*. New York: United Nations, Department of Economic and Social Affairs, Population Division.

Weiss, C. H. (1979). The many meanings of research utilization. *Public Administration Review, 39*(5), 426–431.

WHO. (2008). *International Public Health Symposium on Environment and health research science for policy, Policy for science: Bridging the gap, Madrid, Spain, 20–22 October 2008. Report*. Copenhagen: WHO Regional Office for Europe.

WHO. (1986). *Ottawa charter for health promotion*. Copenhagen: World Health Organization European Regional Office.

WHO. (1988). *Health 21, the Health for all framework for the European region*. Copenhagen: World Health Organisation Regional Office for Europe.

WHO. (2009). *International Public Health Symposium on Environment and health research science for policy, Policy for science: Bridging the gap, Madrid, Spain, 20–22 October 2008*. Copenhagen: WHO Regional Office for Europe.

WHO. (2016). *Action plan to strengthen the use of evidence, information and research for policy-making in the WHO European Region. Regional Committee for Europe 66th Session, 12–15 September*. Copenhagen: WHO.

Wright, J., Parry, J., & Mathers, J. (2007). What to do about political context? Evidence, synthesis, the New Deal for Communities and the possibilities for evidence-based policy. *Evidence and Policy, 3*(2), 253–269.

Chapter 33
Using Conceptual Models to Shape Healthy Sustainable Cities

George Morris, Brigit Staatsen, and Nina van der Vliet

33.1 Introduction

Many features characterise a healthy urban environment in today's world. A healthy city is self-evidently one where physical hazards are kept to a minimum but also where healthy and sustainable lifestyles are promoted and inhabitants are stimulated to become involved with the places they live. These and other criteria can only be fulfilled using an integrated approach that connects health goals with the protection of natural resources and ecosystems while stimulating multisectoral and multidisciplinary collaboration.

In this chapter, we identify the urban environment as a key battleground in what is arguably the twenty-first century's greatest public health challenge—the need for humanity to reflect in all its decisions and actions and the interconnection between humans and the natural processes and systems on which they rely for health, wellbeing and, ultimately, survival. The term 'ecological public health' is increasingly used to encapsulate both this challenge and the required response. Recent international developments, notably the United Nations 2030 Agenda for Sustainable Development, offer a favourable policy platform and an unprecedented opportunity to pursue meaningful change in a coordinated way. While applauding the emergence of this improved policy context, no one should underestimate the complexity of a task that demands, among other things, a rethink of how we live, move and consume in our urban centres. Our purpose here is to briefly describe the urban environment and its challenges but, more especially, to promote the utility and importance of conceptual models as tools with which to think and guide the dialogue among stakeholders to shape the urban policies for a healthier, more sustainable and more equal world.

G. Morris (✉)
European Centre for Environment and Human Health, University of Exeter, Exeter, UK

B. Staatsen · N. van der Vliet
National Institute for Public Health and the Environment, Bilthoven, The Netherlands
e-mail: brigit.staatsen@rivm.nl; nina.van.der.vliet@rivm.nl

M. Nieuwenhuijsen, H. Khreis (eds.), *Integrating Human Health into Urban and Transport Planning*, https://doi.org/10.1007/978-3-319-74983-9_33

33.2 Ecological Public Health

The recognition that health, wellbeing, disease and social patterning in these outcomes are emergent products of a complex interaction of factors operating at the level of society with characteristics specific to the individual has been hugely influential across the field of public health for over 40 years.

What were typically described as 'ecological' or 'socioecological' perspectives in public health gained prominence in the 1970s, stimulated in large part by concerns over the burgeoning and unsustainable costs of healthcare. Through their writings and schematic representations, proponents of a 'new public health' (see, e.g. Lalonde 1974; Evans and Stoddart 1990; Dahlgren and Whitehead 1991) elegantly expressed complexity in the determinants of health in a helpful and policy-relevant way. Among the benefits was a renewed policy focus on the wider structural determinants of health and of health-relevant behaviours and an implicit challenge to public health to abandon intra- and interdisciplinary silos and reach out to other constituencies of policy and practice. Moreover, socioecological perspectives helped drive research which created a much richer understanding of the role of physical environment in health and wellbeing. Going beyond narrow compartmentalised and hazard-focused environmental concerns, the environment came, increasingly, to be recognised as contributory in a wide variety of high-profile health and health-related outcomes. These included non-communicable diseases, mental health and wellbeing, health inequalities and, later, the obesity epidemic. Spurred in part by socioecological perspectives but also moral indignation over health inequity, scientific research has gradually revealed the true importance of the environment for health and wellbeing in the modern era. An important finding is, for example, that high-quality environments, especially pleasant green and natural spaces, can promote better, more equal health (see, e.g. Hartig et al. 2014). There is also growing acceptance of a psychosocial dimension in the relationship between people and the physical context in which they live which contributes to mental and physical wellbeing (Gee and Payne-Sturges 2004).

Despite describing significant complexity in the determinants of health, wellbeing and equity, much of the rhetoric and many of the schematics initially proposed to support the 'ecological' perspectives of the late twentieth century (Evans and Stoddart 1990; Dahlgren and Whitehead 1991) now appear deficient in one very significant respect. Specifically, they seem to disregard or downplay the dynamic interaction and interdependence between human beings and the natural environment and the fact that human health and wellbeing is a product of this interdependence (Coutts et al. 2014; Morris et al. 2015). Fortunately, as the twenty-first century has progressed, references to the ecological perspectives and the concept of 'ecological public health' permeating the literature reflect not only human social complexity in the determinants of health but also the interconnectedness of humans with their natural environment. 'Ecological public health', as it is now construed, recognises the absurdity of any model that regards human beings, their social and economic activities and their health and wellbeing as somehow separate and distinct from natural systems. Indeed, there is a growing acceptance, given the damage now being wrought on planetary processes and systems by human activity that it may well be impossible to deliver health and wellbeing, healthcare or equity in any of these areas in the medium to long term without thinking in a much more focused way about the environment (Rayner and Lang 2012; Morris et al. 2015; Morris and Saunders 2017).

For most of the last 10,000 years, the Earth's systems and processes have been essentially stable, presenting conditions favourable to human health and development, and, despite gross inequity in their distribution, abundant resources have also been available to support human life. However, over the space of less than two centuries, humans have moved to become the principal drivers of change at planetary level. The grave implications for humanity are cogently expressed in the concept of 'planetary boundaries'. Refining earlier work (Rockström et al. 2009), Steffen and colleagues proposed nine specific limits or 'planetary boundaries' relating to the biophysical subsystems and processes which ensure the Earth's habitability for humans (Steffen et al. 2015). Yet, already by 2015, four of these limits—those relating to the addition of phosphorus and nitrogen to crops and ecosystems, biodiversity, land use changes (including deforestation) and, arguably most alarmingly, climate change—had all been breached and, with them, what the authors termed a 'safe operating space for humanity'. The risks are amplified by the likelihood that the systems, which are in various ways connected, may now respond unpredictably and in a non-linear fashion. Citing this and the outcomes of similar research, the World Health Organisation is just one among a host of commentators now emphasising that a 'business as usual' approach is no longer a viable option to secure health and wellbeing and to improve equity (see, e.g. WHO 2017, Morris and Saunders, 2017; Rayner and Lang 2012). In an insightful commentary, Rayner and Lang (2012) have argued that *ecological public health* is *the twenty-first century's unavoidable task.* While different terminology may be used across the public health literature to encapsulate the threat, there is broad agreement concerning the need, henceforth, to build population health on ecological principles and around the huge scale of the challenge. The fact that wherever we live in the world, our health will in future be profoundly influenced by the changes to planetary processes and systems, which we as individuals influence confers a distinctly global dimension to everyday public health practice in the twenty-first century. Nonetheless, an important frontline in the battle is likely to remain the towns, cities and rural locations in which we move, consume and live out our daily lives. Working in partnership, policymakers, public agencies, the private sector and indeed citizens themselves carry responsibility for identifying and delivering the policies and actions necessary for a healthy, equitable and sustainable future. Here, we argue that relatively simple conceptual models of health and its determinants already exist which have capacity to guide these policies and actions at national and subnational level.

33.3 Policy Platform

Healthy and sustainable urban development has been included as a key objective in various recent international agreements such as UN 2030 Agenda for Sustainable Development, the New UN Urban Agenda and the EU Urban Agenda.

The UN 2030 Agenda for Sustainable Development with its Sustainable Development Goals (SDGs) might be regarded as the political embodiment, not only of the quest for global sustainability and equity but also modern ecological thought in public health. The SDGs and in particular SDG11 '*Make cities inclusive, safe, resilient and sustainable*', can be seen as an overarching framework for policy to improve the environment and health in cities (UN 2015). SDG11, however, is not the only SDG with an urban dimension. The 17 SDGs and 169 related indicators aim to eradicate poverty and inequality, create inclusive economic growth, preserve the planet and improve population health. Many have an environmental and health dimension that, as well as addressing climate change, holds potential for significant public health improvements particularly in cities (WHO 2017; Carmichael et al. 2017; Pruss Ustun et al. 2016).

Drawing on ideas and developments from the Dutch context, Agenda Stad (2016), and championed by the Netherlands during its European Union presidency, the 2016 EU Urban Agenda European Commission (2015) also recognises both the challenge and opportunity present in the urban context. It aims to strengthen the urban dimension of European policies, to create better regulations and promote knowledge exchange. The EU Agenda focuses on sectors relevant to environment, health and equity, with pilot partnerships established to address four of these themes—air quality, housing, inclusion of migrants and refugees and urban poverty—over the next 2–3 years (Urban Agenda for the EU 2015). Importantly, it promotes vertical and horizontal coordination of policies, impact assessment, and knowledge exchange. A major objective is to contribute to SDG11.

A further component of the international policy platform is represented by the New Urban Agenda (NUA). This was adopted by the UN Habitat III Conference in 2016 and emphasises the key role of cities in the pursuit of sustainable development. The NUA reiterates the commitment to interlink social, environmental and economic principles and to rethink how we build, manage and live in the urban context. Of particular relevance to the theme of this chapter and its emphasis on tools to think with at a local level, the NUA recognises that national governments are leaders in the definition and implementation of inclusive and effective urban policies. The contribution of subnational and local government, civil society and other relevant stakeholders is no less important in order to achieve satisfactory outcomes (United Nations 2016).

33.4 Why Is the Urban Environment So Central to Ecological Public Health?

The process of urbanisation is seemingly unremitting and most of the World's population already lives in cities. In Europe, for example, from 1950 to 2011, the percentage of the population living in urban areas increased from 51% to 73%, and this has been predicted to reach 83% in 2050 (EEA 2015). However, the importance of urbanisation for health wellbeing and the future of humanity is founded on much more than numbers and scale. How cities are planned and managed can profoundly affect citizen health and wellbeing, both directly and indirectly.

Self-evidently, cities can provide access to services—such as energy, water, housing, green spaces, transport and climate protection. They can also be the source of multiple economic and social benefits, which enhance the mental and physical wellbeing of those who live there.

In contrast, badly planned, poorly managed and ill-regulated cities expose citizens to threats from air pollution, noise, waste and vehicle-related hazards. Cities can produce social isolation and shape the life choices of their citizens in ways which can be damaging to their health and wellbeing—perhaps promoting or sustaining unhealthy and sedentary lifestyles implicated in a global epidemic of non-communicable diseases and obesity (Carmichael et al. 2017; Staatsen et al. 2017).

Health inequalities by many measures are all too evident between and within countries, and it is often the urban context that throws up the most striking examples of poor health and health inequity. Differences in the quality, availability and maintenance of urban infrastructures and services (such as housing, water and sanitation, the work environment, transport systems, green infrastructure and food shops) frequently manifest in health inequity in many countries. Some of the poorest life circumstances are endured by the urban poor (FRESH 2015a; Carmichael et al. 2017).

For these and other reasons, the urban context must remain a key focus for public health activity (WHO 2017). However, it is the recognition that cities are also a key battleground in the pursuit of global sustainability that catapults them to the forefront of the modern public health challenge.

Put simply, through influencing and managing change to how people live, move and consume, cities can offer among the best health-promoting and sustainable opportunities. For example, a shift from motorised transport to active transport (providing enough safe biking and walking routes as well as public transport) provides health and wellbeing benefits. Promoting active transport and clean fuel use will lead to a reduction of greenhouse gas emissions.

Scientific evidence is growing on the important role played by urban green and blue spaces in delivering considerable health and wellbeing benefits, promoting biodiversity as well as protecting from the impacts of climate change (heavy rainfall, temperature increases). Urban gardening can contribute to more sustainable and healthy food consumption. Improvement of housing quality and use of renewable energy sources may reduce fuel poverty and respiratory complaints, as well as mitigate climate change.

Urban planning decisions play an important role in protecting and promoting the health and wellbeing of its residents, as well as enhancing urban ecosystems. Figure 33.1 offers an interesting typology for modern cities. It describes for example features of a 'clean' city, 'green' city or a 'social city'. In a 'clean' city, the focus is on the protection against classic environmental risk and the provision of basic services. In a 'green 'or 'physically active' city, the focus is on the health promoting features of a city. This model has been developed by a consortium of Dutch architects and health professionals and is used in dialogues with stakeholders to help determine ambitions and focus (Platform Healthy Design 2013).

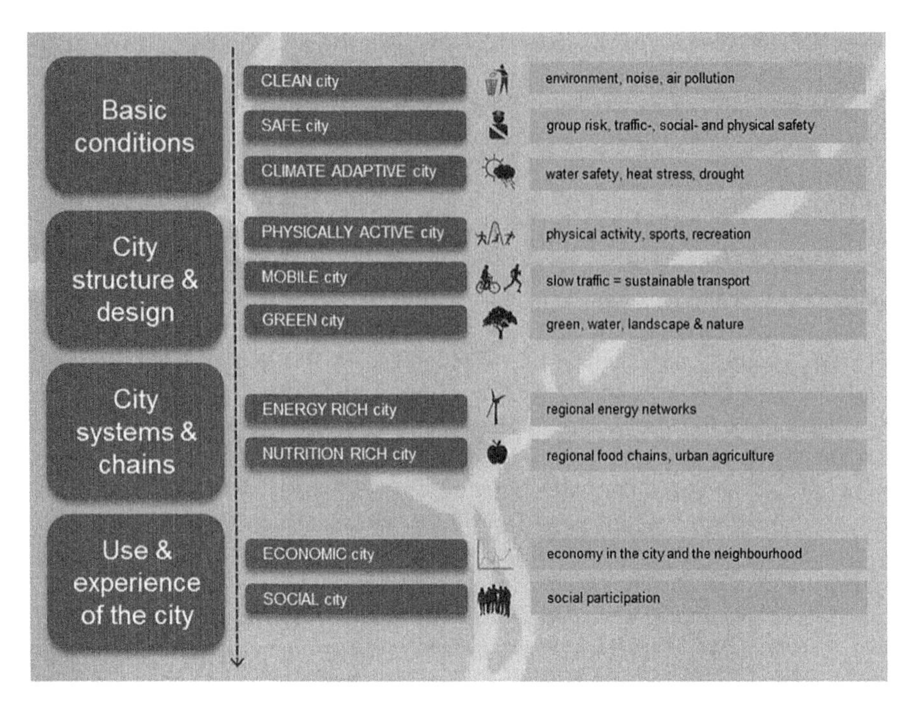

Fig. 33.1 Ten typologies of cities (Source: Platform Healthy Design 2013)

It is abundantly clear that the urban environment represents an extremely dynamic context for public health activity. Those in positions of responsibility in cities must anticipate and respond in a timely fashion to an evolving situation. Demographic changes are creating a rapidly ageing population, many with pre-existing illness and enhanced vulnerability. A changing climate may fundamentally alter the physical context of urban life in ways that are important for health, wellbeing and equity. Together, these and other developments significantly challenge policymakers and demand a detailed understanding of the forces that act to shape the physical and social environment in cities and how citizens interface with them. Conceptual models have potential to enable policy and other decision-makers to navigate in complex and changing circumstances to identify effective policies.

33.5 Conceptual Models in Public Health

Driven by advances in science and technology, but also by wider societal changes over two centuries, there has been a clear evolution in ideas about the determinants of health and disease. Shifts in what are sometimes called 'health paradigms' have delineated different eras in public health and enabled useful configuring frameworks within which to pursue better, more equitable population health (Susser and Susser 1996).

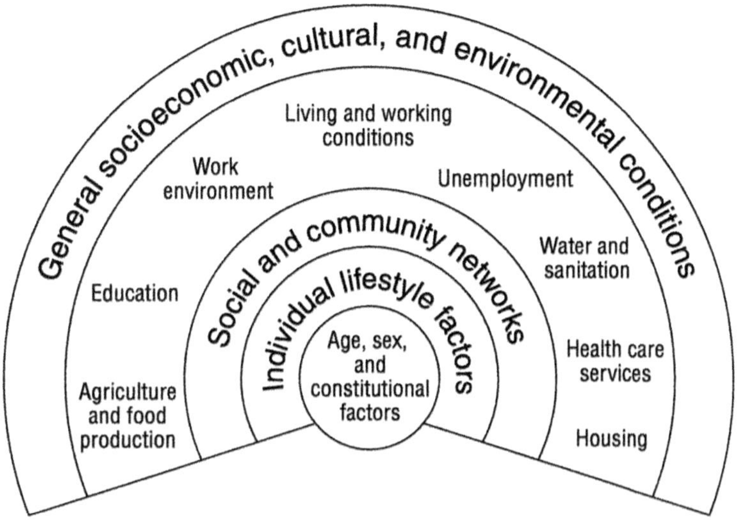

Fig. 33.2 The Social Model of Health. Source: Dahlgren and Whitehead (1991)

In 1974, a Canadian government white paper, ‘A New Perspective on the Health of Canadians’, was launched in response to concerns over potentially bankrupting healthcare costs in that country (Lalonde 1974). In seeking a more inclusive and preventative approach to tackling the burden of disease, the white paper argued that any health problem could be traced to one or more of four elements, termed ‘health fields’. These were lifestyle, human biology, environment (physical and social) and healthcare organisation. This more inclusive and preventive perspective made explicit the many different constituencies of policy and practice that influence health outcomes. Despite its specific Canadian focus, the white paper had international impact, driving the socioecological perspectives that would come to dominate the public health rhetoric in Western society. However, the white paper might also be seen as important in nurturing interest in conceptual modelling in public health. Specifically, the ‘health fields’ offered a tool for the analysis of health problems and allowed the significance and interaction of various influences to be explored in a policy-relevant way. The capacity to generate ‘maps of the health territory’ for different issues was emphasised although the white paper was not intended to present actual evidence. Rather, its aim was to present an approach which could elucidate causal pathways (Lalonde 1974; Morris et al. 2006).

In hindsight, it seems counterintuitive that the Canadian white paper did not go as far as to present a schematic or visual representation of the four health fields, yet what was described was, by any interpretation, a ‘conceptual model of health’. It was to become a creating precedent for much that would follow. Emerging somewhat later, the iconic ‘Social Model of Health’ (Dahlgren and Whitehead 1991) (see Fig. 33.2) and Socioecological Model of Health (Evans and Stoddart 1990) (see Fig. 33.3) added depth and breadth to the socioecologi-

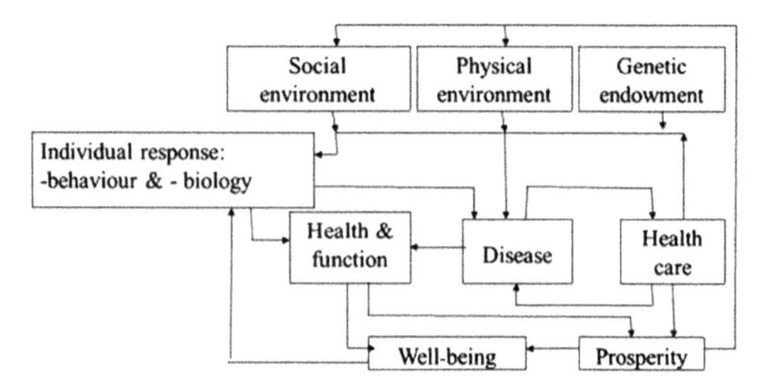

Fig. 33.3 The Socioecological Model of Health. Source: Evans and Stoddart (1990)

cal perspectives of Lalonde and further developed the idea of conceptual models as tools for analysis in public health policy and action. Citing the white paper as direct precedent, in 1990, Evans and Stoddart presented their 'Socioecological Model of Health' as a framework for the assembly of evidence 'in a way that would make its implications more apparent' and added rather insightfully that 'there is much more to policy than evidence' (Morris et al. 2006).

Although it has its own distinct origins and trajectory, the development of conceptual modelling in environmental health policy, right up until the present day, is unquestionably influenced by these earlier attempts to model the determinants of health.

33.6 Conceptual Models in Environmental Health

For much of the post-World War II period in public health, the operational challenge around environment was invariably presented as one of identifying, monitoring and controlling a limited set of toxic or infectious agents in environmental carriers such as food, air, water and soil (Morris and Saunders 2017). Expressed conceptually, the challenge was about how policy and action might impact beneficially on a linear causal pathway leading from hazardous environmental state(s) to group or individual environmental exposure, and on to a (usually negative) health effect. This might be seen as a 'State Exposure-Effect' sequence. From a policy perspective, this representation has proved useful but is also limiting, not least in its failure to represent the (frequently anthropogenic) driving forces that, in turn, create the pressures which act upon and modify the environmental state: a 'Driver-Pressure-State' sequence.

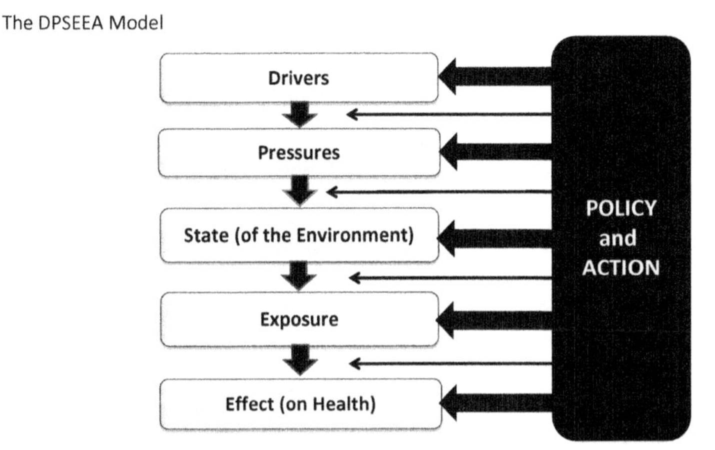

Fig. 33.4 The DPSEEA model. Source: Adapted and simplified from Corvalán et al. (2000)

33.6.1 The DPSEEA Model and Its Derivatives

Although their initial purpose was to create an organising framework for the development and use of environmental health indicators, the importance for policy of integrating 'Driver-Pressure-State' with 'State-Exposure-Effect' sequences to create a single full chain approach was first articulated by Kjellstrom and Corvalan (1995) and Briggs et al. (1996). These co-workers subsequently presented a simplified version of their Drivers-Pressures-State-Exposure-Effect-Actions model as a configuring framework for decision-making in environmental health (Corvalán et al. 2000).

The DPSEEA model offered, for the first time, a much more integrated and policy-relevant conceptualisation of the relationship among human activity, the physical environment and health. Unsurprisingly perhaps, the DPSEEA model has formed the basis for development of a family of conceptual models now finding application in the field of environmental health (Fig. 33.4).

33.6.2 The 'Modified DPSEEA Model'

In 2006, Morris and colleagues developed a modified version of the DPSEEA model (Morris et al. 2006); see Fig. 33.5. The modified or mDPSEEA model was developed as the configuring framework for a new policy approach to environment and health in Scotland. The policy 'Good Places Better Health' (Scottish Government 2008, 2012) was intended to embody socioecological perspectives, taking environmental health activity beyond narrow, compartmentalised and hazard-focused roots, exploring the contribution of environment, in combination with other societal influences and characteristics of the individual in promoting and perpetuating health.

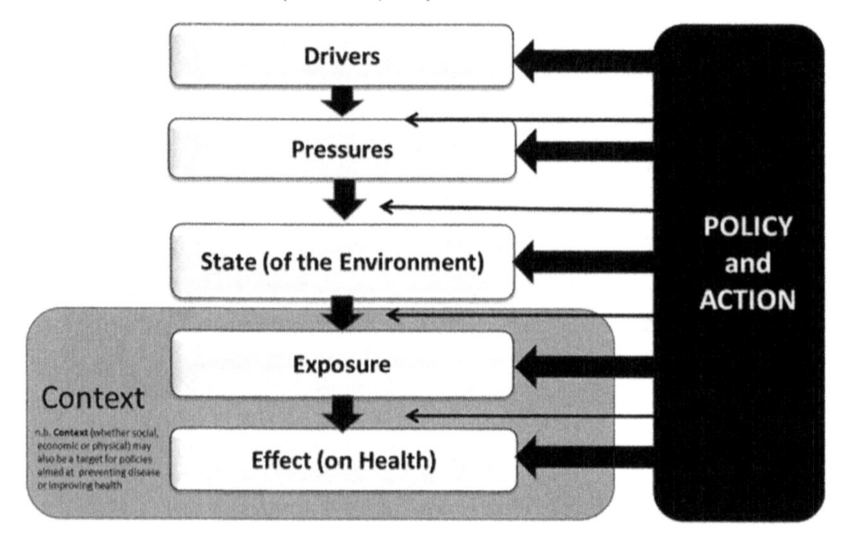

Fig. 33.5 The modified DPSEEA model. Source: Morris et al. (2006)

In this way, it was hoped to identify and address more effectively the environmental contribution to complex health challenges such as obesity, diminished mental health and wellbeing and inequalities in health by a variety of measures. Moreover, the policy sought to reflect a much richer interpretation of the role of environment by recognising the potential of high-quality environments to nurture better, more equal health and wellbeing. The mDPSEEA model differs from the original DPSEEA model (Corvalán et al. 2000) by incorporating a 'contextual bubble' indicating that whether an environmental state (positive or negative) translates into an exposure for an individual is, in every case, critically influenced by a set of interacting contextual factors. These might include demographic, socioeconomic and behavioural influences. In addition, whether an exposure translates into a negative or positive health effect for the individual is influenced by a further set of contextual influences, such as age, gender, pre-existing health status or even the presence of other environmental stressors.

From a policy perspective, it is important to note that existing and proposed interventions (actions) can be directed towards the DPSEEA chain itself and towards elements of the context. A key finding of the prototype phase of Good Places Better Health was that conceptual models like mDPSEEA, when applied to specific issues, can be a means to configure different categories of evidence, a framework for data collection, a vehicle for identifying gaps in knowledge (generating a research agenda) and a tool for policy analysis. Furthermore, the process and product of populating the model for a specific issue or problem is highly effective in engaging the multiple and diverse stakeholders whose inputs or omissions may be critical influence on health outcomes. Perhaps the most useful overarching outcome was the affirmation of the observation by McIntosh et al. (2007) that conceptual models can be useful 'tools to think with'.

This was especially evident through the use of the mDPSEEA model as an issue framing tool in facilitated stakeholder workshops in Scotland.

33.6.3 The 'Ecosystem-Enriched' (eDPSEEA) Model

Cities can usefully be viewed as urban metabolisms and complex systems of flow management and the result of resource allocation, distribution and deployment through time. Conceptual models have a role in explaining the synergies which occur in cities between the environment, human activity and human health, making more explicit the connections among society, the economy, the environment and health and wellbeing and highlighting the importance of biodiversity in both human and planetary health (Carmichael et al. 2017). Reis et al. (2015) further developed the mDPSEEA model to produce the ecosystem (e)-enriched model, or 'eDPSEEA model'. In Fig. 33.6, we present an expanded version of the eDPSEEA model as originally conceived. This reveals its component elements in greater detail. As with mDPSEEA, the model is conceived as a tool to think with in a policy context, but

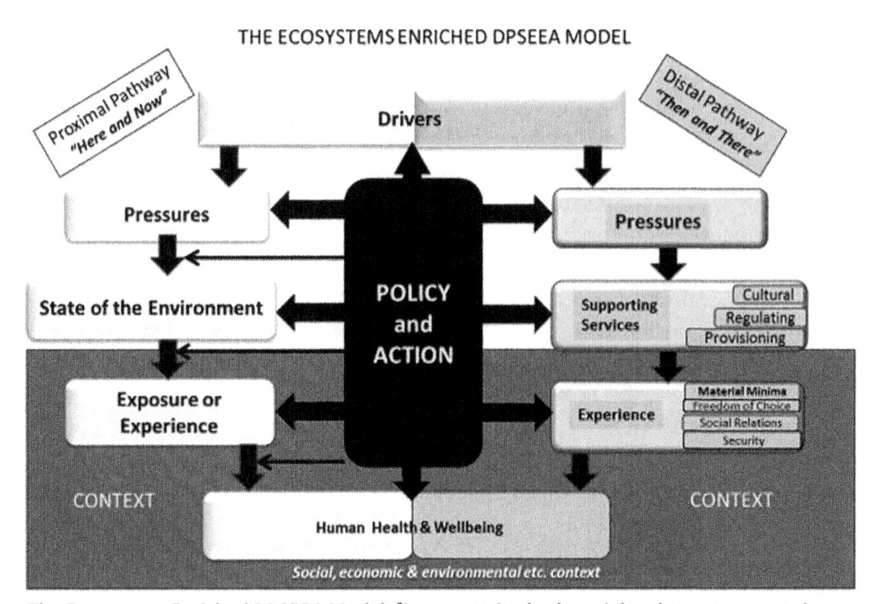

The Ecosystems Enriched DPSEEA Model [incorporating both social and ecosystem services dimensions (Reis, S, Morris G, Beck, S. et al, 2015) as subsequently expanded (EEA, 2015)]

Fig. 33.6 The ecosystem-enriched or eDPSEEA model. Source: Reis et al. (2015)

critically one which recognises that human impacts on global processes and systems are now damaging health, wellbeing and equality within and between communities. The model allows the environmental threat to health and wellbeing to be modelled on the vastly extended temporal and spatial scale necessary if mankind is to build health and wellbeing henceforth on ecological principles. Expressed in another way, it is a tool with which to operationalise the concept of ecological public health described in the introduction of this chapter.

The 'ecosystem-enriched DPSEEA model' continues to capture the dimensions of the environment and health relationship represented by the mDPSEEA model where driving forces create pressures which change health-relevant characteristics of environmental state near, or 'proximal' to, a community. However, eDPSEEA also represents another pathway from driving forces to the health and wellbeing of that community. This so-called 'distal' pathway from driving forces to health and wellbeing effects is concerned with damage to global systems, processes and the consequent changes to environmental state in places remote from the community whose activities generate the harm (Morris et al. 2015). Phenomena such as environmentally driven mass migration or food insecurity mean that any sense of remoteness from the impacts we cause is likely to prove illusory. The eDPSEEA model draws on the insights of the Millennium Ecosystem Assessment (2005) to link ecosystem damage to human health and wellbeing through considering its impact on ecosystem services. Ecosystem services are the benefits human beings receive from the natural environment and are expressed in terms of *provisioning services* (food, fuel, fibre, etc.), *regulating services* (waste treatment, climate regulation, etc.) and *cultural services* (the non-material benefits such as tourism or inspiration) all of which are underpinned by *supporting services* such as soil processes, photosynthesis, etc. When ecosystem services are interupted, human wellbeing can be undermined in five distinct ways. These are the (a) denial of material goods (so-called material minima), (b) disruption of social relationships, (c) diminished security, (d) reduced freedom of choice and (e) direct damage to mental and physical health (Millennium Ecosystem Assessment 2005).

While it is entirely distinct in its origins from the DPSEEA family of models, another useful inclusion within a toolkit of models to aid policy and practice in the urban context is the Egan model (see Fig. 33.7). A product of the 'Egan Review' of 2004, it encapsulates the notion that sustainable communities should address the diverse needs of current and future residents and their children by offering a choice: communities should make effective use of natural resources, enhance and embrace the environment, strengthen social cohesion and increase the economic outlook of communities and the residents.

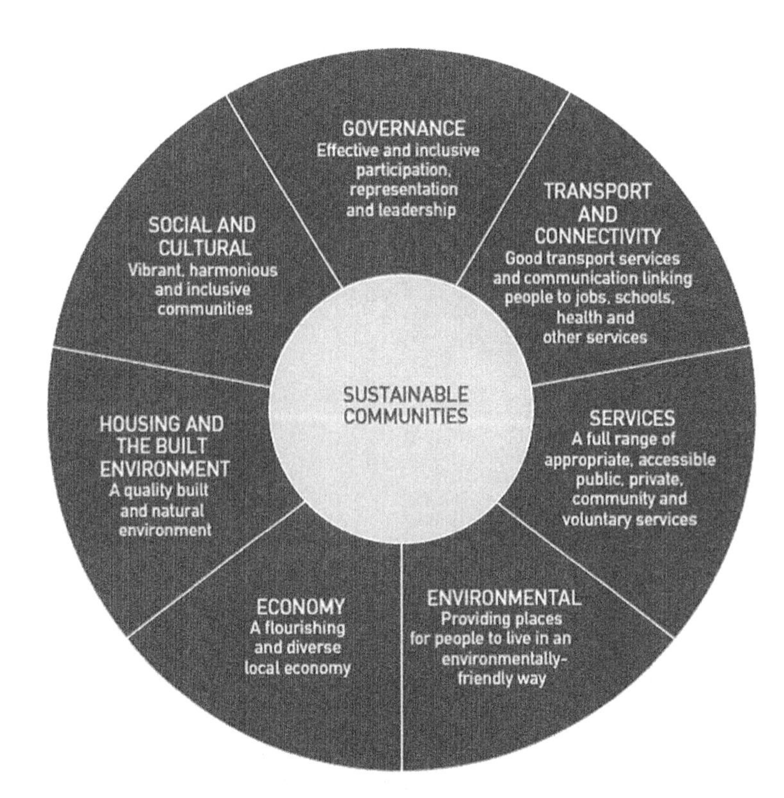

Fig. 33.7 The Egan model (Source: Egan 2004)

33.6.4 *The Ecosystem-Enriched, Behaviourally Enhanced DPSEEA or 'INHERIT* Model'

At the time of writing, the INHERIT model (Staatsen et al. 2017) (see Fig. 33.8) represents the most recent addition in the development of DPSEEA models. Ostensibly quite different in appearance, the model shares key common elements with the eDPSEEA model (most notably in representing two pathways from multiple interacting drivers to health and wellbeing outcomes).[1] The INHERIT model serves to support the development of the Horizon 2020-funded INHERIT project (www.inherit.eu/) which seeks to identify effective intersectoral policies and interventions that promote sustainable and healthy lifestyles and behaviour in the areas of living, moving and consuming.

The INHERIT model emphasises that behaviour should be taken into account when developing and choosing policies and interventions to improve people's health

[1] Whereas the distal and proximal pathways were on both sides of the policy and action box in the eDPSEEA model, the INHERIT model, as expanded from the eDPSEEA model, shows both pathways on the right side of the model.

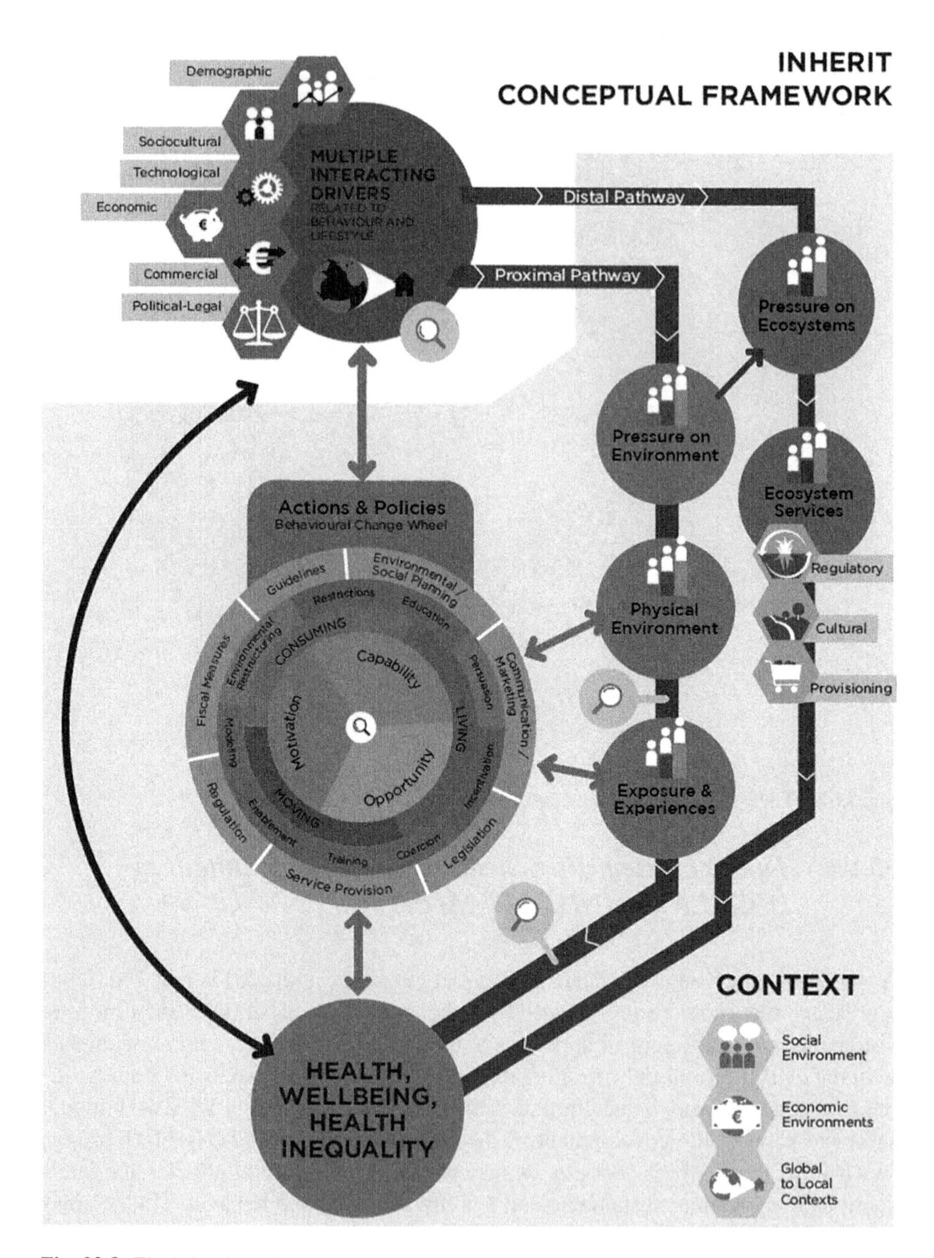

Fig. 33.8 The behaviourally enhanced, ecosystem-enriched DPSEEA model, or INHERIT model (Staatsen et al. 2017)

and environmental sustainability and reduce inequalities and health inequities and thereby achieve a 'triple win'. Accordingly, the first significant difference from other members of the 'DPSEEA family' is the integration of the Behavioural Change Wheel (BCW). The BCW (Michie et al. 2011) is a theory and evidence-based tool with which interventions and policies can be selected/developed that change behaviour. The core of the wheel consists of three behavioural components: COM-B, capability (to be physically or psychologically capable of performing certain behaviour), opportunity (all factors outside individual that can make a certain behaviour possible or not) and motivation (brain processes that direct behaviour, such as conscious decision-making, but also habits). These components interact and together influence behaviour. Different intervention functions (e.g. restriction or education) can influence (one or more) different behavioural components. These interventions can be enabled by different types of policy (such as legislation). Another essential difference with previous versions of the DPSEEA model is that, throughout the model, magnifying glasses are used to represent behavioural hotspots: parts of the causal process that are heavily influenced by human behaviour and thus important parts towards which actions may be directed.

A further feature of the INHERIT model lies in its attention to the issue of inequalities. These are represented throughout the model to emphasise that driving forces, pressures and the resultant environmental state will all differ significantly according to location, creating and sustaining inequalities between individuals and different communities. Expressed in another way, driving forces exist in many combinations and do not create the same pressures for everyone: some populations are affected more than others. Using the example of the food environment in disadvantaged neighbourhood, often a proliferation of fast food outlets varies significantly creating more obesogenic environments for disadvantaged populations. Behaviour is one of the elements in the equation leading to these differences, as people may have different motivations and capabilities that lead them to perform different (health-related) behaviours and interact differently with their environments.

As in the eDPSEEA model (see Fig. 33.6), the INHERIT model shows both a proximal pathway and a distal pathway, the latter demonstrating a Driver-Pressure-Ecosystem Services-Health impacts sequence. This takes account of the fact that multiple interacting and predominantly anthropogenic driving forces which exist today are creating pressures which are impacting Western European populations in the here and now but also future and distant populations. Most obviously, the impacts of global warming, with extreme weather events damaging crops, reducing food supplies, seriously affect health and wellbeing in other parts of the world.

33.7 Applying Conceptual Models to Urban Environmental Health Challenges

To illustrate the effectiveness of conceptual models in an urban context and to identify those actions that will address the environment, health and equity challenges in cities in an effective way, we describe the use of the ecosystem-enriched model in a pilot exercise.

In 2015 a workshop was organised with experts from the network of the European Environmental Agency (EEA). The primary purpose of the workshop was to provide participants with an insight into how the conceptual models might be 'populated' and used for a dialogue with policymakers, scientists, city planners and citizens. The topics selected for further discussion were healthy transport, healthy housing and climate. Using discussion methods like 'knowledge-café' and 'brainwriting', the conceptual framework was discussed, relevant indicators and future challenges and barriers were identified, as well as options for actions.

While presented here as a proof of concept rather than a comprehensive analysis, Fig. 33.9 illustrates the application of the eDPSEEA model in the case of urban transport. It shows how identical drivers such as (lack of) investments in public transport and infrastructure, urban and transport planning, cultural norms and individual preference (the demand for convenience, speed and comfort) influence health through direct pathways (proximal, near in space and time) and more indirectly through land use and ecosystem impacts (distal route).

Drivers put pressure on the proximal urban environment (state) by influencing land use, the availability of (green) space, walking or cycling networks and traffic density. As a result, individuals may be subjected to air pollution, more noise and higher temperatures or have (negative or positive) experiences in relation to exercising or leisure, depending on the availability of cycling networks and parks. The way these changes to the proximal environment in cities affect individual or community health and wellbeing is however dependent on interacting contextual factors such as their stage of life and socioeconomic circumstances. Air pollutants from motorised transport may also have an effect on climate change. Although, for European cities, these changes may appear distal in that they appear to be happening elsewhere or seem to be a concern for future generations, they are of course real and 'proximal' threats to the people in the places affected. Moreover, in a world connected economically, socially and environmentally, Europeans are never isolated from the environmental, social and health changes occurring now and later elsewhere in the world (Staatsen et al. 2017; FRESH Consortium 2014a, b; Carmichael et al. 2017).

The eDPSEEA conceptual framework, and especially the process of populating the model for an issue in multiple stakeholder groups, helps frame the issue and identify (often hidden) associations, as well as co-benefits and unintended side consequences. Many policies in the field of neighbourhood planning, housing and transport offer co-benefits. For example, policies that promote cycling benefit health and wellbeing in a variety of ways while potentially reducing greenhouse gas emissions. However, other policies intended to produce benefits to environment, health

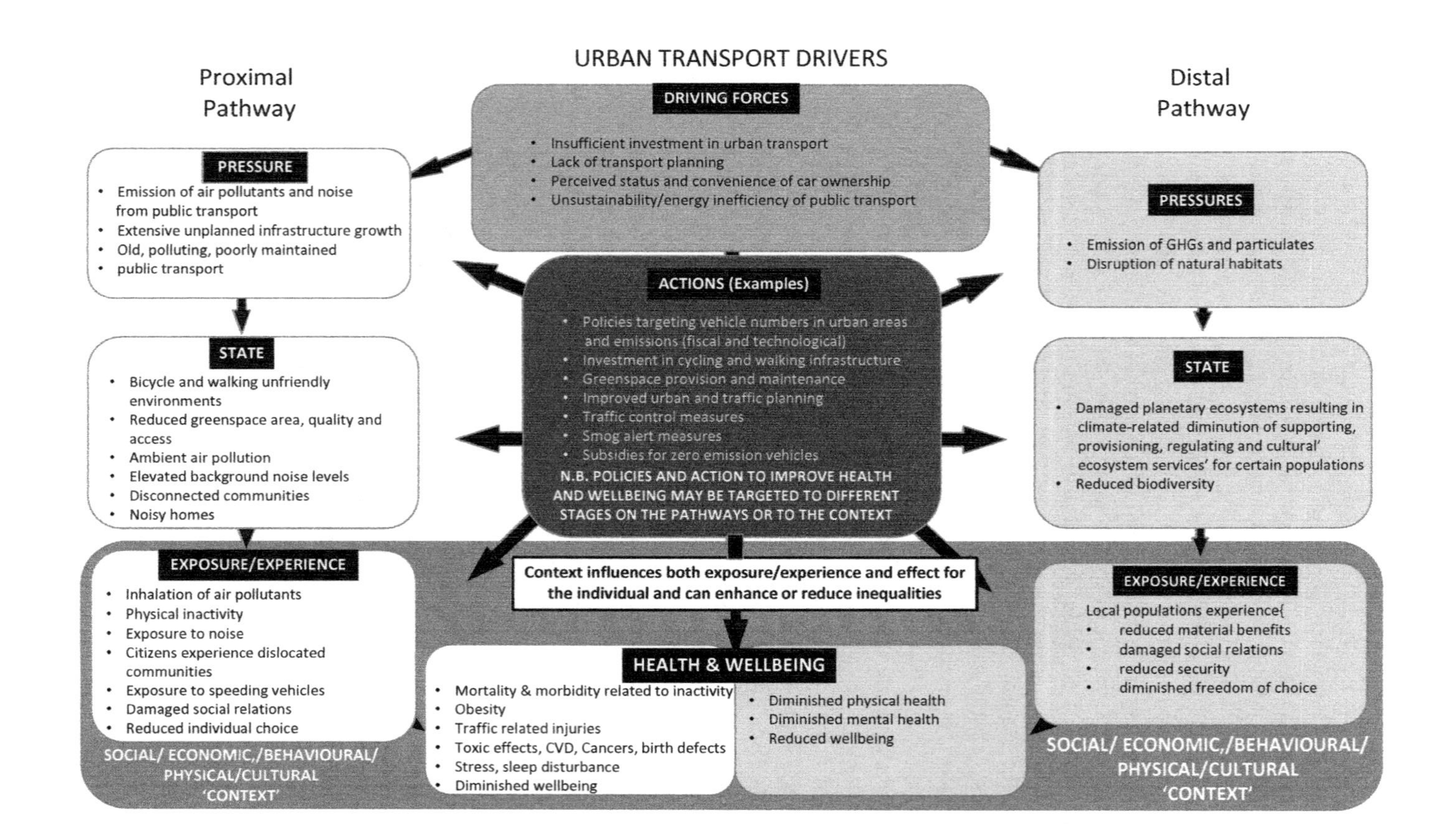

Fig. 33.9 Addressing the effects and actions of transport through DPSEEA models. Source: FRESH (2015b) and FRESH Transport leaflet (FRESH 2015c). Modelling approach derived from Reis et al. (2015)

and wellbeing may have unintended consequences which are damaging to health. For example, measures which have encouraged the widespread use of diesel cars with the intention of reducing CO_2 emissions have added to background concentrations of particulates which are damaging to respiratory and cardiovascular health.

The eDPSEEA conceptual framework also offers a structured approach to identify useful indicators for evaluating and monitoring the impact of policies on the environment, health and wellbeing. The correct indicators, presented in a relational way, are central to understanding problems and measuring progress in addressing proximal and distal impacts (Fig. 33.10). Thus, models like eDPSEEA can offer a useful configuring framework around which to build an information system.

In many cities, there are experiments involving the promotion of active mobility, electric driving, free public transport or banning private cars from city centres. Information is not always available for these experiments, and there is a need to improve the evaluation and exchange of such information. For example, the evidence base for stimulating people to shift from car use to active transport is based mostly on cross-sectional studies. There is a lack of long-term measurements addressing the proximal and distal impacts of a modal shift (Staatsen et al. 2017). The development and application of a common set of indicators to evaluate this type of intervention (such as those which have been developed in FRESH and are being developed in INHERIT), would be extremely useful.

An analysis of Sustainable Urban Mobility Plans (SUMP) shows a wide variation in the way climate change and equity are addressed in cities (Arsenio et al. 2016 in Staatsen et al. 2017). The elaborated INHERIT model holds promise as vehicle for raising awareness of equity issues in transport and critically the role of human behaviour in influencing outcomes.

Also of direct relevance to the urban context, the eDPSEEA model has also been applied to the domain of housing (see Fig. 33.11). Housing construction, maintenance, the location and occupation of houses and access to services have an impact on the health and wellbeing of the occupants (Braubach et al. 2011). Important drivers in the area of healthy housing include population density, age profile, economy/income (increasing socioeconomic inequalities), lifestyles, spatial planning and policies regarding housing materials and construction standards, consumer products and energy saving. There is a variety of indoor chemicals (VOCs, CO, NO_2, tobacco smoke, asbestos), microbiological (mould, dust) and physical agents in houses and building blocks associated with a range of health impacts (i.e. respiratory and cardiovascular diseases). The populated model also shows that policies to secure energy efficiency might reduce ventilation rates and by extension GHG emissions. However, these very same policies might profoundly reduce indoor air quality to the detriment of occupant health and wellbeing. The potential for conflict between housing policies motivated by energy conservation or other building-related concerns and environmental health should be considered at an early stage of planning in order to minimise negative health impacts. Large social inequalities in housing between and within EU member states are also evident (FRESH 2015a; Staatsen et al. 2017).

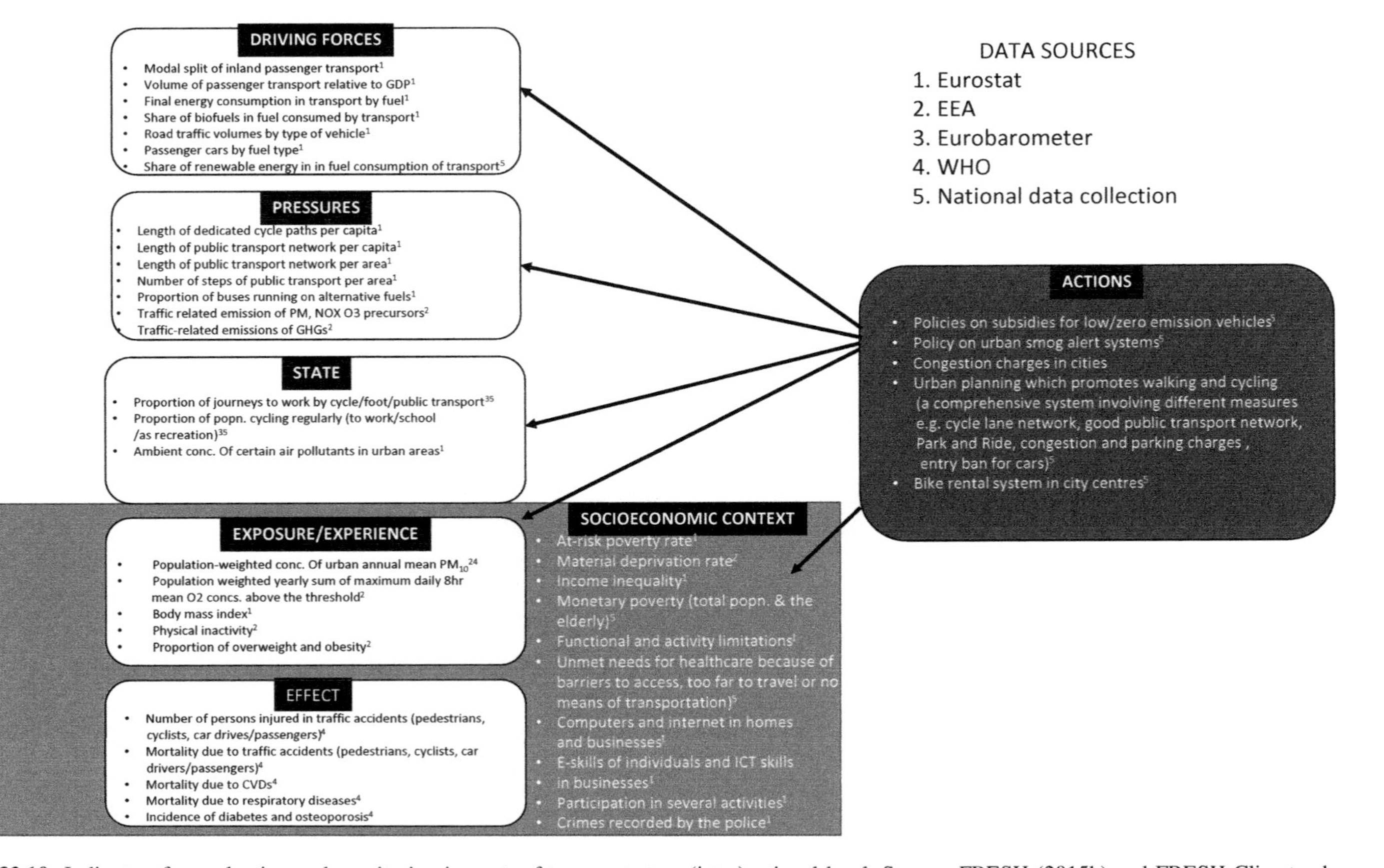

Fig. 33.10 Indicators for evaluating and monitoring impacts of transport at an (inter)national level. Source: FRESH (2015b) and FRESH Climate change leaflet (FRESH Consortium 2014a, b)

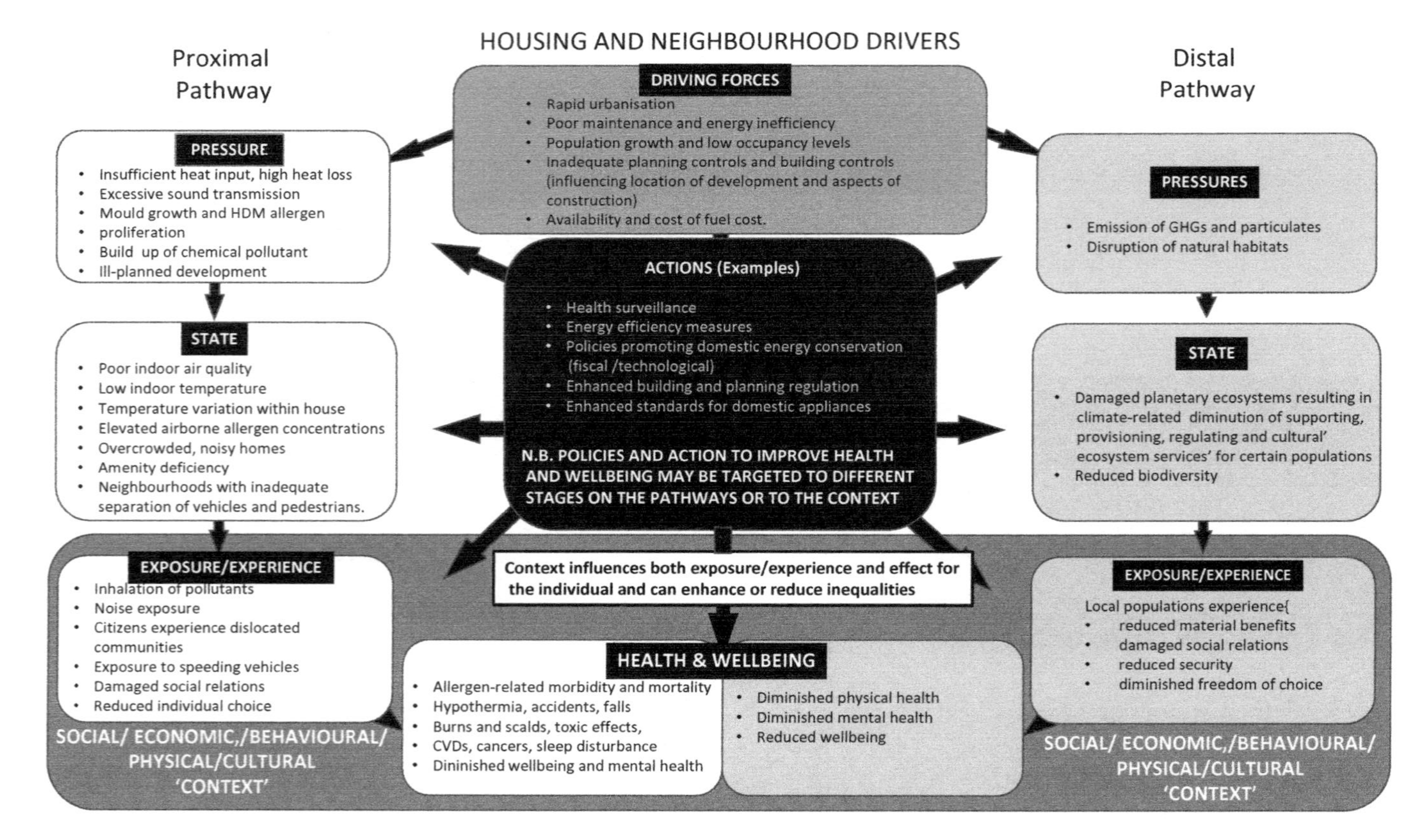

Fig. 33.11 Addressing the effects and actions of housing through the DPSEEA model. Source: FRESH Housing leaflet (FRESH Consortium 2014b)

Effective design and governance of healthy and sustainable cities relies on the awareness of these complex interactions, feedback loops and the trade-offs involved. This can be achieved, by securing more stakeholder involvement, to inform cross-cutting analysis and synthesis and to reflect on implications.

The integrated conceptual models described above imply that urban and transport planners and architects need to work together with environmental scientists, public health specialists, local communities and those from many other disciplines to make sense of the complexities of the urban metabolism and inform policymakers about actions to regulate human activities, encourage changes in behaviour and promote equity (Carmichael et al. 2017, p. 51; Staatsen et al. 2017).

33.8 Discussion and Conclusions

In this chapter, we have presented a number of different conceptual models. The very existence of so many models implies that many researchers and policymakers find them useful in understanding the complex interacting factors that shape our world and our health. However, those creating the models we have illustrated would readily subscribe to the maxim that 'all models are wrong but some are useful' (Box 1979).

The Dahlgren and Whitehead and Evans and Stoddart models (see Figs. 33.2 and 33.3) each offer representations of the relationship between societal and individual factors in creating and destroying health and wellbeing. We highlighted a limitation common to each of these models in that both fail to fully represent the interdependence of human health and that of natural systems and processes. However, it would be remiss not to recognise the seminal influence of both models in challenging siloed thinking and work practices in public health. In particular, the models were and remain very valuable tools for analysing the socially complex 'proximal' context, which many policymakers and practitioners encounter in the here and now.

Coming a little later, and following something of a societal awakening around the threat to the natural systems on which we rely, the Egan Model is particularly effective in illustrating the many dimensions of a sustainable community and has clear application as a tool for community engagement in pursuing that aspiration.

We have presented above the evolution of the 'DPSEEA family of models' (DPSEEA, mDPSEEA, eDPSEEA and INHERIT) each of which places the relationship between physical environment and health at its core. It would be a mistake to assume however that each subsequent iteration of the model represents a progression towards a more sophisticated and, by implication, 'better' understanding. It is true that with each iterations add further dimensions and increased scope are added. Yet it is important to note that each form of the model was developed to serve a particular purpose and earlier versions are not necessarily superseded or rendered obsolete by those developed later. All are useful inclusions within a toolkit of conceptual models that can be deployed by policymakers and practitioners when seeking to engage with others and to produce beneficial change.

The product and process of populating models for different issues has proved useful for engaging multiple stakeholders: providing an understanding of the contribution of others and communicating messages to a wider constituency. Models have great potential as configuring frameworks for information systems and for highlighting in gaps whether in knowledge, information or policy. At their simplest, conceptual models are 'tools to think with' in an increasingly complex world and can be applied at differing scales from the international to the very local.

We are attracted by the idea that a conceptual model is analogous to a camera. The user initially has a number of cameras from which to choose. Having chosen, he or she may then adjust the focus, zoom in or out (scale), make a film (time) and think about different perspectives (Schram-Bijkerk et al. 2016). A user can focus on health (and use, e.g. Dahlgren and Whitehead's model (1991)) and sustainability (the Egan model, life cycle analysis, etc.) or opt for an integrated approach and use the eDPSEEA or INHERIT model described above.

In this chapter, our focus has been on the application of conceptual models in the cities of the twenty-first century. Our justification is simple. Wherever we live in the world, our health and wellbeing, our pursuit of equity and our future as individuals and as a species are inextricably bound up with the natural environment. This is both the message and the challenge of ecological public health. It is a global challenge which demands insightful public health thinking on a markedly extended temporal and spatial scale. However, arguably the most important frontline in the battle to achieve the triple win of health and wellbeing, equity and sustainability remains, as it has so often been throughout history the urban environment. Cities are the theatres where society can make the greatest progress in meeting the ecological public health challenge and where risks that attend failure will be felt most acutely. Impacts on city level depend on and interact with processes and impacts on neighbourhood, district and national scale, which is difficult to capture in one-dimensional models. The aspect of time is also important to consider. We need integrated models like the eDPSEEA and INHERIT model that help us selecting those actions that are healthy and sustainable in the future, as well as inclusive.

If we are to begin to identify and gain support for the actions which simultaneously protect both ecosystems and human health and wellbeing, in ways which are socially inclusive, sustainable and equitable, globally and across multiple generations, we need tools to work with.

References

Agenda Stad. (2016). *Verslag Werkconferentie Agenda Stad*. Retrieved from http://agendastad.nl/verslag-werkconferentie-agenda-stad/

Arsenio E, Martens K, Di Ciommo F. (2016) Sustainable urban mobility plans: Bridging climate change and equity targets? Res Transp Econ (55):30-9. In: Staatsen B, van der Vliet N, Kruize H, et al. (2017) INHERIT: Exploring triple-win solutions for living, moving and consuming that encourage behavioural change, protect the environment, promote health and health equity. EuroHealthNet, Brussels

Box, G. E. P. (1979). Robustness in the strategy of scientific model building. In R. L. Launer & G. N. Wilkinson (Eds.), *Robustness in statistics* (pp. 201–236). New York: Academic Press.

Braubach, M., Jacobs, D. E., & Ormandy, D. (2011). *Environmental burden of disease associated with inadequate housing*. Copenhagen: WHO Regional Office Europe. Retrieved from http://www.euro.who.int/__data/assets/pdf_file/0003/142077/e95004.pdf.

Briggs, D., Corvalan, C., & Nurminen, M. (1996). *Linkage methods for environmental health analysis: General guidelines*. Geneva: World Health Organisation.

Carmichael, L., Racioppi, F., Calvert, T., & Sinnett, D. (2017). *Environment and health for European cities in the 21st century: Making a difference*. Geneva: WHO.

Corvalán, C., Briggs, D., Briggs, D. J., & Zielhuis, G. (2000). *Decision-making in environmental health: From evidence to action*. New York: Taylor & Francis.

Coutts, C., Forkink, A., & Weiner, J. (2014). The portrayal of natural environment in the evolution of the Ecological Public Health Paradigm. *International Journal of Environmental Research and Public Health, 11*(1), 1005–1019.

Dahlgren, G., & Whitehead, M. (1991). *Policies and strategies to promote social equity in health*. Stockholm (Mimeo): Institute for Future Studies.

Egan, J. (2004). *The Egan review: Skills for sustainable communities*. London: Office of the Deputy Prime Minister.

European Environmental Agency (EEA). (2015). *The European Environment - State and Outlook (SOER) 2015. A comprehensive assessment of European environment's state, trends, prospects, in a global context*. Copenhagen: EEA.

European Commission. (2015). *The EU Urban Agenda* [Factsheet]. Retrieved from http://ec.europa.eu/futurium/en/content/factsheet-eu-urban-agenda

European Commission. (2016). *Urban agenda for the EU 'Pact of Amsterdam'*. Brussels: EU. Retrieved from http://ec.europa.eu/regional_policy/sources/policy/themes/urban-development/agenda/pact-of-amsterdam.pdf.

Evans, R. G., & Stoddart, G. L. (1990). Producing health, consuming health care. *Social Science & Medicine, 31*(12), 1347–1363.

FRESH Consortium. (2014a). *Climate change leaflet*. Copenhagen: EEA. Retrieved from https://www.eea.europa.eu/articles/a-europe-to-thrive-in.

FRESH Consortium. (2014b). *Housing leaflet*. Copenhagen: EEA. Retrieved from https://www.eea.europa.eu/articles/a-europe-to-thrive-in.

FRESH Consortium. (2015a). *FRESH brochure*. Copenhagen: EEA. Retrieved July, 2017, from https://www.eea.europa.eu/articles/aeurope-to-thrive-in/fresh-brochure/view.

FRESH Consortium. (2015b). *A Europe to thrive in – Environment, health and well-being*. Copenhagen: EEA. Retrieved from https://www.eea.europa.eu/articles/a-europe-to-thrive-in.

FRESH Consortium. (2015c). *Transport leaflet*. Copenhagen: EEA. Retrieved from https://www.eea.europa.eu/articles/a-europe-to-thrive-in.

Gee, G. C., & Payne-Sturges, D. C. (2004). Environmental health disparities: A framework integrating psychosocial and environmental concepts. *Environmental Health Perspectives, 112*(17), 1645–1653.

Hartig, T., Mitchel, R., de Vries, S., & Frumkin, H. (2014). Nature and health. *Annual Review of Public Health, 35*, 207–228.

Kjellstrom, T., & Corvalan, C. (1995). Framework for the development of environmental health indicators. *World Health Statistics Quarterly, 48*, 144–154.

Lalonde, M. (1974). A conceptual framework for health. *RNAO News, 30*(1), 5–6.

McIntosh, B. S., Seaton, R. A., & Jeffrey, P. (2007). Tools to think with? Towards understanding the use of computer-based support tools in policy relevant research. *Environmental Modelling and Software, 22*(5), 640–648.

Millennium Ecosystem Assessment. (2005). *Ecosystems and human well-being: Synthesis*. Washington: Island Press.

Morris, G. P., Beck, S. A., Hanlon, P., & Robertson, R. (2006). Getting strategic about the environment and health. *Journal of Public Health, 120*(10), 889–903.

Morris, G., Reis, S., Beck, S., Fleming, L. E., Adger, W. N., Benton, T. G., et al. (2015). *Health Climate Change impacts report card technical paper No.10. Climate change and health in the UK. Scoping and communicating the longer-term "distal" dimensions*. Swindon: Natural Environment Research Council. Retrieved January 5, 2017, from http://www.nerc.ac.uk/research/partnerships/ride/lwec/report-cards/health-source10/.
Morris G, Saunders P (2017) The environment in health and well-being. Oxford Research Encyclopedia of Environmental Science. Retrieved January 5, 2017, from http://environmentalscience.oxfordre.com/view/10.1093/acrefore/9780199389414.001.0001/acrefore-9780199389414-e-101?print=pdf.
Michie, S., et al. (2011). The behaviour change wheel: A new method for characterising and designing behaviour change interventions. *Implementation Science, 6*, 42.
Platform Healthy Design/Platform Gezond Ontwerp. (2013). *Magazine Gezond Ontwerp*. Eindhoven: Technische Universiteit Eindhoven.
Pruss Ustun, A., Wolf, J., Corvalan, C., et al. (2016). Diseases due to unhealthy environments: An updated estimate of the global burden of disease attributable to environmental determinants of health. *Journal of Public Health*, 1–12. http://www.canberra.edu.au/research/repository/file/67130a67-4143-4df3-8d10-aa2e53814e66/1/full_text_published.pdf
Rayner, G., & Lang, T. (2012). *Reshaping the conditions for good health*. Oxford: Routledge Publishers.
Reis, S., Morris, G., Fleming, L. E., et al. (2015). Integrating health and environmental impact analysis. *Journal of Public Health, 129*(10), 1383–1389.
Rockström, J., Steffen, W., Noone, K., et al. (2009). A safe operating space for humanity. *Nature, 461*(7263), 472–475.
Steffen, W., Richardson, K., Rockström, J., et al. (2015). Planetary boundaries: Guiding human development on a changing planet. *Science, 347*(6223), 1259855.
Schram-Bijkerk, D., Kruize, H., Staatsen, B., & van Kamp, I. (2016). Modellen plaveien de weg naar een gezonde stad. *Tijdschrift Milieu-Dossier, 22*(6), 43–47. Retrieved from http://www.vvm.info.
Staatsen, B., van der Vliet, N., Kruize, H., et al. (2017). *INHERIT: Exploring triple-win solutions for living, moving and consuming that encourage behavioural change, protect the environment, promote health and health equity*. Brussels: EuroHealthNet.
Scottish Government. (2008). *Good places, better health. A new approach to environment and health in Scotland*. Edinburgh: Scottish Government.
Scottish Government. (2012). *Good places better health for Scotland's children*. Edinburgh: Scottish Government.
Susser, M., & Susser, E. (1996). Choosing a future for epidemiology: I. Eras and paradigms. *American Journal of Public Health, 86*(5), 668–673.
United Nations. (2015). *Sustainable development goals*. New York: United Nations. Retrieved from http://www.un.org/sustainabledevelopment/sustainable-development-goals/.
United Nations (UN). (2016). *The New Urban agenda. Adopted at the United Nations Conference on Housing and sustainable urban development on 20 October 2016 (Habitat III)*. Nairobi: Habitat. Retrieved from http://habitat3.org/wp-content/uploads/NUA-English.pdf
World Health Organization (WHO). (2017). *Environment and health in Europe: Status and perspectives*. Copenhagen: WHO Regional Office Europe. Retrieved from http://www.euro.who.int/__data/assets/pdf_file/0004/341455/perspective_9.06.17ONLINE.PDF?ua=1.

Chapter 34
Integrating Human Health into the Urban Development and Transport Planning Agenda: A Summary and Final Conclusions

Mark Nieuwenhuijsen and Haneen Khreis

What is clear is that there are links between urban planning and land use, transport planning and behaviour and environmental exposures and morbidity and premature mortality and that many improvements to urban and transport planning exist. These improvements could reduce morbidity and premature mortality in cities, but are not implemented for various reasons.

34.1 Introduction

This book has brought together people from different sectors and disciplines including urban and transport planning, environment, public health and social sciences to document how urban and transport planning and environmental exposures affect health and what solutions can create cities where citizens are less adversely affected by urban and transport planning practices and related exposures. Although the language and terminology used at times differ substantially, the main messages that come across are that improvements in urban and transport planning are essential and possible to improve public health.

What is clear is that there are links between urban planning and land use, transport planning and behaviour and environmental exposures and morbidity and premature mortality and that many improvements to urban and transport planning exist

M. Nieuwenhuijsen (✉)
ISGlobal, Barcelona, Spain
e-mail: Mark.nieuwenhuijsen@ISGlobal.org

H. Khreis
ISGlobal, Barcelona, Spain

Texas Transportation Institute, Texas A&M, College Station, TX, USA
e-mail: H-Khreis@tti.tamu.edu

M. Nieuwenhuijsen, H. Khreis (eds.), *Integrating Human Health into Urban and Transport Planning*, https://doi.org/10.1007/978-3-319-74983-9_34

that could reduce morbidity and premature mortality in cities, but are not implemented for a large variety of reason.

34.2 A Summary

34.2.1 The Current State of Affairs

In this book, Giles-Corti et al. (2018) described the influence of city planning on transport mode choice, access to open space, walkability and other characteristics of the built environment on chronic diseases and their risk factors—particularly physical activity through walking—and on environmental sustainability. They mentioned that numerous systematic reviews of (mainly) cross-sectional evidence have confirmed earlier reviews that physical activity—principally walking for transport—is associated with three main built-environment features, higher residential density, mixed land use or access to local destinations required for daily living and connected street networks (either measured individually or combined into a composite 'walkability' index), but that there is a complex relationship between these variables (Giles-Corti et al. 2018). For example, density *alone* is unlikely to encourage physical activity. Higher-density development with few local destinations or little public transport—as is now being built in some cities—is simply high-rise sprawl and continues to foster motor-vehicle dependency and traffic congestion. Rather, the relationship between higher-density development and walking is apparent because density is generally a proxy for other environmental characteristics (including demographics; car ownership; access to local destinations, employment, shops and services; frequent public transport; connected street networks that make destinations more proximate) that directly influence choice of transport mode and hence levels of physical activity (Giles-Corti et al. 2018).

Findings are similar to a recent systematic review of 18 studies that found that residential density, safety from traffic, recreation facilities, street connectivity and high-walkable environment were associated with physical activity (Malambo et al. 2018) and a review by Wang et al. (2016) that described a whole range of built environment and other measures that promote walking and cycling including the availability and suitability of design of facilities; reducing individuals' opportunities to undertake activities that can substitute walking and cycling activities; shortening the distance; increasing accessibility; improving personal security; improving personal safety; improving transport safety; reducing fear of injury, accident and dog attack; improving aesthetics appearance, natural sceneries and environmental quality; and increasing comfort level and provision of supporting facilities. High-walkable environments were also associated with lower blood pressure, body mass index, diabetes mellitus and metabolic syndrome (Malambo et al. 2018).

Furthermore, in this book, van Wee (2018) described how land use policies are related to transport choices and may affect health. Several policy categories can influence the impact of the transport system on health. Regulations for new road vehicles have an impact on emission and exposure levels and on road safety of both people using these vehicles and people experiencing the risk of being hit by these vehicles. Pricing policies (e.g. subsidies on public transport, levies on fuels, taxes on cars) and parking policies influence mode choice and therefore exposures and health effects. Infrastructure policies influence the (un)attractiveness to travel to distinguished destinations, via influencing travel times, travel costs and effort. Specific public transport policies (such as those having an impact on the services offered and tariffs) influence mode choice and the intensity of using public transport (number of trips, distances travelled). Land use policies influence which activities are located where and next in multiple ways influence travel behaviour and next health (van Wee 2018). Schepers et al. (2018) also noted that dense and diverse land use and development oriented towards public and active transport are associated with high levels of road safety. The relationship between urban form and road safety is indirect via mobility. The aforementioned urban form characteristics are likely to contribute by creating favourable preconditions for road safety (and public health in general).

Poor urban and transport planning practices have led to extensive sprawl, a move towards the use of motorized vehicles and high levels of air pollution and noise, heat island effects and lack of green space, which are all detrimental to health. Hoffmann (2018) described that among environmental risk factors, ambient air pollution is the most important cause of disease, leading to more than 4 million premature deaths annually and more than 100 million DALYs, across the world. Important health effects of ambient air pollution include, but are not limited to, cardiovascular and cerebrovascular disease, chronic respiratory disease and infections and lung cancer, with ischemic heart disease being responsible for most of the estimated annual premature deaths. Anthropogenic sources of air pollution include combustion products from energy production, motorized traffic and household heating with wood, coal or oil, waste incineration, tire and break wear, industrial emissions and emissions from surrounding agricultural areas. Among those, traffic is the most important determinant of within-city exposure contrasts, since people live, work and commute in close proximity to traffic.

Another important traffic-related environmental exposure is noise. Lercher (2018) documented that in 2011, a WHO assessment concluded that the largest health effects of noise in terms of population spread are due to severe sleep loss and severe annoyance (each year more than 1 million healthy life years lost across larger agglomerations in Europe). Also cardiovascular health (myocardial infarction, hypertension) is affected with a 5–10% higher risk for those living in noisier areas. Effects of noise on cognition and performance follow. Updated WHO reviews in 2017 add further noise-associated diseases such as stroke and diabetes to the list. More critical is, however, that the threshold at which severe health effects of noise occur were lowered by about 5–10 dBA.

Basagaña (2018) described that exposure to extremely hot or cold temperature is a well-recognized health threat and increases mortality and hospital admissions. Results from an international study using mortality data from 384 locations in 13 countries estimated that 0.42% of all deaths can be attributed to heat every year, even if no heat waves occur (Gasparrini et al. 2015). Basagaña (2018) reviewed what is known about the effects of temperatures on health, how the heat island effect can intensify health effects, and how several studies have shown that measures to reduce the heat island phenomenon can be translated into health benefits. And finally, Dadvand and Nieuwenhuijsen (2018) described the lack of green space in many cities and how natural environments, including green spaces, have been associated with improved mental and physical health and wellbeing and are increasingly recognized as a mitigation measure to buffer the adverse health effects associated with urban living.

Besides the environmental risk factors, there are other factors related to urban and transport planning that are detrimental to health. Stigendal (2018) described how transport and urban planning measures have deepened a community severance between the socially included and the socially excluded by favouring the more profitable transport modes. He went on to state that a social cohesion of cities on current conditions is not a solution because that would aggravate and preserve the problems causing inequalities. Therefore, the causes of inequality should be combated and that should be done in ways where the people affected by the causes are seen as potentials and where these potentials (in particular their experience and knowledge) are taken advantage of. The social cohesion to strive for in cities, thus, is a collective empowerment of people who want to combat these causes and thereby transform society.

Borrell et al. (2018) addressed the health inequalities in cities. Social health inequalities are differences in health which are 'unnecessary and avoidable but, in addition, are also considered unfair and unjust'. People of working class and immigrants of poor countries have worse health and higher mortality rates. Women have worse self-perceived health although their life expectancy is larger than men's. The WHO Commission on Social Determinants of Health concluded that social inequalities in health arise from inequalities in the conditions of daily life and the fundamental drivers that give rise to them: inequities in power, money and resources. These social and economic inequalities underpin the determinants of health—the range of interacting factors that shape health and wellbeing.

Finally, some of the worst conditions can be found in informal settlements. Corburn and Sverdlik (2018) described the issues in informal settlements or slums that dominate the urban landscape and that in 2016 were the home to almost 900 million people. Global health is now slum health. The determinants of health for slum dwellers are multifactorial and can vary from country to country and even within the same city. They reviewed the definitions of informal settlements or slums and highlighted the key drivers of health, disease and wellbeing in urban informal settlements, including spatial segregation, insecure residential status, poverty and employment, housing structural quality, water and sanitation, energy, transport, violence, climate change and services. They also highlighted the evidence supporting the links between these key factors and infectious and noncommunicable diseases (NCDs).

34.2.2 How to Improve Current Conditions

So how can we improve the current situation and what is needed? Crowhurst Lennard (2018) argued that a major barrier preventing us achieving these goals is over-reliance on the economic model of city making. There are two competing value systems at work. The gross domestic product (GDP) model sees the city as an economic engine, and its function is to fuel economic growth and increase wealth. The second is based on quality of life. In this model, the function of the city is the 'care and culture' of human being and of the earth. She promotes the Quality of Life model for shaping cities through the Principles of True Urbanism, which includes facilitating community social life, facilitating contact with nature, facilitating independent mobility and creating a hospitable built environment.

For this, Tsouros (2018) argued that the Healthy Cities as a movement embodied a number of key features that proved crucial in its success to promote health in cites including a strong emphasis on values, political commitment, partnership-based approaches, democratic governance, strategic thinking and networking. It combined the discipline of a well-defined project involving committed cities with mechanisms of inclusiveness and engagement of all interested cities. Healthy Cities in Europe thrived at the cutting edge of public health, continuously broadening and adapting its agenda to new knowledge, global and regional developments and emerging local needs.

Further to this, Lawrence (2018) argued that an interdisciplinary approach based on the generic principles of human ecology can improve our understanding of the consequences of large-scale urban development for health and wellbeing. This knowledge should be the foundation of urban planning and building construction. The advantage of applying principles of human ecology stems from its integrated conceptual framework of the multiple relations between human groups and all the components of their natural and built environments.

However, Grant (2018) noted that having to a large extent understood how to design-out communicable disease in urban areas, we seem to lack knowledge about how to build human habitats without risk of noncommunicable disease. We don't deliberately build places to support healthy lifestyles and reduce health inequity. Planners, urban designers, transport practitioners and public health specialists are waking up to this reality. The evidence base is building, and there is relevant experience in the European Healthy City Network and other cities. The solution lies in better understanding what elements of urban form support health and why cities develop in ways that undermine health and then changing the ways we manage, renew and build urban environments. We need to form coalitions for healthier places, using a health lens across planning, transport and all other urban policy. Land use pattern, transport, green space and urban design are key for manipulation and design to better support urban health and health equity, and this includes planetary health.

For this, Hermansen et al. (2018) placed emphasis on cities as *human* habitats, underlining the importance of reintroducing, or perhaps introducing, a human-centric approach to urban design. They used a working definition of health-promoting human habitats to mean well-designed, built environments that foster strong social cohesion as well as individual mental and physical well-being. Building on this definition, they uncovered the interconnectedness that exists between *people* and *place*.

Further to this, Stevenson and Gleeson (2018) argued that there is a growing need for sustainable urbanization in the form of compact cities, cities of short distances in which there are higher residential and population densities, greater mixed land use and an urban design amenable to walking and cycling. Importantly, research is pointing to enhanced health outcomes associated with compact cities as a consequence of energy efficiency and reduced emissions as residents are more likely to live closer to amenities and walk, cycle or use public transport. However, although urban density is an important dimension of urban form and structure, it acts in a complex way with other social and morphological features to shape cities, human behaviour and population health.

Some more specific measures were described to improve our urban environment and people's health. Rueda (2018) suggested that Superblocks are a suitable solution for addressing the main dysfunctions and challenges that urban systems face today. The Superblock, an urban cell, when repeated creates a mosaic that extends throughout the city. It is an urban model that is used for new developments and regeneration, and it is currently applied in Barcelona. Nieuwenhuijsen et al. (2018a) described the rationale, prerequisites, barriers, facilitators and strategies for car-free cities as a mean to improving the urban environment. They described nine prerequisites to facilitate the transition towards becoming car-free.

Solutions can also be specifically found through transport planning and measuring progress and indicators for performance. Zietsman and Ramani (2018) stated that sustainable transportation principles provide a desirable framework for the development of an urban transportation planning agenda that addresses environmental, economic and societal goals. Several transportation planning activities around the world use some form of a sustainable transportation definition in the framing of their higher-level goals and priorities, with indicators to reflect the same. May (2018) drew on recent European guidance designed to help cities to select suitable measures and package them more effectively to achieve a wide range of objectives or overcome identified transport-related problems. In doing so, he considered the range of measures and the evidence on them, the design of packages, the development of strategies and the constraints on implementing specific measures and ways of overcoming them. He then introduced a knowledgebase, KonSULT, whose Measure Option Generator and Policy Guidebook are designed to assist cities with this crucial task. Subsequently he considered the contribution of the range of available policy measures to the specific objective of enhanced public health.

Furthermore, Khreis et al. (2018) provided a review of the ways transport urban climate action provides direct and more immediate benefits—in climate terms, 'co-benefits'—to public health. They focused on the impacts of five key transport policy measures which have been established to yield significant greenhouse gas reductions and substantial economic benefits. These were (1) compact land use planning to reduce motorized passenger travel demand, (2) passenger modal shift and improving transit efficiency, (3) electrification and passenger vehicle efficiency, (4) freight logistics and (5) freight vehicle efficiency and electrification. They showed that these measures have great potential to improve public health in urban areas whilst mitigating climate change and provide arguments that in some cases these benefits may rival, or exceed, benefits to the economy and climate from these actions.

Finally, Ferreira (2018) argued that transport planners need a much more embodied and inner engagement with their work. If embodied engagement is critical to understand a given reality or problem and to find ways to solve complex issues, it becomes then clear that transport planners need to become physically active individuals as much as they can, if they are to effectively promote active modes of transport and healthy built environments.

What is also important in all of this is also the involvement of citizens. Verlinghieri (2018) described the connections between citizens' participation and health showing its potential and limits in an increasingly complex world with an example of the Ringland initiative. Ringland is a crowd-brained and crowd-funded infrastructure project for a 6-billion-euro investment that has been completely initiated and developed bottom-up by local citizens. It consists of a road tunnelling project that, differently from the originally proposed new ring road, would mitigate the health impacts associated with the ring road. Furthermore, Haklay and Eleta (2018) explored the potential of community-led air quality monitoring. Community-led air quality monitoring differs from top-down monitoring in many aspects: it is focused on community needs and interests and a local problem. They suggested that accessible and reliable community-led air quality monitoring can contribute to the understanding of local environmental issues and improve the dialogue between local authorities and communities about the impacts of air pollution on health and urban and transport planning.

34.2.3 Conceptual Frameworks and Models

To further our understanding and guide our solutions, we need conceptual frameworks and models. Nieuwenhuijsen and Khreis (2018) provided a simplified framework for urban and transport planning, environment and health showing the relationships between urban planning, travel behaviour, environmental exposures and morbidity and premature mortality. Morris et al. (2018) discussed conceptual models to support policy and action in environmental health but, especially, to deliver ecological public health in the urban context. One 'family' of conceptual models, DPSEEA model and its derivatives, is particularly suited to this role as it offers an integrated and policy-relevant conceptualisation of the relationship between human activity, the physical environment and health. They illustrated a number of conceptual models and emphasized their utility as tools to think, communicate and deliver when addressing public health and societal challenges of unprecedented complexity. They stated that only by developing and applying the correct tools can we confront the challenges of the urban context and design policies that create a healthier, more sustainable and more equal world.

Bettencourt (2018) stated that the modelling of cities as complex systems across all these interdependent dimensions is relatively new. He presented a primer to different modelling approaches to cities and urbanization, pointing out their realms of applicability as well as critical areas where they are currently insufficient. Specifically, he gave a short introduction to five main traditions—agent-based, spatial equilibrium, contagion, life-course and growth models—and discussed how they are becoming increasingly articulated and interdependent to create more realistic and more predictive models of cities. From this perspective, he summarized a research agenda for urban theory and modelling focused on understanding issues of growth and development across scales of organization from individuals to neighbourhoods, cities and urban systems.

Nieuwenhuijsen et al. (2018b) described Health Impact Assessment (HIA) as an important tool to integrate evidence in the decision-making process and introduce health in all policies. In urban and transport planning, HIAs have been used generally to qualitatively assess urban interventions rather than offering more useful/powerful estimations to stakeholders through quantitative approaches. HIAs could answer various pressing questions such as: what are the best and most feasible urban and transport planning policy measures to improve public health in cities? Also the process on how to get there is often as important as the actual output of the HIA, as the process may provide answers to important questions as to how different disciplines/sectors can effectively work together and develop a common language, how to best incorporate citizen and stakeholder views, how different modelling and measurement methods can be effectively integrated, and whether a public health approach could make changes in urban and transport planning. Furthermore, Rojas-Rueda (2018) described Health Impact Assessment of active transportation as a valuable tool to promote, assess and maximize the benefits of active transportation for health promotion and prevention. Borrell et al. (2018) presented policies to tackle health inequalities in urban with the example of Urban HEART as a useful tool.

34.2.4 Barriers and Enablers

But what are the barriers and enablers for these solutions? Riley and de Nazelle (2018) discussed the barriers and enablers to ensuring that urban and transport planning decision-making effectively integrates health evidence. These barriers were discussed under the themes of (1) differing understandings of health between sectors, (2) differing understandings of evidence and difficulties around evidence translation, (3) governance and politics and (4) institutional context. They went on to put forward solutions to overcoming these barriers and suggested that enabling factors reside in improving communication and collaboration across sectors and disciplines. Such collaboration is likely to be facilitated by changes to the institutional context in which decisions are made as well as in the way research is developed and communicated.

Grant and Davis (2018) further explored the tensions, the arguments and the possible solutions for those involved in this struggle, be they researchers or practitioners. They stated that the city is the laboratory for change and the subject. But its complexity and its adaptability make it a laboratory like no other. And as a subject, it responds to our interventions with unpredictability. They argued that we need a new transdisciplinary science, not the business as usual of built environment and transport, and beyond the traditional evidence hierarchy of the public health world. New actors and new approaches are needed in the research arena. That we need strong advocacy to support good evidence, and that we need to blend tactical urbanism with action research. And that for the sake of future population health, city leadership, from many quarters, needs to learn how to collaborate for cogeneration of new knowledge that will make a difference to people's lives.

34.3 In Conclusion

What has become apparent is that to a large extent, we know the problems, we know the solutions and we have adequate tools and methods to quantify and communicate them. However, many of the paradigms and tools that are useful and relevant are coming from different disciplines and are not in the form of a holistic and comprehensive framework that can be used by or presented to city leaders and planners. Furthermore governance structures are not in place to bring health into the decision process and play an important part in maintaining the current state of affairs.

There is an urgent need to bring together researchers and practitioners from the different sectors and disciplines (e.g. urban and transport planning, environment, public health, social sciences) to develop a common framework and model for healthy urban and transport planning. This is not a new concept, but one that has slowly been gathering pace over the last few decades after seeing the many health problems related to poor urban and transport planning or the lack thereof.

We need a change because the current urban and transport planning practices are detrimental to health, whilst we should and can create cities that promote health. And for a change to happen, we need a number of prerequisites:

- A crisis
- Knowledge
- Technology
- Partnership
- Vision
- Leadership

In this book we addressed a number of these prerequisites, but what is lacking is a clear vision of healthier future cities and a governance structure that allows good leadership and partnership for health. We need to build long-term visioning of a healthy urban future and involve many stakeholders. We need to address embedded issues such as if we want cities for people or cities for cars. How should healthy, sustainable, liveable, vibrant, equitable, happy, economically viable cities look like? We need urgent leadership and partnerships to address these issues, something which, to a great extent, has been lacking.

As Grant (2018) mentioned, we need to form coalitions for healthier places, using a health lens across planning, transport and all other urban policy. Land use pattern, transport, green space and urban design are key to better support urban health and health equity. We hope that this book has provided further impetus and leads to better and healthier cities.

Across the different disciplines, we have a lot of knowledge of what is good for health, as described in this book. Yet, further research is still needed in many areas, and it would be desirable to develop a good research agenda that brings together all the different sector and disciplines involved. It is beyond the scope of this chapter to spell out the possible items for the research agenda, but throughout the book, many suggestions have been made. We should and can transform our urban environments that are detrimental to health into places that promote health and make cities not only the engines of innovation and wealth creation but also of health promotion.

References

Basagaña, X. (2018). Heat islands/temperature in cities: Urban and transport planning determinants and health in cities. In M. Nieuwenhuijsen & H. Khreis (Eds.), *Integrating human health into urban and transport planning*. Cham: Springer International Publishing.

Bettencourt, L. M. A. (2018). Complex systems modeling of urban development: Understanding the determinants of health and growth across scales. In M. Nieuwenhuijsen & H. Khreis (Eds.), *Integrating human health into urban and transport planning*. Cham: Springer International Publishing.

Borrell, C., Gotsens, M., & Novoa, A. M. (2018). Establishing social equity in cities – A health perspective. In M. Nieuwenhuijsen & H. Khreis (Eds.), *Integrating human health into urban and transport planning*. Cham: Springer International Publishing.

Corburn, J., & Sverdlik, A. (2018). Informal settlements and human health. In M. Nieuwenhuijsen & H. Khreis (Eds.), *Integrating human health into urban and transport planning*. Cham: Springer International Publishing.

Crowhurst Lennard, S. H. C. (2018). Livable cities: Concepts and role in improving health. In M. Nieuwenhuijsen & H. Khreis (Eds.), *Integrating human health into urban and transport planning*. Cham: Springer International Publishing.

Dadvand, P., & Nieuwenhuijsen, M. (2018). Green space and health. In M. Nieuwenhuijsen & H. Khreis (Eds.), *Integrating human health into urban and transport planning*. Cham: Springer International Publishing.

Ferreira, A. (2018). (Un)healthy bodies and the transport planning profession: The (Im)mobile social construction of reality and its consequences. In M. Nieuwenhuijsen & H. Khreis (Eds.), *Integrating human health into urban and transport planning*. Cham: Springer International Publishing.

Gasparrini, A., Guo, Y., Hashizume, M., Lavigne, E., Zanobetti, A., Schwartz, J., Tobias, A., Tong, S., Rocklöv, J., Forsberg, B., Leone, M., De Sario, M., Bell, M. L., Guo, Y.-L. L., Wu, C., Kan, H., Yi, S.-M., de Sousa Zanotti Stagliorio Coelho, M., Saldiva, P. H. N., Honda, Y., Kim, H., & Armstrong, B. (2015). Mortality risk attributable to high and low ambient temperature: a multicountry observational study. *The Lancet, 386*, 369–375. https://doi.org/10.1016/S0140-6736(14)62114-0.

Giles-Corti, B., Gunn, L., Hooper, P., Boulange, C., Zapata Diomedi, B., Pettit, C., & Foster, S. (2018). Built environment and physical activity. In M. Nieuwenhuijsen & H. Khreis (Eds.), *Integrating human health into urban and transport planning*. Cham: Springer International Publishing.

Grant, M. (2018). Planning for healthy cities. In M. Nieuwenhuijsen & H. Khreis (Eds.), *Integrating human health into urban and transport planning*. Cham: Springer International Publishing.

Grant, M., & Davis, A. (2018). Translating evidence into practice. In M. Nieuwenhuijsen & H. Khreis (Eds.), *Integrating human health into urban and transport planning*. Cham: Springer International Publishing.

Haklay, M., & Eleta, I. (2018). On the front line of community-led air quality monitoring. In M. Nieuwenhuijsen & H. Khreis (Eds.), *Integrating human health into urban and transport planning*. Cham: Springer International Publishing.

Hermansen, B., Werner, B., Evensmo, H., & And Graphics Works By Nota, M. (2018). The human habitat: My, our and everyone's city. In M. Nieuwenhuijsen & H. Khreis (Eds.), *Integrating human health into urban and transport planning*. Cham: Springer International Publishing.

Hoffmann, B. (2018). Air pollution in cities: Urban and transport planning determinants and health in cities. In M. Nieuwenhuijsen & H. Khreis (Eds.), *Integrating human health into urban and transport planning*. Cham: Springer International Publishing.

Khreis, H., Sudmant, A., Gouldson, A., & Nieuwenhuijsen, M. J. (2018). Transport policy measures for climate change as drivers for health in cities. In M. Nieuwenhuijsen & H. Khreis (Eds.), *Integrating human health into urban and transport planning*. Cham: Springer International Publishing.

Lawrence, R. J. (2018). Human ecology in the context of urbanization. In M. Nieuwenhuijsen & H. Khreis (Eds.), *Integrating human health into urban and transport planning*. Cham: Springer International Publishing.

Lercher, P. (2018). Noise in cities: Urban and transport planning determinants and health in cities. In M. Nieuwenhuijsen & H. Khreis (Eds.), *Integrating human health into urban and transport planning*. Cham: Springer International Publishing.

Malambo, P., Kengne, A. P., De Villiers, A., Lambert, E. V., & Puoane, T. (2018). Built environment, selected risk factors and major cardiovascular disease outcomes: A systematic review. *PLoS One, 11*(11), e0166846. https://doi.org/10.1371/journal.pone.0166846.

May, A. D. (2018). Measure selection for urban transport policy. In M. Nieuwenhuijsen & H. Khreis (Eds.), *Integrating human health into urban and transport planning*. Cham: Springer International Publishing.

Morris, G., Staatsen, B., & Van Der Vliet, N. (2018). Using conceptual models to shape healthy sustainable cities. In M. Nieuwenhuijsen & H. Khreis (Eds.), *Integrating human health into urban and transport planning*. Cham: Springer International Publishing.

Nieuwenhuijsen, M. J., Bastiaanssen, J., Sersli, S., Waygood, E. O. D., & Khreis, H. (2018a). Implementing car free cities: Rationale, requirements, barriers and facilitators. In M. Nieuwenhuijsen & H. Khreis (Eds.), *Integrating human health into urban and transport planning*. Cham: Springer International Publishing.

Nieuwenhuijsen, M. J., Khreis, H., Verlinghieri, E., Mueller, N., & Rojas-Rueda, D. (2018b). The role of health impact assessment for shaping policies and making cities healthier. In M. Nieuwenhuijsen & H. Khreis (Eds.), *Integrating human health into urban and transport planning*. Cham: Springer International Publishing.

Nieuwenhuijsen, M. J., & Khreis, H. (2018). Urban and transport planning and health. In M. Nieuwenhuijsen & H. Khreis (Eds.), *Integrating human health into urban and transport planning*. Cham: Springer International Publishing.
Riley, R., & De Nazelle, A. (2018). Barriers and enablers of integrating health evidence into transport and urban planning and decision making. In M. Nieuwenhuijsen & H. Khreis (Eds.), *Integrating human health into urban and transport planning*. Cham: Springer International Publishing.
Rojas-Rueda, D. (2018). Health impact assessment of active transportation. In M. Nieuwenhuijsen & H. Khreis (Eds.), *Integrating human health into urban and transport planning*. Cham: Springer International Publishing.
Rueda, S. (2018). Superblocks for the design of new cities and renovation of existing ones. Barcelona's case. In M. Nieuwenhuijsen & H. Khreis (Eds.), *Integrating human health into urban and transport planning*. Cham: Springer International Publishing.
Schepers, P., Lovegrove, G., & Helbich, M. (2018). Urban form and road safety: Public and active transport enable high levels of road safety. In M. Nieuwenhuijsen & H. Khreis (Eds.), *Integrating human health into urban and transport planning*. Cham: Springer International Publishing.
Stevenson, M., & Gleeson, B. (2018). Complex urban systems: Compact cities, transport and health. In M. Nieuwenhuijsen & H. Khreis (Eds.), *Integrating human health into urban and transport planning*. Cham: Springer International Publishing.
Stigendal, M. (2018). Aiming at social cohesion in cites to transform society. In M. Nieuwenhuijsen & H. Khreis (Eds.), *Integrating human health into urban and transport planning*. Cham: Springer International Publishing.
Tsouros, A. D. (2018). Healthy cities: A political movement which empowered local governments to put health and equity high on their agenda. In M. Nieuwenhuijsen & H. Khreis (Eds.), *Integrating human health into urban and transport planning*. Cham: Springer International Publishing.
Van Wee, B. (2018). Land-use policy, travel behavior and health. In M. Nieuwenhuijsen & H. Khreis (Eds.), *Integrating human health into urban and transport planning*. Cham: Springer International Publishing.
Verlinghieri, E. (2018). Participating to health: The healthy outcomes of citizen participation in urban and transport planning. In M. Nieuwenhuijsen & H. Khreis (Eds.), *Integrating human health into urban and transport planning*. Cham: Springer International Publishing.
Wang, Y., Chau, C. K., Ng, W. Y., & Leung, T. M. (2016). A review on the effects of physical built environment attributes on enhancing walking and cycling activity levels within residential neighborhoods. *Cities, 50*, 1–15.
Zietsman, J., & Ramani, T. (2018). Advancing health considerations within a sustainable transportation agenda – Using indicators and decision-making. In M. Nieuwenhuijsen & H. Khreis (Eds.), *Integrating human health into urban and transport planning*. Cham: Springer International Publishing.

Index

M. Nieuwenhuijsen, H. Khreis (eds.), *Integrating Human Health into Urban and Transport Planning*, https://doi.org/10.1007/978-3-319-74983-9

H

I

V

W

Printed by Printforce, the Netherlands